COMPSTAT 2002

COMPSTAT

Proceedings
in Computational Statistics

15th Symposium Held in Berlin,
Germany, 2002

Edited by
Wolfgang Härdle
Bernd Rönz

ebook

With 205 Figures
and 89 Tables

Physica-Verlag
A Springer-Verlag Company

Professor Dr. Wolfgang Härdle
Humboldt-Universität zu Berlin
CASE – Centre of Applied Statistics and Economics
Spandauer Straße 1
10178 Berlin, Germany

Professor Dr. Bernd Rönz
Humboldt-Universität zu Berlin
Wirtschaftswissenschaftliche Fakultät
Institut für Statistik und Ökonometrie
Spandauer Straße 1
10178 Berlin, Germany

ISBN 3-7908-1517-9 Physica-Verlag Heidelberg New York

Library of Congress Cataloging-in-Publication Data applied for
Die Deutsche Bibliothek – CIP-Einheitsaufnahme
Proceedings in Computational Statistics:
15th Symposium Held in Berlin, Germany, 2002; with 89 Tables /
COMPSTAT. Ed. by Wolfgang Härdle; Bernd Rönz. –
Heidelberg; New York: Physica-Verl., 2002
ISBN 3-7908-1517-9

Physica-Verlag Heidelberg New York
a member of BertelsmannSpringer Science + Business Media GmbH

http://www.springer.de

Cover design: Erich Kirchner, Heidelberg

SPIN 10886369 88/2202-5 4 3 2 1 0 – Printed on acid-free papaer

Preface

This COMPSTAT 2002 book contains the Keynote, Invited, and Full Contributed papers presented in Berlin, August 2002. A companion volume including Short Communications and Posters is published on CD.

The COMPSTAT 2002 is the 15th conference in a serie of biannual conferences with the objective to present the latest developments in Computational Statistics and is taking place from August 24th to August 28th, 2002.

Previous COMPSTATs were in Vienna (1974), Berlin (1976), Leiden (1978), Edinburgh (1980), Toulouse (1982), Prague (1984), Rome (1986), Copenhagen (1988), Dubrovnik (1990), Neuchâtel (1992), Vienna (1994), Barcelona (1996), Bristol (1998) and Utrecht (2000).

COMPSTAT 2002 is organised by CASE, Center of Applied Statistics and Economics at Humboldt-Universität zu Berlin in cooperation with Freie Universität Berlin and University of Potsdam.

The topics of COMPSTAT include methodological applications, innovative software and mathematical developments, especially in the following fields: statistical risk management, multivariate and robust analysis, Markov Chain Monte Carlo Methods, statistics of E-commerce, new strategies in teaching (Multimedia, Internet), computerbased sampling/questionnaires, analysis of large databases (with emphasis on computing in memory), graphical tools for data analysis, classification and clustering, new statistical software and historical development of software.

A conference like COMPSTAT is not possible without an active scientific input from many high class researchers. The contributions were classified by the Scientific Program Commitee (SPC): Prof. Wolfgang Härdle (Berlin), Prof. Antony Unwin (Augsburg), Prof. Jaromir Antoch (Prague), Prof. Adrian Bowman (Glasgow), Prof. Michel Delecroix (Paris), Prof. Wenceslao Gonzalez Manteiga (Santiago de Compostela), Prof. Junji Nakano (Tokyo), Prof. Michael Schimek (Graz) and Prof. Peter van der Heijden (Utrecht).

The papers were refereed by the SPC and we would like to thank our colleagues for all their help. The resulting proceedings provide a broad overview of the current research areas in statistical computing.

We are also very grateful to the members of the Local Organizing Committee: Dr. Yasemin Boztug, Prof. Dr. Herbert Büning, Prof. Dr. Wolfgang Härdle, Prof. Dr. Lutz Hildebrandt, Dr. Sigbert Klinke, Prof. Dr. Uwe Küchler, Prof. Dr. Bernd Rönz, Dr. Peter Schirmbacher, Dipl.-Ing. Benjamin Schüler, Prof. Dr. Vladimir Spokoiny, Prof. Dr. Hans Gerhard Strohe, Prof. Dr. Jürgen Wolters and Uwe Ziegenhagen.

We would also like to thank Patricia Ahrend and Luise Richter who together with Dr. Klinke did a great job in managing the papers.

Berlin, August 2002:

W. Härdle

B.Rönz

Contents

Invited Papers

An Implementation for Regression Quantile Estimation

Thomas W. Yee,
Department of Statistics, University of Auckland, Private Bag 92019,
Auckland, New Zealand

Abstract. Of the many methods that have been developed for quantile regression (Wright and Royston, 1997) the LMS method (Cole and Green, 1992) can be easily understood, is flexible and based on splines. The basic idea is that, for a fixed value of the covariate, a Box-Cox transformation of the response is applied to obtain standard normality. The three parameters are chosen to maximize a penalized log-likelihood. One unpublished extension by Lopatatzidis and Green is to transform to a gamma distribution, which helps overcome a range-restriction problem in the original formulation. This paper proposes a new method based on the Yeo and Johnson (2000) transformation. It has the advantage that it allows for both positive and negative values in the response. R/S-PLUS software written by the author implementing the LMS method and variants are described and illustrated with data from a New Zealand workforce study.

Keywords. age-reference centile analysis, LMS method, penalized likelihood, quantile regression, R, S-PLUS, vector generalized additive models

1 Introduction

The classical problem of estimating quantiles (also known as centiles and percentiles) is a well-known one: given a random sample $y_1, \dots, y_n$ from some distribution or from a reference population, estimate the α quantile, e.g., $\alpha = 0.25$ is the lower quartile. An extension of this problem is concerned with scatter plot data (x_i, y_i), $i = 1, \dots, n$, where it is desired to obtain the conditional quantile curve $\alpha(x)$ of the response y given x. A second extension is involves handling a *vector* of explanatory variables $\boldsymbol{x}_i$. For example, for many clinical and anthropometric variables, the major influence is age, followed by the gender of the individual. Often one of the variables in $\boldsymbol{x}_i$ is of special interest so we can write $\boldsymbol{x}_i = (x_i, \boldsymbol{x}_i^{*T})^T$ where x will be called the 'primary' covariate. If the other explanatory variables are fixed at $\boldsymbol{x}_i^* = \boldsymbol{x}_0^*$, say, then it is possible to plot $\hat{\alpha}(x, \boldsymbol{x}_0^*)$ versus x. The software described in this paper has capabilities that handle this second extension to a limited degree.

Quantile regression has been widely used in many fields, e.g.,

1. In medical statistics, age-centile curves result from X = age, and common responses Y include height, weight, and body mass index (BMI = weight $\div$ height^2; in kg/m^2, a measure of obesity). Many researchers have proposed gestational-age-specific centile charts and tables for a variety of relevant measurements, e.g., Chitty et al. (1994).
2. Clinical chemistry and clinical epidemiology, e.g., in population screening for heart disease and stroke according to risk factors such as serum cholesterol concentration and blood pressure. Harris and Boyd (1995) gives a

useful overview of methods applied to data sets from these areas of research and clinical practice.
3. Economics, e.g., it has been used to study determinants of wages, discrimination effects, and trends in income inequality. See Buchinsky (1998) for examples.
4. Education, e.g., the performance of students in public schools on standardized exams as a function of socio-economic variables such as parents' income and educational attainment. Other variables are policy variables, e.g., class size, teacher qualifications. See Koenker and Hallock (2001) for examples.

The purpose of this paper is to present vector generalized additive models (VGAMs) as a unifying class of models for which LMS quantile regression models naturally fall in. A new variant based on the Yeo and Johnson (2000) transformation is proposed, which offers some advantage over the 'classical' LMS method of Cole and Green (1992). Software written by T. Yee implementing several variants of the LMS method is described. The author's code is under continual development, therefore it is likely that future changes will occur. What is presented here uses version 4 of the S language (Chambers, 1998); in particular, it was run under S-PLUS 6.0 and R 1.4.1, under UNIX and LINUX. It follows the modelling approach described by Chambers and Hastie (1993).

2 Vector Generalized Linear and Additive Models

The unifying theory behind this article is the VGLM/VGAM framework in which the prominant ideas are iteratively reweighted least squares (IRLS), vector splines, modified backfitting and constraints-on-the-functions. This gives rise to a class of maximum penalized likelihood models (see, e.g., Green and Silverman (1994) for an overview of these ideas). We briefly review the ideas in this section. See Yee and Wild (1996) for more details.

Suppose now that a response $\boldsymbol{y}$ is a q-dimensional vector. VGLMs are defined as a model for which the conditional distribution of $\boldsymbol{Y}$ given $\boldsymbol{x}$ is of the form

$$f(\boldsymbol{y}|\boldsymbol{x};\mathbf{B}) \;=\; h(\boldsymbol{y},\eta_1,\ldots,\eta_M,\phi)$$

for some known function $h(\cdot)$, where $\mathbf{B} = (\boldsymbol{\beta}_1\,\boldsymbol{\beta}_2\,\cdots\,\boldsymbol{\beta}_M)$ is a $p \times M$ matrix of unknown regression coefficients, ϕ is an optional scale parameter, and

$$\eta_j \;=\; \eta_j(\boldsymbol{x}) \;=\; \boldsymbol{\beta}_j^T\boldsymbol{x} \;=\; \beta_{(j)0}+\beta_{(j)1}\,x_1+\cdots+\beta_{(j)p}\,x_p, \quad j=1,\ldots,M, \tag{1}$$

is the jth linear predictor. VGLMs are thus like GLMs, but can have more than one linear predictor. Importantly, they cover models outside the confines of the exponential family. One can write

$$\boldsymbol{\eta}(\boldsymbol{x}_i) \;=\; \begin{pmatrix} \eta_1(\boldsymbol{x}_i) \\ \vdots \\ \eta_M(\boldsymbol{x}_i) \end{pmatrix} \;=\; \boldsymbol{\eta}_i \;=\; \boldsymbol{\beta}_0 + \mathbf{B}^T\boldsymbol{x}_i \;=\; \boldsymbol{\beta}_0 + \begin{pmatrix} \boldsymbol{\beta}_1^T\boldsymbol{x}_i \\ \vdots \\ \boldsymbol{\beta}_M^T\boldsymbol{x}_i \end{pmatrix}$$

where $\boldsymbol{\beta}_0 = (\beta_{(1)0},\cdots,\beta_{(M)0})^T$ is the vector of intercepts. The η_j of VGLMs may be applied directly to parameters of a distribution rather than just to means for GLMs. For example, in the negative binomial distribution (with parameters p and k), $\eta_1 = p$ and $\eta_2 = k$ would be possible with VGLMs, though, because $0 < p < 1$ and $k > 0$, it would be better to have $\eta_1 = \text{logit}\, p$ and $\eta_2 = \log k$, say.

VGAMs provide additive-model extensions to VGLMs, that is, (1) is generalized to

$$\eta_j(\boldsymbol{x}) \;=\; \beta_{(j)0} + f_{(j)1}(x_1) + \cdots + f_{(j)p}(x_p), \quad j = 1, \ldots, M,$$

a sum of smooth functions of the individual covariates, just as with ordinary GAMs. Once again, the $f_{(j)k}(x_k)$ are fitted using smoothers—indeed, $\boldsymbol{f}_k = (f_{(1)k}(x_k), \ldots, f_{(M)k}(x_k))^T$ are estimated simultaneously using *vector smoothers*. VGAMs are thus a visual data-driven method that is well suited to exploring data. They retain the simplicity of interpretation that GAMs possess. Like GLMs, the VGAM class has the fortunate property that the IRLS algorithm can fit them all. This unified method of solution makes it easier for practitioners too.

In practice, it is very useful to put "constraints-on-the-functions" because the functions in $\boldsymbol{f}_k$ may be related. For example, specify one parameter to be independent of a covariate X_1:

$$\begin{aligned}\eta_1 &= \beta_{(1)0} + \beta_{(1)1} X_1 \\ \eta_2 &= \beta_{(2)0}.\end{aligned}$$

This constraint, which is useful in quantile regression, can be enforced with the constraints-on-the-functions framework. For VGAMs, we have more generally

$$\begin{aligned}\boldsymbol{\eta}(\boldsymbol{x}) &= \boldsymbol{\beta}_0 + \boldsymbol{f}_1(x_1) + \cdots + \boldsymbol{f}_p(x_p) \\ &= \mathbf{H}_0\, \boldsymbol{\beta}_0^* + \mathbf{H}_1\, \boldsymbol{f}_1^*(x_1) + \cdots + \mathbf{H}_p\, \boldsymbol{f}_p^*(x_p)\end{aligned}$$

where $\mathbf{H}_0, \mathbf{H}_1, \ldots, \mathbf{H}_p$ are known full-column rank *constraint matrices*, $\boldsymbol{f}_k^*$ is a vector containing a possibly reduced set of component functions and $\boldsymbol{\beta}_0^*$ is a vector of unknown intercepts. With no constraints at all, $\mathbf{H}_0 = \mathbf{H}_1 = \cdots = \mathbf{H}_p = \mathbf{I}_M$ ($M \times M$ identity matrix) and $\boldsymbol{\beta}_0^* = \boldsymbol{\beta}_0$ ($= \boldsymbol{\eta}_0$). Like the $\boldsymbol{f}_k$, the $\boldsymbol{f}_k^*$ are centered for uniqueness. For VGLMs, the $\boldsymbol{f}_k$ are linear so that

$$\mathbf{B}^T \;=\; \begin{pmatrix}\mathbf{H}_0\boldsymbol{\beta}_0^* & \mathbf{H}_1\boldsymbol{\beta}_1^* & \cdots & \mathbf{H}_p\boldsymbol{\beta}_p^*\end{pmatrix}.$$

For models with a log-likelihood $\ell = \sum \ell_i$, VGAMs can be estimated by applying a modified vector backfitting algorithm to an adjusted dependent vector $\boldsymbol{z}_i = \boldsymbol{\eta}_i + \mathbf{W}_i^{-1}\boldsymbol{d}_i$, with weights $\mathbf{W}_i$, where $\boldsymbol{d}_i = \partial \ell_i / \partial \boldsymbol{\eta}$, and $\mathbf{W}_i = -\partial^2 \ell_i / (\partial \boldsymbol{\eta}\, \partial \boldsymbol{\eta}^T)$ or its expected value.

3 The LMS Method

The LMS method gets its name from the first letters of the Greek letters, λ, μ and σ, which are the three parameters. The idea behind the LMS method is that a Box-Cox power transformation of the y_i, given x_i, has some parametric distribution such as the standard normal, that is,

$$Z = \begin{cases} \dfrac{\left(\dfrac{Y}{\mu(x)}\right)^{\lambda(x)} - 1}{\sigma(x)\,\lambda(x)}, & \lambda(x) \neq 0; \\[2ex] \dfrac{1}{\sigma(x)} \log\left(\dfrac{Y}{\mu(x)}\right), & \lambda(x) = 0, \end{cases}$$

is $N(0,1)$ (Cole, 1988). Because σ must be positive, it is wise to use $\boldsymbol{\eta}(x) = (\lambda(x), \mu(x), \log(\sigma(x)))^T$, which is the default in VGAM. Given $\hat{\boldsymbol{\eta}}$, the $100\alpha\%$ quantile (e.g., $\alpha = 0.5$ for median) can be estimated by inverting the Box-Cox power transformation at the $100\alpha\%$ quantile of the parametric distribution.

Cole and Green (1992) advocated estimation by maximizing a penalized log-likelihood using splines to estimate the three functions, thus allowing a great deal of flexibility. Of the three functions, it is often a good idea to allow $\mu(x)$ to be more flexible because the functions $\lambda(x)$ and $\sigma(x)$ usually vary more smoothly with x. In fact, it is sometimes a good idea to set λ and σ to be an intercept term only—something which can be achieved by constraints-on-the-functions. These preferences are illustrated in the example below. Cole and Green (1992) estimated $\boldsymbol{\eta}$ iteratively, whereas VGAM fits the functions simultaneously using a vector smoothing spline (see Yee (1998)). Consequently there may be a little difference in estimates.

The LMS method can handle vector $\boldsymbol{x}_i$, and VGAM's methods function `qtplot.lmscreg()` can plot the fitted quantiles against the 'primary' covariate if the other covariates are fixed at some value.

One problem with the above formulation is that $1+\lambda(x)\,\sigma(x)\,\Phi^{-1}(\alpha/100) > 0$ is required in order for the centiles to be available. Hence a disadvantage is that the range of transformation depends on λ. To overcome this range-restriction problem in its parameters, Lopatatzidis and Green proposed replacing the normal distribution by a gamma distribution.

3.1 The Gamma Distribution Version

The transformed variable $W = (Y/\mu)^{\lambda}$ is assumed gamma with unit mean and variance $\lambda^2\sigma^2$. Then the $100\alpha\%$ quantile of Y at x is $\mu(x)\, W_\alpha^{1/\lambda(x)}$ where W_α is the equivalent deviate of size α for the gamma distribution with mean 1 and variance $\lambda(x)^2\sigma(x)^2$. In the case $\lambda(x) = 0$, the distribution of W at x reduces to

$$\log\left(\frac{y}{\mu(x)}\right) \sim N(0, \sigma(x)^2),$$

giving $100\alpha\%$ quantile of Y at x as $\mu(x)\exp\{\sigma(x) Z_\alpha\}$.

Although the gamma model is not traditional in quantile regression it avoids a range of problems. Lopatatzidis and Green showed it has finite expectations of the required derivatives of the likelihood function, which can prove to be an important problem for the normal version, particularly when σ is small. In such situations it appears that the penalized likelihood can go to infinity and thus the algorithm cannot converge. In such cases the gamma version of the LMS method can give an acceptable solution. Further technical details of the gamma version and a comparison between the normal and gamma distributions is given in Lopatatzidis and Green.

3.2 VGAM Software for the LMS method

The LMS method has at least three software implementations. T. Cole has a stand-alone FORTRAN program, and V. Carey has S-PLUS code at http://biosun1.harvard.edu/~carey/index.ssoft.html; there is a link to "lmsqreg".

T. Yee has written S-PLUS and R functions implementing the class of VGLMs and VGAMs, called the VGAM library. It is freely available and can be downloaded starting from http://www.stat.auckland.ac.nz/~yee. Detailed documentation is also provided. Currently it runs under the latest UNIX and LINUX versions of S-PLUS and R. In the author's software, each model is fully described by a VGAM "family function"—a small piece of code containing specific information about that distribution or model. Users of VGAM are encouraged to write and document their own family functions, and to make them available for others to use.

In practice, the LMS method involves using *approximate* derivatives. In fact, if too many iterations are performed, the solution may diverge and fail! Also, good initial values are usually required.

At this stage the family functions `lms.bcn()`, `lms.bcg()`, and `lms.yjn()` have been written to implement quantile regression. The "bc" stands for Box-Cox, "yj" for Yeo-Johnson, "n" for normal, and "g" for gamma. Two methods functions, `qtplot()` and `deplot()`, have also been written, which produce quantile and density plots of fitted models.

A typical VGAM family function for quantile regression has the following arguments.

```
> args(lms.bcn)
function(percentiles = c(25, 50, 75), zero = NULL, link.sigma =
        c("loge", "identity"), init.lambda = NULL, init.sigma = NULL)
NULL
```

Hence, in a typical usage such as

```
fit0 = vgam(BMI ~ s(age, df=4), lms.bcn(zero=c(1,3)), data=wfmen)
```

the corresponding output will be something like

```
> fitted(fit0)[1:2,]
        25       50       75
1 23.30000 25.23782 27.48002
2 24.68716 26.74023 29.12546
```

That is, the fitted values are a n-row matrix of quantiles. The default percentiles are 25%, 50% and 75%, but the user can specify alternative values, e.g., `lms.bcn(percentiles=c(50,95))` for 50% and 95%. Thus `fitted(fit0)` do not contain actual fitted means such as for GLMs.

The `zero` argument can be assigned a vector taking the values 1, 2, and/or 3. If set to a value j, it forces η_j to be modelled as a intercept only. By default, `zero=NULL`, meaning all the η_j are modelled as functions of the predictors. Thus `zero` is a convenient way of implementing a certain type of constraints-on-the-functions.

4 Extension to the Yeo and Johnson (2000) Transformation

In this section we extend the LMS method to use the Yeo and Johnson (2000) transformation. This power transformation is potentially useful for improving normality, and has the advantage over the Box-Cox transformation in that it is well defined on the whole real line. It is given by

$$\psi(\lambda, y) = \begin{cases} \{(y+1)^\lambda - 1\}/\lambda & (y \geq 0, \lambda \neq 0), \\ \log(y+1) & (y \geq 0, \lambda = 0), \\ -\{(-y+1)^{2-\lambda} - 1\}/(2-\lambda) & (y < 0, \lambda \neq 2), \\ -\log(-y+1) & (y < 0, \lambda = 2). \end{cases}$$

Subsequently, the corresponding inversion formulae is given by

$$\psi^{-1}(\lambda, \psi) = y(\lambda, \psi) = \begin{cases} \{\psi\lambda + 1\}^{1/\lambda} - 1 \\ \exp(\psi) - 1 \\ 1 - \{-(2-\lambda)\psi + 1\}^{1/(2-\lambda)} \\ 1 - \exp(-\psi). \end{cases}$$

Derivatives are needed for the treatment below; these are

$$\psi^{(k)} = \begin{cases} [(y+1)^\lambda \{\log(y+1)\}^k - k\psi^{(k-1)}]/\lambda & (y \geq 0, \lambda \neq 0), \\ \{\log(y+1)\}^{k+1}/(k+1) & (y \geq 0, \lambda = 0), \\ -[(-y+1)^{2-\lambda}\{-\log(-y+1)\}^k - k\psi^{(k-1)}]/(2-\lambda) & (y < 0, \lambda \neq 2), \\ \{-\log(-y+1)\}^{k+1}/(k+1) & (y < 0, \lambda = 2). \end{cases}$$

where $\psi^{(k)} = \partial^k \psi(\lambda, y)/\partial\lambda^k,\ k \geq 1,\ \psi^{(0)} = \psi(\lambda, y)$.

4.1 The Yeo-Johnson Transformation Version

With the log-likelihood $\ell = \sum_{i=1}^n \ell_i$, then

$$\ell_i = -\frac{1}{2}\log(2\pi) - \log(\sigma_i) - \frac{1}{2\sigma_i^2}(\psi_i - \mu_i)^2 + (\lambda_i - 1)\,\mathrm{sgn}(y_i)\,\log(|y_i| + 1).$$

Using the above transformation it is straightforward to derive the score vector and (observed) Hessian matrix, however, Newton-Raphson is unsuitable because the observed Hessians are not positive-definite in general. Fisher scoring helps alleviate this problem. The expected value of three elements of the information matrix are tractable:

$$-E\left[\frac{\partial^2 \ell_i}{\partial\sigma_i^2}\right] = \frac{2}{\sigma_i^2}, \quad -E\left[\frac{\partial^2 \ell_i}{\partial\mu_i^2}\right] = \frac{1}{\sigma_i^2}, \quad -E\left[\frac{\partial^2 \ell_i}{\partial\mu_i\,\partial\sigma_i}\right] = 0.$$

Unfortunately there are three elements whose expected values are untractable, viz.,

$$-\frac{\partial^2 \ell_i}{\partial\lambda_i^2} = \frac{1}{\sigma_i^2}\left[\left(\frac{\partial\psi_i}{\partial\lambda_i}\right)^2 + (\psi_i - \mu_i)\frac{\partial^2\psi_i}{\partial\lambda_i^2}\right], \tag{2}$$

$$-\frac{\partial^2 \ell_i}{\partial\lambda_i\,\partial\sigma_i} = -\frac{2}{\sigma_i^3}\frac{\partial\psi_i}{\partial\lambda_i}(\psi_i - \mu_i), \quad -\frac{\partial^2 \ell_i}{\partial\lambda_i\,\partial\mu_i} = -\frac{1}{\sigma_i^2}\frac{\partial\psi_i}{\partial\lambda_i}. \tag{3}$$

The difficulty in obtaining formulae for the $\mathbf{W}_i$ is usually the case for the LMS method—see Cole and Green (1992) for the classical LMS method.

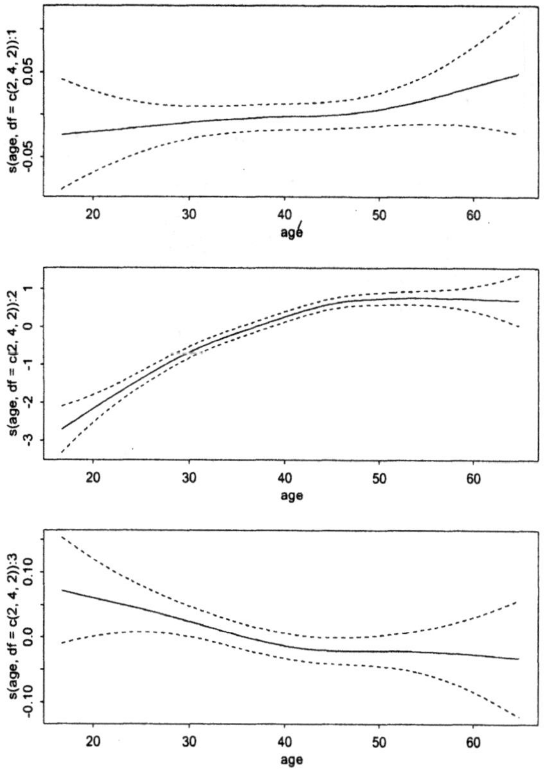

Fig. 1. Fitted functions $\hat{\eta}_1 = \hat{\lambda}(x)$, $\hat{\eta}_2 = \hat{\mu}(x)$, $\hat{\eta}_3 = \log(\hat{\sigma}(x))$ of the LMS method fitted to BMI data.

In the author's `lms.yjn()`, the three elements in (2) can be approximated by taking a weighted mean of those quantities, for example, using the property that the first derivatives have zero mean,

$$-E\left[\frac{\partial^2 \ell_i}{\partial \lambda_i \, \partial \sigma_i}\right] \approx \frac{2}{\sigma_i} \text{weighted.mean}\{\text{sgn}(y_i) \log(|y_i| + 1)\}.$$

This often (but not always!) makes the $\mathbf{W}_i$ positive-definite usually, but convergence is rather slow. A refinement is to smooth the quantities against the covariate x with a limited degrees of freedom—this is also implemented.

The $\alpha(x)$-quantile function is $\psi^{-1}(\lambda(x), \mu(x) + \sigma(x) Z_\alpha)$, and the density of Y at $X = x_0$ is

$$g_Y(y) = \frac{1}{\sigma(x_0)} \phi\left(\frac{\psi(y, \lambda(x_0)) - \mu(x_0)}{\sigma(x_0)}\right) \cdot (|y| + 1)^{\text{sgn}(y) \cdot (\lambda(x_0) - 1)},$$

where $\phi(\cdot)$ is the density function of a standard normal.

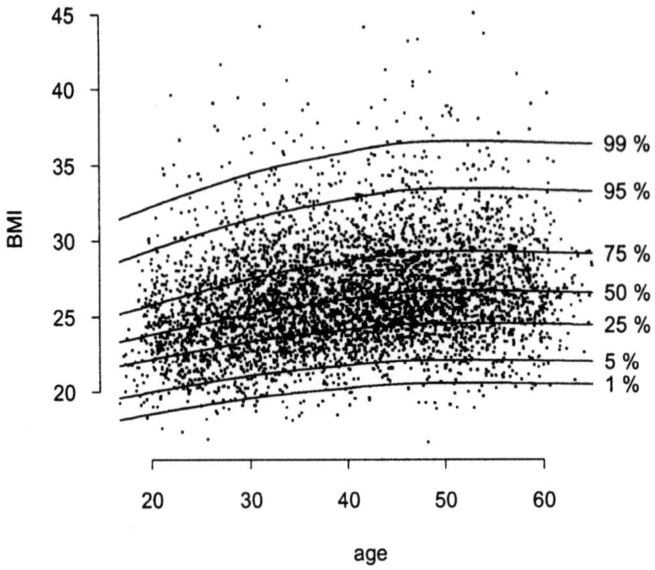

Fig. 2. Fitted quantile curves from `fit2`.

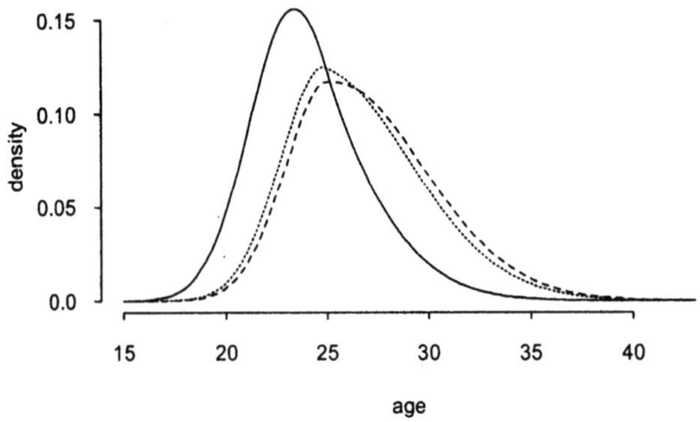

Fig. 3. Density plots of BMI from `fit2` at ages 20, 40 and 60.

4.2 Example

From a cross-sectional workforce study described by Yee and Wild (1996) we look at the ages (in years) and body mass indexes (BMI) of 4786 European-type male New Zealanders aged between 16 and 65. They can be thought of as an approximate random sample from the adult working population. Because BMIs are affected by age, gender and race, healthy values can range anywhere from about 20 to 30 for most such adults.

A scatter plot of the data can be seen from Fig. 2. A simple fit and plots are obtained from

```
fit = vgam(BMI ~ s(age, df=c(2,4,2)), lms.yjn(yoffset=-25), data=wfmen)
par(mfrow=c(3,1), mar=c(6,5,2,1)+0.1, las=0)
plot(fit, se=T, col=1, cex=1.0, rug=F)
```

This results in Fig. 1. Here, the first and third functions appear linear, but the second ($\hat{\mu}(x)$) is clearly nonlinear. Like GAMs, pointwise ± 2 SE bands are useful as they give a rough indication of how much to trust the fitted functions. Note that the linearity of the fit could also be investigated by plotting the derivatives on the same scale (not done here), e.g.,

```
plot(fit, col=1, deriv=1, scale=.2, cex=1.0, rug=F)
```

A more formal summary of `fit` is provided by

```
> summary(fit)
Call:
vgam(formula = BMI ~ s(age, df = c(2, 4, 2)), family = lms.yjn(yoffset =
        -25), data = wfmen)
Number of linear predictors:    3
Names of linear predictors: lambda, mu, log(sigma)
Dispersion Parameter for lms.yjn family:   1
Log-likelihood: -7998.61 on 14347.15 degrees of freedom
Number of Iterations:  4
DF for Terms and Approximate Chi-squares for Nonparametric Effects

                          Df Npar Df Npar Chisq  P(Chi)
            (Intercept):1  1
            (Intercept):2  1
            (Intercept):3  1
s(age, df = c(2, 4, 2)):1  1     0.9      1.701 0.16730
s(age, df = c(2, 4, 2)):2  1     3.0     54.652 0.00000
s(age, df = c(2, 4, 2)):3  1     1.0      3.170 0.07264
```

The P-values for testing linearity of the functions match the visual inspections of the plots. Let's try an even simpler model:

```
fit2 = vgam(BMI ~ s(age, df=4), lms.yjn(yoffset=-25, zero=c(1,3)),
            data=wfmen)
```

which models λ and σ as an intercept only. Then

```
par(mfrow=c(1,1), mar=c(5,5,1,3)+0.1, bty="L", las=1)
more = qtplot(fit2, per=c(1,5,25,50,75,95,99), ylab="BMI",
              pcex=0.25, pch=15, lwd=2)
```

results in Fig. 2. For fixed ages, the distribution of BMI is positively skewed. These results show that BMI generally increases with age until about 50, and then stabilize. Sometimes data like this decreases past a certain point—this is probably a selection bias because people with high BMIs are more prone to diseases (particular cardiovascular types) and hence die younger.

Density plots can be obtained by

```
par(mfrow=c(1,1), mar=c(5,5,.1,.1)+0.1, bty="L", las=1)
at = seq(15, 43, by=0.25)
par(mfrow=c(1,1), lwd=2)
deplot(fit2, x0=20, at=at, xlab="age")
deplot(fit2, x0=40, at=at, add=T, lty=2, col=1)
deplot(fit2, x0=60, at=at, add=T, lty=3, col=1) -> a
```

This gives Fig. 3. There is a little difference in the distribution of BMIs between the 40 and 60 year olds, but they both are subtantially higher compared to 20 year olds. The positive skew is also evident.

Both `qtplot()` and `deplot()` deposit useful computed quantities in the slot `@post`. For example, in the above, we have

```
> more@post$qtplot$percentiles
[1]  1  5 25 50 75 95 99
> more@post$qtplot$fitted[1:2,]
         1        5       25       50       75       95       99
1 19.56188 21.05564 23.34619 25.18737 27.51502 31.42792 34.43878
2 20.57540 22.12450 24.57543 26.75066 29.32948 33.46696 36.59444
> a@post$deplot$newdata
  age
1  60
> a@post$deplot$at[1:5]
[1] 15.00 15.25 15.50 15.75 16.00
> a@post$deplot$y[1:5]
[1] 7.061750e-07 1.326083e-06 2.443224e-06 4.417287e-06 7.838161e-06
```

4.3 Adjusting for Gender

In theory the LMS method can operate with more than one 'primary' covariate (usually age), e.g., adjust for other variables such as gender and race. To plot the results using `qtplot.lmscreg()`, however, is not easy, but possible by using its `newdata` argument. This allows prediction to occur in that data frame. Note that plotting the centiles against the primary variable only makes sense if the non-primary variables have a common value.

To illustrate the LMS method with a vector $\boldsymbol{x}$, we fit a simplified model to both male and female data. We will use `wf`, which consists of a random sample of 350 men and 350 women, and there is another variable, `wf$sex`, which one adjusts for—it has values 0 for men and 1 for women. The 'primary' variable is `wf$age`.

```
fit3 = vgam(BMI ~ s(age, df=c(2,4,2)) + sex,
            fam=lms.yjn(percentile=50, yoffset=-25), data=wf)
nn = 80
Age = seq(18, 64, len=nn)   # Predict at these values
half = split(wf, wf$sex)

par(bty="l", mar=c(5,4,1,3)+0.1, xpd=T, las=1)
plot(wf$age, wf$BMI, xlab="age", ylab="BMI", type="n")  # Set up axes

points(half[["0"]]$age, half[["0"]]$BMI, pch=15, cex=0.5)
newdat = data.frame(age=Age, sex=rep(0, nn))
men = qtplot(fit3, newdata=newdat, add=T)

points(half[["1"]]$age, half[["1"]]$BMI, pch="o")
newdat = data.frame(age=Age, sex=rep(1, nn))
women = qtplot(fit3, newdata=newdat, add=T, llty=3, lcol=2)
```

The plot is given in Fig. 4. Not surprising, men are more solidly built than women of the same age. Indeed,

```
> (men@post$qtplot$fitted - women@post$qtplot$fitted)[1:4]
[1] 2.001926 2.001926 2.001926 2.001926
```

shows that the median BMI of men is about 2.0 kg/m^2 higher than women regardless of age. Of course, this is a no-interaction model which would be an unreasonable assumption if infants were included in the data set. The absence of an age–gender interaction would need checking.

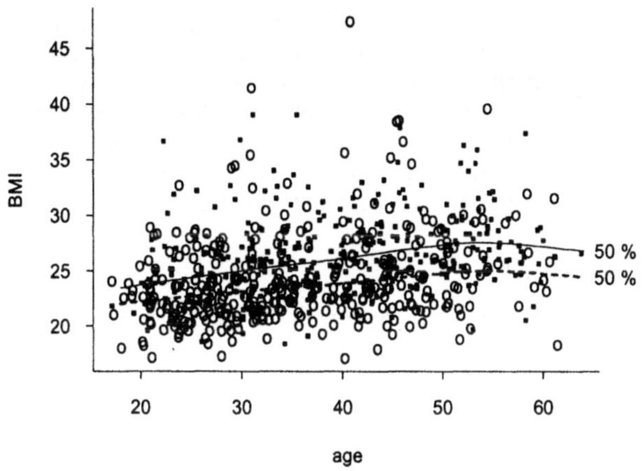

Fig. 4. Output of `qtplot()` when there are non-primary variables. The solid line and squares are for men, and the dashed line and circles are for women.

5 Discussion

Quantile regression is an interesting and useful type of regression that allows modelling the entire distribution of a response rather than just through its mean function. It is a large topic and it has a wide range of applications. In this paper we have focussed on estimating quantiles given a random sample from a reference population. However, another type of related data that is often collected is extreme value data where the R_i largest values of some phenomenon is measured at each value of a covariate x_i. Here are some examples.

1. For each year from 1931 to 1981 the 10 highest daily sea levels (cm) at Venice are recorded. These values correspond to the 10 worse floodings each year.
2. The top 10 runners in each age group (integer valued) in a school are used to estimate the 99 percentile of running speed as a function of age.

Compared to the simple random sampling scheme assumed in this paper, this type of data consists of the top portion of the data set only and omitting the vast majority of the data. This censoring often occurs from the literature because tables only include the most extreme values, e.g., worse disasters, largest insurance claims, most exceptional people. Fortunately, the VGAM framework based on penalized likelihood can accommodate such problems too! Such extreme value data can be modelled using the Gumbel distribution, which is a generalized extreme value distribution with $\xi = 0$. Then one can apply a penalized likelihood argument similar to the LMS method to obtain extreme centiles (e.g., $\alpha = 0.99$) based on such data. See Rosen and Cohen (1996), who extended the work of Smith (1986). VGAM family functions have been written to implement their method, with a methods function for `qtplot()` applicable as well.

The ability to handle extreme value data illustrates some of the capability of the VGAM framework: its sheer breadth enables many models to be encompassed, and therefore can be fitted using a function such as `vglm()` and `vgam()`, often with only the need to write a single VGAM family function.

One disadvantage of the LMS method is that it is difficult for it to handle bimodal data. For example, Hyndman et al. (1996) show that, during the summer in Melbourne, the distribution of temperature for a given day, given the previous day's temperature, is bimodal. This requires further investigation. The ability for `qtplot()` to plot derivatives is also worth attempting.

Acknowledgements

The author wishes to thank Professor Peter Green for providing a copy of the Lopatatzidis and Green manuscript.

References

Buchinsky, M. (1998). Recent advances in quantile regression models: a practical guide for empirical research, *The Journal of Human Resources*, **33**, 88–126.

Chambers, J.M. (1998). *Programming with Data: A Guide to the S Language.* New York: Springer-Verlag.

(eds.) Chambers, J.M. and Hastie, T.J. (1993). *Statistical Models in S.* Pacific Grove, CA.: Wadsworth and Brooks/Cole.

Chitty, L.S., Altman, D.G., Henderson, A. and Campbell, S. (1994). Charts of fetal size: 2, head measurements, *Br. J. Obstetr.*, **101**, 35–43.

Cole, T.J. (1988). Fitting smoothed centile curves to reference data. *J. R. Statist. Soc.* A, **151**, 385–406.

Cole, T.J. and Green, P.J. (1992). Smoothing reference centile curves: the LMS method and penalized likelihood. *Statist. Med.*, **11**, 1305–1319.

Green, P.J. and Silverman, B.W. (1994). *Nonparametric Regression and Generalized Linear Models.* London: Chapman & Hall.

Harris, E.K. and Boyd, J.C. (1995). *Statistical Bases of Reference Values in Laboratory Medicine.* New York: Dekker.

Hyndman, R.J., Bashtannyk, D.M. and Grunwald, G.K. (1996). Estimating and visualizing conditional densities. *J. Comp. Graph. Statist.*, **5**, 315–336.

Koenker, R. and Hallock, K. F. (2001). Quantile regression: an introduction. *Journal of Economic Perspectives*, **15(4)**, 143–156.

Lopatatzidis, A. and Green, P.J. (unpublished manuscript). Semiparametric quantile regression using the gamma distribution.

Rosen, O. and Cohen, A. (1996). Extreme percentile regression. In: Härdle, W. and Schimek, M.G. (eds.), *Statistical Theory and Computational Aspects of Smoothing: Proceedings of the COMPSTAT '94 Satellite Meeting held in Semmering, Austria, 27–28 August 1994*, pp. 200–214, Heidelberg: Physica-Verlag.

Smith, R.L. (1986). Extreme value theory based on the r largest annual events. *J. Hydrology*, **86**, 27–43.

Wright, E.M. and Royston, P. (1997). A comparison of statistical methods for age-related reference intervals. *J. R. Statist. Soc.* A, **160**, 47–69.

Yee, T.W. (1998). On an alternative solution to the vector spline problem, *J. Roy. Statist. Soc.* B, **60**, 183–188.

Yee, T.W. and Wild, C.J. (1996). Vector generalized additive models. *J. Roy. Statist. Soc.* B, **58**, 381–493.

Yeo, I.-K. and Johnson, R.A. (2000). A new family of power transformations to improve normality or symmetry. *Biometrika*, **87**, 954–959.

Computational Methods for Time Series Analysis

Genshiro Kitagawa[1,2] Tomoyuki Higuchi[1,2] and Seisho Sato[1]

[1] The Institute of Statistical Mathematics
4-6-7 Minami-Azabu, Minato-ku, Tokyo 106-8569 Japan
[2] Graduate University for Advanced Study

Abstract. By the progress of fast computing facilities and various computing technologies, it becomes realistic to apply computer intensive methods to statistical analysis. In time series analysis, sequential Monte Carlo methods was developed for general state space models which enables to consider very complex nonlinear non-Gaussian models.

In this paper, we show algorithms, implementations and parameter estimation for Monte Carlo filter and smoother. Various ways of the use of parallel computer are also discussed. The usefulness of the general state space modeling is illustrated with several examples.

Keywords. General state space model, nonlinear filtering, sequential Monte Carlo method, self-organizing state space model, parallel computation

1 Introduction

In statistical time series analysis, a model is built based on the data and prior knowledge on the object and the objective of the analysis. By using a proper model, it is possible to combine various knowledge on the object and the information from other and current data sets, and thus can extract essential information from the data. This is the main feature of statistical information processing procedures.

In time series analysis, the prior knowledge on the dynamics of the signal process and the mechanism of the observation process can usually be combined into state space model form. Therefore, many important problems in time series analysis can be solved by using the ordinary state space model, and various nonstationary time series models have been developed by using the state space models (Harrison and Stevens 1976, Harvey 1989, Kitagawa and Gersch 1996). However, there are many situations where the ordinary time series model is insufficient and a more general nonlinear model or non-Gaussian model is required (Kitagawa 1987). By the recent progress of computing ability, the use of nonlinear or non-Gaussian time series model is becoming realistic.

In this paper, we first review a sequential Monte Carlo based non-Gaussian filtering and smoothing method for the estimation of the state of the nonlinear non-Gaussian state space model is shown (Gordon et al. 1993, Kitagawa 1996, Doucet et al. 2001). In this method, arbitrary distributions of the state and the noise are expressed by many particles. Then it is possible to develop a recursive filter and smoother for very complex nonlinear non-Gaussian state space models. To mitigate the difficulty associated with the use of Monte Carlo filter and the non-Gaussian filter, a self-organizing state space model has been developed. We then consider the use of parallel computation for the sequential Monte Carlo filtering and the parameter estimation for the general state space models.

2 General State Space Model and Sequential Monte Carlo Filtering

2.1 General State Space Model

Consider the linear Gaussian state space model for the time series y_n,

$$x_n = F_n x_{n-1} + G_n v_n \tag{1}$$
$$y_n = H_n x_n + w_n, \tag{2}$$

where x_n is an unknown state vector, v_n and w_n are the system noise and the observation noise with $v_n \sim N(0, Q_n)$ and $w_n \sim N(0, R_n)$, respectively. The initial state x_0 is assumed to be distributed as $x_0 \sim N(x_{0|0}, V_{0|0})$. (1) and (2) are called the system (or state) model and the observation model, respectively. It is well-known that this model has been used in the modeling of various types of time series, such as the nonstationary time series in the mean, in the variance and in the covariance (Harrison and Stevens 1976, Harvey 1989, Kitagawa and Gersch 1996).

This model can be generalized to a nonlinear version

$$x_n = F_n(x_{n-1}, v_n) \tag{3}$$
$$y_n = H_n(x_n, w_n), \tag{4}$$

where the densities for the system noise and the observation noise v_n and w_n are given by $q_n(v)$ and $r_n(w)$, respectively. $F_n(x, v)$ and $H_n(x, w)$ are possibly nonlinear functions of the state and the noise inputs. Nonstationary time series with abrupt structural changes, outliers, non-Gaussian observation distribution, and nonlinear dynamics can be treated with this model (Kitagawa and Gersch 1996).

The above nonlinear non-Gaussian state space model specifies the conditional density of the state given the previous state, $q(x_n|x_{n-1})$, and the that of the observation given the state, $r(y_n|x_n)$. This is the essential features of the state space model, and it is sometimes convenient to express the model in this general form based on conditional distributions

$$x_n \sim Q_n(\,\cdot\,|x_{n-1}) \tag{5}$$
$$y_n \sim R_n(\,\cdot\,|x_n). \tag{6}$$

With this model, it is possible to treat a discrete process or a discrete state model, such as the inhomogeneous Poisson process or the hidden Markov model.

2.2 State Estimation and Prediction

Many important problems in time series analysis such as the likelihood computation for parameter estimation, the prediction, the interpolation and the decomposition of a time series can be solved via estimating the state. Therefore, the essential problem in state space modeling is the estimation of the state x_n from the observations.

The problem of state estimation can be formulated as an evaluation of the conditional density $p(x_n|Y_t)$, where Y_t is the set of observations $\{y_1, \ldots, y_t\}$.

Corresponding to the three distinct cases, $n > t$, $n = t$ and $n < t$, the state estimation problem can be classified into three categories where the conditional density $p(x_n|Y_t)$ is called the predictor, the filter and the smoother, respectively.

For the standard linear-Gaussian state space model, each density can be expressed by a Gaussian density and its mean vector and the variance-covariance matrix can be obtained by computationally efficient recursive formulas such as the Kalman filter and smoothing algorithms (Anderson and Moore 1979).

2.3 Non-Gaussian Filtering

For nonlinear non-Gaussian state space models or general state space models, however, the conditional distributions are non-Gaussian and various types of approximations to or assumptions on the densities are used to obtain recursive formulas for state estimation. Some of the examples are the extended Kalman filter (Anderson and Moore 1979), the Gaussian-sum filter (Alspach and Sorenson 1972), the dynamic generalized linear model (West et al. 1985).

However, the following non-Gaussian filter and smoother can yield an arbitrarily precise posterior density.

[Non-Gaussian Filter]

$$\begin{aligned} p(x_n|Y_{n-1}) &= \int p(x_n|x_{n-1})p(x_{n-1}|Y_{n-1})dx_{n-1} \\ p(x_n|Y_n) &= \frac{p(y_n|x_n)p(x_n|Y_{n-1})}{p(y_n|Y_{n-1})}, \end{aligned} \tag{7}$$

where $p(y_n|Y_{n-1})$ is the predictive distribution of y_n and is defined by $\int p(y_n|x_n)p(x_n|Y_{n-1})dx_n$.

[Non-Gaussian Smoother]

$$p(x_n|Y_N) = p(x_n|Y_n)\int \frac{p(x_{n+1}|x_n)p(x_{n+1}|Y_N)}{p(x_{n+1}|Y_n)}dx_{n+1}. \tag{8}$$

These formula can be realized by using the step function or piecewise linear function approximation of the related densities and the numerical integration (Kitagawa 1987). This method can yield very precise However, the direct implementation of the formula requires computationally very costly numerical integration and can be applied only to lower dimensional state space models.

2.4 Monte Carlo Filtering

In the Monte Carlo filtering, we approximate each density function by many particles which can be considered as realizations from that distribution (Gordon et al. 1993, Kitagawa 1996). Specifically, assume that each distribution is expressed by using m (m=10,000, say) particles as follows:

$$\begin{aligned} &\{p_n^{(1)},\ldots,p_n^{(m)}\} \sim p(x_n|Y_{n-1}) && \text{Predictor} \\ &\{f_n^{(1)},\ldots,f_n^{(m)}\} \sim p(x_n|Y_n) && \text{Filter} \\ &\{s_{n|N}^{(1)},\ldots,s_{n|N}^{(m)}\} \sim p(x_n|Y_N) && \text{Smoother} \end{aligned}$$

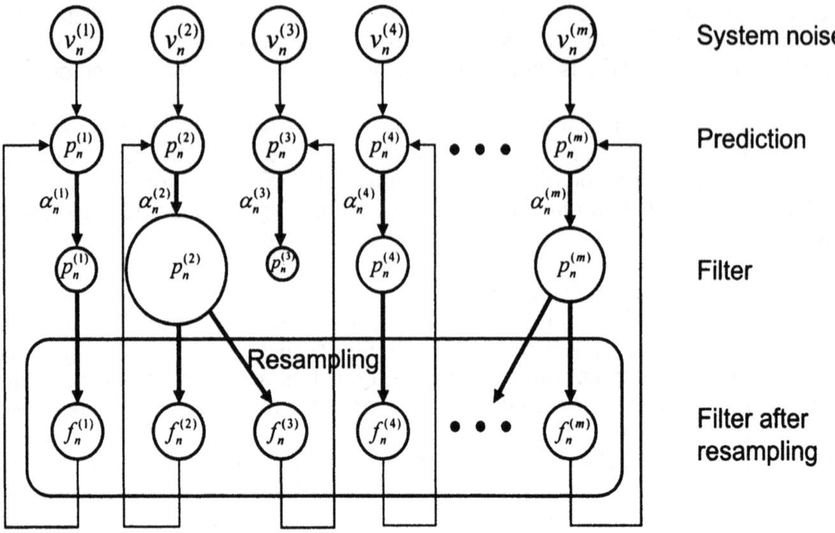

Fig. 1. Monte Carlo Filter

Namely, we approximate the distributions by the empirical distributions determined by m particles. This means that $p(x_n|Y_{n-1})$ is approximated by the probability function

$$\Pr(x_n = p_n^{(j)}|Y_{n-1}) = \frac{1}{m}, \qquad \text{for } j = 1, \cdots, m. \tag{9}$$

Then it can be shown that a set of realizations expressing the one step ahead predictor $p(x_n|Y_{n-1})$ and the filter $p(x_n|Y_n)$ can be obtained recursively as follows.

[**Monte Carlo Filter**]

1. *Generate a random number* $f_0^{(j)} \sim p_0(x)$ *for* $j = 1, \ldots, m$.
2. *Repeat the following steps for* $n = 1, \ldots, N$.
   (a) *Generate random numbers* $v_n^{(j)} \sim q(v)$, *for* $j = 1, \ldots, m$.
   (b) *Compute* $p_n^{(j)} = F(f_{n-1}^{(j)}, v_n^{(j)})$, *for* $j = 1, \ldots, m$.
   (c) *Compute* $\alpha_n^{(j)} = p(y_n|p_n^{(j)})$ *for* $j = 1, \ldots, m$.
   (d) *Generate* $f_n^{(j)}$, $j = 1, \ldots, m$ *by the resampling of* $p_n^{(1)}, \ldots, p_n^{(m)}$.

2.5 Likelihood of the Model and Parameter Estimation

The state space model usually contains several unknown structural parameters such as the variances of the noises and the coefficients of the functions F_n and H_n. The vector consisted of such unknown parameters is hereafter denoted by θ. Then $p(x_n|Y_{n-1}, \theta)$ defines the density of the one-step-ahead predictor. On the other hand, in general, the likelihood of time series model specified by the parameter θ is obtained by

$$L(\theta) = p(y_1, \cdots, y_N|\theta) = \prod_{n=1}^{N} p(y_n|Y_{n-1}, \theta), \tag{10}$$

where $p(y_n|Y_{n-1}, \theta)$ is the conditional density of y_n given Y_{n-1} and is obtained by

$$p(y_n|Y_{n-1}, \theta) = \int p(y_n|x_n, \theta)p(x_n|Y_{n-1}, \theta)dx_n. \tag{11}$$

When we used the Monte Carlo filter, the conditional density in the right hand side of (11) can be approximated by

$$p(y_n|Y_{n-1}, \theta) = \frac{1}{m}\sum_{j=1}^{m} \alpha_n^{(j)}. \tag{12}$$

The maximum likelihood estimate of the parameter θ is then obtained by maximizing the log-likelihood

$$\ell(\theta) = \log L(\theta) = \sum_{n=1}^{N} \log p(y_n|Y_{n-1}, \theta). \tag{13}$$

If there are several candidate models, we can evaluate the goodness of the models and find the best one by finding the model with smallest value of AIC defined by

$$\text{AIC} = -2\text{max}\ell(\theta) + 2\#(\theta) \tag{14}$$

where $\#(\theta)$ denotes the dimension of the parameter θ.

2.6 Self-Organizing Representation

If the non-Gaussian filter is implemented by the Monte Carlo filter, the sampling error due to the approximation (13) sometimes renders the maximum likelihood method impractical. Also, in the actual time series modeling, structural parameters are not constant over time and we sometimes need to consider that the parameter $\theta = \theta_n$ evolves with time n. In these cases, instead of estimating the parameter θ by the maximum likelihood method, we consider a Bayesian estimation by augmenting the state vector as

$$z_n = \begin{bmatrix} x_n \\ \theta_n \end{bmatrix}. \tag{15}$$

The state space model for this augmented state vector z_n is given by

$$\begin{aligned} z_n &= F^*(z_{n-1}, v_n) \\ y_n &= H^*(z_n, w_n) \end{aligned} \tag{16}$$

where the nonlinear functions $F^*(z, v)$ and $H^*(z, w)$ are defined by $F^*(z, v) = [F(x, v), \theta_n]^T$, $H^*(z, w) = H(x, w)$.

Assume that we obtain the posterior distribution $p(z_n|Y_N)$ given the entire observations $Y_N = \{y_1, \cdots, y_N\}$. Since the original state vector x_n and the parameter vector θ_n are included in the augmented state vector z_n, it immediately yields the marginal posterior densities of the parameter and of the original state.

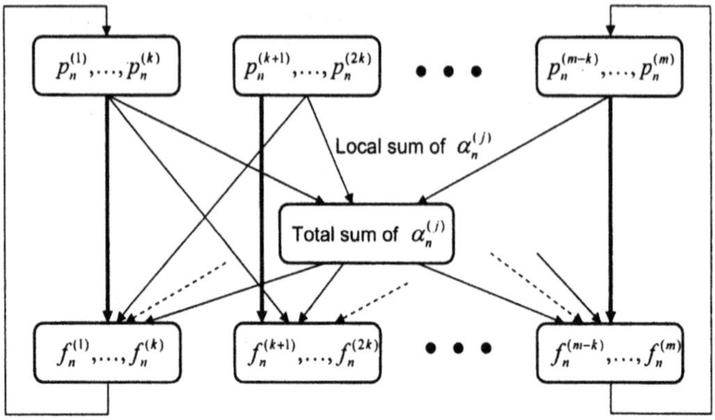

Fig. 2. Parallel implementation of Monte Carlo filter

3 Parallel Computation

3.1 Direct Parallelization of Monte Carlo Filter

The Monte Carlo filter and smoother can be parallelized. In particular, in the prediction step, particles can be generated independently in each node (Figure 2). However, in the resampling of the filtering step, many communications between nodes occur. Therefore, for usual Monte Carlo filter, where the amount of computations for one particle is small, a direct parallelization will deteriorate the scalability of the computation.

One exception occurs when the computation for one step is very heavy. Takahashi and Sato (2000) used parallel implementation of the Monte Carlo filter and smoother for estimating the term structure of interest rates for which explicit evaluation of the Bayesian weight $\alpha_n^{(j)}$ is difficult and they used a Monte Carlo simulation method for approximating it.

3.2 Parallel Monte Carlo Filter with Crossover

It is also possible to run many Monte Carlo filters in parallel. It that case, the distributions obtained from these filters approximate the same true distribution. Therefore, it will be natural to crossover the Monte Carlo filters by exchanging the particles of two Monte Carlo filters that are selected at random. Another possible way of crossovering Monte Carlo filters are to add small portion of the particles of other Monte Carlo filters to each Monte Carlo filter (Figure 3). Further research is necessary to evaluate whether this method accutally has a merit in computational effciency or accuracy compared with a signle Monte Carlo fitler.

3.3 Parallelization of Gaussian-sum Filter

Gaussian-sum filter and smoother were developed to approximate the non-Gaussian or nonlinear filter/smoother for higher order state space models (Alspach and Sorenson 1972, Harrison and Stevens 1976 and Anderson and Moore 1979). In Kitagawa (1994), it was shown that such a Gaussian-sum filter can be easily derived from the non-Gaussian filter by using Gaussian mixture approximations to the related densities. Specifically, the following

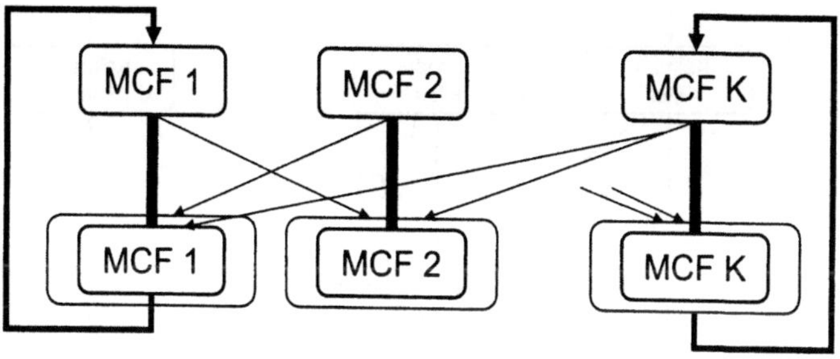

Fig. 3. Parallel Monte Carlo filter

approximations were used:

$$p(v_n) = \sum_{i=1}^{K_v} \alpha_i \varphi_i(v_n), \qquad p(w_n) = \sum_{j=1}^{K_w} \beta_j \varphi_j(w_n)$$

$$p(x_n|Y_{n-1}) = \sum_{k=1}^{L_n} \gamma_{kn} \varphi_k(x_n|Y_{n-1}), \qquad p(x_n|Y_n) = \sum_{\ell=1}^{M_n} \delta_{\ell n} \varphi_\ell(x_n|Y_n) \tag{17}$$

Here φ_i denotes a properly defined Gaussian density. Substituting these into (7), we immediately obtain the following algorithm for the Gaussian-sum filtering.

One step ahead prediction:

$$p(x_n|Y_{n-1}) = \sum_{i=1}^{K_v} \sum_{\ell=1}^{M_{n-1}} \gamma_{i\ell,n} \varphi_{i\ell}(x_n|Y_{n-1}) \equiv \sum_{k=1}^{L_n} \gamma_{k,n} \varphi_j(x_n|Y_{n-1}). \tag{18}$$

Here $\gamma_{i\ell,n} = \alpha_i \delta_{\ell,n-1}$, $L_n = K_v M_{n-1}$ and $\varphi_{i\ell}(x_n|Y_{n-1})$ is the one-step-ahead predictor of x_n obtained by the Kalman filer assuming that filter distribution of x_{n-1} is $\varphi_\ell(x_{n-1}|Y_{n-1}) \sim N(x^\ell_{n-1|n-1}, V^\ell_{n-1|n-1})$ and that $p(v_n) = \varphi_i(v_n) \sim N(0, Q_i)$. Therefore, $\varphi_{i\ell}(x_n|Y_{n-1})$ is also Gaussian and its mean and covariance can be obtained by the ordinary Kalman predictor:

$$\begin{aligned} x^{i\ell}_{n|n-1} &= F_n x^\ell_{n-1|n-1} \\ V^{i\ell}_{n|n-1} &= F_n V^\ell_{n-1|n-1} F^T + G_n Q_i G_n^T. \end{aligned} \tag{19}$$

Filtering:

$$p(x_n|Y_n) \propto \sum_{j=1}^{K_w} \sum_{k=1}^{L_n} \delta_{jk,n} \varphi_{jk}(x_n|Y_n) \equiv \sum_{\ell=1}^{M_n} \delta_{\ell n} \varphi_\ell(x_n|Y_n) \tag{20}$$

Here, $\delta_{jk,n} = \beta_j \gamma_{kn} \varphi_{jk}(y_n|Y_{n-1})$, $M_n = K_w L_n$ and $\varphi_{jk}(x_n|Y_n) \sim N(x^{jk}_{n|n}, V^{jk}_{n|n})$ is the filter of x_n obtained under the assumption that the one-step-ahead predictor of x_n is $N(x^k_{n|n-1}, V^k_{n|n-1})$ and $p(w_n) = \varphi_j(w_n) \sim N(0, \sigma_j^2)$. Therefore,

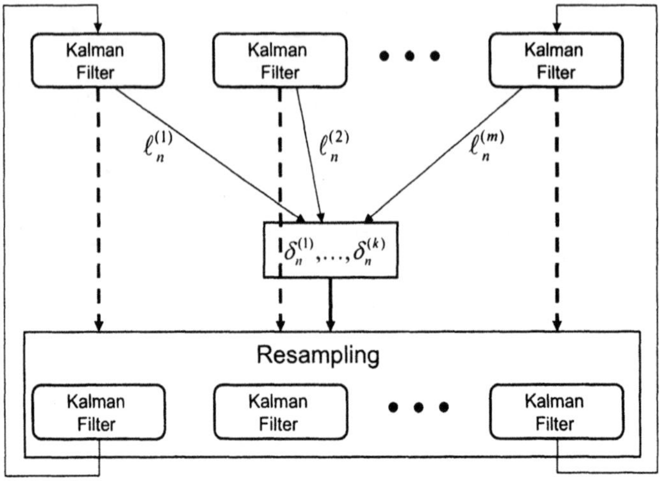

Fig. 4. Parallel implementation of Gaussian-sum filter

it can be obtained by the following Kalman filter:

$$\begin{aligned} K_n^{jk} &= V_{n|n-1}^k H_n^T (H_n V_{n|n-1}^k H_n^T + \sigma_j^2)^{-1} \\ x_{n|n}^{jk} &= x_{n|n-1}^k + K_n^{jk}(y_n - H_n x_{n|n-1}^k) \\ V_{n|n}^{jk} &= (I - K_n^{jk} H_n) V_{n|n-1}^k. \end{aligned} \tag{21}$$

The Gaussian-sum filter/smoother is suited for parallel computation, since 10 to 100 Gaussian components are usually sufficient. In the Gaussian-sum filters developed so far, in the corrupsing stage namely the stage that reduces the number of Gaussian components, various ad hoc methods were used (Figure 4). However, this step can be realized by the similar method as the resampling in the Monte Carlo filter.

3.4 Parallel Grid Search for Hyper-Parameters of the Model

Even when we use the self-organizing state space model for simultaneous estimation of the state and the unknown parameters, it still remains to determine the hyper-parameter(s) which controls the smoothness of the time change of the parameters. However, our experiences suggest that the fine-tuning of this hyper-parameter is unnecessary because a small change of the hyper-parameter will not yield a significant change of the state estimates. Therefore, we determine the best value of the hyper-parameter by a coarse grid search. Since no communications are necessary except for the parameter value and the log-likelihood, the computation for the grid search can be implemented to parallel computation very efficiently (Figure 5 and Higuchi 2001).

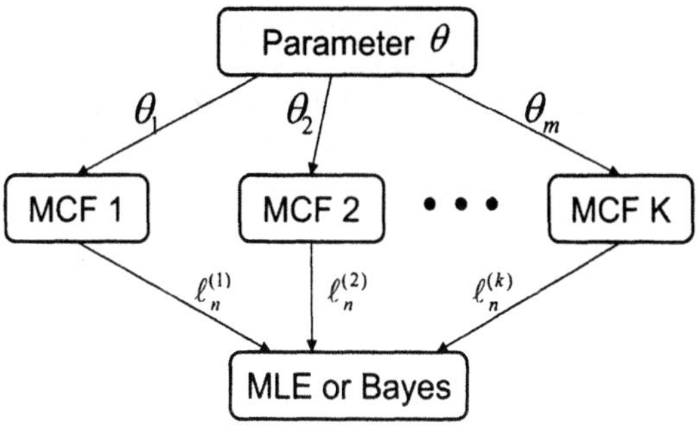

Fig. 5. Parallel parameter estimation

4 Conclusion

General state space model is useful tool for expressing and computing for Bayesian time series modeling. Recursive filtering and smoothing formula corresponding to the Kalman filter and smoother for linear Gaussian model can be obtained even for the general state space model, and can be realized by a sequential Monte Carlo approximation. The Monte Carlo filter also enables the realization of the self-organizing state space model which can simultaneously estimate the unknown state vector and the structural parameters of the general state space model.

There are several ways of parallelizing the Monte Carlo filter/smoother and the self-organizing state space model. In this paper, we considered the direct parallelization of the Monte Carlo filter, parallel Monte Carlo filter with crossover, parallelization of Gaussian-sum filter/smoother and parallel grid search for the hyper-parameters of the general state space model.

References

Alspach, D. L., and Sorenson, H. W., (1972). Nonlinear Bayesian Estimation Using Gaussian Sum Approximations. *IEEE Transactions on Automatic Control*, **AC-17**, 439–448.

Anderson, B. D. O., and Moore, J. B., (1979). *Optimal Filtering.* Prentice-Hall, New Jersey.

Andrede Netto, M. L., Gimeno, L. and Mendes, M. J., (1978). On the optimal and suboptimal nonlinear filtering problem for discrete-time systems, *IEEE Trans. Automat. Control*, **23**, 1062–1067.

Doucet, A., de Freitas, N. and Gordon, N., (2001). *Sequential Monte Carlo Methods in Practice.* Springer-Verlag, New York.

Gordon, N. J., Salmond, D. J. and Smith, A. F. M., (1993). Novel approach to nonlinear/non-Gaussian Bayesian state estimation, *IEE Proceedings–F*, **140**, 107–113.

Harrison, P.J. and Stevens, C.F. (1976). Bayesian Forcasting, *Journal of the Royal Statistical Society*, Series B, **38**, 205–247.

Harvey, A.C. (1989). *Forecasting structural time series models and the Kalman filter*, Cambbridge University Press, Victoria, Australia.

Higuchi, T. (1997). Monte Carlo filter using the Genetic algorithm operators, *Journal of Statistical Computation and Simulation*, **59**, 1–23.

Higuchi, T. (2001). Evolutionary Time Series Model with Parallel Computing, *Proceedings of the 3rd Japan-US Joint Seminar on Statistical Time Series Analysis*, Kyoto, Japan, June 18-22, 2001, 183–190.

Kitagawa, G., (1987). Non-Gaussian state-space modeling of nonstationary time series (with discussion), *Journal of the American Statistical Association*, **82**, 1032–1063.

Kitagawa, G., (1991). A nonlinear smoothing method for time series analysis, *Statistica Sinica*, **1**, 371–388.

Kitagawa, G., (1994). The two-filter formula for smoothing and an implementation of the Gaussian-sum smoother, *Annals of the Institute of Statistical Mathematics*, **46**, 605–623.

Kitagawa, G., (1996). Monte Carlo filter and smoother for non-Gaussian nonlinear state space model, *Journal of Computational and Graphical Statistics*, **5**, 1–25.

Kitagawa, G., (1998). Self-Organizing State Space Model, *Journal of the American Statistical Association*, **93**, 1203–1215.

Kitagawa, G. and Gersch, W., (1996). *Smoothness Priors Analysis of Time Series*, Springer-Verlag, New York.

Ljung, L., (1979). Asymptotic behavior of the extended Kalman filter as a parameter estimator for linear systems, *IEEE Transactions on Automatic Control*, **AC–24**, 36–50.

Solo, V., (1980). Some aspect of recursive parameter estimation, *International Journal of Control*, **32**, 395–410.

Takahashi, A. and Sato, S. (2001). A Monte Carlo filtering approach for estimating the term structure of interest rates, *Annals of the Institute of Statistical Mathematics*, **53**, 50–62.

West, M., Harrison, P. J. and Migon, H. S., (1985). Dynamic generalized linear models and Bayesian forecasting (with discussion), *Journal of the American Statistical Association*, **80**, 73–97.

Forecasting PC-ARIMA Models for Functional Data

Mariano J. Valderrama, Francisco A. Ocaña and Ana M. Aguilera

Department of Statistics and Operations Research, 18071-Granada, SPAIN

Abstract. This paper introduces an improvement on the forecasting models previously developed by the authors for continuous time series based on the PCA of the stochastic process by cutting series in seasonal periods. The new approach consists of modelling principal components as ARIMA processes and then to formulate a mixed PC-ARIMA model for the time series. This methodology is then applied to the climatic phenomenon known as *El Niño*.

Keywords. Principal components, functional data, ARIMA, *El Niño*.

1 Introduction

In many statistical applications the sample information about a continuous time series is given by discrete time observations not always evenly spaced. Because of this we are interested in developing a forecasting model for such a time series that takes into account the continuous nature of data and avoids some of the restrictive hypotheses imposed by classical approaches.

The model proposed in this paper is based on functional principal component analysis (FPCA) of the sample curves obtained by cutting the series into pieces and on classical ARIMA modelling of the associated principal components that can be seen as a set of discrete time series. This mixed PC–ARIMA model allows not only to forecast a continuous time series in a whole future interval but also to reconstruct it between the discretization time points in the past.

FPCA (Ramsay and Silverman, 1997) is a generalization of the classic principal component analysis (PCA) of a sample of data vectors for the reduction of dimension of a set of sample curves (realizations of a stochastic process). Alternative models for continuous time series based on FPCA are the autoregressive Hilbertian models (ARH) introduced by Bosq (1991) and the PCP models proposed in Aguilera *et al.* (1997, 1999a, 1999b) and Valderrama *et al.* (2000).

At the end of this paper a mixed PC–ARIMA model will be applied to model and forecast a climatic time series describing *El Niño* natural phenomenon. The forecasting performance of the adjusted mixed PC-ARIMA model will be finally compared with classical ARIMA models.

[1]This research was supported by Project No. BFM2000-1466 of *Dirección General de Investigación, Ministerio de Ciencia y Tecnología*, Spain

2 The problem settings

Let $\{x(t) : t \in \mathbb{R}_0^+\}$ be a continuous time series which is only observed at discrete times,

$$\tau_1 < \tau_2 < \ldots < \tau_P \leq T_f\,, \tag{1}$$

until an instant T_f. As it has said before, the objective of this paper is to forecast that continuous time series in the future, $\{x(t) : t > T_f\}$.

As in the case of many real time series, let us assume that $\{x(t)\}$ shows a seasonal period $T > 0$.

In order to formulate a forecasting model based on functional principal component analysis, we propose to cut the time series in periods of amplitude T (see Figures 1 and 2) and to obtain a set of sample paths of the following continuous time process defined by rescaling as follows:

$$X(w,s) \overset{def}{=} x(wT + s) \quad \left\{ \begin{array}{c} \forall\, w = 0,1,2,\ldots \\ \forall\, s \in (0,T] \end{array} \right. \tag{2}$$

As we will consider $T_f = (N+1)T$ for simplicity, we can establish that the continuous time process X has been observed in sampled paths ranging from $w = 0,\ldots,N$.

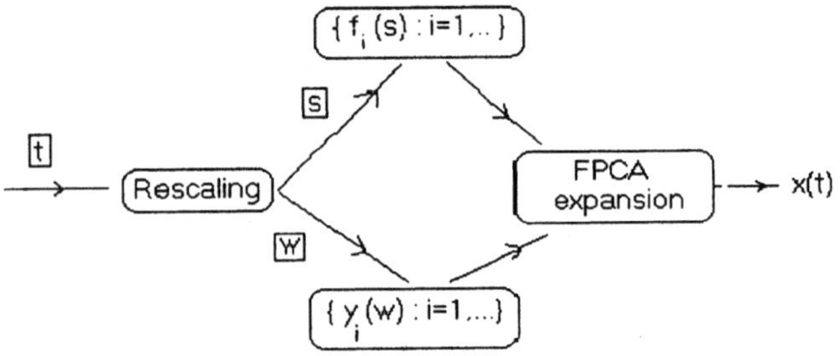

Figure 1: Sketch of functional PCA approach to continuous time series description.

Let us consider that the process obtained by cutting the series into pieces, $\{X(w,s) : s \in (0,T], w = 0,1,\ldots\}$, is a second order and quadratic mean continuous random process whose sample–paths are square integrable functions over the interval $(0,T]$.

By analogy with the multivariate case, the principal components of the process $X(s)$ are defined as uncorrelated generalized linear combinations of the process variables having maximum variance. It can be shown (Ramsay and Silverman, 1997) that the estimation of the *ith* principal component associated with the process $\{X(s)\}$ is obtained as

$$y_i(w) \stackrel{def}{=} \int_0^T \widetilde{X}(w,s)\, f_i(s)\, ds\,, \qquad \forall\; w = 0, \ldots, N\,,$$

where f_i, called the *ith* principal factor (p.f.), is the normalized eigenfunction corresponding to the *ith* largest eigenvalue λ_i (*ith principal value*) of the sample covariance operator $\mathcal{R}$ defined as

$$\mathcal{R}(f)(s_1) = \int_0^T R(s_1, s_2)\, f(s_2) ds_2\,, \quad \forall f \in L^2(0, T],$$

whose kernel is the sample covariance function

$$R(s_1, s_2) \stackrel{def}{=} \frac{1}{N+1} \sum_{w=0}^{N} \widetilde{X}(w, s_1) \widetilde{X}(w, s_2)\,,$$

where $\widetilde{X}(w,s) \stackrel{def}{=} X(w,s) - \mu(s)$ and $\mu(s) \stackrel{def}{=} \frac{1}{N+1} \sum_{w=0}^{N} X(w,s)$, $\forall\; s \in (0,T]$. Therefore, the principal factors are the solutions of the Fredholm second kind integral equation

$$\int_0^T R(s_1, s_2) f_i(s_2)\, ds_2 \;=\; \lambda_i\, f_i(s_1)\,, \quad \forall s_1\,,$$

verifying that they are orthonormal, $\int_0^T f_i(s) f_j(s) ds = \delta_{ij}$, where the associated eigenvalues form a decreasing sequence of positive real numbers

$$\lambda_1 \geq \lambda_2 \geq \ldots \geq 0\,.$$

As a consequence of PCA definition, the principal components are centered and uncorrelated random variables whose variances are the associated eigenvalues. Thus the variance of process X explained by the *ith* principal component is $V_i = \lambda_i / V$, with $V = \sum_{j \geq 1} \lambda_j$ being the total variance of the process in the interval $(0, T]$.

The functional PCA of process X yields to an expansion of the continuous time series $\{x(t) : 0 \leq t \leq T_f\}$ as follows

$$X(w,s) = x(t) \;=\; \mu(s) + \sum_{i \geq 1} y_i(w)\; f_i(s)\,, \quad \left\{ \begin{array}{c} \forall\; w = 0, 1, \ldots, N, \\ \forall\; s \in (0, T], \\ t = wT + s\,. \end{array} \right. \tag{3}$$

Through this expansion the variability of the time series has been then expanded in terms of the seasonal period (see Figure 1). The principal factors (f_i) account the variability in the seasonal period $(0, T]$ and the principal components account the variability among seasonal periods $\{(wT, (w+1)T] \,:\, w = 0, \ldots, N\}$.

Moreover, if we only consider the q first p.c.'s in Equation (3), an approximating expansion of $x(t)$ can be obtained by truncating this series as follows

$$x^{(q)}(t) \;=\; \mu(s) + \sum_{i=1}^{q} y_i(w)\; f_i(s)\,, \quad \left\{ \begin{array}{c} \forall\; w = 0, 1, \ldots, N, \\ \forall\; s \in (0, T], \\ t = wT + s\,. \end{array} \right. \tag{4}$$

In the stochastic process settings, it is known that this truncated expansion is the best q-dimensional linear model for X. We can thus consider the error term for such a truncated expansion in the time series settings as follows:

$$\epsilon^2 \stackrel{def}{=} \int_0^{T_f} \left[x(t) - x^{(q)}(t)\right]^2 dt \;=\; (N+1)\sum_{i>q} \lambda_i \,.$$

If we define $m : (0, T_f] \longrightarrow \mathbb{R}$ as the periodic extension of μ, it can be defined and proved that

$$\mathrm{Var}[x] \stackrel{def}{=} \int_0^{T_f} [x(t) - m(t)]^2 \, dt = (N+1)\sum_{i\geq 1} \lambda_i \qquad \text{and}$$

$$\mathrm{Var}[x^{(q)}] \;\stackrel{def}{=}\; \int_0^{T_f} \left[x^{(q)}(t) - m(t)\right]^2 dt = (N+1)\sum_{i=1}^{q} \lambda_i \,.$$

Therefore, we have

$$\mathrm{Var}[x] \;=\; \mathrm{Var}[x^{(q)}] + \epsilon^2 \,,$$

where the variability of $x(t)$ is decomposed as a sum of the variability of the truncated series $x^{(q)}$ and the squared error term ϵ^2. The rate of variability explained by the truncated series is then obtained by the variance explained by the first q p.c.'s as follows

$$\frac{\mathrm{Var}[x^{(q)}]}{\mathrm{Var}[x]} \;=\; \frac{\sum_{i=1}^{q} \lambda_i}{\sum_{i\geq 1} \lambda_i} \;=\; \sum_{i=1}^{q} V_i \,.$$

In practice, the original continuous time series $\{x(t)\}$ is only observed at discrete values, as it was said in Equation (1). So we only have discrete observations of each sample path $X(w, *)$, this means that information on the sample continuous curves is only provided by discrete observations. In order to estimate FPCA we propose to define an approximating time series given by rescaling the stochastic process whose sample curves are defined by continuous approximations based on numerical techniques, as for example natural cubic spline interpolation (Aguilera *et al.*, 1996).

3 PC-ARIMA forecasting model

Let us observe that FPCA provides an orthogonal expansion of the continuos process $\{X(w, s)\}$ in terms of a set of deterministic functions (principal factors) and a set of random variables (principal components). As it has been remembered in previous section, for the principal components the following conditions hold:

- $\displaystyle \frac{1}{N+1}\sum_{w=0}^{N} y_i(w) = 0,$

- $\frac{1}{N+1}\sum_{w=0}^{N} y_i(w)y_l(w) = \lambda_i\,\delta_{i,l}$.

It is obvious that in our case the sample paths of the stochastic process X are not independent observations, which is an unusual assumption in classic FPCA. Thus each principal component, $\{y_i(w) : w = 0,\ldots,N\}$, can be considered as a sequence of correlated values with w as independent parameter, what yields to consider $\{y_i(w)\}_w$ as a discrete time series with w ranging in seasonal periods.

We propose to model each principal component by an ARIMA model (Box and Jenkins, 1970). The general expression of an ARIMA(p_i, d_i, q_i) model for the *ith* principal component y_i is given by

$$\Phi_i(B)(1-B)^{d_i}y_i(w) = \theta_i(B)\epsilon_i(w), \tag{5}$$

where B is the backward shift operator, $\Phi_i(B)$ is an autoregressive operator defined as

$$\Phi_i(B) = 1 - \phi_{i,1}B - \phi_{i,2}B^2 - \ldots - \phi_{i,p_i}B^{p_i},$$

$\Theta_i(B)$ is the moving average operator given by

$$\Theta_i(B) = 1 - \theta_{i,1}B - \theta_{i,2}B^2 - \ldots - \theta_{i,q_i}B^{q_i},$$

and $\epsilon_i(w)$ is a white noise process for each i.

Roughly speaking, given the continuous time series

$$\{x(t) : 0 \leq t \leq T_f = (N+1)T\} \qquad (w = 0,\ldots,N)$$

with seasonal period T, we try to forecast $\{x(Tw_* + s) : \; s \in (0,T]\}$, for $w_* = N+1, N+2, \ldots$

The prediction model proposed in this paper is based on ARIMA forecasting of each of the first q principal components selected for reconstructing the continuous time series as in Expansion (4). In Figure 2 the PC-ARIMA modelling approach is sketched, when two p.c.'s ($q = 2$) are considered in Expansion (4).

After estimating in the usual form these q ARIMA models formulated in Equation (5), one for each p.c., we will be able to obtain one forecast for each principal component in the future period w_*, denoted by $\widehat{y_i}(w_*)$.

Finally, the original continuous time series $x(t)$ is predicted in the whole future interval as follows

$$\widehat{x}^{(q)}(w_*T + s) \;=\; \mu(s) \;+\; \sum_{i=1}^{q} \widehat{y_i}(w_*)\; f_i(s)\,, \quad \forall\, s \in (0,T]\,. \tag{6}$$

4 Application to climatic data

El Niño[2] is a climatic phenomenon consisting of a disruption of the ocean-atmosphere system in the tropical Pacific having important consequences for

[2] `http://www.pmel.noaa.gov/tao/elnino/el-nino-story.html`

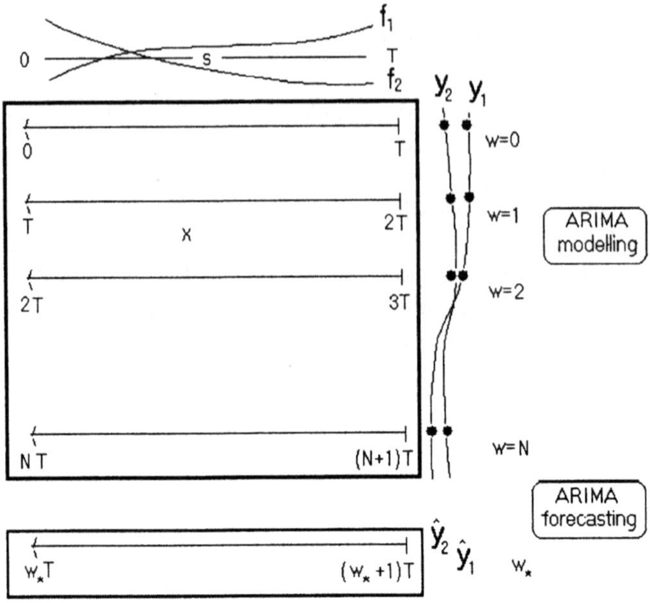

Figure 2: Sketch of PC-ARIMA modelling with q=2.

weather around the globe, such as the increased rainfall across the southern tier of US and in Peru, which has caused destructive flooding, and drought in the West Pacific, sometimes associated with devastating brush fires in Australia. *El Niño* can be seen in sea surface temperature in the Equatorial Pacific Ocean.

In this study data from the *El Niño* series [3] has been monthly considered from January 1950 to July 2001. We can then consider that we have discrete observations given by $\{x(t) : t = 1, 2, \ldots, 619\}$, which can bee seen in Figure 3. A more detailed look of data can be obtained in Figure 4

The seasonal behaviour of *El Niño* series is stated in Figure 5, where the estimated spectral density is plotted against several periods. The spectral density confirms that *El Niño* series exhibits an annual cycle, as is usually assumed for this climatic series. This means that we can consider $T = 12$.

In this section we firstly aim to fit a functional PC-ARIMA models from (discrete) monthly data ranging from 1950 to 1999 (50 years). Thus the adjust of functional PC-ARIMA model is carried out by the discrete observations given by $\{x(t) : t = 1, 2, \ldots, 600\}$. Secondly, with the fitted functional PC-ARIMA model, we are then going to continuously forecast *El Niño* series along the year 2000 and the first seven months of year 2001 (until July 2001). Finally, the forecasting results from the adjusted PC-ARIMA model will be compared to the one obtained by the Box-Jenkins methodology over the knots where we can carry out the comparisons, which are only the discrete instants given by

[3]`http://www.cpc.ncep.noaa.gov/data/indices/`

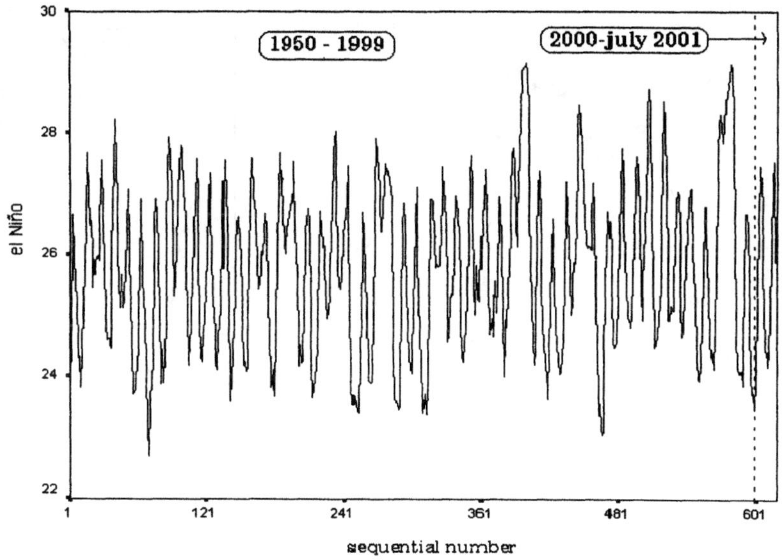

Figure 3: *El Niño* monthly series from 1950 to july 2001. The model fitting is obtained from data ranging 1950 to 1999. Series from 2000 to july 2001 is only used for comparing.

$\{x(t) : t = 601, \ldots, 619\}$.

The fitting of the functional PC-ARIMA model begins by selecting a numerical technique for approximating estimators in the functional PCA. In our case, natural cubic splines interpolating sample paths has been considered for defining the approximating stochastic process X, where as it was established before $T = 12$. The functional PCA of X is then estimated, where the explained variances of p.c.'s appear in Table 1 and the estimated p.f.'s are plotted in Figure 6. As we can see in Table 1, the 96.85% of the total variability of *El Niño* series, $\{x(t)\}$, is explained by the truncated continuous time series $\{x^{(4)}(t)\}$, which is given by the discrete series $\{y_i(w)\}_w$, for $i = 1, \ldots, 4$. The continuous time reconstruction availability for the truncated series $x^q(t)$ (see Equation (4)) is illustrated in Figure 7 for $q = 1$ and $q = 4$, where they have been plotting with the approximated continuous time series $x(t)$ in 1993.

Once the functional PCA has been performed, we have considered four discrete time series $\{y_i(w)\}_w$. Hence, to these series we can apply any forecasting methodology for discrete time series. In our case, we have adjusted the following seasonal ARIMA models:

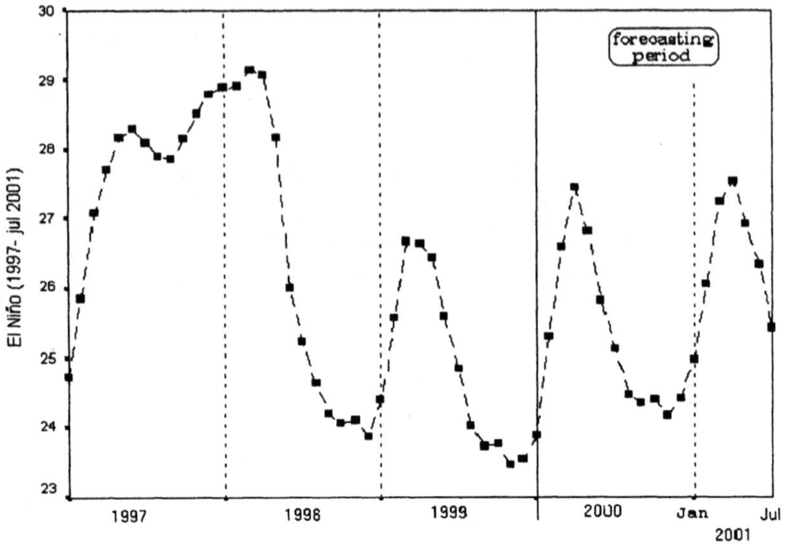

Figure 4: *El Niño* monthly series from 1997 to July 2001.

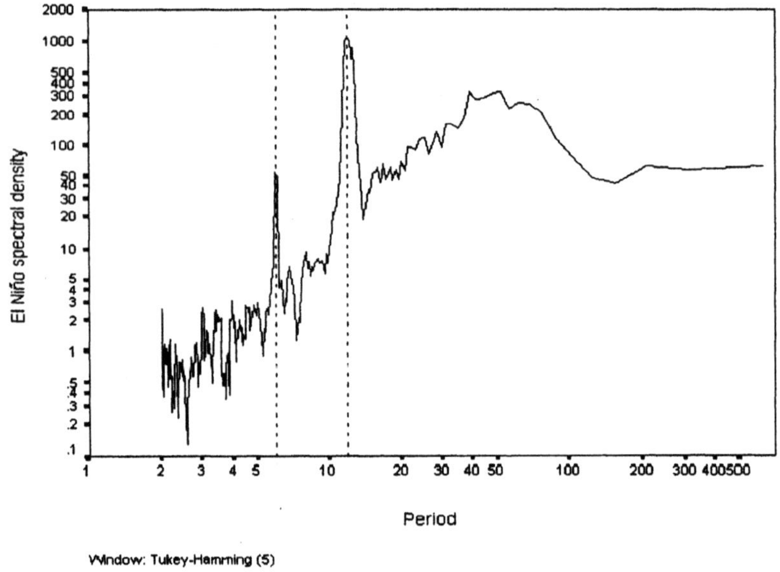

Figure 5: Estimated spectral density of *El Niño* series based on data from 1950 to 1999.

# p.c. (i)	Estimated variance (λ_i)	% Estimated variance (V_i)	% Explained variance $(\sum_{l=1}^{i} V_l)$
1	6.183	68.605	68.605
2	2.064	22.896	91.501
3	0.305	3.383	94.884
4	0.178	1.971	96.855
5	0.122	1.353	98.208
6	0.059	0.652	98.860
7	0.032	0.360	99.220
⋮	⋮	⋮	⋮
12	0.006	0.068	100.000
TOTAL	9.013		100.000

Table 1: Estimated principal component variances of *El Niño* series (1950-1999) based on natural cubic splines interpolation.

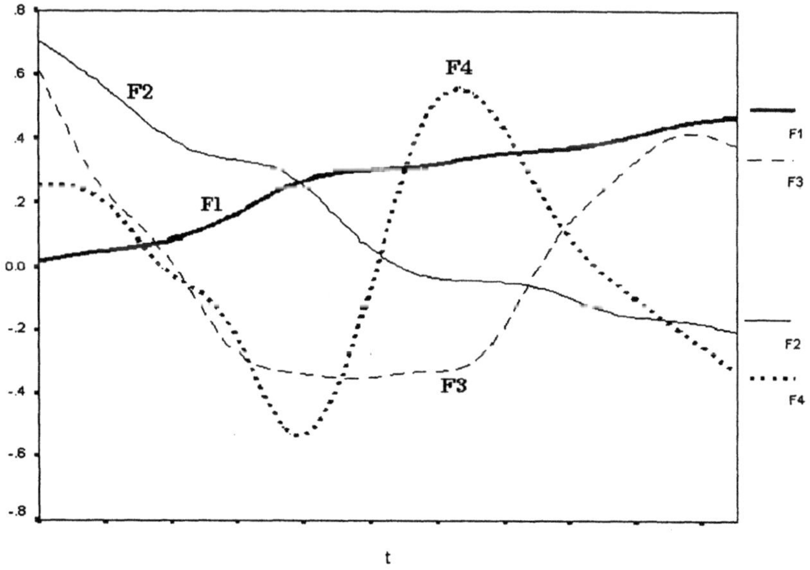

Figure 6: The first four estimated principal factors based on data from 1950 to 1999.

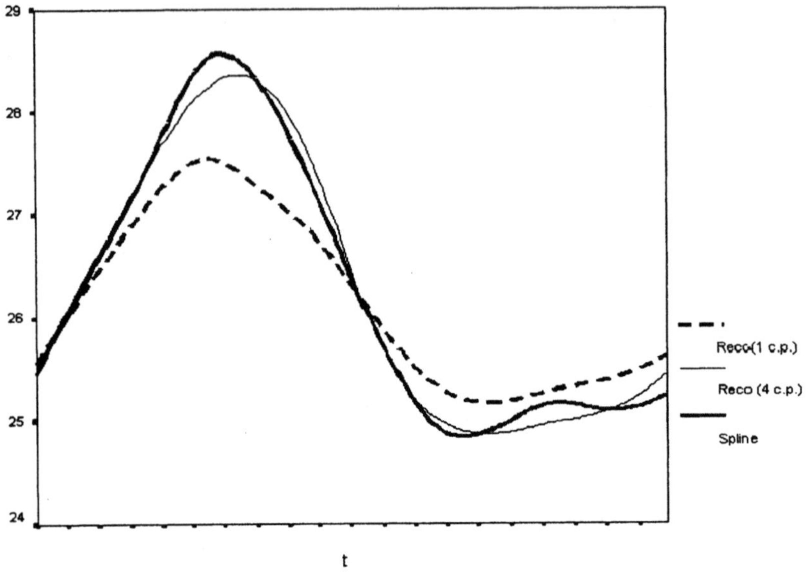

Figure 7: Reconstructions of *El Niño* series in 1993 by the truncated FPCA expansion with $q = 1$ and $q = 4$ and the interpolating spline.

$$\{y_1(w)\}_w \quad \leadsto \quad \text{SARIMA}(0,0,2) \times (0,1,2)_5$$

$$\{y_2(w)\}_w \quad \leadsto \quad \text{SARIMA}(0,0,1) \times (0,1,1)_5$$

$$\{y_3(w)\}_w \quad \leadsto \quad \text{ARIMA}(1,0,1)$$

$$\{y_4(w)\}_w \quad \leadsto \quad \text{SARIMA}(0,1,1) \times (0,1,1)_{11}$$

The previous fitting has been performed from monthly observations in fifty years (1950-1999). The principal components ARIMA models let to forecast the first four p.c.'s in the years 2000 and 2001. Thus it follows from Equation (6) the *El Niño* series forecasting in the interval given by year 2000 and 2001. In Figure 8 the continuous time PC-ARIMA forecasting of *El Niño* series and the real discrete observations are plotted. For comparing the proposed forecasting approach, we have applied the classical Box-Jenkins methodology to *El Niño* series in years 1950-1999,

$$\{x(t) : t = 1, ..., 600\}_t \leadsto \text{SARIMA}(0,1,1) \times (0,1,2)_{12},$$

and we have also plotted the discrete forecasts in the forecasting interval in Figure 8.

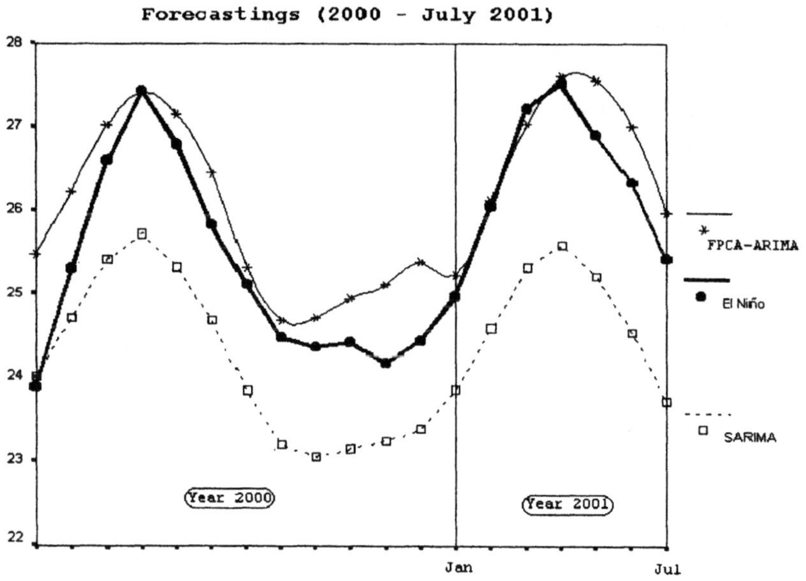

Figure 8: Forecastings of *El Niño* series from January 2000 to July 2001 by the mixed FPCA-ARIMA model and by the classical seasonal ARIMA approach.

References

Aguilera, A.M., Gutiérrez, R. and Valderrama, M.J. (1996). Approximation of estimators in the PCA of a stochastic process using B-splines. *Communications in Statistics (Simulation and Computation)* **25**, 671-691.

Aguilera, A.M., Ocaña, F.A. and Valderrama, M.J. (1997). An approximated principal component prediction model for continuous time stochastic processes. *Applied Stochastic Models and Data Analysis* **13**, 61-72.

Aguilera, A.M., Ocaña, F.A. and Valderrama, M.J. (1999a). Forecasting time series by functional PCA. Discussion of several weighted approaches. *Computational Statistics* **14**, 443-467.

Aguilera, A.M., Ocaña, F.A. and Valderrama, M.J. (1999b). Forecasting with unequally spaced data by a functional principal component approach". *Test* **8**, 233-254.

Besse, P. (1991). Approximation spline de l'ánalyse en composantes principales d'une variable aléatoire hilbertienne. *Annales de la Faculté des Sciences de Toulouse* **12**, 329-346.

Bosq, D. (1991). Modelization, non-parametric estimation and prediction for continuous time processes. In: *Nonparametric functional estimation and related topics*, 509-529, NATO ASI Series.

Box, G.E.P. and Jenkins, G.M. (1970). *Time series analysis forecasting and control.* San Francisco: Holden Day.

Ramsay, J.O. and Silverman, B.W. (1997). *Functional Data Analysis.* New York: Springer-Verlag.

Valderrama, M.J., Aguilera, A.M. and Ocaña, F.A. (2000). *Predicción dinámica mediante análisis de datos funcionales. Introducción a los modelos PCP.* Madrid: La Muralla-Hespérides.

KyPlot as a Tool for Graphical Data Analysis

Koichi Yoshioka

Department of Biochemistry and Biophysics, Graduate School of Allied Health Sciences, Tokyo Medical and Dental University, 1-5-45 Yushima, Bunkyo-ku, Tokyo 113-8519, Japan

Abstract. KyPlot is a software package intended to provide an integrated environment for data analysis and visualization. It offers a broad range of statistical procedures on a spreadsheet interface and versatile graphing tools. KyPlot has the functionality of running a screen show to present the created graphs and schemes. It also supports an interactive system of graph fitting, including nonlinear regression, interpolation and a variety of smoothing methods for nonparametric regression and density estimation.

Keywords. Data visualization, graph fitting, nonparametric regression, density estimation, smoothing methods

1 Introduction

Graphical visualization is an important aspect of data analysis. Graphs allow us to grasp overall patterns of data or find detailed structures of data. Graphs also allow us to view mathematical models fitted to data and check the validity of the models. Iterative exploration is critical for graphical visualization. Graphing needs to be iterative because graphs can assist to discover unidentified aspects of the data, and once we notice them, we can often formulate a new question to the data.

Graphing is also important for data communication, when we want to present information to others. Interactive graphing is necessary in producing high-quality graphs for presentation and publication. Because it is difficult to determine all the specifications of a desired graph beforehand, graphs need to be fully customizable after they are created.

There is thus an increasing need for a software system of data analysis and visualization that is easy to use, capable of creating customizable graphs and has the functionality of interactive graph fitting. Although statistical software systems currently available usually include graphing tools, they may not be adequate in terms of versatility and interactivity.

KyPlot is a software package intended to provide an integrated environment for data analysis and visualization. It offers a broad range of statistical procedures on a spreadsheet interface and versatile tools for creating graphs to visualize the results. Overall functionality of version 2 of KyPlot was described elsewhere (Yoshioka, 2002). In this paper the general design of a new version (version 3) of the software is presented in section 2 and the functionality of graph fitting, particularly, smoothing, is described in section 3.

2 Structure of KyPlot

KyPlot is a stand-alone Windows application with menus, toolbars and tabbed dialog boxes (Figure 1). Within the main window, it contains two types of child windows. One is "`spread`" containing one or more spreadsheets where one can enter, process and analyze data. The functionality of the spreadsheets is similar to that of other spreadsheet applications, supporting cell formulae and formatting. The syntax of cell formulae is compatible with Microsoft Excel and data can be moved about and cross-referenced similarly. As most computer users are familiar with a spreadsheet, they will easily learn how to use the spreadsheet of KyPlot. Furthermore, KyPlot provides a variety of functions for data processing and statistical procedures, including parametric and nonparametric tests, procedures for regression analysis, multivariate analysis, survival analysis and time series analysis (Yoshioka, 2002).

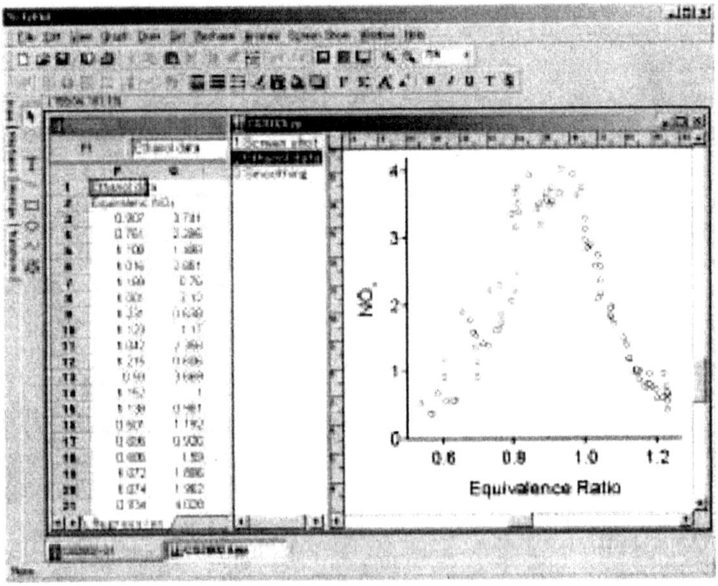

Figure 1. Screen shot of KyPlot showing a spread window containing the ethanol dataset and a scatter plot of the data in a page of a figure window.

The other type of windows is "`figure`" containing one or more pages for drawing graphs and schemes. Figure pages are made object-oriented. The classes of drawing objects supported are line, rectangle, ellipse, B-spline, text, picture and graph. KyPlot has a versatile drawing functionality to create technical drawings: one can move and resize objects with a "select and drag" operation with a mouse; one can cut, copy and paste them and apply operations such as grouping and layering. KyPlot supports many types of 2- and 3-dimensional graphs. The versatility in creating many types of graphs and in customizing graph components is one of the most significant features of KyPlot.

On-screen presentation using a personal computer has been increasingly popular recently. In version 3 of KyPlot, multiple pages can be created in a figure window and the user can perform a screen show of these pages in a full screen mode within the application.

3 Graph fitting functionality of KyPlot

A variety of graph fitting methods are available in KyPlot: nonlinear regression for user-given models based on least squares or maximum likelihood methods, polynomial regression, B-spline fits, piecewise polynomial interpolation and surface interpolation. KyPlot also offers smoothing methods for nonparametric regression and density estimation. KyPlot supports three major methods for smoothing: smoothing splines or penalized likelihood methods, kernel-based or local likelihood methods and basis function methods such as wavelets and Fourier series.

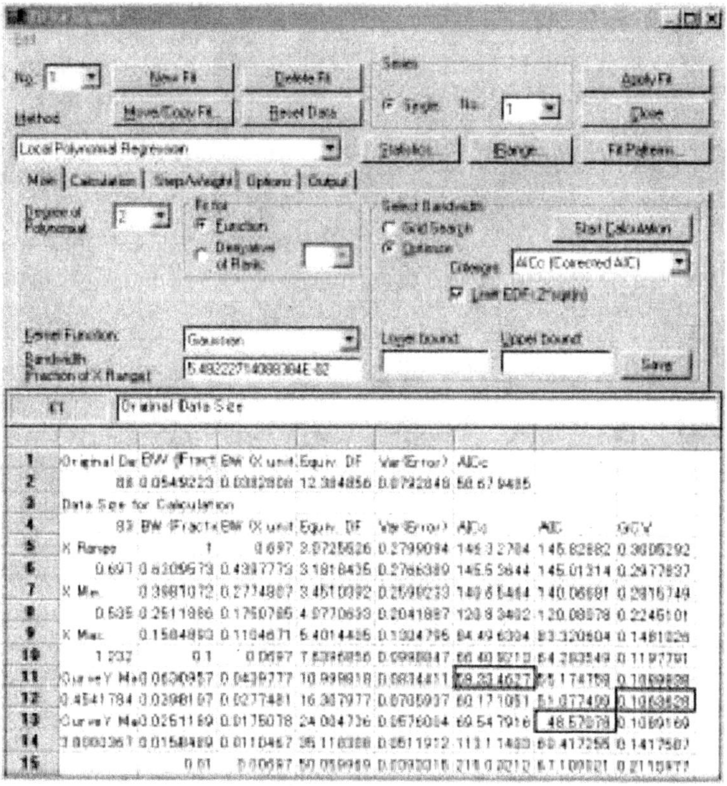

Figure 2. Fit dialog box for local polynomial regression.

Figure 3 shows the results of applying local polynomial least squares regression to the ethanol dataset (Simonoff 1996; Loader 1999). The procedures to perform graph fitting are as follows. The first step of graph fitting is to create a graph of the data: the user selects the range of a dataset in a spreadsheet and opens the [Create Graph] menu, which pops up a dialog box to specify how the data is formatted and then another dialog box to select the graph type and other graph options. After finishing these operations, a graph is created in a figure page (Figure 1). The user then opens the [Graph] – [Fit] menu, which pops up a dialog box for fitting, and selects "Local Polynomial Regression" from the "Method" combo box (Figure 2). As shown in Figure 2, settings for fitting procedures can be done by selecting radio button options, marking check boxes and entering values in text boxes so that no programming or macro is necessary: the user can perform fitting by clicking with mouse and minimal amount of typing.

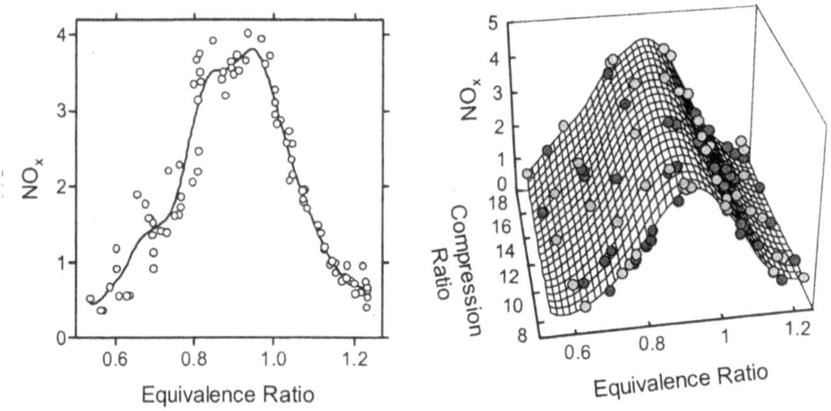

Figure 3. Univariate (A) and bivariate (B) local quadratic fits to the ethanol dataset. For the bivariate fit, a spherically symmetric Gaussian kernel was used; the data points under the fitting surface are darkened. The bandwidths of the local regressors were determined by AICc.

In the Fit dialog box, the user can specify parameters of fitting, such as the degree of local polynomial (2 in this example), type of kernel function (Gaussian in this example), and the value of the bandwidth. Then the user clicks the "Apply Fit" button to start calculation and the fitting curve is drawn on the graph (Figure 3). To determine the value of the bandwidth, selection criteria such as GCV (generalized cross validation), AIC (Akaike's information criterion) and AICc (corrected AIC) are available. The user can perform a grid search for the criteria: in the "Select Bandwidth" frame in the dialog box, selecting the "Grid Search" option and clicking the "Start Calculation" button give a range of bandwidths and the values of these criteria in the spreadsheet of the dialog box (Figure 2). When the "Optimize" option is selected, clicking the Start Calculation button activates an iterative routine to optimize one of the criteria. In Figure 3 the bandwidths of both the univariate and bivariate local quadratic estimates were determined by AICc (Hurvich et al., 1998). These procedures can be repeated until a satisfactory fit is obtained. The

Fit dialog box thus serves as a "toolbox" for interactive graph fitting.
KyPlot provides flexible options for fitting: the dataset used for fitting is the one given in the spreadsheet in the Fit dialog box so that the user may add, exclude or change values of the data for fitting without changing the data for graphing; weights can be assigned to individual data points; and by clicking the "`New Fit`" button, multiple fits can be applied to the same graph.
KyPlot also supports penalized least squares regression: smoothing splines for univariate regression and thin plate splines for bivariate regression. Figure 4 shows a thin plate spline interpolant (A) and smoother (B) to the ore-bearing layer dataset (Green and Silverman, 1996). The smoothing parameter of the splines can be determined in a similar manner to the bandwidth of local polynomial estimators.

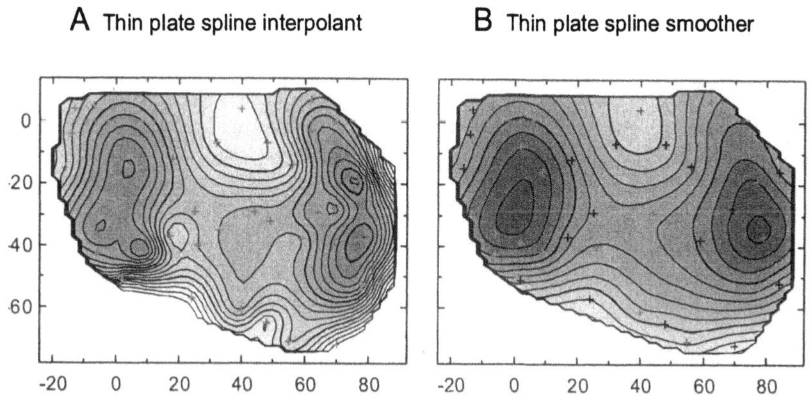

Figure 4. Contour plots of thin plate interpolant (A) and smoother (B) to the ore-bearing layer dataset (positions of 38 data sites and the "true width" of the ore-bearing layer measured at each site) (Green and Silverman, 1996).

Wavelet transforms are a device for representing functions in a way that is localized in both time and frequency domains. It has been demonstrated that wavelet-based smoothing methods can adapt readily to spatially inhomogeneous curves (Donoho et al., 1995; Bruce and Gao, 1996). KyPlot provides wavelet regression methods by thresholding (shrinkage). In Figure 5 wavelet thresholding is applied to the MNR spectrum dataset (Bruce and Gao, 1996). The threshold level can be determined by the methods of Donoho et al. (1995), Nason (1996) or Hurvich and Tsai (1998).
Local or penalized least squares methods are generalized to likelihood methods when response variables have a non-Gaussian distribution (Green and Silverman, 1996; Loader, 1999). KyPlot supports local and penalized likelihood regressions with a variety of families (binomial, Poisson, gamma, negative binomial, Pearson and Huber) and link functions. In Figure 6, bivariate nonparametric logistic regression is applied to a binary dataset. A contour plot of local logistic regression estimate is drawn for the MBA grade dataset (Simonoff, 1996). The data are the score on the Graduate Management Admission Test (GMAT) and first-year grade point average for 61 second-year MBA students at New York University's Stern School of Business in 1995, with data values for men marked by x and those for women

marked by O. The bold contour is the 0.5 level, which can be used as the classification boundary for men and women.

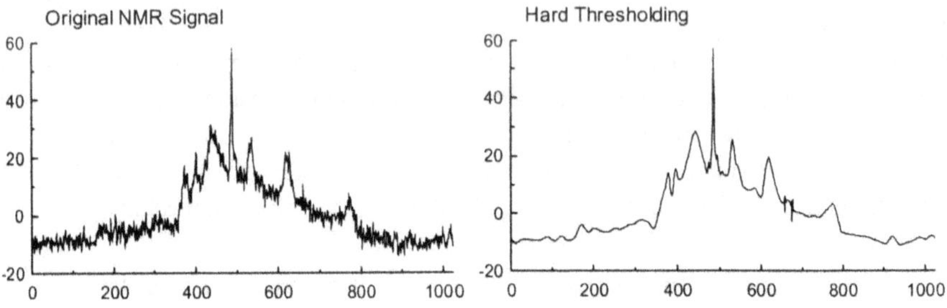

Figure 5. Wavelet regression by thresholding. Left panel: original NMR spectrum signal; right panel: result of wavelet hard thresholding. Symmlet of order 6 was used and the thresholding level was determined by the cross-validation criterion of Nason (1996).

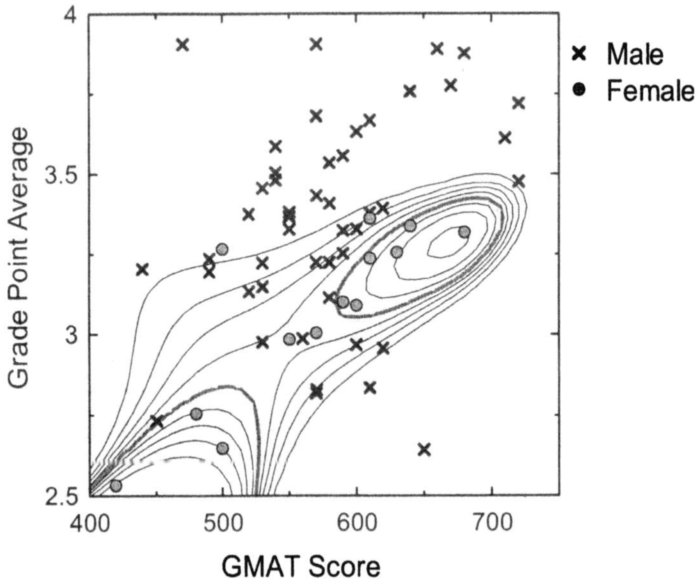

Figure 6. Contour plot of local logistic regression estimate for the MBA grade dataset. Local quadratic logistic regression was applied to the data and the result was plotted as a contour plot. The bold contour is the 0.5 level. Gaussian kernel was used and the bandwidth was determined by AICc.

When the errors have a long tailed distribution, least squares estimates can be excessively sensitive to extreme observations or outliers. Robust regression methods attempt to remedy this by identifying and down-weighting influential observations. A popular robust nonparametric procedure is LOWESS, which reduces the influence of outliers by an iterative reweighted least square scheme (Cleveland, 1979). Another approach is to apply the local likelihood or penalized likelihood method with a long tailed distribution for errors (Loader, 1999). For this purpose, KyPlot supports the Pearson family of distributions that includes Cauchy and t-distributions and Huber function family. The dataset shown in Figure 7 is generated by adding noise with a long tailed distribution to a sine curve. As shown in Figure 7A, the smoothing spline fits with three different values of smoothing parameter are considerably deviated from the sine curve. By contrast, the penalized likelihood estimates with Cauchy and Huber families are successful in identifying the sine curve (Figure 7B).

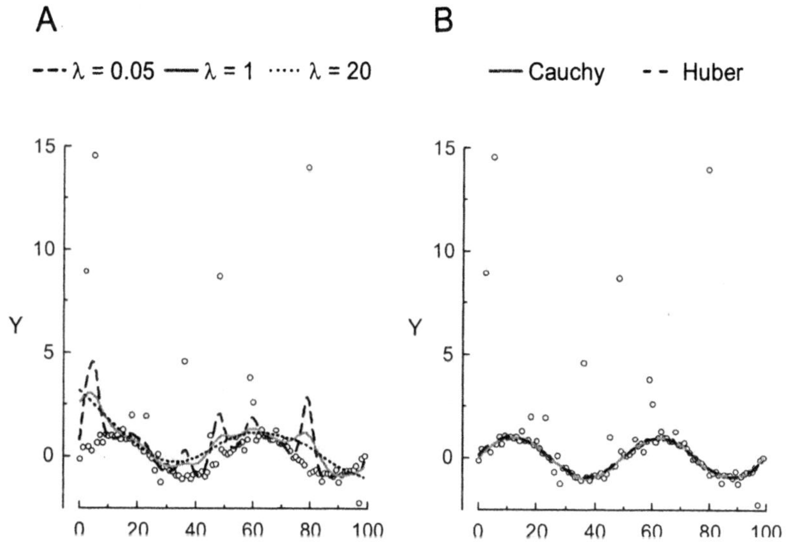

Figure 7. A. Cubic smoothing spline fits with three different values of the smoothing parameter (λ) to an artificial dataset. B. Penalized likelihood fits with Cauchy and Huber family distributions to the same dataset. The values of λ were determined by AICc.

KyPlot also offers nonparametric density estimation methods. In the left panel of Figure 8, the density of the Old Faithful geyser dataset, which contains durations of 107 eruptions, was estimated by the local likelihood method of Loader (1996) and the penalized likelihood method of Silverman (1982). The smoothing parameters of both estimates were determined by AIC (O'Sullivan, 1988; Loader, 1999). When data are given as binned data with a sufficiently small bin width, the density can be estimated by applying nonparametric Poisson regression (Simonoff, 1996; Loader, 1999). A histogram with the bin width of 0.03 was constructed from the geyser data and the local likelihood and penalized likelihood Poisson regressions were applied with the smoothing parameters determined by AICc. As shown in the right panel of Figure 8, the fit using local or penalized Poisson regression is comparable to that of local or penalized likelihood density estimation method. Density of bivariate data can be similarly estimated (Figure 9).

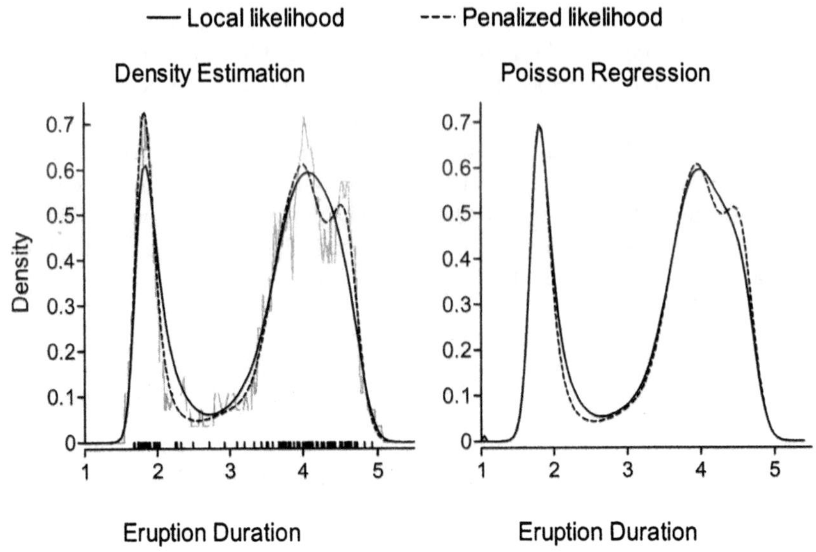

Figure 8. Density estimation for the Old Faithful geyser dataset. Left panel: local likelihood density estimate (solid line; local quadratic with Gaussian kernel) and penalized likelihood estimate (dashed line; with 3rd order derivative penalty). The grey jaggy line is a kernel estimate using uniform kernel. Right panel: density estimates for binned data using local likelihood (solid line; local quadratic with Gaussian kernel) or penalized likelihood (dashed line; with 3rd order derivative penalty) Poisson regression. The smoothing parameters of density estimates were determined by AIC and those of Poisson regression fits were determined by AICc.

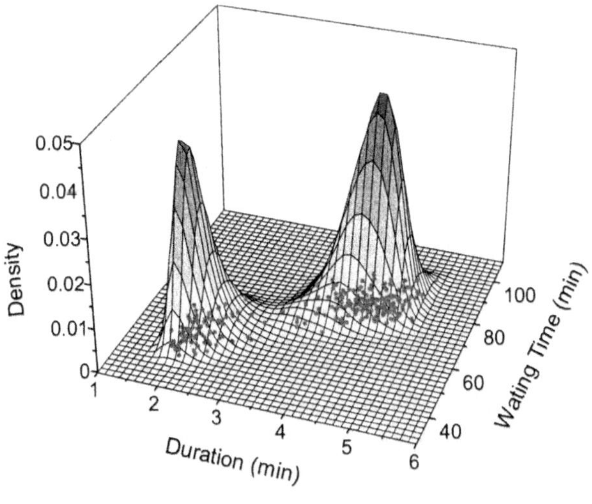

Figure 9. Local likelihood (local quadratic with Gaussian kernel) density estimate for the bivariate Old Faithful geyser dataset. The bandwidth was determined by AIC.

The hazard rate function for survival time data can be estimated nonparametrically with a slight modification of the density estimation methods (O'Sullivan, 1988; Loader, 1999). The upper panel in Figure 10 shows the local likelihood and penalized likelihood estimates for the hazard rate of the Stanford heart transplant data. These estimates suggest an initial high hazard followed by a sharp drop after the operation. In the lower panel, the estimates of survival function are plotted, showing that both the likelihood estimates are very close to the Kaplan-Meier estimate.

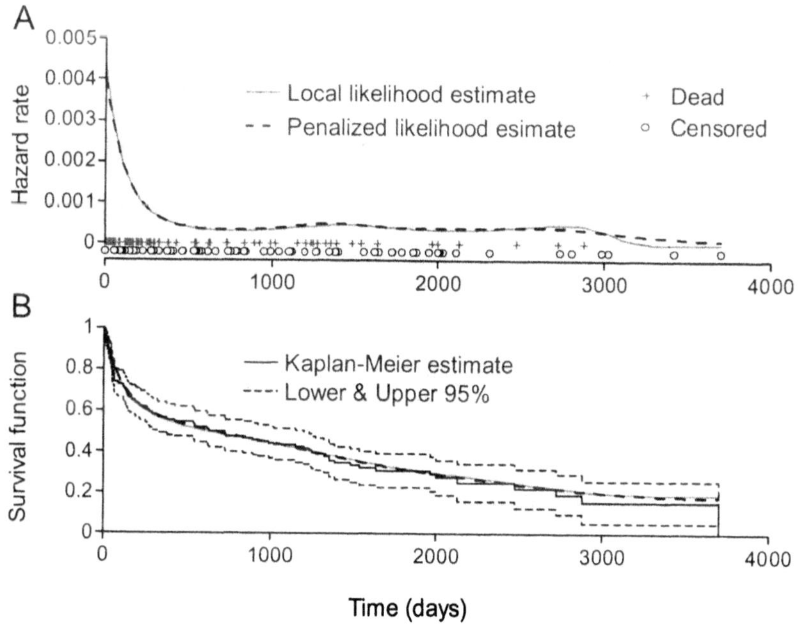

Figure 10. A. Local likelihood and penalized likelihood hazard rate estimates of heart transplant data. B. Kaplan-Meier estimate and its 95% confidence bands with the local and penalized likelihood estimates.

4 Discussion

Development of the software started from the necessity for my work. I am a neurophysiologist, studying how the brain works. I have been analyzing experimental data by applying various analytical methods. I needed software tools to perform such analyses and tools to create graphs and schemes. Because existing computer programs made for these purposes did not satisfy me, I began making programs by myself, gradually added many kinds of routines and integrated them into a software package, KyPlot. There exist still many plans for the project of KyPlot. I plan to add more statistical procedures and more graphical tools. For smoothing, procedures to be added are methods for multivariate regression such as additive models and adaptive regression splines, semiparametric models and neural networks.

References

Bruce, A. and Gao, H.-Y (1996). *Applied Wavelet Analysis with S-Plus*. New York: Springer-Verlag.

Cleveland, W.S. (1979). Robust locally weighted regression and smoothing scatter plots. *Journal of American Statistical Association*, **74**, 829-836.

Donoho, D.L., Johnstone, I.M., Kerkyacharian, G. and Picard, D. (1995). Wavelet shrinkage: asymptopia? (with discussion) *Journal of the Royal Statistical Society, Series B*, **57**, 362-366.

Green, P.J. and Silverman, B.W. (1994). *Nonparametric Regression and Generalized Linear Models*. London: Chapman & Hall.

Hurvich, C.M., Simonoff, J.S. and Tsai, C.-L. (1998). Smoothing parameter selection in nonparametric regression using an improved Akaike information criterion. *Journal of the Royal Statistical Society Series B*, **60**, 271-293.

Hurvich, C.M. and Tsai, C.-L. (1998). A cross-validatory AIC for hard thresholding in spatially adaptive function estimation. *Biometrika*, **85**, 701-710.

Loader, C.R. (1996). Local likelihood density estimation. *Annals of Statistics*, **24**, 1602-1618.

Loader, C.R. (1999). *Local Regression and Likelihood*. New York: Springer.

Nason, G.P. (1996). Wavelet shrinkage using cross-validation. *Journal of the Royal Statistical Society, Series B*, **58**, 463-479.

O'Sullivan, F. (1988). Fast computation of fully automated log-density and log-hazard estimators. *SIAM Journal on Scientific and Statistical Computation*, **9**, 363-379.

Silverman, B.W. (1982). On the estimation of a probability density function by the maximum penalized likelihood method. *Annals of Statistics*, **10**, 795-810.

Simonoff, J.S. (1996), *Smoothing Methods in Statistics*. New York: Springer.

Yoshioka, K. (2002). KyPlot - a user-oriented tool for statistical data analysis and visualization, *Computational Statistics*, To appear.

Mice and Elephants Visualization of Internet Traffic

J. S. Marron[1], Felix Hernandez-Campos[2] and F. D. Smith[2]

[1] School of Operations Research and Industrial Engineering, Cornell University, Ithaca, NY, 14853, USA and Department of Statistics, University of North Carolina, Chapel Hill, NC 27599-3260, USA

[2] Department of Computer Science, University of North Carolina, Chapel Hill, NC 27599-3175, USA

Abstract. Internet traffic is composed of flows, sets of packets being transferred from one computer to another. Some visualizations for understanding the set of flows at a busy internet link are developed. These show graphically that the set of flows is dominated by a relatively few "elephants", and a very large number of "mice". It also becomes clear that "representative sampling" from heavy tail distributions is a challenging task.

Keywords. Heavy Tail Distributions, HTTP Flows, Visual Representation, Zooming Graphics.

1 Introduction

Internet traffic is composed of a large number of small "packets" flowing from one computer to another. These packets are organized into flows that represent blocks of data, such as file transfers. In this paper, a visualization is developed for studying the nature of the population of flows passing through a given internet link.

An example of the visualization is shown in Figure 1. The behavior of flows over time is shown by representing each flow as a horizontal line segment, whose left end point is the start time of the flow (i.e. the time of the first packet), and whose right endpoint is the end time of the flow (the time of the last packet). Thus each line segment represents the full time period where the flow is "on". For convenient visual separation of a reasonable number of flows, a random height is added. This follows the spirit of the jitter plot idea of Tukey and Tukey (1990). See also pages 121-122 of Cleveland (1993).

An interesting feature of the flows shown in Figure 1 is a very large number of very short flows, and a few very long flows. The terminology "mice and elephants" provides a useful metaphor for understanding this characteristic of internet traffic: there are relatively few elephants, and a large number of mice. Another way of thinking of mice and elephants is that the distribution of lengths of the flows is heavy tailed.

The data shown in Figure 1 are HTTP (Web browsing) response times. They were collected during a four hour period on a Thursday afternoon in April of 2001. This time period was chosen to represent a "heavy traffic" time. An HTTP response time is defined here as the time between the first and last packets of a single HTTP data transfer. For more details on the data collection and processing methods, see Smith, Hernandez-Campos, Jeffay and Ott (2001).

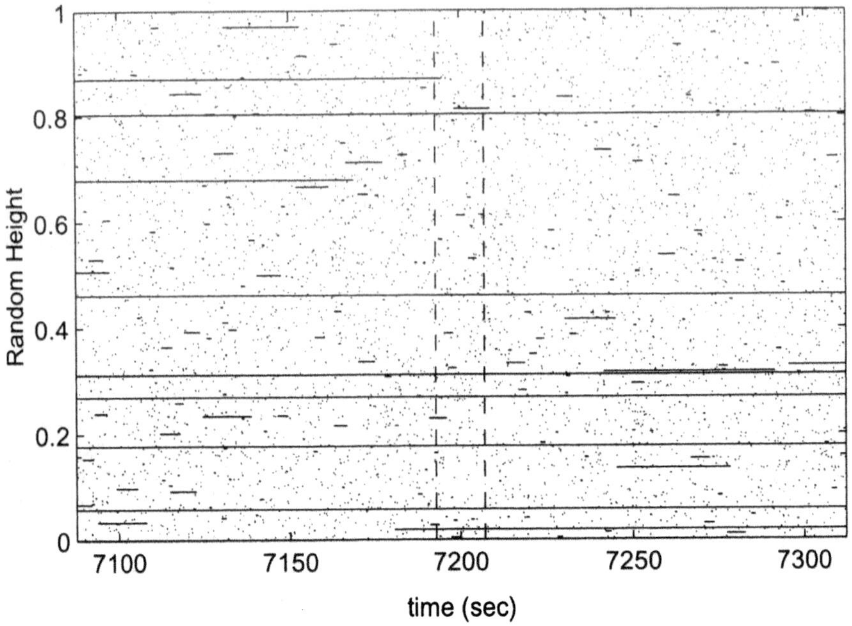

Fig. 1. Mice and Elephant display, showing HTTP flows. This is a random sample of 5000 out of 104,839 flows overlapping a 3.75 minute time interval.

During the four hour time window, there were 6,870,022 total HTTP responses. If all of these were shown in Figure 1, then the image would be useless, because it would be totally black. Sub-sampling is a natural approach to this problem.

One way to sub-sample is to reduce the time window. This is done in Figure 1, where only flows which spend at least part of their life times in the shown interval are considered. The interval has length 3.75 minutes = 4 hours / 4^3 (the reason for this choice will become apparent in Section 2), and is centered in the full 4 hour range of the data. The vertical dashed bars indicate the window of a further time restriction, that will be discussed in Section 2. When the full 6,870,022 responses were sub-sampled to the window of Figure 1, there were still *104,839* left. This is still far too many to show all of the mice and elephants, without heavy overplotting.

Another approach to reducing the number that are shown is random sampling. This was also done in Figure 1, where only a random sample of 5000 is shown. The number 5000 was the result of some experimentation with the goal of pleasing visual effect. Larger numbers made some of these plots "too busy", and a smaller number seemed inefficient in terms of not showing enough of the data.

Random sampling has very broad appeal as a means of finding a "representative subsample". Indeed this notion lies at the heart of a large number of statistical procedures in routine use. However, random sampling has some very serious drawbacks in the present case, because the distribution of lengths is heavy tailed. In particular, since there are relatively quite few elephants, their chance of appearing in the random sample can be quite negligible, which means they may not be properly represented in the visualization. This problem is investigated in Section 2. The problem becomes very clear from looking across a range of different "scales" (time windows). An alternative visualization is discussed in Section 2.

This mice and elephants visualization also provides visual insight into the causes of the interesting and compelling theory that has been built up to explain phenomena that have been observed in internet traffic. This is discussed in Section 3.

2 Deeper Look at Sampling Issues

In this section, sampling issues are carefully investigated. The key insights come from the construction of mice and elephant plots over a wide range of different time scales. An interesting sequence of views of this type may be found following the link "Zoomed-in Views of Mice & Elephants in Thursday Afternoon Trace" from the web page

`http://www-dirt.cs.unc.edu/marron/MiceElephants/.`

To save space, only a carefully selected few are shown as figures in this paper.

Figure 2 show the mice and elephants view over the full four hour time interval. This can be thought of as "zooming out in time scale". Once again only a random sample of 5000 is displayed (to avoid massive overplotting). The parallel bars near the center show the 3.75 minute time interval that is used in Figure 1 (thus giving a visual impression of the amount of "zooming out" that has been done). Figure 2 gives the impression that the traffic is all mice, and there are no elephants, which is clearly different from the impression of Figure 1. In particular, Figure 1 shows several elephants which span the full time interval, while none of these are visible over the same time interval in Figure 2.

The differences between Figures 1 and 2 are explained by the sampling process. In particular, as noted by a number of authors (referenced in Section 3), this distribution is quite "heavy tailed", which means that one may expect some very large values, but there will be only a very few of them. In particular, a careful look showed that there were 116 flows (in the full data set), that covered the full time range in Figure 1. Several of these show up in Figure 1, because there 5000 flows were randomly selected from the 104,839 that intersected that time interval, i.e. the subsample was about 5% of the total. Hence it is not surprising that about 6 of the 116 flows that cross the full interval appear in Figure 1. But the sampling situation is far different in Figure 1, where the 5000 flows were drawn from the full population of *6,870,022*, i.e. the subsample is less than 0.1% of the total. Hence it is not surprising that none of the 116 flows that completely cross the time interval of Figure 1, appear in Figure 2.

Figure 3 show the effect of "zooming in" (from the view shown in Figure 1) with respect to the time scale. Here the time interval is reduced to only the rather short 14 second interval between the vertical dashed bars in Figure 1. Once again, the view is radically different from that shown in Figure 1. This time the visual impression is of far more elephants. This view even conveys the impression that mice are a fairly negligible component of the population!

As above, this mistaken impression of the full population is the result of random sampling. This time 5000 flows were randomly selected from the 6910 that intersect this time interval, i.e. about 83%. This time interval is

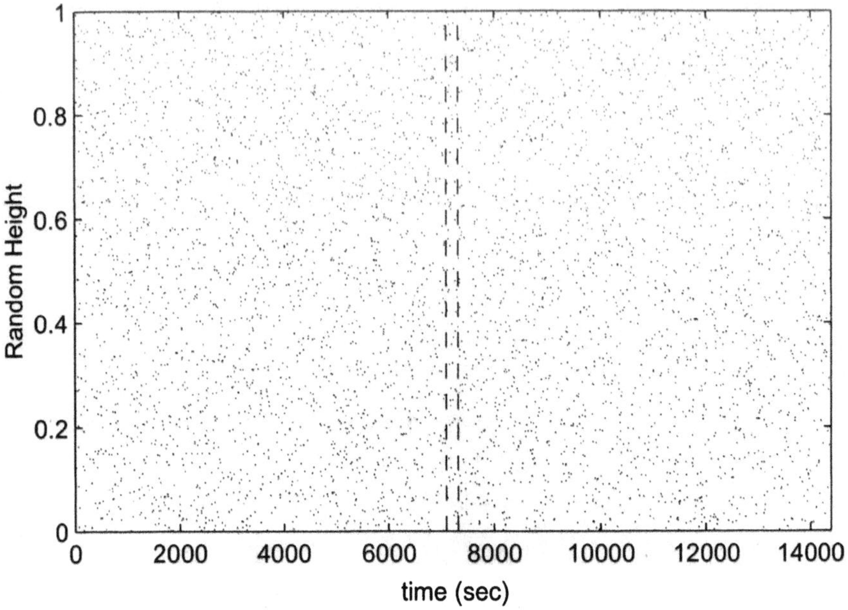

Fig. 2. Mice and Elephant display, over a long 4 hour time interval. This is a random sample of 5000 out of 6,870,022 total flows.

completely crossed by 163 flows (this might be viewed as surprisingly few, since 116 flows crossed the 16 times longer interval shown in Figure 1). Since most of these appear in the sample shown, the visual impression is dominated by these elephants. Indeed a reasonable opinion is that overplotting has occurred in Figure 3.

Figure 4 shows a sense in which Figure 3 is rather representative of the data. In Figure the full 4 hour time interval is shown, but instead of random sampling, the longest 5000 flows are shown. This is called the corresponding "elephant plot". This provides another way of seeing that this distribution of flow lengths is very heavy tailed. As noted above, this is less than 0.1% of the total data, and yet the smallest flows shown appear as dots (no visible length at this time scale). This means that the other 99.9% of the data would also appear as dots (if they could all be usefully plotted).

It is also interesting to view modifications of Figure 4, which zoom through smaller values of the threshold k, where the largest k flows are shown (thus $k = 5000$ is shown in Figure 4). This can be viewed following the link "Elephants-only Views for Thursday Afternoon Trace" from the above web page.

The Thursday afternoon time period shown here was chosen to represent a peak traffic time. Similar plots were also constructed for the off peak time period of Sunday morning.. The mice and elephants plots, as in Figures 1, 2 and 3, can be found following the link "Zoomed-in Views of Mice & Elephants in Sunday Morning Trace", and some elephant plots, as in Figure 4, can be found following the link "Elephants-only Views for Sunday Morning Trace"

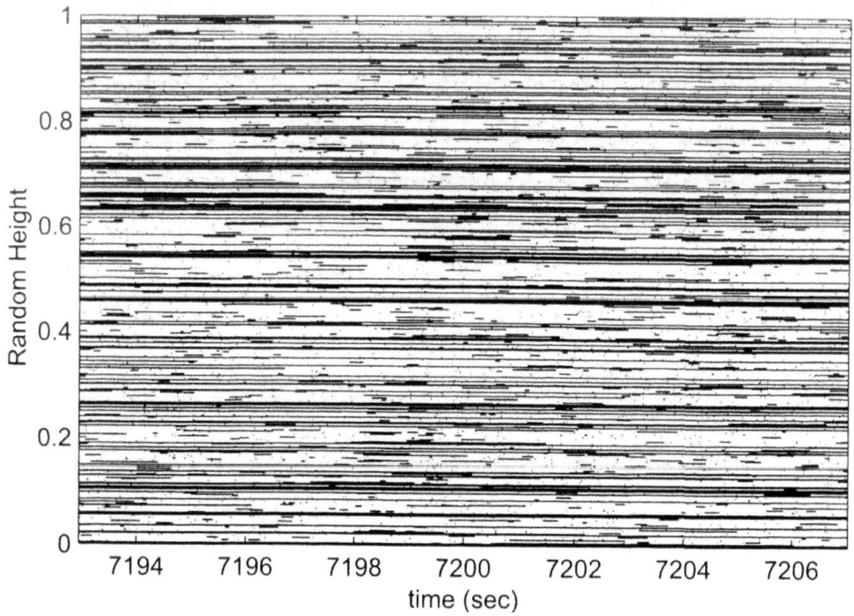

Fig. 3. Mice and Elephant display, showing HTTP flows over a very narrow time window. This is a random sample of 5000 out of 6910 flows overlapping a 14 second time intervals.

from the above web page.

A clear lesson is that random sampling is not an effective method for choosing a "representative sample" for internet data, or indeed data from heavy tail distributions that appear elsewhere, e.g. financial and environmental contexts. This is an interesting exception to the usual lesson taught in elementary statistics courses, in the special case of heavy tail distributions. An interesting open problem is how can a representative sample be chosen, for this type of visual impression, in heavy tailed contexts? There are related open sub-sampling problems (with perhaps different answers), for other purposes. For example, internet traffic data sets in generally tend to be large, and often need to be subsampled to efficiently store for later analysis and comparison. How can such a sample be made "representative"?

3 Related Theory

The mice and elephants plots presented in this paper provide visual insight into some important theoretical results often applied to Internet traffic data. The main idea is that heavy tailed flow length distributions give rise to long range dependence in the aggregated traffic.

The heavy tail of the flow length distribution has been observed by Garrett and Willinger (1994), Paxson (1994), Crovella and Bestavros (1996). See Downey (2001) for an alternative view, and some apparently contradictory ideas. The controversy raised by Downey was resolved by Gong, Liu, Misra

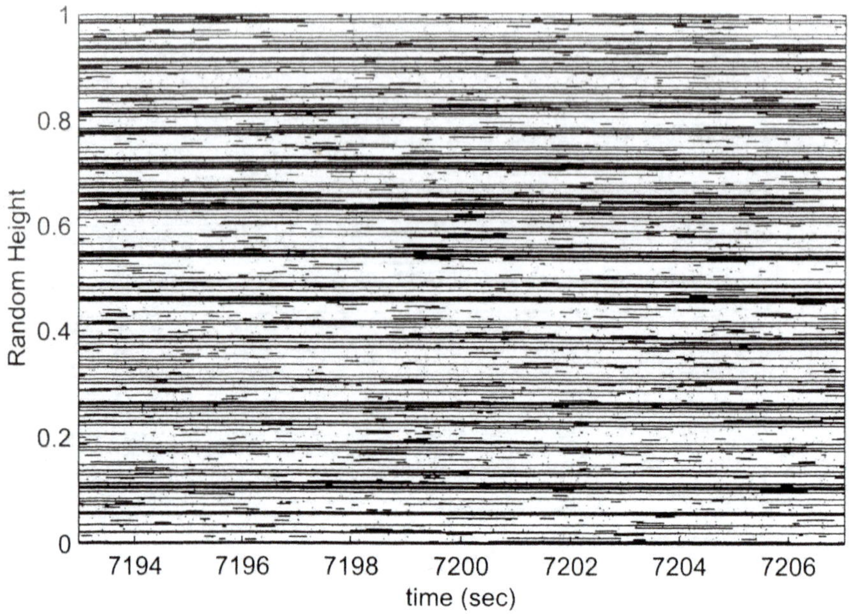

Fig. 4. Elephants display, showing the 5000 largest HTTP flows, over the longer 4 hour time window.

and Towsley (2001) and by Hernandez-Campos, Marron, Samorodnitsky and Smith (2002). The latter paper is a particularly detailed study, with more than usual attention paid to statistical issues, such as careful accounting for sampling variability.

Beautiful mathematics showing how heavy tailed flow lengths lead to long range dependence can be found in Mandelbrot (1969), Cox, D. R. (1984), Taqqu and Levy (1986), Leland, Taqqu, Willinger and Wilson (1994), Crovella and Bestavros (1996), Heath, Resnick and Samorodnitsky (1998), Resnick and Samorodnitsky (1999). A recent and general version of this can be found in Hernandez-Campos, Marron, Samorodnitsky and Smith (2002).

Plots such as Figure 1 give a clear intuitive view of how a few elephants create long range dependence. The corresponding short range situation is illustrated in Figure 5. Figure 5 is based upon a simulated set of flows, whose starting times are the same as those in the original full data set ($n = 6,870,022$). However, the lengths of the flows are simulated from an exponential distribution, where the mean is taken to be the sample mean from the mean lengths. The exponential flow length distribution is the foundation of classical queueing theory, and is well known to result in short range dependent aggregated traffic. Although the mean flow length is the same as in Figure 1, the visual impression of the flows in Figure 5 is radically different: there are no elephants. How can the mean lengths be the same when there are no elephants? This happens because the mice in Figure 1 are smaller than those in Figure 5. This is not visible here, but can be seen by zooming in further, follow the link "Zoomed-in Views of Mice & Elephants in

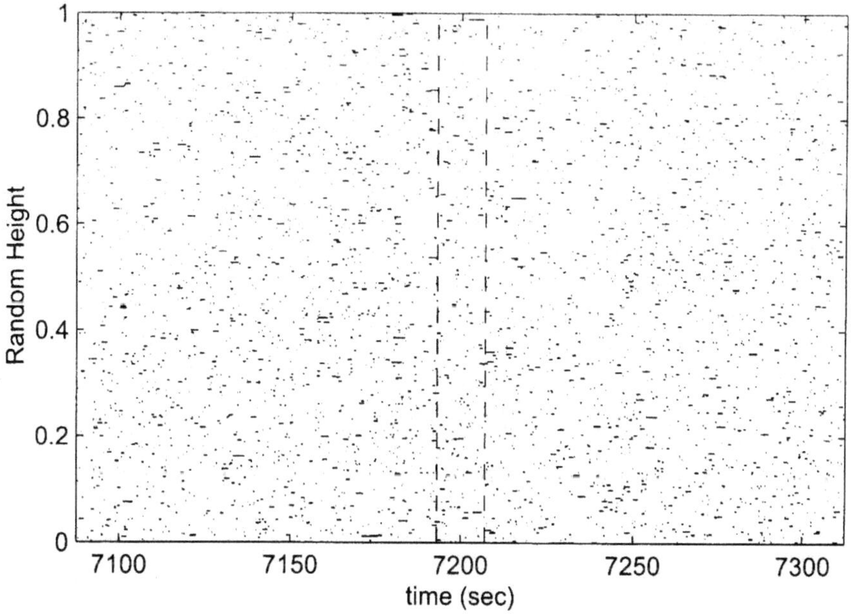

Fig. 5. Mice and Elephant display, showing simulated Exponentail (with the same mean) HTTP flows over the same time window as Figure 1.

Synthetic Version of Thursday Afternoon Trace" from the above web page.

Because there are no elephants in Figure 5, when the traffic is aggregated into bins, the correlations between the counts fall off quickly. However, when a few elephants, such as shown in Figure 1 enter the picture, the correlations between bin counts die off far more slowly, creating long range dependence.

An aspect of the distribution of flows not studied here is the distribution of the flow starting times. Recent analysis of these may be found in Cleveland, Lin and Sun (2000), Nuzman, Saniee, Sweldens and Weiss (2002) and Marron, Hernandez-Campos and Smith (2001).

References

Cleveland, W. S. (1993) *Visualizing Data*, Hobart Press, Summit, New Jersey, U.S.A.

Cleveland, W. S., Lin, D. and Sun, D. X. (2000) IP packet generation: statistical models for TCP start times based on connection-rate superposition, *Performance Evaluation Review: Proc. ACM Sigmetrics 2000*, 28, 166-177.

Cox, D. R. (1984) Long-Range Dependence: A Review, in *Statistics: An Appraisal, Proceedings 50th Anniversary Conference.* H. A. David, H. T. David (eds.). The Iowa State University Press, 55-74.

Crovella, M. E. and A. Bestavros, A. (1996) Self-similarity in world wide web traffic evidence and possible causes, *Proceedings of the ACM SIGMETRICS 96*, pages 160–169, Philadelphia, PA.

Downey, A. B. (2001) Evidence for long tailed distributions in the internet, ACM SIGCOMM Internet Measurement Workshop, November 2001. Internet available at `http://rocky.wellesley.edu/downey/longtail/`.

Garrett, M. W. and Willinger, W. (1994). Analysis, Modeling and Generation of Self-Similar Video Traffic, *Proc. of the ACM Sigcom '94*, London, UK, 269-280

Gong, W., Liu, Y., Misra, V. and Towsley, D. (2001). On the tails of web file size distributions, *Proceedings of 39-th Allerton Conference on Communication, Control, and Computing.* Oct. 2001. Internet available at: `http://www-net.cs.umass.edu/networks/publications.html`.

Hannig, J., Marron, J. S., Samorodnitsky, G. and Smith, F. D. (2001) Log-normal durations can give long range dependence, unpublished manuscript, web available at `http://www.stat.unc.edu/postscript/papers/marron/NetworkData/LogNorm2LRD/`

Heath, D., Resnick, S. and Samorodnitsky , G. (1998) Heavy tails and long range dependence in on/off processes and associated fluid models, *Mathematics of Operations Research*, 23, 145-165.

Hernandez-Campos, F., Marron, J. S., Samorodnitsky, G. and Smith, F. D. (2002) Variable Heavy Tailed Durations in Internet Traffic, unpublished manuscript, web available at `http://www.stat.unc.edu/postscript/papers/marron/NetworkData/VarHeavyTails/`.

Marron, J. S., Hernandez-Campos, F. and Smith, F. D. (2001) A SiZer analysis of IP Flow start times, unpublished manuscript.

Leland, W. E., Taqqu, M. S., Willinger, W. and Wilson, D. V. (1994). On the Self-Similar Nature of Ethernet Traffic (Extended Version), *IEEE/ACM Trans. on Networking*, 2, 1-15.

Mandelbrot, B. B. (1969) Long-run linearity, locally Gaussian processes, H-spectra and infinite variance, *International Economic Review*, 10, 82-113.

Nuzman, C., Saniee, I., Sweldens, W. and Weiss, A. (2002) A compound model for TCP connection arrivals for LAN and WAN applications, unpublished manuscript.

Paxson, V. (1994) Empirically-Derived Analytic Models of Wide-Area TCP, Connections. *IEEE/ACM Transactions on Networking*, 2, 316–336.

Resnick, S. and Samorodnitsky, G. (1999) Activity periods of an infinite server queue and performance of certain heavy tailed fluid queues, *Queueing Systems*, 33, 43-71.

Smith, F. D., Hernandez, F., Jeffay, K. and Ott, D. (2001) "What TCP/IP Protocol Headers Can Tell Us About the Web", *Proceedings of ACM SIGMETRICS 2001/Performance 2001*, Cambridge MA, June 2001, pp. 245-256.

Taqqu, M. and Levy, J. (1986) Using renewal processes to generate LRD and high variability, in: *Progress in probability and statistics*, E. Eberlein and M. Taqqu eds. Birkhaeuser, Boston, 73-89.

Tukey, J., and Tukey, P. (1990). Strips Displaying Empirical Distributions: Textured Dot Strips. Bellcore Technical Memorandum.

Relativity and Resolution for High Dimensional Information Visualization with Generalized Association Plots (GAP)

Shun-Chuan Chang[1], Chun-houh Chen[2], Yueh-Yun Chi[3], and Chih-Wen Ouyoung[4]

[1] Office of Statistics, Department of Health, Taiwan, R.O.C., Taipei 100, Taiwan
[2] Institute of Statistical Science, Academia Sinica, Taipei 115, Taiwan
[3] Department of Biostatistics, The University of North Carolina, Chapel Hill, NC 27599, USA
[4] Department of Statistics, National ChengChi University, Taipei 116, Taiwan

Abstract. Generalized association plots (GAP) (Chen, 1996; 1999; 2002) is an information visualization environment for high dimensional data structure without dimension reduction. There is no limit for sample size and variable number. Three matrix maps for raw data matrix, object proximity matrix, and variable proximity matrix are created for visually extracting grouping structures for objects and variables and the interaction information between object-clusters and variable-groups. Seriation algorithms are developed to permute objects and variables such that rows and columns with similar profiles are arranged at closer positions. Categorical generalized association plots (cGAP) (Chen, 1999; Chen et al., 2002) is an extension of GAP adapted for visualizing high dimensional categorical data structure. Optimal scaling (multiple correspondence analysis) is applied to compute the proximity matrices for objects as well as for variables and to obtain colors for coding all categories in the raw data matrix. Relativity and resolution are two related critical issues in conducting efficient GAP and cGAP analyses. This article discusses possible solutions when standard procedures fail in generating satisfactory relativity and resolution for GAP and cGAP.

Color version of figures presented in this article together with software packages for GAP and cGAP can be obtained from our website at http://gap.stat.sinica.edu.tw/.

Keywords. Categorical data, color-coding, data mining, matrix map, multiple correspondence analysis, seriation

1 Continuous and categorical generalized association plots

Originally developed for analysing psychosis disorder data (Lin et al., 1998; Hwu et al., 2002), generalized association plots (GAP) is a statistical visualization and clustering environment for retrieving important pieces of high dimensional information structure embedded in many different kinds of raw data matrices and proximity matrices. Illustrated in Figure 1 are four major steps in constructing a GAP analysis for a data set with ninety-five objects and fifty variables (see Chen and Chen (2000) and Chen (2002) for descriptions of the data set).

Given a raw data matrix in step 1, GAP first computes two proximity matrices, one for objects and the other for variables. These two proximity matrices together with raw data matrix are then projected through selected color spectrums to create three matrix maps. In step 2, suitable seriations for objects and variables are obtained to permute three matrices (maps) such that the concept of relativity (to be discussed in Section 2) is satisfied. These three permuted matrices (maps) are further partitioned into sub-matrices (sub-maps) in step 3 to identify clusters of objects and groups of variables. Finally a sufficient graph is created in step 4 to summarize information embedded in these three permuted and partitioned matrices (maps).

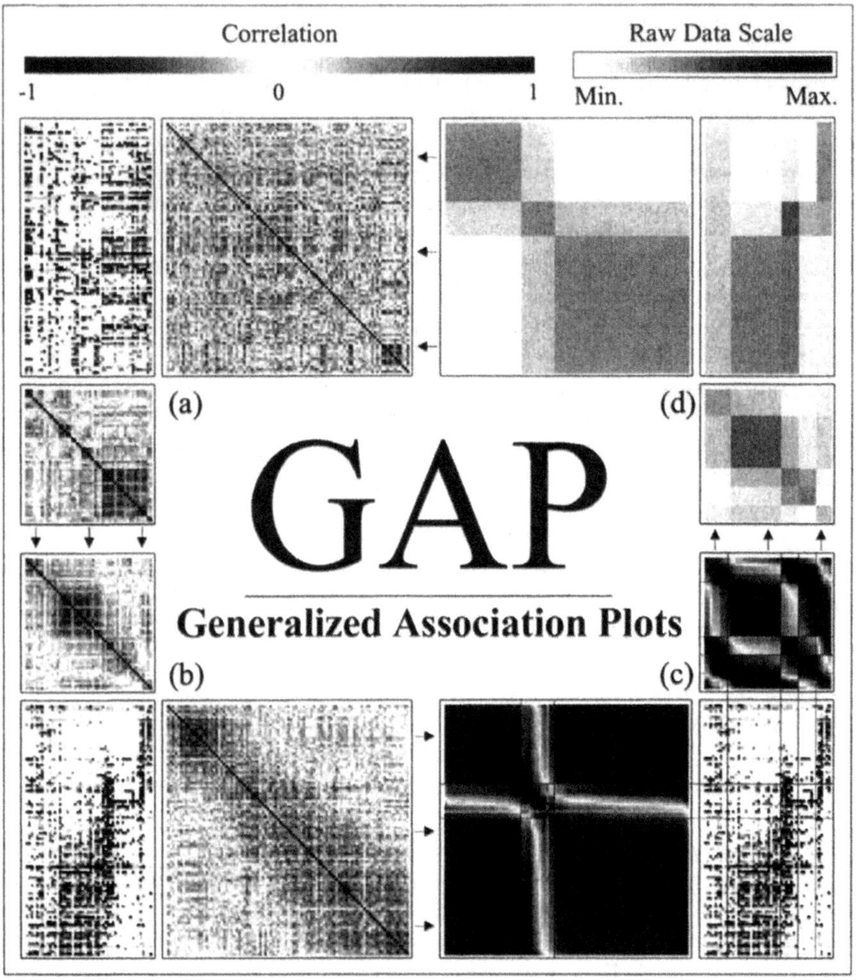

Fig. 1. Generalized Association Plots (GAP): (a) Raw data map and proximity maps with suitable color projections; (b) Sorted raw data maps and proximity maps with the concept of relativity; (c) Partitioned raw data map and proximity maps with clustering analyses; (d) Sufficient graphs with summarized information.

1.1 Categorical generalized association plots (cGAP)

The most commonly used statistical graphic tools for visualizing contingency table type data structure are correspondence analysis (Hill, 1974; Benzecri, 1976; Nishisato, 1980) and the mosaic display (Hartigan and Kleiner, 1981; Friendly 1994). For visualizing higher dimensional contingency tables, there are multiple correspondence analysis (MCA) (Gifi, 1990; Michailidis and de Leeuw, 1999) and extending mosaic displays (Friendly, 1999). However, they are not designed for visualizing very high dimensional data set. Categorical generalized association plots (cGAP) (Chen, 1999; Chen et al., 2002), an extension of GAP, is a categorical information visualization environment designed to simultaneously visualize and extract information embedded in thousands of categorical variables with thousands of objects. In step 0 of a cGAP analysis, a multiple correspondence analysis (MCA) with a three-dimensional solution space is utilized to scale the input categorical raw data matrix into the three matrices in step 1 of an ordinary GAP analysis. The rest of the procedures are identical to the four steps of an ordinary GAP analysis. A three-dimensional solution space for Hartigan's Hardware data (Hartigan, 1975) is illustrated in Figure 2.

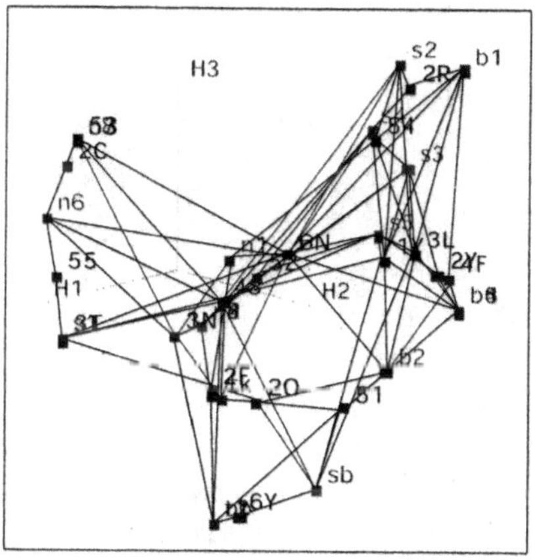

Fig. 2. Multiple correspondence analysis (MCA) for Hardware data (Hartigan, 1975).

2 Concept of relativity

In step 2 of a GAP analysis, Chen (2002) proposed the concept of relativity of a statistical graph. The basic idea is to place similar (distinct) objects at positions close to (far away from) each other in a plot for representing the association structure. In graphic tools for continuous variables, this concept of relativity always holds because the nature of a continuous variable plot is to represent the metric structure of objects geometrically in a display. Figures 3a and 3b depict the relativity concept of continuous variables for Iris data (Fisher 1936). Two iris flowers (represented by red and blue dots) with similar petal widths and petal lengths are displayed next to each other automatically in a histogram and a scatter-plot. On the other hand, the relativity concept does not necessarily exist in a categorical type graph except for correspondence analysis type of displays where categorical data sets are scaled into quantitative ones. Illustrated in Figures 3c and

3d are pie chart and mosaic display for Hardware data (Hartigan 1975). Two categories arranged next to each other do not necessary share similar properties. Friendly (1994) suggested a general method that orders the categories in a mosaic display according to their scores on the largest dimension in a correspondence analysis of residuals.

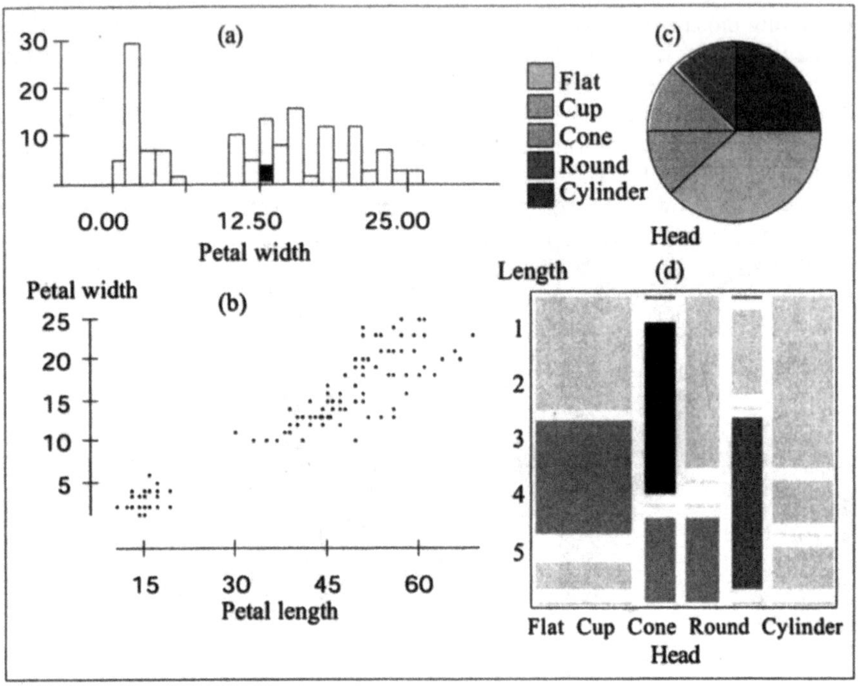

Fig. 3. Relativity of a statistical graph: (a) Histogram and (b) Scatter-plot for continuous Iris data (Fisher, 1936) with the concept of relativity; (c) Pie chart and (d) Mosaic display for categorical Hardware data (Hartigan, 1975) without the concept of relativity.

2.1 Concept of relativity for GAP

In a GAP analysis with one raw data matrix map and two proximity matrix maps, this concept of relativity has to be forced in with appropriate seriations (permutations). Without suitable seriations to arrange these matrix maps such that rows (columns) with similar (distinct) properties are placed at closer (farther) positions, these matrix maps are useless. One more important aspect of the relativity concept in a GAP type analysis is the selection of color spectrums for projecting the numerical values (raw data and proximity) into pixels with suitable colors. An equally spaced gray scale (Figure 1) or a rainbow spectrum is suitable for representing single-directional numerical values while spectrums with double hue scales are necessary for displaying bi-directional intensities. Correlation coefficients with a blue-white-red spectrum in Figure 1 and gene expression ratios with a green-black-red spectrum for cDNA microarray data (Eisen et al., 1998; Iyer et al., 1999) are two such bi-directional examples. When variables (columns) or objects (rows) have different scales, it may be necessary to transform (normalization, standardization, rank, etc.) them before color projection is applied such that color version of the relativity concept can hold. That is to apply comparable colors to represent similar numerical values within a single column (row) and amongst neighboring columns (rows).

2.2 Concept of relativity for cGAP

There are also three corresponding aspects of the concept of relativity in a categorical GAP analysis. Objects with similar category-profiles should be placed in closer rows while variables with identical object-distributions are arranged in columns next to each other. Categories sharing similar (different) object-distributions arc also assigned with comparable (distinct) colors to represent their characteristics. Given a categorical input data matrix, it is necessary to define the proximity measures for objects as well as for variables before any seriation algorithm can be applied to search for good permutations such that the relativity concept is valid. Color version of the relativity concept is the most difficult problem in creating a cGAP display. In a conventional pie chart (Figure 3c) or cartogram for a single categorical variable, especially nominal variable, colors are only used as labels for readers to visually distinguish different categories in the display. Canonical colors (red, green, blue, cyan, magenta, yellow, black, and white) can be assigned sequentially for each category presents in the variable for this purpose. Color in a cGAP display is utilized to represent high dimensional association structure embedded in multi-way contingency table for a categorical data matrix. Each category in every variable needs a specific color to represent its relative distribution structure among objects. Categories share similar object distribution should be assigned comparable colors. Colors cannot be assigned to categories in each variable separately. A systematic scheme is required to simultaneously assign colors to every category in all variables for preserving color version of the concept of relativity.

A complete cGAP analysis consists of the following seven components: (a) appropriate proximity measurement for objects; (b) appropriate proximity measurement for variables; (c) suitable color assignment for categories; (d) good seriation algorithm for objects; (e) good seriation algorithm for variables; (f) proper clustering analysis for objects; and (g) proper clustering analysis for variables. An MCA is first carricd out in stcp 0 of a cGAP analysis to identify the first three components. These three components are then fed into an ordinary GAP analysis (Figure 1) to carry out the rest four major steps.

When all seven goals are achieved, cGAP users can easily extract different levels of information embedded in the input categorical data matrix: (I) clustering structure (within cluster and between clusters) of objects; (II) grouping pattern (within group and between groups) of variables; (III) association structure between object-clusters and variable-groups; (IV) extreme objects (outliers) with rare categories.

2.3 Rotation of the Color Cube

One issue related to the color version of relativity in a cGAP analysis is to rotate the Red/Green/Blue color space in corresponding to the three-dimensional MCA solution space. Categorical GAP converts three MCA dimensions into red, green, and blue hues respectively by default. There are six possible permutations for assigning three MCA dimensions to three dimensions of the RGB color cube.

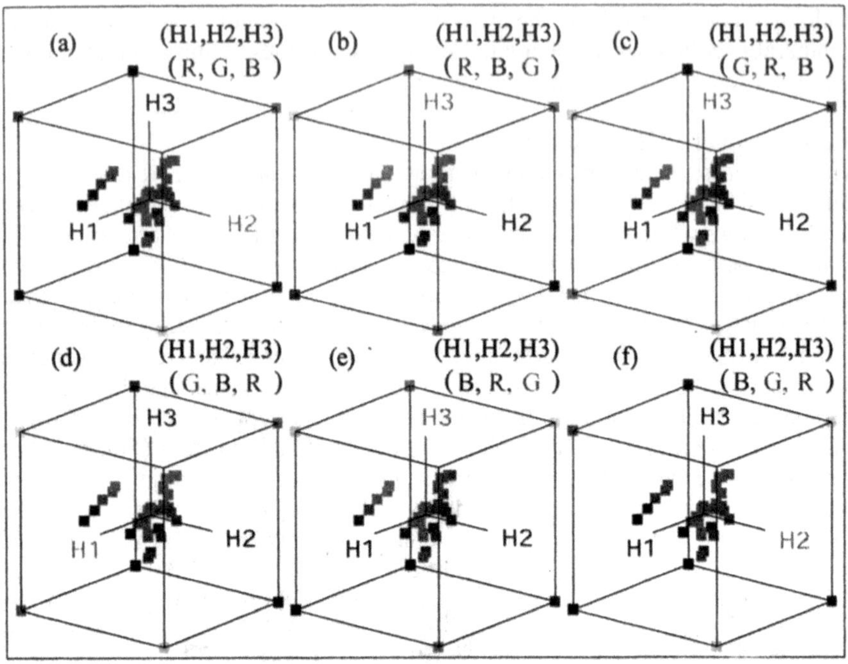

Fig. 4. Rotation of color cube for all six Red-Green-Blue permutations for three MCA dimensions, H1-H2-H3, Hardware data (Hartigan, 1975).

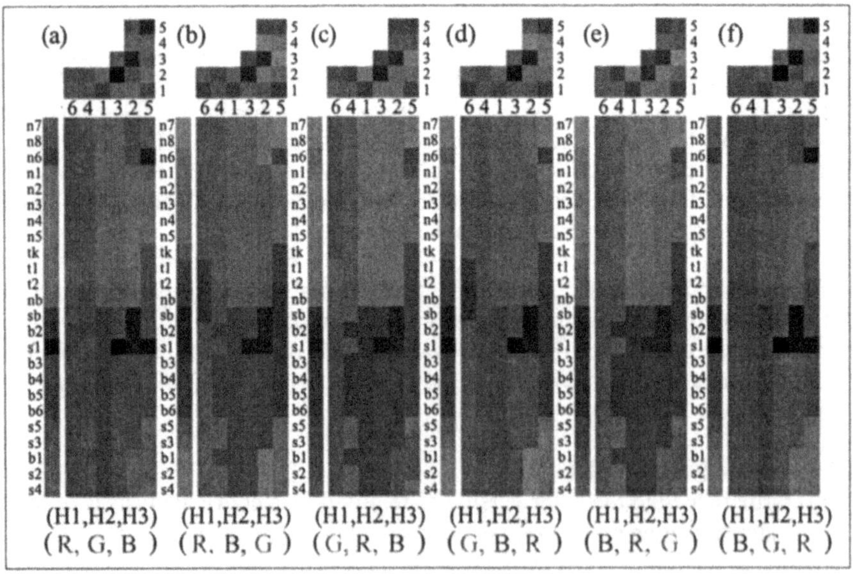

Fig. 5. Categorical GAP displays with all six Red-Green-Blue permutations for three MCA dimensions, H1-H2-H3, Hardware data.

Figure 4 has these six color cubes for Hardware data. These six permutations result in six cGAP raw data matrix maps, Figure5. A cGAP display is not unique in this sense. The color relativity concept is guaranteed by the MCA procedure. Similar categories are placed in closer configuration points in an MCA solution space so they are assigned similar colors sharing similar triplets in the solution space no matter which R/G/B combination is chosen. Computations of two proximity matrices for object and variable are unique since rotation of color cube only affects color assignment of the raw data matrix maps, not the relative configuration structure.

3 Resolution of a statistical graph

Illustrated in Figure 6a is a two-dimensional scatter-plot of data points with an outlier. Since this outlying observation is located far away from the main body of the data cloud the only feature can be identified in Figure 6a is the relative structure of this outlier to the rest of the data points. It is necessary to exclude this outlying observation to release resolution of the scatter-plot for extracting information embedded in the main data structure (Figure 6b). Transformation (logarithmic, rank, etc.) of variables (Wilkinson 1999) can also achieve better resolution of data structure without removing outlying observations, Figure 6c. Similar resolution problems exist in all types of continuous data displays.

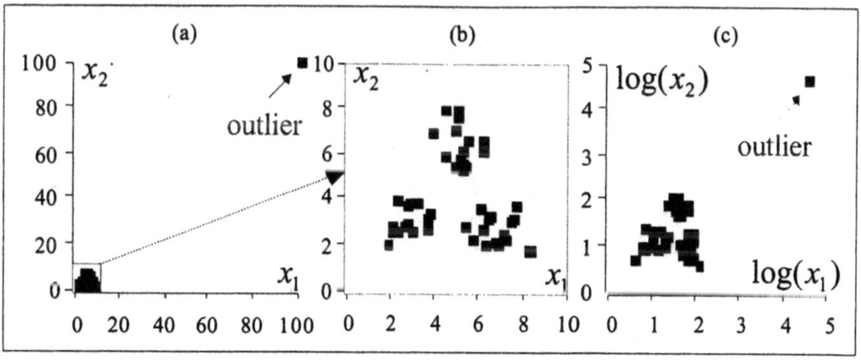

Fig. 6. Resolution for a continuous statistical graph (simulation data): (a) Scatter-plot with poor resolution due to presence of outliers; (b) Zoomed-in scatter-plot with better resolution. (c) Scatter-plot of variables with logarithmic transformation.

3.1 Resolution for GAP

GAP encounters similar resolution problem when there are outliers. Outliers may exhaust hues in a GAP color spectrum. Figures 7abc illustrate series of GAP raw data matrix maps for the same data set used in Figures 6abc. In Figure 7a with an outlier, only hues to two ends of the rainbow spectrum are used to represent relative structure of outlier to the main data cloud. All hues in the middle of the spectrum are wasted and important structure embedded in the main data cloud is completely lost. Figures 7b and 7c give GAP raw data matrix maps without the outlier and with logarithmic transformation respectively.

Proximity matrix map in a GAP analysis is essentially a direct visualization of all pair-wise geometric relationships in conventional statistical graphs (histogram, scatter-plot, etc.). Multidimensional scaling (MDS) (Borg and Groenen (1997), Chen and Chen (2000), Cox and Cox (2002)), on the contrary, tries to represent pair-wise measurements in a proximity matrix as geometric relationships of points on a low-dimensional Euclidean space. GAP plots pair-wise two-dimensional Euclidean distances for points in Figures 6a and 6b as color pixels in matrix maps in Figures 8a and 8b. Geometrical resolution difficulties are now converted into color resolution problems. Figure 8c displays the proximity matrix map with logarithmic transformed Euclidean distances for retrieving better color resolution.

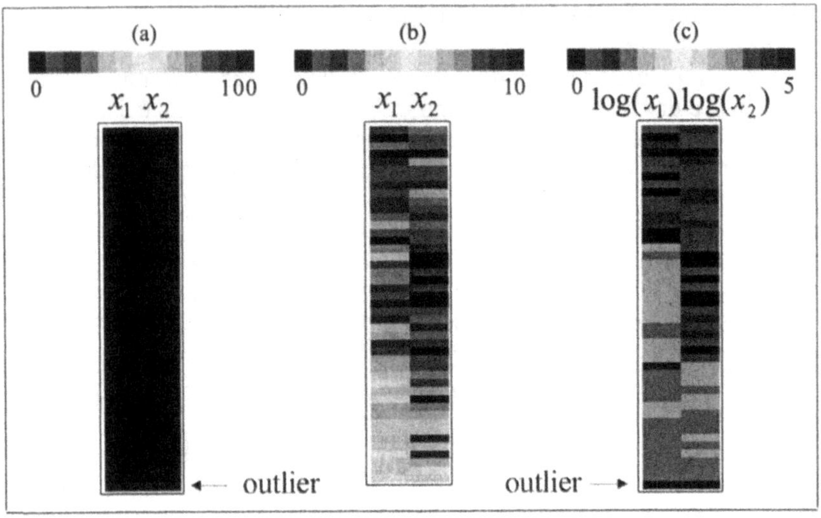

Fig. 7. Resolution for GAP raw data matrix maps (same data as in Figure 6): (a) Raw data matrix map with poor color resolution with an outlier; (b) Good color resolution without the outlier; (c) Raw data matrix map with logarithmic transformation.

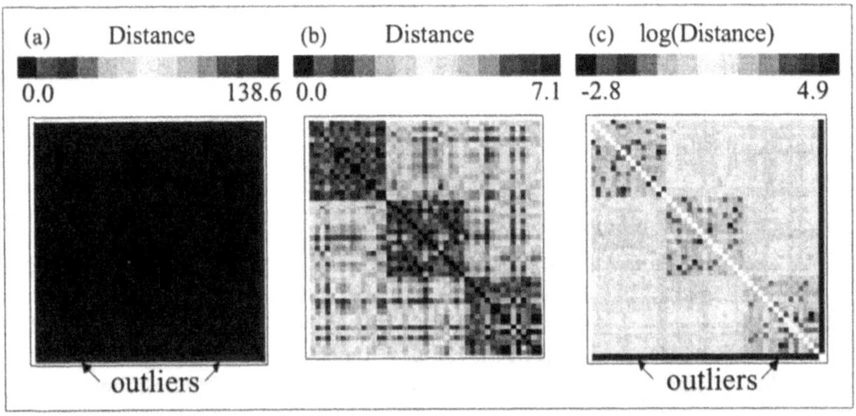

Fig. 8. Resolution for GAP proximity matrix maps (Euclidean distance matrix for points in Figure 6): (a) Poor color resolution due to outlier distances; (b) Good color resolution without outliers; (c) Proximity matrix map with logarithmic transformation.

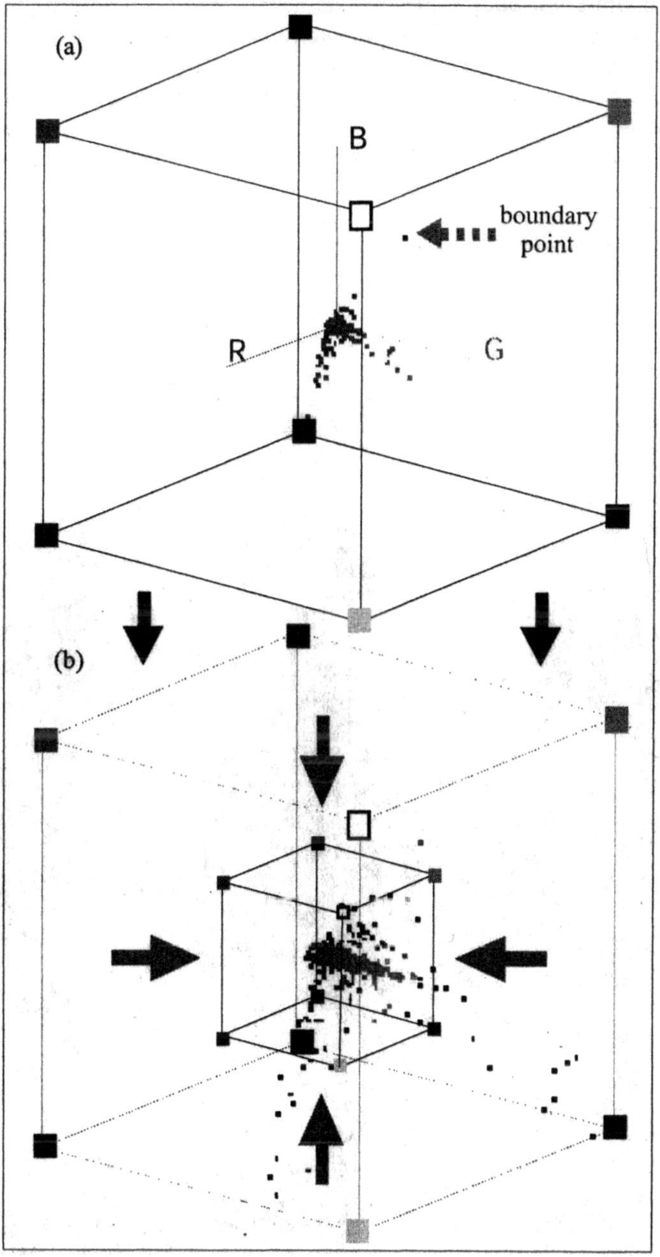

Fig. 9. Resolution problem for cGAP (simulation data): (a) RGB cube with poor color resolution due to the presence of outliers (boundary points); (b) Zoomed-in RGB cube to release color resolution (colors outside the RGB cube are recoded in black).

3.2 Resolution for cGAP

Similar to a GAP analysis, a cGAP analysis also has possible resolution difficulties for color-coding of raw data matrix map and two proximity matrix maps. Color resolution difficulties for proximity matrix maps in a cGAP display can be remedied using identical procedures as in a GAP analysis. Resolution problem for color-coding of each category in every variable is a much more difficult issue.

Colors for categories in a cGAP raw data matrix display are automatically assigned by a three-dimensional MCA solution configuration of the input categorical raw data matrix. Association structure in the multi-way contingency table of the data set and algorithms used in an MCA analysis determine the solution configuration and the color-coding scheme. Three-dimensional MCA configuration points for the most common categories (categories with largest proportions) reside near the origin while configuration points for the most rare categories (categories with smallest proportions, outlying category) occupy the outer most positions. Categorical GAP converts these two extremes into colors near the center (grey) of the RGB cube and colors on the boundaries (brightest colors) of the RGB cube.

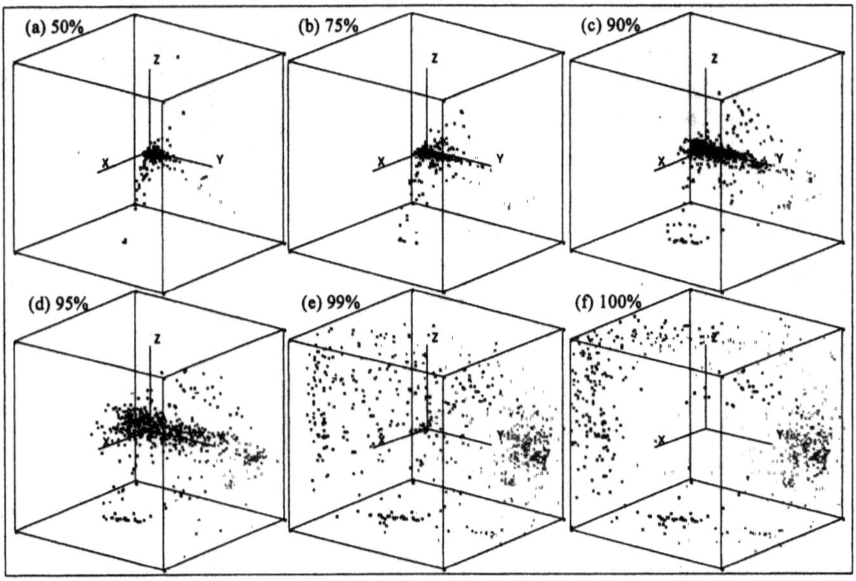

Fig. 10. Release of color space with zoomed-in colors projected onto the boundaries of RGB cube (same simulation data as in Figure 9): (a) 50% zoomed-in; (b) 75% zoomed-in; (c) 90% zoomed-in; (d) 95% zoomed-in; (e) 99% zoomed-in; (f) 100% zoomed-in.

If a category only applies to a single object, the configuration points for the object and that category will coincide. Objects with a unique profile (outliers) will be pushed further away from the origin and assigned brightest color. Relative structure of these distinct objects and categories to the main data cloud of objects and categories exhausts both geometric resolution and color resolution. Figure 9a illustrates one such simulation with most categories and objects located near the origin and are assigned colors near pure grey. It is necessary to release geometric resolution for these near origin categories and objects to obtain better color assignments.

The easiest approach is to rerun an MCA analysis with these outlying objects and categories (variables) excluded. A complete new color-coding scheme is obtained which may not be compatible with original colors for those excluded objects and categories. A second approach is to zoom-in the color cube until certain criteria are satisfied. Categories and objects outside the color cube are assigned with certain colors. Figure 9b adopts such an approach with categories and objects outside the cube coded in black. One variant of the second approach zooms-in the RGB cube to a specific proportion and projects categories and objects onto boundaries of the cube. That is to color these categories and objects with brightest possible colors with at least one component of the RGB triplet reaches its maximum. Figure 10 displays this approach applied to the same data in Figure 9a with various zoomed-in proportions.

4 Conclusions

GAP and cGAP are very powerful visualization and clustering tools for extracting information structure embedded in high dimensional continuous and categorical data sets. Both versions have many potential applications in fields ranging from survey sampling, data mining, education, psychometrics, to biomedical studies, functional genomics, and astronomy. This article discusses possible procedures in generating satisfactory relativity and resolution for GAP and cGAP.

It is possible to modify algorithms used in multiple correspondence analysis directly in several aspects to make cGAP more robust with outliers presented in the data set for better relativity and resolution. Michailidis and de Leeuw (2000) discuss a homogeneity analysis (MCA) by alternating least absolute deviations with various normalization procedures (*1-norm*, *∞-norm*, etc.). These MCA variants can be integrated into cGAP for searching of better relativity and resolution.

Only three-dimensional hue (R/G/B) is used in representing the first three MCA configuration dimensions in current version of cGAP. It is also possible to use brightness and saturation of colors to introduce information contained in fourth and fifth dimensions of MCA into the framework of cGAP. There are many other potential aspects that can be improved and extended (Chen 1999) to make GAP and cGAP better and more powerful information visualization environments.

References

Benzecri, J. P. (1976). Histoire et préhistoire de l'Analyse des données. *Cah. Anal. Données*, **1**, No. 1-4 (published in book form: 1982, Dunod, Paris).

Borg, I., Groenen, P. (1997). *Modern Multidimensional Scaling: Theory and Applications*. Springer-Verlag, New York.

Chen, C. H. (1996). The properties and applications of the convergence of correlation matrices. In: *1996 Proceedings of the Section on Statistical computing*, 49-54, American Statistical Association.

Chen, C. H. (1999). Extensions of generalized association plots (GAP). In: *1999 Proceedings of the Section on Statistical Graphics*, 111-116, American Statistical Association.

Chen, C. H. (2002). Generalized association plots: information visualization via iteratively generated correlation matrices. *Statistica Sinica*, **12**, 7-29.

Chen, C. H., Chen, J. A. (2000). Interactive diagnostic plots for multidimensional scaling with applications in psychosis disorder data analysis. *Statistica Sinica*, **10**, 665-691.

Chen, C. H., Chang, S. C., Chen, J., Chi, Y. Y., Li, Y. S., Ouyoung, C. W., and Peck, K. (2002). Information visualization for high dimensional categorical data structure with categorical generalized association plots (cGAP), *Technical Report, Academia Sinica, Taiwan, Institute of Statistical Science.*

Cox, T. F., Cox, M. A. A. (2001). *Multidimensional Scaling*. Chapman & Hall, London.

Eisen, M. B., Spellman, P. T., Brown, P. O., and Botstein, B. (1998). Cluster analysis and display of genome-wide expression patterns. *Proc. Nat'l. Acad. Sci. U. S. A.*, **95**, 14863-14868.

Fisher, R. A. (1936). The Use of Multiple Measurements in Axonomic Problems. *Annals of Eugenics*, 7, 179-188.

Fisher, R. A. (1938). The precision of discriminant functions. *The Annals of Eugenics*, **10**, 442-429.

Friendly, M. (1994). Mosaic displays for multi-way contingency tables. *Journal of the American Statistical Association*, **89**, 190-200.

Friendly, M. (1999). Extending Mosaic Displays: Marginal, Partial, and Conditional Views of Categorical Data. *Journal of Computational and Graphical Statistics*, **8**, 373-395.

Gifi, A. (1990). *Nonlinear multivariate analysis*. Wiley, Chichester.

Hartigan, J. A. (1975). *Clustering Algorithm*. New York: Wiley.

Hartigan, J. A., and Kleiner, B. (1981). Mosaics for contingency tables. In W. F. Eddy (Ed.), Computer Science and Statistics: *Proceedings of the 13th Symposium on the Interface*. New York: Springer-Verlag.

Hill, M. O. (1974). Correspondence analysis: a neglected multivariate method. *Applied Statistics*, **23**, 340-354.

Hwu, H. G., Chen, C. H., Hwang, T. J., Liu, C. M., Cheng, J. J., Lin, S. K., Liu, S. K., Chen, C. H., Chi, Y. Y., Ouyoung, C. W., Lin, H. N., and Chen, W. J. (2002). Symptom Patterns and Subgrouping of Schizophrenic Patients: Significance of Negative Symptoms Assessed on Admission. Paper accepted for publication. *Schizophrenia Research.*

Iyer, V. R., Eisen, M. B., Ross, D. T., Schuler, G., Moore, T., Lee, J. C., Trent, J. M., Staudt, L. M., Hudson Jr., J., Boguski, M. S., Lashkari, D., Shalon, D., Botstein, D., Brown, P. O. (1999). The transcriptional program in the response of human fibroblasts to serum, *Science*, Jan 1 1999, 83-87.

Lin A. S., Chen, C. H., Hwu H. G., Lin H. N., and Chen J. A. (1998). Psychopathological Dimensions in Schizophrenia: A Correlational Approach to Items of the SANS and SAPS. *Psychiatry Research*, **77**, 121-130.

Michailidis, G, and de Leeuw, J. (1998). The Gifi system for descriptive multivariate analysis. *Statistical Science*, **13**, 307-336.

Michailidis, G. and de Leeuw, J. (2000). Homogeneity analysis by alternating least absolute deviations. *Technical report, Department of Statistics, The University of Michigan.*

Nishisato, S. (1980). *Analysis of categorical data. Dual scaling and its applications*. University of Toronto Press.

Wilkinson, L. (1999). *The grammar of graphics*. Springer-Verlag, New York.

Supervised Learning from Microarray Data

Trevor Hastie[1], Robert Tibshirani[2], Balasubramanian Narasimhan[3] and Gilbert Chu[4]

[1] Statistics Department and Department of Health, Research and Policy, Stanford University, CA 94305, USA. hastie@stanford.edu
[2] Department of Health, Research and Policy, Stanford University, CA 94305, USA. tibs@stat.stanford.edu
[3] Statistics Department, Stanford University, CA 94305, USA. naras@stanford.edu Policy, Stanford University, CA 94305, USA. tibs@stat.stanford.edu
[4] Departments of Biochemistry and Medical Oncology, Stanford University, CA 94305. chu@cmgm.stanford.edu

Abstract. Gene expression arrays pose challenging problems for most traditional supervised learning techniques. We present a discussion of some of the issues involved. We then propose a simple approach to class prediction for DNA microarrays, based on a enhancement of the nearest centroid classifier. Our technique uses soft-thresholded class centroids as prototypes for each class. The shrinkage improves significantly prediction performance, and identifies a subset of the genes most responsible for class separation. The method performs as well or better than competitors from the literature, and is easy to understand and interpret. We illustrate the technique on data from three studies: small round blue cell tumors, leukemia and breast cancer.

Keywords. Microarrays, expression, classification, discriminant analysis.

1 Introduction

The statistical problem of class prediction is an important one, and has recently received a great deal of attention in the context of DNA microarrays. Here the task is to classify and predict the diagnostic category of a patient, based on his/her gene expression profile.

The microarray problem is special in that

- there are a large number p of inputs (genes) from which to predict classes,
- a relatively small number n of samples, and
- it is especially important to identify which genes contribute towards the classification.

Typically p is in the (tens of) thousands, while n is of the order of 100. This poses a statistical dilemma, since the impetus in data mining in recent years was prompted by huge sample sizes rather than a huge number of predictors. Despite this imbalance, there have been a wide variety of exotic techniques proposed to solve this difficult classification problem. For example, in a headlined article in *Nature Genetics*, Khan et al. (2001) proposed using a neural network to solve the classification problem. Their paper includes complex network diagrams, learning curves and requires thousands of fits of bootstrapped fits of a neural network.

It is clear that even the simplest linear methods for classification will require heavy regularization to work in this setting. In this paper we report on a very simple approach to class prediction, based on nearest centroids, that performs well and is easy to understand and interpret (Tibshirani et al. 2002). We have investigated regularized linear discriminant analysis and logistic regression as well; in the interest of space, we report on the former briefly and the latter elsewhere.

As our motivating example, we consider the data from Khan et al. (2001), which is available on their web site. These consist of expression measurements on 2308 genes from small,round blue cell tumors (SRBCT) of childhood. These tumors are classified into BL (Burkitt lymphoma), EWS (Ewing), NB (neuroblastima) and RMS (rhabdomyosarcoma). There are 63 training samples and 25 test samples, although five of the latter were not SRBCTs. Using their complex neural network approach, Khan et al. (2001) achieve a test error of 0%, and are able to identify 96 genes that are most important to the classification.

The starting point for our method is nearest centroid classification. Figure 1 (light grey bars) shows the training-set centroids for each of the 4 classes, represented as contrasts from the overall centroid.

To apply nearest centroid classification, we take the gene expression profile of the test sample (array), and compute its squared distance from each of the 4 class centroids. The predicted class is the one whose centroid is closest to the expression profile of the test sample. This procedure makes 5 errors on the 20 test samples (2 on the training), and has the major drawback that it uses all 2308 genes.

Here we propose a simple modification of the nearest-centroid method, "nearest prototype", which uses de-noised versions of the centroids as prototypes for each class. The shrunken centroids are the red bars in Figure 1. Classification is then made to the nearest prototype. The resulting procedure has 0 test and 0 training errors. In addition, only 94 genes have a non-zero red bar for one or more classes in Figure 1, and hence are the only ones that contribute towards the classification. The amount of shrinkage is determined by cross-validation.

The nearest prototype method can also be used in unsupervised problems as well. For example, it is by now standard to use hierarchical clustering methods on expression arrays to discover clusters in the samples (Eisen et al. 1998). The methods described here can identify subsets of the genes that succinctly characterize each cluster.

2 Details of centroid shrinkage

Let x_{ij} be the expression for genes $i = 1, 2, \ldots p$ and samples $j = 1, 2, \ldots n$. We have classes $1, 2, \ldots K$, and let C_k be indices of the n_k samples in class k. The ith component of the centroid for class k is $\bar{x}_{ik} = \sum_{j \in C_k} x_{ij}/n_k$, the mean expression value in class k for gene i; the ith component of the overall centroid is $\bar{x}_i = \sum_{j=1}^{n} x_{ij}/n$.

In words, we shrink the class centroids towards the overall centroids. However, we first normalize by the within class-standard deviation for each gene. Let

$$d_{ik} = (\bar{x}_{ik} - \bar{x}_i)/s_i, \tag{1}$$

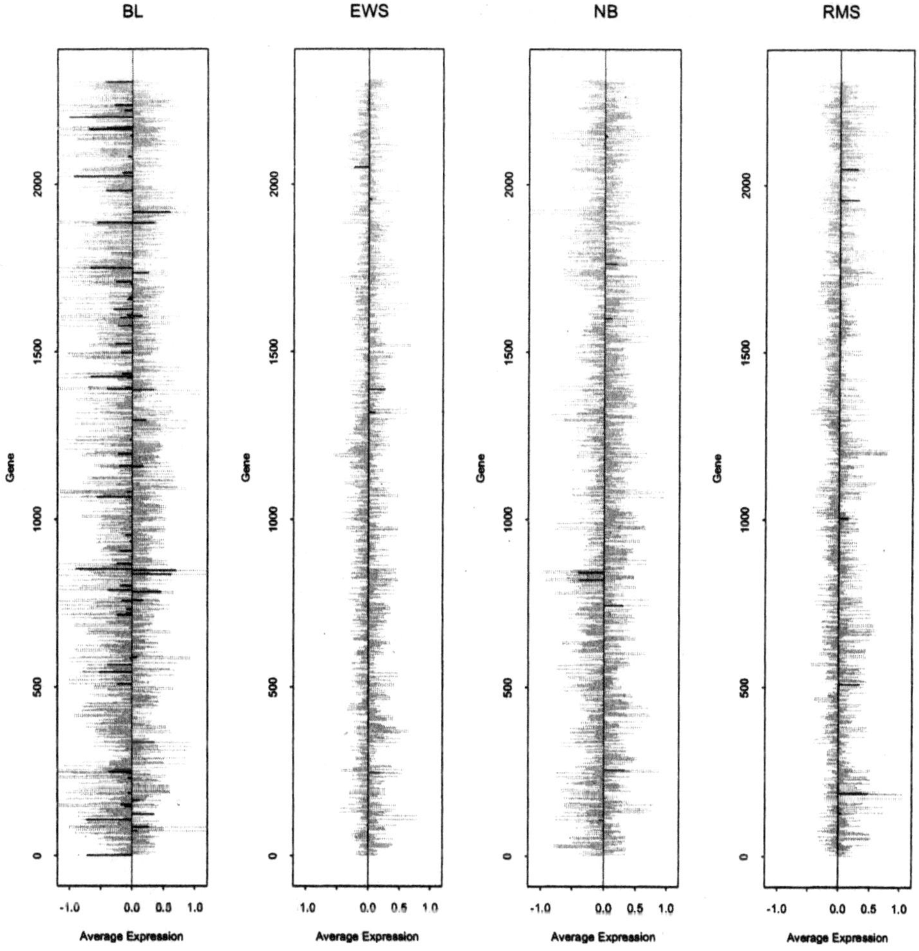

Fig. 1. *Centroids (grey) and shrunken centroids (red) for the SRBCT dataset. Each centroid has the overall centroid subtracted, and hence what we see are contrasts.*

where s_i is the pooled within class standard deviation for gene i:

$$s_i^2 = \frac{1}{n-K}\sum_k \sum_{i \in C_k} (x_{ij} - \bar{x}_{ik})^2. \tag{2}$$

This d_{ik} is a kind of t-statistic for gene i, comparing class k to the other classes. We can write

$$\bar{x}_{ik} = \bar{x}_i + s_i d_{ik} \tag{3}$$

Our proposal shrinks each d_{ik} towards zero, giving d'_{ik} and new shrunken centroids or prototypes

$$\bar{x}'_{ik} = \bar{x}_i + s_i d'_{ik} \tag{4}$$

The shrinkage we use is called *soft-thresholding*: each d_{ik} is reduced by an amount Δ in absolute value, and is set to zero if it's absolute value is less than zero. Algebraically, this is expressed as

$$d'_{ik} = \text{sign}(d_{ik})(|d_{ik}| - \Delta)_+ \tag{5}$$

where + means *positive part* ($t_+ = t$ if $t > 0$, and zero otherwise). Since many of the $\bar{x}_{ik}$ will be noisy and close to the overall mean $\bar{x}_i$, soft-thresholding produces "better" (more reliable) estimates of the true means (Donoho & Johnstone 1994, Tibshirani 1996). It has the nice property that many of the components (genes) are effectively eliminated as far as class prediction is concerned, if the shrinkage parameter Δ is large enough. We choose Δ by cross-validation, as illustrated below.

Note that the standardization by s_i above has the effect of giving higher weight to genes whose expression is stable within samples of the same class. This same standardization is inherent in other common statistical methods, such as linear discriminant analysis.

3 Class probabilities and discriminant functions

We classify test samples to the closest shrunken centroid, again standardizing by s_i. We also make a correction for the relative abundance of members of each class.

Suppose we have a test sample (a column vector) with expression levels $x^* = (x_1^*, x_2^*, \ldots x_p^*)$. We define the *discriminant score* for class k

$$\delta_k(x^*) = \sum_{i=1}^{p} \frac{(x_i^* - \bar{x}'_{ik})^2}{s_i^2} - 2 \log \pi_k \tag{6}$$

The first part of (6) is simply the standardized squared distance of x^* to the kth shrunken centroid. The second part is a correction based on the class *prior probability* π_k, where $\sum_{k=1}^{K} \pi_k = 1$. This prior gives the overall proportion of class k in the population. The classification rule is then

$$C(x^*) = \ell \text{ if } \delta_\ell(x^*) = \min_k \delta_k(x^*) \tag{7}$$

If the smallest distances are close and hence ambiguous, the prior correction gives a preference for larger classes, since they potentially account for more errors. We usually estimate the π_k by the *sample priors* $\hat{\pi}_k = n_k/n$. If the sample prior is not representative of the population, then more realistic priors can be used instead, or even uniform priors $\pi_k = 1/K$.

We can use the discriminant scores to construct estimates of the class probabilities, by analogy to Gaussian linear discriminant analysis:

$$\hat{p}_k(x^*) = \frac{e^{-\frac{1}{2}\delta_k(x^*)}}{\sum_{\ell=1}^{K} e^{-\frac{1}{2}\delta_\ell(x^*)}} \tag{8}$$

Figure 2 displays these probabilities for the training data (top) and test samples (bottom). All samples have at least 25% separation between the highest and next highest probability.

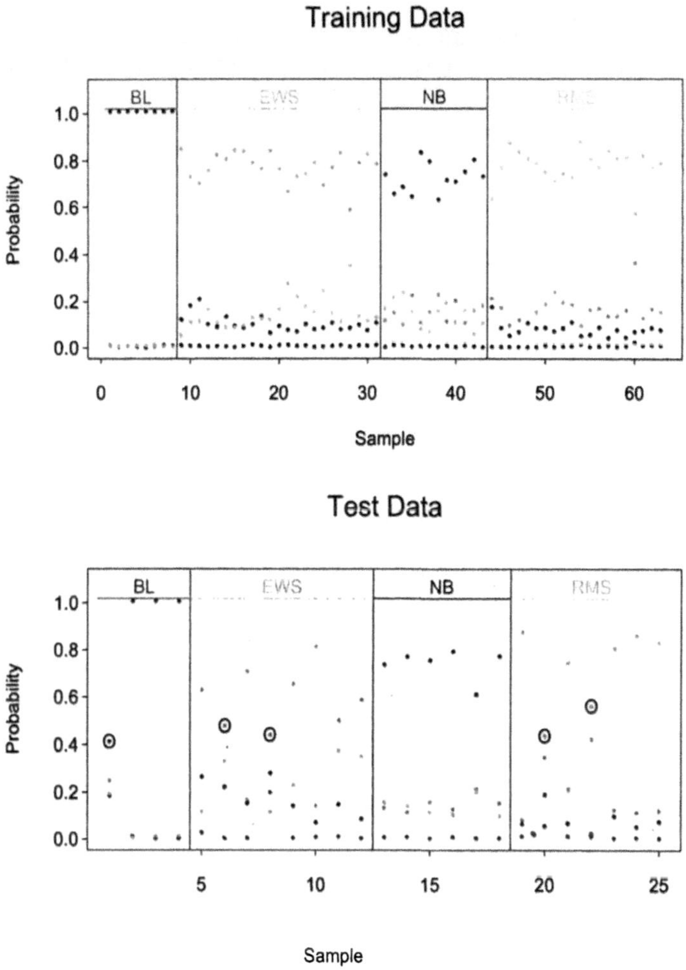

Fig. 2. Estimated probabilities for the training data (top) and test data (bottom). Probabilities are partitioned by the true class in the top panel, and the predicted class in the bottom panel. All 20 of the SRBCT (small round blue cell tumors) test samples are correctly classified. Five of the test samples are non-SRBCT, and hence should not be classified as such. The maximum estimated probability for these five samples is marked with a circle: they are (slightly) below the maximum probabilities for the other test samples in each class.

The discriminant scores (6) are similar to those used in linear discriminant analysis (LDA), which arise from using the *Mahalanobis* metric in computing distance to centroids:

$$\delta_k^{LDA}(x^*) = (x^* - \bar{x}_k)^T W^{-1}(x^* - \bar{x}_k) - 2\log \pi_k. \tag{9}$$

Here we are using a vector notation, and W is the pooled, within-class covariance matrix. With thousands of genes and tens of samples ($p \gg n$), W is huge and any sample estimate will be singular (and hence its inverse is undefined) Our scores can be seen to be a heavily restricted form of LDA, necessary to cope with the large number of variables (genes). The differences are that

- we assume a diagonal within-class covariance matrix for W; without this LDA would be ill-conditioned and fail.
- we use shrunken centroids rather than centroids as a prototype for each class.
- as Δ increases, an increasing number of genes will have *all* their $d'_{ik} = 0,\ k = 1, \ldots, K$ due to the soft-thresholding in (5). Such genes contribute no discriminatory information in (6), and in fact cancel in (8)

We discuss other ways of restricting LDA in Section 6.

Both our scores (6) and (9) are effectively *linear* in the x_i^*. If we expand the square in (6), discard the terms involving x_i^{*2} (since they are independent of the class index k and hence do not contribute towards class discrimination), and multiply by -2, we get

$$\tilde{\delta}_k(x^*) = \sum_{i=1}^{p} \frac{x_i^* \bar{x}'_{ik}}{s_i^2} - \frac{1}{2} \sum_{i=1}^{p} \frac{\bar{x}'^{\,2}_{ik}}{s_i^2} + \log \pi_k, \tag{10}$$

which is linear in the x_i^*. Because of the sign change, our rule classifies to the largest $\tilde{\delta}_k(x^*)$. Likewise the LDA discriminant scores have the equivalent linear form

$$\tilde{\delta}_k^{LDA}(x^*) = x^{*T} W^{-1} \bar{x}_k - \frac{1}{2} \bar{x}'^{\,T}_k W^{-1} \bar{x}'_k + \log \pi_k \tag{11}$$

This linear version of our score (10) makes for easier comparisons with other linear scoring proposals in the literature (see (Tibshirani et al. 2002)).

4 Choosing the amount of shrinkage

Figure 3 shows the training, test, and cross-validation error as the shrinkage parameter Δ is varied. The "Size" axis at the top of the plot indicates the number of genes retained (for the training data) at that particular threshold.

We used balanced 10-fold cross-validation, ensuring that the classes were distributed evenly among the 10 folds. The left end of the figure represents no shrinkage, while the right end represents complete shrinkage. Both the cross-validated and test error are minimized near $\Delta = 1.02$, which is the value we used to produce the red curves in Figure 1. The upper axis shows the number of *active* genes with at least one non-zero component d'_{ik}, as Δ is varied. At $\Delta = 1.02$ there are 94 active genes.

5 The Genes that Matter

Figure 4 shows the shrunken differences d_{ik} for only the 94 genes having at least one non-zero difference. We have ordered the genes from bottom to top, with non-zero differences for classes EWS, NB, RMS in the bottom three panels. In the large top panel are the genes with non-zero differences only in BL.

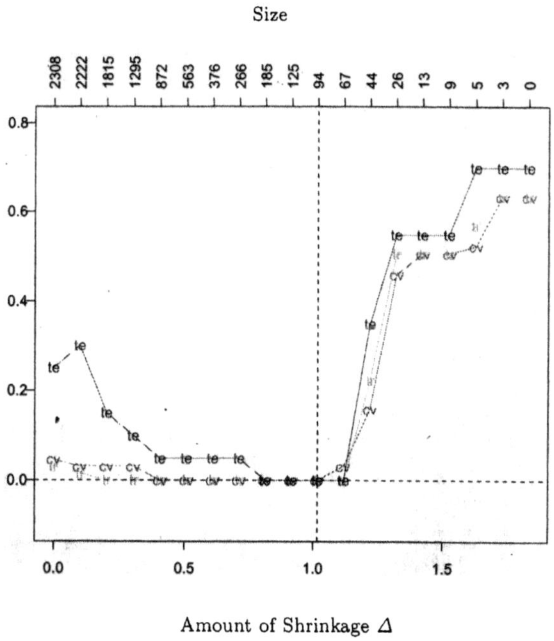

Fig. 3. *Training, test, and cross-validation error as the threshold parameter Δ is varied. The value $\Delta = 1.02$ is chosen, resulting in a subset of 94 selected genes.*

Comparing these genes to the 96 genes reported by Khan et al. (2001), the two lists have 59 genes in common.

Figure 5 shows the heat-map of the chosen 94 genes. Within each of the horizontal partitions, we have ordered the genes by hierarchical clustering, and similarly for the samples within each vertical partition.

Now we look back at the $2308 \times 4 = 9232$ unshrunken t-statistics d_{ik}. The 94th largest in absolute value is only 1.965. This value would not be declared significant in the context of 9232 t-statistics. Similarly, we can perform a separate one-way anova for each gene. The 94th smallest p-value is 0.16, and only 36 genes make the 5% cut-off. This demonstrates how this procedure differs from the more common approach of screening genes by significance of their individual t-statistics or p-values. Our procedure uses soft-thresholding rather than screening, and focuses on misclassification error. The latter is a more relevant measure for class prediction problems.

6 Other approaches

We saw in Section 3 that our nearest prototype scores (6) are a restricted form of the LDA scores (9)

$$\delta_k^{LDA}(x^*) = (x^* - \bar{x}_k)^T W^{-1}(x^* - \bar{x}_k).$$

Since $p \gg n$, the within-class covariance W is singular, and so this solution is undefined. Nearest shrunken centroids makes W diagonal which solves the singularity problem; in addition it shrinks the $\bar{x}_k$.

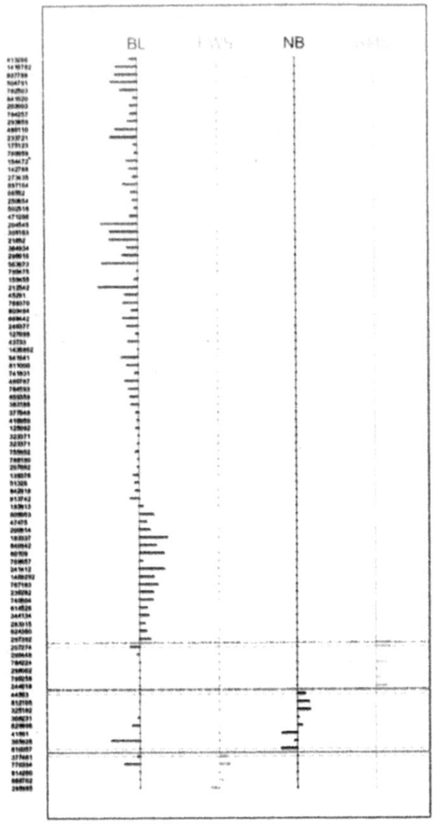

Fig. 4. *Shrunken differences d_{ik} for the 94 genes having at least one non-zero difference. We have ordered the genes from bottom to top, with non-zero differences for classes EWS, NB, RMS in the bottom three panels.*

Regularized Discriminant Analysis (Friedman 1989) leaves the centroids alone, and modifies the covariance matrix in a different way:

$$\delta_k^{RDA}(x^*) = (x^* - \bar{x}_k)^T (W + \lambda I)^{-1} (x^* - \bar{x}_k), \tag{12}$$

where λ is a parameter (like our Δ.) The fattened $W + \lambda I$ is nonsingular, and as λ gets large, this procedure approaches the nearest centroid procedure (with no variance scaling, nor centroid shrinking). A slightly modified version uses $W + \lambda D$, where $D = \text{diag}(s_1^2, s_2^2, \ldots, s_p^2)$. As λ gets large, this approaches the variance weighted nearest centroid procedure. In practice, we normalize this regularized covariance by dividing by $1 + \lambda$, leading to the convex combination $(1 - \alpha)W + \alpha D$, where $\alpha = \lambda/(1 + \lambda)$. Although the relative distances do not change, this is important when making the adjustment for the class priors.

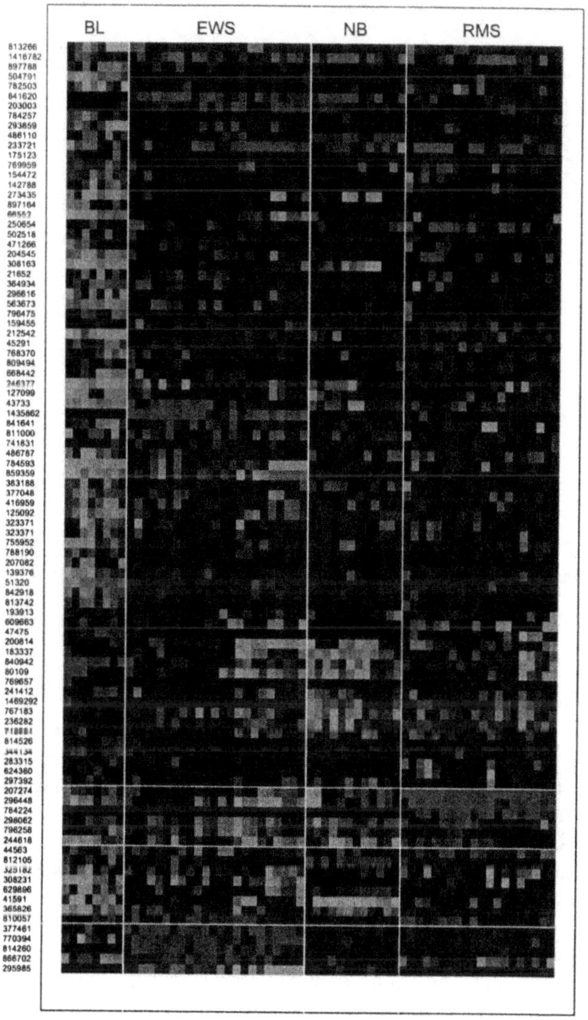

Fig. 5. *Heat-map of the chosen 94 genes. Within each of the horizontal partitions, we have ordered the genes by hierarchical clustering, and similarly for the samples within each vertical partition.*

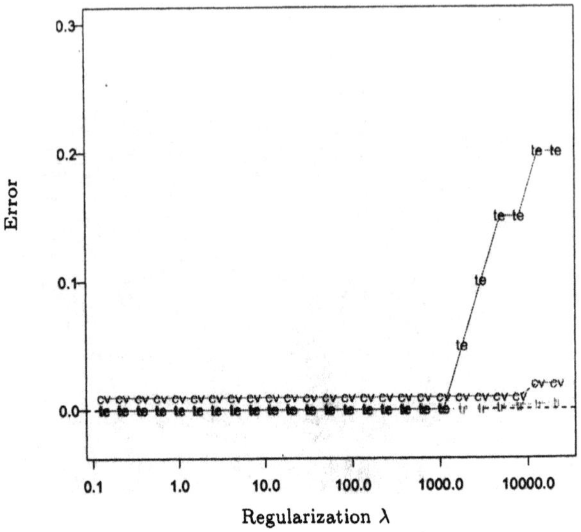

Fig. 6. *SRBCT data. Training, CV and test error curves for RDA, as a function of the regularization parameter* λ.

Figure 6 shows the training, CV and test error curves for the SRBCT data, as a function of λ. Although for a wide range of values for λ, RDA achieves zero training and test error, the CV curve in this example would not select the best model (if we follow the convention of selecting the most restricted model with minimum CV).

Although RDA shows some promise, it is more complicated than our nearest shrunken centroid procedure. Furthermore, in the process of its regularization, it does not select a subset of genes as the shrunken centroid procedure does. We are considering other hybrid approaches of RDA and nearest centroids in ongoing research projects.

The neural network approach of Khan et al. (2001) can also be interpreted as a form of regularized linear discriminant analysis. In that paper they actually use a *linear* network, using the first 10 principal components (eigengenes). In statistical parlance, this is known as principal components regression, and requires no iterative learning procedure and learning curves. Principal component regression is a hard-thresholded version of ridge regression, which in turn has strong connections with the RDA procedure described above (Hastie et al. 2001). The authors do use a model-averaging procedure, similar to bagging (Breiman 1996) but based on three-fold cross-validation, to regularize the procedure further. Although their procedure also produced zero training and test errors on the SRBCT data, it is far more complex than either nearest shrunken centroids or RDA. With so many genes and so few samples, it is very likely that restricted versions of standard statistical routines will suffice, as is the case with our procedure.

7 Software and computational details

The computations involved in the shrunken nearest centroid method are quite straightforward. One important detail—in the denominator the the statistics d_{ik} in (1) we add (the same) positive constant s_0 to each of the s_i values. This guards against the possibility of large d_{ik} values arising by chance, from genes at very low expression levels. We set s_0 equal to the median value s_i over the set of genes. A similar strategy was used in the SAM methodology of Tusher et al. (2001). We are currently developing a program similar to SAM, to implement nearest centroid classification.

The computations in RDA look horrendous, since it seems we have to invert a $p \times p$ matrix $W + \lambda I$. However, we can exploit the fact that W has rank $n - K$, and using the singular value decomposition of the expression matrix X adjusted for the class centroids, the computations can be performed efficiently.

References

Breiman, L. (1996), 'Bagging predictors', *Machine Learning* **26**, 123–140.

Donoho, D. & Johnstone, I. (1994), 'Ideal spatial adaptation by wavelet shrinkage', *Biometrika* **81**, 425–455.

Eisen, M., Spellman, P., Brown, P. & Botstein, D. (1998), 'Cluster analysis and display of genome-wide expression patterns', *Proc. Natl. Acad. Sci., USA.* **95**, 14863–14868.

Friedman, J. (1989), 'Regularized discriminant analysis', *Journal of the American Statistical Association* **84**, 165–175.

Hastie, T., Tibshirani, R. & Friedman, J. (2001), *The Elements of Statistical Learning; Data mining, Inference and Prediction*, Springer Verlag, New York.

Khan, J., Wei, J., Ringner, M., Saal, L., Ladanyi, M., Westermann, F., Berthold, F., Schwab, M., Antonescu, C., Peterson, C., & Meltzer, P. (2001), 'Classification and diagnostic prediction of cancers using gene expression profiling and artificial neural networks', *Nature Medicine* **7**, 673–679.

Tibshirani, R. (1996), 'Regression shrinkage and selection via the lasso', *J. Royal. Statist. Soc. B.* **58**, 267–288.

Tibshirani, R., Hastie, T., Narasimhan, B. & Chu, G. (2002), 'Diagnosis of multiple cancer types by shrunken centroids of gene expression', *Proceedings of the National Academy of Sciences* .

Tusher, V., Tibshirani, R. & Chu, C. (2001), 'Significance analysis of microarrays applied to transcriptional responses to ionizing radiation', *Proc. Natl. Acad. Sci. USA.* **98**, 5116–5121.

Teaching Statistics with Electronic Textbooks

Jürgen Symanzik[1] and Natascha Vukasinovic[1]

[1] Utah State University, Department of Mathematics and Statistics, 3900 Old Main Hill, Logan, UT 84322–3900, USA

Abstract. In the first part of this paper, we will provide a general overview on electronic "Introductory Statistics" textbooks and summarize their individual strengths and weaknesses. Examples of such electronic textbooks are the CD–ROM–based ActivStats, the Web–based CyberStats, and the CD–ROM– and Web–based MM*Stat,

In the second part of this paper, we will present our experiences from teaching an undergraduate "Introductory Statistics" course at Utah State University, using CyberStats. We compare this Web–based course with similar textbook–based courses and report student and instructor viewpoints.

Keywords. ActivStats, CyberStats, MM*Stat, Introductory Statistics

1 Introduction

Learning statistics requires students to develop a variety of skills including quantitative and graphical insights along with mathematical and analytical abilities (Härdle, Klinke & Marron 1999). In addition, it is sometimes difficult for students to relate statistical concepts presented in class to real world problems and everyday situations. Therefore, introductory statistics courses are commonly considered difficult and intrinsincly boring by students.

The idea to use computers for statistical teaching dates back to times before the Web became widely popular and before everyone had almost immediate access to a personal computer equipped with a CD–ROM drive. Snell & Peterson (1992), for example, state that "the computer can be used profitably in almost any elementary mathematics course, but it is especially helpful in teaching probability and statistics. [...] We have identified three broad areas in which the computer is helpful: reducing the need for lengthy manual calculations, facilitating graphical data analysis, and illustrating statistical concepts by means of simulation experiments." As it turns out, modern electronic textbooks, combined with an underlying statistical software package, show some of their strengths in exactly these fields. Interactivity, hands–on exercises, colorful applets, well–documented real–life examples, and the possibility to analyze even larger data sets quickly and effectively may be the most attractive features of statistical teaching software (or teachware).

An increasing level of computer literacy among students and the availability of necessary equipment in classrooms have stimulated developing and using statistical teaching software in introductory statistics courses. Using statistical teachware in introductory statistics courses has almost become a standard over the past years. Using software when teaching statistics varies from occassional interactive exercises and in–class software demonstrations to completely Web–based courses without any use of "traditional" teaching tools or even pen and paper, as in Kent L. Norman's courses (`http://cognitron.umd.edu/cognitron.html`).

Chromiak & Rossman (1992) describe how to use HyperCard, a tool that was freely available for Macintosh computers, to teach statistics. Their ideas

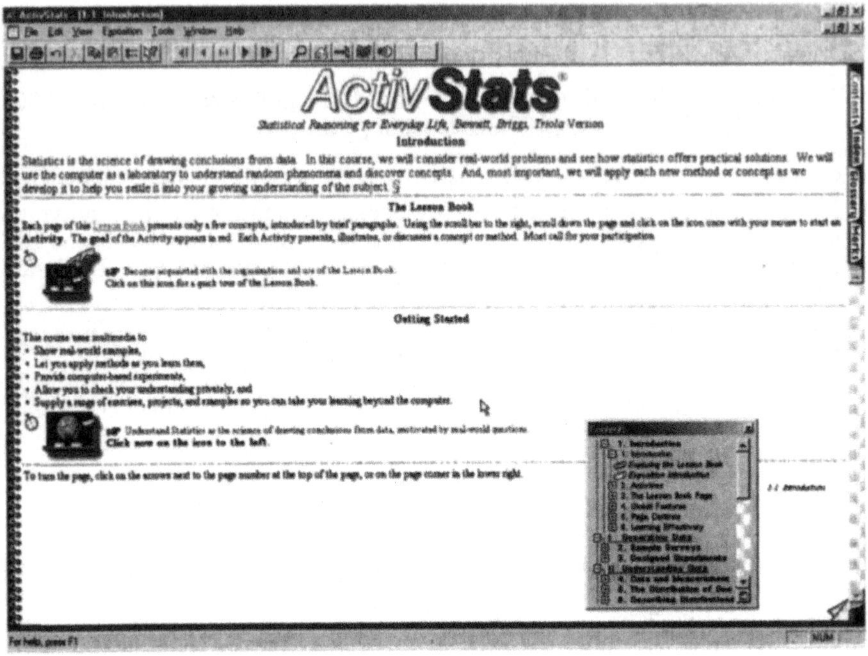

Fig. 1. Contents Page of ActivStats.

are to get the students involved, give students immediate feedback, present the same idea in many different ways, and focus on concepts rather than on calculations. Lock (1997) provides a general overview on Internet resources for teaching statistics. Currall (1997) and Rönz (1997) discuss general ideas of computer–aided teaching in statistics. Symanzik (1998) summarizes additional resources related to Web–based teaching.

The two main purposes of statistical teaching software can be identified as follows:

1. A teaching software should enable better and easier understanding of (rather abstract) underlying statistical concepts through interactive examples and exercises.
2. Using teaching software should provide a more interesting and active learning environment. Engaging into hands–on activites can help capturing attention and interest especially of weaker and less motivated students.

There are many statistical teachware packages and electronic textbooks available. Some of them are freely accessible via the Internet, such as the GASP educational procedures (West, Ogden & Rossini 1998, West & Ogden 1998*a*), accessible at `http://www.stat.sc.edu/rsrch/gasp/`, Hyperstat by David M. Lane (Lane 1999), accessible at `http://davidmlane.com/hyperstat/`, and the UCLA Statistics e–book by Jan de Leeuw (de Leeuw 1997), accessible at `http://www.stat.ucla.edu/textbook/`, to mention only a few. Others such as Seeing Statistics (`http://www.seeingstatistics.com/`) or CyberStats (`http://www.cyberk.com`) are commercial packages that al-

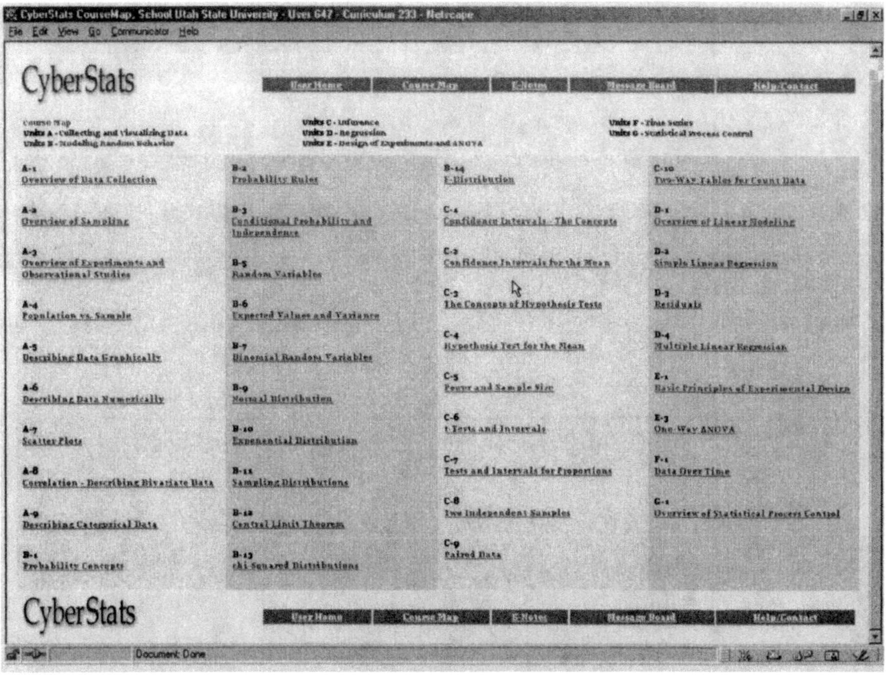

Fig. 2. Contents Page of CyberStats.

low Internet access upon registration. Some of the commercial teachware packages are distributed on a CD–ROM in addition to Internet access e.g., MM*Stat (`http://www.quantlet.com/mdstat/mmstat.html`), or on a CD–ROM only, e.g., ActivStats (`http://www.datadesk.com/ActivStats`).

In the first part of this paper, we have comparatively reviewed three popular commercial electronic textbooks: ActivStats, CyberStats, and MM*Stat. In the second part of the paper, we report about our own experiences in teaching an introductory statistics course to undergraduate students at Utah State University in the Fall 2001 semester using CyberStats.

2 Comparative Review: ActivStats, CyberStats, and MM*Stat

The three electronic textbooks ActivStats, CyberStats, and MM*Stat are commercial products that can be purchased by students and organizations. All three packages exist for several years and have been actively used at a variety of schools.

The oldest package is ActivStats by Paul Velleman, whose first version was released around 1996 by Adison Wesley, Inc. and Data Description, Inc. The current version of ActivStats (2001–2002 release) is available for purchase at a price of \$51 and \$225 for individual and commercial users, respectively. New releases of ActivStats appear for each academic year to keep up with advances in software and operating systems and to add new material. The package is distributed on a CD–ROM accompanyed by a 30–page "User's Guide" booklet. Macnaughton (1998) and Maatta (1999) provide thorough

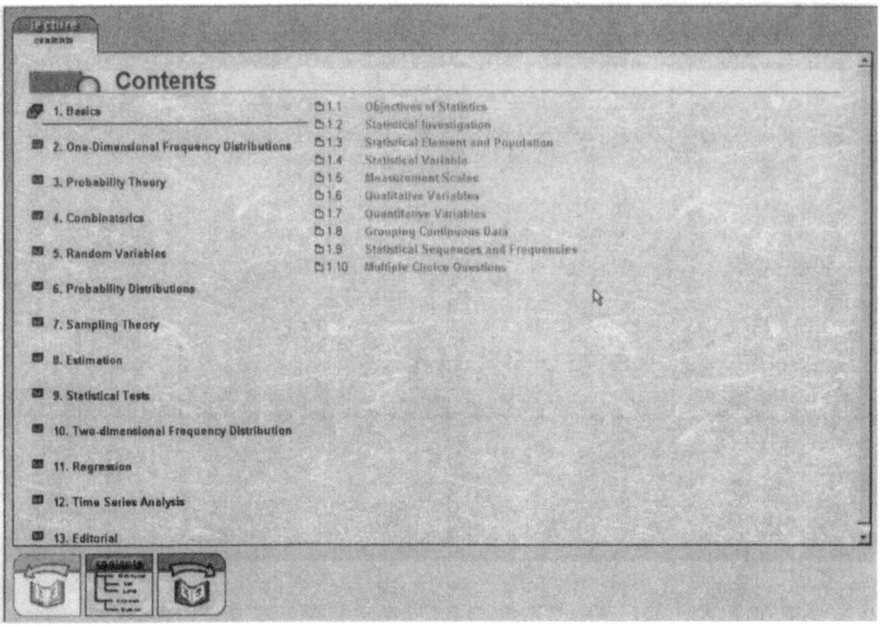

Fig. 3. Contents Page of MM*Stat.

reviews of ActivStats.

CyberStats is a product of CyberGnostic, Inc. The first version of CyberStats appeared around 1999 and was released to interested individuals free of charge exclusively for test purposes. Since January 1, 2001, CyberStats 2.0 has been offered as a commercial product. CyberStats is a fully Web-based package and access is granted for about one semester upon registration and payment of $30 per individual user.

The multimedia project MM*Stat (Rönz, Müller & Ziegenhagen 2000, Härdle, Lehmann & Rönz 2001, Rönz 2001) has been developed at Humboldt–University at Berlin and has been released as a commercial product in 2001. The German version of the package is distributed by Springer–Verlag on a CD–ROM accompanied by a short manual booklet at a price of EUR 17.20 and can be ordered from `http://www.springer.de/cgi-bin/search_book.pl?isbn=3-540-14893-0`. The English version and other foreign language versions of this package are Web-based and are currently freely accessible at `http://www.quantlet.com/mdstat/mmstat.html`. The German version was previously also available on the Web from a password protected site. The MM*Stat developers are closely collaborating with Springer–Verlag in Germany, a major publisher of electronic books. It is planned that in the near future, readers/users will gain access to the Web–based version of MM*Stat and other electronic books after payment of an access fee.

All three packages can be considered complete electronic textbooks. They possess all the features of a traditional statistics textbook: coverage of statistical concepts with text, graphics, and exercises, including interactive table of contents, index, glossary, and self–assessment tools. In addition, CyberStats provides all tools necessary for complete course management, such as instructor's and students' course Web pages, electronic submission of homework assignments, a message board, and a chat room for students.

Both ActivStats and CyberStats target undergraduate students, especially non–science majors, and emphasize concept understanding, data vizualization, and data analysis. Less emphasis is put on formulas and mathematical components of statistics. Figure 1 shows the start page of ActivStats with the opened content window. In ActivStats, the sequence of topics follows closely the "standard" sequence for introductory courses set by one of the currently most popular introductory textbooks (Moore & McCabe 1999): data and variables, univariate distributions, regression and correlation, probability and randomness, random variables, sampling distributions, and statistical inference. ActivStats is a multimedia package that, in addition to "traditional" examples and exercises, also contains narrated lectures and mini–videos that enormously influence students' perception of the topic. Each of the 24 lectures includes several activities in form of interactive exercises, data analysis exercises, and concept–review exercises. Also, each lecture is followed by a set of homework questions and a set of projects in which students have to actually collect data and analyze them using the built–in data analysis package.

In CyberStats, the course content is divided into seven major units: collecting and visualizing data, modeling random behavior, inference, regression, design of experiments and ANOVA, time series, and statistical process control. Each unit is divided into several subunits, as presented in Figure 2. Each subunit represents a separate lecture that begins with a "Summary", followed by a set of motivational questions related to the topic ("Think first"). The actual lecture is presented in form of "Three Keys": "Basics", where the basic concepts are presented and illustrated by real–life examples and interactive exercises; "Uses", where a student has a chance to work through a set of examples and exercises; and "Warning", where potential dangers of wrongly used statistical concepts are discussed. Each of these three keys is accompanied by plenty of examples and exercises. An additional set of exercises is provided in "Examples", that is followed by a self–assessment test. In each exercise set, a student can submit the answers to the system, which are automatically recorded on the CyberStats server and can be accessed by the instructor. A student actively aquires knowledge of the topic by following the pages step–by–step and doing the interactive exercises.

As opposed to ActivStats and CyberStats, MM*Stat targets students in science and engineering, using a more abstract level of presentation and a different order of topics. Although it contains many real–life examples and interactive exercises, the emphasis is on definitions, formulas, and underlying mathematical concepts. The topics include: basics, one–dimensional frequency distributions, probability theory, combinatorics, random variables, probability distributions, sampling theory, estimation, statistical tests, two–dimensional frequency distributions, regression, and time series analysis, as shown in Figure 3. Each topic is divided into several subtopics that are presented as separate lectures. The text of each lecture appears on the screen in a scrollable format. Additional information and detailed examples can be accessed by clicking on the corresponding icons at the bottom of the screen. "Interactive" icons provide access to interactive exercises based on XploRe quantlets. A student can open several windows, or "cards", at the same time. Each topic ends with a set of multiple choice questions. Student's answers can be evaluated with or without indicating errors.

Each of the three electronic textbooks is tightly linked to a data analysis software package that can be used from inside the textbook for an immediate presentation of examples and interactive exercises, but also for analyzing one's own data. The data analysis package integrated into ActivStats is the award–winning Data Desk 6.1 (Data Description, Inc.). ActivStats is also available in versions that teach JMP, Minitab, SPSS, and Excel. CyberStats is equipped with Data Tools, a clone of the popular WebStat software (West et al. 1998, West & Ogden 1998*b*). MM*Stat incorporates the well–known XploRe software (Härdle, Klinke & Müller 2000). In terms of using the data analysis package, ActivStats seems to be most convenient, because the data analysis software is tightly integrated into the package and is directly accessible from the CD–ROM. If a student clicks on an activity or exercise that uses Data Desk, Data Desk is automatically launched and loaded with step–by–step instructions and the data set to be analyzed. On the other hand, CyberStats and MM*Stat lack a direct integration of the data analysis software. The accompanying software must be loaded from the server, which is very time demanding and might be disturbing for a student when the Internet connection is slow.

3 Teaching Experiences with CyberStats

In the Fall 2001 semester, Utah State University (USU) ran two sections of its introductory statistics course "Stat 2000". One section was a regular textbook–based section, using Moore & McCabe (1999). The other section was a Web–based section based on CyberStats. Students had a choice which section they wanted to attend. The regular section had 17 students and the Web–based section had 14 students. One student that was initially enrolled in the Web–based section changed to the regular section within the first few lectures, claiming that he did not want to participate as a guinea pig in an untested version of this course. There were 42 lectures at 50 minutes each.

The reason to run two sections of this course (instead of just the Web–based section) where usually only one section has been offered had practical reasons — the largest available computer classroom could not hold the expected number of students for this course. Although the results described below are only the outcome of an observational study and not of a controlled experiment, we have made some interesting observations. These may be of interest not only for instructors using CyberStats but also for instructors interested in working with other electronic textbooks.

Table 1 shows the grade distribution (in full letter grades) for the regular and Web–based sections in relation to some of the previous offerings. No major difference can be noticed for the A, B, C, and D grades. The relatively high percentage of A grades (5 out of 14, i.e., 36%) in the Web–based section can be easily explained by chance, recalling that there were only 14 students in that section. However, the lack of any F grades in both sections is noticeable. Confounding factors such as smaller class size (previous offerings had between 32 and 48 students in a section), closer interaction between instructor and students, and overall higher attendance may lead to this outcome.

Grade	Text (Fall 1999)	Text (Spring 2000)	Text (Fall 2001)	Web (Fall 2001)
A	25%	22%	18%	36%
B	31%	38%	47%	29%
C	25%	16%	35%	29%
D	9%	5%	0%	7%
F	9%	19%	0%	0%

Table 1. Grade distribution during different offerings of Stat 2000.

Due to the different arrangement of lecture topics in Moore & McCabe (1999) and CyberStats, it was not possible to provide identical exams for the two sections. However, about 50% of the questions in all exams were identical for both sections. There was no significant difference in the performance between the two sections.

Homework assignments (a maximum of 300 points) consisted of textbook/ CyberStats questions, old exam questions, and other questions. There was a noticeable difference in homework performance in both sections as shown in the following summary statistics:

```
Summary: Regular Section
   Min.  1st Qu.  Median  Mean  3rd Qu.  Max.
   196    290      299     289    300     300
Summary: Web-based Section
   Min.  1st Qu.  Median  Mean  3rd Qu.  Max.
    38    269      290     262    299     300
```

Students from the regular section did far better in homeworks than students from the Web-based section. A breakdown of the homework questions of the Web-based section into CyberStats and non-CyberStats questions reveals some interesting results. Students from the Web-based section did similarly well for the CyberStats questions (which had to be submitted electronically) as the students from the regular section for the textbook questions. But the weaker students from the Web-based section often did not turn in the non-CyberStats questions (which could be e-mailed, faxed, or turned in during class).

3.1 CyberStats Course Material

Prior to teaching the Web-based section of "Stat 2000", it was planned to cover material that is also covered in the regular section. This would have been units A-1 through A-8, B-1 through B-9, B-11, B-12, C-1 through C-4, C-6, and D-1 through D-3 in CyberStats (see Figure 2). Additional material to be discussed in both sections covers use and misuse of statistical graphics, scatterplots and the idea of linked and geographic scatterplot brushing, micromaps, and mosaic plots.

However, it was not possible to discuss all material that was previously planned in CyberStats. Units B-11, C-1 through C-4, C-6, and mosaic plots have not been discussed in CyberStats. There are several reasons for these omissions:

- Interactive teaching using the Web in class requires considerably more time than teaching the same material on the board. The instructor has to make sure that each student arrives at the intended Web page that contains the interactive applet that is being used in class. At the beginning of a class, it often took 5 to 7 minutes before each student arrived at a designated Web page (going to CyberStats first, logging on, navigating from the main menu to a unit, and finding and starting the interactive in a unit).
- Technical problems, related to the USU computer system (and not to CyberStats) required some spontaneous reorganization of lectures. During four lectures, the instructor's computer was not able to access the Web, during another four lectures, students did not have Web access (instructor's computer and student computers were differently routed). This considerably affected the flow of those eight lectures, having to work on the board and demonstrating the most useful interactives in a future lecture.
- Incompatibilities between different settings with respect to Web browsers and Data Tools/WebStat caused major delays in class. Due to the restricted availability of the computer classroom, it was not possible to

test all features of CyberStats in the classroom. Instead, a laptop and the instructor's office computer have been used for testing. Usually, everything worked as expected. Unfortunately, long delays occured in the classroom because features for incorporating data (copy/paste, load from URL, etc.) worked differently (or did not work at all) under Netscape and Internet Explorer for Data Tools and WebStat. After a complete failure in one lecture to load data, it took half of the next lecture before every student was using the correct Web browser and the correct approach to incorporate the data.

- Additional depth of some of the material in CyberStats compared to the regular textbook resulted in more time spent on some of the topics than previously planned.

Overall, working with CyberStats requires a careful reevaluation what can be taught in one semester due to unavoidable delays when using technology in class and problems with computers, Web access, and related technology that are likely to occur over the course of a semester.

3.2 Teaching Style with CyberStats

Regarding the teaching style, several alternatives have been tried. In the beginning of the semester, where topics such as data collection, sampling, experiments and observational studies had been discussed, the course format was mostly discussion–based (with no extra handouts), using examples and the self assessment questions from CyberStats as basis. Later in the semester, when talking about linear regression, a more hands–on approach was used, with some comments and sketches on the board and more work with the interactives and Data Tools/WebStat. There was little focus on hand–calculations and formulas, although formulas have been briefly explained. Eventually, topics such as variance/standard deviation, binomial random variables, and the normal distribution had been discussed using handouts adopted from the regular section and the interactives within CyberStats. The main task for an instructor when using CyberStats in a classroom setting is to integrate the given technology, in particular the interactives and Data Tools/WebStat, into the overall course material. Finding an optimal teaching style that incorporates technology probably requires more efforts than designing a good regular course.

Finally, it should be stated that the instructor should not expect that many students work on interactives prior to a lecture unless credit is given. When asking students to work on a particular interactive for the next lecture (without providing credit for doing so) with the intention of discussing the students' observations and results during that lecture, there was rarely any student that did the work prior to class. Eventually, all interactives worth discussing have been discussed in class or have been assigned as a homework assignment with credit.

3.3 Course Evaluation and CyberStats Extra Questionnaire

Sixteen students from the regular section and 12 students from the Web–based section filled out the official teacher/course evaluation during the last week of the semester. Based on a scale ranging from 6 (excellent) down to 1 (very poor), the "Overall quality of the course" obtained a mean of 4.3 (with a standard deviation of 0.86) in the regular section and a mean of 4.2 (0.83) in the Web–based section. The "Instructor's effectiveness" obtained a mean of 4.1 (0.81) in the regular section and a mean of 4.2 (1.03) in the Web–based section. Obviously, there is no significant difference in these two criteria between the two sections. When compared with the results from previous offerings of this course, the Fall 2001 results were identical to the median and also to the mean outcomes over the previous years. Also, none of the other 18 criteria on the official evaluation showed major differences between the two sections. The highest difference was observed for "Course organization" with a 4.6 (1.26) for the regular section and 4.0 (1.04) for the Web–based section. This difference may be attributed to the fact that it was not always possible to teach the intended material due to the reported problems to access CyberStats and the required reorganization of course material, which resulted in jumps from one topic to another.

In addition, students could comment on aspects of the teaching or content of this course that were especially good and suggest changes to improve the teaching or the content. In the Web–based section, four students listed CyberStats (and the interactives) and its combined use with handouts in a lecture format as "especially good". In the "changes" question, six students of the Web–based section had no suggestion at all. No suggested change was listed more than once and only one student suggested not to use CyberStats.

In addition to the official teacher/course evaluation, an additional questionnaire with extra questions related to CyberStats has been passed out in the Web–based section. Thirteen (of the 14) students answered this questionnaire. Note that some students listed more than one feature. The questions and most frequently given answers follow below:

1. *What did you like* **most** *in CyberStats?*
   Eleven students cited the interactives, three students liked the possibility to submit homework online, and two students liked the examples best, in particular the ones related to current problems such as AIDS and social issues. Five students listed different advantages of electronic books such as the self–assessment with answers to the problems, the glossary and the ease of finding definitions, and the overall user–friendliness.
2. *What did you like* **least** *in CyberStats?*
   Four students complained that CyberStats or the access to it often did not work. Two students stated that they had no computer access at home and therefore could not work on homework assignments at home. Two students did not like the time it takes to access CyberStats and find information and that it is impossible to mark information. Several other topics have been listed once, including the price. Only one student did not like CyberStats at all.
3. *Are there any particular problems you would like to see fixed in CyberStats?*
   Four students would like to see a fix of the timeout issue, such that they are no longer being kicked out of CyberStats after 15 minutes without reloading a new page. Three students would like to see fixes related to the speed of the server, internet access, or that they simply do not have to depend on the Web anymore. Several other topics have been listed once. Three students did not have any suggestions.
4. *Overall, did you enjoy working with CyberStats? Circle one of the answers:*
   – *Yes, a lot.* [5 students]
   – *A little bit.* [7 students]
   – *Not very much.* [0 students]
   – *No, not at all.* [1 student]

5. *What did you like* **most** *in the way this course was taught?*
   Five students liked the open book exams most, which allowed to use CyberStats during the exam and to look up definitions and examples within CyberStats. Four students mentioned the combination of CyberStats and additional handouts. Four students liked the fact that this course was taught by a good instructor (and not only through the Web without instructor) and using technology as a tool. A few other topics have been listed once. Only one student did not have any suggestion.
6. *What did you like* **least** *in the way this course was taught?*
   Three students liked least when CyberStats or the access to it did not work and that it was hard to do homework assignments with no computer access at home. Two students thought that there was too much emphasis on computers and did not like that almost everything was online. Otherwise, one student did not like the combination of CyberStats and additional handouts and would have preferred either CyberStats or handouts only. Three students had specific comments regarding exams. Three students did not suggest any topic they liked least.
7. *Do you have any suggestions how to improve the teaching of this course when using CyberStats?*
   While some suggestions have been given once, nine students did not suggest any improvements or just indicated that some of the problems with CyberStats encountered early in the semester have been worked out.
8. *Overall, what would you suggest for future Stat 2000 classes here at USU? Ultimately, we can either user CyberStats* **or** *a standard textbook, but we have to decide for one of these two options. Circle one of the answers:*
   - *Use CyberStats for all future Stat 2000 classes.* [6 students]
   - *Fix the problems mentioned before, then do another experimental class with CyberStats and only decide then whether to use CyberStats or a textbook.* [6 students]
   - *Do not use CyberStats and immediately decide to work with a standard textbook for all future Stat 2000 classes.* [1 student]

3.4 Discussion of Evaluation and Questionnaire

Even though there was no major difference in the overall course evaluation of the Web–based and the regular sections, it is obvious that the majority of students who took the Web–based section overall enjoyed it, in particular the interactives, the possibility to submit homeworks electronically, and the access to CyberStats during the exams. However, as previously stated, the comparison between the two sections was not a designed experiment because students could decide themselves which section to attend.

Problems that were mentioned all over again relate to the nature of a Web–based textbook only: problems to access it at all, slow access, problems to do homeworks if no computer is available at home, difficulties to find something in the electronic version, and no possibility to mark and highlight important facts in an electronic book. One of the cited problems was immediately fixed during the semester: students are no longer automatically logged out after 15 minutes of idle time.

Based on the experience with the Web–based section and the students' suggestions, one should definitely consider CyberStats (or any other electronic textbook) for future use in Utah State's "Stat 2000" class. However, it should be investigated how students who have not decided themselves to take the Web–based course react to an electronic textbook before permanently adopting this electronic textbook.

Even though the additional questionnaire suggests that CyberStats was popular among the students, it can be noticed that the overall performance of students and the overall course/instructor evaluation was within the range of previous courses. Only subjectively, students had the impression that CyberStats is better than a regular textbook. Nevertheless, as stated in the Introduction, electronic textbooks are important to better motivate students and teach statistical ideas and concepts in a more convenient manner.

4 Discussion

Textbooks that are available in non–standard form, i.e., on the Web or on CD, have considerable advantages but also disadvantages compared to a regular, printed textbook. For a Web–only textbook, the main disadvantage are access–related problems such as slow connections and servers that are down. Also, students that live further away from the university and do not have a computer at home have a clear disadvantage as they have to commute to the university to access the system and read the material and work on homework problems.

Not having a printed version of the textbook also turned out to be inconvenient for the instructor. It is very difficult to determine on the computer screen which exercises to assign, in particular when there is more than just one set of exercises for a unit or when a homework assignment consists of questions related to several units. Eventually, all exercises (and solutions) have been printed out to assign selected exercises. Also, looking up a definition or an example in a Web–only textbook is time consuming. Finding such information in a printed textbook (or on a CD) is much faster. Recently, a printed version of CyberStats has been made available upon request.

Clearly, a Web–based textbook has an immediate advantage: errors and typos can be fixed immediately. It does not take several months (or years) before the next updated edition of a textbook on CD or in print is being published. As experiences with MM*Stat show (Rönz et al. 2000), about 48% of the students prefers the CD–ROM version, 40% prefer the Web version, and 12% have no preference. The CD has the advantage that it is independent from a network connection and can be run on virtually any computer. Servers that are down, slow connections, a telephone line for a modem that is shared by three or four roommates — these are features the CD does not depend on.

Another main advantage of a Web–based textbook, compared to a CD or print version, is the possibility to upload homeworks to the main server and download the answers after the submission deadline. After some initial technical difficulties with timeout problems and the access to the server, the homework upload was well accepted among the CyberStats students. As described before, weaker students from the Web–based section used to work on the CyberStats questions but often did not turn in the additional homework exercises on paper. CyberStats offers a possibility to give students an entirely electronic exam: the "Test Bank" allows instructors to write electronic exams that include the instructor's own questions, CyberStats questions, and questions from other instructors who use CyberStats. A nice addition would be if each instructor could also add individual questions to the unit exercises and ask students to answer all questions electronically.

To summarize, the main advantage of electronic textbooks on the Web (or on CD) are the interactives and additional features such as narrative texts and videos. Also, the ability to work on homework problems on the computer and upload answers to a server from which they can be collected after the due date is a major advantage for students and the instructor.

Ideally, an electronic textbook (Web-based or on CD) should not be used alone, but accompanied by a printed book and optionally with the other electronic format. An additional printed textbook has several benefits. In particular, it is faster to "access" a textbook and find the desired information. The textbook can be read while commuting in buses or trains, it can be taken back home over the weekend, and it can be read during short breaks at the university when no computer lab is at hand.

Acknowledgements

The authors want to thank Dan Coster for providing information on previous "Stat 2000" courses, Palyne Gaenir for technical assistance with CyberStats, and Wolfgang Härdle, Alex Kugushev, and Paul Velleman for their helpful comments regarding the electronic textbooks discussed in this paper.

References

Chromiak, W. & Rossman, A. (1992), Using HyperCard to Teach Statistics, *in* F. Gordon & S. Gordon, eds, 'Statistics for the Twenty-First Century (MAA Notes, Number 26)', The Mathematical Association of America, Washington, D.C., pp. 243–257.

Currall, J. (1997), 'Computer Aided Statistics Teaching: Real Advance or Technological Fashion?', *Computing Science and Statistics* **29**(2), 321–327.

de Leeuw, J. (1997), The UCLA Statistics Textbook and Modules, *in* 'Bulletin of the International Statistical Institute, 51st Session Istanbul 1997, Proceedings Book 2', pp. 55–58.

Härdle, W., Klinke, S. & Marron, J. S. (1999), 'Connected Teaching of Statistics', *Statistical Computing and Statistical Graphics Newsletter* **10**(1), 12–20.

Härdle, W., Klinke, S. & Müller, M. (2000), *XploRe Learning Guide*, Springer, Berlin.

Härdle, W., Lehmann, H. & Rönz, B. (2001), MM*Stat — Eine interaktive Einführung in die Welt der Statistik, Exponat auf der CeBIT 2001, Discussion Paper, No. 4/2001, Humboldt-Universität zu Berlin, Sonderforschungsbereich 373. `http://sfb.wiwi.hu-berlin.de`.

Lane, D. M. (1999), 'The Rice Virtual Lab in Statistics', *Behavior Research Methods, Instruments & Computers* **31**(1), 24–33.

Lock, R. H. (1997), 'Internet Resources for Teaching Statistics', *Computing Science and Statistics* **29**(2), 339–343.

Maatta, J. (1999), 'ActivStats 2.0 and The Active Practice of Statistics', *The American Statistician* **53**(1), 85–86.

Macnaughton, D. B. (1998), 'Review of ActivStats 2.0', *The Statistics Teacher Network*. `http://www.matstat.com/teach/p0030.htm`.

Moore, D. S. & McCabe, G. P. (1999), *Introduction to the Practice of Statistics (Third Edition)*, W. H. Freeman and Company, New York, NY.

Rönz, B. (1997), 'Computer-aided Teaching in Statistics — Some Experiences', *Computing Science and Statistics* **29**(2), 319–320b.

Rönz, B. (2001), MM*Stat — A Multimedia Tool for Teaching of Statistics, Discussion Paper, No. 85/2001, Humboldt-Universität zu Berlin, Sonderforschungsbereich 373. `http://sfb.wiwi.hu-berlin.de`.

Rönz, B., Müller, M. & Ziegenhagen, U. (2000), The Multimedia Project MM*Stat for Teaching Statistics, *in* J. G. Bethlehem & P. G. M. van der Heijden, eds, 'COMPSTAT 2000: Proceedings in Computational Statistics', Physica-Verlag, Heidelberg, pp. 409–414.

Snell, J. L. & Peterson, W. P. (1992), Does the Computer Help us Understand Statistics?, *in* F. Gordon & S. Gordon, eds, 'Statistics for the Twenty-First Century (MAA Notes, Number 26)', The Mathematical Association of America, Washington, D.C., pp. 167–188.

Symanzik, J. (1998), 'Current Internet Technology and Statistics — Blessing or Curse?', *Computing Science and Statistics* **30**, 500–509.

West, R. W. & Ogden, R. T. (1998*a*), 'Interactive Demonstrations for Statistics Education on the World Wide Web', *Journal of Statistics Education* **6**(3). `http://www.amstat.org/publications/jse/v6n3/west.html`.

West, R. W. & Ogden, R. T. (1998*b*), 'WebStat: An Environment for Statistical Analysis on the the World Wide Web', *Computing Science and Statistics* **29**(1), 307–310.

West, R. W., Ogden, R. T. & Rossini, A. J. (1998), 'Statistical Tools on the World Wide Web', *The American Statistician* **52**(3), 257–262.

Trans-Dimensional Markov Chains and their Applications in Statistics

S. P. Brooks

Statistical Laboratory, Centre for Mathematical Sciences, Wilberforce Road, Cambridge, CB3 0WB, UK

Abstract. We discuss the application of trans-dimensional Markov chains to both Bayesian and Classical model discrimination. We discuss how to measure and improve the efficiency of the chains and illustrate the ideas via a classical variable selection problem and a Bayesian recovery-recapture analysis.

Keywords. Model Discrimination; Variable Selection, Recovery-Recapture, Markov Chain Monte Carlo, Reversible Jump MCMC, Convergence Assessment.

1 Introduction

Monte Carlo methods have long been used as a means of numerically approximating complex functions. These methods typically involve generating observations from some distribution and using this sample to approximate the function of interest. Markov chain Monte Carlo (MCMC) methods provide a very effective set of tools for simulating from extremely complex distributions and have revolutionised the field of Bayesian statistics in particular since it's introduction to the Statistical literature a little over 10 years ago.

Similar methods exist within the classical literature too. For example, the idea of simulated annealing (SA; Brooks and Morgan 1995) involves simulating Markov chains on a sequence of target distributions which converge to a point mass at the MLE. Thus, though MCMC methods have had a far greater impact on the Bayesian literature than the classical, they should not be regarded as exclusively Bayesian tools.

An important innovation in the MCMC literature was the introduction of trans-dimensional (TD)MCMC, and in particular reversible jump (RJ)MCMC (Green 1995), which allows Markov transitions between states of different dimension. This is particularly useful in the context of model selection or exploration. Within the Bayesian context we may be interested in simulating observations from a posterior distribution defined jointly over both parameter and model space. With models of different dimension (i.e., with different numbers of parameters) a between-model move involves adding or deleting parameters and TDMCMC methods allow us to perform these moves. However, as we shall see SA algorithms can be similarly extended to allow automatic classical model selection by optimising an objective function (such as the AIC) across model and parameter space simultaneously. Again, with models of different dimension, this involves the use of trans-dimensional jumps.

In this paper we examine the application of trans-dimensional Markov chains within the context of both Bayesian and classical model exploration. We begin with an introduction to SA and MCMC methods and explain how they be extended, using TDMCMC updates, to tackle model discrimination problems. We discuss the difficulty in implementing these algorithms in an efficient manner and suggest an adaptive updating scheme which automatically selects proposal parameter values to improve the efficiency of between-model moves. We also discuss problems associated with convergence assessment for TDMCMC algorithms and provide extensions to several methods recently developed to tackle this problem. Finally, we provide two applications of trans-dimensional Markov chain methodology. The first, uses RJMCMC to look at model discrimination in a Bayesian analysis of capture-recapture data. The second uses trans-dimensional (TD)SA to tackle a classical variable selection problem.

2 Markov Chains for Bayesian and Classical Inference

2.1 The Posterior Distribution

Given a sample $\boldsymbol{x}$ from a distribution with likelihood $L(\boldsymbol{x}|\boldsymbol{\theta})$ and a prior density for $\boldsymbol{\theta} \in \boldsymbol{\Theta} \subseteq \mathbb{R}^p$ given by $p(\boldsymbol{\theta})$, Bayes' Theorem relates the posterior $\pi(\boldsymbol{\theta}|\boldsymbol{x})$ to the prior via the formula

$$\pi(\boldsymbol{\theta}|\boldsymbol{x}) \propto L(\boldsymbol{x}|\boldsymbol{\theta})p(\boldsymbol{\theta}).$$

This posterior distribution is the basis for inference within the Bayesian paradigm and represents out beliefs about the model parameters having observed data, $\boldsymbol{x}$. Our posterior beliefs are often summarised in the form of posterior means and variances for example, but the calculation of these summary statistics requires the integration of this (often complex) distribution over the parameter space. Explicit evaluation of these integrals is generally not possible but we can calculate empirical estimates of statistics of interest by drawing samples from this distribution. MCMC methods provide a mechanism for obtaining draws from almost arbitrarily complex posterior distributions by simulating a Markov chain with $\pi(\cdot)$ as its stationary distribution. An initial burn-in period is discarded and subsequent states are used as a dependent sample from the posterior upon which inference can be based (Brooks 1998).

This approach can be extended to allow inference on both parameter and model space. Suppose that we have a collection of models $M_1, M_2, \ldots$ indexed by the parameter $m \in \mathcal{M}$ and that under each model we have an associated parameter vector $\boldsymbol{\theta}_m \in \boldsymbol{\Theta}_m \subseteq \mathbb{R}^{n_m}$ with corresponding likelihood function $L_m(\boldsymbol{x}|\boldsymbol{\theta}_m)$. Then our posterior distribution can be extended to model space by taking

$$\pi(\boldsymbol{\theta}_m, m|\boldsymbol{x}) \propto L_m(\boldsymbol{x}|\boldsymbol{\theta}_m)p(\boldsymbol{\theta}_m|m)p(m),$$

where $p(\boldsymbol{\theta}_m|m)$ denotes the prior for $\boldsymbol{\theta}_m$ under model m. As in the fixed-model case, we may draw samples from this posterior distribution to obtain empirical estimates of quantities of interest such as posterior means or the marginal distribution of the models. However, since each model may have a different number of parameters, between-model moves may require a Markov transition between states of different dimension and for this we require TDMCMC updates.

2.2 Maximising the Likelihood

Classical inference is based upon maximising the likelihood function $L(\boldsymbol{x}|\boldsymbol{\theta})$ with respect to the model parameters, $\boldsymbol{\theta}$. This maximisation is usually done numerically, but many numerical optimisation techniques are liable to get "stuck" in local rather than global maxima. One way to avoid this problem is to introduce a stochastic element to the maximisation algorithm and one such technique is known as simulated annealing, which mimics the natural process by which crystalline solids cool to a state of minimum energy.

Given an objective function $f(\boldsymbol{\theta}) = -L(\boldsymbol{x}|\boldsymbol{\theta})$ which we wish to minimise over the p-dimensional vector of parameters $\boldsymbol{\theta}$, we define the density $\pi_T(\boldsymbol{\theta}) \propto \exp[-f(\boldsymbol{\theta})/T]$. Clearly, as $T \to 0$, this density converges to a point mass at the minimum of f. The SA algorithm begins by simulating a Markov chain with stationary density $\pi_T(\boldsymbol{\theta})$ at some initial temperature $T = T_0$, allowing the chain to reach equilibrium. The temperature is then decreased and the chain continued until equilibrium is achieved in the new temperature. We continue to repeat this process until we reach a temperature sufficiently low that the system essentially freezes. This point corresponds to the minimum of f (Brooks and Morgan 1995) and hence the maximum of $L(\boldsymbol{x}|\boldsymbol{\theta})$.

Clearly, the SA and MCMC algorithm are closely linked in that both require the simulation of Markov chains with some pre-determined stationary distribution. In fact the SA scheme is essentially a series of MCMC schemes on a converging sequence of target distributions. Thus, the SA algorithm may also be extended to consider the problem of model selection as follows. Suppose we choose to discriminate between competing models using the AIC statistic. Then we can take $f(\boldsymbol{\theta}_m, m) = -2logL_m(\boldsymbol{\theta}_m) + 2n_m$ and define

$$\pi_T(\boldsymbol{\theta}_m, m) \propto \exp(-f(\boldsymbol{\theta}_m, m)/T).$$

Again, we can use simulated annealing to obtain the minimum of f, but this involves constructing a series of Markov chains with stationary distribution $\pi_T(\cdot)$ and, as with the Bayesian posterior above, since each model may have a different number of parameters, between-model moves will again require Markov transitions between states of different dimension and so we need TDMCMC updates.

2.3 Simulating a Fixed-Dimensional Markov Chain

The Metropolis Hastings algorithm provides the most general scheme for the simulation of a Markov chain with pre-determined stationary density $\pi(\boldsymbol{\theta})$.

The algorithm proceeds in two stages. Given the state of the chain $\boldsymbol{\theta}^t$ at time t, we propose a new state $\boldsymbol{\phi}$ which is drawn from some (essentially arbitrary) proposal distribution $q(\boldsymbol{\phi}|\boldsymbol{\theta}^t)$. This proposal is then accepted as the new state of the chain (so that $\boldsymbol{\theta}^{t+1} = \boldsymbol{\phi}$) with probability $\alpha = \min(1, A)$ where

$$A = \frac{\pi(\boldsymbol{\phi})q(\boldsymbol{\theta}^t|\boldsymbol{\phi})}{\pi(\boldsymbol{\theta}^t)q(\boldsymbol{\phi}|\boldsymbol{\theta}^t)},$$

otherwise $\boldsymbol{\theta}^{t+1} = \boldsymbol{\theta}^t$.

Various special cases of the general Metropolis Hastings scheme arise with different choices for the proposal q, including the single update Metropolis Hastings scheme (in which each component of the vector $\boldsymbol{\theta}^t$ is updated individually), the Gibbs sampler (in which the elements of $\boldsymbol{\theta}^t$ are updated in blocks with their posterior conditional distribution as a proposal), the Langevin algorithm (which uses gradient information to move towards areas of high posterior mass) and the slice sampler (which augments the state space so that updates essentially reduce to draws from a Uniform distribution). Any of these methods can be used either alone or in combination to obtain a Markov chain with the desired stationary distribution (Brooks 1998).

2.4 Trans-Dimensional Markov Chains

Though alternative methods are available (Stephens 2000), in order to simulate a trans-dimensional chain with stationary distribution $\pi(\boldsymbol{\theta}_m, m)$ we shall use a generalisation of the Metropolis Hastings algorithm known as reversible jump MCMC (Green 1995).

Suppose that we wish to perform a TD jump moving from $\boldsymbol{\psi} = (\boldsymbol{\theta}_k, k)$ to $\boldsymbol{\psi}' = (\boldsymbol{\theta}_{k'}, k')$, where $n_k < n_{k'}$ and $k, k' \in \mathcal{M}$. Though there are a wide variety of implementations, the simplest RJMCMC update proceeds as follows. Suppose that we choose a move of type $r \in \mathcal{R}$ to take the chain from $\boldsymbol{\psi}$ to $\boldsymbol{\psi}'$. We can do this by generating a random vector $\boldsymbol{u} \sim q_{r,k,k'}(\boldsymbol{u})$ of length $n_{k'} - n_k$ and choose as our proposed state $\boldsymbol{\psi}' = (g(\boldsymbol{\theta}_k, \boldsymbol{u}), k')$ for some function g. This proposal would then be accepted with probability $\alpha_r(\boldsymbol{\psi}, \boldsymbol{\psi}') = \min(1, A)$, where

$$A = \frac{\pi(\boldsymbol{\psi}')p_{r'}(\boldsymbol{\psi}')}{\pi(\boldsymbol{\psi})p_r(\boldsymbol{\psi})q_{r,k,k'}(\boldsymbol{u})}\left|\frac{\partial g(\boldsymbol{\theta}_k, \boldsymbol{u})}{\partial(\boldsymbol{\theta}_k, \boldsymbol{u})}\right|. \tag{1}$$

Here, $p_r(\boldsymbol{\phi})$ denotes the probability of proposing move type r in state $\boldsymbol{\psi}$ and move type r' is the opposite move type to move r.

Note that in this case, the move from $\boldsymbol{\psi}'$ to $\boldsymbol{\psi}$ is accomplished by using the inverse transformation. This means that the proposal is deterministic in this direction. There are more general implementations for the RJMCMC procedure and we shall briefly discuss these at the end of the next section.

Thus, for model determination problems we can construct the necessary chains using a combination of MCMC updates for within-model moves and reversible jump MCMC for between-model-moves.

3 Efficient Proposals for Trans-Dimensional Chains

Though the proposal distributions for both the within- and between-model moves are essentially arbitrary, the performance of the resulting Markov chain depends critically upon this choice. For within-model moves it is fairly easy to choose efficient proposals that both lead to high acceptance rates and rapid movement around the state space, this is considerably harder for the TD moves because there is no Euclidean structure between models to guide proposal choice. In order to obtain an efficient algorithm, proposals are typically "tuned" on the basis of a short pilot simulation of the chain. However, in the model determination context in which the space of models is large, the computational expense associated with pilot tuning quickly becomes prohibitive and automatic and/or adaptive methods are required.

Brooks *et al* (2002b) suggest a method based upon the construction of a Taylor Series expansion of the acceptance ratio for certain canonical jumps to select the most appropriate parameters for a pre-specified proposal distribution. Without loss of generality, let us suppose that we need only generate a scalar value u in order to move between two models M_k and $M_{k'}$, so that $n_{k'} = n_k + 1$. Let us suppose also that the proposal density q is parameterised by one parameters s.

Now, in order to maximise movement around model space, we would like the value of A in (1) to be as close to 1 as possible (since values higher than 1, mean that the reverse move becomes unlikely). Thus, we might pick a particular value of u, and set $A = 1$ at this point by choosing an appropriate value for the proposal parameter, s. Suppose there exists some value u^* such that $L(\boldsymbol{x}|g(\boldsymbol{\theta}, u^*)) = L(\boldsymbol{x}|\boldsymbol{\theta})$, then we call this the weak non-identifiability "centering point". Essentially, this value of u identifies a point in the higher dimensional space whose corresponding likelihood contribution is equal to that of the original state in the smaller-dimensional model. Thus, as far as the model is concerned, the new state is equally "desirable" as the original state and this seems an appropriate point about which to centre. (This is somewhat akin to the random walk Metropolis algorithm where updates are centred around the current point.)

Suppose, for example, that we take g to be the identity, centre at $u = 0$, take independent and identical $N(0, \sigma_\theta^2)$ priors for all elements of $\boldsymbol{\theta}_m$ and for all m, and a $N(o, \sigma_q^2)$ proposal for u. Then, the Jacobian term in (1) is 1, the likelihood terms in the posterior cancel, as do all but one of the prior terms (leaving one in the numerator) and, setting $A = 1$, we obtain

$$\sigma_q^2 = \sigma_\theta^2 \left(\frac{p_{r'}(g(\boldsymbol{\theta}, 0), k')}{p_r(\boldsymbol{\theta}, k)} \right)^2$$

This so-called "zeroth order" method may be extended to the case where the proposal depends upon two or more parameters by simultaneously setting $A = 1$ and one or more derivatives of A (with respect to u) to zero, thereby ensuring that A not only takes the value 1 at $u = u^*$ but also takes values "close" to 1 in a neighbourhood around u^*. The more derivatives we take, the larger this neighbourhood.

The main advantage of these methods over the obvious alternative, in which the proposal parameters are fixed throughout the simulation at values chosen on the basis of an initial pilot run of the algorithm, is that these methods allow the proposal parameters to "adapt" to the current state of the chain. In this way, the proposal distribution is tailored directly to the current state of the chain, rather than being fixed at some value which "performs well on average". This adaptation is particularly important in the context of the TDSA algorithm, where TD jumps become increasing difficult as the temperature decreases (Brooks *et al* 2001a,b).

The development of these methods as described in Brooks *et al* (2002b) focusses upon transitions for which the move to a lower dimensional model is deterministic. However, these methods can be directly extended. Suppose that we wish to move from some model k with parameter vector $\boldsymbol{\theta}_k$ to some new model k' and parameter vector $g(\boldsymbol{\theta}_k, \boldsymbol{u})$ for some n-dimensional $\boldsymbol{u} \sim q_{k,k'}(\boldsymbol{u})$. Suppose also that the reverse move would be performed by generating an $(n + n_k - n_{k'})$-dimensional $\boldsymbol{u}' \sim q_{k',k}(\boldsymbol{u}')$ and that $g'(\boldsymbol{\theta}_{k'}, \boldsymbol{u}') = \boldsymbol{\theta}$. Then this move (from k to k') would be accepted with probability $\min(1, A)$, where

$$A = \frac{\pi(\boldsymbol{\theta}', k'), p_{r'}(\boldsymbol{\theta}_{k'}, k')q(\boldsymbol{u}')}{\pi(\boldsymbol{\theta}, k), k), p_r(\boldsymbol{\theta}, k)q(\boldsymbol{u})} \left| \frac{\partial(\boldsymbol{\theta}_{k'}, \boldsymbol{u}')}{\partial(\boldsymbol{\theta}_k, \boldsymbol{u})} \right|.$$

The method of Brooks *et al* (2002b) can be then be extended to this general case by choosing appropriate centering values $\boldsymbol{u}^*$ and $\boldsymbol{u}'^*$ and setting $A = 1$ and any necessary derivatives of A (with respect to $[\boldsymbol{u}, \boldsymbol{u}']$) to zero to obtain the parameters for the proposal distributions $q_{k,k'}$ and $q_{k',k}$. Note that in the special case where $n_k = n_{k'}$ (i.e., we essentially have a within-model move) an infinite order method is always available (given an appropriate choice of q) – the Gibbs sampler.

4 Convergence Assessment for TDMCMC

One of the drawbacks of using Markov chain-based methods is that inference can only be gained once the chain has converged to a state of equilibrium. In the context of model determination, convergence monitoring is particularly difficult since both the length and interpretation of the state vector may change from iteration to iteration.

One way around this problem is to parameterise the model in such a way as to allow as many model parameters as possible to retain their interpretation as the sampler moves from model to model. If such a parameter (or set of parameters) can be identified then it may be monitored and used to provide an indication as to how well the MCMC sampler is performing. Brooks and Giudici (2000) suggest a method for monitoring simulations with this property, based upon a two-way ANOVA-type decomposition of the simulation output which essentially extends the popular method of Gelman and Rubin (Gelman and Rubin 1992; Brooks and Gelman 1998) to the TDMCMC case.

Though various ANOVA decompositions of the variance are possible, Brooks and Giudici (2000) focus upon one particular decomposition of the variance of a scalar statistic $h(\boldsymbol{\theta})$ which proves to be particularly sensitive to differences in the mean of h between different replications. Let h^i_{mj} denote the value of $h(\theta)$ the ith time that model m is visited in chain j ($m =\in \mathcal{M}$, $j = 1, ..., J$ and $i = 1, ..., n_{mj}$). Then we can monitor

$$B = \sum_{m \in \mathcal{M}} (\bar{h}^{\cdot}_{m\cdot} - \bar{h}^{\cdot}_{\cdot\cdot})^2 / (|\mathcal{M}| - 1)$$

$$\text{and } B_w = \sum_{j=1}^{J} \sum_{m \in \mathcal{M}} (\bar{h}^{\cdot}_{mj} - \bar{h}^{\cdot}_{\cdot j})^2 / (|\mathcal{M}| - 1)J,$$

both of which estimate the variation in the value of h between chains (the first overall, and the second averaging the within-chain variation between models). Here, $\bar{h}^{\cdot}_{mj}$ denotes the empirical mean of the h-values over the n_{mj} occurrences of model m in chain j and the other terms are defined similarly.

Brooks and Giudici (2000) argue that convergence of these two terms to the same value is a necessary condition for convergence to have been achieved and suggest monitoring the values (or their ratio, which should tend to 1) over the length of the simulation to ensure that this is the case. Of course comparisons of other variance decompositions could also be monitored, but we shall focus only upon these variance terms here. This diagnostic approach can be extended to the multivariate case, by taking $\boldsymbol{h} = (h_1(\boldsymbol{\theta}), ..., h_p(\boldsymbol{\theta}))^T$ where each of the $h_i(\boldsymbol{\theta})$ are scalar quantities whose meaning remains constant across all models. Now, if we wish to monitor the vector $\boldsymbol{h}$, the within- and between-chain variances become covariance matrices and convergence assessment is then made on the basis of comparisons between the corresponding B and B_w matrices. This is most easily measured by looking at properties of the matrix $B^{-1}B_w$ such as the largest eigenvalue, for example (generalising the multivariate diagnostics of Brooks and Gelman 1998 to the trans-dimensional context). A straightforward extension of their Lemma 4.3 shows that the largest eigenvalue of $B^{-1}B_w$ bounds above any of the ratios based upon the corresponding scalar statistics $h_1, .., h_p$. Thus, by monitoring this eigenvalue until it converges to a value around 1, we know that the corresponding ratios based upon the constituent scalar statistics will also be close to 1. This means that we need only monitor this single statistic and greatly improves the interpretability of the diagnostic.

One drawback with this approach is the identification of scalar statistics that retain a consistent interpretation between models. In addition, the convergence of the ratio B/B_w is a necessary but not sufficient condition for convergence. In order to be confident that convergence has indeed been achieved it is better to monitor a variety of convergence measures.

An alternative approach suggested by Brooks *et al* (2002a) focusses upon a comparison of the empirical distribution of the "model indicator" between multiple replications. Though various measures are suggested, we shall discuss only their chi-squared diagnostic here. If the output of J chains have each converged, then they should all have similar estimates of the posterior model probabilities. Let N_{mj} denote the number of times that model m is visited by chain J within a fixed number of iterations. We can therefore construct an $|\mathcal{M}| \times J$ matrix whose (m, j)th entry is N_{mj} corresponding to the observed number of visits to each model within each chain. Under the assumption that the chains are homogeneous with respect to the model probabilities, we can estimate the corresponding expected number of counts E_{mj} by

$$E_{mj} = \frac{N_{m\cdot}}{J}, \quad \forall j = 1, ..., J,$$

where $N_{m\cdot} = \sum_{j=1}^{J} N_{mj}$. The chi-squared test for between-chains homogeneity is therefore based upon the following statistic:

$$\chi^2(T) = \sum_{m \in \mathcal{M}} \sum_{j=1}^{J} \frac{(N_{mj} - E_{mj})^2}{E_{mj}}$$

and, under the homogeneity hypothesis, $\chi^2(T)$ is asymptotically distributed as a chi-squared variate with $(J-1)(|\mathcal{M}|-1)$ degrees of freedom. Of course, this distribution is based upon the assumption of independence between observations, which will not be the case with MCMC output. One way to achieve approximate independence is to proceed by sub-sampling each chain, by taking every λth observation, where λ is a thinning parameter to be tuned. Of course, the thinning parameter cannot be taken as large as possible, as its increase has two effects: on one hand the output is less dependent; but on the other hand, the accuracy of the asymptotic approximations employed in the tests decreases. Thus, we require some means by which to determine the most efficient thinning rate.

Raftery and Lewis (1992) propose a convergence assessment technique based upon the decomposition of a continuous Markov chain into a two state sub-Markov chain. They then estimate the convergence rate of the sub-Markov chain and use this as an (under)estimate of the convergence rate of the original chain. In the context of the convergence assessment of TDMCMC simulations we already have a very natural $|\mathcal{M}|$-state sub-Markov chain obtained by monitoring the model indicator through time. After T iterations have been completed, a transition matrix for the model indicator may be estimated with (i,j)th element given by $P_{ij} = R_{ij}/T$. Here R_{ij} denotes the number of times a move from model i to model j is observed. Typically, this matrix will consist of a very large proportion of zeroes even with large T since many transitions will not be possible under any given TDMCMC updating scheme. Once this matrix has been obtained the rows may be permuted in order to place any null rows at the bottom, and the second largest eigenvalue may then be computed. As long as at least two models are visited during the simulation, this eigenvalue will be real and positive and will provide both an estimate for the convergence rate of the sub-Markov chain and of the original chain.

In order to choose the thinning parameter λ, we use the following rationale. If the chain is uniformly ergodic, then the distance between the t-step transition density and the stationary density is bounded above by $C\rho^t$, where C is an unknown constant and ρ is the convergence rate estimated above. Suppose also that there exists an integer $k < T$, such that $\rho^k \simeq 0.01$. This means that, after k iterations, the chain is 100 times "closer" to convergence than it as at the beginning. Alternatively, we can say that the dependence between two iterations spaced k apart is 100 times smaller than that between successive observations. This implies that a thinning value of around k would seem sensible, using an error level of 0.01. Therefore, given an estimate of ρ, we obtain λ as $\lambda = \frac{log\, 0.01}{log\, \rho}$ and this can be used to thin our output.

5 Examples

5.1 Bayesian Analysis of Recovery-Recapture Data

Catchpole *et al* (1998) (and subsequently Brooks *et al* 2002c) analyse nine years of mark-recapture-recovery data on shags ringed on the Isle of May in the Firth of Forth, Scotland. In total, 8,637 birds were marked in the study, comprising 1087 breeding adults and 7550 young. The model they chose to describe the data has year-dependent first-year survival probabilities, $\{\phi_{1,t}\}$; a separate survival probability ϕ_{imm} for birds aged 1 and 2 years; a survival probability ϕ_a associated with all birds over the age of 3; age-dependent recapture probabilities p_i, $i \in \{0,1,2\}$; a year-dependent adult recapture probability, $\{p_{a,t}\}$; and a year-dependent recovery probability, $\{\lambda_t\}$. This results in a model with 32 identifiable parameters in total.

Here, we focus on the problem of determining the the dependence (or otherwise) of each of these six groups of parameters upon sex. The sex covariate takes one of three possible values, 0 (unknown), 1 (male) or 2 (female). The "unknown" code refers primarily to birds ringed as chicks which were not later observed breeding. Thus we can expect these birds to have smaller estimated survival and recapture probabilities. This has been found to be the case, and in the analysis we therefore focus on the effects of the male and female codes, 1 and 2 respectively.

As an illustration, for a model with only sex-dependent adult survival we write ϕ_a, in self-explanatory notation, as logit $(\phi_a) = \mu$, for birds of unknown sex, and

$$\text{logit}\, \phi_{a,M} = \mu + \tau_M; \qquad \text{logit}\, \phi_{a,F} = \mu + \tau_F$$

for birds of known sex, where $\tau_M \sim N(0, \sigma_\tau^2)$, $\tau_F \sim N(0, \sigma_\tau^2)$, and $\sigma_\tau^2 \sim \Gamma^{-1}(a_\tau, b_\tau)$. Thus, τ_M and τ_F are essentially random effects.

With a vague prior on μ and similar constructions for the remaining parameters, we can use an MCMC simulation scheme to obtain posterior means and variances for any model of interest. Similarly, RJMCMC updates can be introduced to explore model space and obtain posterior probabilities of sex dependence for each of the parameters. In this case, we have a total of 2^6 competing models (since each parameter is either sex-dependent or not). The RJMCMC algorithm quickly explores this moderate model space providing the following summary outputs.

Parameter	ϕ_1	ϕ_{imm}	ϕ_a	p	p_a	λ
$\mathbb{P}$(Sex effect)	0.005	0.017	0.736	0.762	0.021	0.114

Clearly there is little evidence to suggest any sex-dependence for the majority of parameters, but there is a high posterior probability of a sex effect for both p and ϕ_a. The result for p is interesting, as the value for young males is larger than that for young females (not shown). We would expect this from the biology, as attempts at early breeding are far more common in males than in females, resulting in the higher value of p for males. Similarly, the dynamics of the adult population would suggest that the adult survival would be higher for males and this is widely believed to be the case. Nonetheless, this is the first statistical analysis to demonstrate and quantify these sex differences.

5.2 Classical Variable Selection

We consider the data of Hosmer and Lemeshow (1989) concerning potential factors associated with low infant birth weight (less than 2.5kg). Birth weights for 189 infants are given, of which 59 are considered to have low birth weight. In addition, nine explanatory variables related to the mother are recorded, and these are thought to explain the variability in birth weights. Following Hosmer and Lemeshow (1989) we fit a logistic regression. Denoting the set of variables to be included in the analysis by $m \subseteq \{0, 1, ..., 9\}$ and using the AIC to determine the best model we base our TDSA algorithm upon

$$\pi_T(\boldsymbol{\theta}, m) \propto e^{-(|m|+1)/T} \prod_{i=1}^{n} \left(\frac{e_i(\boldsymbol{\theta}, m)}{1 + e_i(\boldsymbol{\theta}, m)} \right)^{y_i/T} \left(\frac{1}{1 + e_i(\boldsymbol{\theta}, m)} \right)^{(1-y_i)/T}$$

where

$$e_i(\boldsymbol{\theta}, m) = \exp\Big(\sum_{j \in m} \theta_j x_{ji}\Big),$$

θ_j denotes the parameter associated with variable j (variable 0 is simply a vector of 1's), x_{ji} denotes the ith value of variable j, and y_i the corresponding dependent variable (Brooks and Friel 2001). Always keeping the constant term, we have a total of 9 explanatory variables from which we must choose the "best" subset. Therefore, there are a total of $2^9 = 512$ possible models.

Lemeshow and Hosmer (1998) use a sequential model selection method in which they begin with the saturated model and sequentially remove variables using a likelihood ratio-based testing procedure. Their solution contains six variables and suggests the following model:

$$\text{logit } p_i = 0.095 - 0.017x_{2i} + 1.264x_{3i} + 0.864x_{4i} + 0.876x_{5i} + 1.231x_{6i} + 1.767x_{7i}.$$

This solution has log-likelihood $= -100.24$ with a corresponding AIC value of 214.48. Using TDSA, Brooks and Friel (2001) take an initial temperature $T_0 = 10$ which we reduce by a factor 0.995 every 1000 iterations. We performed 1400 temperature reductions to give a final temperature of approximately 0.01. Our simulation using TDSA finds the model

$$\begin{aligned}\text{logit } p_i = -0.110 - 0.016x_{2i} + 1.313x_{3i} + 0.831x_{4i} + 0.874x_{5i} + \\ 1.152x_{6i} + 1.885x_{7i} + 0.769x_{8i}\end{aligned}$$

with log-likelihood -98.94 and corresponding AIC value of 213.87. Thus, our TDSA method provides a better model, giving a lower AIC value than that of Lemeshow and Hosmer (1998). This demonstrates the advantage of a mechanism for properly searching model space over the commonly-used sequential methods that simply add or delete parameters and measure improvement.

6 Discussion

In this paper, we have demonstrated the utility of trans-dimensional Markov chains in both Bayesian and classical inference. The advantage of these techniques for model exploration is obvious and in more complex examples than we consider here (for which we may have of the order of millions of competing models), these provide the only feasible methods for model exploration, see King and Brooks (2001). Exhaustive enumeration is simply impractical in such cases.

As well as an area active in terms of applications, we are still in urgent need of fundamental methodological work. For example, convergence assessment for trans-dimensional chains is an open and active area. Perhaps the most challenging area concerns proposal choice. Though we have shown in this paper how to tune parameters given a particular form of proposal distribution, there are no results which help guide the choice of q in the first place. Questions relating to the construction of appropriate centering points, particularly between non-nested models is also an area full of interesting research possibilities.

References

Brooks, S. P. (1998), Markov Chain Monte Carlo Method and its Application. *The Statistician* **47**, 69–100

Brooks, S. P. and N. Friel (2001), Model Selection via Simulated Annealing. Technical report, University of Cambridge

Brooks, S. P. and A. Gelman (1998), Alternative Methods for Monitoring Convergence of Iterative Simulations. *Journal of Computational and Graphical Statistics* **7**, 434–455

Brooks, S. P. and P. Giudici (2000), MCMC Convergence Assessment via Two-Way ANOVA. *Journal of Computational and Graphical Statistics* **9**, 266–285

Brooks, S. P., P. Giudici and A. Philippe (2002a), On Non-Parametric Convergence Assessment for MCMC Model Selection. *Journal of Computational and Graphical Statistics* in press

Brooks, S. P., P. Giudici and G. O. Roberts (2002b), Efficient Construction of Reversible Jump Proposal Distributions (with discussion). *Journal of the Royal Statistical Society, Series B* accepted for publication

Brooks, S. P. and B. J. T. Morgan (1995), Optimisation using Simulated Annealing. *The Statistician* **44**, 241–257

Brooks, S. P., B. J. T. Morgan, E. A. Catchpole and M. Harris (2002c), Bayesian Methods for Analysing Ringing Data. *Journal of Applied Statistics* in press

Catchpole, E. A., S. N. Freeman, B. J. T. Morgan and M. P. Harris (1998), Integrated recovery/recapture data analysis. *Biometrics* **54**, 33–46

Gelman, A. and D. B. Rubin (1992), Inference from Iterative Simulation using Multiple Sequences. *Statistical Science* **7**, 457–511

Green, P. J. (1995), Reversible Jump MCMC Computation and Bayesian Model determination. *Biometrika* **82**, 711–732

Hosmer, D. and S. Lemeshow (1989), *Applied Logistic Regression*. New York: Wiley

King, R. and S. P. Brooks (2001), Bayesian Estimation of Census Undercount. *Biometrika* **88**, 317–336

Lemeshow, S. L. and D. W. Hosmer (1998), Logistic Regression. In P. Armitage and T. Colton (eds.), *Encyclopedia of Biostatistics*, pp. 2311–2327, Chichester: Wiley

Raftery, A. E. and S. M. Lewis (1992), How Many Iterations in the Gibbs Sampler? In J. M. Bernardo, A. F. M. Smith, A. P. Dawid and J. O. Berger (eds.), *Bayesian Statistics 4*, pp. 763–774, Oxford University Press

Stephens, M. (2000), Bayesian Analysis of Mixture Models with an Unknown Number of Components: An Alternative to Reversible Jump Methods. *Annals of Statistics* **28**, 40–74

Full Papers

A Bayesian Model for Compositional Data Analysis

Mark J Brewer[1], Chris Soulsby[2] and Sarah M Dunn[3]

[1] Biomathematics and Statistics Scotland, Macaulay Institute, Aberdeen, AB15 8QH, UK
[2] Department of Geography and Environment, University of Aberdeen, Aberdeen, AB24 3UF, UK
[3] Macaulay Institute, Aberdeen, AB32 6RE, UK

Abstract. We introduce a Bayesian hierarchical model for the analysis of compositional data, where the source data are observed as indicator measurements rather than direct proportions. We focus on an application of the model to water quality data, where we apportion stream water to three sources.

Keywords. Compositional data, Bayesian model, MCMC

1 Introduction

The problem of apportioning composite material to original sources is a well-established yet difficult problem. In this paper, we propose a model based on modelling log-ratios of mixing proportions which takes data in the form of measurements rather than compositions, and which simultaneously models source and mixture data.

The situation studied in this paper differs from some other applications of compositional analysis, such as the classic examples of Aitchison (1986); we observe the concentrations of two chemical markers—nitrates and silicates. Our model reflects three sources of uncertainty: (a) in the measurement or applicability of the source samples; (b) in the mixing proportions; and (c) in the measurements of the mixture samples.

1.1 A Water Quality Example

We consider data arising from a study at the Macaulay Institute, Aberdeen, into water quality, where three sources of water were identified for the water found in a particular stream. The task was to identify the composition of the stream water with regard to the three source components from water chemistry data taken at various times defined by "events", i.e. periods of heavy rainfall. The data were first analysed by Soulsby *et al.* (2002).

Data were presented in the form of measurements of two "indicator" variables—concentrations (in mg/l) of nitrates and silicates—for both source and mixture (stream) samples. The sources are assumed to be, in this water quality example: ground water (GW), sub-surface flow (SSF) and overground flow (OF). One such data set is illustrated by Figure 1, where we also impose a triangle connecting the sample means of the three sources. Note that purely deterministic methods of obtaining compositions will fail here since a number of mixture data points fall outside this triangle—what is required is an analysis using a model which accounts for variability in the source means *in addition* to variation in the contributions from each source which make up the mixture samples. In the next Section we define such a model in general terms, and go on to apply it to water quality data in Section 3.

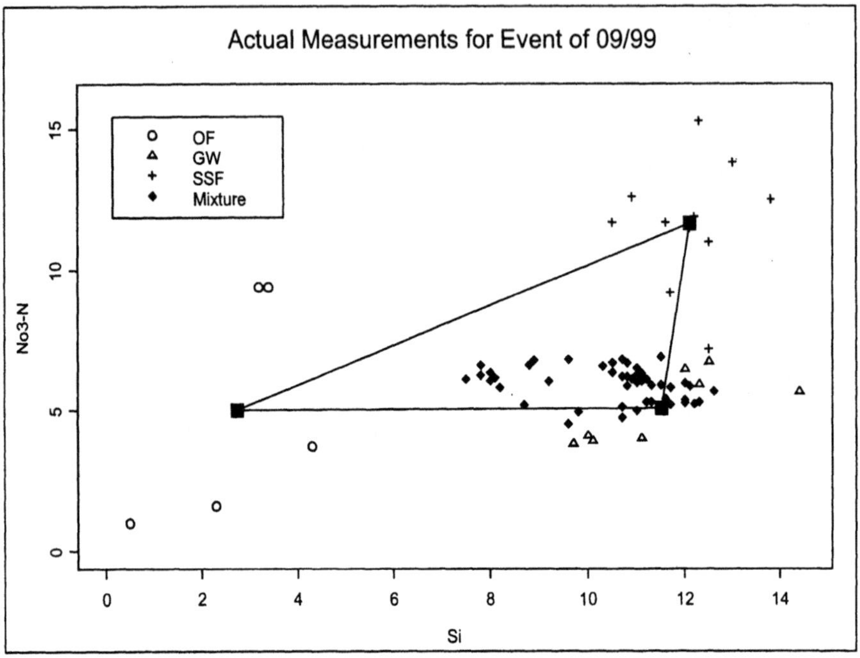

Fig. 1. Plot of indicator measurements for all three sources and event data for the September 1999 rainfall event. The filled squares represent the source means

The method of Genereux (1998) has become widely-used in hydrology for estimating compositions. While this method accounts for "uncertainty" in sources, it can lead to interval estimates of proportions which extend beyond the range $[0, 1]$. Analysis using our Bayesian model produces credible intervals which will necessarily lie wholly within that range. Finally here, we note that Billheimer (2001) and Billheimer *et al.* (2001) also use a Bayesian model for compositional analysis, but deal with data observed directly as compositions; the "compositional algebra" developed in those papers is not required here.

2 A Bayesian Hierarchical Model

We suppose that two kinds of data are available. Firstly, there are training data available on each source s, $s = 1, \ldots, N$, of interest; in our examples, $N = 3$. The data are in the form of A indicator measurements, where we require $A \geq N-1$ in order to distinguish between all N sources. Assume that, for each source, the data on the concentration of the indicator measurement, $x_{a,s}$, $a = 1, \ldots, A$, has a multivariate Normal distribution with unknown mean and variance. Letting $\boldsymbol{x}_s = (x_{1,s}, \ldots, x_{A,s})^T$ be the A-dimensional vector of concentrations of indicators in source s, we can write this situation as

$$\boldsymbol{x}_s^i \sim \mathrm{MVN}_A(\boldsymbol{\mu}_s, \boldsymbol{\Sigma}_s), \qquad s = 1, \ldots, N, \quad i = 1, \ldots, n_{x,s}, \tag{1}$$

where $n_{x,s}$ is the number of training samples for source s, $\boldsymbol{\mu}_s$ is an A-vector of means for source s, and $\boldsymbol{\Sigma}_s$ is the corresponding $(A \times A)$-dimensional covariance matrix. Note that our water quality examples have $A = 2$.

Secondly, there are n_z samples from a mixture of the above sources, all drawn from the sample population of possible mixtures, yielding indicator data denoted by the A-dimensional vector of concentrations $\boldsymbol{z}^j = (z_1^j, \ldots, z_A^j)^T$, $j = 1, \ldots, n_z$.

Each mixed sample j has unknown mixing proportions p_s^j of each source s. The contribution of each source to the indicator measurements in the sample will be $p_s^j \boldsymbol{y}_s^j$, where $\boldsymbol{y}_s^j$ follows the pure sample distribution for source s in (1). Summing these contributions over sources gives a compositional equation for each sample, namely

$$\boldsymbol{z}^j = \sum_{s=1}^{N} p_s^j \boldsymbol{y}_s^j + \boldsymbol{e}^j, \qquad j = 1, \ldots, n_z,$$

where

$$\boldsymbol{y}_s^j \sim \mathrm{MVN}_A(\boldsymbol{\mu}_s, \boldsymbol{\Sigma}_s), \qquad s = 1, \ldots, N, \quad j = 1, \ldots, n_z,$$

$$\sum_{s=1}^{N} p_s^j = 1 \qquad \text{where } 0 \leq p_s^j \leq 1, \quad j = 1, \ldots, n_z,$$

and $\boldsymbol{e}^j = (e_1^j, \ldots, e_A^j)$ is an A-dimensional vector of i.i.d. measurement errors ($j = 1, \ldots, n_z$) with common Normal distribution having mean $\mathbf{0}_A$ (an A-vector of zeros) and variance $\boldsymbol{\Sigma}_e$.

We assume further that, for each mixed sample j, the log-ratios of the p_s^j with respect to p_N^j follow a multivariate Normal distribution, thus:

$$\begin{aligned} q_s^j &\equiv \log(p_s^j / p_N^j), \qquad s = 1, \ldots, N-1, \quad j = 1, \ldots, n_z, \\ \boldsymbol{q}^j &= (q_1^j, \ldots, q_{N-1}^j)^T, \\ \boldsymbol{q}^j &\sim \mathrm{MVN}_{N-1}(\boldsymbol{\mu}_q, \boldsymbol{\Sigma}_q) \end{aligned}$$

where $\boldsymbol{\mu}_q$ is the mean location vector on the log-ratio scale, and $\boldsymbol{\Sigma}_q$ the associated variance matrix.

We complete the specification of the model by defining the following prior distributions: $\boldsymbol{\mu}_q \sim \mathrm{MVN}_{N-1}\left(\mathbf{0}_{N-1}, \boldsymbol{\tau}_q^{-1}\right)$, where $\mathbf{0}_{N-1}$ is an $(N-1)$-vector of zeros and $\boldsymbol{\tau}_q$ is an $(N-1) \times (N-1)$ diffuse precision matrix, typically taken to be diagonal; $\boldsymbol{\mu}_s \sim \mathrm{MVN}_A\left(\boldsymbol{m}_s, \boldsymbol{\tau}_s^{-1}\right)$, where $\boldsymbol{m}_s$ is an A-vector often taken as the sample means of the A indicators for each source $s = 1, \ldots, N$, and $\boldsymbol{\tau}_s$ is an $A \times A$ diffuse precision matrix as before (here accounting for source measurement error); $\boldsymbol{\Sigma}_q^{-1} \sim \mathrm{Wishart}\left(\boldsymbol{\Omega}_q, d_q\right)$ where $\boldsymbol{\Omega}_q$ is a diagonal $(N-1) \times (N-1)$ matrix and d_q is a scalar quantity of value at least $N-1$; finally $\boldsymbol{\Sigma}_e$ and $\boldsymbol{\Sigma}_s$ ($s = 1, \ldots, N$) which are defined in a similar manner to $\boldsymbol{\Sigma}_q$.

Analysis of the above model was performed via Markov chain Monte Carlo (MCMC) using the freely-available WinBUGS package (Lunn *et al.*, 2000). The main focus will be on estimates of: the overall composition proportions $\boldsymbol{p}$ (obtained from the simulated $\boldsymbol{\mu}_q$); and the individual observation proportions $\boldsymbol{p}^j$ ($j = 1, \ldots, n_z$). We plot the latter against time to observe how the composition of water in the stream changes during the event, but note that the model does not explicitly take time into account.

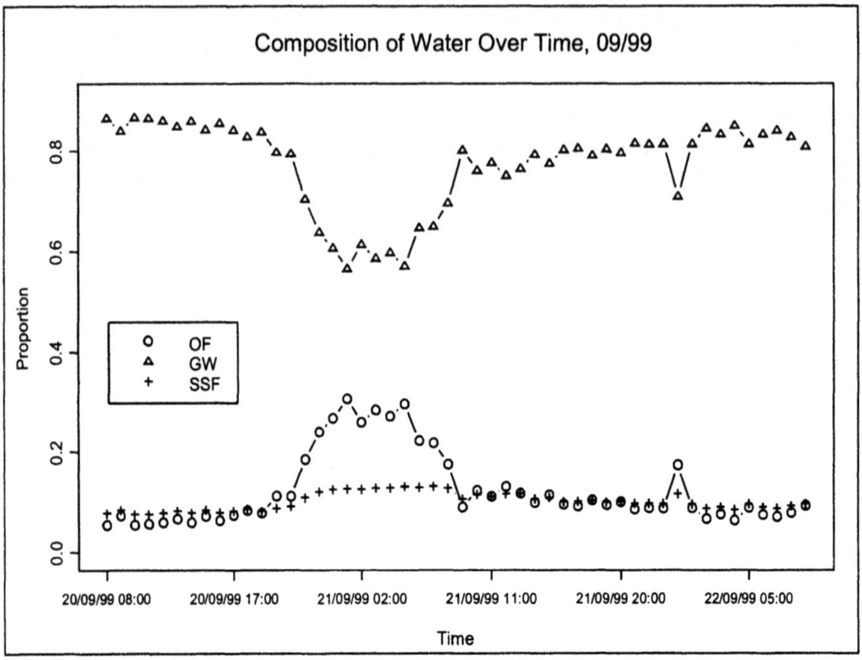

Fig. 2. Trace of individual estimated component proportions against time

Source	Estimated Proportion	95% CI Lower	95% CI Upper
GW	0.805	0.706	0.909
OF	0.096	0.025	0.167
SSF	0.099	0.036	0.180

Table 1. Estimated source component proportions

3 Results

Figure 2 shows the plots of the estimated compositions over time for a rain event in north-east Scotland during September 1999, with $n_z = 50$. The rain itself fell mainly during 20/09/99 and the early part of 21/09/99, i.e. in the period covered by the left-hand side of the plot. This event followed a relatively dry period; as such, the initial ground water (GW) contribution (as a proportion of the total) is fairly high. Some time after the start of the event, we can see the GW (proportion) dipping somewhat, with a corresponding rise in overground flow (OF). The SSF contribution rises slightly during this period also, and a day or so after the event the system has settled back to (approximately) its original state. Table 1 shows the estimated overall component proportions averaged over time for the three sources with 95% credible interval (CI) bounds. At first sight, it may seem strange that the estimated proportion for OF is lower than that for SSF—it may appear from Figure 2 that the OF contribution is higher than the SSF due to the OF's peak. However, this apparent conflict is simply a consequence of the skewness induced by the log-ratio transformation.

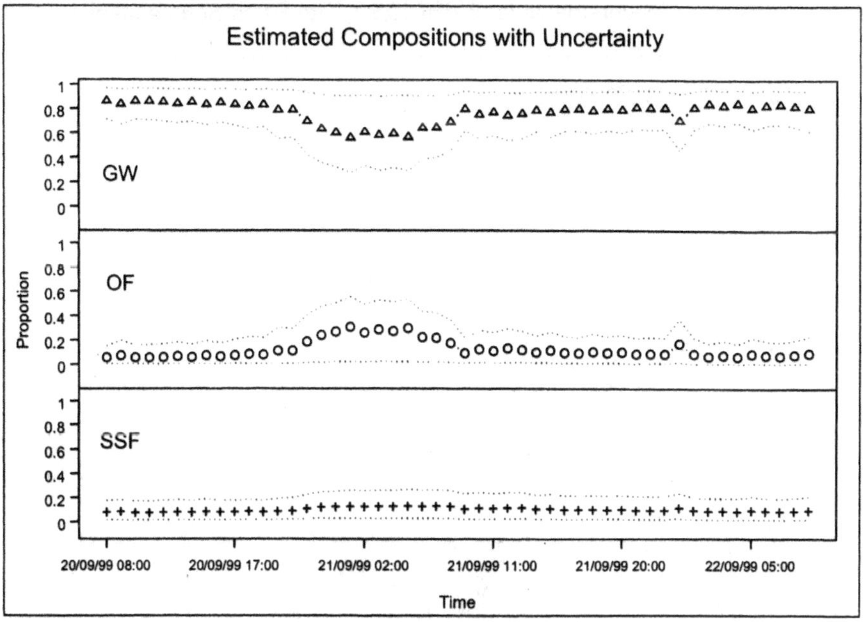

Fig. 3. Traces of individual estimated component proportions with uncertainty in the form of 95% credible intervals

To illustrate the uncertainty in the estimates of the individual component proportions, we study Figure 3. Here we notice that the uncertainty of the estimates for GW and OF is greatest during the GW dip/OF peak period, where the credible intervals overlap for a time.

Finally, we consider data for another event, in October 1999, shown in Figure 4. Here we simply note the difference between the fitted source mean (shown as an empty square) and the ordinary sample mean (filled square) for OF. Source data for overground flow is difficult to obtain—which is why there are only 5 data points. The points are also very widely scattered; our analysis of these data has used information from the *mixture* sample in order to help estimate the *source* means. On the basis of this, the OF source mean for this particular event (or time) has been estimated to have a higher nitrate concentration than the source sample mean would have suggested. Correspondingly, more weight has been placed on the two sample points with higher nitrate concentration, and this seems reasonable given the shape of the cloud of mixture data points, for example. In this way, insight can be gained into the suitability of individual source samples for analysis of any particular rainfall event.

4 Discussion

The model proposed has allowed us to determine the overall source composition of stream water and to study the progression of component proportions over time, in this case. The results obtained are explicable, and uncertainty can be quantified in a coherent manner. Analysis of the current data might be improved by relaxing the assumption of independence on the mixture sam-

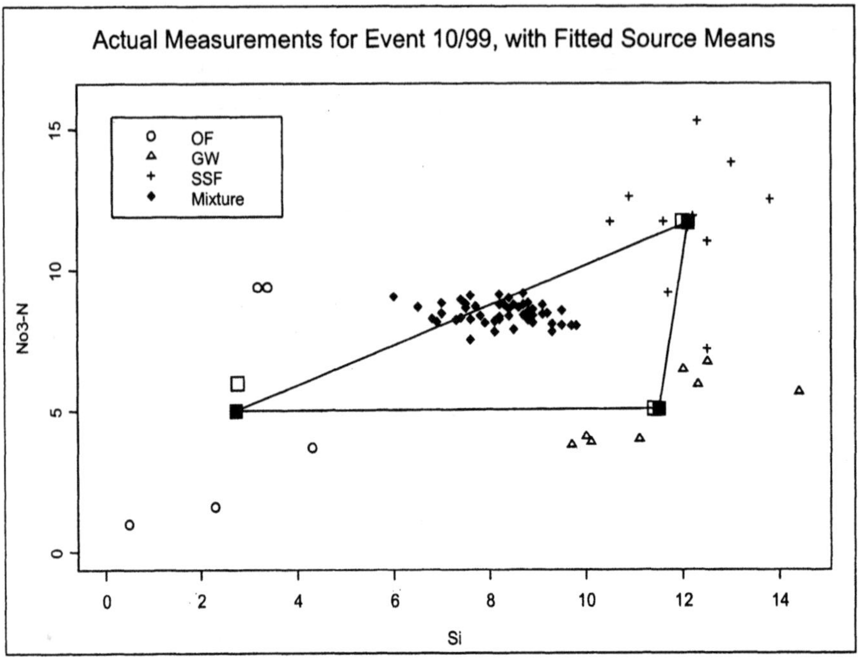

Fig. 4. Plot of indicator measurements for all three sources and event data for the October 1999 rainfall event. The filled squares represent the source means; the empty squares show the source means estimated from the model

ples, perhaps by modelling time itself directly or by inclusion of a suitable surrogate for time (such as stream flow rate) in the model.

References

Aitchison, J. (1986) *The Statistical Analysis of Compositional Data.* Chapman and Hall, London.

Billheimer, D. (2001) Compositional Receptor Modelling. *Environmetrics*, **12**, 451–467.

Billheimer, D., Guttorp, P., and Fagan, W.F. (2001) Statistical Interpretation of Species Composition. *Journal of the American Statistical Association*, **96**, 1205–1214.

Genereux, D. (1998) Quantifying Uncertainty in Tracer-Based Hydrograph Separations. *Water Resources Research*, **34**, 915–919.

Lunn, D.J., Thomas, A., Best, N. and Spiegelhalter, D. (2000) WinBUGS - A Bayesian Modelling Framework: Concepts, Structure and Extensibility. *Statistics and Computing*, **10**, 325–337.

Soulsby, C., Petry, J., Brewer, M.J., Dunn, S.M., Ott, B. and Malcolm, I.A. (2002) Identifying and Assessing Uncertainty in Hydrological Pathways: A Novel Approach to End Member Mixing Analysis in a Scottish Agricultural Catchment. *Journal of Hydrology* (Submitted).

A Comparison of Marginal Likelihood Computation Methods

Charles S. Bos[1]

[1] Faculty of Economics and Operations Research, *Vrije* Universiteit *Amsterdam*, De Boelelaan 1105, NL-1081 HV Amsterdam, The Netherlands, cbos@feweb.vu.nl

Abstract. In a Bayesian analysis, different models can be compared on the basis of the expected or marginal likelihood they attain. Many methods have been devised to compute the marginal likelihood, but simplicity is not the strongest point of most methods. At the same time, the precision of methods is often questionable.

In this paper several methods are presented in a common framework. The explanation of the differences is followed by an application, in which the precision of the methods is tested on a simple regression model where a comparison with analytical results is possible.

JEL classification. C11, C52, C63
Keywords. Marginal likelihood, Bayesian analysis

1 Introduction

In Bayesian inference, the development of sampling methods has come far since the introduction of the Metropolis-Hastings algorithm (Metropolis, Rosenbluth, Rosenbluth, Teller & Teller, 1953). With the resulting posterior samples, it is often of interest to contrast two competing models, to judge which model fits the data best. The (related) concepts of the marginal likelihood, the Bayes factor and the posterior odds can be used for this purpose, see e.g. Aitkin (1991). Kass & Raftery (1995) even summarize a range of computational methods for obtaining estimates of the Bayes factor, with a theoretical indication of the accuracy that can be expected.

The marginal likelihood and corresponding Bayes factor and posterior odds are used in a number of recent articles, but no extensive analysis of the practical precision of the methods has been performed yet. In this paper, first an overview of the available estimation methods is given in section 2. For each of the methods an indication of the precision is obtained in section 3. The precision is evaluated in a small simulation exercise, on a simple regression model where analytical results are available.

2 Marginal likelihood and its computation

2.1 The concept of marginal likelihood and posterior odds

When models are estimated in a classical manner, they can be compared on the basis of the likelihood they attain. The likelihood function is evaluated at the point indicated by the parameter estimates, often at the location of maximum likelihood. In a Bayesian framework, there is not one parameter vector characterizing the fit of the model. Instead, based on the likelihood and the prior, the full posterior distribution of the parameters is derived. A characteristic for the fit of a model M is in this case the expected or marginal likelihood $m(Y|M)$, where the expectation is taken over the likelihood $\mathcal{L}(Y;\theta|M)$ with respect to the prior distribution $\pi(\theta|M)$ of the parameters,

$$m(Y|M) = \int_\theta \mathcal{L}(Y;\theta|M)\pi(\theta|M)\,\partial\theta. \tag{1}$$

The marginal likelihood is the major ingredient for statistics like the Bayes factor BF$= m(Y|M_1)/\ m(Y|M_2)$, comparing the evidence in favour of two competing models. The posterior odds (PO) ratio is again based on the BF, with PO$=$ BF$\times\pi(M_1)/\pi(M_2)$, and relates the posterior evidence of the models.

Though other ways exist to compute the Bayes factor or the posterior odds (see Verdinelli & Wasserman (1995) for the (generalized) Savage-Dickey density ratio), the method using the marginal likelihoods is conceptually the simplest. The next section focuses on computational methods for the marginal likelihoods.

2.2 Computational methods

Only in very special cases, most notably for the exponential likelihood with conjugate priors, can the marginal likelihood m be calculated analytically as the integrating constant of the posterior kernel.[1] In other cases, numerical methods are needed. Table 1 summarizes a range of methods. Details can be found in Bos (2001) and the references therein.

The following remarks concerning the methods in table 1 can be made: The brute-force integration method suffers from the curse of dimensionality. When it is not viable, a simulation method may help. The method m_{IS} is not operational without a choice for the importance sampling density $\pi^*(\theta)$, approximating the prior density. Using the prior density as importance function leads to weights $w_i \equiv 1$ as in m_{Prior}, but many drawings will fall in low-likelihood regions. Sampling from the posterior density gives more drawings in the correct region, but leads to an estimate of m which may not have a finite variance (Newton & Raftery 1994). Intermediate positions, such as $\pi^*(\theta) = \delta\pi(\theta) + (1-\delta)p(\theta|Y)$ can be chosen: This gives a consistent estimate with better convergence behaviour (see also Newton & Raftery 1994). A more recent solution for stabilizing the harmonic mean estimate, utilizing a technique for lowering the dimension of the problem, is given in Satagopan, Newton & Raftery (2000), and is not discussed here.

m_{LP} and m_{Kern} are special versions of m_{App}. The first essentially fits a normal density to the mode of the posterior. The method can be expected to work well in cases where the posterior is highly peaked, less well in multimodal cases. m_{Kern} applies a kernel smoother, and therefore also suffers to some extent from the curse of dimensionality, as more drawings $\theta^{(i)}$ are needed when the dimension increases, in order to get a good approximation to the posterior density.

Finally, the Gibbs sampling method can be applied in cases where the conditional densities are available, and may be the only method when data augmentation (Gelfand & Smith 1990) is used. Implementing may prove to be

---

[1] As we are calculating the marginal likelihood of a specific model M on a fixed data set Y in this section, dependence of m on these quantities is suppressed in the notation.

Table 1. Computational methods for marginal likelihoods

$m_{\text{Anal}} = \int \mathcal{L}(Y,\theta)\pi(\theta)\partial\theta$	
	The exact solution, available only in special cases.
m_{Num}	Using brute-force numerical integration, m can be computed for low-dimensional problems.
$m_{\text{Prior}} = \mathsf{E}_{\pi(\theta)}\mathcal{L}(Y,\theta) \approx \frac{1}{n}\sum \mathcal{L}(Y,\theta^{(i)}); \theta^{(i)} \sim \pi(\theta)$	
	Simulating from the prior, m is the average likelihood. A very unstable and inefficient estimate.
$m_{\text{IS}} = \frac{1}{\sum w_i}\sum w_i\mathcal{L}(Y,\theta^{(i)}); \theta^{(i)} \sim \pi^*(\theta), w_i = \pi(\theta^{(i)})/\pi^*(\theta^{(i)})$	
	Sampling from an importance density with higher probability mass around the posterior mode improves on m_{Prior}.
$m_{\text{HM}} = \left(\frac{1}{N}\sum \frac{1}{\mathcal{L}(Y,\theta^{(i)})}\right)^{-1}; \theta^{(i)} \sim p(\theta \mid Y)$	
	Using the posterior density as importance sampling density in m_{IS} leads to using the harmonic mean of the likelihood values as estimator.
$m_{\text{App}} = \mathcal{L}(Y,\theta)\pi(\theta)/p(\theta \mid Y)$	
	In general, relating the height of the posterior kernel to the height of an approximating posterior density in a high density point θ can give an estimate of m.
$m_{\text{LP}} = (2\pi)^{k/2}\lvert\tilde{\Sigma}\rvert^{\frac{1}{2}}\mathcal{L}(Y,\tilde{\theta})\pi(\tilde{\theta})$	
	This is m_{App} using a quadratic expansion of the posterior kernel around the posterior mode $\tilde{\theta}$.
$m_{\text{Kern}} = \mathcal{L}(Y,\theta)\pi(\theta)/p_{\text{Kern}}(\theta \mid Y)$	
	This is m_{App} using a kernel smoother to approximate the posterior density at a high density point θ.
$m_{\text{Gibbs}} = \mathcal{L}(Y,\theta)\pi(\theta)/\left[p(\theta_1 \mid Y)\prod p(\theta_i \mid \theta_1,..,\theta_{i-1},Y)\right]$	
	Chib (1995) proposed to use $k-1$ samples from separate Gibbs chains to approximate $p(\theta_i \mid \theta_1,..,\theta_{i-1},Y) \approx \sum p(\theta_i \mid \theta_1,..,\theta_{i-1},\theta_{i+1}^{(j)},..,\theta_k^{(j)},Y)/N$, where N samples are drawn for the missing elements $\theta_{i+1},..,\theta_k$ from the conditional densities.

hard, and for each element of the vector of parameters a separate Gibbs chain has to be run, leading to a rather hefty computational burden, as compared to the other methods.

3 Marginal likelihood in practice: A comparison

3.1 Data, model and analytical marginal likelihood

In Brownlee (1965, page 454) a data-set concerning the oxidation of ammonia for producing nitric acid is examined. Explanatory variables are the air flow in the tower, the temperature of cooling water and the acid concentration which is produced. The dependent variable is the so-called stack-loss, which is 10 times the percentage of ammonia which escapes from the process without having been converted into nitric acid.

This data-set was analysed often (e.g. Justel & Peña 1996) using regression models, with the focus on recognizing outliers among the 21 observations on the 4 variables. For the purpose of displaying the differences of the computational methods for marginal likelihoods, it is sufficient to limit ourselves to the pure regression model

$$y = X\beta + \epsilon, \qquad \epsilon = \mathcal{N}(0, \sigma^2 \mathcal{I}_n). \tag{2}$$

For simplicity, we assume that the variance σ^2 is known and fix it at its least squares estimate.

With a normal prior $\pi(\beta)$, with expectation β_0 and covariance matrix $\Sigma_0 = \sigma_0^2 \mathcal{I}_k$, the posterior $p(\beta|Y)$ is normal as well, with covariance $\tilde{\Sigma} = (\hat{\Sigma}^{-1} + \Sigma_0^{-1})^{-1}$ and expectation $\tilde{\beta} = \tilde{\Sigma}^{-1}(\hat{\Sigma}^{-1}\hat{\beta} + \Sigma_0^{-1}\beta_0)$, where $\hat{\beta}$ and $\hat{\Sigma}$ are the least squares estimates. Computing the marginal likelihood from the quotient of the posterior kernel $\mathcal{L}(Y;\theta)\pi(\theta)$ and the posterior density $p(\theta|Y)$ gives the expression

$$m_{\text{Anal}} = (2\pi)^{-\frac{n}{2}} \sigma^{-n} |\Sigma_0|^{-\frac{1}{2}} |\tilde{\Sigma}|^{\frac{1}{2}} \exp\left(-\frac{1}{2}\left[\frac{y'y}{\sigma^2} + \beta_0 \Sigma_0^{-1} \beta_0 - \tilde{\beta}\tilde{\Sigma}^{-1}\tilde{\beta}\right]\right) \tag{3}$$

Note that all terms containing β drop out of this equation, as should be the case for the integrand of the posterior kernel over β.

3.2 Estimation results

Table 2 displays the main estimation results for the regression model on the stack-loss data. For the prior, an expectation of zero and a large standard deviation of the elements was chosen, such that the prior hardly influences the results. Indeed, the difference between the least squares and posterior estimates is minimal.

Table 2. Prior, least squares and posterior moments of the parameters

	Prior		Least squares		Posterior	
Variable	β_0	σ_0	$\hat{\beta}$	$\hat{\sigma}$	$\tilde{\beta}$	$\tilde{\sigma}$
Air flow	0	20	0.7968	0.166	0.7969	0.166
Water temperature	0	20	1.1114	0.456	1.1109	0.456
Acid concentration	0	20	-0.6250	0.085	-0.6249	0.085

Note: Non-zero correlations between the LS and posterior parameters are not reported.

The methods explained in section 2 were used to construct the figures in table 3. For each of the methods indicated in the first column, the table reports an estimate of the logarithm of marginal likelihood in the second column.

With numerical integration, the result of the analytical computation can be found if enough time/effort is put into the integration routine. For this model, no difference is found indeed. However in practice, for more elaborate models, it is often difficult to find sensible bounds for the integration region, and either the tails of the posterior density might be missed or too much effort is put into regions with little mass, leading to large computational efforts being wasted.

Table 3. Logarithm of the marginal likelihood of the regression model for Stack-loss data

Method	Log m	Repeated simulation Mean	s.d.	Min	Max	%Time
m_{Anal}	-74.271					
m_{Num}	-74.271					
m_{Prior}	-110.710	-759.827	558.03	-3141.87	-81.90	11
m_{HM}	-61.281	-61.111	0.57	-64.18	-60.16	12
m_{LP}	-74.271	-74.280	0.04	-74.38	-74.18	0
m_{Kern}	-74.166	-74.109	0.08	-74.30	-73.80	1
m_{Gibbs}	-74.354	-74.304	0.49	-75.30	-72.92	76

Note: Results are reported for one large simulation of 50,000 drawings and for 100 repeated simulations of sample size 1,000. Timings indicate the percentage of time spent computing the measures in the repeated simulation.

Both the m_{Prior} and m_{HM} methods are special versions of m_{IS}. They both suffer from instability: even after more than 45,000 drawings,[2] the estimates of $\log m_{\text{Prior}}$ and $\log m_{\text{HM}}$ can be seen to change.

The LaPlace method uses a normal density to approximate the normal posterior density: As the approximation is perfect, the correct solution is found. In general cases the estimate will not be perfect, but practice indicates that this method works rather well, especially for highly peaked, unimodal posterior densities.

Approximating the posterior with a kernel smoothing density (a multivariate Gaussian kernel smoother with automatic bandwidth selection) is simple, but gives good results as well. The Gibbs method gives an answer which is slightly more precise, but is more costly in terms of computing time and in complications in the estimation procedure.

Columns 3–7 of table 3 report results for replicating the estimation 100 times, using sample sizes of 1,000. The prior method proves to be totally unreliable, with an enormous standard deviation in column 4. The harmonic mean method does not converge to the correct estimate either.[3] For the LaPlace method, the estimate was now computed at the mean of the parameters sampled from the posterior density, with the covariance of the sample as covariance estimate. It is seen that this very simple method is quite good (at least for this model). Also the kernel and Gibbs methods perform well, with more variation in results for the latter.

The last column gives an indication of the respective durations of the calculations. The m_{Prior} and m_{HM} methods use rather costly function evaluations, leading to computation times increasing with the size of the sample. The method applying Gibbs sampling needs to collect new samples from the

[2] For m_{Prior}, 50,000 values are sampled from the prior density. For m_{HM}, and likewise for m_{Kern}, a sample of 50,000 *accepted* drawings from a Metropolis-Hastings chain with a multivariate Student-t candidate with 4 degrees-of-freedom was collected. At an acceptance rate of 73%, this gave a total sample of size 68,071 from the posterior density.

[3] A separate computation, continuing for 10^9 drawings from the posterior, continued to display jumps in the cumulative estimate of m_{HM} even at the end of the run.

conditional densities, through several runs of the Gibbs chain. In the present implementation it is taking about 7 times as long as either m_{Prior} or m_{HM}. Depending on the sample size, the dimension of the problem and the integration method used, it can be quicker than the brute-force integration method. For the results in column 2, on the larger sample, the Gibbs computation finished in roughly one third of the time needed for computing the integral using Gaussian quadrature.

4 Concluding remarks

In this paper, a range of computational methods for obtaining an estimate of the marginal likelihood of a model are reviewed briefly, and applied in a small application. This way, the qualitative impression of the precision of the methods available until now was checked in a quantitative manner.

The main finding is that both the method of computing the marginal likelihood as the expectation of the likelihood under the prior, as the method of computing the harmonic mean of likelihood values when sampling from the posterior, are not trustworthy. Though work is under way to stabilize m_{HM} (Satagopan, Newton & Raftery 2000), the methods applying either a LaPlace or kernel approximation to the posterior density in a high-density location prove to be more useful. The Gibbs sampling method of Chib (1995) can give good results as well, though the computational costs are high especially with increasing dimension of the vector of parameters.

References

Aitkin, M. (1991), 'Posterior Bayes factors', *Journal of the Royal Statistical Society, Series B* **53**(1), 111–142.

Bos, C. S. (2001), Time Varying Parameter Models for Inflation and Exchange Rates, PhD thesis, Tinbergen Institute, Erasmus University Rotterdam. **TI 256**.

Brownlee, K. A. (1965), *Statistical Theory and Methodology in Science and Engineering*, 2 edn, Wiley, New York.

Chib, S. (1995), 'Marginal likelihood from the Gibbs output', *Journal of the American Statistical Association* **90**(432), 1313–1321.

Gelfand, A. E. & Smith, A. F. M. (1990), 'Sampling-based approaches to calculating marginal densities', *Journal of the American Statistical Association* **85**(410), 398–409.

Justel, A. & Peña, D. (1996), 'Gibbs sampling will fail in outlier problems with strong masking', *Journal of Computational & Graphical Statistics* **5**(2), 176–189.

Kass, R. E. & Raftery, A. E. (1995), 'Bayes factors', *Journal of the American Statistical Association* **90**(430), 773–795.

Metropolis, N., Rosenbluth, A. W., Rosenbluth, M. N., Teller, A. H. & Teller, E. (1953), 'Equations of state calculations by fast computing machines', *Journal of Chemical Physics* **21**, 1087–1091.

Newton, M. A. & Raftery, A. E. (1994), 'Approximate Bayesian inference by the weighted likelihood bootstrap', *Journal of the Royal Statistical Society, Series B* **3**, 3–48.

Satagopan, J. M., Newton, M. A. & Raftery, A. E. (2000), Easy estimation of normalizing constants and Bayes factors from posterior simulation: Stabilizing the harmonic mean estimator, Unpublished manuscript.

Verdinelli, I. & Wasserman, L. (1995), 'Computing Bayes factors using a generalization of the Savage-Dickey density ratio', *Journal of the American Statistical Association* **90**(430), 614–618.

A Hotelling Test Based on MCD

G. Willems[1], G. Pison[1], P.J. Rousseeuw[1] and S. Van Aelst[2]

[1] Department of Mathematics and Computer Science, University of Antwerp (UIA), Universiteitsplein 1, B-2610 Wilrijk, Belgium
[2] Department of Applied Mathematics and Computer Science, Ghent University, Krijgslaan 281 S9, B-9000 Ghent, Belgium

Abstract. Hypothesis tests and confidence intervals based on the classical Hotelling T^2 statistic can be adversely affected by outliers. Therefore, we construct an alternative inference technique based on a statistic which uses the highly robust MCD estimator of Rousseeuw (1984) instead of the classical mean and covariance matrix. The distribution of this new statistic differs from the classical one. Similarly to the classical T^2 distribution, we obtain a multiple of a certain F-distribution. It is shown through a Monte Carlo study that this approximation is very accurate, both at the normal model and at contamination models.

Keywords. Hotelling, Minimum Covariance Determinant, Robustness, One-sample hypothesis test

1 Introduction

Hotelling's T^2 statistic is the standard tool for inference about the center of a multivariate normal distribution. An example is the one-sample T^2 hypothesis test for a sample $X_n = \{\boldsymbol{x}_1, \ldots, \boldsymbol{x}_n\}$ from $N_p(\boldsymbol{\mu}, \boldsymbol{\Sigma})$ where $\boldsymbol{\mu}$ and $\boldsymbol{\Sigma}$ are unknown. The hypothesis $H_0 : \boldsymbol{\mu} = \boldsymbol{\mu}_0$ is rejected at the level α if

$$T^2 := n(\overline{\mathbf{x}} - \boldsymbol{\mu}_0)'\mathbf{S}^{-1}(\overline{\mathbf{x}} - \boldsymbol{\mu}_0) > \frac{(n-1)p}{(n-p)} F_{p,n-p,1-\alpha} \qquad (1)$$

where $\overline{\mathbf{x}} = \frac{1}{n}\sum_{i=1}^{n} \boldsymbol{x}_i$ and $\mathbf{S} = \frac{1}{n-1}\sum_{i=1}^{n}(\boldsymbol{x}_i - \overline{\mathbf{x}})(\boldsymbol{x}_i - \overline{\mathbf{x}})'$ are the mean and covariance of the sample (see e.g. Johnson and Wichern 1988). Other applications of the T^2 statistic include simultaneous confidence intervals and two-sample hypothesis tests.

However, these tests and confidence intervals are based on the classical mean and covariance, hence the results can be heavily influenced by outliers. Therefore, we propose to use robust estimators of location and scatter instead of the classical mean and covariance in the expression for T^2 given by (1). In this paper we propose to use a reweighted version of the Minimum Covariance Determinant estimator (MCD) of Rousseeuw (1984) which is a highly robust estimator of location and scatter. The MCD estimator will be summarized in Section 2.

The distribution of the test statistic based on MCD differs from the classical one. Therefore, the key problem is to find a good approximation for this distribution. In Section 3 we construct an approximate distribution using ideas that are similar to the construction of the classical T^2 distribution.

Based on a Monte Carlo study, in Section 4 we construct functions which yield values for the constants that are necessary to obtain the approximate distribution of the robust T^2 statistic. The resulting method gives an accurate approximation as will be shown in Section 5.

2 Minimum Covariance Determinant Estimator

Given n data points in $\mathbb{R}^p$ the MCD is determined by the subset of size $h = \lfloor \gamma n \rfloor$ (where $0.5 \leq \gamma \leq 1$) whose covariance matrix has the smallest determinant. The MCD location estimate T is defined as the mean of that subset, and the MCD scatter estimate C is a multiple of its covariance matrix. The multiplication factor consists of a consistency factor and a finite-sample correction factor.

To increase the efficiency of the MCD we compute the reweighted MCD. The reweighted estimates are the weighted mean $T^1 = (\sum_{i=1}^n w_i \boldsymbol{x}_i)/\sum_{i=1}^n w_i$ and weighted covariance $C^1 = c_{\delta,p} d_{n,p} (\sum_{i=1}^n w_i (\boldsymbol{x}_i - T^1)(\boldsymbol{x}_i - T^1)')/\sum_{i=1}^n w_i$ with weights based on the robust distances (Rousseeuw and Van Zomeren 1990) of the observations. Observations with robust distance below a cutoff value q obtain weight 1, while the other observations obtain weight 0. The factor $c_{\delta,p}$ is the consistency factor for the normal model and can be found in Croux and Haesbroeck (1999). It depends on δ which here denotes the fraction of observations that obtained weight 0 and have a robust distance smaller than $2q$ (observations that have a robust distance larger than $2q$ are considered to be far outliers). The constant $d_{n,p}$ is a correction factor which makes the estimator unbiased at small samples (see Pison et al. 2002).

The MCD estimator is intuitively appealing and has interesting properties like affine equivariance and asymptotic normality. It also has a bounded influence function and a positive breakdown value depending on γ . The reweighted MCD estimator inherits each of these properties while achieving a higher efficiency (Lopuhaä 1999, Croux and Haesbroeck 1999, Lopuhaä and Rousseeuw 1991). In this paper we will use $\gamma = 0.75$ as a good compromise between efficiency and breakdown (25%). To compute the reweighted MCD we use the FAST-MCD algorithm (Rousseeuw and Van Driessen 1999) which makes the MCD available for routine use, even for large data sets.

3 The robust T^2 statistic

To obtain robust inference techniques for the mean of a multivariate normal distribution we construct a new statistic by replacing the classical estimators in Hotelling's T^2 by the reweighted MCD estimators (T^1, C^1):

$$T_R^2 := n(T^1 - \boldsymbol{\mu})'(C^1)^{(-1)}(T^1 - \boldsymbol{\mu}).$$

The finite-sample distributions of the reweighted MCD location and scatter are unknown, but it turns out they have properties that are similar to the classical estimators. That is, for a sample of size n from $N_p(\boldsymbol{\mu}, \boldsymbol{\Sigma})$ the reweighted MCD estimators approximately satisfy the properties:

1. There exists some κ such that $T^1 \sim N_p(\boldsymbol{\mu}, \kappa \frac{1}{n} \boldsymbol{\Sigma})$.

2. There exist m and c such that $mc^{-1}C^1 \sim W_p(\boldsymbol{\Sigma}, m)$ (Wishart) and $\mathrm{E}[C^1] = c\boldsymbol{\Sigma}$.

3. T^1 and C^1 are independent.

Property 2 was formulated in Hardin and Rocke (2001). These properties allow us to obtain an approximate F-distribution for T_R^2, analogous to the F-distribution of the classical T^2. From Mardia et al. (1995), theorem 3.5.2, we obtain that

$$T_R^2 \approx \kappa c^{-1} \frac{mp}{m-p+1} F_{p,m-p+1} \tag{2}$$

Instead of determining values for the constants κ, c and m we will use a more direct approach. First we rewrite (2) as

$$T_R^2 \approx dF_{p,q} \tag{3}$$

which has the advantage that only two constants , d and q, have to be determined. These constants, which both depend on sample size n and dimension p, will be obtained by matching the mean and variance of the distribution given by (3) such that we obtain a very close approximation to the exact distribution of T_R^2. Since the MCD is affine equivariant it follows that the T_R^2 statistic is affine invariant. Therefore it suffices to determine values of d and q for samples from the standard Gaussian distribution $N_p(\mathbf{0}, \mathbf{I}_p)$.

From the definition of the F-distribution we obtain the following expressions for the constants d and q:

$$d \approx \mathrm{E}[T_R^2]\frac{q-2}{q} \tag{4}$$

$$q \approx \left(\frac{\mathrm{Var}[T_R^2]}{\mathrm{E}[T_R^2]^2}\frac{p}{2} - 1\right)^{-1}(p+2)+4 \tag{5}$$

Since the mean and variance of the T_R^2 distribution can not be obtained analytically, they will be approximated by simulation. It will be shown that this approach results in a very accurate approximation of the true distribution of T_R^2. Moreover, we will construct functions which yield approximations for $\mathrm{E}[T_R^2]$ and $\mathrm{Var}[T_R^2]$ for all n and p, so there will be no need for further simulation in practice.

Note that we do know the exact asymptotic distribution of T_R^2 since the asymptotic normality of T^1 and C^1 implies that

$$T_R^2 = n(T^1 - \mu)'(C^1)^{-1}(T^1 - \mu) \underset{n\to\infty}{\to} \kappa\chi_p^2$$

where κ is the asymptotic variance of a component of T^1. This asymptotic variance depends on p and can be derived from the influence function of the reweighted MCD location given in Lopuhaä 1999. It follows that

$$\mathrm{E}[T_R^2] \underset{n\to\infty}{\to} \kappa p \quad \text{and} \quad \mathrm{Var}[T_R^2] \underset{n\to\infty}{\to} \kappa^2 2p \tag{6}$$

These results will be incorporated in the functions that determine finite-sample values of d and q.

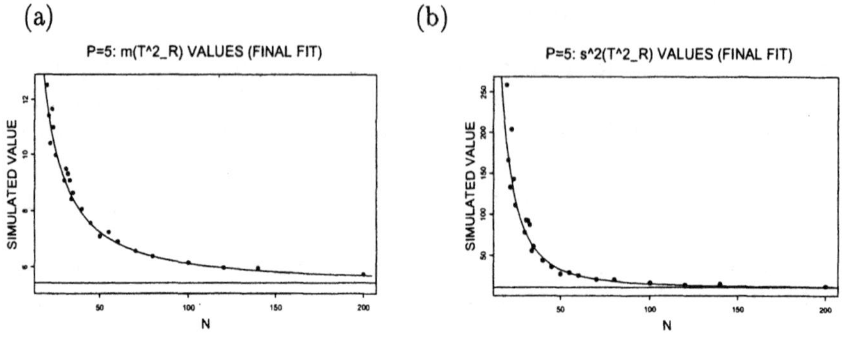

Fig. 1. Simulated values for $p = 5$ of (a) $\mathrm{m}(T_R^2)$; (b) $\mathrm{s}^2(T_R^2)$.

4 Monte Carlo simulations

For several sample sizes n and dimensions p, we generated $m = 3000$ samples $X_j; j = 1, \ldots, m$ from a standard Gaussian distribution. For each sample X_j we computed the reweighted MCD location and scatter estimates and the corresponding values of T_R^2. Then the empirical mean $\mathrm{m}(T_R^2)$ and variance $\mathrm{s}^2(T_R^2)$ were computed as an approximation to $\mathrm{E}[T_R^2]$ and $\mathrm{Var}[T_R^2]$. Figure 1a shows some values of $\mathrm{m}(T_R^2)$ versus the sample size n for $p = 5$ while Figure 1b shows the corresponding $\mathrm{s}^2(T_R^2)$ values. These plots show a smooth pattern, hence we determined smooth functions to fit these points. On the plots we also added a horizontal line indicating the asymptotic value given by (6). For p fixed ($p = 1, \ldots, 8$ and 10) we fitted the values of $\mathrm{m}(T_R^2)$ using the following regression model:

$$f_p(n) = \kappa p + \frac{\alpha_p}{n^{\beta_p}}$$

Hence, once we have determined the parameter values α_p and β_p for a given dimension p, then for any sample size n the value of $f_p(n)$ is an approximation of $\mathrm{E}[T_R^2]$. However, this would require simulations for all dimensions p. Therefore, for $p \geq 3$ we fitted the values $f_p(rp^2) - \kappa p$ for $r = 7$ and $r = 10$ as a function of the dimension p. These two fits can then be considered as a system of equations yielding values $\widehat{\alpha_p}$ and $\widehat{\beta_p}$ for any p. Finally, for any n an approximation for $\mathrm{E}[T_R^2]$ is given by $\widehat{f_p}(n) := \kappa p + \frac{\widehat{\alpha_p}}{n^{\widehat{\beta_p}}}$. Similarly a function that returns values of $\mathrm{Var}[T_R^2]$ is derived. Overall these procedures yield good approximations for the actual simulated values of $\mathrm{E}[T_R^2]$ and $\mathrm{Var}[T_R^2]$, as is illustrated by superimposing the resulting curves for $p = 5$ in Figure 1.

5 Results

In the previous section we performed a simulation study to obtain estimates for $\mathrm{E}[T_R^2]$ and $\mathrm{Var}[T_R^2]$. They are to be used in equations (4) and (5) which in turn yield the appropriate parameter values for the approximative distribution given by (3). Now we verify the accuracy of this approximation. For this purpose, for several sample sizes n and p we performed simulations with

$m = 3000$ data sets $Z_j; j = 1, \ldots, m$ generated from the standard Gaussian distribution. For each data set Z_j we computed the T_R^2 statistic. In general the empirical distribution of these T_R^2 values compared very well to our approximate distribution. Because we are specifically interested in the one-sample hypothesis test, let us focus here on the 95% and 99% quantiles.

Table 1. Percentage above the 5% cutoff for the F-distribution

$p \setminus n$	10	20	30	50	70	100	140	200	1000
2	6.6	5.8	5.1	5.2	5.4	5.0	4.9	4.8	4.5
3	7.3	4.7	4.5	4.9	5.0	4.8	5.0	5.0	4.6
5		6.4	4.9	4.8	5.1	5.1	5.1	5.3	5.2
7		3.8	6.5	5.5	5.3	4.9	5.2	5.3	4.5
10			4.8	5.5	4.5	4.4	4.8	5.1	4.8

Table 2. Percentage above the 1% cutoff for the F-distribution

$p \setminus n$	10	20	30	50	70	100	140	200	1000
2	1.8	0.9	0.8	1.1	1.0	1.0	0.9	0.8	1.0
3	2.2	1.1	0.9	0.9	1.1	0.9	1.2	1.1	0.9
5		1.8	1.0	0.7	1.1	0.8	1.1	1.0	1.0
7		0.8	1.6	1.0	1.2	0.9	1.1	0.9	0.9
10			0.8	1.0	0.8	0.8	1.1	1.1	0.8

For several values of n and p Table 1 shows the actual percentage of simulated T_R^2 values above the 95% quantile of the approximate distribution given by (3), while Table 2 contains the results for the 99% quantile. We clearly see that in all cases the difference between the actual percentage and the nominal value is very small. Hence, by using this approximate distribution the hypothesis test based on T_R^2 will give good Type I-error probabilities for the normal model.

Other simulation results, that are not reported here, suggest that the loss in power compared to the classical Hotelling test is acceptable. (Note that the classical test has maximal asymptotic power.)

The main advantage of the MCD based Hotelling test is its robustness against outliers in the data. To investigate this, we performed simulations with contaminated data sets. For these data sets 90% of the observations were generated from the standard Gaussian distribution $N_p(\mathbf{0}, \mathbf{I}_p)$ and the other 10% were taken from $N_p(\boldsymbol{\mu}, \boldsymbol{\Sigma})$ with $\boldsymbol{\mu} = (5+p, \ldots, 5+p)'$ and $\boldsymbol{\Sigma} = \text{diag}(0.1, \ldots, 0.1)$. Table 3 lists the percentage of tests in the simulations that rejected the hypothesis $H_0 : \boldsymbol{\mu} = \mathbf{0}$ at the 5% level. We see that the actual percentages for the test based on T_R^2 are still close to the nominal values, and are a big improvement compared to the results of the classical T^2 test. Hence we can conclude that the approximate T_R^2 distribution proposed in this paper offers a robust and reliable alternative to the classical Hotelling test and other inference techniques based on T^2.

Table 3. 10% far outliers: Percentage of erroneous rejections of H_0

$p \setminus n$		10	20	30	50	70	100	140	200
2	T_R^2	4.7	5.0	4.4	5.1	5.4	5.9	5.4	5.0
	T^2	2.4	5.4	11.1	40.0	78.8	99.2	100.0	100.0
5	T_R^2		3.9	3.6	4.7	5.0	5.9	6.3	6.7
	T^2		6.1	12.0	23.3	42.2	88.3	100.0	100.0
10	T_R^2			1.3	2.8	3.9	5.3	7.2	6.8
	T^2			10.6	19.2	28.2	57.1	91.4	100.0

We implemented an S-Plus function that determines for every n and p the approximate T_R^2 distribution given by (3) and which uses the functions derived in Section 4. The function returns a desired quantile of the approximate T_R^2 distribution or the p-value corresponding to the value of the T_R^2 statistic. The S-Plus function is available from our website http://win-www.uia.ac.be/u/statis/.

References

Croux, C. and Haesbroeck, G. (1999). Influence and Efficiency of the Minimum Covariance Determinant Scatter Matrix Estimator. *Journal of Multivariate Analysis*, **71**, 161-190.

Hardin, J. and Rocke, D.M. (2001). The Distribution of Robust Distances. Technical Report, Univ. of California at Davis.

Johnson, R.A. and Wichern, D.W. (1988). *Applied Multivariate Statistical Analysis*. Prentice-Hall, Inc., Englewood Cliffs, New Jersey.

Lopuhaä, H.P. (1999). Asymptotics of Reweighted Estimators of Multivariate Location and Scatter. *The Annals of Statistics*, **27**, 1638-1665.

Lopuhaä, H.P. and Rousseeuw, P.J. (1991). Breakdown Points of Affine Equivariant Estimators of Multivariate Location and Covariance Matrices. *The Annals of Statistics*, **19** 229-248.

Mardia, K.V., Kent J.T. and Bibby J.M. (1995). *Multivariate Analysis*. Academic Press Ltd., London.

Pison, G., Van Aelst, S. and Willems, G. (2002). Small Sample Corrections for LTS and MCD. *Metrika*, to appear.

Rousseeuw, P.J. (1984). Least Median of Squares Regression. *Journal of the American Statistical Association*, **79**, 871-880.

Rousseeuw, P.J. and Van Driessen, K. (1999). A Fast Algorithm for the Minimum Covariance Determinant Estimator. *Technometrics*, **41**, 212-223.

Rousseeuw, P.J. and Van Zomeren, B.C. (1990). Unmasking Multivariate Outliers and Leverage Points. *Journal of the American Statistical Association*, **85**, 633-651.

A Resampling Approach to Cluster Validation

Volker Roth, Tilman Lange, Mikio Braun, and Joachim Buhmann
University of Bonn, Institute of Computer Science III,
Römerstrasse 164, D-53117 Bonn, Germany,
{*roth, lange, braunm, jb*}@*cs.uni-bonn.de*

Abstract. The concept of *cluster stability* is introduced as a means for assessing the validity of data partitionings found by clustering algorithms. It allows us to explicitly quantify the quality of a clustering solution, without being dependent on external information. The principle of maximizing the cluster stability can be interpreted as choosing the most *self-consistent* data partitioning. We present an empirical estimator for the theoretically derived stability index, based on imitating independent sample-sets by way of resampling. Experiments on both toy-examples and real-world problems effectively demonstrate that the proposed validation principle is highly suited for model selection.

Keywords. Unsupervised learning, cluster validation, model selection

1 Introduction

Unsupervised learning or clustering aims at extracting hidden structure in a data set. This goal requires to find an answer to some fundamental questions: (i) what structure are we interested in? (ii) what kind of clustering model provides a suitable method for extracting the desired structure? (iii) what is the "correct" number of clusters? The first question can be viewed as a problem definition, the last two ones are usually referred to as the problem of cluster validation. In general, validating clustering solutions means evaluating results of cluster analysis in a quantitative and objective fashion. Such an evaluation can be based on two types of criteria: (i) External criteria: a clustering solution is matched to a priori information, i.e. external information that is not contained in the dataset. (ii) Internal criteria: the quality measure is exclusively based on the data themselves. Internal criteria can roughly be subdivided into those methods which assess the fit between the data and the expected structure, and techniques which focus on the stability of the solution.

In this work we present a resampling-based method for cluster validation. According to the above taxonomy, it is an internal criterion which requires no prior information. The central ingredient of the validation method is the notion of *cluster stability*. Stability measures the variability of solutions, which were built on two independent samples from the same data source. The stability concept has a clear theoretical interpretation as choosing the most self-consistent data partitioning. Clustering is treated as the problem of identifying the *hypothetical* supervisor, that explains the data in the most consistent way. Among all possible labeling of the data (i.e among all possible supervisors), we select the one which provides us with the smallest expected number of misclassifications on a test set, given a fixed grouping algorithm.

Seen from a technical viewpoint, our resampling method refines related approaches initially described in [1] and later generalized in different ways in

[4] and [5]. From a conceptual viewpoint, however, it extends these heuristic approaches by providing a clear theoretical background for model selection in unsupervised clustering problems.

2 Stability and risk minimization

Applying the stability concept to unsupervised clustering problems is only meaningful, if we think about clustering as partitioning the *whole object space.* This view of the problem requires the user to specify a predictive inference mechanism for data-grouping methods. While this may be viewed as a very strong requirement, we believe that otherwise any general concept of purely data-dependent (internal-) cluster validity would be inherently ill-defined.

Contrary to classification problems, algorithms for clustering data cannot be stated in a purely discriminative fashion. The missing labeling information requires an additional modeling step: given a set of physical measurements (or *features*) describing the objects to be partitioned, we must specify how these features relate to the structure we are interested in. This is usually done by specifying a grouping *principle*, such as inter-cluster compactness criteria or intra-cluster separation measures. Based on such a principle, different grouping algorithms can be formulated. In spite of the fact that the stability framework also applies to this general setup, in the following we will restrict ourselves to the more intuitive situation in which we have already specified an algorithm. The open question within this framework is the "correct" number of clusters, a problem which is usually called the *model order selection problem.*

We may formalize this by considering potential supervisors, each of which labels the n objects contained in a training set by a different number $2 \leq k < n$ of clusters (we do not consider the degenerate case $k = 1$ which of course always has perfect stability). A grouping algorithm α assigns labels to objects, and in this sense it defines a potential supervisor. For this supervisor, α can then be viewed as a classifier with zero empirical risk. More precisely, we have the following situation: a data generator produces a series of i.i.d. sets of n objects

$$O_1^n, \dots, O_l^n, \dots \quad (1)$$

If the set O_i^n is presented to a fixed clustering algorithm α, we obtain a set of hypothetical labels

$$Y^n := \{\alpha_i(O_i^n)\}, \quad (2)$$

where the variable Y takes values in the range of $\{1, \dots, k\}$. In the following we will always denote by α_i a fixed algorithm trained on the sample O_i^n. Despite the clustering algorithm is considered fixed, the actual partition of the object space still depends on the training set. The principle of cluster stability estimates the degree of this dependence with respect to differences between partitions. Presenting a second set of objects O_j^n to the algorithm, we define (in-)stability as the empirical risk of the rule α_j on the test-set O_i^n, assuming the true labels are those defined in (2). Note that this is a measure of the generalization ability of α_j, since both sets O_i^n and O_j^n are drawn *independently* from the same data source. Denoting with $\mathbf{1}\{\cdot\}$ an indicator function, our instability functional is defined as the empirical probability of false predictions (y_l are the ones defined in equation (2))

$$d'' [\alpha_i(O_i^n), \alpha_j(O_i^n)] = \frac{1}{n} \sum_{l=1}^{n} \mathbf{1}\{y_l \neq \alpha_j(o_l)\}, \text{ with } O_i^n = \{o_1, \dots, o_n\}. \quad (3)$$

It is important, that the setting is asymmetric in the sense that the rule α_j is used in a predictive sense, whereas α_i is evaluated on its training set. If we would have tested both rules on a *third* set, it is possible to measure high stability in overfitting situations where both rules cannot generalize at all: consider for example an algorithm that only memorizes the objects seen in the training set, and always predicts label i for unseen objects. The same problem occurs, if we measure the stability based on a *common fraction* of the learning sets. The latter also explains, why stability measures are only meaningful with respect to generalization ability, if the grouping algorithm can be used in a predictive sense.

Selecting a grouping solution with high stability corresponds to identifying a hypothetical supervisor, for which the chosen learning algorithm provides on average a small risk on a test set of size n (we averaged over many tuples of learning- and training sets). Since by assumption all solutions have zero empirical training risk, the principle of favoring stable solutions can be viewed as selecting the most self-consistent labeling.

Due to the inherent permutation symmetry of clustering algorithms, however, the following problem arises in equation (3): even if exactly the same objects are grouped together, the labels may be arbitrarily permuted. To overcome this problem, we consider the loss incurred for all permutations π of the set of predicted labels $\alpha_j(O_i^n)$, and choose the one with smallest risk:

$$d'_i(\alpha_i, \alpha_j) := \min_{\pi} d'' \left[\alpha_i(O_i^n), \pi(\alpha_j(O_i^n))\right]. \tag{4}$$

Fortunately, there is no need to explicitly examine all $k!$ possible matches. We can rather use the *Hungarian method* for solving *minimum weighted perfect bipartite matching* problems with computational complexity of $\mathcal{O}(k^3)$ ([6], p. 248).

It holds that $d'_i(\alpha_i, \alpha_j) \leq 1 - 1/k$. To see this let α, β be two labelings, and let π be drawn uniformly from the set of all permutations of k indices. Then, straightforward calculations show that $E_\pi d''(\alpha, \pi(\beta)) = 1 - 1/k$ and therefore $\min_\pi d''(\alpha, \pi(\beta)) \leq 1 - 1/k$. This bound is tight: Let ρ be the *random predictor* which assigns labels uniformly at random. It can be shown that the instability of ρ converges to $1 - 1/k$ as $n \to \infty$.

In the non-asymptotic case, we cannot neglect the influence of the minimization over permutations. The estimator is biased towards higher stability. However, we can easily estimate the random stability by sampling. In order to use the (in)stability value for selecting the "correct" number of clusters k, the dependency of $d'_i(\rho_i, \rho_j)$ on k requires us to normalize the measurements. Taking expectations with respect to different pairs of sets (O_i^n, O_j^n), the final estimator that ensures fair comparisons between different values of k then reads:

$$\hat{d}(\alpha) = \frac{E_{i,j}[d'_i(\alpha_i, \alpha_j)]}{E_{i,j}[d'_i(\rho_i, \rho_j)]} \tag{5}$$

In practice, different pairs of sets (O_i^n, O_j^n) are hardly available (note that the size of the individual sets n should be large), but we may emulate these sets by iteratively splitting the total set of objects into two disjoint halves.

3 The Algorithm in Practice

With the theoretical derivation of the stability concept in the last section, a practical algorithm can now be outlined as follows:

1. Split the object set of size $2n$ into two sets of equal size, O_1^n and O_2^n.
2. Present the first dataset to the algorithm. The result is the mapping α_1 of each of the objects in O_1^n to one of k clusters.
3. Apply α to the second set O_2^n. Use α_2 to predict the cluster membership of all objects contained in the first set.
4. Set O_1^n now has two different labelings. Find the correct permutation of labels by using the Hungarian method for *minimum weighted perfect bipartite matching*. The costs for identifying labels i and j are the number of miss-classifications with respect to the labels $Y^n = \alpha_1(O_1^n)$ (these are assumed correct).
5. Normalize with respect to the random stability.
6. Iterate the whole procedure from step 1 to 5, average over assignment costs and compute the expected (in-)stability value.
7. Iterate the whole procedure for each k to be tested.

4 Experiments

In a first **toy example**, we demonstrate the model selection performance of the proposed stability estimator for synthetically generated datasets. We have drawn data samples from 3 different sources forming concentric rings in the 2-dimensional plain. A suitable clustering method for line-shaped data of this kind is *Path-based Clustering*, [2]. Figure 1 shows the estimated (in-)stability curve for this grouping model. We can clearly identify the solution for $k = 3$ as the most stable partition with an estimated risk of $\hat{d}(\alpha) < 0.1\%$. This model significantly outperforms all other choices of k.

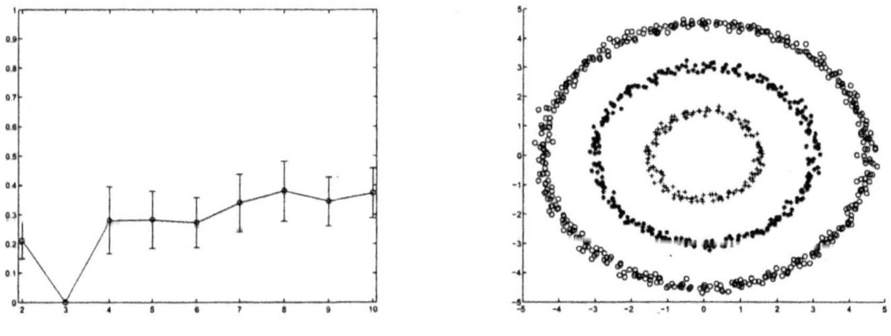

Fig. 1. Averaged instability curve and standard deviations for the 3 rings toy-example (left panel). The most stable solution with $k = 3$ (right panel).

The **Iris Dataset** is perhaps the best known database to be found in the pattern recognition literature [3]. The data report four characteristics (sepal width, sepal length, pedal width and pedal length) of three species of Iris flower. Each class contains 50 instances. One class is linearly separable from the other two, the latter are not linearly separable from each other. These observations are important because we use k-means for clustering, which generates linear class boundaries. Thus, we expect a two-cluster solution to two be very stable on different subsets. Estimating three clusters should also be possible with high stability. Our experiments nicely coincide with these expectations: we have randomly split the data 30 times in two subsets, applied a k-means clustering algorithm (optimized by deterministic annealing), and measured the matching costs between the two partitions. The resulting instability curve is depicted in figure 2. The fact that the most stable number of clusters is two and not three reflects from our design decision to link cluster validity to stability only and to discard improvements of model fit by more complex clustering solutions. Two clusters yield a larger distortion for the Iris data than three clusters but the stability is lower for three clusters since linear separability is violated.

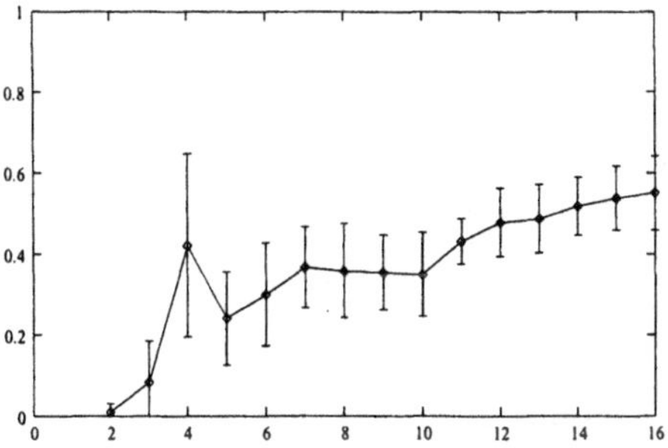

Fig. 2. Estimated instability for the Iris dataset vs number of classes, and standard deviations. The solution for $k = 2$ is extremely stable (risk $\hat{d}(\alpha) \approx 0.1\%$, the one for $k = 3$ still has a relatively low estimated risk $\hat{d}(\alpha) \approx 8\%$.

The **yeast cell cycle dataset** [7] shows the fluctuation of expression levels of yeast genes over cell cycles. Using periodicity and correlation algorithms, (Spellman *et al.*) identified 800 genes that meet an objective minimum criterion for cell cycle regulation. By observing the time of peak expression, (Spellman *et al.*) crudely classified the 800 genes into 5 different groups.

In our experiments, we investigated both the validity of clustering solutions and their accordance with the classification proposed. From the complete dataset (available at `http://cellcycle-www.stanford.edu`), we used the 17 *cdc28* conditions for each gene, after log-transforming and normalization to zero mean and unit variance. We grouped the 17-dimensional data by minimizing the k-means cost function using a deterministic annealing strategy. The estimated instability curve over the range of $2 \leq k \leq 20$ is shown in figure 3. For each k, we averaged over 20 random splits of the data. A dominant peak occurs for $k = 5$, with an estimated misclassification risk of $\hat{d}(\alpha) \approx 19\%$.

In order to compare a 5 cluster solution with the labeling proposed by (Spellman *et al.*), we again used the bipartite matching algorithm to break the inherent permutation symmetry. The averaged agreement rate over all 5 groups is $\approx 52\%$. This may be viewed as a rather poor performance. However, the labels should not be considered as the "ground truth", but as a "crude classification with many disadvantages" as stated by the authors. Moreover, a closer view on the individual agreement rates per group shows, that at least two groups could be matched with more than 70% agreement.

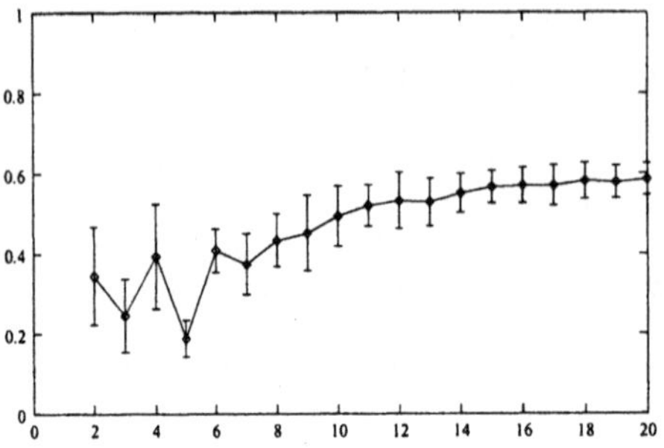

Fig. 3. Estimated instability for the Yeast Cell-Cycle dataset vs. number of classes.

5 Conclusions

We have introduced the concept of *cluster stability* as a means for solving the model order selection problem in unsupervised clustering. Given two independent object-sets from the same source, we recast grouping problems as a supervised classification task. In this scenario, the first grouping solution identifies a hypothetical supervisor, for which it provides a classifier with zero empirical risk. Presenting the second object-set to the fixed grouping algorithm, the (in-)stability functional measures the empirical probability of misclassifications with respect to the labeling by the identified supervisor. Taking expectations over tuples of object-sets, and normalizing by the stability of a random predictor, we derive a stability measure that allows us to compare solutions for different numbers of clusters in a fair and objective way. In order to estimate the cluster stability in practical applications, we introduced an empirical estimator that emulates independent samples by way of iteratively splitting the total object-set. Unless many other validation indices proposed in the literature, the estimated instability has a clear interpretation in terms of misclassification risk. The results presented in section 4 effectively demonstrate that cluster stability is a suitable measure for estimating the most self-consistent data partitioning.

References

1. J. Breckenridge. *Replicating cluster analysis. Method, consistency and validity* Multivariate Behavioral research, 1989.
2. B. Fischer, T. Zöller, J. Buhmann. *Path Based Pairwise Data Clustering with Application to Texture Segmentation.* In: Energy Min. Meth. in Computer Vision and Pattern Recognition, LNCS 2134, 235-250, Springer, 2001.
3. R. A. Fisher. *The use of multiple measurements in taxonomic problems*, Ann. Eugen., 1936.
4. J. Fridlyand & S. Dudoit. *Applications of resampling methods to estimate the number of clusters and to improve the accuracy of a clustering method.* Stat. Berkeley Tech Report. No. 600, 2001.
5. E. Levine, E. Domany. *Resampling Method for Unsupervised Estimation of Cluster Validity.* Neural Computation 13: 2573-2593, 2001.
6. C.H. Papadimitriou & K. Steiglitz. *Combinatorial Optimization, Algorithms and Complexity*, Prentice-Hall, Englewood Cliffs, NJ, 1982.
7. P.T. Spellman, G. Sherlock, MQ. Zhang, V.R. Iyer, K. Anders, M.B. Eisen, P.O. Brown, D. Botstein, B. Futcher. *Comprehensive Identification of Cell Cycle-regulated Genes of the Yeast Saccharomyces cerevisiae by Microarray Hybridization.* Molecular Biology of the Cell 9, 3273-3297, 1998.

A Self Documenting Programming Environment for Weighting

Wilfried Grossmann and Petra Ofner

Department of Statistics and Decision Support Systems, University Vienna, A-1010 Vienna, Universitätsstrasse 5, Austria

Abstract. The paper presents a proposal for a statistical computing environment for weighting, which processes statistical data and description of data simultaneously. The developed model is of interest also for other types of computing in official statistics.

Keywords. Programming environments, Weighting, Metadata

1 Introduction

Weighting plays an important role in survey computing, mainly for improving quality of estimates with respect to bias and variance but also for adjusting of estimates in such way that they match controls. Various weighting algorithms have been developed, which need besides the actual data additional information about the data, in particular information about underlying populations and information about the sampling design, stored frequently separate from survey data. Hence, only data with computed weights are delivered outside the production environment. Consequently, in case of reuse of the data it is difficult to evaluate the adequacy of existing weights, in some cases even impossible.

The paper presents a proposal for an environment, which supports weighting as well as structured documentation of the weighting process. Besides weighting the proposed approach seems to be useful for integration of processing and documentation in other areas of statistical computing. Section 2 discusses the requirements on the environment from the statistical point of view. In section 3 we consider a data model fulfilling the requirements outlined in section 2 and section 4 describes processing inside the environment in case of weighting.

2 Statistical requirements for a weighting environment

Let $y = (y_1, y_2, \dots, y_N)'$ be the vector of the values of a variable Y of interest in a finite population and denote by $y_s = (y_{1s}, y_{2s}, \dots, y_{ns})'$ the observed values of the variable in a sample s. One of the primary goals in survey computing is calculation of a linear estimate $\hat{\Theta}(Y) = \sum w_{is} y_{is}$ for population totals or means. The coefficients w_{is} are called *weights* for the observations y_s. Various methods are known for calculation of the weights, which may be classified into the following three approaches.

The first one is the design-based approach assuming known sampling probabilities for the observation units. Based on the sampling probabilities π_{is} for the observed units one can define the weights proportional to $1/\pi_{is}$. Such weights are often called *base weights* valid for all variables in the survey dataset and the corresponding estimator is known as Horvitz-Thompson estimator.
The second approach closer to traditional statistical modeling is the model-based approach. It assumes that the values of *auxiliary variables* $X = (X_1, X_2, \ldots, X_p)$ are available for all population units prior to sampling. Based on the auxiliary information one defines a linear model for the target variable Y of the following form: $E[Y] = X\beta,\ Var(Y) = V$. Using the observations y_s one can estimate the parameter vector β from the sample and define an optimal estimator for the quantity of interest (cf. Valliant et al. (2000)). Usually these weights are specific for each target variable Y.
A third alternative is the model-assisted approach, which tries to bridge the two alternatives by using information from the sampling design for calculation of base weights and auxiliary variables for modification of base weights. These new weights are calibrated in such sense that the weighted sums of population totals for the auxiliary variables reproduce the known population totals. Depending on the type of the auxiliary variables a number of computational procedures are available. Well known examples are generalized regression estimates (GREG), based on a universal regression model for all target variables, or calibration weights, defined by minimizing a predefined distance between base weights and the new calibrated weights (Deville and Särndal (1992)). In case of qualitative auxiliaries a rather simple calculation method known as raking can be used (cf. Kalton et al. (1998) for an introductory survey). Usually these weights are considered as universal for the whole data set.
Although the model based approach seems to be more favorable from a statistical point of view, the appealing fact of the model-assisted approach is that it can be integrated into statistical databases rather easily. The property of reproduction of population totals for summary tables produced with the calibration weights is sometimes called numerical consistency (cf. also Renssen et al. (2001) for analyzing numerical consistency in connection with output databases produced from more than one survey).
From the exposition it is obvious that calculation of weights presupposes the combination of data from a number of different sources: besides survey data we need also data for the sampling design and data for auxiliary variables from a population database. These data sources are often not fully compatible with respect to their structure, for example case by variates survey data may be used together with a sampling plan defined in a table of selection probabilities for strata and auxiliary variables represented in tables of marginal counts for the population. Consequently, calculation of weights needs (as usual in applied statistics) a number of tedious preprocessing steps for aligning the data according to the requirements of the algorithm. Hence, an integrated repository for all data occurring in connection with survey data, which supports preprocessing and computation, may be useful. From the data user perspective it seems important to

have access not only to the weights but also to the documentation of the weighting process.

3 Data Structures for a weighting environment

In order to fulfill the requirements stated at the end of the previous section we need a data structure which allows a unified and integrated representation of different types of data (survey data, sampling data, population data) occurring in different formats (case data, summary tables). Moreover for supporting also preprocessing we need not only representation of the data itself (*first order data*), but also a data representation of the description of these data (*second order data* or *metadata*). A rather simple data structure fulfilling both requirements is a so-called *container structure* (cf. Denk et al (2002) for a theoretical foundation). Basically a container consists of a *bucket* carrying first order data as a relational table and a *bucket schema* containing the description of the data bucket. In agreement with the main statistical data structures we discern two *bucket formats*: case type buckets for case level data and summary type buckets for dimensional data (tables or more general multidimensional cubes). All buckets occurring in connection with a data set are pooled together into a *composite*. The integrated representation necessary for a data source is schematically given in figure 3.1. Bullet nodes in the graph correspond to directories and circled nodes correspond to physical datasets with B denoting the buckets and S the corresponding bucket schemas.

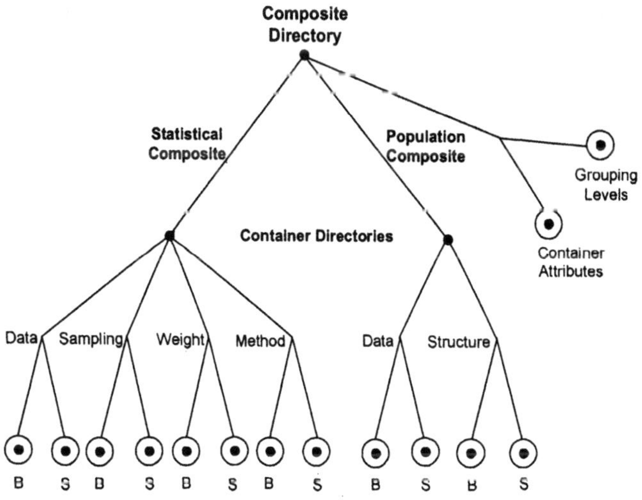

Figure 3.1. Composite structure for statistical data

Corresponding to the requirement analysis we discern *statistical composites* carrying all information about survey data and *population composites* for population data. Inside the statistical composite we have containers of four different types: a *data container* for survey data, a *sampling container* for sampling information, a *weight container* for the weights available for the dataset and a *method container* for the results of the calculations applied to the data (cf. section 4). The following figure shows an example of a bucket schema for a data container.

Container variable	Variable reference	Grouping level
id (key)	V1	-
sex	V2	0
education	V3	1
income	V4	1

Fig 3.2. Bucket schema for a data container

The column "variable reference" in the examples refers to the *container attributes* listing all variables used inside the composite together with a number important properties. Besides the usual specification for statistical variables like quantitative, qualitative or string variables it gives also a reference to the admissible ranges for the variable. These ranges are stored in the *grouping level* sets and the selected range is indicated in the grouping level column of the bucket schema. The grouping levels support a number of predefined recoding and transformation operations for variables (cf. Papageorgiou et al. (2001) for details in a similar model). For our purpose the most important descriptive element for variables is the *role specification* of a variable, informing about the (statistical) meaning of the variable inside the composite. Typical roles in a data bucket may be *identifying* variable (key inside the bucket), *auxiliary* variable or *observation* variable. In the sampling bucket possible roles are *selection probability* or *stratum count.* Figure 3.3 shows an example for a bucket in a sampling container using the qualitative variable V5 for defining the strata and for the other variables in the column headings roles are indicated.

V5	V6 (stratum count)	V7 (selection probability)
Stratum 1	880	0,25
Stratum 2	144	0,30

Fig 3.3. Bucket for a sampling container

The roles of variables in a weight bucket correspond to the weighting method and inside a method bucket the roles are defined according to the applied transformation. A typical role in the context of weighting may be variance of the estimate for the population total. Note that these roles are global roles inside the composite. Besides these global roles we have to consider also roles of variables in the context of compassion, for example explanatory variables in a regression model.

The population composite is structured similar to the data composite: the data bucket is reserved for population data available either as a register (bucket format case) or marginal counts (bucket format summary). The *structure bucket* informs about the population structure, usually defined by stratification.

The implementation of the structure could be done in different environments. In order to be fully compatible with some standard in statistical computing the prototype was implemented in the SAS environment, i.e. all containers as well as the mentioned directories are realized as SAS datasets. Such an implementation is convenient for performing statistical computing with the data but rather cumbersome with respect to data administration inside the structure.

4 Transformation of composites

Using the structure outlined in section 3 computation of weights (as any other statistical computation) can be interpreted as a sequence of transformations T producing from a number of input composites $C_1, C_2, \ldots, C_k$ a new output composite $C_{new} = T(C_1, C_2, \ldots, C_k)$. Description of the transformation encompasses besides the input and output composites the definition of an operator together with specification of operator parameters, a variable list to which the operator applies and a variable list which is generated by the operator. Based on these specifications the transformation is executed by the following main processing activities:

(i) Feasibility check for the transformation
(ii) Definition of a detailed computation plan
(iii) Statistical computation according to the plan
(iv) Generation of the output composite and documentation

The feasibility check infers whether the required computation is possible using only second order data, i.e. information contained in the bucket schemas and the description of container attributes, in particular the roles of the variables inside the composite. Checking presupposes a careful analysis of a family of transformations from a statistical point of view. In case of weighting we have to analyze the data requirements for different types of weighting. Let us consider two examples:

Calculation of base weights is feasible, provided that the variables describing the sampling design in the sampling schema are available also in the data container and the grouping levels are in both containers compatible. For instance in the example given by figures 3.2 and 3.3 base weights are not feasible, because variable V5 is not part of the bucket schema in the data container.

Checks for calibration weights have to take into account the type of the auxiliary variables. Consider for instance the case of two categorial auxiliaries and a desired adjustment according to the marginals with respect of these two variables (defined in the parameter specification of the transformation). Calibration is only feasible if the two auxiliaries are used in the data container in compatible form and calculation of base weights can be done in advance.

From computer science point of view such checks can be treated formally using the theory of semistructured data sets. Details in the context of this model may be found in Denk et al (2002).

Based on the results of the feasibility check one can define a detailed computation plan by specifying a sequence of conventional statistical algorithms, together with

a data structure obtained as a view on the input composites. For example, in the above sketched case of calibration weighting the plan encompasses following steps: Calculation of base weights using data contained in the sampling container, calculation of calibration factors using data from the population composite and determination of weights in a form compatible to the structure of the survey data in the statistical composite.
Execution of this plan by statistical computation and generation of the new output composite is rather straight forward, provided that the various numerical procedures are accessible.
Documentation of the transformations is given in the transformation repository, which is shown schematically in figure 4.1. *Trafo history* informs about the applied transformation, details about composites and variables are kept in *Origin Composite* and *Origin Variables*, structural information about the operators are contained in the *Operator Directory*. Also these structures are implemented as SAS data sets, the outlined processing steps are realized as SAS macros.

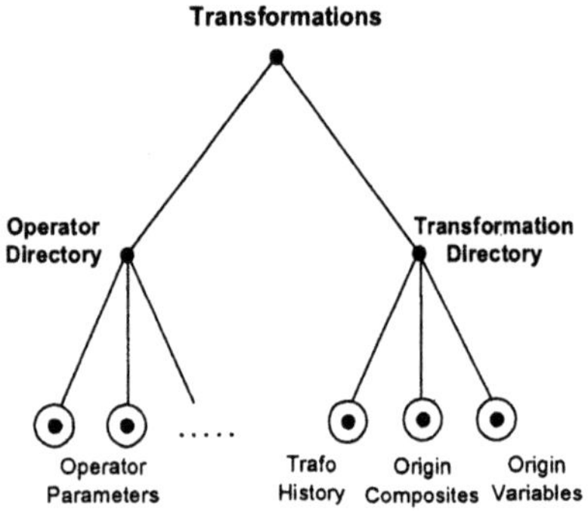

Fig 4.1. Structure of the transformation repository

References

Denk, M., Froeschl, K.A., Grossmann, W. (2002). Statistical Composites: A Transformation-bound Representation of Statistical Data. Submitted to Statistical and Scientific Databases 2002, Edinburgh.

Deville, J.C., Särndal, C.E. (1992). Calibration estimators in survey sampling. J. American Statistical Association, 87, 376-382.

Kalton, G. Flores-Cervantes I. (1998). Weighting Methods. In: New Methods in Survey Research (A. Westlake et al. eds.), 77-93. Association for Survey Computing, Chesham Bucks, UK.

Papageorgiou, H. Pentaris, F. Theodorou, E. Vardaki, M. Petrakos, M. (2001). A statistical metadata model for simultaneous manipulation of both data and metadata. J. of Intelligent Information Systems 17, 169 – 192.

Renssen, R.H., Kroese, A.H., Willeboordse, A.J. (2001). Aligning Estimates by Repeated Weighting. Working paper, Statistics Netherlands (Project no: RSM-70110).

Valliant, R., Dorfmann, A.H., Royall, R.M. (2000). Finite Population Sampling and Inference. Wiley Series in Probability and Statistics, New York.

A State Space Model for Non-Stationary Functional Data [1]

Mónica Ortega-Moreno[2], Mariano J. Valderrama[3] and Juan Carlos Ruiz-Molina[4]

[2] Department of General Economics and Statistics, Faculty of Economics Sciences, University of Huelva, 21071 Huelva, Spain
[3] Department of Statistics and Operations Research, Faculty of Pharmacy, University of Granada, 18071 Granada, Spain
[4] Department of Statistics and Operations Research, Faculty of Social Sciences, University of Jaén, 23071 Jaén, Spain

Abstract. A time dependent state space model with minimal dimension is introduced in this paper by approximating the stochastic process of continuous time nature by means of spline interpolation of its sample paths and then by differentiating its Karhunen-Loève expansion. A comparative study of forecasting, using the Kalman-Bucy filter, with simulated data is presented from a well known non-stationary process, the Brownian motion, discussing its advantages.

Keywords. Karhunen-Loève expansion, Kalman-Bucy filter, B-splines, Brownian motion

1 Introduction

In the field of stochastic system, non stationary processes have been deeply considered in several theoretic problems, but in practice the problem of building one of them from sample-paths of the process is much more difficult of solving, because of the difficulty to identify and estimate the optimal structure and moreover its computational implementation.

For solving this problem we present a time dependent state-space model that introduces flexibility in the formulation. The method is based on a decomposition of the process in a finite series and a recurrent relation that allows us to define a state space model and estimate the process by mean of Kalman-Bucy filtering. The most usual decomposition is the Karhunen-Loève orthogonal expansion that has good statistical properties although it presents two problems: there are not general procedures to obtain the random variables and eigenfunctions of the decomposition and, on the other hand, it is not usual in practice to know the covariance of the process. Valderrama et al. (2000) solve these problems by interpolating the sample-paths of the original process with B-spline functions. The recurrent relation between these functions makes us possible to relate the interpolated process and its derivatives and to derive a state space model dependent on the time with minimal dimension.

The new model is easily implementable from a computational point of view and we have not to identify and estimate the parameters. Furthermore their

[1] This research was supported in part by Project No. BFM2000-1466 of Dirección General de Investigación, Ministerio de Ciencia y Tecnología.

performance is acceptable in several situations including when the original process is non-stationary.

The paper is structured as follows; the Section 2 contains a brief summary about the formulation of the approximated PCA of a process by spline interpolation. Then, in Section 3 is proposed the time variable state-space model. Finally, in the Section 4, we include an application with simulated data from non-stationary process, the Brownian motion, discussing its advantages.

2 Approximated PCA by Spline Interpolation

Let us consider a second order and quadratic-mean continuous stochastic process $\{X(t), t \in [0,T]\}$ whose sample-paths belong to a functional space $L^2[0,T]$. Although these are continuous functions, in practice they can only be observed at the times of a partition of the interval: $\pi \equiv \{0 = t_0 < t_1 < \ldots < t_{n-1} < t_n = T\}$, whose knots can be unequally spaced. Therefore, each sample-path is reduced to a vector in $\Re^{n+1}$. In our aim of taking into account the real nature of the stochastic process and dealing with continuous and differentiable sample-functions, we interpolate the knots by means of cubic B-splines, so that each sample-path ω is interpolated at points $(t_l, X(t_l))$ and thanks to the recursive relation between B-splines functions we obtain:

$$IX_\omega(t) = \sum_{i=-3}^{n-1} \alpha_i B_i^{(3)}(t). \tag{1}$$

where $\{B_{-3}^{(3)}(t), B_{-2}^{(3)}(t), \ldots, B_{n-2}^{(3)}(t), B_{n-1}^{(3)}(t)\}$ is a basis of the (n+3)-dimensional space of cubic splines on the partition π and α_i represents the coefficients of interpolation.

We reduce the expression (1) to a linear system. To guarantee an unique solution for this system it is necessary to impose that all the sample-paths have second order derivative in the extremes of the definition interval. We also assume that these derivatives are null because it is not common to observe such values in each sample and it is not a restrictive hypothesis as prove Valderrama *et al.* (1998).

Once the sample-paths have been interpolated we also approximate the stochastic process by means of its Karhunen-Loève expansion:

$$X_\omega(t) = \sum_{i=1}^{\infty} \xi_i f_i(t),$$

where ξ_i denote the principal components and each eigenfactor $f_i(t)$ is projected on the space of B-splines so that it can be expressed as follows:

$$f_j(t) = \sum_{i=-3}^{n-1} \gamma_{ji} B_i^{(3)}(t) + \eta_j(t),$$

where $\eta_j(t)$ is the interpolation error.

3 Derivation of the State-Space Model by Interpolation of Stochastic Process

In order to derive the linear state-space model we define the two-dimensional state vector: $Z(t) = [\,IX(t) \quad \frac{\partial IX(t)}{\partial t}\,]^T$ and due to the relation between the derivatives of the B-spline functions it is possible to write the following relation:

$$\frac{\partial Z(t)}{\partial t} = F(t)Z(t) + LU(t), \tag{2}$$

where: $F(t) = \begin{bmatrix} 0 & 1 \\ 6 & -2t \end{bmatrix}$, $L(t) = [0\ 1]^T$ and $U(t)$ depends on the coefficients α_i and on the used knots.

On the other hand, let us suppose that the observation of the process $IX(t)$ is disturbed by a white noise $\{N(t)\}$ with variance r. Then, if we define the matrix $H = [\,1\ 0\,]$, the measurement equation is given by

$$Y(t) = IX(t) + N(t) = HZ(t) + N(t) \tag{3}$$

Both equations (2) and (3) define a state space model which is introduced nature by means of B-spline interpolation. Let us observe that the model is not invariant on the time so that improves the one introduced by Ruiz et al. (1995) when we are describing from the data a non-stationary process.

Once the parameters of the system and the initial conditions are specified it is possible to obtain the Kalman-Bucy estimate which verifies the differential equation:

$$\frac{\partial \widehat{Z}(t/t)}{\partial t} = F(t)\widehat{Z}(t/t) + Lu(t) + K(t)\left[Y(t) - H\widehat{Z}(t/t)\right],$$

for $t \geq t_0$, where $\widehat{Z}(t_0/t_0) = 0$. The gain matrix is determined by the relation:

$$K(t) = P(t/t)H^T(t)r^{-1}$$

where $P(t/t)$ is the estimate of the variance-covariance matrix of the estimation error and verifies the Riccati equation:

$$\frac{\partial \widehat{Z}(t/t)}{\partial t} = F(t)P(t/t) + P(t/t)F^T(t) - P(t/t)H^T r^{-1} HP(t/t),$$

for $t \geq t_0$ with the initial conditions: $P(t_0/t_0) = E[Z(t_0)Z^T(t_0)]$.

The general solution for a differential equation with input as the one given in (2) is:

$$Z(t) = \phi(t, t_0)Z(t_0) + \int_{t_0}^{t} \phi(t,\tau)L(\tau)u(\tau)d\tau. \tag{4}$$

where $Z(t_0)$ is a initial value of the state vector $Z(t)$ and $\phi(t,\tau)$ is a matrix known as transition matrix.

There is not a general procedure for obtaining the transition matrix when the elements of the involved matrixes are not constants and in almost all the situations numerical algorithms are required. In our situation, from the structure of matrix $F(t)$, we need to solve the following second order differential equation with time-variable coefficients:

$$\frac{\partial^2 x(t)}{\partial t} + 2t\frac{\partial x(t)}{\partial t} - 6x(t) = U(t). \tag{5}$$

Once we have solved the equation and the transition matrix has been obtained for two solutions of the equation (5), we are going to formulate the discrete approach of the model Then, the integral equation (4) is converted in a difference equation so that we can derive the discrete state-space model and provides predictions of the process from noisy observations recorded in discrete times.

4 Application to the Brownian Motion

In order to evaluate from a practical point of view the behaviour of the proposed model on a non-stationary stochastic process we are going to simulate a sample-path of a standard Brownian motion on $[0, 1]$ (i.e. with variance equal to one). Then an approximated expansion of this sample function from a finite basis of cubic B-splines defined on $[0, 1]$ with sixty-five unequally spaced observations is developed. These observation times have been obtained from simulations of an uniform distribution in $[0, 1]$. The obtained approximation is considered as the real sample-path of the process and it is the one used for testing the behaviour of the estimates on the interval.

The observed sample-path has been obtained by disturbing the real one with an unity variance white noise, who has been used to start the Kalman filter equations and provides the necessary information at each step of the algorithm.

Figure 1 shows the results of the optimal filter obtained with the proposed state-space model given by equations (3) and (4) superposed with the original sample-path. In fact they almost agree along the interval.

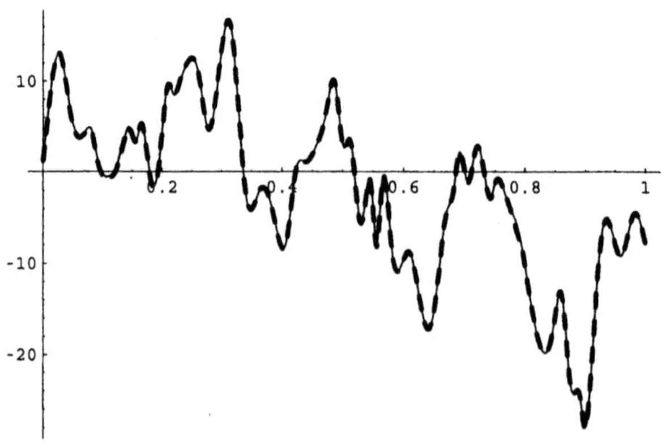

Fig. 1. Real (continuous line) and estimated (discontinuous line) sample path of the Brownian motion.

Figure 2 and 3 contain the simple and partial autocorrelation functions respectively of the model innovations and prove the white noise character of such innovations.

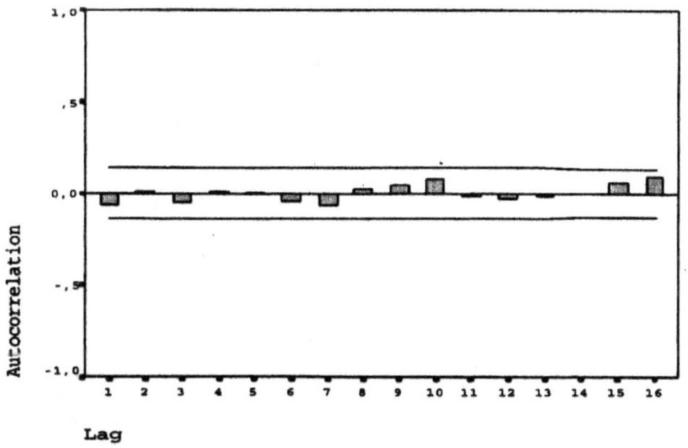

Fig. 2. Estimated simple autocorrelation function of innovations of the model.

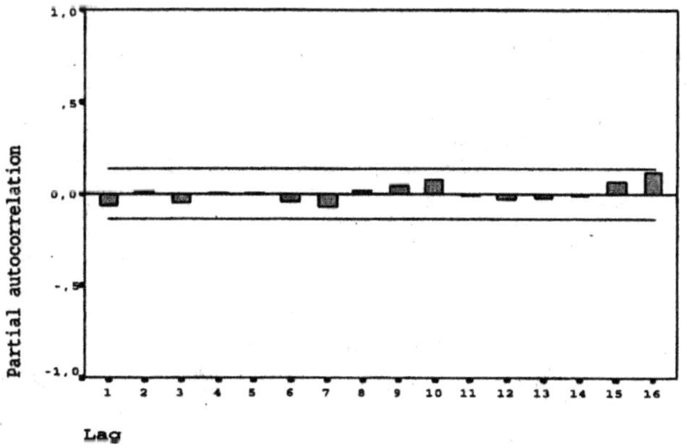

Fig. 3. Estimated partial autocorrelation function of innovations of the model.

Finally, Figure 4 presents the graphics of the elements of the covariance matrix of estimation error for the Brownian motion. We observe that $lim_{t\to\infty}P(t) = 0$, so that all their terms are decreasing. This means that also the gain matrix verifies $lim_{t\to\infty}K(t) = 0$ and the filter pays more attention to previous estimates of $z(t)$ and less to the new information provided by the data.

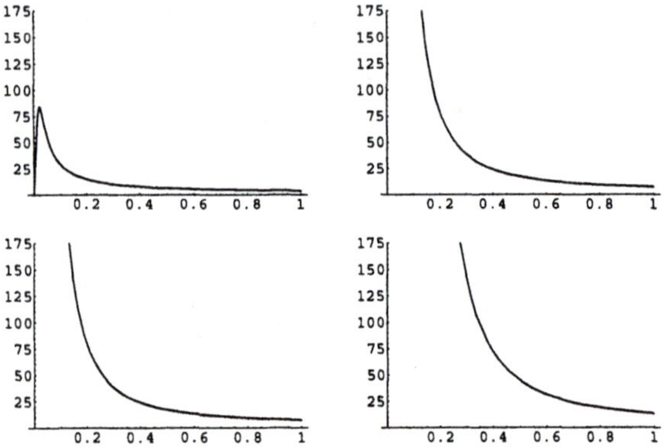

Fig. 4. Terms of the covariance matrix of the estimation error for the Brownian motion.

Although this is only a result for simulated data, the auto-explicative character of this model, on the basis of its Kalman gain matrix, allows to guess that this behaviour will be general for every phenomenon that we take into account, with the great advantage of starting with a previously identified two-dimensional state-space model.

References

Ruiz, J.C., Valderrama, M.J., and Gutiérrez, R. (1995). Kalman Filtering on Approximative State Space Models. *Journal of Optimization Theory and Applications*, **84**(2), 415-431.

Valderrama, M.J., Aguilera, A.M. and Ocaña, F.A. (2000). *Predicción Dinámica mediante Análisis de Datos Funcionales*. Hespérides-La Muralla, Madrid.

Valderrama, M.J., Aguilera, A.M. and Ruiz, J.C. (1998). Time Series Forecasting by Principal Component Methods. In: *COMPSTAT98 Proceedings in Computational Statistics*, 137-146. Heidelberg: Physica-Verlag.

A Wildlife Simulation Package (*WiSP*)

Walter Zucchini[1], Martin Erdelmeier[1] and David Borchers[2]

[1] Institut für Statistik und Ökonometrie, Georg-August-Universität Göttingen, Platz der Göttinger Sieben 5, Göttingen, Germany

[2] Research Unit for Wildlife Population Assessment, Mathematical Institute, North Haugh, University of St Andrews, St Andrews, Fife, KY16 9SS, Scotland

Abstract. ***WiSP*** is an ***R*** library of functions designed as a teaching tool to illustrate methods used to estimate the abundance of closed wildlife populations. It enables users to generate animal populations having realistically complex spatial and individual characteristics, to generate survey designs for a variety of survey techniques, to survey the populations and to estimate the abundance. It can be used to assess properties of estimators when the model assumptions are violated.

Keywords. ***WiSP***, statistical software, wildlife, abundance estimation

1 Introduction

The problem of estimating the abundance of wildlife populations has generated an enormous literature describing a rich variety of methods (see e.g. Seber, 1982) necessitated by divergent physical, behavioural and demographic characteristics of animals (and plants) and considerations of the costs and practicability of surveying populations. Borchers, Buckland and Zucchini (2002) illustrate that, despite their superficial diversity, most of the standard methods are fundamentally similar and can be understood as variations on a single theme. The *Wildlife Simulation Package* (***WiSP***) was primarily designed to illustrate the ideas described in that book.

The main objective in developing ***WiSP*** was to create a convenient environment for students to experiment with different survey designs in an artificial but realistic setting, namely using generated populations with versatile spatial and animal-level characteristics. The advantage of course is that the true abundance is known and it is therefore possible to study how the properties of estimators are affected by the design parameters (survey effort, observer skill, etc.) and by specific violations of the model assumptions.

WiSP is a library of functions in ***R*** (Ihaka and Gentleman, 1996) that is available as free software under the terms of the Free Software Foundation's GNU General Public License (Hornik, 2002). Important advantages of ***R*** in our application are that it is easy for users of ***WiSP*** to modify the existing functions (e.g. change an estimator), to add new functions (e.g. a new survey design) and to utilize the full range of ***R*** tools for interactive data manipulation, calculation and graphical display. It has a standardized LaTeX-like documentation format for generating *help* documentation. A most convenient, if not critical, feature of ***R***, from the point of view of the system design for ***WiSP,*** is that it is an object-oriented language.

This paper is organized as follows: Section 2 outlines the main concepts associated with the estimation of animal abundance, and the methods that are currently implemented in ***WiSP***. Section 3 sketches the features of ***WiSP*** with special emphasis on the classes of objects that are defined in ***WiSP***. Section 4 offers some concluding remarks.

2 Building Blocks

The model describing the state of a population, that is the spatial distribution and characteristics of the individual animals is called the ***state model***. The term *population* is more complex in the context of estimating animal abundance than it is in standard statistical investigations. That is because many abundance estimators are based on assumptions about the state model, of which the real population is regarded as one realization. In some cases information regarding specific characteristics of the individual animals must be collected during the survey. The individuals might congregate in groups of varying sizes; each animal has a size, gender, age, and so on. Such properties, as well as behavioural factors, will affect the probability that a given animal will be detected or captured. Some techniques are based on assumptions regarding the spatial distribution of the individuals or groups in the survey region. Thus for the purposes of comparing survey strategies or estimators in a simulated setting and (especially) for assessing the sensitivity of abundance estimators to violations in model assumptions, the software must be able to generate populations that are realistically complex in their spatial and animal-level characteristics.

The ***observation model*** is the probabilistic process used to describe which animals are detected or captured, given the search region, the positions and other characteristics of the animals as well as features of the survey (visibility conditions, survey effort, etc.) This model applies to the ***covered region*** (the sub-region that is searched), which is chosen using a set of rules, called the ***survey design***, and usually involves some randomisation. The surveyor has control of the survey design, e.g. the survey effort (number of traps, observers, how long they search, etc.), the shape of the survey units (rectangular, circular, strips, etc.) and the method to select them (simple random sampling, systematic sampling on a grid, etc.). ***WiSP*** currently implements the following methods:

Plot sampling covers methods such as quadrat sampling and strip sampling, which, although different in their implementation, are effectively identical from a statistical point of view. They differ only in the shape of the survey units; it is assumed that animals in the covered region are detected with certainty.

Removal, catch-effort and change-in-ratio methods require at least two survey occasions. The animals captured on each occasion are counted and removed from the population. The *removal* method assumes that capture probabilities are equal for all animals, that they remain the same on every survey occasion and that the captures are independent. These assumptions are relaxed in *catch-effort* models that let the capture probability to depend on some measure of the catch-effort on each occasion; the estimates are based on changes in the catch-per-unit-effort rather on changes in the catch. *Change-in-ratio* methods constitute an additional refinement; the capture probability is allowed to depend on animal-level variables (size, age, sex, etc.) that can be recorded in the surveys. This set of methods and the next make no assumption regarding the state model.

Mark-recapture methods are similar to the removal methods but the captured animals are not removed; they are marked and replaced in the population. The methods vary in complexity depending on the model assumptions. In the simplest case one assumes that on each occasion each animal is equally likely to be captured; other models allow for various kinds of heterogeneity, for example that the capture probabilities depend on certain (observable) features of the captured animals, on their capture history and so.

Distance sampling methods estimate the relationship between the detection probability and the distance of detected animals to the centreline of strip of given width (line transect sampling) or to the centre of a circle of given radius (point transect sampling). It is assumed that detection is certain if the distance is zero. (This assumption is unrealistic for some animals (e.g. whales) and can be relaxed by using ***double platform methods*** in which two observers independently record their detections and the distance of the animals from the centreline.) Only a single survey is required; randomisation is usually applied to select the strips (or circles) that make up the covered region. Here too the methods vary in complexity depending on the assumptions made and the type of information collected.

It is clear that both the nature and credibility of the assumptions underlying the above methods differ substantially; consequently the properties of the estimators will vary enormously. Of course that does not imply that the models with the least restrictive assumptions will perform best because, among other things, their performance depends on the sample size, which, if small, reverses the benefits gained by relaxing suspect assumptions. That is why it is important to compare the properties of estimators when the assumptions are violated.

Currently ***WiSP*** can compute maximum likelihood point estimates and at least one type of interval estimate (non-parametric/parametric bootstrap, asymptotic normal-based or profile likelihood confidence intervals) for all simple cases, and for some of the more complex versions of the above models.

3 Design and object-classes in *WiSP*

We now outline the main components of ***WiSP,*** its design, object-classes and a few functions, relying on examples to illustrate some features of the library. The ***WiSP*** library is designed to simulate the abundance estimation process in its entirety. The main steps are specifying the survey region and the properties of the state model, generating a population, specifying the survey design and the observation model, conducting the survey and, finally, computing point and interval abundance estimates. These steps, which are illustrated in Figure 1, determine the main classes of objects defined in ***WiSP***, namely *region, population, design, sample, estimate* objects and a number of parameter objects that specify the properties the state model, the observation model, the survey design and the method of estimation.

In the current implementation a region is a simply a rectangle with given height and width. The parameters of the state model are specified using a function *setpars.population* that creates an object in the class *pars.population*. Objects in that class contain details of the probability distribution of the group sizes, the distribution of exposure (catchability/detectibility), the number and distribution of user-specified types (e.g. female/male, size class, etc.) and options regarding the number of groups to be generated. Two of the required attributes of *population* objects are themselves objects, namely the *region* and the *density.population;* the latter controls the spatial distribution of the groups. Special functions are available to create and to modify *density.population* objects: *generate.density* creates a linear density surface over the region, *add.hotspot* increases/decreases the density around a given point; *set.stripe* changes the density in a rectangular sub-region. These functions enable the user generate populations having realistically complex spatial distributions (Figure 2).

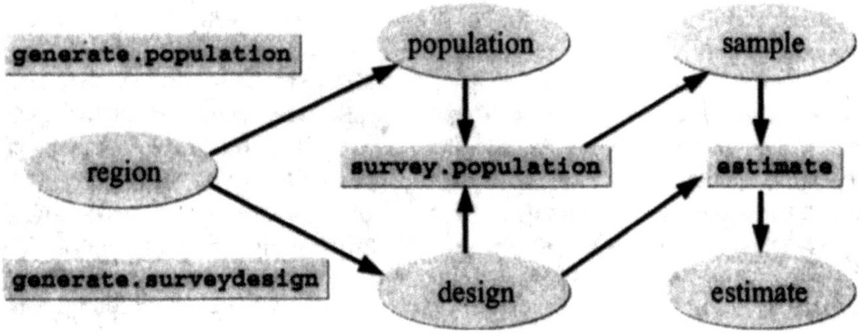

*Figure 1: Generic functions (rectangles) and objects (ovals) in **WiSP**.*

The observation model and the survey design are related because, *inter alia*, the latter determines where we look and how hard we look, but there is also an important distinction between them. The former determines the detection/capture probabilities, which can depend on properties of the population that are not taken into account by the survey design. It is precisely this distinction that enables us to assess the performance of the estimators when the model assumptions are violated. Thus two different classes of objects are needed: *design* objects (whose properties are defined by *design.parameter* objects created by appropriate functions) and *sample* objects created as a result of surveying the population using the given design. However, some properties of the observation process are controlled by parameter objects in a class *pars.survey*. Figure 3 illustrates this for a line transect example.

In practice the information available after completion of the survey comprises the survey design and the sample information because the latter depends on the survey method used. In plot sampling one merely has the counts in each plot; line transect samples must specify the distances of each detected group to the centreline and, possibly, other information about animal-level variables (e.g. group size) and/or survey-related variables (e.g. visibility conditions). Thus *sample* objects contain both design and sample information; they are passed to the appropriate estimation functions to create *estimate* objects that contain the relevant results, namely point and interval estimates, graphical displays, diagnostics, etc. Figure 4 illustrates this for the example in Figure 3.

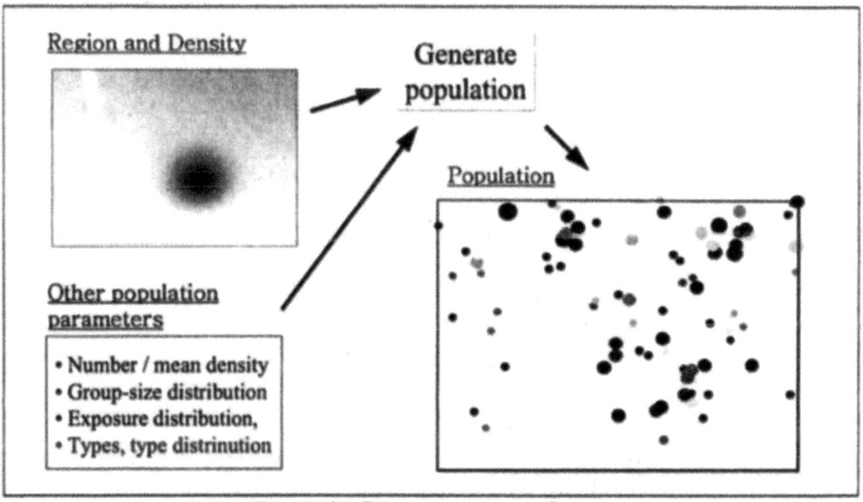

Figure 2: Generating a population.

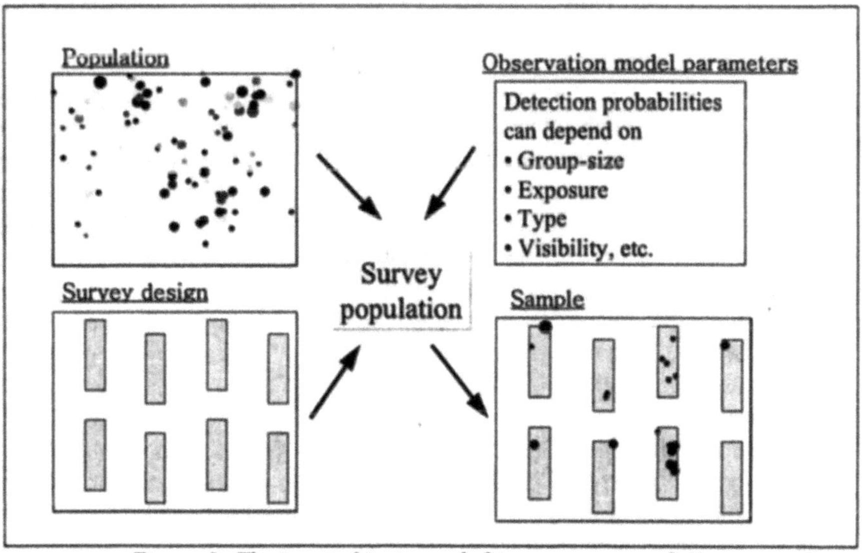

Figure 3: The input objects needed to survey a population.

A very convenient feature of R is the facility to define generic functions for standard commands such as *print, plot, and summary*. One defines which properties of objects in a given class are to be printed, plotted or summarized after which the command *plot(object)*, for example, will generate the plot that is appropriate for the object. Similarly, the command as.population(<appropriate information>) will to coerce information into an object of the class *population* and *is.population* (*object*) checks if a given object is indeed in the class *population*.

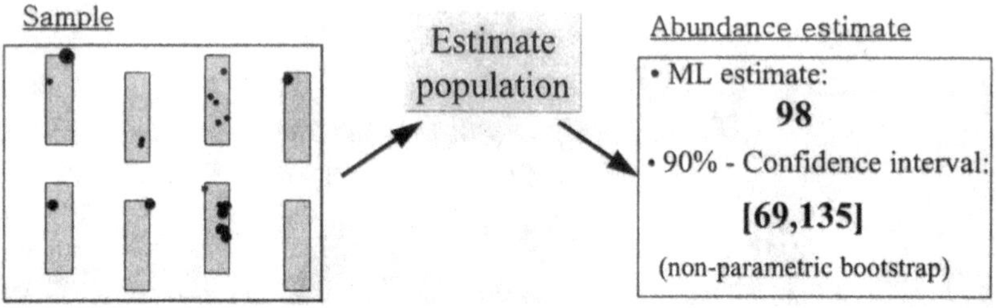

Figure 4: Estimates are computed using sample objects which must contain information about the survey design.

4 Concluding remarks

WiSP is primarily intended as a teaching tool to illustrate the basic principles underlying the most popular (closed population) abundance estimation techniques and to study their strengths and weaknesses. In its current form it does not have the functionality of the more comprehensive software packages available for the analysis of field data. (Useful lists of links can be found at www.phidot.org/software and www.mbr.nbs.gov/software.html.)

The free availability of ***R*** was an important consideration in selecting it as the language for ***WiSP*** but the other advantages that ***R*** offers were even more important in terms of future development of the ***WiSP*** library. The code can be made available in text form which makes it conveniently accessible to users who wish to modify it define new functions to cover a greater range of survey designs, estimators, or types of complexities in the state process. The object-orientated design of ***WiSP*** that is facilitated by ***R*** renders the task of making such modifications and enhancements incomparably less forbidding than would be the case if a language without this feature had been used.

References

Borchers, D.L. Buckland, S.T. & Zucchini W. (2002). *Estimating Animal Abundance: closed populations*, Springer-Verlag (in press).

Hornik (2002), "The R FAQ", http://www.gnu.org/copyleft/gpl.html

Ihaka, Ross and Gentleman, Robert (1996), R: A Language for Data Analysis and Graphics, *Journal of Computational and Graphical Statistics*, 5(3), 299-314.

Seber, G.A.F. (1982), *The estimation of animal abundance and related* information *parameters: Second Edition*. Griffin, London.

Algorithmical and Computational Procedures for a Markov Model in Survival Analysis

Juan Eloy Ruiz-Castro[1], Rafael Pérez-Ocón[1] and Delia Montoro-Cazorla[2]

[1] Departamento de Estadística e Investigación Operativa, Universidad de Granada, SPAIN

[2] Departamento de Estadística e Investigación Operativa, Universidad de Jaén, SPAIN

Abstract. A methodology for studying the homogeneous time-continuous Markov processes that allow the usual quantities of interest in survival studies to be expressed in a well-structured form is considered. This is performed by introducing the phase-type distributions, which enable algorithmic expressions for the performance measures used in survival analysis. A Markov model with several absorbent states is studied, incorporating the phase-type distributions. Following this procedure, we show that Markov models can be applied in a direct and tractable way. In this general model covariates are introduced and applied to analysing the behaviour of breast cancer. The computational implementation can be developed from the formulae proposed in a more convenient way than previous ones and this has been performed using the *Matlab* program.

Keywords. Markov model, Survival data, Type-phase distribution, Matlab

1 Introduction

In the biostatistical literature, homogeneous Markov processes have frequently been used. Many authors have studied multistate Markov processes in the modelling of diseases (Kalbfleisch and Lawless, 1985; Kay, 1986; Pérez-Ocón, Ruiz-Castro and Gámiz-Pérez, 1998). All these papers considered an absorbent state and a finite number of transient states. When the number of absorbent states is greater than one, the literature is not extensive. In this line, Beck (1979) developed a stochastic survival model with covariates using competing-risk theory. Chiang et al. (1989) extended Beck's paper, introducing time-dependent covariates.

Most of the authors calculated mathematical expressions for survival measures following classical methods, these calculations becoming complicated when stochastic processes are involved. As a result, the transition probability functions are usually calculated considering the eigenvectors and the eigenvalues of the infinitesimal-generator process or solving the Kolmogorov equations in a direct way. Moreover, in these works several measures of interest such as survival function, hazard rate function and average times are determined by applying numerical methods. For this, it is suitable to introduce algorithmic procedures for the computational implementation.

We consider a homogeneous time-continuous Markov model with an absorbent state, whereupon the phase-type distributions (Neuts, 1981) arise in the analysis of survival times in a natural way. This type of distribution has shown its utility in queue theory, and has the advantage of an algorithmic treatment. The calculation of the mathematical expressions can be presented in a closed form that allows algebraic treatment, and the corresponding computational implementation in an appropriate way.

When the evolution of a disease is being studied, it is informative to analyse the effect of several factors that can affect its behavior. Some works have introduced these factors through a covariate vector (Lu and Stitt, 1994; Pérez-Ocón et al., 1998 among others). In general, these works consider a classical methodology in their development. In the present work the covariates and the multi-censoring data mechanism are introduced into the model. We have developed some performance measures following this procedure, thereby showing the advantage of the new methodology; moreover, new measures determined using tractable computations, are suggested.

Applying the methodology of phase-type distributions, we can incorporate new models with several absorbent states. The mathematical expressions are very similar to the case with one absorbent state and in some cases the expressions are identical after replacing a vector by a matrix. Thus, it is possible to consider more complex models without excessive complication in the calculations.

The paper is organized as follows. In Section 2 the model is introduced and several algorithmic results are presented for calculating some measures of interest in survival analysis. In Section 3 the results above are applied to a breast cancer data set. In Section 4 some conclusions are presented.

2 Algorithmization of a Markov Model with several absorbent states

A Markov process with several absorbent states is considered. In Section 2.1 the type-phase distributions are introduced, and its structure is used in Section 2.2 to obtain different measures in matrix form in survival analysis. Covariates are introduced in the quantities.

2.1 The model

Let $X \equiv \{X(t);\ t \geq 0\}$ be a Markov process with m transient states $W = \{1, 2, \ldots, m\}$ and k absorbent ones $F = \{m+1, m+2, \ldots m+k\}$. The generator of the process can be expressed by blocks as follows:

$$\mathbf{Q} = \left|\begin{array}{c|c} \mathbf{T} & \mathbf{T^0} \\ \hline \mathbf{0} & \mathbf{0} \end{array}\right|,$$

where $\mathbf{T}$ contains the transitions among transient states, $\mathbf{T^0}$ the transitions from a transient state up to an absorbent one and the other two blocks are filled with zeros. The orders of $\mathbf{T}$ and $\mathbf{T^0}$ are $m \times m$ and $m \times k$, respectively. The initial vector of the process is given by (α, α^0), with α being the initial vector from a transient state and α^0 from an absorbent state. Let us assume α^0 is a zero vector.

The time up to the absorption for one absorbent state follows a PH-distribution with representation $(\alpha, \mathbf{T})$. The distribution function is given by

$$F(t)=1-\alpha\exp(\mathbf{T}t)\mathbf{e},\ t \geq 0,$$

where **e** is the m-dimensional column vector of 1's. The hazard-rate function has the following algebraic expression,

$$r(t) = \frac{\alpha \exp(\mathbf{T}t)\mathbf{T}^0}{\alpha \exp(\mathbf{T}t)\mathbf{e}}, t \geq 0,$$

and the *ith* moment is

$$\mu_i = (-1)^i i!\, (\alpha \mathbf{T}^{-i}\mathbf{e}),\ i \geq 0.$$

Other results can be seen in Neuts (1981).

For this model, several measures of interest in survival analysis are developed algorithmically in algebraic and matrix form. In Section 2.2 absorption for an absorbent class and survival probabilities are calculated, while, in Section 2.2.1 sojourn average times are calculated.

2.2 Absorption

The transition-probability functions of the model can be determined solving the Kolmogorov equations. Considering the structure of the PH-distributions, these functions can be expressed algorithmically in the following way. We define the following matrices. The matrix

$$\mathbf{T}(t)=\exp(\mathbf{T}t),\ t \geq 0$$

contains the transition probabilities among transient states at time t, and

$$\mathbf{R}(t)=\mathbf{T}^{-1}(\exp(\mathbf{T}t)-\mathbf{I})\mathbf{T}^0,\ t \geq 0,$$

the transition probabilities from a transient state up to an absorbent one, with **I** being the identity matrix with order m × m. The entry (i, j) of this matrix is the distribution function of the first time by the absorbent state j, given that the process initiates in a transient state i.

Taking limits for $t \to \infty$ in this matrix expression gives the absorption probabilities between transient and absorbent states in the process, which is given by the matrix

$$\mathbf{R} = -\mathbf{T}^{-1}\mathbf{T}^0,$$

where the entry (i, j) of this matrix contains the absorbing probability by the state j given that the process begins in state i.

Given an initial probability vector α from the transient states, of order 1 × m, the h*th* component of the function vector of order 1 × k given by

$$\mathbf{S}(t) = \mathbf{e}' - \alpha\mathbf{T}^{-1}(\exp(\mathbf{T}t)-\mathbf{I})\mathbf{T}^0,\ t \geq 0$$

represents the probability that the process at time t will occupy a transient state or an absorbent one other than m + h, h = 1, 2,..., k, (where the vector **e'** is the transpose vector of **e**). When $t \to \infty$, $S_h(t)$ tends to the absorption probability for a state other than m + h, since the absorbing occurs with probability one. Thus,

$$\mathbf{S} = \lim_{t\to\infty} \mathbf{e}' - \alpha\mathbf{T}^{-1}(\exp(\mathbf{T}t) - \mathbf{I})\mathbf{T}^0 = \mathbf{e}' + \alpha\mathbf{T}^{-1}\mathbf{T}^0.$$

The h*th* component of the probability vector **S**(t) is a defective probability distribution and can be divided in two parts. The first is the probability of staying in a transient state at time t, and then the probability of being absorbed at time t by an absorbing state different from h*th*. This can been expressed in the following way:

$$\mathbf{S}_h(t) = \alpha\exp(\mathbf{T}t)\mathbf{e} + \mathbf{a}_h(\alpha\mathbf{T}^{-1}(\exp(\mathbf{T}t)-\mathbf{I})\mathbf{T}^0),\ t \geq 0,$$

where $\mathbf{a}_h$ is the k-dimensional vector column of 1's except for the h*th* component that is a zero. Taking limits when $t \to \infty$ is obtained

$$S_h = -a_h \alpha T^{-1} T^0.$$

2.3 Sojourn average times

For this model, it is well-known that the mean staying time in a state j within a period of time [0,t], given that the process initiates in a state i, can be calculated by integrating the transition probability functions. If the transient states are considered, this mean time is found by integrating the function **T**(t), and the matrix expression is given by

$$\mathbf{U}(t) = \int_0^t \mathbf{T}(u)du = \mathbf{T}^{-1}(\exp(\mathbf{T}t) - \mathbf{I}),\ t \geq 0.$$

Since the absorption occurs with probability one, the mean time in the transition state j up to the absorption, given that the initial state is i, is the entry (i, j) of the matrix $-\mathbf{T}^{-1}$. Thus, the mean time of staying in state j up to the absorption is given by the entry *jth* of the vector $-\alpha\mathbf{T}^{-1}$, and the mean time up to the absorption is $-\alpha\mathbf{T}^{-1}\mathbf{e}$, the expected value of the PH-distribution with representation (α, **T**).

2.4 Covariates

In this section, covariates are introduced into the model. A column covariate vector, **z**, is introduced into the model in the same way as in the proportional-hazards model. The infinitesimal generator has the same structure as without covariates,

$$\mathbf{Q}(\mathbf{z}) = \left|\begin{array}{c|c} \mathbf{C}(\mathbf{z}) & \mathbf{C}^0(\mathbf{z}) \\ \hline \mathbf{0} & \mathbf{0} \end{array}\right|.$$

The blocks of the generator are

$$C_{ij}(\mathbf{z}) = C_{ij} \exp(\mathbf{z}'\beta_{ij}),\ i \neq j,\quad i, j \in W$$

$$C_{ii}(\mathbf{z}) = -\sum_i C_{ij}(\mathbf{z}),\ i, j \in W$$

$$C_{ij}^0(\mathbf{z}) = C_{ij}^0 \exp(\mathbf{z}'\beta_{ij}),\ i \in W,\ j \in F$$

where C_{ij} and C_{ij}^0 are the baseline transition intensities and β_{ij} the regression-parameters vector associated to the transition $i \to j$.

Similar matrix algorithmic expressions can be formulated for calculating the measures described above, considering several risk groups.

3 Application to Breast Cancer

The methodology described above has been applied to censored breast cancer data set for modeling the influence of treatments in relapse and survival times. The subjects were 300 patients who were submitted for surgical treatment. The follow-up of these patients dates from 1970 up to December 1995 and they were seen longitudinally. The homogeneous Markov process considered has three states: no relapse, relapse and death. The treatments (chemotherapy, radiotherapy and hormonal therapy) are introduced as covariates by means of transition intensities, thus providing three covariates. More detail can be found in Pérez-Ocón, Ruiz-Castro and Gámiz-Pérez (1998).

The likelihood function is built from the dataset considering the matrices **T**(t) and **R**(t), and it has been implemented with the *Matlab* program. Thus, we have obtained the estimated parameters given in Pérez-Ocón et al. (1998) considering this new methodology without having to solve Kolmogorov equations.

We have built some subroutines, m files, for the *Matlab* program to calculate some interesting measures in survival analysis, implementing in this program the algorithmical results developed in this work. For instance, the following subroutine calculates the survival probability up to a time s given the representation of the distribution (a, T),

```
function out=surv(a, T, s)
m=length(T);
out = a*expm(T*s)*ones(m,1);
```

Survival curves from the model for different treatments have been calculated by considering the type-phase distributions. The time until death is the type-phase distribution with representation (α, **C**(**z**)), with $\alpha = (1, 0)$ given that every patient begins the study in the first state. Figure 1 shows the survival function for patients treated with radiotherapy and chemotherapy.

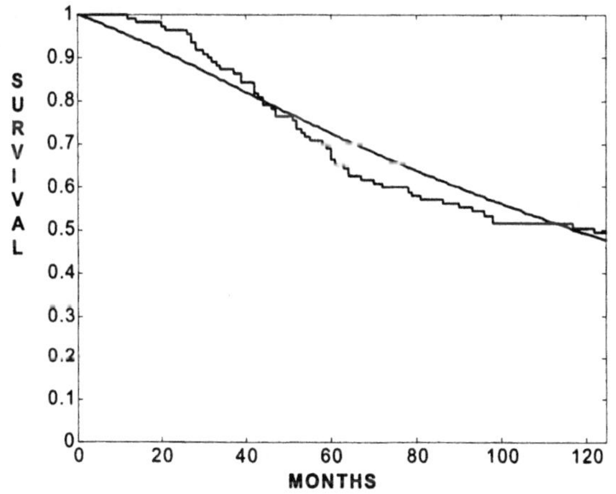

Figure 1. Survival of breast cancer for patients treated with Radiotherapy and Chemotherapy (the empirical curve is the step function and the estimated curve is the continuous line)

Also, considering the algorithmic matrix expressions, we have calculated life tables, transition-probability functions and staying mean times for different risk groups. Table 1 shows the average sojourn time in any transient state until the last death time registered according to the different risk groups .

These survival measures have been compared with the empirical ones and the goodness fit of the model has been tested.

z	Last death time: t	Average sojourn time: αU(t, z)e
RT, HT, CT	92	72.9600 (72.0155)
RT,CT	122	89.2046 (88.0170)
RT, HT	144	85.7997 (91.9054)
HT, CT	206	97.8188 (84.2500)
RT	101	69.6341 (63.6832)
HT	62	48.2991 (46.8551)
CT	59	45.3124 (48.3862)
Without treatment	112	84.2491 (51.5892)

Table 1. Average lifetime until last death time for different risk groups (empirical average lifetime in parentheses)

4 Conclusions

In the present paper, we have developed the methodology of phase-type distributions, and it has been applied to breast cancer. We have studied the cohort considered in Pérez-Ocón *et al.* (1998) using classical Markov methods under this new methodology, which is more elegant and more efficient in terms of computational time than the classical one.

Acknowledgement

The first and the second author gratefully acknowledge the financial support by Ministerio de Ciencia y Tecnología, Proyecto BFM2001-3802, España.

References

Beck, G.J. (1979). Stochastic Survival Models with Competing Risks and Covariates. *Biometrics*, **35**, 427-438.

Chiang, Y.K.; Hardy, R.J.; Hawkins, C.M. and Kapadia, A.S. (1989). An Illness-Death Process with Time-Dependent Covariates. *Biometrics*, **45**, 669-681.

Kalbfleisch, J.D. and Lawless, J.F. (1985). The Analysis of Panel Data under a Markov Assumption. *Journal of the American Statistical Association*, **80**, 392, 863-871

Kay, R. (1986). A Markov Model for Analysing Cancer Markers and Disease States in Survival Studies. *Biometrics*, **42**, 855-865.

Lu , Y. and Stitt, F.W. (1994). Using Markov Processes to Describe the Prognosis of HIV-1 Infection. *Medical Decision Making*, **14**, 266-272.

Neuts, M.F. (1981). *Matrix Geometric Solutions in Stochastic Models. An Algorithmic Approach.* Johns Hopkins Univ Press.

Pérez-Ocón, R.; Ruiz-Castro, J.E. and Gámiz-Pérez, M.L. (1998). A Multivariate Model to Measure the Effect of Treatments in Survival to Breast Cancer. *Biometrical Journal*, **40**, 6, 703-715.

An Algorithm for the Construction of Experimental Designs with Fixed and Random Blocks

P. Goos[1], A.N. Donev[2], M. Vandebroek[1]

[1] Department of Applied Economics, Katholieke Universiteit Leuven, 3000 Leuven, Belgium.

[2] Department of Probability and Statistics, University of Sheffield, Sheffield, S3 7RH, U.K.

Abstract. This paper is concerned with the design of experiments where the relationship between a response variable of interest and a number of quantitative variables is studied. The observations have to be divided into blocks. Some of the block effects are regarded as random while others as fixed. An algorithmic approach to the construction of such designs is shown to be very successful.

Keywords. BLKL exchange algorithm, correlated observations, D-optimality

1 Introduction

One of the important techniques of designing efficient experimental designs is blocking. The observations are divided in groups (blocks) of homogeneous units. This allows for obtaining estimates of the parameters of interest adjusted for the difference between the blocks. Depending on the experiment, the block effects can be modeled as random or fixed.

There is a considerable amount of literature on the design of experiments where the block effects can be treated as fixed. For example, see Atkinson and Donev (1989, 1992) and Cook and Nachtsheim (1989). However, in many experiments the blocks of observations are sampled from a large population for which an inference about the parameters of interest is required. The blocks in such studies are regarded as random and this should be taken into account when the experiment is designed. Goos and Vandebroek (2001) discuss the problem where a single random blocking variable exists. However, in practice there are often several blocking variables and some of them can be treated as random.

For example, consider a clinical trial where a number of treatments for asthma have to be compared. The treatments are applied repeatedly over several time periods to the patients. These can be regarded as random blocks of observations taken from the population of patients that can receive the treatments. On the other hand, the effect of each treatment would depend on the atmospheric conditions and the weather as these are know to affect the condition of the asthma sufferers. Choosing the times of administering the treatment at random brings a second random block variable, time or weather conditions. Hence, random effects for both the experimental units and the time are needed in the model.

The algorithmic approach to the construction of experimental designs has been proved effective and flexible in situations where the experimental situation is complex. For example, the BLKL exchange algorithm of Atkinson and Donev (1989) can be used to obtain designs when the observations have to be divided in blocks and the block effects can be regarded as fixed. There is however no available solution to the problem described in this paper. Here we present an algorithm that extends the BLKL algorithm to generate designs that can be used in experiments with more than one (random or fixed) block variable.

In the following section we discuss what models would be appropriate to use when both types of blocks exist as well as the estimation of these models. We also discuss some design issues and present the criterion of D-optimality. The new algorithm is described in Section 3. It generates designs that are D-optimum. Its usefulness is illustrated with an example in Section 4. The paper concludes with a discussion and some suggestions for further work.

2 Models and their estimation

We are interested in the case where the response variable of interest, Y, can be explained by regressors being functions of, say, m continuous and l qualitative variables. The observations taken for each combination of the levels of the qualitative variables form blocks of observations which differ only due to the levels of the continuous explanatory variables. Depending on the experimental conditions, some of these block effects may be regarded fixed while the others may be considered random. Hence, the model is

$$\begin{aligned}\mathbf{y} &= \mathbf{Z}\boldsymbol{\alpha} + \mathbf{X}\boldsymbol{\beta} + \boldsymbol{\varepsilon} \\ &= \mathbf{Z}\boldsymbol{\alpha} + \mathbf{W}\boldsymbol{\gamma} + \mathbf{F}\boldsymbol{\delta} + \boldsymbol{\varepsilon},\end{aligned} \tag{1}$$

where $\mathbf{y}$ is a vector of n observations, $\boldsymbol{\alpha}$, $\boldsymbol{\gamma}$ and $\boldsymbol{\delta}$ are vectors of u, b and q parameters corresponding to the random block effects, the fixed block effects and the remaining regressors, respectively. **Z**, **W** and **F** are the corresponding design matrices, with rows $\mathbf{z}_i^{\mathrm{T}}$, $\mathbf{w}_i^{\mathrm{T}}$ and $\mathbf{f}_i^{\mathrm{T}}$, $i = 1, 2, \ldots, n$, corresponding to the individual observations. Hence, $\boldsymbol{\beta}$ is the vector of all p, $p = b + q$, fixed parameters $\boldsymbol{\beta} = \{\boldsymbol{\gamma}^{\mathrm{T}} \quad \boldsymbol{\delta}^{\mathrm{T}}\}^{\mathrm{T}}$ with corresponding design matrix **X**. In (1) the vector $\boldsymbol{\varepsilon}$ contains the errors of the observations, which we assume to be independent, normally distributed with zero mean and variance σ_ε^2. We denote the variance of the ith random block effect σ_i^2 and assume the random block effects to be independent from each other and from $\boldsymbol{\varepsilon}$, i.e.

$$\mathrm{var}(\boldsymbol{\alpha}) = \mathbf{G} = diag\{\sigma_1^2, \ldots, \sigma_1^2, \sigma_2^2, \ldots, \sigma_2^2, \ldots, \sigma_u^2, \ldots, \sigma_u^2\}.$$

Hence

$$\mathrm{var}(\mathbf{y}) = \mathbf{V} = \mathbf{Z}\mathbf{G}\mathbf{Z}^{\mathrm{T}} + \sigma_\varepsilon^2 \mathbf{I}_n, \tag{2}$$

where $\mathbf{I}_n$ is an identity matrix of size n. Equation (2) can also be written as

$$\mathbf{V} = \left(\mathbf{Z}\mathbf{H}\mathbf{Z}^{\mathrm{T}} + \mathbf{I}_n\right)\sigma_\varepsilon^2,$$

where

$$\mathbf{H} = diag\{\eta_1, \ldots, \eta_1, \eta_2, \ldots, \eta_2, \ldots, \eta_u, \ldots, \eta_u\}$$

and

$$\eta_i = \frac{\sigma_i^2}{\sigma_\varepsilon^2}, \qquad i = 1, 2, \ldots, u.$$

For example, if there is one blocking variable acting at l_1 levels generating blocks whose effects are considered fixed, and two random blocking variables acting at u_1 and u_2 levels, respectively, generating random blocks so that $\mathbf{Z} = (\mathbf{Z}_1 \quad \mathbf{Z}_2)$, the model becomes

$$\mathbf{y} = \mathbf{Z}_1\alpha_1 + \mathbf{Z}_2\alpha_2 + \mathbf{W}\gamma + \mathbf{F}\boldsymbol{\delta} + \boldsymbol{\varepsilon}. \tag{3}$$

Then,

$$\begin{aligned} \text{var}(\mathbf{y}) &= \mathbf{Z}_1\mathbf{Z}_1^{\mathrm{T}}\sigma_1^2 + \mathbf{Z}_2\mathbf{Z}_2^{\mathrm{T}}\sigma_2^2 + \mathbf{I}_n\sigma_\varepsilon^2 \\ &= \left(\eta_1\mathbf{Z}_1\mathbf{Z}_1^{\mathrm{T}} + \eta_2\mathbf{Z}_2\mathbf{Z}_2^{\mathrm{T}} + \mathbf{I}_n\right)\sigma_\varepsilon^2. \end{aligned}$$

The parameters in equation (1) that have to be estimated are $\boldsymbol{\alpha}$, $\boldsymbol{\beta}$ ($\boldsymbol{\gamma}$ and $\boldsymbol{\delta}$), although the main interest is usually in $\boldsymbol{\delta}$ or in linear combinations of the elements of $\boldsymbol{\delta}$. Customarily, σ_ε^2 and σ_i^2, i = 1, 2, …, u, are not known and also need to be estimated.

Model (1) can be estimated using the maximum likelihood, the residual maximum likelihood and the iterative generalised least squares methods. The parameter estimates are

$$\hat{\boldsymbol{\beta}} = \left(\mathbf{X}^{\mathrm{T}}\mathbf{V}^{-1}\mathbf{X}\right)^{-1}\mathbf{X}^{\mathrm{T}}\mathbf{V}^{-1}\mathbf{y}$$

and

$$\hat{\boldsymbol{\alpha}} = \mathbf{G}\mathbf{Z}^{\mathrm{T}}\mathbf{V}^{-1}(\mathbf{y} - \mathbf{X}\boldsymbol{\beta}).$$

The statistical properties of the parameters estimates when different methods for estimation are used are well studied - for example, see Brown and Prescott (1999).

We will be mainly concerned with the construction of D-optimum experimental designs, i.e. designs for which $\det\left(\mathbf{X}^{\mathrm{T}}\mathbf{V}^{-1}\mathbf{X}\right)$ is maximized. The confidence ellipsoid for the fixed effects in model (1) for designs optimum under this criterion of optimality has a minimum volume. The algorithm presented in the next section allows for choosing combinations of the levels of the continuous explanatory variables for a pre-specified block structure for which this is achieved.

3 An algorithm

In order to compute D-optimum designs for model (1), we have developed a point exchange algorithm. The input to the algorithm is a description of the design problem and includes the specification of the polynomial model in the explanatory variables, the numbers of random and fixed blocking variables and the number of their levels, and the number of observations possible for each combination of the levels of the blocking variables. The number of runs for each combination of the levels of the blocking variables need not be the same. Also, the user has to specify expected values for η_i, i=1,2,...,u, and the number of times the algorithm is to be run. The latter is done in order to minimize the probability of finding a local optimum. Each of these runs is called a *try*. By default, the algorithm assumes a cubic design region for the quantitative variables and computes a standard grid of candidates points as in the BLKL algorithm. However, the user can also specify the set of candidate points. This is a very useful feature when constraints are imposed on the values that the explanatory variables can take during the experiment. The output of the algorithm contains the optimal levels of the explanatory variables in each observation, as well as the best possible assignment of the observations to the blocks in the experiment.

The structure of the algorithm is as follows:

- Step1. Generate a non-singular starting design with the required block structure. Some observations are randomly selected from a list of candidate points and then the design is completed by sequentially adding those candidate points whose prediction variance is largest and assigning them randomly to blocks that have not already been filled in with the required number of observations yet.
- Step2. Improve the starting design by considering exchanges of pairs of design points from different blocks and exchanges of design points with points from the list of candidates. The most beneficial exchange is made and this step is repeated until no further improvement of the design is possible.
- Step 3. Repeat steps 1 and 2 the required number of *tries* and store the experimental design that is currently best with respect to the D-optimality criterion.

Similarly to nearly all design construction algorithms available in the literature, the algorithm described here cannot guarantee that the best possible D-optimal design will be found. Our experience suggests that this problem is common when large designs with complicated block structures are generated. However, the probability of finding the global optimum increases with the number of tries. Therefore, this problem becomes less important with the ever-increasing speed of the modern computers.

4 Example

We have used the algorithm described in the previous section to compute D-optimum designs for various experimental situations. In order to allow for a graphical representation, we restrict our attention in this paper to the case of two quantitative explanatory variables, say x_1 and x_2. Suppose that 12 runs are available for estimating a full quadratic model in x_1 and x_2 and that two blocking variables are present, each one acting at 2 levels. As a result, 4 blocks are available. We also assume that the block sizes are homogeneous, i.e. we assume that 3 runs can be performed for each combination of the levels of the blocking variables (i.e. in each block).

D-optimum designs for this situation were computed under different assumptions about the blocking variables. Two interesting designs are displayed in Figure 1.

The design shown in panel (a) is optimum in the following cases:

- blocking variable 1 is fixed and blocking variable 2 is random with η_2 close to zero, e.g. η_2=0.1, or
- both blocking variables are random with η_1 close to zero and η_2 not too small,

e.g. η_1=0.1 and η_2=1, or

- both blocking variable are random with η_1 and η_2 close to zero, e.g. $\eta_1=\eta_2=0.1$.

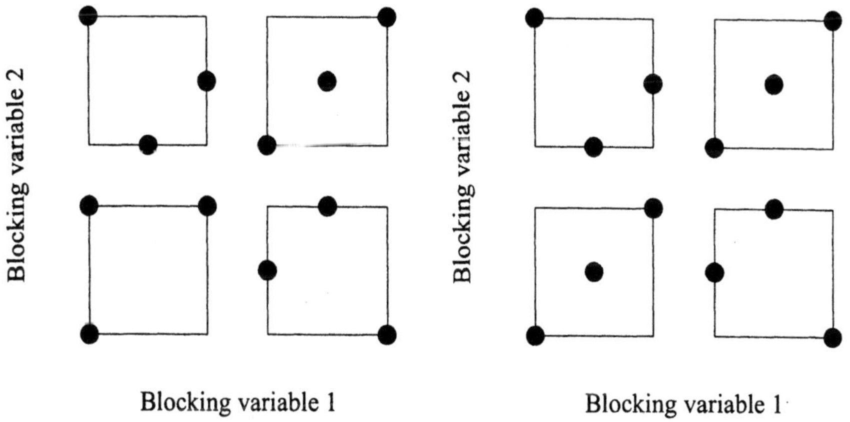

(a) **Design 1** (b) **Design 2**

Fig. 1. Optimum designs found by the algorithm

The design in panel (b) is D-optimum when

- both blocking variables are fixed, or
- both blocking variables are random with η_1 and η_2 not close to zero, e.g. $\eta_1=\eta_2=1$.

If both blocking variables are random, a contour plot can be drawn to show for what values of η_1 and η_2 each of the designs in Figure 1 is optimal. This is done in Figure 2 for $10^{-2}\leq\eta_i\leq10^2$ (i=1,2).

This small example clearly demonstrates that the structure of the D-optimum design depends on the nature of the blocking variables. It can be verified that the projections obtained by ignoring the blocks are different for both designs. This implies that not only the assignment of the design points to the blocks is important, but also their choice.

From the example, it can also be seen that when it comes to computing a D-optimum design, there is no difference between a fixed blocking variable and a random blocking with large η_i. A similar result was obtained by Goos and Vandebroek (2001) for the case of one blocking variable.

5 Conclusions and further work

The algorithm presented in this paper allows experimenters to design complex experiments involving more than one blocking variables. It was shown that the D-optimality of the designs depends on the assumptions regarding the blocking variables.

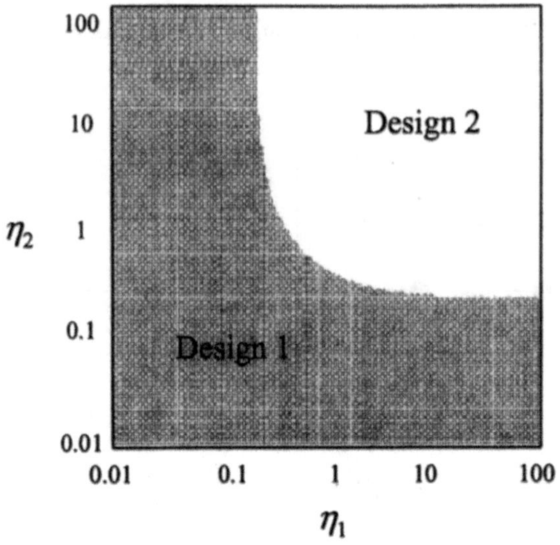

Fig. 2. Optimum designs for different values of η_1 and η_2

An interesting problem that needs to be addressed in future work is the choice of the optimal number of levels for the blocking variables and the number of observations at each of these levels.

Acknowledgement

The research that led to this paper was carried out while Dr. Peter Goos was a Postdoctoral Researcher of the Fund for Scientific Research - Flanders (Belgium).

References

Atkinson, A.C. and Donev, A.N. (1989) The construction of exact D-optimum experimental designs with application to blocking response surface designs, *Biometrika*, **76**, 515-526.

Atkinson, A.C. and Donev, A.N. (1992) *Optimum Experimental Design*, Oxford: Clarendon Press.

Brown, H. and Prescott, R. (1999). *Applied Mixed Models in Medicine*. Wiley.

Cook, R.D. and Nachtsheim, C.J. (1989) Computer-aided blocking of factorial and response-surface designs, *Technometrics*, **31**, 339-346.

Goos, P. and Vandebroek, M. (2001) Optimal response surface designs in the presence of random block effects, *Computational Statistics and Data Analysis*, **37**, 433-453.

An Algorithm to Estimate Time Varying Parameter SURE Models under Different Type of Restrictions

Susan Orbe[1], Eva Ferreira[1] and Juan Rodríguez-Póo[2]

[1] Dpto de Econometría y Estadística, Universidad del País Vasco, Spain
[2] Dpto de Economía, Universidad de Cantabria, Spain

Keywords. Nonparametric methods, constrained estimators, seasonal restrictions, cross restrictions

1 Introduction

A system of *seemingly unrelated regression equations* (SURE) is considered where constraints on the coefficients are allowed. These possible constraints can be of two different types, either single or cross equation restrictions. An example of the first type are seasonal restrictions (see Hylleberg, 1986). The second type appears on empirical specifications of production and consumption systems. In production theory, for example, the estimation of the structural parameters is subject to restrictions implied by homogeneity, product exhaustion, symmetry and monotonicity (see Jorgenson, 2000). Also in consumption, when demand functions are specified, the assumption of linear budget constraints lead to adding up and homogeneity restrictions. Furthermore, additional restrictions are sometimes suggested by the theory, as the Slutsky symmetry condition and the Engle condition (see Deaton and Muellbauer, 1980).

Motivated by the previous analysis, a nonparametric method is proposed to estimate time varying coefficients subject to these types of restrictions in a SURE framework. A closed form for the time varying coefficient estimators is provided so the coefficients can be estimated without the need of iterative procedures.

2 A general time-varying parameter SURE model

Let us consider the m-th equation in the SURE model given by

$$y_{mt} = \beta_{m1t}x_{m1t} + \beta_{m2t}x_{m2t} + \ldots + \beta_{mp_mt}x_{mp_mt} + u_{mt} \tag{1}$$

for $m = 1, \ldots, M \quad t = 1, \ldots, T$; where the endogenous variable in the m-th equation, y_m, is explained by a set of p_m explanatory variables, $x_{m1}, x_{m2}, \ldots, x_{mp_m}$. Thus, the total number of explanatory variables is $P = \sum_{m=1}^{M} p_m$. The vector of unknown coefficients needed to be estimated is $\beta_{mt} = (\beta_{m1t}, \beta_{m2t}, \ldots$ $\ldots, \beta_{mp_mt})'$. For the error terms, u_{mt}, it is assumed that: $E(u_{mt}) = 0$, $E(u_{mt}u_{m't}) = \sigma_{mm't}$ and $E(u_{mt}u_{m't'}) = 0, \forall m, t$. This covariance structure allows for heteroskedasticity and also for a time varying contemporary correlation structure. The following assumption imposes smoothness on the behavior of the parameters along time. It prevents abrupt changes and it is required for identification of the model of interest.

(A.1) $\beta_{mit} = f_{mi}(t/T)$ is a smooth function such that $f_{mi}(\cdot) \in C^2[0,1]$ for all $i = 1, 2, \ldots, p_m$ and $m = 1, 2, \ldots, M$.

The estimator for the coefficients set in each equation, $\beta_m = (\beta'_{m1} \cdots \beta'_{mT})$, is obtained minimizing the smoothed residual sum of squares for each m,

$$S_{mr}(\beta_{mr}) = \sum_{t=1}^{T} K_{mrt}\left(y_{mt} - \beta_{m1r}x_{m1t} - \beta_{m2r}x_{m2t} - \ldots - \beta_{mp_m r}x_{mp_m t}\right)^2.$$

where $K_{mrt} = (Th_m)^{-1}K((r-t)/Th_m)$ is a kernel function.

(A.2) The function $K(\cdot)$ is a second order kernel with compact support $[-1,1]$. Assume that $\int_{-1}^{1} u^4 K(u)du$ and $\int_{-1}^{1} K^4(u)du$ are finite and strictly positive.

The parameter h_m is the bandwidth that will regulate the smoothness degree in the m-th equation (Härdle, 1990). Note that the function kernel $K(\cdot)$ is univariate, it depends only on the time index and it is the same for all equations and variables. Therefore, there is no curse of dimensionality in this context.

The proposed estimator is derived by minimizing $S_{mr}(\beta_{mr})$ for all equations together and taking into account the time varying cross restrictions and the seasonal restrictions. For the first type restrictions they are represented through a general linear equation $G_t\beta_t = g_t$ $t = 1, 2, \ldots, T$ where G_t is a matrix of order $Q_t \times P$, g_t a vector of length Q_t and $\beta_t = (\beta_{1t} \ldots \beta_{Mt})^T$. That is, Q_t is the number of restrictions considered at each time t, and these restrictions can relate coefficients associated to the same equation or different equations. In a matrix form, the cross restrictions be written as $G\beta = g$ where $G = (G_1\ G_2 \ldots G_T)'$ and $g = (g_1\ g_2 \ldots g_T)'$ are a $Q \times TP$ order matrix and a Q order vector, respectively ($Q \equiv \sum_{t=1}^{T} Q_t$). The submatrix G_r and the subvector g_r describe the set of linear restrictions imposed among the coefficients corresponding to the r-th time index.

To account for the seasonal restrictions, define s_T as the number of the observations per season, such that s_T/T remains constant as T increases. For a simpler notation, the subscript in s_T will be dropped from now on. The seasonal constraints will be introduced through the inequalities

$$T^{-1}\sum_{t=s+1}^{T}\left(\beta_{mit} - \beta_{mi(t-s)}\right)^2 \leq \rho_{mi} \quad m = 1, 2, \ldots M \quad i = 1, 2, \ldots, p_m \quad (2)$$

where ρ_{mi} controls the degree of seasonality for each sequence of coefficients in the system. Given (A.1), the difference between β_{mit} and $\beta_{mi(t+1)}$ can not be excessively large and so $T^{-1}\sum_{t=s+1}^{T}\left(\beta_{mit} - \beta_{mi(t-s)}\right)^2$ will always be bounded by some nonnegative constant $\rho^{\star}_{mi}$. Then, if $\rho_{mi} \geq \rho^{\star}_{mi}$, no seasonality is imposed at all, but as far as $\rho_{mi} < \rho^{\star}_{mi}$, a seasonal pattern is imposed, that becomes stronger as ρ_{mi} decreases. The seasonal measure in

(2) can be written in matrix notation as $\beta' A \beta$, where $A = \bar{R}'\bar{R}$ and

$$\bar{R} = \begin{bmatrix} 1 & 0 & \overbrace{0\ldots0}^{sP-2} & -1 & 0 & \ldots & \ldots & 0 \\ 0 & 1 & 0 & \overbrace{0\ldots0}^{sP-2} & -1 & 0 & \ldots & 0 \\ \vdots & \ddots & & & \ddots & & & \vdots \\ 0 & \ldots & \ldots & \ldots & 1 & \overbrace{0\ldots0}^{sP-1} & -1 & \end{bmatrix}. \qquad (3)$$

Two types of restriction can be considered in a system such as (1). Note that the difference between seasonal and cross restrictions is that in the first case, the sum of squares of the seasonal differences is constrained, whereas in the second case an equality constraint is imposed.

In this context, a global estimation method of the time varying parameters will be derived that simultaneously takes into account all the equations and all the previous sources of restrictions. The new criterion function will be

$$\tilde{S}(\beta) = (Y - X\beta)' W^{1/2} \Omega^{-1} W^{1/2} (Y - X\beta) + \beta' A_\lambda \beta + 2\lambda_L'(G\beta - g), \quad (4)$$

where

- $\mathbf{Y} = \mathbf{i_T} \otimes \mathbf{Y}$, where i_T a T-order unit vector and Y is a vector of order MT^2 defined as $(y_1' \, y_2' \ldots y_M')'$.
- $\mathbf{X} = \mathbf{I_T} \otimes \mathbf{X}$, where I_T is a T-order identity matrix and X is the data matrix of order $MT \times TP$, defined as $diag(\{X_m\}_{m=1}^{M})$ where X_m denotes the $T \times p_m$ data matrix for the m-th equation. Thus, $\mathbf{X}$ is a matrix of order $T^2M \times TP$
- β is TP-order vector and $\beta = (\beta_{11}' \ldots \beta_{M1}' \ldots \beta_{1T}' \ldots \beta_{MT}')'$.
- W is a squared matrix of order T^2M defined as $diag(\{W_t\}_{t=1}^{T})$, with $W_t = diag(\{W_{mt}\}_{m=1}^{M})$ and $W_{mr} = diag(\{K_{mrt}\}_{t=1}^{T})$.
- Ω is a squared matrix of order T^2M defined as $\text{diag}\left(\{\Omega_t\}_{t=1}^{T}\right)$ where $\Omega_t = \Sigma_t \otimes I_T$ and $\Sigma_t = \{\sigma_{mm't}\}_{m,m'=1}^{M}$.
- $A_\lambda = \lambda^\star A$, $\lambda^\star = I_T \otimes diag(\lambda)$ and $\lambda = (\lambda_{11} \ldots \lambda_{1p_1} \ldots \lambda_{M1} \ldots \lambda_{Mp_M})$.

For the existence and uniqueness of the minimizer of (4) a sufficient condition is:

(A.3) $\text{Rank}\left(\Omega_r^{-1/2} W_r^{1/2} X\right) = P < T$, for $r = 1, \ldots, T$,

The role of each λ_{mi} is to control the degree of seasonality imposed over $\{\beta_{mit}\}_{t=1}^{T}$, being proportionally inverse to ρ_{im}. That is, large values of λ_{im} (small values of ρ_{im}) impose stronger seasonal patterns and vice versa. The vector λ_L represents the Lagrange multipliers for the cross restrictions and, as usual, measures the cost of imposing untrue restrictions. Let us denote by $\widehat{\beta}$ the solution to the optimization problem $\widehat{\beta} \equiv \text{argmin}_\beta \tilde{S}(\beta)$ and consider for a moment that the smoothing parameters $h_1, \ldots, h_M$, the vector λ and the covariance matrix Ω are given. From the first order conditions of (4), the solution for β can be written as

$$\widehat{\beta} = \left[(\mathbf{X}'\mathbf{W}^{1/2}\Omega^{-1}\mathbf{W}^{1/2}\mathbf{X}) + \mathbf{A}_\lambda - \mathbf{G}'\left[\mathbf{G}(\mathbf{X}'\mathbf{W}^{1/2}\Omega^{-1}\mathbf{W}^{1/2}\mathbf{X})^{-1}\mathbf{G}'\right]^{-1} \right.$$
$$\left. \times\, G(\mathbf{X}'\mathbf{W}^{1/2}\Omega^{-1}\mathbf{W}^{1/2}\mathbf{X})^{-1}\mathbf{A}_\lambda\right]^{-1}(\mathbf{X}'\mathbf{W}^{1/2}\Omega^{-1}\mathbf{W}^{1/2}\mathbf{X})\hat{\beta}_{\mathbf{hR}}, \qquad (5)$$

where the vector $\hat{\beta}_{hR}$ is defined as the solution to the optimization problem taking $\lambda \equiv 0$.

3 The estimation procedure

The expression given in (5) provides a closed matrix form for the estimators and the structure involves the inverse of matrices of order T^2M (Ω) and of order TP The purpose of this section is to reduce this computational cost. To attain this objective, the main idea is to use the system of normal equations that comes just before the expression (5) and exploit the particular structure of the matrices involved. A brief description follows. The system of normal equations can be written as $\mathcal{Z}\beta = \Upsilon$, where a $\mathcal{Z}$ is a symmetric matrix of order TP and β and Υ are TP order vectors, P is the number of parameters at any time and T is the sample size. This system will be divided into a set of systems of smaller order. Each system will provide the solution for a subvector of the TP order vector $\widehat{\beta}$, that will be solved in terms of another subvector. Thus, these systems will be solved using a recursive method that gives to the algorithm its recursive character. The explicit solution of one of these subvectors will be explicit and, once it is obtained, the rest will be computed. All together will provide the components of the complete estimator given by (5).

The system of normal equations is given by,

$$\left[(\tilde{\mathbf{X}}'\tilde{\mathbf{X}}) + \mathbf{A}_\lambda - \mathbf{G}'\left[\mathbf{G}(\tilde{\mathbf{X}}'\tilde{\mathbf{X}})^{-1}\mathbf{G}'\right]^{-1}\mathbf{G}(\tilde{\mathbf{X}}'\tilde{\mathbf{X}})^{-1}\mathbf{A}_\lambda\right]\widehat{\beta} = (\tilde{\mathbf{X}}'\tilde{\mathbf{X}})\hat{\beta}_{\mathbf{hR}}, \quad (6)$$

where $\tilde{\mathbf{X}}$ is defined as $\Omega^{-1/2}W^{1/2}\mathbf{X}$. The system (6) can be rewritten as $\mathcal{Z}\widehat{\beta} = \Upsilon$ where the matrix $\mathcal{Z}$ and the vector Υ are defined according to the following algebraic equality:

$$\begin{bmatrix} \mathcal{Z}_{11} & \mathcal{Z}_{12} & \dots & \mathcal{Z}_{1T} \\ \mathcal{Z}_{21} & \mathcal{Z}_{22} & \dots & \mathcal{Z}_{2T} \\ \vdots & \vdots & & \vdots \\ \mathcal{Z}_{T1} & \mathcal{Z}_{T2} & \dots & \mathcal{Z}_{TT} \end{bmatrix} \cdot \begin{bmatrix} \widehat{\beta}_1 \\ \widehat{\beta}_2 \\ \vdots \\ \widehat{\beta}_T \end{bmatrix} = \begin{bmatrix} \Upsilon_1 \\ \Upsilon_2 \\ \vdots \\ \Upsilon_T \end{bmatrix} \qquad (7)$$

where a general subvector $\widehat{\beta}_r$ contains all the estimated coefficients corresponding to the r-th moment. The symmetric TP order matrix $\mathcal{Z}$ contains the T^2 submatrices of order P that have been denoted by $\{\mathcal{Z}_{rr'}\}_{r,r'=1}^{T}$. In order to describe the structure of each of these submatrices, let us define Δ_λ as the diagonal matrix of order P containing the seasonal parameters $(\lambda_{11} \dots \lambda_{1p_1} \dots \lambda_{M1} \dots \lambda_{Mp_M})$.

Define for each $r = 1, \ldots, T$, $\triangle_{\lambda,r} = (I_P - H_r)\triangle_\lambda$ where I_P is the identity matrix of order P and H_r is

$$H_r = G_r' \left[G_r(\tilde{\mathbf{X}}_\mathbf{r}'\tilde{\mathbf{X}}_\mathbf{r})^{-1}\mathbf{G}_\mathbf{r}'\right]^{-1} G_r(\tilde{\mathbf{X}}_\mathbf{r}'\tilde{\mathbf{X}}_\mathbf{r})^{-1}; \quad \tilde{\mathbf{X}}_\mathbf{r} = \Omega_\mathbf{r}^{-1/2}\mathbf{W}_\mathbf{r}^{1/2}\mathbf{X},$$

where G_r is the matrix for the cross restrictions. Taking into account that s is the seasonal period and the special structure of the matrix $\bar{R}$, the expressions for the diagonal submatrices $\mathcal{Z}_{rr}$ and nondiagonal $\mathcal{Z}_{rr'}$ are:

$$\mathcal{Z}_{rr} = \begin{cases} \tilde{\mathbf{X}}_\mathbf{r}'\tilde{\mathbf{X}}_\mathbf{r} + \triangle_{\lambda,\mathbf{r}} & \text{for } r = 1, \ldots, s \\ \tilde{\mathbf{X}}_\mathbf{r}'\tilde{\mathbf{X}}_\mathbf{r} + \mathbf{2}\triangle_{\lambda,\mathbf{r}} & \text{for } r = s+1, \ldots, T-s \\ \tilde{\mathbf{X}}_\mathbf{r}'\tilde{\mathbf{X}}_\mathbf{r} + \triangle_{\lambda,\mathbf{r}} & \text{for } r = T-s+1, \ldots, T \end{cases} \tag{8}$$

$$\mathcal{Z}_{rr'} = \mathcal{Z}_{r'r} = \begin{cases} -\triangle_{\lambda,r} & \text{for } r' = r+s \\ 0 & \text{for } r' \neq r+s \end{cases} \tag{9}$$

Each subvector Υ_r of the TP order vector Υ is given by

$$\Upsilon_r = \tilde{\mathbf{X}}_\mathbf{r}'\tilde{\mathbf{Y}}_\mathbf{r} - \mathbf{G}_\mathbf{r}' \left[\mathbf{G}_\mathbf{r}(\tilde{\mathbf{X}}_\mathbf{r}'\tilde{\mathbf{X}}_\mathbf{r})^{-1}\mathbf{G}_\mathbf{r}'\right]^{-1} (\mathbf{G}_\mathbf{r}\widehat{\beta}_\mathbf{h} - \mathbf{g}_\mathbf{r}); \quad \tilde{\mathbf{Y}}_\mathbf{r} = \Omega_\mathbf{r}^{-1/2}\mathbf{W}_\mathbf{r}^{1/2}\mathbf{Y}$$

Hence the algorithm uses (8) and (9) to compute the estimators as described in the following steps.

Estimation Algorithm

STEP 1 Given the seasonal parameters and the variances matrix, build $\triangle_{\lambda,r} = H_r\triangle_\lambda$. For $r = 1, 2, \ldots T$, compute $\mathcal{Z}_r$ and Υ_r.

STEP 2 For $j = 1, 2, \ldots, s$, compute $a_j = \mathcal{Z}_j^{-1}$
For $i = 1, 2, \ldots, T/s - 1$, compute
$a_{is+j} = \left[\mathcal{Z}_{is+j} - \triangle_{\lambda,is+j} \cdot a_{(i-1)s+j} \cdot \triangle_{\lambda,(i-1)s+j}\right]^{-1}$

STEP 3 For $j = 1, 2, \ldots, s$, compute $b_j = a_j \cdot \Upsilon_j$
For $i = 1, 2, \ldots, T/s - 1$, compute
$b_{is+j} = a_{is+j} \cdot \Upsilon_{is+j} + a_{is+j} \cdot \triangle_{\lambda,is+j} \cdot b_{(i-1)s+j}$

STEP 4 For $j = 1, 2, \ldots, s$, compute
$\widehat{\beta}_{(T-s)+j} = c_{(T-s)+j} = b_{(T-s)+j}$
For $k = (T-s)/s - 1, \ldots, 1, 0$, compute
$\widehat{\beta}_{(k-1)s+j} = a_{(k-1)s+j} \cdot \triangle_{\lambda,(k-1)s+j} \cdot c_{ks+j} + b_{(k-1)s+j}$
$\vdots$
$\widehat{\beta}_j = c_j = a_j \cdot \triangle_{\lambda,j} \cdot c_{s+j} + b_j$

Note that the dimensions of the matrices to be inverted, a_{is+j}, are of order P and therefore the computational cost is low. The whole procedure considered that the smoothing and the seasonal parameters are given. In practice, these parameters are unknown and they must be selected using some criterion. Orbe, Ferreira and Rodríguez-Póo (2001) provide a data driven method to select the seasonal parameters that presents a good performance in a simulation study.

4 Conclusions

SURE models are very useful in Econometrics. However, the estimation of the coefficients can lead to a cumbersome problem if time varying parameters are allowed. Of course, the difficulty increases if restrictions in the parameters are considered. In this paper, a method to estimate time varying parameters in a very general SURE model, where seasonality and cross restrictions are considered, has been developed. One of the main advantage of the proposed methodology is that it provides a general expression for the estimator which is valid for the case of several covariates, cross restrictions and seasonal effects at the same time. An algorithm is derived to compute the estimator at a reasonable computational cost, without the need of any iterative method.

References

Chow, G.C. (1984). Random and changing coefficients models. In: *Z. Griliches and M.D. Intriligator (Ed) Handbook of Econometrics Vol. 2*, 1213-1245. North Holland, New York.

Deaton, A. and J. Muellbauer. (1980) *Economics and consumer behavior.* Cambridge: Cambridge University Press.

Härdle, W. (1990). *Applied nonparametric estimation.* Cambridge University Press.

Hylleberg, S. (1986). *Seasonality in regression.* London: Academic Press.

Jorgenson, D.W. (2000). *Econometric modeling of producer behavior.* Cambridge: MIT Press.

Orbe, S., E. Ferreira and J.M. Rodríguez-Póo. (2001). Nonparametric estimation of time varying parameters under shape restrictions. *Working paper Biltoki 2001.02* Department of Econometrics and Statistics. University of the Basque Country. Spain.
http://www.et.bs.ehu.es/biltoki/dt200102.shtml

Analyzing Data with Robust Multivariate Methods and Diagnostic Plots

Greet Pison[1] and Stefan Van Aelst[2]

[1] Department of Mathematics and Computer Science, Universitaire Instelling Antwerpen (UIA), Universiteitsplein 1, B-2610 Wilrijk, Belgium.

[2] Department of Applied Mathematics and Computer Science, Ghent University, Krijgslaan 281 S9, B-9000 Gent, Belgium.

Abstract. Principal Component Analysis, Canonical Correlation Analysis and Factor Analysis (Johnson and Wichern 1998) are three different methods for analyzing multivariate data. Recently robust versions of these methods have been proposed by Croux and Haesbroeck (2000), Croux and Dehon (2001) and Pison et al. (2002) which are able to resist the effect of outliers. Influence functions for these methods are also present. However, there does not yet exist a graphical tool to display the results of the robust data analysis in a fast way. Therefore we now construct such a diagnostic tool based on empirical influence functions. These graphics will not only allow us to detect the influential points for the multivariate statistical method but also classify the observations according to their robust distances. In this way we can identify regular points, good (non-outlying) influential points, influential outliers, and non-influential outliers. We can downweigh the influential outliers in the classical estimation method to obtain reliable and efficient estimates of the model parameters. Some generated data examples will be given to show how these plots can be used in practice.

Keywords. Diagnostic plot, Influence functions, Robustness, Outliers, Robust Distances, Cutoff values

1 Introduction

In this paper we construct a diagnostic plot to analyze the results of a robust multivariate method. The main idea can be used for principal component analysis, canonical correlation analysis and principal factor analysis. Here however, we restrict us to principal component analysis because of page limit. Throughout the paper we assume that these methods are based on the correlation matrix (see e.g. Johnson and Wichern 1998). The diagnostic plots will be based on robust estimates of the model parameters and empirical influence functions.

Croux and Haesbroeck investigate the robustness of principal components by means of the influence function. Besides the asymptotical influence function approach, their paper also presents empirical versions of the influence function which can be used in the finite sample case to determine the influence of each of the observations on the estimates. However, since these multivariate models contain several (high dimensional) parameters, we obtain several empirical influence functions of the same dimensions. Hence, a graphical tool to represent this information in a comprehensive way will be useful. The goal of these plots is to show what influence each of the observations has on the estimated model parameters.

It is well-known that not all influential points are necessarily (bad) outliers that have to be removed from the data. This is very similar to regression analysis (Rousseeuw and van Zomeren 1990) where a huge outlier in the cariers can have a very small standardized residual and therefore is called a "good leverage point". It is not recommendable to delete such points from the data because these observations make the estimation more precisely. In regression analysis a diagnostic tool helps us to identify very quickly the outliers, the good leverage points and the bad leverage points. Similarly, we now also construct such a graphical tool for robust multivariate methods.

In section 2 we define principal component analysis. Section 3 contains the theoretical influence functions of this method together with the definition of the empirical influence function. The construction of the diagnostic plot is given in Section 4. In Section 5 some simple generated data will be used to explain the interpretation of the plots.

2 Principal Component Analysis

Given a set of variables $X_1, \ldots, X_p$, in principal component analysis we construct a set of new variables Y_i with $i = 1, \ldots, p$ which are uncorrelated linear combinations of the original variables with maximum variance (Johnson and Wichern 1998). Principal component analysis is a powerful tool which is often used to represent the original p-dimensional data in a low dimensional subspace of dimension k spanned by the principal components that have the k largest variances. Denote $\lambda_1 \geq \lambda_2 \geq \ldots \geq \lambda_p$ the eigenvalues of the (theoretical) correlation matrix R and $e_1, e_2, \ldots, e_p$ the corresponding eigenvectors. Let $X = [X_1, X_2, \ldots, X_p]$, then the principal components are given by

$$Y_i = e_i^t X \tag{1}$$

and

$$Var(Y_i) = \lambda_i \tag{2}$$

$$Corr(Y_i, Y_j) = 0 \; i \neq j \tag{3}$$

Hence, the principal components are determined by the eigenvalue-eigenvector pairs of the correlation matrix R.

3 Influence Functions

The influence functions of the eigenvalues $l_i(F)$ and eigenvectors $v_i(F)$ corresponding to the classical correlation matrix $R(F)$ at $F = N_p(\mu, \Sigma)$ can e.g. be derived from (Critchley 1985) and are given by

$$IF(x, l_j, F) = \tilde{z}_j^2 - \lambda_j e_j^t D_{\tilde{x}} e_j \tag{4}$$

$$IF(x, v_j, F) = \sum_{\substack{k=1 \\ k \neq j}}^{p} (\tilde{z}_k \tilde{z}_j - \frac{\lambda_k + \lambda_j}{2} e_j^t D_{\tilde{x}} e_k) \frac{e_k}{\lambda_j - \lambda_k} \tag{5}$$

with $\tilde{x} = \Sigma_D^{-1/2}(x - \mu)$, $\tilde{z}_j = e_j^t \tilde{x}$ and $D_{\tilde{x}} = diag(\tilde{x}\tilde{x}^t)$.

Note that the influence function of a functional depends on the underlying distribution. We consider F to be multivariate normal with mean μ and covariance matrix Σ. In the finite sample case, to obtain the empirical influence function, this underlying distribution is replaced by an empirical distribution $F(\hat{\mu}, \hat{\Sigma})$ where $\hat{\mu}$ and $\hat{\Sigma}$ are estimates of the parameters μ and Σ (see also Jaupi and Saporta 1993). This empirical influence function (EIF) can then be evaluated in each of the data points to assess their influence.

To detect the most influential observations, we propose to substitute robust estimates for the unknown parameters μ and Σ in the influence functions of the classical estimators (functionals) for the multivariate model. This kind of empirical influence function was first proposed by Pison et al (2002) where it is shown that this is indeed the only empirical influence function that detects all the influential points. This can be explained as follows. Since the robust estimates of μ and Σ are not attracted by the outliers, they are close to the true parameter values that were to be estimated. Therefore, the influence of outliers on the classical estimates for the multivariate model can be investigated by the empirical influence functions of the classical estimators based on these robust location/scatter estimates.

To robustly estimate the unknown location and scatter parameters we propose to use the one-step reweighted MCD estimator of Rousseeuw (1984) or S-estimators (Davies 1987, Rousseeuw and Leroy 1987) . These are highly robust estimators with bounded influence function and positive breakdown value that are easy to compute.

To summarize the influence of the observations on the classical estimates, we propose to compute the overall influence of each of the observations on the first k eigenvalues and eigenvectors of interest. The overall influence of an observation is obtained by computing the norm of its empirical influences on each of the k components. We scale this norm to avoid increasing values with increasing value of k. Formally, the overall influence of observation x_i on the first k eigenvalues and eigenvectors is given by

$$EIF_k(x_i, l) = \sqrt{\frac{1}{k}\sum_{j=1}^{k} EIF(x_i, l_j)^2} \tag{6}$$

$$EIF_k(x_i, v) = \sqrt{\frac{1}{kp}\sum_{j=1}^{k}\sum_{l=1}^{p} EIF(x_i, v_{jl})^2} \tag{7}$$

where $v_j = (v_{j1}, \ldots, v_{jp})^t$ for $j = 1, \ldots, p$.

4 Diagnostic Plot

The diagnostic plot can now be constructed as follows. We plot the overall empirical influence of the observations versus their robust distance which is defined by

$$RD(x_i) = \sqrt{(x_i - \hat{\mu})^t(\hat{\Sigma})^{-1}(x_i - \hat{\mu})}.$$

To detect influential points we also need a cutoff value. Therefore we generate 100 datasets from the same size as the original one from a multivariate normal distribution with covariance matrix equal to the robust scatter estimate of the original data. For each of these datasets we compute the empirical influences by substituting the classical sample correlation matrix and the corresponding eigenvalues and eigenvectors in expressions (4) and (5) to obtain the empirical influences. We now use the classical estimates because the data were generated without outliers such that these estimates are more accurate and can also be computed faster. Finally, we take the 95 % percent quantile of these influences as the cutoff value.

This value is added as a horizontal line. The vertical line in the plot corresponds with the square root of the 97.5% of a Chi-square distribution with p degrees of freedom since it is commonly assumed that the squared distances are (approximately) Chi-square distributed. In this way, the diagnostic plot contains four different areas. In the lower left quadrant we find the regular points which are close to the center of the data and have a small influence on the classical estimates. Non-outlying points that have a high influence on the estimates are visible in the upper left quadrant. The outliers are identified in the right quadrants of the plots. Outliers that have a high influence on the estimates are shown in the upper right quadrant while outliers that only have small influence are found in the lower right quadrant. If you remove the latter from the data, then a classical principal component analysis (so a reweighted one) can be carried out which will yield more efficient estimation of the parameters.

5 Generated data sets

Let us look at 2 generated examples which are extreme cases that allow us to further clarify these diagnostic plots. In the first example we generated 40 points from a bivariate Gaussian distribution with correlation 0.7 between the two variables. We then moved four objects (observations 10, 11, 39 and 40) in the direction of the first principal component. The diagnostic plots of the eigenvalues and eigenvectors based on the reweighted MCD with breakdown value $r = 0.75$ are shown in Figure 1. From Figure 1(a) we immediately see

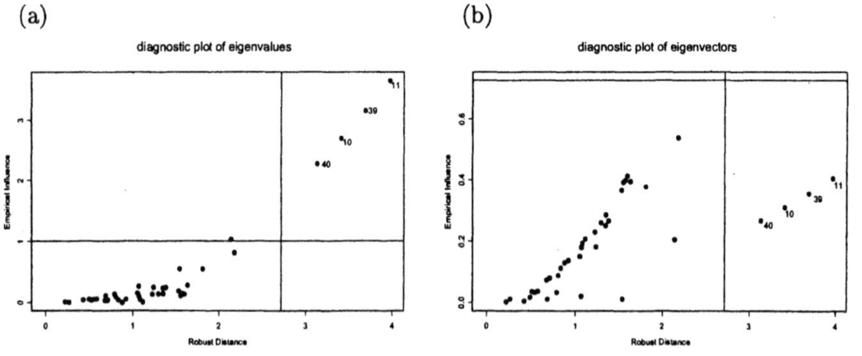

Fig. 1. Diagnostic plot based on reweighted MCD for generated data with $p = 2$ and $n = 40$; (a) of the eigenvalues; (b) of the eigenvectors.

that the four outliers lie in the upper right quadrant which means that they have a high influence on the classical estimates of the eigenvalues. However, from Figure 1(b) it is obvious that those points have little influence on the eigenvectors. Intuitively, this was expected because we moved the points in the direction of the first principal component. Hence, these outliers do not change the directions of the principal components. However, the variance of the first principal component will become larger due to the four outliers which is reflected by the high influences in figure 1(a).

In the second example we start from the same 40 observations as in the first example but now we moved observations 10, 11, 39 and 40 in the direction of the second principal component. The scatter plot of this dataset is given in Figure 2. The corresponding diagnostic plots based on S estimates with

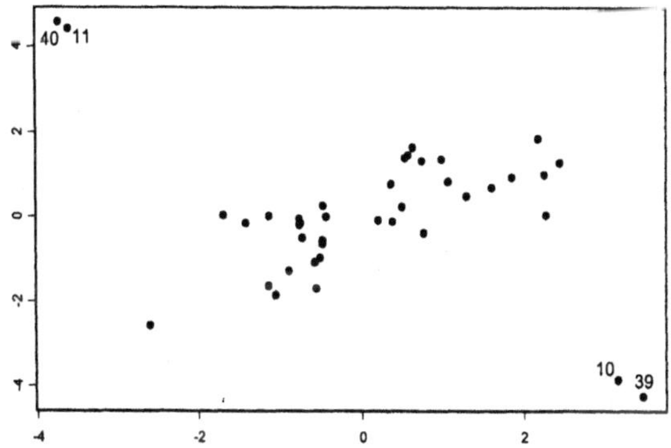

Fig. 2. Scatter plot of the generated data with $n = 40$ and p=2

breakdown value $r = 0.5$ are given in Figure 3. From both plots we see that the outliers have a high influence on both the estimates of eigenvalues and eigenvectors. Also here this is intuitively clear. From Figure 2 we see that the four outliers increase the variance in the direction of the second principal component of the uncontaminated data such that it becomes larger than the variance in the direction of the first original principal component. Hence, the outliers do not only influence the eigenvalues but also drastically affect the eigenvectors as reflected in Figure 3(b). Therefore we conclude that it is recommendable to downweigh these points when analyzing these data using classical principal component analysis.

References

Critchley, F. (1985), "Influence in Principal Component Analysis", *Biometrika*, 72, 627–636.

Croux, C. and Haesbroeck, G.(2000), "Principal Component Analysis based on Robust Estimators of the Covariance or Correlation Matrix: Influence Functions and Efficiencies," *Biometrika*, 87, 603-618.

Croux, C. and Dehon, C. (2001), "Analyse Canonique Basée sur des Estimateurs Robustes de la Matrice de Covariance," La Revue de Statistique

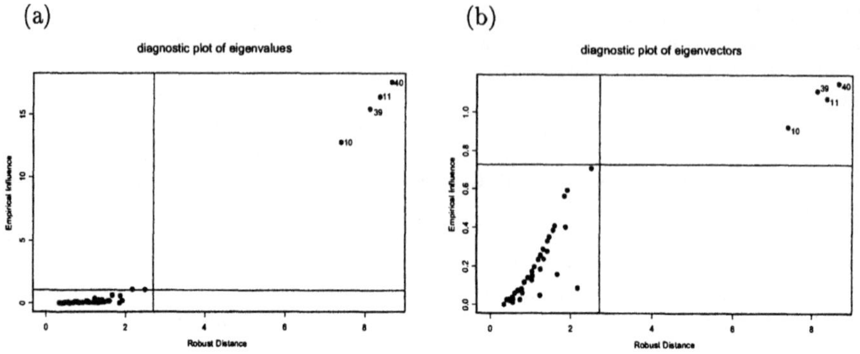

Fig. 3. Diagnostic plot based on S estimates for generated data with $p = 2$ and $n = 40$; (a) of the eigenvalues; (b) of the eigenvectors.

Appliquee, to appear.

Davies, L. (1987), "Asymptotic Behavior of S-estimators of Multivariate Location Parameters and Dispersion Matrices," *The Annals of Statistics*, 15, 1269-1292.

Hampel, F.R., Ronchetti, E.M., Rousseeuw, P.J., and Stahel, W.A. (1986), *Robust statistics: the approach based on influence functions,* John Wiley and Sons, New York.

Jaupi, L., and Saporta, G. (1993), "Using the Influence Function in Robust Principal Components Analysis," in *New Directions in Statistical Data Analysis and Robustness*, eds. S. Morgenthaler, E. Ronchetti, and W.A. Stahel, Basel: Birkhäuser, 147–156.

Johnson, R.A. and Wichern, D.W. (1998), *Applied Multivariate Statistical Analysis*, Fourth Edition, Prentice Hall, New Jersey.

Pison G., Rousseeuw P.J., Filzmoser P. and Croux C. (2002), "Robust Factor Analysis," *Journal of Multivariate Analysis*, to appear.

Rousseeuw, P.J. (1984), "Least Median of Squares Regression," *Journal of the American Statistical Association*, 79, 871–880.

Rousseeuw, P.J. (1985), "Multivariate Estimation with High Breakdown Point," in *Mathematical Statistics and Applications*, Vol. B, eds. W. Grossmann, G. Pflug, I. Vincze and W. Wertz, Dordrecht: Reidel, 283–297.

Rousseeuw, P.J. and Leroy, A.M. (1987), *Robust Regression and Outlier Detection*, Wiley-Interscience, New york.

Rousseeuw, P.J. and van Zomeren, B.C. (1990), "Unmasking Multivariate Outliers and Leverage Points," *Journal of the American Statistical Association*, 85, 633-651.

Application of "Aggregated Classifiers" in Survival Time Studies

A. Benner
Biostatistik, DKFZ, 69120 Heidelberg, Germany

Abstract. A gradient-descent boosting algorithm is presented for survival time data, where the individual additive components are regression trees.

Keywords. Survival Trees, Bagging, Boosting, Loss functions

1 Tree-structured survival analysis

In general two approaches are used to build survival trees. One use a measure of within-node homogeneity asking how similar are the survival experiences in a node. Examples are given by Gordon & Olshen, 1985, who measure the distance between Kaplan-Meier estimates of the survival curves, or LeBlanc & Crowley, 1992, whose measure is based on Poisson deviance residuals. The other approach is to use a measure of between-node separation by using a test statistic to distinguish between survival experiences. Examples are Segal, 1988, who use the Tarone-Ware statistics, and LeBlanc & Crowley, 1993, who use the logrank test.

2 Aggregated classifiers

Unstable classifiers like neural nets or trees can have low bias but are characterized by their high variance. Combining multiple versions of unstable classifiers is variance reducing. Several researchers already demonstrated that aggregating multiple versions of unstable "classifiers" like trees or neural nets results in reduced test set error (e.g. Dietterich, 2000). In the following only regression trees will be considered.

2.1 Bagging and Boosting

Bagging (an acronym for bootstrap aggregating, Breiman, 1996) fits many (large) trees to bootstrap-resampled versions of the training data, and builds an aggregated predictor by averaging. Growing the tree on a learning sample $\mathcal{L} = \{(y_i, x_i), i = 1, ..., n\}$ provides a predictor $\phi(x, \mathcal{L})$ for the response y. Using a sequence of learning samples $\{\mathcal{L}_k\}$, who are drawn from the same underlying distribution as $\mathcal{L}$, allows to replace $\phi(x, \mathcal{L})$ by the average of $\phi(x, \mathcal{L}_k)$,

$$\phi_A(x) = E\phi(x, \mathcal{L})$$

Bootstrapping lets us imitate the process leading to $\phi_A(x)$ by using B bootstrap samples $\mathcal{L}_1, ..., \mathcal{L}_B$ from $\mathcal{L}$ and computing

$$\phi_B(x) = ave_{b=1}^{B}\phi(x, \mathcal{L}_b)$$

Boosting (Freund & Schapire, 1996) fits many (large or small) trees to reweighted versions of the training data. The most popular boosting algorithm, `AdaBoost` (adaptive boosting; Freund & Schapire, 1996), is applied to a dichotomous response $y_i \in \{-1, 1\}$.

Friedman et al. (2000) describe a generalized version of AdaBoost in which the classifier returns a class probability estimate rather than the classifications themselves.

While the goal of bagging is variance reduction, boosting is primarily a bias reduction procedure. Therefore several proposals were made to combine the two methods (e.g. arcing, which is the acronym for adaptively resampling and combining, Breiman, 1998).

2.2 Gradient boosting

Friedman et al. (2000) show that both AdaBoost algorithms fit an additive logistic regression model. The algorithms optimize an exponential criterion which is second order equivalent to the binomial log-likelihood criterion as used in linear or additive logistic regression analyses. Friedman et al. (2000) and Friedman (2001) extend these adaptive boosting algorithms towards a theoretically based function estimation for classification and regression. The target is to find an estimate $\hat{F}(x)$ of the function $F^*(x)$, which maps x to y,that minimizes the expected value of a loss function $\psi(y, F(x))$,

$$F^* = argmin_F \; E_{y,x} \; \psi(y, F(x))$$

The focus here is on additive models, $F(x; \beta_m, a_m) = \sum_{m=1}^{M} \beta_m h(x, a_m)$. Numerical optimization is done by a steepest descent algorithm, where the negative gradient defines the steepest descent direction. This is *gradient boosting* as described in Friedman (2001). The generic algorithm of gradient boosting is shown in Fig. 1.

Initialize $F_0(x) = argmin_\rho \sum_{i=1}^{n} \psi(y_i, \rho)$ and repeat for $m = 1, ..., M$:

1. Compute the negative gradient as "pseudo-response",
   $\tilde{y}_i = - \left[\frac{\partial \psi(y_i, F(x_i))}{\partial F(x_i)} \right]_{F(x)=F_{m-1}(x)}$
2. Use a fitting criterion that estimates conditional expectation (given x) to estimate the negative gradient by h(x, a).
3. Compute the gradient step by
   $\rho_m = arg\ min_\rho \sum \psi \left[(y_i, F_{m-1}(x_i)) + \rho\ h(x_i, a_m) \right]$
4. Update the estimate, $F_m(x) = F_{m-1}(x) + \rho_m\ h(x, a_m)$

Fig. 1. The generic algorithm of gradient boosting

The generic algorithm estimates a function $h(x, a)$ which is most correlated with the negative gradient. $h(x, a)$ is a parameterized function of x, characterized by parameter vector a. In gradient boosting $h(x, a)$ is typically a classification or regression tree.

3 Interpretability

Among the purported biggest advantages of single regression trees is interpretability, whereas aggregated trees are thought to lack this feature.

Aggregated trees can be interpreted (at least approximately) using partial dependence plots (Friedman, 2001) in conjunction with a measure of the relative importance of input variables as described in Breiman et al., 1984.

3.1 Partial dependence

The aim of this plots is to show the dependence of the response on input variables by graphical displays of $\hat{F}(x)$ as a function of its arguments. Let z_l be a subset of the input variables and z_l^c be the complement subset. Now $\hat{F}(x) = \hat{F}(z_l, z_l^c)$. Conditioning on specific values for z_l^c $\hat{F}(x)$ is approximated by $\hat{F}(z_l|z_l^c)$ and a summary of the partial dependence is then

$$\bar{F}(z_l) := E_{z_l^c}[\hat{F}(x)] = \frac{1}{n}\sum_{i=1}^{n} \hat{F}(z_l, z_l^c(i)) = ave_{z_l^c}[\hat{F}(z_l, z_l^c]$$

which is the contribution of z_l to $\hat{F}(x)$ after accounting for z_l^c.

For regression trees this function can be computed by a weighted traversel: At the root a weight value of one is assigned. For each non-terminal node visited, if its split variable is in z_l the appropriate left or right daughter node is visited and the weight is not modified. If the node's split variable is a member of z_l^c, then both daughters are visited and the current weight is modified by the fraction of observations that went left or right at that node. Each terminal node visited is assigned the current weight value. When the tree traversel is complete, the value of $\bar{F}(z_l)$ is the corresponding weighted average of the $\hat{F}(x)$ values over those terminal nodes visited during the traversel. For a collection of B regression trees, the results for the individual trees are averaged.

4 Loss functions in survival

Let the positive random variable T denote the time from a starting point $t = 0$ to an event. The marginal probability of being event-free up to time t is given by $S(t) = P(T > t)$. Given a vector of covariates, x, the estimate of patient-specific survival probabilities $S(t|x)$ is denoted by $\hat{\pi}(t|x)$.

Define $Y = I(T > t^*)$ for a fixed time point t^*. For demonstrating gradient boosting for survival data we use the Brier score loss function for measuring inaccuracy (Graf et al., 1999),

$$\psi(Y, \hat{\pi}) = (Y - \hat{\pi}(t^*|x))^2$$

Using the empirical version of the Brier score for censored data at a fixed time point t^* gives the loss function

$$\psi(y, F) = (0-F(x))^2 I(T \le t^*, \delta = 1)(1/\hat{G}_T) + (1-F(x))^2 I(T > t^*)(1/\hat{G}_{t^*}).$$

The resulting boosting algorithm is shown in Fig.2.

Initialize $F_0(x) = \hat{\pi}(t^*|x)$
Repeat for $m = 1, ..., M$:

1. $\tilde{y}_i = (0 - F_{m-1}(x_i))I(T \le t^*, \delta = 1)(1/\hat{G}_T)$
   $+ (1 - F_{m-1}(x_i))I(T > t^*)(1/\hat{G}_{t^*})$, G denotes the Kaplan-Meier estimate of the censoring distribution.
2. Compute $\{R_{lm}\}$, an L-terminal node tree of $(\{\tilde{y}_i, x_i\})$
3. Compute for each terminal node $l = 1, .., L$,
   $\gamma_{lm} = \sum_{x_i \in R_{lm}} \tilde{y}_i / \sum_{x_i \in R_{lm}} [I(T \le t^*, \delta = 1)(1/\hat{G}_T) + I(T > t^*)(1/\hat{G}_{t^*})]$
4. Update $F_m(x) = F_{m-1}(x) + \sum_{l=1}^{L} \gamma_{lm} 1(x \in R_{lm})$

Fig. 2. A boosting algorithm using the Brier score for censored data

5 Example

Between October 1990 and August 1998,325 patients with chronic lympocytic leukemia from a single institution were enrolled in a clinical study and followed with regard to survival (Doehner et al., 2000). A Cox regression model was used to identify possible prognostic factors. Beside others input variables used were age,Binet stage,white-cell count (WBC),serum lactate dehydrogenase (LDH), and presence or absence of genomic aberrations (e.g. deletion in 17p, deletion in 11q). As a reference we compute the survival tree using the tssa program of Segal (Segal, 1988). The resulting tree is

```
node), split, (n, failures), km-median, split-statistic
      * denotes terminal node, choice is Mantel-Haenzel

 1) root (322,112) 60.5 8.04
   2) LDH.at.study<248.5 (268,72) 72.0 7.22
     4) Binet.at.study<2.5 (237,50) 72.0 5.28
       8) age.at.study<73.5 (205,37) 85.5 5.04
        16) X11q<1.5 (175,23) 85.5 *
        17) X11q>1.5 (30,14) 47.5 *
       9) age.at.study>73.5 (32,13) 47.5 *
     5) Binet.at.study>2.5 (31,22) 21.5 *
   3) LDH.at.study>248.5 (54,40) 31.5 *
```

In contrast to this result the boosting procedure using the rpart algorithm (Therneau & Atkinson, 1997) and the Brier score for 3 years survival does not include age or 11q deletion as relevant variables (Fig. 3).

On the other side LDH seems to be an important prognostic factor for three year survival (Fig. 4).

But note that these results depend strongly on the choice of t^*. For short term survival ($t^* = 12$ months) deletion 17p was identified as most relevant prognostic factor whereas age was most important for long term survival. If we do not want to restrict to one time point t^* Graf et al. consider to average loss over time, that is to use an integrated version of the Brier score for $t \in [0, t^*]$.

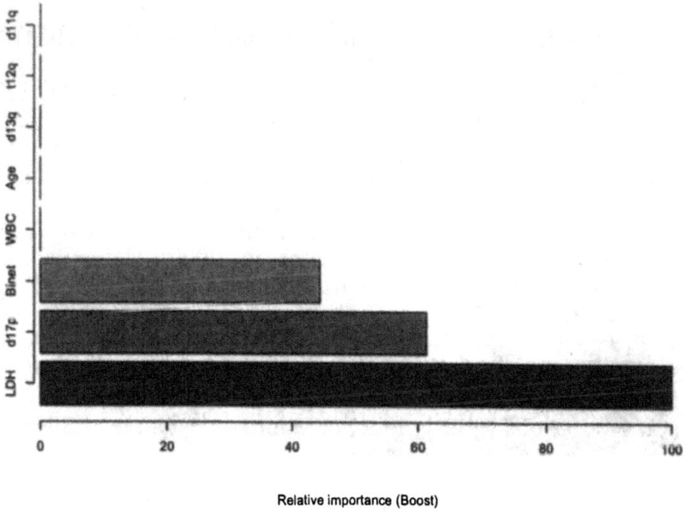

Fig. 3. The relative importance of input variables as computed from the boosting procedure using the Brier score for 3 years survival.

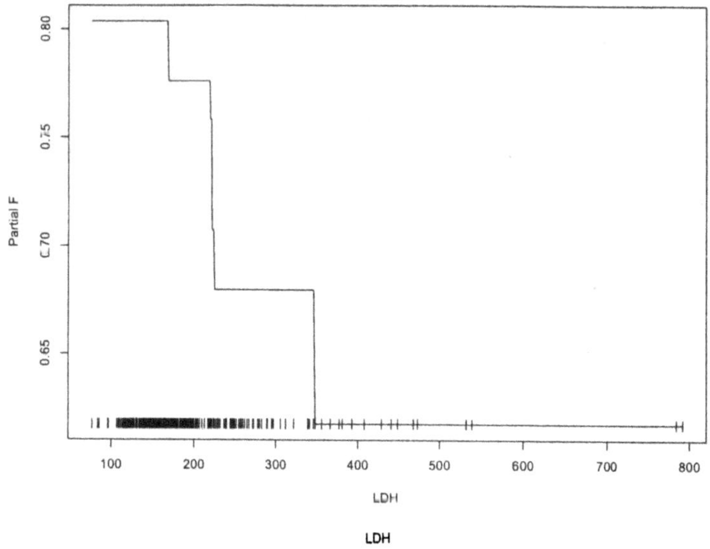

Fig. 4. The partial dependence plot for LDH for the boosting procedure using the Brier score for 3 years survival.

References

Breiman, L., Friedman, J.H., Olshen, R.A. and Stone, C.J. (1984). *Classification and Regression Trees.* Wadsworth, CA

Breiman, L. (1996). Bagging predictors. *Machine Learning*, **24**, 123-140.

Breiman, L. (1998). Arcing classifiers. *Annals of Statistics*, **26**, 801-849.

Dietterich, T.G. (2000). An experimental comparison of three methods for constructing ensembles of decision trees: Bagging, boosting, and randomization. *Machine Learning*, **40**, 139-157.

Doehner, H., Stilgenbauer, S., Benner, A., Leupolt, E., Kroeber, A., Bullinger, L., Doehner, K., Bentz, M. and Lichter, P. (2000). Genomic aberrations and survival in chronic lymphocytic leukemia. *New England Journal of Medicine*,**343**, 1910-1916.

Freund, Y. and Schapire, R.E. (1996). Experiments with a new boosting algorithm. in In: *Proceedings of the 13th International Conference on Machine Learning*, 148-156. San Francisco, CA: Morgan Kaufmann.

Friedman, J.H. (2001). Greedy function approximation: A gradient boosting machine. *Annals of Statistics*, **29**.

Friedman, J.H., Hastie, T. and Tibshirani, R. (2000). Additive logistic regression: A statistical view of boosting. *Annals of Statistics*, **28**, 337-407.

Gordon, L. and Olshen, R.A.(1985). Tree-structured survival analysis (with discussion). *Cancer Treatment Reports*, **69**, 1065-1068.

Graf, E., Schmoor, C., Sauerbrei, W. and Schumacher, M. (1999). Assessment and comparison of prognostic classification schemes for survival data. *Statistics in Medicine*, **18**, 2529-2545.

LeBlanc, M. and Crowley, J. (1992). Relative risk trees for censored survival data. *Biometrics*, **48**, 411-425.

LeBlanc, M. and Crowley, J. (1993). Survival trees by goodness of split. *Journal of the American Statistical Association*, **88**, 457-467.

Segal, M.R. (1988). Regression trees for censored data. *Biometrics*, **44**, 35-47.

Therneau, T.M. and Atkinson, E.J. (1997). An Introduction to Recursive Partitioning Using the RPART Routines. *Mayo Clinic Section of Biostatistics technical report #61*

Application of Hopfield-like Neural Networks to Nonlinear Factorization*

D. Husek[1], A.A. Frolov[2], H. Rezankova[3] and V. Snasel[4]

[1] Institute of Computer Science, Academy of Sciences of the Czech Republic, Prague, 182 00, Czech Republic

[2] Institute of Higher Nervous Activity, Russian Academy of Sciences, Moscow

[3] Department of Statistics and Probability, University of Economics, Prague, 130 67, Czech Republic

[4] Department of Computer Science, Mining University, Ostrava, 701 00, Czech Republic

Abstract. The problem of binary factorization of complex patterns in recurrent Hopfield-like neural network was studied by means of computer simulation. The network ability to perform a factorization was analyzed depending on the number and sparseness of factors mixed in presented patterns. Binary factorization in sparsely encoded Hopfield-like neural network is treated as efficient statistical method and as a functional model of hippocampal CA3 field.

Keywords. Binary factorization, Hopfield network, sparse encoding

1 Information redundancy problem

The idea of redundancy is intuitively clear. In everyday experience, we deal with temporally and spatially regular information structure, and this regularity tends to be universal through individual life and even from specie to specie. Such regularity forms common informational space and implies existence of some lexicon of features that guarantee this regularity. The universality of this lexicon is provided by existence of co-occurring features that form invariant clusters, further termed as factors. Clusterization of features is provided by evolutionary increase of complexity of the matter, supposing that the entropy of the environment that surrounds a human is substantially bigger than that of random one. The idea of evolutionary structuring can be ranged from gene level to macro scale level of organic matter. Since "knowledge" of this structuring is implicitly encoded in brain organization by means of natural selection information storage mechanism used in brain is adopted for elimination of this redundancy. Thus, every incoming sensory pattern can be represented as a nonlinear combination of intrinsically existing invariant factors. We shall only consider spatial

* This work was supported by grants Grant Agency of Czech Republic No. 201/01/1192 and 201/00/1031.

redundancy, i.e. redundancy caused by co-occurrence of several factors in one sensory pattern.

In his influential papers Marr (1970, 1971) raised this problem and noted that redundant information may be stored efficiently if some collection of features that commonly occur were extracted from input pattern and added to the vocabulary of brain's experience as a new entity (concept). Through this vocabulary, the brain later interprets and records its experience. Marr also noted, that memory task is close to clusterization process.

Problem of efficient information storage mentioned above could be resolved by means of proper network organization, special learning rules and dynamics and will be addressed later.

2 Statistical point of view

As was shown in the previous chapter information redundancy is the main source of information compression and thus its elimination is a prerequisite of efficient information storage. Furthermore, redundancy elimination allows one to reveal the underlying structure that is not evident from the row data. This widely occurring problem is conventionally solved by means of various statistical algorithms. Revealing the underlying structure consists in determining factors (clusters) and memberships of variables in these factors. There is a problem of clustering variables.

The basic algorithm for redundancy elimination is a *principal component analysis* (PCA), which allows extracting the main relations in the data of high dimensionality. A common way to find the principal components of a data set is by calculating the eigenvectors of the data correlation matrix. For categorical data, PCA with optimal scaling can be used. In comparison with standard PCA, categorical (nonlinear) PCA allows variables to be scaled at different levels. The weighted form of PCA is called *correspondence analysis*. It is appropriate for frequency data. For our purpose multiple correspondence analysis (MCA) must be used. It is sometimes called *homogeneity analysis*.

The most popular method for clustering of numeric variables is *factor analysis*. This technique assumes linear relationships between variables. The cluster membership of the certain variable is not hard (disjoint) and overlapping clusters can be found. The special technique for clustering (both variables and cases) is *cluster analysis*. There are different algorithms for this purpose. Disjoint (*k*-means) techniques attempt to identify relatively homogeneous groups of variables based on selected characteristics. Some other techniques are called hierarchical. They are based on proximity matrices of dissimilarity or similarity measures for all pairs of variables. The third important group is made by overlapping (fuzzy) clustering algorithms. A proximity matrix is also a base for further technique – *multidimensional scaling* (MDS). This method can be considered to be an alternative to factor analysis, and it is typically used as an exploratory method. With MDS the user can analyze not only correlation matrices but also any kind of similarity or dissimilarity matrix.

Some constrains exist for some methods in statistical packages. For optimal scaling methods in statistical package SPSS categories must be coded from value 1. Further, SPSS can analyze only up 100 variables by multidimensional scaling.

There have been recently developed many neural network based algorithms that perform a function of redundancy elimination both in linear and nonlinear case (independent component analysis, PCA network,). Most of them utilize artificially constructed learning rules, multi-layered networks with nonlinear transformation function or simply mapped nonlinear case onto linear etc. that allow taking into account second-order cross-correlations of incoming pattern.

In this paper we propose a factor analysis method for sufficiently nonlinear case that naturally follows from Hebbian-like learning rule and properties of Hopfield-like neural network dynamics. Furthermore, this conceptual mechanism seems plausible in frame of memory storage and processing in brain.

3 Factorization

We define factorization as decomposition of a complex vector signal into a set of simple factors based on correlation between components of the former. In the case of binary factorization a complex vector signal (pattern) has the form of a logical sum of weighted binary factors: $X^0 = \bigvee_{l=1}^{L} \alpha_l \mathbf{f}^l$. This case of factorization allows an interpretation in terms of attractor neural networks with binary activity. The central idea is that network can easily learn cross–correlations that underlie in incoming complex pattern using Hebbian learning rule. Hebbian rule forms connections matrix as a covariance matrix for the set of learned patterns. Neurons that tend to fire together (represent one common factor) will be more correlated and corresponding connection strengths will be larger in respect to those neurons that belong to different factors. Hence, each group of neurons that forms a factor might correspond to the attractor of the network dynamics. This paper is devoted to investigation of conditions under which factors actually form attractors in sparsely encoded Hopfield-like neural network by means of computer simulation.

4 Model description

Detailed theoretical and computational analysis of sparsely encoded Hopfield-like neural networks was given elsewhere (Amari (1988), Frolov (1993), Frolov (1997), Perez-Vincente (1989)). In contrast to these works we trained the network by set of complex patterns (binary superposition of factors). Thus, on the learning stage, fully connected network of N binary neurons was trained by a set of M patterns of the form $X^m = \bigvee_{l=1}^{L} \alpha_l^m \mathbf{f}^l$, where $\mathbf{f}^l \in \mathrm{B}_p^N$ (pN is maintained constant) are L factors and $\alpha^m \in B_{p_f}^L$ are factor scores[1]. Both factors and factor scores were

---

[1] $\mathrm{B}_p^N = \{\mathbf{X} \mid X_i \in \{0,1\}, P\{X_i = 1\} = p \quad \forall i = 1,\ldots,N\}$

chosen statistically independent. Here p and p_f are sparsenesses (ratio of number of active elements to the total number of elements) of factors with respect to neurons and patterns with respect to factors. In a limit $p_f \to 0$ patterns become pure factors that correspond to ordinary Hopfield case.

Connections matrix J was formed using the correlation Hebbian rule:

$$\mathbf{J}_{ij} = \frac{1}{A}\sum_{m=1}^{M}(X_i^m - q\{X^m\})(X_j^m - q\{X^m\}) \; i \neq j, \; \mathbf{J}_{ii} = 0,$$

where $q\{X^m\} = \sum_{i=1}^{N} X_i^m / N$ is the total activity of the pattern.

On the recall stage, on presentation of an initial pattern, the network was let to evolve until it stabilized in some attractor. The evolution of the network's state is determined by the synchronous dynamics equation for activity X in time:

$$X_i(t+1) = \Theta(h_i(t) - T(t)), \; X_i(0) = f_i^l, \; i = 1,\ldots,N$$

where

$$h_i(t) = \sum_{j=1}^{N} \mathbf{J}_{ij} X_j(t)$$

is synaptic excitation, Θ – step function, and $T(t)$ – activation threshold. The threshold $T(t)$ is chosen at each time step in such a way that sparseness of the network activity is kept constant and equal to p. Thus, on each step $n = pN$ "winners" (neurons with greatest synaptic excitation) are chosen and only they are active on the next step. This procedure ensures that attractors are only point and cyclic of length two. The stable pattern (point attractor) or first pattern of cyclic attractor was taken as a resulting pattern of the recall process. In order to check whether factors do form attractors we took pure factors as initial network states.

5 Measures

The closeness between the recalled factor and the final pattern was measured by their overlap

$$m^f = m(X^{in}, X^f) = \frac{1}{N}\sum_{i=1}^{N} X_i^{in} X_i^f .$$

As a measure of the relative informational loading we use $\alpha \approx Lh(p)/N$, where $h(p)$ is the Shannon function. The informational capacity of the network is α_{cr}, which is a maximum α, for which stable states in the vicinities of stored factors still exist. Average number of active factors in patterns $C = Lp_f$ we used as a measure of problem *complexity*. The simplest case $C = 1$ corresponds to the standard Hopfield model.

6 The results of the simulation

In order to analyze the dependency of α_{cr} on parameters of the network the model was simulated on the computer. Calculations were performed for N from 200 to 4000, for p = 0.5, 0.1, 0.02. The program generated random factors, mixed them randomly (regarding p and p_f) into the set of M patterns, trained the net with this set, and tested the net with factors. The distribution of final overlaps m^f has two distinct modes: for $m^f \approx 1$ and $m^f \approx 0$, that correspond to stabilization of the network in true and spurious attractors respectively. The threshold value $m^f{}_{thr}$ used for separation was determined as a point of a minimum in case of balanced distribution between two modes. The probability of existence of stable attractors in the vicinities of factors was estimated by the probability that m^f belongs to the "true" mode.

For each network size the probability P fast drops when the network loading increases and as N increases the curve of the function $P(\alpha)$ approaches step-like function. A point at which the drop occurs was taken as α_{cr}. The dependence of α_{cr} on C is shown in Fig 1.

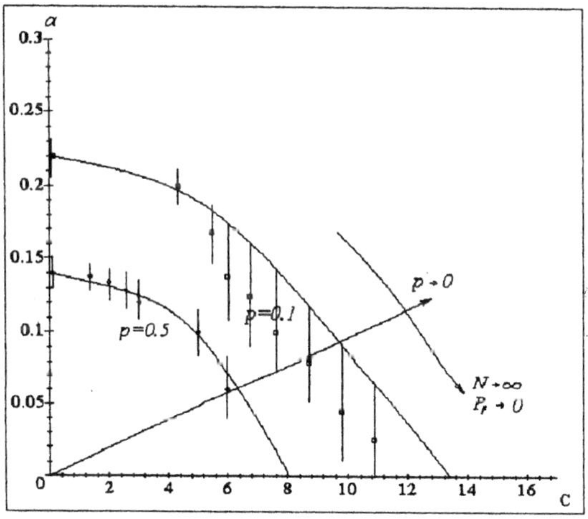

Fig 1. Critical curves in (α,C) - plane

For each p a curve in (α,C)-plane separates phases of possibility and impossibility of factorization. As C increases the curve approaches abscise, and one may estimate the value C_{max} – maximum complexity of patterns, for which factorization is still possible. As sparseness increases (p decreases) C_{max} increases.

7 Conclusions

Thirty years ago Marr developed conceptual model of hippocampus. His functional idea is close to our idea of factorization. Marr proposed that the hippocampus plays a role of the processor of the complex incoming pattern with its subsequent transfer to the neocortex as a set of "classificatory units". A role of temporal storage of information was assigned to a field CA3, which due to an extensive system of recurrent collaterals is a natural autoassociator. We suggest that due to presented ability of autoassociator to perform factorization, this field can carry out a function of decomposition of complex information into elementary factors.

Computer simulation proved the validity of this idea. Recurrent neural network happened to be capable of extracting factors from the complex patterns structure learned using Hebbian rule. Our investigation has not yet covered the internal structure of attraction basins, number of spurious attractors and so on. This will be the aim of future research. Factorization plausibility turned out to be dependent on the absolute complexity of the patterns. It hints the idea to extend the network parameters to the real ones ($N \approx 10^5, p \approx 0.04, P\{\mathbf{J}_{ij} \neq 0\} \approx 0.02$) and use more biologically plausible neural dynamics to estimate such critical macroparameters that could be compared with their analogues in a behavioral experiment.

References

Amit, D.J., Gutfreund, H. & Sompolinsky, H. (1987). Statistical mechanics of neural networks near saturation. *Ann. Phys.*, **173**, 30-67.

Amari, S. & Maginu, K. (1988). Statistical neurodynamics of associative memory. *Neural Networks*, **1**, 63-73.

Amari, S. (1989). Characteristics of sparsely encoded associative memory. *Neural Networks*, **2**, 451-457.

Buzsaki, G. (1996). Hippocampo-neocortical dialogue. *Cerebral Cortex*, **6**, 81-92.

Frolov, A.A. & Muraviev, I.P. (1993). Informational characteristics of neural networks capable of associative learning based on Hebbian plasticity. *Network*, **4**, 495-536.

Frolov, A.A., Husek, D. & Muraviev, I.P. (1997). Informational capacity and recal quality in sparsely encoded Hopfield-like neural network: Analytical approaches and computer simulation. *Neural Networks*, **10**, 845-855.

Gifi, A. (1990). *Nonlinear Multivariate Analysis.* New York: John Wiley & Sons.

Hopfield, J.J. (1982). Neural network and physical systems with emergent collective computational abilities. *Proc. Natl. Acad. Sci. USA*, **79**, 2544-2548.

Marr, D. (1970). A theory of cerebral neocortex. *Proc R Soc Lond*, B, **176**, 161-234.

Marr, D. (1971). Simple memory. A theory of archicortex. *Phil Trans R Soc Lond*, B, **262**, 24-81.

McDonald, R.P. (1985). Factor Analysis and Related Methods. New Jersey: Lawrence Erlbaum Associates, Publisher.

Perez-Vicente, C.J. & Amit, D. (1989). Optimized network for sparsely encoded patterns. *J. of Physics A: Math. Gen.*, **22**, 559-569.

Bagging Tree Classifiers for Glaucoma Diagnosis

Torsten Hothorn and Berthold Lausen

Institut für Medizininformatik, Biometrie und Epidemiologie
Friedrich-Alexander Universität Erlangen-Nürnberg,
Waldstraße 6, D-91054 Erlangen, Deutschland

Abstract. The aggregation of multiple unstable classifiers leads to substantial reduction of misclassification error in many applications and bench mark problems. We focus on the problem of classifying eyes as normal or glaucomatous based on measurements derived from laser scanning images of the optic nerve head. The performance of various aggregated classifiers is investigated for a clinical training sample and for a simulation model of eye morphologies.

Keywords. Discriminant analysis, method selection bias, error rate estimation

1 Introduction

The quest for an efficient and robust classifier is an important issue in clinical research. The possibilities to use diagnostic image data or other high throughput measurements are an increasing part of the clinical routine work and a challenge for modern data analysis in general. In our application, predictors derived from laser scanning images of the eye background are used to classify eyes as normal or glaucomatous.

Breiman (1996a, 1998) suggests bootstrap aggregating (acronym: bagging) of classifiers to improve an unstable classification method. Another idea is to use P-value adjusted classification trees as introduced in Lausen et al. (1994). Bagging classification trees allows substantial reduction of misclassification error in many applications and bench mark problems (cf. Breiman, 1998).

The selection of a classifier with respect to minimal estimated misclassification error using a learning sample only leads to over-fitting and bias in both model selection and error rate estimation. We suggest two strategies to deal with the problem: first, a simulation model which is partially based on parameters estimated from the study population is used to select an "optimal" classifier for glaucoma classification. The error rate estimation for the selected method, e.g. using the .632+ estimator by Efron & Tibshirani (1997), does not suffer a method selection bias. As a second procedure we use a combination of linear discriminant analysis (LDA) and tree classifiers which combines the advantages of both methods.

2 Data and Simulation Model for Laser Scanning Images

Glaucoma is an ocular disease that causes progressive damages in the optic nerve fibres and leads to visual field loss. Laser scanning images of the eye background are used to detect a loss of retinal nerves. The images taken by the Heidelberg Retina Tomograph (HRT, Heidelberg Engineering, 1997) are used to derive measurements for the loss of retinal nerves.

A learning sample of 98 HRT examinations of normal eyes and 98 examinations of glaucomatous eyes is available. To classify the observations as normal or glaucomatous, 62 predictors derived from the HRT images are used. In the following, we give a short description of the simulation model, cf. Hothorn & Lausen (2002).

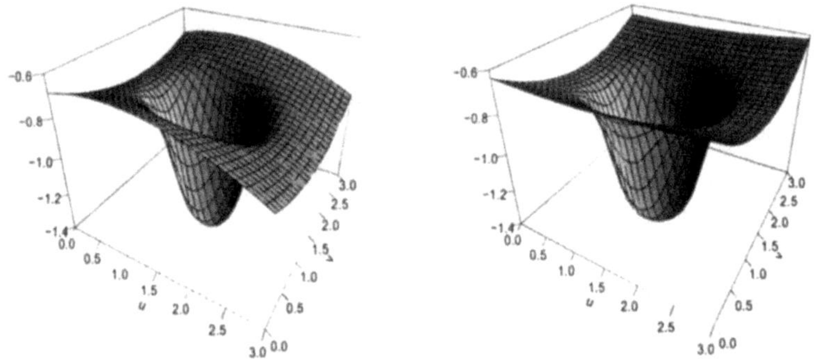

normal eye glaucomatous eye

Fig. 1. Surface of the optic nerve head model for normal (left, $\delta = 0$) and glaucomatous (right, $\delta = 1$) eyes with parameters estimated from the study population. The excavation of the papilla of the glaucomatous eye is wider and the border is flat compared to the normal papilla.

The model of the surface of the optic nerve head introduced by Swindale et al. (2000) is used to describe the morphology of the eye background in normals and patients. For a realistic setup, the parameters of the model are estimated from the given study population. Figure 1 shows the model surface for a normal and glaucomatous eye. Progression of retinal nerve fibre loss is modeled by a weighted average of the two surfaces computed on a 40×40 grid where $\delta \in [0, 1]$ is the coefficient of the weighted average. We simulate the measurement error of the HRT with normal errors with expectation null and variance σ^2. The errors are added to the surface at each point of the grid.

The first setup simulates the measurement error of the HRT itself for repeated examination of one normal and one glaucomatous eye. Three levels of progression for the glaucomatous eye are used: $\delta = 0.1, 0.5$ and 0.9. The measurement error is fixed at $\sigma = 0.2$.

The second setup is more suitable for comparing the performance of classifiers in a human population with varying morphology of the optic nerve head and simulates a population with three subgroups with respect to the size of the optic disc (small: 1.72 mm^2, medium: 2.61 mm^2 and large: 4.01 mm^2). Again, the measurement error is fixed at $\sigma = 0.2$ and the level of progression varies for $\delta = 0.1, 0.5$ and 0.9.

Finally, we study the influence of the measurement error of the HRT on the classification of glaucoma for a clinical population. In the third setup we partition every subgroup with respect to the size of the optic disc into three equally sized subgroups with respect to the level of progression $\delta = 0.2, 0.5$ and 0.8. The measurement error of the HRT varies for $\sigma = 0.1, 0.2$ and 0.3 in this situation.

3 Bagging a Combination of LDA and Classification Trees

A classifier that combines two methods and therefore avoids the method selection process is most desirable. We suggest "Double-Bagging" for a combination of LDA and classification trees.

Let $\mathcal{L} = \{(y_i, \mathbf{x}_i), i = 1, \dots, N\}$ denote a learning sample of N independent observations consisting of p-dimensional vectors of predictors $\mathbf{x}_i = (x_{i1}, \dots, x_{ip}) \in \mathbb{R}^p$ and class labels $y_i \in \{1, \dots, J\}$. The observations in the learning set are a random sample from some distribution function F

$$(y_1, \mathbf{x}_1), \dots, (y_N, \mathbf{x}_N) \stackrel{\text{iid}}{\sim} F.$$

A classifier $C(\mathbf{x}, \mathcal{L})$ predicts future y-values for a vector of predictors $\mathbf{x}$ based on a learning sample $\mathcal{L}$. The aggregated classifier C_A is given by

$$C_A(\mathbf{x}) = \mathrm{E}_F C(\mathbf{x}, \mathcal{L}),$$

where the expectation is over learning samples $\mathcal{L}$ distributed according to F. Bagging as introduced by Breiman (1996a) estimates the aggregated rule $C_A(\mathbf{x})$ using the bootstrap by

$$\hat{C}_A(\mathbf{x}) = \mathrm{E}_{\hat{F}} C(\mathbf{x}, \mathcal{L}^*),$$

where $\mathcal{L}^*$ is a random sample from the empirical distribution function $\hat{F}$

$$(y_1^*, \mathbf{x}_1^*), \dots, (y_N^*, \mathbf{x}_N^*) \stackrel{\text{iid}}{\sim} \hat{F}.$$

Drawing a random sample of size N from the empirical distribution. a bootstrap sample of size N covers approximately 2/3 of the observations of the learning sample. The observations, which are not in the bootstrap sample, are called out-of-bag sample and may be used for estimating the misclassification error or for improved class probability estimates, see Breiman (1996a,b).

In our framework, the out-of-bag sample is used to perform a LDA. The results are the coefficients of a $J-1$ dimensional linear transformation of the original variables called linear discriminant or canonical variables (for example Ripley, 1996, Chapter 3.1.). The discriminant variables are computed for the bootstrap sample and a new classifier is constructed using the original variables as well as the discriminant variables. Double-Bagging is performed as follows:

1) Draw B random samples $\mathcal{L}*^{(1)}, \dots, \mathcal{L}*^{(B)}$ with replacement from $\mathcal{L}$ and let $X*^{(b)}$ denote the matrix of predictors $\mathbf{x}_1^{*(b)}, \dots, \mathbf{x}_N^{*(b)}$ from $\mathcal{L}*^{(b)}$.

2) Compute a LDA using the out-of-bag sample $\mathcal{L} \setminus \mathcal{L}*^{(b)}$, that gives a $p \times (J-1)$ matrix $Z^{(b)}$, where the columns are the coefficients of the linear discriminant functions.

3) Construct the classifier C using the original variables as well as the discriminant variables of the bootstrap sample $(\mathcal{L}*^{(b)}, X*^{(b)} Z^{(b)})$.

4) Iterate steps 2) and 3) for all B bootstrap samples.

A new observation $\mathbf{x}$ is classified by majority voting using the predictions of all classifiers

$$C\left((\mathbf{x}, \mathbf{x}Z^{(b)}), (\mathcal{L}*^{(b)}, X*^{(b)} Z^{(b)})\right) \text{ for } b = 1, \dots, B.$$

Using the out-of-bag sample for the LDA, the coefficients of the discriminant function are estimated by an independent sample, thus avoiding over-fitted discriminant variables in the tree growing process.

The package `rpart` (version 3.1-3, see Therneau & Atkinson, 1997) in the `R` programming environment (version 1.4.0, `cran.r-project.org`, see Ihaka & Gentlemen, 1996) is used for the construction of classification trees. The tree building continues until the nodes are pure. We used 50 bootstrap replications for both CTREE-Bagging and Double-Bagging.

4 Results

We investigate the three different setups to compare the performance of LDA, CTREE, CTREE-Bagging and Double-Bagging of CTREE and LDA. The simulated misclassification errors for all three setups are given in Table 1. The classifiers are trained using a learning sample of size 200 (100 normal and 100 glaucoma surfaces) and tested using a test sample of the same size. We use 1000 Monte-Carlo replications.

Setup 1

δ	LDA	CTREE	CTREE-Bagging	Double-Bagging
0.1	0.376	0.470	0.452	0.431
0.5	0.027	0.223	0.158	0.031
0.9	0.000	0.110	0.063	0.000

Setup 2

δ	LDA	CTREE	CTREE-Bagging	Double-Bagging
0.1	0.443	0.459	0.440	0.437
0.5	0.246	0.216	0.155	0.142
0.9	0.147	0.137	0.077	0.059

Setup 3

σ	LDA	CTREE	CTREE-Bagging	Double-Bagging
0.3	0.337	0.340	0.298	0.288
0.2	0.278	0.271	0.216	0.206
0.1	0.178	0.147	0.101	0.101

Table 1. Model of glaucoma classification: Simulated error rates in three setups for linear discriminant analysis (LDA), classification trees (CTREE), bagged classification trees (CTREE-Bagging) and Double-Bagging of LDA and CTREE.

In the first setup LDA performs best. CTREE-Bagging suffers a much higher error rate but improves the even larger error rate of CTREE. It may be assumed that the two classes are separable by linear combinations of the predictors. Note, that for progression $\delta = 0.9$ LDA classifies all observations correctly while CTREE-Bagging has a misclassification error rate of 6.3%.

CTREE outperforms LDA in the second setup. Classification trees are able to identify the subgroups with respect to the size of the optic disc. CTREE-Bagging reduces the simulated misclassification error of CTREE significantly.

The subgroups with respect to both the size of the optic disc and progression are identified by classification trees in the third setup. On the one hand, this situation is most realistic for analysing the performance of the three classifiers for glaucoma. Therefore, one would prefer bagged classification trees for glaucoma classification. On the other hand, LDA has much lower misclassification error especially in the first setup. In all situations, Double-Bagging performs at least comparable to the best of the other procedures. Therefore, we can avoid a method selection bias for our dataset by using Double-Bagging as combination of the two methods that performed best in different setups.

The 10-fold cross-validated misclassification error of Double-Bagging for the learning data is 0.157.

In summary, simulation results indicate that, depending on the specific situation, LDA or bagged classification trees perform best for glaucoma classification. Double-Bagging is at least comparable to the best one of both methods. Additionally, the estimated misclassification error of the combined classifier does not suffer a method selection bias.

Acknowledgments

Financial support from Deutsche Forschungsgemeinschaft, grant SFB 539-A4/C1, is gratefully acknowledged.

References

Breiman, L. (1996a). Bagging predictors. *Machine Learning*, **24**(2), 123–140.

Breiman, L. (1996b). *Out-Of-Bag Estimation.* Tech. rep., Statistics Department, University of California Berkeley, Berkeley CA 94708.

Breiman, L. (1998). Arcing classifiers. *The Annals of Statistics*, **26**(3), 801–824.

Efron, B. & Tibshirani, R. (1997). Improvements on cross-validation: The 632+ bootstrap method. *Journal of the American Statistical Association*, **92**(438), 548–560.

Heidelberg Engineering (1997). *Heidelberg Retina Tomograph: Bedienungsanleitung Software version 2.01.* Heidelberg Engineering GmbH, Heidelberg.

Hothorn, T. & Lausen, B. (2002). Bagging combined classifiers. In *Studies in Classification, Data Analysis, and Knowledge Organization*, Heidelberg. Proceedings IFCS (accepted), Springer.

Ihaka, R. & Gentlemen, R. (1996). R: A language for data analysis and graphics. *Journal of Computational and Graphical Statistics*, **5**, 299–314.

Lausen, B., Sauerbrei, W., & Schumacher, M. (1994). Classification and regression trees (CART) used for the exploration of prognostic factors measured on different scales. In Dirschedl, P. & Ostermann, R. (eds.), *Computational Statistics*, pp. 483–496, Heidelberg. Physica-Verlag.

Ripley, B. D. (1996). *Pattern Recognition and Neural Networks.* Cambridge, UK: Cambridge University Press.

Swindale, N. V., Stjepanovic, G., Chin, A., & Mikelberg, F. S. (2000). Automated analysis of normal and glaucomatous optic nerve head topography images. *Investigative Ophthalmology and Visual Science*, **41**(7), 1730–42.

Therneau, T. & Atkinson, E. (1997). *An Introduction to Recursive Partitioning using the rpart Routine.* Tech. Rep. 61, Section of Biostatistics, Mayo Clinic, Rochester.

Bayesian Automatic Parameter Estimation of Threshold Autoregressive (TAR) Models using Markov Chain Monte Carlo (MCMC)

Esmail Amiri

Department of Mathematics & Statistics, Imam Khomeini International University, P.o.Box 288, Ghazvin, Iran.

Abstract. In nonlinear time series analysis a Threshold Autoregressive model (TAR) is considered as an approximation to a Nonlinear autoregressive process(NLAR). A TAR model is a piece-wise linear model over the state space. It is linear in the space of the thresholds.

In this article we confine our studies to a class of TAR models which is called Self-Exciting Threshold Autoregressive (SETAR).

The first step in fitting a SETAR model to a data-set is to decide the number of regimes. The second step is to identify the delay, the threshold(thresholds) and the order of each regime. Our approach is to estimate automatically, the number of regimes, the delay, the threshold(thresholds), the order and parameters of each regime for this class of threshold models via a Bayesian approach using MCMC methods. Model selection is done through calculating mean of the root mean square error (MRMSE) of forecasts.

Keywords. Non linear Time Series, Threshold Autoregressive models, Markov Chain Monte Carlo (MCMC), Gibbs Sampler.

1 Introduction

In nonlinear time series analysis a Threshold Autoregressive model (TAR) is considered as an approximation to a *nonlinear autoregressive process*(NLAR). A TAR model is a piece-wise linear model over the state space. It is linear in the space of the thresholds. TAR models are applicable to a variety of time series such as Biological, Physical, Economical and etc. (see Bhansali,1980).

In this article we confine our studies to a class of TAR models which is called Self-Exciting Threshold Autoregressive (SETAR). SETAR models were proposed by Tong(1978) as an alternative model for describing periodic time series. The full details regarding these models can be obtained from (Tong, 1990).

The first step in fitting a SETAR model to a data-set is to decide the number of regimes which will be denoted by l. The second step is to identify the delay, the threshold(thresholds) and the order of each regime, which the first two in this step, play a key role in the nonlinear nature of the model. The initial step in identification of the thresholds is the selection of the delay parameter which sometimes is called threshold lag and will be denoted by d(a positive integer). In the literature Akaike's Information Criterion (AIC) is used, to select values for l, d and the order in each regime. Our approach is to estimate automatically, the values of l, d, the threshold(thresholds) which we denote it by r and the order of each regime for this class of threshold models in a Bayesian approach via MCMC methods. Model selection is done through calculating mean of the root mean square error of forecasts(MRMSE). The order of each regime will be denoted by $k_j (j = 1, ..., l)$.

Recent Bayesian studies of the d and r for this class of models in the literature are Geweke and Terui(1993) and McCulloch and Tsay(1994). Chen and Lee(1995) in a Bayesian simulation study focused on the posterior distribution of a two-regime SETAR model and applied MCMC methods to estimate all parameters of the model given $k_j (j = 1, 2)$. We employ a hybrid Gibbs sampler to generate samples from the joint posterior density of d, $r = \{r_1, r_2, ..., l-1\}$ and $k = \{k_1, k_2, ..., k_l\}$. The prior densities for d, $r_i \in r \ \ (i = 1, 2, ..., l-1)$ and $k_j \in k, \ \ (j = 1, 2, ..., l)$ are assumed uniform densities.

In section 2 SETAR models are introduced and their Bayesian analysis is discussed. In section 3 the algorithm of the automatic parameter estimation is illustrated.

2 Bayesian analysis of Self-exciting Threshold Autoregressive models(SETAR)

A time series y_t is a SETAR process if it follows the model

$$y_t = \alpha_0^{(j)} + \sum_{i=1}^{k_j} \alpha_i^{(j)} y_{t-1} + \epsilon_t^{(j)} \quad \text{if} \quad r_{j-1} < y_{t-d} \leq r_j \qquad (1)$$
$$j = 1, 2, ...l \qquad -\infty = r_0 < r_1 < .... < r_{l-1} < r_l = \infty$$

where $\alpha_i^{(j)}, r_j \in R$(set of real numbers) and d is a positive integer($d \in \{1, 2, ..., \max k_j\}$). We assume $\epsilon_t^{(j)}$ is normal with mean zero and standard deviation σ_j and denote a SETAR with l regimes as SETAR($l; k_1, ..., k_l$). If $l = 2$ then r is a scalar, for $l > 2$, r is a vector.

Assume $y^0 = \{y_0, y_{-1}, ..., y_{1-\max\{k_1, k_2, ..., k_l\}}\}$ as an initial known condition. Let d, r and $k_j (j = 1, 2, ..., l)$ be fixed. We divide the data set $y = \{y_1, y_2, ..., y_T\}$ (T is the actual number of time series) into subsets $Z_1, Z_2, ..., Z_l$ such that

$$Z_1 = \{y_t; y_{t-d} \leq r_1\}, Z_2 = \{y_t; r_1 < y_{t-d} \leq r_2\}, ..., Z_l = \{y_t; y_{t-d} > r_{l-1}\} \qquad (2)$$

where the number of observations in $Z_1, Z_2, ..., Z_l$are $T_1, T_2, ..., T_l$,respectively. $T = T_1 + T_2 + ... + T_l$ and also the elements in the subsets are ordered.

The $SETAR(l; k_1, k_2, ..., k_l)$ can be written in the following matrix form

$$\begin{cases} Z_1 = X_1 \alpha_1 + \epsilon_1 \\ Z_2 = X_2 \alpha_2 + \epsilon_2 \\ \cdots\cdots\cdots\cdots\cdots\cdots \\ \cdots\cdots\cdots\cdots\cdots\cdots \\ Z_l = X_l \alpha_l + \epsilon_l \end{cases} \qquad (3)$$

where $X_j (j = 1, 2, ..., l)$ is a $T_j \times k_j + 1$ matrix of lagged variables corresponding to $Z_j (j = 1, 2, ..., l)$: if row i of Z_j contains y_t, then row i and column $p + 1$ of X_j contains y_{t-p} with 1 in column 1. $\epsilon_j (j = 1, 2, ..., l)$ is a vector $T_j \times 1$ whose elements are corresponding innovations of each y_t in(1). Suppose $\tau = (d, r, k)$, $\alpha_j = (\alpha_0^{(j)}, \alpha_1^{(j)}, ..., \alpha_{k_j}^{(j)})$ and the prior distribution of parameters and variances conditional on τ is improper Jeffrey's prior,

$$p(\alpha_j, \sigma_j, j = 1, 2, ..., l|\tau) \propto 1/(\sigma_1 \sigma_2 ... \sigma_l). \qquad (4)$$

Then the posterior density of $(\alpha_j, \sigma_j, j = 1, 2, ..., l)$ conditional on τ and data is

$$\pi(\alpha_j, \sigma_j, j = 1, 2, ..., l|y, \tau) \propto \prod_{j=1}^{l} \sigma_j^{-(T_j+1)} (2\pi)^{-(T_j/2)} \exp\{-\frac{(Z_j - X_j\alpha_j)'(Z_j - X_j\alpha_j)}{2\sigma_j^2}\} \quad (5)$$

which can be written in the following form

$$\pi(\alpha_j, \sigma_j, j = 1, 2, ..., l|y, \tau) \propto \prod_{j=1}^{l} \sigma_j^{-(T_j+1)} (2\pi)^{-(T_j/2)} \exp\{-\frac{\nu_j s_j^2 + (\alpha_j - \hat{\alpha}_j)'X_j'X_j(\alpha_j - \hat{\alpha}_j)}{2\sigma_j^2}\} \quad (6)$$

where $\hat{\alpha_j} = (X_j'X_j)^{-1} X_j'Z_j$, $s_j^2 = (Z_j - X_j\hat{\alpha}_j)'(Z_j - X_j\hat{\alpha}_j)/\nu_j$, $\nu_j = T_j - k_j - 1$. Integrating over $\sigma = (\sigma_1, \sigma_2, ..., \sigma_l)$ on (6), we obtain

$$\pi(\alpha_j, j = 1, 2, ..., l|y, \tau) \propto \prod_{j=1}^{l} \sigma_j^{-(T_j/2-1)} (2\pi)^{-(T_j/2)} \Gamma(\tfrac{T_j}{2}) \{\nu_j s_j^2 + (\alpha_j - \hat{\alpha}_j)'X_j'X_j(\alpha_j - \hat{\alpha}_j)\}^{-(T_j/2)} \quad (7)$$

In applying the Gibbs sampler for estimation of the parameters of SETAR$(l; k_1, k_2, ..., k_l)$, conditional distributions of all elements of $(\alpha_j, \sigma_j, j = 1, 2, ..., l)$ are necessary. Hence, when the values of $k_1, k_2, ..., k_l$ are large, the convergence of the simulated chains to the posterior distribution will take a along time. In this case applying the *multi-state Gibbs sampler* will improve the convergence.

Carter and Kohan(1994) and Fruhwirth-Schnatter(1994) have independently suggested a direct implementation of multi-state sampling for time series models. The implementation is based on the identity:

$$\pi(\alpha_j|\sigma_j, y, \tau) = \pi(\alpha_{jk}|\sigma_j, y, \tau)\pi(\alpha_{jk-1}|\alpha_{jk}, \sigma_j, y, \tau) \pi(\alpha_{jk-2}|\alpha_{jk}, \alpha_{jk-1}, \sigma_j, y, \tau).......\pi(\alpha_{j0}|\alpha_{j,-(0)}, \sigma_j, y, \tau) \quad (8)$$

where $\alpha_{j,-(k)}$ is all of the elements of vector α_j except α_{jk}. It can easily be seen that

$$\alpha_j|\sigma_j, y, \tau \sim N(\hat{\alpha_j}, \sigma_j^2(X'X)^{-1}) \quad (9)$$

$$\sigma_j|\alpha_j, y, \tau \sim IG(T_j/2, \frac{\nu_j s_j^2 + (\alpha_j - \hat{\alpha}_j)'X_j'X_j(\alpha_j - \hat{\alpha}_j)}{2}) \quad (10)$$

where IG denotes an inverse Gamma density.

The joint posterior density of d, r, k conditional on data(for simplicity we suppress y^0) can be presented as follows. A uniform prior is chosen as prior for the orders.

$$f(d, r, k|y) \propto \prod_{j=1}^{l} 2^{-(\nu_j/2+1)} (\pi)^{-(\nu_j/2)} \Gamma(\frac{\nu_j}{2})(\frac{\nu_j s_j^2}{(\nu_j/2)} |X_j'X_j|^{-1/2} \quad (11)$$

where the vector $k = \{k_1, k_2, ..., k_l\}$ contains the order of regimes from 1 to l, respectively. The aim is to apply MCMC methods and to estimate values of d, r, k. Closed form of the conditional posterior density of r given other parameters is not available. So it is not possible to apply the Gibbs sampler, directly. Closed form of the conditional posterior probability function of each order or the delay given other parameters is available. Therefore the sampling strategy should be based on a hybrid Gibbs sampler(see Smith and Roberts,1993). The probability functions involved in this sampling strategy are:
(i) The conditional posterior density of r

$$f(r|d,k,y) \propto \prod_{j=1}^{l} 2^{-(\nu_j/2+1)}(\pi)^{-(\nu_j/2)}\Gamma(\frac{\nu_j}{2})(\frac{\nu_j s_j^2}{2})^{-(\nu_j/2)}|X_j' X_j|^{-1/2} \quad (12)$$

where $k = (k_1, k_2, ..., k_l)$ is the vector of orders.
(ii) The conditional posterior probability function of d

$$f(d|r,k,y) = \frac{\prod_{j=1}^{l} \Gamma(\frac{\nu_j}{2})(\frac{\nu_j s_j^2}{2})^{-(\nu_j/2)}|X_j' X_j|^{-1/2}}{\sum_{d=0}^{I}\prod_{j=1}^{l} \Gamma(\frac{\nu_j}{2})(\frac{\nu_j s_j^2}{2})^{-(\nu_j/2)}|X_j' X_j|^{-1/2}} \quad (13)$$

where $d = 0, 1, ..., I$(I is an integer, denotes the maximum number of the delay).
(iii) The conditional joint posterior probability function of orders

$$f(k|r,d,y) = \frac{\prod_{j=1}^{l} \Gamma(\frac{\nu_j}{2})(\frac{\nu_j s_j^2}{2})^{-(\nu_j/2)}|X_j' X_j|^{-1/2}}{\sum_{k_1=0}^{n_1}\sum_{k_2=0}^{n_2} \cdots \sum_{k_l=0}^{n_l}\prod_{j=1}^{l} \Gamma(\frac{\nu_j}{2})(\frac{\nu_j s_j^2}{2})^{-(\nu_j/2)}|X_j' X_j|^{-1/2}} \quad (14)$$

In practice we assume $0 \le k_1 \le n_1$, $0 \le k_2 \le n_2, ..., 0 \le k_l \le n_l$ where $n_1, n_2, ..., n_l$ are maximum order of each regime and their values are less than the actual number of the time series in each regime. It worth to mention that we derive the joint posterior probability function of orders, because it will make the convergence of the sampler faster in comparison with using the full conditional of each order.

2.1 Model Selection

With the improper priors on the parameters, model selection criterion such as Bayes factor, that involve ratios of predictive densities, lack interpretability. A solution to this problem is through the use of cross-validation predictive density and the forecasting predictive density(see Gelfand and Dey, 1993). We take a simple approach and compute mean of the root mean square error(MRMSE) for each model, using a preserved portion of data after time T which is not involved in the estimation.

If we update the model with an additional observation and generate the next h step forecast, and repeat this exercise n times, then we may estimate the mean square error (RMSE) for the lead period h as

$$RMSE(h) = \sqrt{n^{-1}\sum_{i=1}^{n}(y_{T+h+i-1} - z_{T+h+i-1})^2} \qquad h = 1, 2, ..., H \quad (15)$$

where z_{T+h} is the forecast corresponding to the actual value y_{T+h} and recursively could be generated (see,Geweke and Terui,1993). The following equation is the proposed model selection criterion

$$MRMSE(h) = \frac{1}{H}\sum_{h=1}^{H}\sqrt{n^{-1}\sum_{i=1}^{n}(y_{T+h+i-1} - z_{T+h+i-1})^2} \qquad (16)$$

where H is the last step of consecutive steps ahead.

3 Algorithm

It has been assumed that the joint prior distribution of d, r, k is flat. A discrete uniform distribution can be chosen as a candidate-generating distribution for r. r can be either a scalar or a vector, if it is a scalar the sampling from the distribution of r is straightforward. If r is a vector with $l-1$ elements($l > 2$), in order to draw samples from the joint density of thresholds, first it can be partitioned into the univariate distributions as

$$f(r_1, r_2, ..., r_{l-1}) = f(r_{l-1}).f(r_{l-2} \mid r_{l-1})......f(r_1 \mid r_2, ..., r_{l-1}) \qquad (17)$$

and then dependent random samples could be drawn by the methods in Devroye(1986).

In brief the algorithm proceeds as follows:

Stage 1

- Step(1): set d, r and$k$ to the reasonable initial values.
- Step(2): draw r from a candidate-generating density and accept or reject it via the Metropolis steps using (12).
- Step(3): draw d from (13) by the inversion method.
- Step(4): draw k from (14).
- Step(5): repeat step(2), step(3) and step(4) for t times, then control the convergence of the chain(chains) by a suitable convergence diagnostics method. If the convergence is diagnosed go to step(6), otherwise go to step(5).
- Step(6): to take m samples from the joint posterior distribution of d, r, k, repeat step(2), step(3) and step(4) m times(some gaps before taking each sample is necessary).

Stage 2

Using multi-state Gibbs sampler idea, instead of sampling from $\sum_{j=1}^{l}(k_j + 2)$ full conditionals, we only sample from $2l$ full conditional distributions. Evaluation of the posterior distribution can be done in the following way,

$$\pi(\alpha, \sigma|y, \tau) = f(\sigma|y, \tau).f(\alpha|\sigma, y, \tau) \tag{18}$$

where

$$\frac{1}{\sigma_j^2}|y, \tau \sim G(\nu_j/2,\ 2/\nu_j s_j^2)$$

and G denotes a Gamma distribution.

To get the estimate of $\sigma_j, \alpha_j, j = 1, 2, ..., l$, the algorithm proceeds as follows:

- Step(7): set $\alpha_j = \alpha_0, \sigma_j = \sigma_0, \quad j = 1, 2, ..., l$, where α_0 and σ_0 are arbitrary values (for each j, these values can be different).
- Step(8): draw samples from $\alpha_j|\sigma_j, y, \tau$ and $\sigma_j|\alpha_j, y, \tau$ for $j = 1, 2, ..., l$ in turn.
- step(9): repeat step(2) for t times and control the convergence by a suitable convergence diagnostic tool, until the convergence is diagnosed.
- step(10): take the desired number of samples in order to estimate the parameters.
- step(11): compute model selection criterion.

The algorithm has been successfully applied to some examples, but we skip these for the sake of page limitation.

A computer code is provided in **R** environment to test the fitness of a SETAR model to a dataset using the proposed algorithm.

References

Bhansali, R.J. (1980). Threshold autoregression, limit cycles and cyclical data (with discussion. *Journal of the Royal Statistical Society, Series B*, **42**, 245-292.

Carter, C.K. and Kohan, R. (1994). On Gibbs sampling for state space models.*Biometrica*, **81**(3), 105-114.

Chen, Cathy W.S. and Lee, J.C. (1995). Bayesian Inference of Threshold Autoregressive Models, *Journal of the time series analysis*, **16**, 5, 483-492.

Devroye, L. (1986). *Non-Uniform Random Variate Generation*. Springer-Verlag, New York, Berlin, Heidelberg, Tokyo.

Fruhwirth-Schnatter, S. (1994). Data Augmentation and dynamic linear models. *Journal of time series analysis*, **15**(2), 183-202.

Gelfand, A. E. and Dey, D. K. (1993). Bayesian Model choice: Asymptotics and exact calculations, *Journal of the Royal Statistical Society, B*,**55**,501-514.

Geweke, J. and Terui, N. (1993). Bayesian Threshold Autoregressive Models for Nonlinear Time Series. *Time Series Analysis*, **14**,5, 441-454.

McCulloch, R.E. and Tsay, R.S. (1994). Bayesian analysis of autoregressive time series via the Gibbs sampler. *Journal of Time Series Analysis*, **15**,2, 235-250.

Smith,A.F.M. and Roberts,G.O. (1993). Bayesian computation via the Gibbs sampler and related Markov Chain Monte Carlo methods. *Journal of the Royal Statistical Society series B*, **55**, 3-24.

Tong,H. (1978). On a threshold model. *In Chan C. H. (ed.),Pattern Recognition and Signal Processing,Netherlands:Sijthoff and Noordhoff, Amsterdam*

Tong, H. (1990). *Non-linear time series a dynamical system approach*, Oxford Science Publications.

Bayesian Semiparametric Seemingly Unrelated Regression

Stefan Lang, Samson B. Adebayo and Ludwig Fahrmeir

Department of Statistics, University of Munich, Ludwigstr. 33,
80539 Munich, Germany.

Abstract
Parametric seemingly unrelated regression (SUR) models are a common tool for multivariate regression analysis when error variables are reasonably correlated.

A weakness of parametric models is that they require strong assumptions on the functional form of possibly nonlinear effects of metrical covariates. In this paper, we develop a semiparametric SUR model based on Bayesian P-splines. Inference is fully Bayesian and uses recent Markov chain Monte Carlo techniques.

Keywords Bayesian semiparametric models, correlated responses, MCMC, P-splines.

1 Introduction

Multivariate regression analysis is needed in many applications. Neglecting existing association between response variables can lead to biased and inefficient estimation of covariate effects. Correspondingly, analysis of correlated response data has received considerable attention and interest. For multivariate Gaussian response, the parametric seemingly unrelated regression (SUR) model (Zellner, 1962) is a standard tool in econometrics.

As in univariate regression, the assumption of a parametric linear predictor for assessing the impact of covariates on responses is often too restrictive in realistically complex situations. For univariate responses, the development of more flexible non- and semiparametric regression models has been a main topic of recent research, see e.g. Fahrmeir and Tutz (2001, ch. 5) for an introductory survey.

Astonishingly, there is a distinct lack of non- and semiparametric regression models for truly multivariate responses, in particular for seemingly unrelated regression. A notable exception is the Bayesian approach of Smith and Kohn (2000).

In this paper, we present an approach to semiparametric seemingly unrelated regression (SSUR) based on Bayesian P-splines. It extends previous work of Lang & Brezger (2001) for univariate additive models to the present situation. Inference is fully Bayesian and relies on efficient MCMC techniques.

2 Bayesian SSUR Regression

2.1 Observation model

Suppose that regression data consist of observations $y_i = (y_{i1}, ..., y_{ik})'$, $i = 1, \ldots, n$, on a multivariate response y and on covariates. We distinguish between a vector $x_{ir} = (x_{ir1}, ..., x_{irp_r})'$ of covariates, whose influence on the r th component of y_i will be modeled nonparametrically, and a further vector $v_{ir} = (v_{ir1}, ..., v_{irq_r})'$ of covariates, whose effect is modeled in the common usual form. For each component y_{ir}, $r = 1, ..., k$, of the response we assume a semiparametric regression model

$$y_{ir} = \eta_{ir} + \varepsilon_{ir}, \quad i = 1, \ldots, n, \tag{1}$$

with additive predictors

$$\eta_{ir} = f_{r1}(x_{ir1}) + \ldots + f_{rp_r}(x_{irp_r}) + v_{ir}'\gamma_r. \tag{2}$$

The functions f_{rj} are possibly nonlinear functions of the covariates. Type and degree of smoothness is controlled by priors described in the following section. The linear combination $v_{ir}'\gamma_r$ corresponds to the usual parametric part of the predictor, including an intercept term. Note that the mean levels of the unknown functions are not identifiable. To ensure identifiability, the functions are constrained to have zero means. The errors $\varepsilon_i = (\varepsilon_{i1}, \ldots, \varepsilon_{ik})'$, $i = 1, \ldots, n$, are assumed to be i.i.d. multivariate Gaussian with mean zero and a covariance matrix Σ, i.e. $\varepsilon_i | \Sigma \sim N(0, \Sigma)$.

2.2 Prior specifications

Priors for the unknown (possibly nonlinear) functions f_{rj} of covariates x_{rj} are based on Bayesian P-splines (Lang and Brezger, 2001). For notational convenience, we omit indices and illustrate the approach for a specific function f of a covariate x.

The basic assumption behind the P-splines approach is that the unknown smooth function f can be approximated by a spline of degree l defined on a set of equally spaced knots $\zeta_0 = x_{min} < \zeta_1 < \ldots < \zeta_{s-1} < \zeta_s = x_{max}$ within the domain of x. It is well known (de Boor, 1978) that such a spline can be written in terms of a linear combination of $m = s + l$ B-spline basis functions B_t, i.e.

$$f(x) = \sum_{t=1}^{m} \beta_t B_t(x).$$

A crucial point with splines is the choice of the number and also the position of the knots. For a small number of knots the resulting function space may be not flexible enough to capture the variability of the data. For a large number of knots estimated curves may tend to overfit the data. As a remedy Eilers and Marx (1996) suggest a moderately large number of knots (usually between 20 and 40) to ensure enough flexibility, and to define a roughness penalty based on differences of adjacent regression coefficients to guarantee sufficient smoothness of the fitted curves. In our Bayesian approach we replace difference penalties by their stochastic analogues, i.e. first or second order random walk models for the regression coefficients

$$\beta_t = \beta_{t-1} + u_t, \quad \beta_t = 2\beta_{t-1} - \beta_{t-2} + u_t \tag{3}$$

with Gaussian errors $u_t \sim N(0, \tau^2)$ and diffuse priors $\beta_1 \propto const$, or β_1 and $\beta_2 \propto const$, for initial values, respectively. A first order random walk penalises abrupt jumps $\beta_t - \beta_{t-1}$ between successive states and a second order random walk penalises deviations from the linear trend $2\beta_{t-1} - \beta_{t-2}$. Note that the priors (3) can equivalently be rewritten in form of a global smoothness prior

$$\beta|\tau^2 \propto exp\left(-\frac{1}{2\tau^2}\beta' K \beta\right), \tag{4}$$

with appropriate penalty matrix K.

The variance parameter τ^2 is the equivalent to the smoothing parameter in a frequentist approach and controls the trade off between flexibility and smoothness. In order to be able to estimate the "smoothing parameter" τ^2 simultaneously with β, a highly dispersed but proper hyperprior is assigned to it. We choose an inverse gamma distribution with hyperparameters a and b, i.e. $\tau^2 \sim IG(a, b)$.

In the absence of any prior knowledge a natural assumption for fixed effects parameters are independent diffuse priors, i.e. $p(\gamma_{rj}) \propto const.$

For the covariance matrix Σ of the errors, we choose an inverse Wishart prior $\Sigma \sim IW(A, B)$ where A is a scalar and B is a k x k symmetric and positive definite matrix. A standard choice for A is 1 and $B = diag(c, \dots, c)$ with a small c, e.g. $c = 0.005$.

We complete the Bayesian model specification by the assumption that priors for function evaluations, fixed effects parameters and for variances are all mutually independent.

3 Bayesian Inference through MCMC

Inference is fully Bayesian and uses MCMC simulation, drawing from full conditionals of single parameters or blocks of parameters given the rest and the data. Some matrix notation will be introduced for deriving full conditionals. Let $y_{.r} = (y_{1r}, \dots, y_{nr})'$ and $\eta_{.r} = (\eta_{1r}, \dots, \eta_{nr})'$ denote the vector on the r th response variable and the corresponding vector of predictors. Then the additive predictors (1) can be written as

$$\eta_{.r} \; = \; \sum_{j=1}^{p_r} X_{rj}\beta_{rj} \; + \; V_r\gamma_r \tag{5}$$

where β_{rj} is the vector of regression parameters for function f_{rj} and X_{jr} is a design matrix whose columns consist of the B-spline basis functions evaluated at the observations. The matrix V_r is the usual design matrix for fixed effects with rows v'_{ir}, i=1,...,n.

Let $\beta = (..., \beta'_{rj}, ...)'$ denote the stacked vector of all regression parameters, $\tau^2 = (..., \tau^2_{rj}, ...)'$ the vector of corresponding variances and $\gamma = (\gamma'_1, ..., \gamma'_r)'$ the stacked vector all fixed effects parameters. Posterior analysis is then based on

$$p(\beta, \tau^2, \gamma, \Sigma | y) \propto \prod_{i=1}^{n} p(y_i | \eta_i, \Sigma) \prod_{r=1}^{k} \prod_{j=1}^{p_r} \left(p(\beta_{rj} | \tau^2_{rj}) p(\tau^2_{rj})\right) p(\Sigma), \qquad (6)$$

where $p(y_i|\eta_i)$ is given by the Gaussian observation model for observation $y_i = (y_{i1}, ..., y_{ik})'$ and predictor $\eta_i = (\eta_{i1}, \ldots, \eta_{ik})'$.

MCMC simulation is carried out by drawing from full conditionals for the blocks β_{rj}, γ_r, τ^2_{rj} and Σ. They are given as follows:

(i) The full conditional for β_{rj} is Gaussian, $\beta_{rj}|\cdot \sim N(\mu_{rj}, P^{-1}_{rj})$, with precision matrix and mean given by

$$P_{rj} = \frac{X'_{rj} X_{rj}}{\sigma^2_{r|-r}} + \frac{K_{rj}}{\tau^2_{rj}}, \quad \mu_{rj} = P^{-1}_{rj} \left(\frac{1}{\sigma^2_{r|-r}} X'_{rj} (y_{.r} - o_{.r}) \right). \qquad (7)$$

In (7), $\sigma^2_{r|-r}$ is the (conditional) variance $\sigma^2_{r|-r} = \sigma^2_r - \Sigma_{r,-r} \Sigma^{-1}_r \Sigma'_{r,-r}$, derived from partitioning Σ into

$$\Sigma = \begin{pmatrix} \sigma^2_r & \Sigma_{r,-r} \\ \Sigma'_{r,-r} & \Sigma_r \end{pmatrix},$$

(after reordering for the r th component of the error variable). The vector $o_{.r}$ in (7) is an offset vector. The i th component o_{ir} of the offset vector $o_{.r}$ is given by

$$o_{ir} = \Sigma_{r,-r} \Sigma^{-1}_r (y_{i,-r} - \eta_{i,-r}) + \tilde{\eta}_{ir}, \qquad (8)$$

where $y_{i,-r}$ and $\eta_{i,-r}$ are obtained from y_i and η_i by omitting the r th components of y_i and η_i, respectively. The working predictor $\tilde{\eta}_{ir}$ is obtained from η_{ir} by deleting the j th effect f_{rj}.

(ii) The full conditional for γ_r is Gaussian, $\gamma_r|\cdot \sim N(\mu_{\gamma_r}, P^{-1}_{\gamma_r})$, with

$$P_{\gamma_r} = \frac{1}{\sigma^2_{r|-r}} V'_r V_r, \quad \mu_{\gamma_r} = (V'_r V_r)^{-1} V'_r (y_{.r} - o_{.r}). \qquad (9)$$

where the offset $o_{.r}$ is defined in (8) and $\tilde{\eta}_{ir}$ now obtained from η_{ir} by deleting the linear fixed effects term $v'_{ir}\gamma_r$.

(iii) Full conditionals for the variance parameters τ^2_{rj} are inverse Gamma distributions with parameters

$$a'_{rj} = a_{rj} + \frac{rank(K_{rj})}{2}, \qquad b'_{rj} = b_{rj} + \frac{1}{2} \beta'_{rj} K_{rj} \beta_{rj}. \qquad (10)$$

(iv) The full conditional for Σ is an inverse Wishart distribution with parameters

$$A' = A + \frac{n}{2}, \qquad B' = B + \frac{1}{2}\sum_{i=1}^{n}(y_i - \eta_i)(y_i - \eta_i)'. \quad (11)$$

All full conditionals involved have known distributions, hence Gibbs sampling can be used to update the parameters of the model. Computational efficiency is guaranteed by Cholesky decompositions for band matrices.

4 Simulation Study

To investigate how well our SSUR method performs, we carried out a simulation study based on a model used also by Smith and Kohn (2000). Our model includes the same four functions with $n = 100$ observations as in Smith and Kohn (2000), see Figure 1 (solid lines) for graphical displays of the functions. Because the covariance matrix $cov(\varepsilon) = \Sigma$ reported in Smith and Kohn (2000) turned out not to be positive definite, we chose

$$\Sigma = \begin{pmatrix} 1 & 0.7 & 0.6 & 0.9 \\ & 1 & 0.7 & 0.9 \\ & & 1 & 0.7 \\ & & & 1 \end{pmatrix}.$$

We generated 250 replications of the model.

Figure 1 a) - d) shows posterior mean estimates of f_1-f_4 averaged over the 250 replications for both multivariate fits (left panels) and the corresponding separate univariate fits (middle panels). The right panels display boxplots of $\log(MSE)$ for multivariate and univariate fits respectively.

As we can see, fitting a $SSUR$ model is worthwhile when the responses are correlated rather than ignoring the existing correlations. There is a significant and pronounced bias in the univariate estimates of all functions, except for f_3 (the linear function). The boxplots of f_3 (Figure 1 c) reveals that $SSUR$ is still more efficient than the corresponding univariate model even though the two point estimates are indistinguishable.

We finally note that estimates of the covariance matrix Σ are more or less unbiased.

References

de Boor (1978): *A Practical Guide to Splines.* Springer-Verlag, New York.

Eilers, P. and Marx, B. (1996): Flexible Smoothing using B-splines and Penalized Likelihood. *Statistical Science*, 11, 89-121.

Fahrmeir, L. and Tutz, G. (2001): *Multivariate Statistical Modelling based on Generalized Linear Models (3rd ed).* Springer, New York.

Lang, S. and Brezger, A. (2001): Bayesian P-Splines. SFB 386 discussion paper 236, University of Munich.

Smith, M. and Kohn, R. (2000): Nonparametric Seemingly Unrelated Regression. *Journal of Econometrics*, 98, 257-281.

Zellner, A. (1962): An Efficient Method of Estimating Seemingly Unrelated Regression Equations and Tests for Aggregation Bias. *Journal of the American Statistical Association*, 57, 500-509.

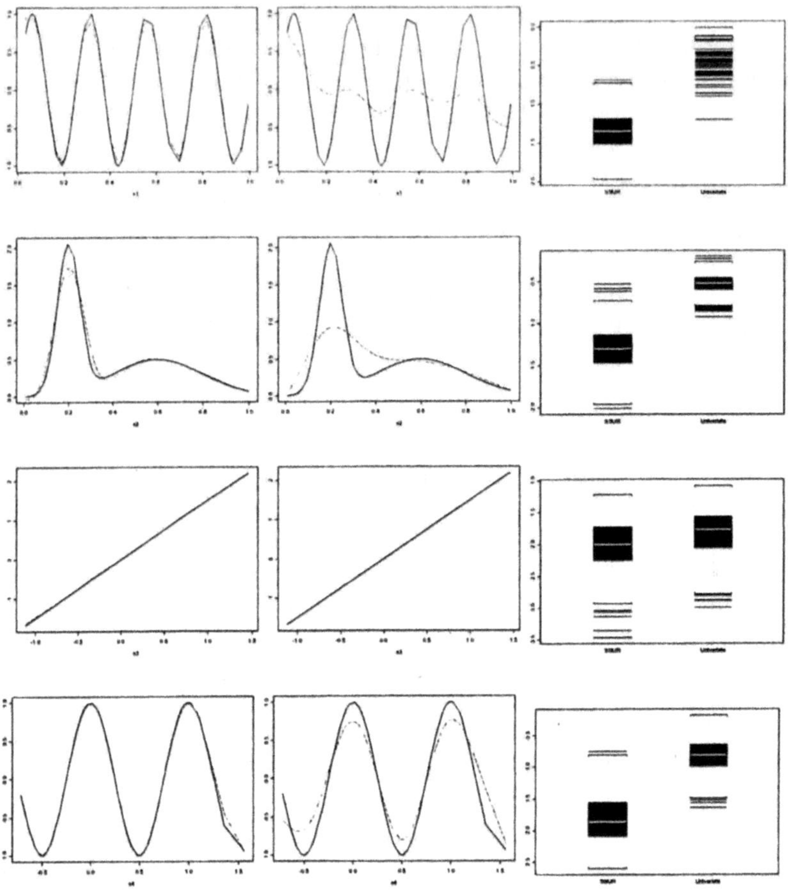

Fig. 1. *Posterior mean estimates (dashed lines) of nonparametric functions f_1-f_4 averaged over the 250 repliactions. Left panels display SSUR estimation results, middle panels the corresponding univariate regressions ignoring correlations. For comparison the true functions are included (solid lines). The right panels display boxplots of log(MSE) for SSUR and univariate regressions, respectively.*

Blockmodeling Techniques for Web Mining

Gabriella Schoier

Dipartimento di Scienze Economiche e Statistiche, Università di Trieste, Piazzale Europa 1, 34127 Trieste, Italy

Abstract. The aim of this paper is to introduce the concept of blockmodeling to Web data (log files), in order to obtain clusters of units (visited pages) which have similar patterns of relationships and to interpret the pattern of relationships among clusters. Doing this we can study the behaviour of the users (sessions) according to the relations among the visited pages. To explain these concepts an application to a Web site is performed.

Keywords. Web Mining, Web Usage Mining, blockmodeling, cluster analysis, log files

1 Introduction

Web Mining is the application of Data Mining techniques to discover meaningful patterns, profiles, trends, from Web resources (see e.g. Srivastava *et al.* (2000)). It is crucial to E-Commerce success. E-intelligence enables businesses to customize Web relationships, to optimise profitability by predicting customer behaviour, to deliver that information to all relevant points of contacts, to provide the tools to efficiently run the Web business operations. Web Mining can be divided into three components: Web Content Mining, Web Structure Mining and Web Usage Mining.
The aim of this paper is to understand how an Internet user goes through the pages of a Web site. This is important both for user personalization and for the insertion of targeted e-mail and banners. For this purpose we will consider Web Usage Mining, which is the process of discovering user access patterns or user habits. Recently referrer logs, which collect information about referring pages for each reference and user registration, have also been included in it (see e.g. Mobasher *et al.* (2000)).
The new idea is the introduction of the concept of blockmodeling to Web data, in particular to log files. Blockmodeling seeks to cluster units which have substantially similar patterns of relationship and to interpret the pattern of relationships among clusters. Usually it is applied in social network analysis, the main goal of which is the detection and interpretation of patterns of relations between actors (see e.g. Batagelj *et al.* (1992), Ferligoj *et al.* (1996), Wasserman and Faust, K., (1994)).
By using this technique in our case we can, not only develop profiles of items with similar characteristics, but also find relations between the items. In this way we can study the behaviour of the users according to the relations among the visited pages and not only consider the page with its characteristics disjoined from the other

pages.
The application regards the site *www.girotondo.com*, which is a site for children.

2 Blockmodeling

Let $E = \{x_1, x_2, ..., x_n\}$ be a finite set of units or *actors*, these are related by binary relations: $R_t \subseteq E \times E$ (for $t = 1,...,r$) which determine a *network* (set of units and relation(s) defined over it) *N*: $N = (E, R_1, R_2, ..., R_r)$.
In the following we will consider a single relation *R* described by a corresponding binary matrix: $\mathbf{R} = \{r_{ij}\}$ where $r_{ij} = \begin{cases} 1. \text{ if } x_i \, R \, x_j \\ \text{otherwise} \end{cases}$

In the case of Web data the actors are the visited pages while the relation between two pages is to share common users.
Given a network, one of the main procedural goals of *blockmodeling* is to identify clusters of units that share structural characteristics defined in terms of a relation *R*. The units within a cluster have the same or similar connection patterns to other units. They form a clustering $C = \{C_1, C_2, ..., C_k\}$ which is a partition of the set *E*:

$$\bigcup_i C_i = E \quad \text{and} \quad \text{for } \forall i \neq j \Rightarrow C_i \cap C_j = \varnothing .$$

A clustering *C* partitions also the relations *R* into blocks each of which is defined by the units belonging to clusters C_i and C_j in terms of the arcs leading from cluster C_i to cluster C_j.
Each partition determines an *equivalence relation* (and vice versa).
The problem of establishing a partition of units in a network in terms of a considered equivalence may be viewed as a special case of a clustering problem, where the optimal value of a criterion function has to be found. The criterion function can be constructed indirectly as a function of compatible (dis)similarity measure between pairs of units, or directly as a function measuring the fit of a clustering to an ideal one with perfect relations within each cluster and between clusters according to the considered types of equivalence. In this paper we have chosen the indirect approach.
Different definitions of equivalence lead to distinct partitions of the units. One of the basic is the so-called *structural equivalence*: equivalent units have the same (or similar) connection patterns to the same neighbour. In terms of sociometrics notation, letting r_{ijt} the presence or absence of a tie from actor *i* to actor *j* on relation R_t, then actors *i* and *j* are structurally equivalent if:

$$r_{iht} = r_{jht} \text{ and } r_{hit} = r_{hjt} \text{ for } \forall h \neq i, j; \quad i,j,h = 1,2,...n; \quad t = 1,2,...,r,$$

in our case $t=1$.
The formal property of structural equivalence can be used to define a partition of actors in a social network into equivalence classes, called positions, each position contains actors who relate in similar ways to and from other actors in the network. We can interpret the result of a positional analysis when it is in the form of a *blockmodel*. A blockmodel consists of two things: a partition of actors into *B*

discrete subsets called positions: $\mathcal{B}_1, \mathcal{B}_2, \ldots, \mathcal{B}_B$ and, for each pair of position, a statement of the presence or absence of a tie b_{klt} $k,l = 1,2,\ldots,B$ and $t = 1,2\ldots,r$, within or between the positions on each of the relations. A blockmodel is also represented by an *image matrix*: $\mathbf{B} = \{b_{klt}\}$ for $k,l = 1,2,\ldots,B$ and $t = 1,2\ldots,r$, Each of the entries b_{klt} of the image matrix is called a *block*. Each block corresponds to a submatrix of the original sociomatrix that contains the relevant interposition or intraposition ties. A block containing a *1* is called *oneblock* while a block containing a *0* is called *zeroblock*. There are several criteria useful for deciding whether a block should be coded as a zeroblock or a oneblock, among these we have used the *α-density criterion* according to which if the observed block density is greater than or equal to α then the block is coded as one block otherwise it is coded as zeroblock, one guideline for choosing α is that it should depend on the density of the relations in the analysis (see Wasserman et. al. (1994)).
An image matrix can be represented by a *reduced graph* in which the nodes correspond to the positions while the lines or arcs correspond to the relations between the positions.

3 An Application

The data, which have been considered, are the log files of a Web site for children, called *Girotondo*, of the NE&T s.r.l., a firm of the Area Science Park of Trieste (Italy). In this site there are seven different sessions: *Bacheca, Corso, Favolando, Giochi, Links, News, Percome* , it has 362 jhtml pages.
The period of observation is from the 29/11/2000 to the 18/01/2001. After the pre-processing of the data the observations, formed by the IP address and by the address of the page, are jointed on the base of the IP address. We obtain a matrix every row of which represents a session and every column of which is associated with a page (the values are one if the page has been visited during the session, zero otherwise) (see Table 3.1 as an example).

Table 3.1. Matrix users by pages

IP ADDRESS	PAG1	PAG2	PAG3	PAG4	PAG5
138.222.202.11	1	0	0	0	0
151.15.169.130	1	0	0	0	0
151.2.15.154	0	0	0	0	0
151.20.111.0	1	0	0	0	0
151.20.143.184	1	0	0	0	0
151.20.9.10	1	0	0	0	0
151.21.44.178	1	0	0	0	0

We have used the program Pajek. It is a free program available at the address: http://vlado.fmf.uni-lj.si/pub/networks/pajek/ .This program has been built for the analysis of large networks which may be found in different areas.

The data of Table 3.1 represent a 2-mode network (a network formed by two distinct set: IP address and pages), we turned it into a 1-mode network, so we obtain a matrix pages by pages (see Table 3.2).

Table 3.2. Matrix pages by pages

	PAG1	PAG2	PAG3	PAG4	PAG5
PAG1	0	1	1	1	0
PAG2	1	0	1	1	1
PAG3	1	1	0	1	1
PAG4	1	1	1	0	1
PAG5	0	1	1	1	0
PAG6	0	0	0	0	0
PAG7	0	0	0	0	0

Having common users represents the relation between two pages, it is a relation that has the properties of the structural equivalence on the base of which the blockmodel is built and the interpretation of the results is made.
The data are compatible with the notion of structural independence, as a measure of it we use, according to Wasserman (1994), the Euclidean distance. It is possible to look at the relations between the units: if a box is black there are at least two users in common between the Web pages (see Table 3.3).

Table 3.3. Initial network

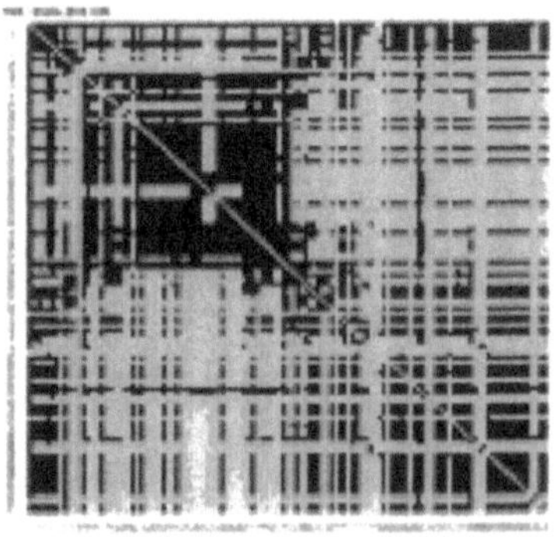

At this point we have applied a cluster analysis, the results suggest to choose six clusters. From that we have constructed the blocks. After applying the permutation, which reflects the distribution of the pages according to the initial configuration, we obtain the figure on the left of Table 3.4. To this subdivision correspond an error equal to 1310 (this means how many times there is a 1 instead of a 0 and vice versa). Now the optimization algorithm has been applied (see Ferligoj *et al.*

(1996)): if in the neighbours (obtained by two different transformations: transitions and transpositions) of the initial partition C there is another partition C' for which the criterion function assumes a small value then this new partition is considered (see the figure on the right of Table 3.4), the final error is 676 (this means how many times there is a 1 instead of a 0 in the matrix respect to the ideal blockmodel).

Table 3.4. Partitions

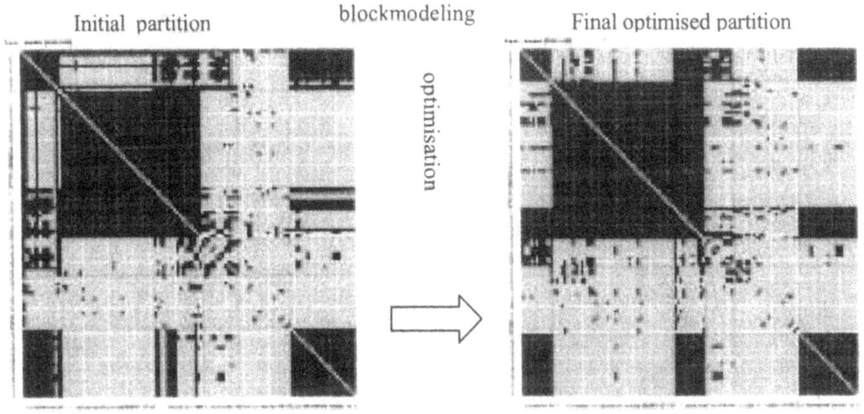

This result may be represented by the image matrix (see Table 3.5) where the positions are the Web pages:

Table 3.5. Image matrix of the final optimised partition

	$\mathcal{B}_1$	$\mathcal{B}_2$	$\mathcal{B}_3$	$\mathcal{B}_4$	$\mathcal{B}_5$	$\mathcal{B}_6$
$\mathcal{B}_1$	1	0	1	1	0	1
$\mathcal{B}_2$	0	1	1	0	0	0
$\mathcal{B}_3$	1	1	1	0	0	1
$\mathcal{B}_4$	1	0	0	0	0	0
$\mathcal{B}_5$	0	0	0	0	0	0
$\mathcal{B}_6$	1	0	1	0	0	1

$\mathcal{B}_1$: this group is formed by the most recent Web pages of the site, there are the last published tale and the last game introduced.

$\mathcal{B}_2$: this is the most numerous cluster, the pages are obtained through a random navigation.

$\mathcal{B}_3$: this group contains the pages, which have the highest number of users in common, in particular there is the highest value of the arcs, here there is also the homepage.
$\mathcal{B}_4$: the pages belonging to this cluster are characterized by having few users in common but a high number of connections with the first block.
$\mathcal{B}_5$: in this cluster the pages have no users in common and the same is for the pages, which are in the other blocks, these are the oldest pages
$\mathcal{B}_6$: in this cluster there are only the pages of the section *Favolando* in which the tales are collected.
The results in Figure 3.1 may be viewed by the reduced graph in which the connections between the clusters are very well observable.

Figure 3.1. Reduced graph

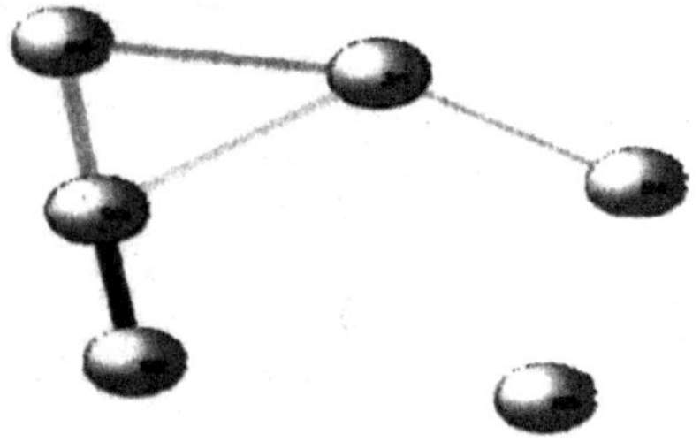

4 Conclusions

As one can see, the results are encouraging as regards the comprehension of the type of behaviour of the users of the site in terms of viewed pages. Future analysis will have to take into consideration the time spent on the singular page and the order of the visits of pages to understand the direction of the navigation through the site.

References

Batagelj, V., Ferligoj, A. and Doreian, P. (1992). Direct and indirect methods for structural equivalence. *Social Networks*, **14**,225-233.

Ferligoj, A., Doreian, P. and Batagelj, V. (1996). Optimizational Approach to Blockmodeling. *Journal of Computing and Information Technology*, **4**,63-90.

Mobasher, B., Doi, H., Luo, T.,Nakagawa, M., Sung, Y. and Wiltshire, Y. (2000). Discovery of Aggregate Usage Profiles for Web Personalization. *Conference on Knowledge Discovery in Databases*, http://www.maya.cs.depaul.edu/~mobasher/personalization.

Srivastava, J., Colley, R., Deshpande, M. and Ton, P. (2000). Web Usage Mining: Discovery and Applications of Usage Patterns from Web Data, http://www.maya.cs.depaul.edu/~mobasher/personalization.

Wasserman, S. and Faust, K. (1994). *Social Network Analysis: Methods and Applications*. New York: Cambridge University Press.

Bootstrapping Threshold Autoregressive Models

John Öhrvik[1] and Gabriella Schoier[2]

[1]Department of Biometry and Informatics, Swedish University of Agricultural Sciences, Box 7013, SE-750 07 Uppsala, Sweden
[2]Dipartimento di Scienze Economiche e Statistiche, Università di Trieste, Piazzale Europa 1, 34127 Trieste, Italy

Keywords. AIC, AIC_c , AIC_u , AR-sieve bootstrap, bootstrap model selection criteria, moving block bootstrap

1 Introduction

Threshold autoregressive (TAR) models have been widely used for periodic time series, as they are nonlinear models relatively simple to handle being linear in different regions of the state space, see e.g. Tong (1990). One of the main problems regards the decision of the selection of the correct order of a TAR model. This problem has been addressed by Tong (1983), Wong and Li (1998) and De Gooijer (2001). Wong and Li proposed the Akaike information criterion (AIC), the biased corrected version AIC_c, the AIC_u which is an approximately unbiased estimate of the Kullback-Leibler information and the Bayesian information criterion (BIC). De Gooijer proposed a cross validation criterion (CU_u^*) corresponding to AIC_u.

The main purpose of this paper is to study how bootstrap selection criteria perform. These criteria are based on a weighted mean of the apparent errors in the sample, and the average error rate obtained from bootstrap samples not containing the point being predicted. We also want to compare these new measures with the traditional ones based on AIC.

Bootstrap methods assume that we resample from independent identically distributed (i.i.d.) observations. This is usually not the case in time series analysis. To overcome this problem we have used different approaches. First a parametric based on the TAR model suggested by AIC_u , then a semiparametric - AR-sieve bootstrap motivated by Buehlmann (1997) and finally moving block bootstrap, see e.g. Kuensch (1989).

2 TAR models and information criteria

Real life systems are usually non linear and some characteristics such as limit cycles asymmetry, amplitude dependent frequency responses, jump phenomena and chaos cannot be captured by linear models. In recent years several non linear time series models have been proposed one of these is the class of threshold autoregressive models (TAR) introduced by Tong and Lim (1980) and Tong (1983). A TAR model is a piecewise linear process whose main aspect is to change the parameters of a linear autoregressive model according to the value of an observable variable. When this variable is a lagged value of the time series the model is called a self-exciting threshold autoregressive model (SETAR)

$$x_t = \phi_{0j} + \sum_{i=1}^{p_j} \phi_{ij} x_{t-i} + \varepsilon_{tj}, \qquad r_{j-1} \le x_{t-d} < r_j \tag{2.1}$$

where d (d>0) is a delay parameter, $-\infty = r_0 < r_1 < \ldots < r_l = +\infty$ are the threshold levels, ϕ_{ij} (j=0,...,l) are constants and $\{\varepsilon_{tj}\}$ (j=1,...,l) are white noise sequences with zero mean and finite variances $\sigma^2_{\varepsilon_j}$. We denote this SETAR(l; p_1,...,p_l).

One of the main problems in the analysis of SETAR models regards selecting the correct order (l; p_1,...,p_l) of the model. In practice the AIC information criteria is usually used, but it has been pointed out by Tong (1990) that one should be cautious in using it because, for this type of models, the exact likelihood function does not exist. Wong and Li have proposed the Akaike information criteria (AIC), the corrected version AIC$_c$, and the unbiased version AIC$_u$.

$$\mathrm{AIC(g)} = \sum_j \{n_j \ln \hat{\sigma}^2_{\varepsilon_j} + 2(p_j+1)\},\ \mathrm{AIC_c(g)} = \sum_j \left\{n_j \ln \hat{\sigma}^2_{\varepsilon_j} + \frac{n_j(n_j+p_j+1)}{n_j-p_j-3}\right\}$$

$$\text{and } \mathrm{AIC_u(g)} = \mathrm{AIC_c(g)} + \sum_j n_j \ln\{n_j/(n_j-p_j-2)\}, \tag{2.2}$$

where n_j is the number of observations in the j:th state and g index the set of candidate SETAR models. Given l states, R candidate combinations of threshold values and a fixed delay d there are $G=R \cdot p_1 \cdot \ldots \cdot p_l$ candidate SETAR models. For a given configuration g the ϕ_{ij}'s could be estimated using conditional least square see e.g. Tong (1990). The best SETAR model with respect to a given criterion is then found by minimizing the criterion over g=1,...,G.

3 Bootstrap criteria

Let $P=(\boldsymbol{\theta},F)$ denote the probability mechanism generating an observed data set x. For SETAR(2;p_1,p_2) models with d known $\boldsymbol{\theta} = (\phi_{10}, \phi_{11}, \ldots, \phi_{1p_1}, \phi_{20}, \phi_{21}, \ldots, \phi_{2p_2}, r)$

and assuming the same error ε_t in both regimes we get $\varepsilon_t \sim F$. We need to estimate the probability mechanism P from the observed data set x in order to draw bootstrap samples $\boldsymbol{x}^*$. Let P_n denote this estimate. The methods used to estimate the probability mechanism P will be described in the next section. Let $\hat{x}_{t,g}$ denote the prediction of x_t from the gth model. The true one step prediction error is then defined as

$$E_P[err_g(x,P)] = E_P[(x_{n+1} - \hat{x}_{n+1,g})^2] = \sigma^2_{\varepsilon,g} \tag{3.1}$$

and the apparent error rate from the sample is

$$err_g(x,P_n) = [n-\max(p_1,p_2)]^{-1} \Sigma_t (x_t - \hat{x}_{t,g})^2 \tag{3.2}$$

which is an estimate of $\sigma^2_{\varepsilon,g}$. The apparent error rate is to "optimistic" in the sense that it underestimates the prediction error. Bootstrap methods can be used to correct for this bias. The ideal bootstrap estimate of the true prediction error is

$$E_{P_n}[err_g(x^*,P_n)] = E_{P_n}\{[n-\max(p_1,p_2)]^{-1} \Sigma_t (x_t - \hat{x}^*_{t,g})^2\} \tag{3.3}$$

Here the bootstrap sample is playing the role of the original sample while the original sample is playing the role of the underlying P. With a finite number B of bootstrap samples we can approximate expression (3.3) by

$$\hat{E}_{P_n}[err_g(x^*,P_n)] = B^{-1} \Sigma_b \left\{[n-\max(p_1,p_2)]^{-1} \Sigma_t (x_t - \hat{x}^{*b}_{t,g})^2\right\} \tag{3.4}$$

changing the order of summation we get

$$\hat{E}_{P_n}[err_g(x^*,P_n)] = [n-\max(p_1,p_2)]^{-1} \Sigma_t \Sigma_b (x_t - \hat{x}^{*b}_{t,g})^2] \tag{3.5}$$

We can now divide the bootstrap samples into those that contain the observation x_t being predicted and those that do not. A bootstrap sample not containing x_t will have a prediction $\hat{x}^{*b}_{t,g}$ not influenced by x_t and can therefore be considered more "true". The idea is to use the average error rate obtained from bootstrap datasets not containing the point being predicted to adjust the apparent error rate. Let $\varepsilon_{0,g}$ denote this error rate. We can estimate $\varepsilon_{0,g}$ by

$$\hat{\varepsilon}_{0,g} = [n - \max(p_1, p_2)]^{-1} \Sigma_t \Sigma_{b\in A_t} (x_t - \hat{x}^{*b}_{t,g})^2 / B_t \tag{3.6}$$

where A_t is the set of indices of bootstrap samples not containing the t:th point and B_t is the cardinality of this set. We suggest the following bootstrap selection criteria, BSC, defined as

$$\text{BSC}(g) = err_g(\mathbf{x},P_n) + (1-e^{-1})[\,\hat{\varepsilon}_{0,g} - err_g(\mathbf{x},P_n)] \tag{3.7}$$

where $[\,\hat{\varepsilon}_{0,g} - err_g(\mathbf{x},P_n)]$ measures the "optimism" in the apparent error rate and the weight factor $(1-e^{-1})$ arises from the fact that when sampling with replacement $P(x_t \in \mathbf{x}^{*b}_g) = 1-(1-n^{-1})^n \rightarrow 1-e^{-1}$, $n \rightarrow \infty$, i.e. we adjust with the proportion of cases where the residual from t:th observation is in the bootstrap sample.

BSC is asymptotically equivalent with the apparent error rate see Efron and Tibshirani (1993) and the asymptotic equivalence between the apparent error rate and the AIC in the context of nonlinear AR order determination has been established by Cheng and Tong (1992). Since AIC has a tendency to overfit for small and moderate sample sizes see e.g. Hjort (1994), it seems reasonable to revise BSC in the same way as AIC has been modified. Hence with BSC(g) given by (3.7) we define the "corrected" and "unbiased" BSC respectively as

$$\text{BSC}_c(g) = [n\text{-}max(p_1,p_2)] \ln [\text{BSC}(g)] - 2\,\Sigma_j (p_j+1) + \Sigma_j \frac{n_j[n_j + p_j + 1]}{n_j - p_j - 3} \tag{3.8}$$

$$\text{BSC}_u(g) = \text{BSC}_c(g) + \Sigma_j n_j \ln[\,n_j/(n_j - p_j - 2)\,] \tag{3.9}$$

The order of the SETAR model and the estimate of r for the different model selection criteria are found by minimizing the respective criterion with respect to g i.e. taking $\text{argmin}_g \text{BSC}(g)$, $\text{argmin}_g \text{BSC}_c(g)$ and $\text{argmin}_g \text{BSC}_u(g)$ respectively.

4 Sampling experiments

In order to study the behaviour of the different bootstrap based and AIC based criteria, a sampling experiment was performed. To achieve a reasonable precision in the results 1000 replicates were used in all cases. The same random numbers were used for each parameter combination in order to reduce the uncertainty when comparing results. The following SETAR(2;1,1) model was used:

$$x_t = \phi_{10} + \phi_{11}x_{t-1} + \varepsilon_t\,, \quad x_{t-1} \le 0 \qquad\qquad (0.\ \text{-}.5\ \ 0.\ .5)$$

$$; \qquad \phi = (1.\ \text{-}.5\ \text{-}.5\ \ .5) \tag{4.1}$$

$$x_t = \phi_{20} + \phi_{21}x_{t-1} + \varepsilon_t\,, \quad x_{t-1} > 0 \qquad\qquad (0.\ \ .2\ \ 0.\ .8)$$

where $\varepsilon_t \sim N(0,1)$ or $\varepsilon_t \sim t(3)$. The sample sizes varied from 64 to 256.

The maximum order (p_1,p_2) from which the orders of the SETAR models were chosen were both set to 4 and r was assumed to belong to the interval $[x_{([.25n]+1)}, x_{(n-[.25n])}]$ were $[\cdot]$ denotes the integer part. d was assumed to be known.

Uniform random deviates were generated using the following multiplicative lagged Fibonacci generator $I_n = I_{n-55} \cdot I_{n-24} \ mod \ 2^{32}$, where the integers I_n were scaled to the interval (0,1) The deviates pass all test of Marsaglia's DIEHARD, a battery of tests of randomness see Marsaglia (1984) except those dependent on the last bit and has a very long period ($\approx 3.868 \cdot 10^{25}$). The $N(0,1)$ variates were generated using the Polar-Marsaglia method and the $t(3)$ variates were then obtained from the normal/independent form of the t-distribution. Three different methods were used for the bootstrap resampling.

In the first, moving block (mbl) we build overlapping blocks of consecutive observations $(x_1,\ldots, x_l)$, $(x_2,\ldots, x_{l+1}),\ldots,(x_{n-l+1},\ldots, x_n)$ where l is the block length parameter. We sample k blocks with replacement from n-l-1 blocks, so that $k \cdot l \approx n$. We do not attempt a reduction to i.i.d. observations, rather we want to keep the probability structure within the resampled units. The blocks should therefore be of sufficient length in order for the dependency to "die" out. This method has the advantage of being more or less "model independent" see Kuensch (1989). The choice of block length l is a trade-off between having enough blocks and not loosing too much of the probability structure. This structure is lost in the points of alignment of the blocks. Carlstein *et al* (1998) has proposed a method to overcome this by aligning with higher likelihood those blocks that match at their ends. We found a block length l=8 to be suitable for our models and sample sizes. No attempt to match the block as suggested by Carlstein *et al* was made.

In the second "best TAR" (tar), the SETAR model chosen by AIC_u is used as fit $\hat{x}_t$ and the estimated residuals $\hat{\varepsilon}_t = x_t - \hat{x}_t$, t = max(p_1, p_2) + 1, …, n are calculated. We then sample n-max(p_1,p_2) observations with replacement from the empirical distribution of the $\hat{\varepsilon}_t$:s $F_{n-\max(p_1,p_2)}$. The bootstrap time series x_t^* is calculated recursively as $x_t^* = x_t$, t = 1, …, max(p_1,p_2) and $x_t^* = \hat{x}_t^* + \hat{\varepsilon}_t^*$, t = max(p_1,p_2) + 1 , …, n, where $\hat{x}_t^*$ is the fit using the bootstrap observations. This method is completely model dependent.

Finally in AR-sieve(ars) an AR(p) model of high order p is used as fit $\hat{x}_t$ and resampling with replacement is made from the estimated residuals $\hat{\varepsilon}_t = x_t - \hat{x}_t$, $t = p + 1$, …, n. The parameters of the AR model has been estimated using the Yule-Walker estimates since this method guarantees a positive definite autocovariance matrix, see Box and Jenkins (1969). The choice of p is often data-driven. Here we have used the AIC criterion suggested by Buehlmann (1997) to find the optimal order p. This method has the advantage over the prewhitening method that it can distinguish between uncorrelated and independent errors ε_t. The method is semi-parametric. Unlike the fully model-based bootstrap approach, like tar above, we here allow for finite sample model misspecification. The AR-sieve procedure uses the linear AR(∞) assumption for the underlying true process, which SETAR models do not fulfil see Buehlmann (1997). However Tsay (1989) suggests that high order AR models could provide reasonable approximation to nonlinear model like SETAR, we have therefore decided to compare the behaviour of this method with the first two.

The results are summarized in Table 4.1 below. It shows the frequency $f(1,1)$ of selecting the correct order (p_1, p_2) = (1,1) over the m =1000 replicates. Approximate 95 % confidence intervals based on the normal approximation and the conservative variance .25*m* are given by the table entry ± 32. From the table we can see that of the AIC based criteria AIC_u is consistently better than the other two, but its performance decreases with increasing sample size contrary of AIC. Comparing with the bootstrap criteria we can see that they are outperforming their

Table 4.1.The frequency $f(1, 1)$ of selecting the correct order.

Error dist	ϕ	Bootstrap Method	Sample Size n	Model selection criteria AIC	AIC_c	AIC_u	BSC	BSC_c	BSC_u
$t(3)$	(0. -.5 0. .5)	mbl	64	417	655	796	555	751	892
			128	462	556	779	432	552	795
		tar	64	422	638	793	480	761	852
			128	485	577	771	429	617	792
		ars	64	405	644	820	457	723	856
			128	508	605	771	373	587	802
$N(0,1)$	(0. -.5 0. .5)	mbl	64	346	578	783	482	693	850
			128	404	526	738	395	518	769
			256	464	512	714	386	453	718
		tar	64	349	572	759	474	719	804
			128	437	546	783	463	665	794
			256	451	492	747	439	535	749
		ars	64	337	561	753	415	701	852
			128	445	540	752	408	593	797
	(1. -.5 -.5 .5)	mbl	64	410	639	818	510	706	894
			128	529	619	811	494	629	823
		tar	64	418	631	819	577	799	841
			128	510	605	792	516	699	801
		ars	64	411	644	803	487	712	845
			128	522	633	823	405	586	782
	(0. .2 0. .8)	mbl	64	352	586	777	528	730	886
			128	429	553	727	456	572	803
		tar	64	363	582	766	511	746	831
			128	428	532	761	432	637	786
		ars	64	342	604	783	460	713	872
			128	423	538	746	408	574	795

corresponding AIC criteria, only AIC is better than BSC for n=128 and n=256. In a few cases also AIC_c is performing better than BSC_c for larger samples while BSC_u is always better then AIC_u except for the parameter combination (1. -.5 -.5 .5), n=128 and the AR-sieve method. The reason for this is probably that SETAR models can not be sufficiently good approximated by a high order AR model when the sample size increases. The model based bootstrap, tar, in our paper is not based on the true model but on the model selected by AIC_u. Thus we can expcct BSC_u to always perform at least as good as AIC_u, but due to wrong order of the model selected by AIC_u the other bootstrap methods ars and mbl can perform better.

5 Conclusions and discussion

We have studied three bootstrap criteria for TAR model selection, BSC, BSC_c and BSC_u. Our sampling experiment indicates that BSC_u outperforms the AIC based criteria. They also show that among these criteria AIC_u is superior. Hence we recommend the use of BSC_u in practice for small to moderate sample sizes. With increasing n the difference between BSC_u and AIC_u seems to vanish. Therefore

AIC_u can be recommended for large samples. All three bootstrap resampling concepts worked quite well, AR-sieve being a little inferior to the other.

The choice of the optimal block length in the block bootstrap is still much of an open question. The problem has much in common with the problem of estimating a density or spectral density. In these cases the choice of the bandwidth is a trade-off between high stability – low variance and high fidelity – low bias. The choice of the block length is a tradeoff between capturing as much as possible of the probability structure in the process – high fidelity and having enough blocks to get stable estimates – high stability. For practical purpose we recommend the method suggested by Hall *et al* (1995) – choose an initial l, subset the original time series into overlapping runs of lengths $s<n$, then slowly increase the block length l_s in these and check for which block lengths the estimate of interest remains relatively stable and close to the estimate for the whole series (i.e. when the mean square error over the n-s subsets with respect to the whole series is small). Then choose one of these block lengths, say l_s^*, get an updated $l^* = (n/s)^{1/k} l_s^*$ (if the quantity of interest is a bias or a variance k=3) and iterate if desired.

References

Box, G. E. P. and Jenkins, G. M. (1969), *Time series Analysis- forecasting and control*, Holden day.

Buehlmann, P. (1997), Sieve Bootstrap for Time Series, *Bernouilli*, **3**, 123-148.

Carlstein, E., Do, K.-A., Hall, P., Hesterberg, T. and Kuensch, H. R. (1998), Matched-block Bootstrap for Dependent Data, *Bernoulli*, **4**, 305-328.

Cheng, B. and Tong, H. (1992), On Consistent Nonparametric Order Determination and Chaos, *Journal of Royal Statistical Society,* **B 54,** 427-449.

De Gooijer, J. G. (2001), Cross-validation Criteria for SETAR Model Selection, *Journal of Time Series Analysis*, **22**, 267-281.

Efron, B. and Tibshirani, R. J. (1993), *An Introduction to the Bootstrap*, Chapman & Hall.

Hall, P., Horowitz, J. L. and Jing, B.-Y. (1995), On Blocking Rules for the Block Bootstrap and Dependent Data, *Biometrika*, **82**, 561-574.

Hjorth, J. S. U. (1994), *Computer Intensive statistical Methods- Validation model selection and bootstrap*, Chapman & Hall.

Kuensch, H. R. (1989), The Jackknife and the Bootstrap for General Stationary Observations, *The Annals of Statistics*, **17**, 1217-1241.

Margasalia, G. (1984), A Current View of Random Number Generators, *The Proceedings of the 16th Symposium on the Interface, Atlanta 1984.* Elsevier Press.

Tong, H. and Lim, K. S. (1980), Threshold Autoregression, Limit Cycles and Cyclical Data, *Journal of Royal Statistical Society,* **B 42,** 245-292.

Tong, H. (1983), *Threshold Models in Non-Linear Time Series Analysis*, Lecture Notes in Statistics, Vol. 21, Springer-Verlag.

Tong, H. (1990), *Non-Linear Time: A dynamical System Approach*, Oxford University Press.

Tsay, R. S. (1989), Testing and Modeling Threshold Autoregressive Processes, *Journal of American Statistical Association*, **84**, 231-240.

Wong, C. S. and Li, W. K. (1998), A Note on the Corrected Akaike Information Criterion for Threshold Autoregressive Models, *Journal of Time Series Analysis* **19**, 113-124.

Canonical Variates for Recursive Partitioning in Data Mining

Carmela Cappelli, Claudio Conversano

Department of Mathematics and Statistics, University of Naples Federico II
e-mail: {carmela.cappelli, conversa}@unina.it.

Abstract. This paper deals with the problem of dimension reduction in the general context of supervised statistical learning, with particular attention to data mining applications. The main goal of the proposed methodology is to improve tree based methods as prediction tool by introducing an alternative approach to data partitioning which is meant to handle large numbers of (possibly correlated) covariates. The key idea is to use suitable combinations of covariates recursively identified.

Keywords. High Dimensional Data Analysis, Canonical Variates, Statistical Learning, Tree-Based Methods.

1 Introduction

During the past decade there has been an explosion in computation and information technology. Vast amounts of data have been collected in a variety of fields such as medicine, biology, finance, and marketing. The goal of understanding these data has led to the development of new tools in the field of statistics, and spawned new areas such as data mining, machine learning and bioinformatics (Friedman et. al 2001). The new challenge for statisticians is to deal with large databases in order to turn data into information.
In the context of supervised classification problems, tree structured methods have proved to be a powerful and effective nonparametric tool for high-dimensional data analysis. Despite their attractiveness and successfulness, they are not always most effective; in fact, the majority of classification tree algorithms are based on axis-orthogonal partitioning of the data and therefore the resulting structure tends to be complex because a number of splits is required to model the underlying relationship between the response variable and the covariates. We deal with the problem of dimension reduction in the context of data mining, where the main emphasis is on generalization, i.e., the ability of a model to make good predictions. In particular, the main goal is to improve the effectiveness of classification trees as data mining tools by introducing an alternative recursive partitioning approach designed to cope with high-dimensionality in the covariates space.
Starting with the milestone work of Brieman et. al (CART, 1984), several methodological proposals and practical applications have been proposed in the literature. Among these, those related in various ways to the dimensionality problem in the covariates and/or in the observation space have been

contributed by Mola and Siciliano (1998), Kim and Loh (2001), and Conversano *et* al. (2001).
The key idea of this paper is to replace subgroups of covariates by their linear combinations. For this purpose, the use of the canonical discriminant function is proposed. By means of using canonical variates the dimension reduction objective is pursued by searching for those linear combinations of the original variables which summarize the relationships among the covariates with respect to a grouping variable.
In our approach, the role of grouping variable is played by the response variable. At each step of the recursive partitioning procedure, the best split of the given node is identified searching over the linear combinations rather than over the original covariates. In this way, without complicating the existing method, a more effective use of the information in the data is achieved.

2 Method

In the following, we briefly describe tree based methods and canonical discriminant function to show how the latter can be cast into the recursive partitioning procedure.

Tree based methods. Suppose a response variable y which takes value in a set of g previously defined classes and a set of p covariates are observed on n objects denoted by $\{(x_{1i}, \ldots, x_{ji}, \ldots, x_{pi}, y_i)\}_{i=1}^{n}$.
A standard tree construction arises from a *divide and conquer* algorithmic strategy which recursively divides the data space into two subregions (subgroups of observations) according to a splitting criterion which allows to select at each step (tree node) the best covariate and cut point along it. The *goodness of split* at a given node is measured in terms of achievable separability among the sibling nodes, i.e., data partitioning is led by the aim of making the distribution of the response variable as "pure" as possible within the nodes. A further step, tree simplification, is usually carried out to avoid overfitting and reduce the size of the tree by discarding away some of the branches. An extensive review and comparative study of pruning methods can be found in Breslow and Awa (1998). In our application, to define the optimally pruned tree we will employ the CART pruning method.
Here, we focus on data partitioning, that is based on a set Q of binary questions (splitting variables) of the form: "Is $x_j \in A$?", so that, if x_j is ordered, Q includes all questions:

$$\text{Is } x_j \leq c? \tag{1}$$

for all c ranging over the domain of x_j.

Canonical discriminat function. In the family of discriminant analysis methods, the canonical model is useful for dimension reduction purposes (McLachlan, 1992) because it is a process of dividing up a p dimensional space into pieces such that the groups (identified by the response variable) are as distinct as possible. This method seeks for a set of vectors that maximizes the ratio of the between-group sum of squares to the (common) within group sum of squares. Let $\mathbf{X}$ denote the $n \times p$ data matrix, $\mathbf{G}$ the $n \times g$ matrix of indicator functions for the g groups (response classes) and $\mathbf{M}$ the $g \times p$ matrix of group means. Then we can define:

- $\mathbf{B} = \frac{(\mathbf{GM}-\mathbf{1\bar{x}})^T(\mathbf{GM}-\mathbf{1\bar{x}})}{g-1}$ as the between group variance matrix;
- $\mathbf{S} = \frac{(\mathbf{X}-\mathbf{GM})^T(\mathbf{X}-\mathbf{GM})}{n-g}$ as the (common) within group covariance matrix.

where $\bar{\mathbf{x}}$ is the vector of covariates means.
The canonical variates are the eigenvectors associated with the eigenvalues of the product matrix $\mathbf{S}^{-1}\mathbf{B}$, and their number is equal to $\min(g-1,p)$. The eigenvector $\mathbf{w}^T$ corresponding to the largest eigenvalue of $\mathbf{S}^{-1}\mathbf{B}$ represents the direction of maximum separation among the groups.
This method enjoys two important properties: robustness to non normality and even to mildly not equal covariance matrices for the groups (Hastie *et* al., 1995; Hand, 1998).

The proposal. The key idea of this paper is to improve the efficiency of the tree based procedure by considering in the (1) functions of the form:

$$\sum_j a_j x_j \tag{2}$$

where a_j are coefficients (some can be zero) to be determined. We propose the use of canonical discriminant analysis so that the problem reduces to create vectors $\mathbf{w}_s^T\mathbf{X}$, where $s = 1, \ldots, min(g-1,p)$. The splitting variables are then defined on the combinations instead of on the original covariates. It is noteworthy that functions (2) are identified recursively at each step. In practice, in our tree growing procedure the canonical discriminant analysis is recursively applied in order to generate new canonical variates which are derived considering the subgroups of observations falling in the current node. In this way, the coefficients of the combinations are updated at each run and, more important, in nodes where only a subgroups of response classes are present the canonical variates are built with respect only to these.
In practical applications, several covariates are collected and often subgroups of these are highly correlated. In cases such as these, the original data matrix can be partitioned with respect to these subgroups and the algorithm finds a reduced number of canonical variates (for a binary response just one) for each subgroups of covariates.
Given that the canonical variates provide the direction in which the groups are best separated, it is convenient to seek the splits along them because they model the relation between the response and the covariates, whereas several axis-orthogonal splits on single variables might be required to approximate a split on a linear combination.

As it will be shown in the applications, this strategy keeps the tree shorter and increases the classification accuracy.
The proposed method shares with popular tools such as Bagging and Boosting (see Dietterich, 2000), the loss of interpretability of the final tree, even if, in the case of the canonical variates, interpretability may be at least partly preserved looking at the values of the coefficients of the combinations.

3 Examples

In order to show how the proposed method works, we have applied it to both real and simulated data sets.
The real data sets are the: 1) wisconsin diagnostic breast cancer data (*wdbc*; 569 instances, 30 real valued covariates, binary response); 2) waveform data (*wave*; 5000 instances, 21 real valued covariates, three class response); 3) wine recognition data (*wine*; 178 instances, 13 real valued covariates, three class response). These datasets have been contributed to the UCI Repository of Machine Learning databases (http://www.ics.uci.edu/ mlearn/MLRepository).
The simulated data (*sim*) are drawn from the model: $y_i = [1 + \exp(k_i)^{-1}]$ where $i = 1, \ldots, 1000$ and k includes 5 covariates linked by the following equation:

$$k_i = \sin(3z_{i1} + 5z_{i2}) - z_{i3} + 2z_{i4} + 4z_{i5} + \epsilon_i$$

with $\epsilon \sim N(0,5)$ and $i = 1, \ldots, 1000$. Each z_j is the additive (linear and/or non linear) combination of 10 covariates randomly drawn from a multivariate normal distribution of the type $N(\mathbf{0}, \Sigma)$ so that the final model has 50 covariates. The data groups have been generated with the goal of obtaining high correlations (positive and/or negative) within each group.
Each data set has been split into a learning set (n^l instances) used to build the trees and a test set (n^{ts} instances) used to estimate the error rate (30%). Table 1 reports the results for the trees built on the original data (T) and using the canonical variates (CVT); we focus on two indicators: accuracy (percentage of misclassifed), and size of both the maximal trees and the best trees [1]. The percentage of misclassified has been evaluated both on the learning set (R^l) and the test set (R^{ts}), the latter providing a more reliable estimate.
We see that for all data sets the proposed approach outperforms remarkably the classical one; in particular, it achieves higher accuracy, both respect to the learning data and the test data, with shorter trees. Only for the wdbc data the the best tree error rate on the learning data is slightly higher with respect to that associated with the classical recursive partitioning, but the tree size is comparatively much smaller.
Actually, providing the recursive partitioning procedure with the relevant information represented by the canonical variates allows the method to reduce

[1] The best trees are the optimally pruned trees defined by means of the CART pruning method.

strongly the error with a few splits. This can be understood by looking at figure 1, that, for the waveform and the simulated data, depicts the error rates produced by the classical and the modified approach, against the tree levels. The curves corresponding to the canonical variate trees show a smaller number of levels (less complex trees) and they are far below those corresponding to the trees built on the original data.

				Maximal			Best		
	n^l	n^{ts}	Trees	$R^l\%$	$R^{ts}\%$	*size*	$R^l\%$	$R^{ts}\%$	*size*
wine	124	54	T	1.61	7.41	6	1.61	7.41	5
			CVT	0.00	0.00	3	0.00	0.00	3
wdbc	300	269	T	1.33	7.81	11	2.00	4.46	4
			CVT	1.33	4.46	4	2.33	2.60	2
wave	3500	1500	T	15.11	24.27	73	16.86	23.53	33
			CVT	4.14	16.40	45	11.77	13.87	18
sim	700	300	T	6.71	36.33	56	16.57	29.33	13
			CVT	2.57	19.33	21	8.86	16.33	8

Table 1. Tree performances indicators

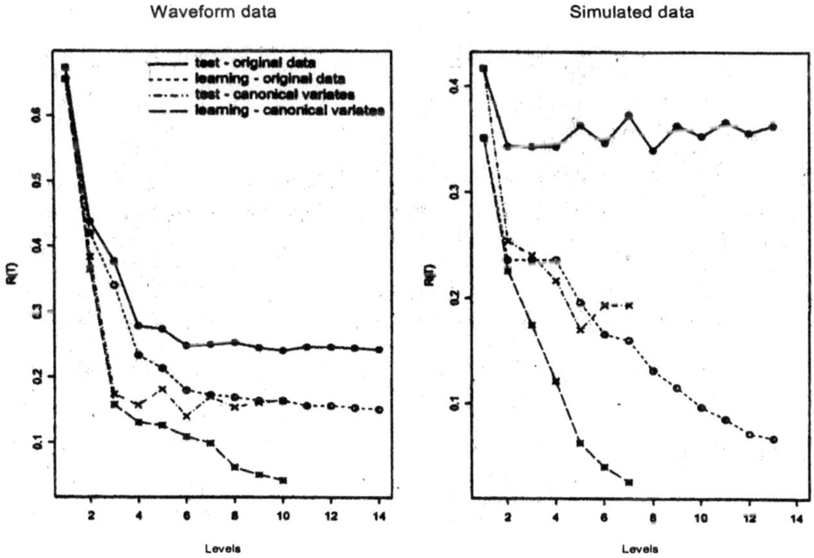

Figure 1. Error rate against tree levels

4 Concluding remarks

Due to the underlying divide and conquer algorithm, trees spend too much effort in modelling all the data and therefore tend to be large and complex.

This problem is particularly relevant in high dimensional data analysis and data mining applications. Our approach is useful because it provides the tree algorithm with the relevant information. In fact, the proposed method is designed for situations in which there is a large number of (highly correlated) covariates. In cases such as these a natural strategy to deal with the high-dimensionality problem is to find a reduced number of combinations of the original covariates. We propose to use canonical variates because they represent (linear) functions of the covariates which are optimal for separating the groups. Applications have shown that these functions recursively defined along with recursive partitioning procedure produce shorter and much accurate trees.

Acknowledgements. Authors wish to thank Prof. R. Siciliano (University of Naples Federico II) for her helpful comments. Research was supported by MURST Funds (PRIN01, coordinator Prof. R. Siciliano).

References

Breiman, L., Friedman, J.H., Olshen, R.A., and Stone, C.J. (1984). *Classification and Regression Trees.* Monterey (CA): Wadsworth & Brooks.

Breslow, L.A., Aha,D.W (1998): Simplifying Decision Trees: a survey. *Knowledge Engeneering Review*, **12**, 1-40.

Conversano C., Mola F. and Siciliano R., (2001): Partitioning Algorithms and Combined Model Integration for Data Mining. *Computational Statistics*, **16**, 323-339.

Dietterich, T. G., (2000). An experimental comparison of three methods for constructing ensembles of decision trees: bagging, boosting, and randomization. *Machine Learning*, **40**(2), 139-158.

Friedman, J. H., Hastie, T. and Tibshirani, R., (2001): *The Elements of Statistical Learning: Data Mining, Inference and Prediction*, Springer.

Hand, D.J. (1998). *Construction and Assessment of Classification Rules.* New York: Wiley.

Hastie, T., Buja, A., Tibshirani, R. (1995): Penalised Discriminat Analysis. *Annals of Statistics*,**23**, 73-102.

Kim, H. and Loh, W. Y. (2001): Classification Trees with Unbiased Multiway Splits. *Journal of the American Statistical Association*, **96**, 589-604.

McLachlan, G.J., (1992). *Discriminant Analysis and Statistical Pattern Recognition.* New York: Wiley.

Siciliano, R. and Mola, F., (1998): Ternary Classification Trees: a Factorial Approach Classification. In: *Visualization of Categorical Data*, 311-323. Academic Press, 223-230.

CAnoVa© a Software for Causal Modeling

Olivia Wüthrich-Martone[1], Christof Nachtigall, Marc Müller, and Rolf Steyer

[1] Institute of Psychology, Friedrich Schiller University, Am Steiger 3, Haus 1, D-07743 Jena, Germany

Abstract. The new impulse given in the last decade to the theory of individual and average causal effects is mostly due to the approach developed by Steyer and others and resulted in valuable theoretical results such as, for example, the link between unconfoundedness and causal unbiasedness. This approach has also allowed for the development of different practical procedures for both testing for confounding and for testing for (average) causal effects. The scope of this contribution is to present CAnoVa, a software for causal modeling that allows a straightforward use of these two methods. CAnoVa is fully compatible with SPSS and allows to test for confounding and for causal effects even in case of designs with multiple treatment variables and multiple confounders.

1 Software for Causal Modeling?

Human interest in elucidating the cause-effect relationship is deeply rooted in the history of mankind and many different approaches to Causality have been developed particularly in the last three centuries. Still Statistics is the discipline that has neglected causality the most, probably because of the intrinsic difficulty of giving a precise mathematical formalization to the basic concept of causality. Another reason of the reluctancy of statisticians to get involved with causality may be traced back in the long misunderstood Fishers dictum that there is no causation outside the randomized experiment. A major turnover in statisticians opinion is due to the development of the theory of individual and average causal effects. The foundations of this theory were set by Rubin (1974), Holland (1986), and Rosenbaum (1984) in a series of papers. This theory has, in this particular formulation (also known as Rubins Causal Model), an approach to causality that is mainly counterfactual and is based on the notion of multiple potential responses, i.e. each unit is supposed to be theoretically exposable to different treatments. Steyer and others (2000) proposed a new formulation of the theory of individual and average causal effect, based on classical probability theory. This formulation allowed a rigorous definition of the multiple potential responses of a unit to different treatments and consequently of the two basic concepts of causal modeling: the individual causal effect (the effect of two different treatments on the same unit) and the average causal effect (the average of different treatments on the whole population).

The causal model underlying randomized experiments is straightforward: once the experiment has been accurately planned and once the experimental conditions have been properly realized, randomization of units to treatments ensures that the experiment outcome is a direct consequence of the treatments considered. In contrast, in nonrandomized experiments a third variable or potential confounder may have induced a bias on the observed treatment means, so that the average causal effects and the treatment mean differences are not equal anymore. In such a case the standard statistical methods, which estimate the treatment means, are no appropriate tool for estimating or testing the average causal effects anymore. Wüthrich-Martone, Nachtigall and Steyer developed different procedures for controlling the bias introduced by potential confounders and for providing useful estimates for the average causal effects even from nonrandomized studies. In particular two methods, one for testing for confounding and the other for testing for (average) causal effects in general linear models with fixed regressors were developed (see Wüthrich-Martone, 2001). Both these methods are based on testing a particular general linear hypothesis and apply to design with qualitative treatment variables, qualitative potential confounders and a continuous response variable. When it became clear that testing for confounding and for average causal effects was possible, at least under certain assumptions, the question then arose which software to use for these purposes? By means of an example (see for example Section 5.4 of Wüthrich-Martone, 2001) it was proved that the methods implemented in the standard routines for the analysis of variance do not test for causal effects, making them useless for the purposes of causal modeling.

The only means of applying the procedures mentioned above to a specific design was to use a software that allowed testing of a general linear hypothesis formulated by the user himself. Only few software offered such a possibility, namely SAS and SYSTAT. Still they required the specification of the coefficient matrix, that is the specification of the matrix of the coefficients of the model parameters appearing in the formulation of a general linear hypothesis. The process of building the coefficient matrix may be quite complicated and definitely time-consuming, especially for designs with a large number of observations or a large number of factor levels. The causal analysis with SAS or SYSTAT of a design with multiple treatment variables and multiple confounder would be even more complicated because of the procedure needed for building the corresponding reduced 2-factor design (see Section 6.2 of Wüthrich-Martone, 2001).

Because of all these difficulties, the authors decided to develop a software for testing for confounding and for causal effects and opted for writing a SCRIPT for SPSS, namely CAnoVa© = **C**ausal **An**alysis **o**f **Va**riance. To date CAnoVa is available in beta version Release 0.95. The only requirements needed for using CAnoVa are a licensed version of SPSS and a copy of CAnoVa itself (available from the authors). CAnoVa may be started directly from an SPSS data file containing the data to analyze. Major practical advantages of CAnoVa are its user-friendly interface and the possibility of getting as output a standard SPSS output window. It is therefore possible to produce in the same output the results of standard statistical procedures (e.g. analysis of variance) and the results of a causal analysis as well. CAnoVa allows to test for confounding and for causal effects even in case of designs with multiple treatment variables and multiple confounders. According to the procedure developed by Wüthrich-Martone, CAnoVa first builds the corresponding reduced 2-factors design and it then performs a causal analysis on the reduced design. The reduced design may be automatically saved in a new SPSS data file.

2 CAnoVa© a Software for Causal Modeling

In order to briefly outline how CAnoVa works the design described in Section 4.5 of Wüthrich-Martone (2001) is analyzed. In this design 500 patients with the same mental disorder are considered. The purpose is to investigate the effects of three different psychotherapies A, B and C on a continuous response variable Y, greater values of which reflect an improvement in the patient's condition. The severity of the disorder is the only potential confounder W considered[1]. Three grades of severity: *high*, *medium* and *low* are distinguished. For each one of the 500 patients, the corresponding severity of the disorder was measured and recorded before treatment. The distribution of W was known and the marginal probabilities $P(W = w_j)$ are

$$P(W = high) = P(W = medium) = 0.4 \quad P(W = low) = 0.2$$

The response variable Y measures the improvement in the patients' condition and was recorded by all patients at an appropriate time after treatment.

2.1 Input

CAnoVa requires that the data are entered in an SPSS file according to the following criteria: factor levels should be coded numerically and no missing observation is allowed. For the considered design the treatment variable has been entered with the name factorx, the confounder as factorw, and the response variable with the name depvary. CAnoVa is launched. The main window of CAnoVa is shown in Figure 1. The variables of the model to analyze are selected. In the options window (not shown), CAnoVa is instructed to use the marginal probabilities $P(W = w_j)$ and not to estimate these probabilities from the cell frequencies (default option).

[1] The assignment of patients to therapies has been performed following the procedure of conditional randomization, which guarantees that the effects of any other confounder have been "screened off".

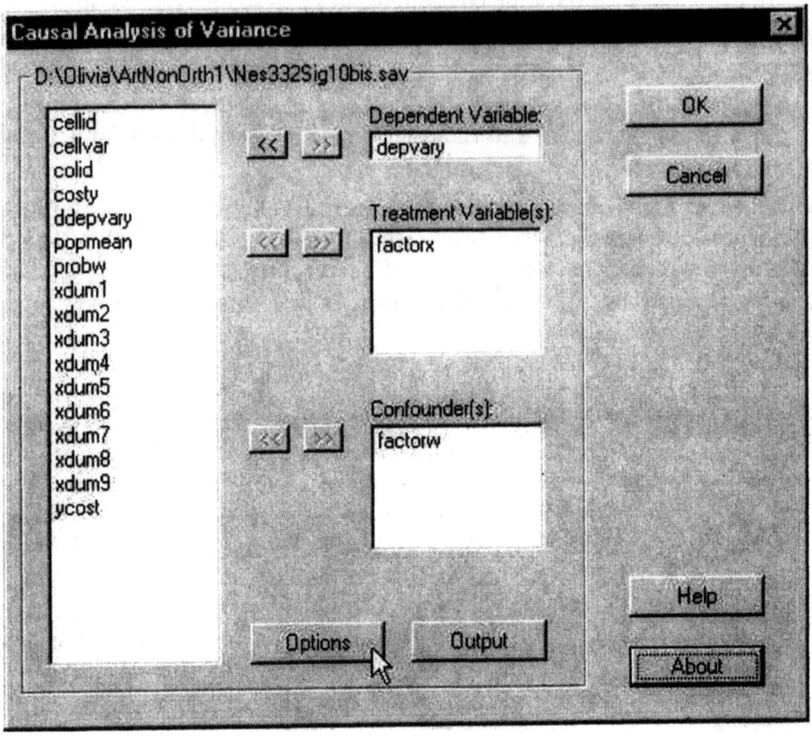

Fig. 1. CAnoVa: Main window

2.2 Output

Two output versions are available for CAnoVa: a long one (printing, among other, the coefficient matrix for the hypothesis of no confounding and the coefficient matrix for the causal null hypothesis[2]) and a short one.

The short version of the output consists of a table with information concerning the data file considered and the options specified, of various summary tables and of the results of the causal analysis of variance, i.e. the estimated treatment means (adjusted for confounding and not) and the tests for confounding and for causal effects.

The most important part of the output is shown in Figure 2. In the first table the estimates of the treatment means (Estim. $\mathrm{E}(Y|X = x)$) and the estimates of the treatment means adjusted for confounding (Estim. $\mathrm{Eadj}(Y|X = x)$) are listed. The last two tables present, in the order, the results of the test for confounding w.r.t. W and of the test for average causal effects[3].

[2] A thorough description of both matrices is given in Chapter 4 of Wüthrich-Martone (2001).

[3] The description of the two tests and corresponding test statistics are given in Chapter 4 of Wüthrich-Martone (2001)

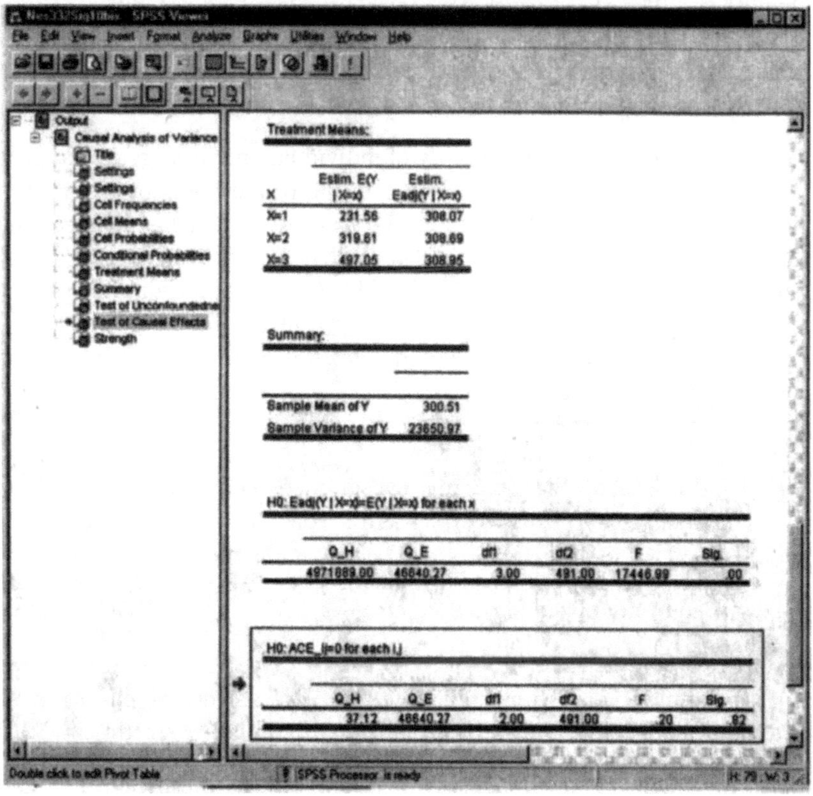

Fig. 2. CAnoVa: Output Part 3

The test for confounding w.r.t. W tests whether the considered confounder (the severity of the disorder) does actually confound the treatment regression considered (that is in our example the treatment regression of patients' improvement on therapies). Since the distribution of the confounder was known, the test statistics follows under the null hypothesis of unconfoundedness an $F_{[3,491]}$ distribution. The p-value corresponding to the test statistics is almost zero, which leads to the rejection of the null hypothesis. It may therefore be concluded that the Severity of the Disorder is an actual confounder for the treatment regression considered.

The last table (the one indicated by an arrow) in Figure 2 presents the result of testing the causal hypothesis (the average causal effects ACE are all zero) $H_0: ACE(i,i') = 0 \; \forall i \; \forall i'$. This hypothesis states that the three therapies do not differ from each other with respect to their (average) causal effects. Since the distribution of the confounder was known, the test statistics follows under the null hypothesis of unconfoundedness an $F_{[2,491]}$ distribution The p-value corresponding to the test statistics is 0.82, which leads to accept the null hypothesis. The three psychotherapies considered do not differ significantly one from each other concerning their average causal effects.

3 Conclusions

The new formulation of the theory of individual and average causal proposed by Steyer has allowed for the developments of methods for testing for confounding and for causal effects in the framework of general linear models. These procedures have been implemented from the authors in CAnoVa, a software fully compatible with SPSS. This compatibility and its user-friendly interface are what makes of CAnoVa, even in the beta version, a practical tool for the experimenter interested in causal modeling and in the analysis of confounding variables.

References

Holland, P. (1986). Statistics and causal inference. *Journal of the American Statistical Association*, **81)**, 945-970.

Rubin, D. B. (1974). Estimating causal effects of treatments in randomized and nonrandomized studies. *Journal of Educational Psychology*, **66(5)**, 688-701.

Rosenbaum, P. and Rubin, D. B. (1984). Estimating the effects caused by treatments. *Journal of the American Statistical Association*, **79(385)**, 26-28.

Steyer, R., Gabler, S., von Davier, A., Nachtigall, C. and Buhl, T. (2000). Causal regression models I: Individual and average causal effects. *Methods of Psychological Research-online*, **5(2)**, http://www.mpr-online.de.

Wüthrich-Martone, O. (2001). *Causal Modeling in Psychology with Qualitative Independent Variables* Aachen: Shaker Verlag.

Classification Based on the Support Vector Machine, Regression Depth, and Discriminant Analysis

Andreas Christmann[1], Paul Fischer[2] and Thorsten Joachims[3]

[1] University of Dortmund, HRZ, 44221 Dortmund, Germany.

[2] University of Dortmund, Department of Computer Sciences, Germany.

[3] Cornell University, Department of Computer Science, USA.

Abstract. The minimum number of misclassifications achievable with affine hyperplanes on a given set of labeled points is a key quantity in both statistics and computational learning theory. We compare the modern approaches the regression depth method and the support vector machine with discrimimant analysis. Summarizing, the regression depth method using currently available algorithms yields often better classifications results for small to moderate data sets, say for sample sizes less than 1000 and dimension up to 10, whereas the support vector machine is often more appropriate for larger or higher dimensional data mining problems.

Keywords. Data mining, Discriminant analysis, Logistic regression, Overlap, Regression depth, Separation, Support vector machine

1 Introduction

The problem to determine the exact minimum number of misclassifications achievable with affine hyperplanes on a given set of labeled points is NP-hard, c.f. Hoeffgen, Simon and van Horn (1995). Hence, there is a need to find reasonable and fast approximation procedures which should work well for small and for large complex data sets, e.g. in data mining. In this paper we compare methods to approximate the minimum number of misclassifications achievable with affine hyperplanes. The regression depth method (RDM) was proposed by Rousseeuw and Hubert (1999). We compare the RDM with the support vector machine proposed by Vapnik (1998) and with linear discriminant analysis.

Commercial statistical software packages are often not able to detect whether a data set has overlap or not. However, the classical maximum likelihood estimate in logistic regression exists if and only if the data set has overlap. A central role concerning the existence and the quality of such estimates plays the quantity n_{co} defined as follows. For a given data set $Z_n = \{(x_{i,1}, \ldots, x_{i,p-1}, y_i);\ i = 1, \ldots, n\} \subset \mathbb{R}^p$ where $y_i \in \{0,1\}$ for $i = 1, \ldots, n$, the quantity n_{co} is the minimum number of misclassifications that any affine hyperplane must incur.

In particular, if $n_{\text{co}} = 0$, the data set is *completely separated* so that there exists a vector $\theta \in \mathbb{R}^p$ such that

$$(\mathbf{x}_i, 1)\theta' > 0 \quad \text{if} \quad y_i = 1 \tag{1}$$

$$(\mathbf{x}_i, 1)\theta' < 0 \quad \text{if} \quad y_i = 0, \qquad i = 1, \ldots, n. \tag{2}$$

Section 5 gives a comparison of the classification results based on the support vector machine and on quadratic discriminant analysis for a two-dimensional example with four response groups.

2 The regression depth approach

Rousseeuw and Hubert (1999) introduced the regression depth method (RDM) for linear regression models. Christmann and Rousseeuw (2001) showed that RDM is useful for binary regression models, too. In the following we will consider the logistic model, although RDM can be used for other binary regression models in an analogous manner. We assume that the design matrix corresponding to the data set Z_n has full column rank and define $\Lambda(z) = 1/[1+e^{-z}]$, $z \in \mathbb{R}$.

Definition 1:
A vector $\theta = (\theta_1, \ldots, \theta_p) \in \mathbb{R}^p$ is called a **nonfit** *to Z_n iff there exists an affine hyperplane V in $\mathbf{x}$−space such that no $\mathbf{x}_i$ belongs to V, and such that the residual $r_i(\theta) = y_i - \Lambda((\mathbf{x}_i, 1)\theta') > 0$ for all $\mathbf{x}_i$ in one of its open halfspaces, and $r_i(\theta) < 0$ for all $\mathbf{x}_i$ in the other open halfspace. The* **regression depth** *of a fit $\theta = (\theta_1, \ldots, \theta_p) \in \mathbb{R}^p$ relative to a data set $Z_n \subset \mathbb{R}^p$ is the smallest number of observations that need to be removed to make θ a nonfit.*

There exists a connection between regression depth and complete separation. Define the horizontal hyperplane defined by $\theta^* = (0, \ldots, 0, 0.5)$. Then θ^* is a nonfit iff $n_{\text{co}} = 0$, and more generally $n_{\text{co}} = \text{rdepth}(\theta^*, Z_n)$. Hence, n_{co} can be computed with algorithms for the regression depth of a given hyperplane. For $p \in \{2, 3, 4\}$ the latter can be computed exactly by the $O(n^{p-1}\log(n))$ time algorithms of Rousseeuw and Hubert (1999) and Rousseeuw and Struyf (1998). For $p \geq 3$, Rousseeuw and Struyf (1998) constructed a fast approximation algorithm based on appropriate projections for the regression depth. The main idea of the algorithm for $p \geq 3$ is to approximate the p−dimensional regression depth by the minimum of certain two-dimensional regression depths. We use

$$n_{\text{co}} = \text{rdepth}(\theta_{opt}, Z_n) = \min_{\theta \in \mathbb{R}^p} \text{rdepth}(\theta, \tilde{Z}_n(\theta)) \tag{3}$$

$$\leq \min_{\theta \in B \subset \mathbb{R}^p} \text{rdepth}(\theta, \tilde{Z}_n(\theta)) =: n_{\text{co}}(B), \tag{4}$$

where $\tilde{Z}_n(\theta) = \{((x_i, 1)\theta', y_i); i = 1, \ldots, n\} \subset \mathbb{R}^2$, $\theta \in \mathbb{R}^p$, and θ_{opt} is an optimal parameter vector. The set B is determined via projections defined by a large number, say 10^4, of random subsamples of the original data set. Details of the algorithm are described in Christmann and Rousseeuw (2001). Of course, this approximation algorithm to compute $n_{co}(B)$ is computer intensive, if the dimensions n or p or the number of samples to be drawn, i.e. $|B|$, are high. Further, drawing random subsamples of the original data set often result in affine hyperplanes for which the number of misclassifications is much higher than for the desired affine hyperplane. There are other determinations of the set B. A naive alternative to $n_{co}(B)$ is to use only one *special* vector b in (4):

$$\begin{aligned} b &= \hat{\theta}_{ML} && \text{if } \hat{\theta}_{ML} \text{ exists} \\ &= \hat{\theta}^{(k)} && \text{otherwise,} \end{aligned}$$

where $\hat{\theta}^{(k)}$ is the last vector computed by the usual Fisher-scoring algorithm to compute the maximum likelihood (ML) estimate in the logistic regression model after stopping due to detection that there is no overlap in the data set. We compute b by the SAS procedure PROC LOGISTIC. This SAS procedure gives a warning if the data set has complete separation or quasicomplete separation, but stores $\hat{\theta}^{(k)}$ and the linear combinations of $(x_i, 1)$ and $\hat{\theta}^{(k)}$. In the same manner, let $s(b)$ be the asymptotic standard error of the ML estimate, if it exists, or the corresponding quantity evaluated for $\hat{\theta}^{(k)}$. Although the naive method $n_{co}(b)$ often gives surprisingly good approximations of n_{co}, it is reasonable to find better approximations of n_{co} in an iterative manner. The heuristic method $n_{co}(h)$ first tries to find a good approximation of n_{co} using the vector b as a starting vector. Secondly a grid search is done where sequentially some of the components of b are set to zero. Then a local search again starting from b is performed based on the following heuristic. We vary the parameter b componentwise in discrete steps by a factor in the interval 0 to 3 taking into account the variability measured by the quantities $s(b)$. If an improvement occurs an additional refinement is made starting from the best solution got so far. Finally the outcomes of all three methods (starting value, grid search, local search) are compared and the best solution is chosen. If during the whole procedure the best possible value of $n_{co} = 0$ is detected the algorithm stops and outputs the current solution. Details of the heuristic method are given in Christmann, Fischer and Joachims (2002).

3 The support vector machine

Vapnik (1998) proposed the support vector machine (SVM), which can be described as follows for the case of pattern recognition. As is usual in the literature on the support vector machine, the responses are recoded as $-1/+1$ instead of 0/1. Decompose θ in the slope part, say $\mathbf{w} = (\theta_1, \ldots, \theta_{p-1})$, and the intercept part θ_p. The aim of the support vector machine is to maximize the width between all possible parallel affine hyperplanes which separate both response groups while penalizing misclassifications by a large positive extra cost C. The support vector machine solves the following quadratic optimization (dual) problem:

$$\begin{aligned} &\text{(D)} \quad \text{minimize (w.r.t. } \alpha\text{):} && \tfrac{1}{2}\alpha'\mathbf{Q}\alpha - \alpha'\mathbf{1} \\ &\phantom{\text{(D)}} \quad \text{subject to} && \alpha'\mathbf{y} = 0 \ \text{ and } \ \mathbf{0} \leq \alpha \leq C\mathbf{1}\ , \end{aligned}$$

where $(\mathbf{Q})_{ij} = y_i y_j x_i' x_j$. Apart from some degenerate cases, the solution of the optimization is unique. It is one advantage of the SVM that the dual problem depends only on inner products between vectors of explanatory variables. Substituting Mercer kernels for the simple dot product allows SVMs to efficiently estimate not only linear but also e.g. kernels with radial basis, cf. Schölkopf and Smola (2002) and Hastie, Tibshirani and Friedman (2001).

4 SVM and RDM

A numerical comparison between RDM and SVM depends on many parameters: algorithms and the actual implementations, tuning constants (number of subsamples for RDM, penalyzing constant C for SVM), definition of the kernel K and the margin for SVM, and of course of the considered data sets.

Joachims (1999) showed that the computation time for the SVM depends heavily on the algorithm and its implementation. His software SVM^{light} is fast even for high dimensional data sets. Schölkopf and Smola (2002, p. 219), demonstrated that different SVM classifiers can yield very different values for the test errors, ranging from 2.6% to 8.9% for the same data set. Hastie, Tibshirani and Friedman (2001, p. 384), compared the SVM using polynomial kernels with degrees d ranging from 1 to 10 with BRUTO and MARS for a simulated data set with 100 observations. The authors showed that the test error of the SVM classifiers may depend critically not only on the kernel but also on the dimensionality of the problem.

Christmann, Fischer and Joachims (2002) compared RDM, SVM and classical linear discriminant analysis (LDA) for various data sets, where some of the data sets are well-known in the field of robust logistic regression. The main criterion was a low number of misclassifications, whereas the computation time was the secondary criterion. Summarizing, the RDM yielded better results for small to moderate sized data sets, say dimension $p \leq 10$ and up to $1,000$ observations. LDA showed the worst performance of these three methods. The misclassification error based on LDA was even larger than the trivial upper bound for n_{co} given by the minimum of the sum of the successes ($y_i = 1$) and the sum of the failures ($y_i = 0$). Especially the heuristic algorithm $n_{\text{co}}(h)$ gave good results. Somewhat surprisingly, there are situations under which the naive algorithm $n_{\text{co}}(b)$ can outperform the algorithm $n_{\text{co}}(B)$. For more complex data sets, the SVM performed well and was very fast. The results for SVM depended on the penalyzing constant C. In general $C = 10^5/(p-1)$ gave better approximations than $C = 10^3/(p-1)$, but the computation time increased. However, SVM sometimes failed to detect that a data set had complete separation, whereas $n_{\text{co}}(b)$ – and $n_{\text{co}}(h)$ if computable in a reasonable time – were able to detect such situations. There do not yet exist algorithms to use RDM for very high dimensional data sets.

5 SVM and QDA

In the following it is shown that the SVM may show an unstable behaviour with respect to training and test errors in certain multi-class classification problems. For simplicity, let us consider a two-dimensional problem with up to four classes without noise, although this phenomenon can also occur for more complex data sets or for data sets with random errors. The explanatory variables x_1 and x_2 are simulated independently from a uniform distribution on the interval $[-1, +1]$. The four possible response classes are constructed as follows, see also Figure 1:

if	$x_1^2 + x_2^2 < 0.15$	then	$y_i = 1$	(ball)
else if	$x_1 - x_2 < -0.75$	then	$y_i = 3$	(triangle) .
else if	$0.15 \le x_1^2 + x_2^2 < 0.6$	then	$y_i = 2$	(ring)
else	$y_i = 0$			(complement)

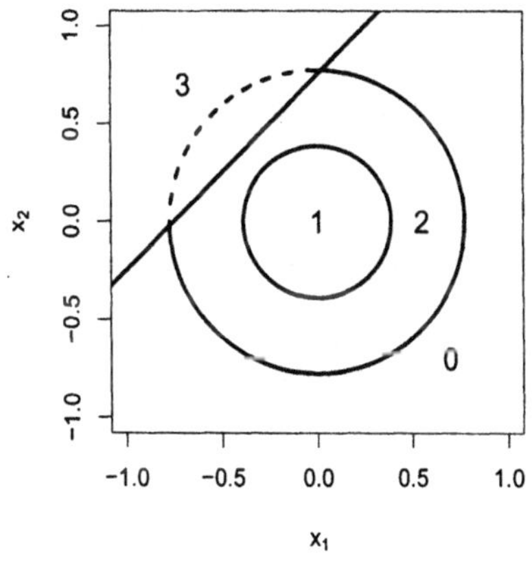

Figure 1: Illustration of simulated patterns.

Three situations are considered: {ball, ring, complement}, {ball, triangle, complement}, and {ball, ring, triangle, complement}. For each situation 250 data sets each with $n = 1,000$ observations were generated. Each data set was randomly splitted into a training data set consisting of 700 observations and a test data set with 300 observations. A SVM with radial basis kernel and cost $C = 10^4$ and a quadratic discriminant analysis (QDA) was done for each training data set and the training error was computed. The R library `e1071` developed by Meyer (2001) was used to train the SVM. The results of training data sets were used to predict the responses of the test data sets to compute the prediction errors for the test data sets, c.f. Table 1. The SVM yields clearly better results than the QDA with respect to training and test errors for both classification problems with 3 response classes. Sometimes even the test errors based on SVM

Table 1: Averaged training and test errors for SVM and QDA.*

Classes	Training error		Test error	
	SVM	QDA	SVM	QDA
(0,1,2)	0.001(0.001)	0.267(0.030)	0.012(0.007)	0.277(0.047)
(0,1,3)	0.001(0.001)	0.275(0.021)	0.011(0.007)	0.278(0.031)
(0,1,2,3)	0.337(0.225)	0.196(0.020)	0.360(0.231)	0.209(0.034)

* Standard deviations are given in parenthesis.

were zero which did not happen for QDA. For the classification problem with 4 response classes the SVM performed often worse than QDA and the prediction errors had a large variability. However, SVM yielded much smaller prediction errors than QDA for *some* of the 250 data sets.

References

Christmann, A., Fischer, P., Joachims, T. (2002). Comparison between the regression depth method and the support vector machine to approximate the minimum number of misclassifications. To appear in: *Computational Statistics.*

Christmann, A. and Rousseeuw, P.J. (2001). Measuring overlap in logistic regression. *Computational Statistics and Data Analysis*, **37**, 65-75.

Hastie, T., Tibshirani, R., Friedman, J. (2001). *The Elements of Statistical Learning. Data Mining, Inference and Prediction.* New York: Springer.

Höffgen, K.U., Simon, H.-U., van Horn, K.S. (1995). Robust Trainability of Single Neurons. *J. Computer and System Sciences*, **50**, 114-125.

Joachims, T. (1999). Making large-Scale SVM Learning Practical. In: *B. Schölkopf, C. Burges, A. Smola (ed.), Advances in Kernel Methods - Support Vector Learning*, MIT-Press.

Meyer, D. (2001). *Support Vector Machines. The Interface to libsvm in package e1071.* Online-Documentation of the package e1071 for "R".

Rousseeuw, P.J. and Hubert, M. (1999). Regression Depth. *J. Amer. Statist. Assoc.*, **94**, 388-433.

Rousseeuw, P.J. and Struyf, A. (1998). Computing location depth and regression depth in higher dimensions. *Statistics and Computing*, **8**, 193-203.

Schölkopf, B. and Smola, A.J. *Learning with Kernels. Support Vector Machines, Regularization, Optimization, and Beyond.* MIT Press, 2002.

Vapnik, V. (1998). *Statistical Learning Theory.* Wiley, New York.

Clockwise Bivariate Boxplots

Aldo Corbellini[1]

[1] Sezione di Statistica, Dipartimento di Economia, University of Parma, Via Kennedy 6, 43100 Parma, ITALY

Abstract. In this paper we suggest a simple way of constructing a robust non parametric bivariate contour based on the rotation of the univariate boxplot which does not necessarily have to use a bivariate generalization of the univariate depth measures. The suggested approach is based on the projection of bivariate data along the round angle. When the angle is a multiple of $\pi/2$ we obtain the traditional univariate boxplot referred to each variable. In all the other cases we obtain univariate boxplots which keep into account in a different way the correlation between the two original variables. We apply the suggested approach to some datasets and exploit the properties of the different choices of defining the inner region and the outer contour. The final result is a simple and easy to construct bivariate boxplot which enables to visualize the location, spread skewness and tails of the data.

Keywords. bivariate boxplot, B-Spline, convex hull, robust centroid

1 Introduction

During the last 10 years there have been some attempts to construct bivariate contours. Goldberg and Iglewicz (1992) suggested a way based on robust elliptic plots and four quarters ellipses matched on their major and minor axes in order to ensure continuity and smoothness in the overall graph. Hyndmann (1996) constructed bivariate contours based on the bivariate density estimated using kernel methods. Zani, Riani and Corbellini (1998) used convex hull peeling and B-spline smoothing. Liu, Parelius and Singh (1999) and Rousseeuw, Ruts and Tukey (1999) developed bivariate contours using the notion of halfspace depth, which is a generalization of the concept of rank to multivariate data. It is well known that there are different generalizations of quantiles for multivariate data (e.g. Chaudhuri, 1996). In this paper we suggest a simple way of constructing a bivariate boxplot based on the rotation of the univariate boxplot which does not necessarily have to use a bivariate generalization of the univariate depth measures. The suggested approach is based on the projection of bivariate data along the round angle. For each of these projections, we construct a univariate boxplot. When the angle is a multiple of $\pi/2$ we obtain the traditional univariate boxplot referred to each variable. In all the other cases we obtain univariate boxplots which keep into account in a different way the correlation between the two original variables.

2 Description of the method

In order to construct a bivariate boxplot we must specify an inner region, (which is the two dimensional extension of the univariate interquartile range of the univariate boxplot), define a robust centroid and devise a method to construct the outer region. In this section we will examine these three steps in detail.

Definition of the inner region. The inner region is the two-dimensional extension of the interquartile range of the univariate boxplot. In one dimension we take the length of the box which contains 50% of the values. In two dimensions a natural way to find the inner region can be based on the univariate projections of the two dimensional data along different directions. In other terms: if we have the coordinates (x_1, x_2) of a point, and we wish to find its new coordinates (x'_1, x'_2) when the axes are rotated clockwise, we can use the following transformation:

$$\begin{pmatrix} x'_1 \\ x'_2 \end{pmatrix} = \begin{pmatrix} \cos\theta & -\sin\theta \\ \sin\theta & \cos\theta \end{pmatrix} \begin{pmatrix} x_1 \\ x_2 \end{pmatrix} \tag{1}$$

In our approach we project the data along different angles, for example,

$$\theta = 0, 2\pi/20, 2\pi/19, \ldots, 2\pi \tag{2}$$

For each projection we construct the univariate boxplot and find the bivariate coordinates of the quartiles. In order to find a convex set of points we can draw a convex hull (Efron, 1965) through them (50%-hull).

As an illustration of the suggested approach we use the data plotted on the left panel of Figure 1. The data are referred to 50 Italian Universities. On the x-axis we have the percentage of students who passed the exams within the prescribed time divided by the total amount of students attending the University, and the y-axis the percentage of courses activated divided by the total amount of teachers. The two variables show a very modest positive correlation ($r = 0.265$). There are a lot of points which seem to depart from an imaginary robust regression line one could draw through the distribution of the data. It is interesting to construct a bivariate non parametric contour which enables us to state what are the atypical Universities. To exemplify our method of construction of the bivariate contour, the right hand panel of Figure 1 shows 3 univariate boxplots. The horizontal and vertical boxplots are based on the projection of the data respectively on the y and x axis. The third boxplot, is based on the projection on the 45° line. For each projection we store the coordinates of the quartiles and we draw a convex hull on these points (left panel of Figure 2). The convex hull of a set of points is the smallest convex polygon containing all the points.

In order to obtain a smooth inner region we can superimpose a B-spline to the 50%-hull. We call inner region the one formed by those units which lie inside or at the boundary of the B-spline superimposed on the 50%-hull.

Remark It is possible to prove that the B-spline will always lie inside the 50%-hull. This implies that the inner region may contain a percentage of points slightly less than 50%. The other theoretical reason why our inner region contains less than 50% of the points is that we take the observations which in every univariate projection always lie inside the interquartile range. The output of the peeling of the univariate boxplots is therefore an inner region which may contain a number of observations smaller than 50%.

The inner region constructed in this way is non parametric (because it does not need parameters to estimate) and adapts to the differing spread of the data in the different directions.

Definition of the robust centroid A robust center can be found as the center of the convex hull superimposed on the points corresponding to the medians of the univariate boxplot in each projection. Alternatively, we can use the

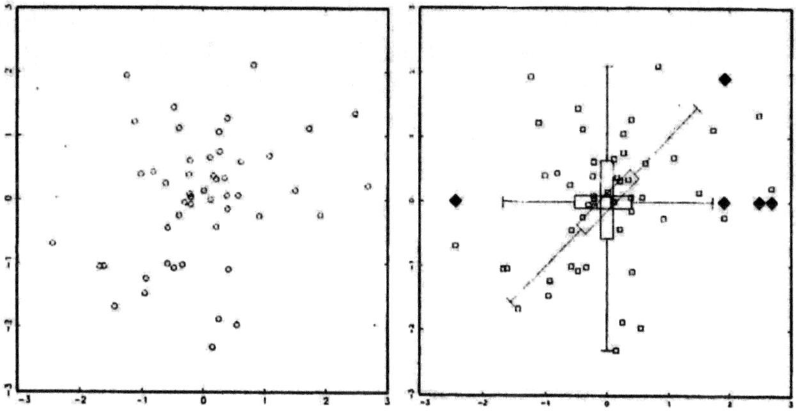

Fig. 1. Plot of percentage of students who passed the exams within the prescribed time versus the number of the activated courses divided by the total number of teachers (left). Univariate boxplots for 3 univariate projection of the data (right).

mean of the points forming the medians in each projection. In this way we can exploit both the efficiency properties of the arithmetic mean and the robustness of the medians. In the left hand panel of Figure 2 we have represented with a cross the center found as mean of the points forming the medians.

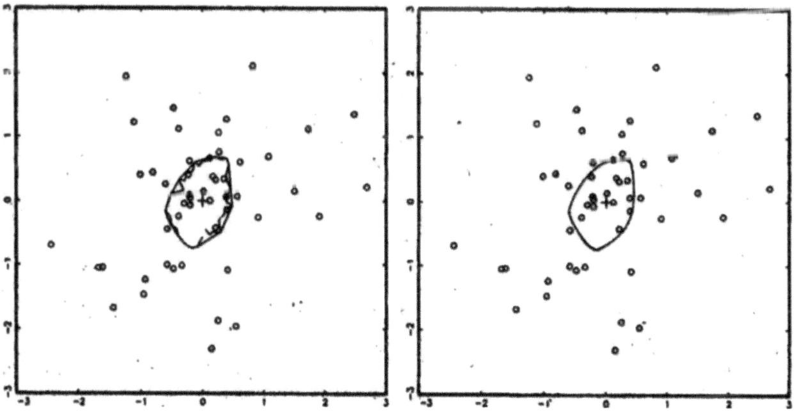

Fig. 2. Plot of the points corresponding to the third quartiles of the univariate projections with superimposed a convex hull (left). *B*-spline superimposed on the convex hull. The points inside or at the boundary of the *B*-spline form the inner region (right).

Definition of the outer region Once we have found a robust bivariate center and a curve representing the inner region we must devise a criterion in order to build an outer contour which discriminates between good and bad observations. In the traditional univariate boxplot we define as outliers those observations which lie outside the interval $[x_{0.25} - 1.5 \times IR, x_{0.75} + 1.5 \times IR]$ where IR is the interquartile range $IR = x_{0.75} - x_{0.25}$. As is well known, if the data come from the normal distribution, the former interval contain approximately 99% of the values. In two dimensions we can use two methods:

- considering a multiple of the distance of the contour which defines the inner region and the robust centroid;
- connecting all the points forming the extremes of the whiskers in the univariate projections.

With the first method the shape of the outer region reflect that of the inner region. The second approach has the advantage that the shape of the outer contour can be completely different from that of the inner region. However, given that the set of points is not necessarily convex, we can draw a convex hull to find a convex set.

Figure 3 shows in the left panel the bivariate boxplot obtained considering a multiple of the distance K of the inner region from the inner region to the robust centroid. The right hand panel shows the outer contour constructed using the whiskers of the projections of the univariate boxplots.

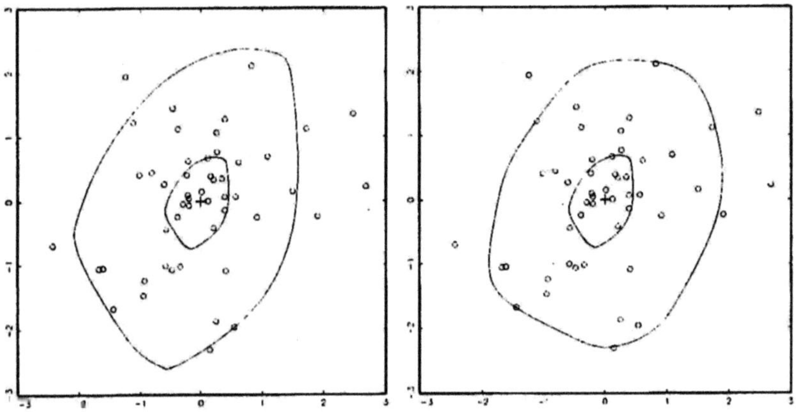

Fig. 3. Bivariate boxplot with the outer contour found using a multiple of the distance from the inner region to the robust centroid (left). Bivariate boxplot with the outer contour found connecting all points forming the upper and lower truncation points of the projections of the univariate boxplots (right).

In this case the contour based on the inner region reflects more the positive correlation present of the points forming the inner region. On the contrary, the outer contour in the right panel is much more elliptical.

3 Examples

In this section we apply our method of constructing the bivariate boxplot to a well known bivariate data set which contains multiple outliers. The data concern the Log light intensity versus Log effective temperature for the 47 stars of the star cluster CYG OB1 (Rousseeuw and Leroy, 1987, pp. 27-28). Once the outliers on the top left panel are removed, the data show a slight positive correlation. Figure 4 shows the bivariate boxplot obtained considering a multiple of the distance from the spline to the robust centroid (left) and the whiskers of the univariate boxplots (right). In both cases the two contours correctly identify the 4 outliers on the top left hand panel. However, the boxplot constructed using the upper whisker seems in same way influence by the presence of the 4 giant stars.

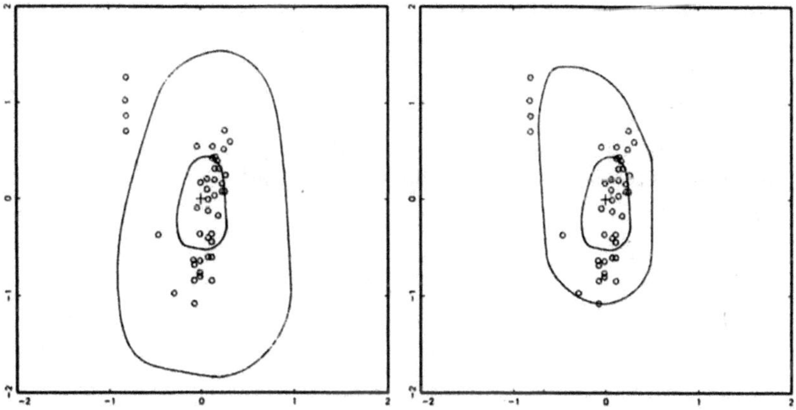

Fig. 4. Plot of Log light intensity and Log effective temperature for the 47 stars of the star cluster CYG OB1. Bivariate boxplot with the outer contour found using a multiple of the distance from the inner region to the robust centroid (left). Bivariate boxplot with the outer contour found connecting all points forming the upper and lower truncation points of the projections of the univariate boxplots (right).

4 Discussion

In this paper we have shown how it is possible to construct a robust bivariate contour based on the projection the univariate boxplot along the round angle. This method produces non parametric contours which adapt to the differing spread of the data in the different directions. In a subsequent paper we will explore the theoretical aspects of the different options of constructing the outer contour. Since the inner contour contains less than 50% of the statistical units, we can therefore consider a different interquantile range, possibly $[x_{0.80} - x_{0.20}]$ or even $[x_{0.85} - x_{0.15}]$. This distance will be reflected on the outer contour plotted with truncation points (whiskers). Another possibility consists in measuring the multiplication factor K of the inner contour with a Monte Carlo approach. Also the B-Spline choice is not mandatory, and some more robust, smoother, second derivative penalized spline could be used instead, as pointed out by Green and Silverman (1994).

References

Chaudhuri, P. (1996). On a geometric notion of quantiles for multivariate data. *J. of the American Statistical Association*, **91**, 862-872.

Efron, B. (1965). The Convex Hull on a Random Set of Points, *Biometrika*, **52**, 331-343.

Green, P.J. and Silverman, B.W. (1994). *Nonparametric Regression and Generalized Linear Models*, Chapman & Hall, London.

Goldberg, K.M., and Iglewicz, B. (1992). Bivariate Extensions of the Boxplot, *Technometrics*, **34**, 307-320.

Hyndman, R.J. (1996). Computing and Graphic Highest Density Regions, *Applied Statistics*, **46**, 295-298.

Liu, R.Y., Parelius, J. M., and Singh, K. (1999). Multivariate Analysis by Data Depth: Descriptive Statistics, Graphics and Inference, *Annals of statistics*, **27**, 783-840.

Rousseeuw, P.J., Ruts, I, and Tukey, W. (1999). The Bagplot: A Bivariate Boxplot, *The American Statistician*, **53**, 382-387.

Rousseeuw, P.J., Leroy, A. (1997). *Robust Regression and Outlier Detection*, Wiley, New York.

Zani, S., Riani, M. and Corbellini, A. (1998). Robust Bivariate Boxpot and Multiple Outlier Detection, *Computational Statistics and Data Analysis*, **28**, 257-270.

Combining Graphical Models and PCA for Statistical Process Control

Roland Fried[1], Ursula Gather[1], Michael Imhoff[2], Melanie Keller[1] and Vivian Lanius[1]

[1] Department of Statistics, University of Dortmund, D-44221 Dortmund, Germany
[2] Surgical Department, Community Hospital Dortmund, D-44137 Dortmund, Germany

Abstract. Principal component analysis (PCA) is frequently used for detection of common structures in multivariate data, e.g. in statistical process control. Critical issues are the choice of the number of principal components and their interpretation. These tasks become even more difficult when dynamic PCA (Brillinger, 1981) is applied to incorporate dependencies within time series data. We use the information obtained from graphical models to improve pattern detection based on PCA.

Keywords. Time series analysis, dimension reduction, online monitoring, pattern detection

1 Introduction

Principal components analysis (PCA) is a frequently used tool for reducing the dimension of high-dimensional data. It is often applied e.g. in statistical process control when many variables have to be monitored simultaneously. The high sampling frequencies, that are possible nowadays for measuring process data, cause both autocorrelations and cross-correlations in the data. Brillinger (1981) suggests a dynamic version of PCA in the frequency domain that takes such correlations at various time-lags into account.

An important problem is the choice of the number of principal components (PC's). Versions of criteria like the eigenvalue criterion, the scree graph and the percentage of explained variability are difficult to use as we have to consider all frequencies. Further problems arise from the interpretation of the PC's. When using methods like Varimax or Procrustes rotation of the eigenvectors we also have to consider all frequencies. In conclusion, this method is computationally expensive when applied to high-dimensional time series.

These problems can be facilitated by partitioning the variables into subsets of closely related variables and extracting PC's for these subsets (MacGregor et al., 1994, Casin, 2001). The subsets need to be specified by background knowledge, that can be derived from prior analysis of historic data. We relate PCA to graphical models (Cox and Wermuth, 1996, Dahlhaus, 2000) and use the latter to enhance the results of pattern recognition based on PCA.

2 Dynamic principal component analysis

Principal component analysis searches those directions in a multivariate data space that capture the largest percentage of variability within the measured data. In the static case, where independent observations of a d-variate random variable $\mathbf{X}$ are observed, the solution is given by an analysis of the eigenvalues and the eigenvectors of the (estimated) covariance matrix $\boldsymbol{\Gamma}$: The j-th principal component $\chi_j = \mathbf{v}_j'\mathbf{X}$, which is defined to have maximum variance given the constraints that $||\mathbf{v}_j|| = 1$ and $\mathbf{v}_j \in \mathbb{R}^d$ is orthogonal to $\mathbf{v}_1, \ldots, \mathbf{v}_{j-1}$, is determined by an eigenvector $\mathbf{v}_j$ for the j-th largest eigenvalue λ_j of $\boldsymbol{\Gamma}$, that is the variance of χ_j. For dimension reduction one uses the first k PC's $\chi = (\chi_1, \ldots, \chi_k)$ instead of the observed data with a suitable $k \leq d$.

PCA as defined above captures relationships only between simultaneous measurements. Brillinger (1981, chapter 9) suggests a dynamic version of PCA for a multivariate time series $\mathbf{x}(1), \ldots, \mathbf{x}(n)$ that takes into account correlations between the components at various time lags assuming that the underlying process $\{\mathbf{X}(t) : t \in \mathbb{Z}\}$ is stationary. For simplicity we further assume that the series has mean zero. While the PC's found by classical ("static") PCA are linear combinations of simultaneous measurements only, the PC's suggested by Brillinger are k-dimensional filtered series

$$\chi(t) = \sum_h \boldsymbol{b}(h)\mathbf{x}(t-h), t \in \mathbb{Z},$$

i.e. linear combinations of past, present and future measurements, where $\{\boldsymbol{b}(h), h \in \mathbb{Z}\}$ is a linear filter of dimension $k \times d$. We get an approximation of the observed time series using another filter of dimension $d \times k$ as

$$\mathbf{x}^\star(t) = \sum_h \boldsymbol{c}(h)\chi(t-h), \quad t \in \mathbb{Z}.$$

The filters minimizing the mean square approximation error are given by

$$\boldsymbol{b}(h) = \frac{1}{2\pi}\int_0^{2\pi} \boldsymbol{B}(\alpha)\exp(ih\alpha)d\alpha, \ \boldsymbol{c}(h) = \frac{1}{2\pi}\int_0^{2\pi} \boldsymbol{C}(\alpha)\exp(ih\alpha)d\alpha$$

$$\boldsymbol{B}(\alpha) = \begin{bmatrix} \overline{\mathbf{v}_1(\alpha)'} \\ \vdots \\ \overline{\mathbf{v}_m(\alpha)'} \end{bmatrix}, \ \boldsymbol{C}(\alpha) = \overline{\boldsymbol{B}(\alpha)}',$$

where $\mathbf{v}_j(\alpha)$ is an eigenvector for the j-th largest eigenvalue $\lambda_j(\alpha)$ of the spectral density matrix $\boldsymbol{f}(\alpha), \alpha \in [0, 2\pi]$, and $\lambda_j(\alpha)$ is the spectral density of the j-th principal component series. The minimal mean square error is

$$E\left([\mathbf{X}(t) - \mathbf{X}^\star(t)]'[\mathbf{X}(t) - \mathbf{X}^\star(t)]\right) = \int_0^{2\pi} \sum_{j=k+1}^{d} \lambda_j(\alpha)d\alpha.$$

The principal component series are uncorrelated at all time lags. Thus, for dynamic PCA we need to perform a classical PCA for each frequency. For the estimation of the spectral density matrix we smooth the periodogram using a Daniell-window.

PCA is popular e.g. in statistical process control since in multivariate data often "redundant" information is contained as there are (approximate) relationships between the variables. A possibility to remove redundancy is to extract those PC's from the data, that contain the relevant information. Typically, one monitors the PC's jointly by Hotelling's T^2 statistic, that is

$\chi(t)'\chi(t)$ as the PC's are orthogonal, and additionally monitors the sum of the squared residuals $Q(t) = (\mathbf{x}(t) - \mathbf{x}^*(t))'(\mathbf{x}(t) - \mathbf{x}^*(t))$. Ideally, significant changes in the former represent a possible change within the same (still valid) relationships, while the latter signal a deviation from these relationships. For a closer inspection one needs to inspect the observed variables then as using T^2 and Q only means a large loss of information. In principle, one could also monitor the PC's individually as they are mutually orthogonal. However, interpretation of the PC's is usually not possible as they are linear combinations of all variables. Sometimes meaningful PC's can be obtained by applying a suitable transformation like Varimax or Procrustes rotation. While the former works automatically but does not always give easily interpretable results, for the latter we have to specify directions using background knowledge. If we were able to derive meaningful PC's in some way we could often specify possible sources of a disturbance directly without looking at the observed variables and avoiding possible problems like T^2 being significant without any of the observed variables showing an unusual pattern.

In a comparative study, Keller (2000) analyses physiological variables measured in intensive care using static PCA, dynamic PCA and multivariate singular spectrum analysis (MSSA, Read, 1993). She finds clinically relevant patterns like outliers and level shifts to be captured best by dynamic PCA, that also needs fewer PC's than the other methods. The essential information contained in 11 physiological variables could usually be compressed into five dynamic PC's. However, problems are noted with respect to a criterion based choice of the number of PC's as dynamic versions of the scree graph or the percentage of explained variability point at fewer PC's in most of the cases. Using too few PC's is not desirable since then the systematic (as opposed to singular or random) variability is partitioned incorrectly into the PC's and the residuals. In consequence we might miss relevant patterns.

This risk is further increased by the fact that PCA has a couple of optimality properties but none of them considers the variables individually (e.g. McCabe, 1984). Particularly, PCA does not guarantee that every variable has a large multiple correlation with the principal components, or, equivalently, that the remaining variability is small for all variables. However, this is interesting in statistical process control. Beale, Kendall and Mann (1967) suggest to base variable selection on this criterion in multivariate analysis. It should guide dimension reduction in the context of online-monitoring, at least if all measured variables are important.

Hence, statistical process control based on any form of PCA may result in unsatisfactory results as the PC's may be difficult to interpret and do not necessarily describe the course of every important variable well. Methods like Procrustes or Varimax rotation do not really solve these problems and are difficult to apply for dynamic PCA since we need to perform the rotation at all frequencies.

3 Graphical models

A *graph* $G = (V, E)$ consists of a finite set of *vertices* V and a set of *edges* $E \subseteq V \times V$, that are ordered pairs of vertices. It can be visualized by drawing a circle for each vertex and connecting each pair a, b of vertices whenever $(a, b) \in E$ or $(b, a) \in E$ by an edge. We focus exclusively on *undirected graphs* where $(a, b) \in E$ implies $(b, a) \in E$. In these graphs all edges represent symmetric associations and are symbolized by lines.

Dahlhaus (2000) introduces *partial correlation graphs* for multivariate time series, which generalize *conditional independence graphs* for independent observations. In these graphs, the vertices represent the measured variables, i.e. the components of the time series. Partial correlation graphs are defined by the *pairwise Markov property*, that states that two components a and b are partially uncorrelated at all time lags given the linear effects of all remaining components if they are not connected by an edge, i.e. $(a,b),(b,a) \notin E$. For independent measurements, the partial correlations given all remaining variables can easily be calculated from the *concentration matrix* $(g_{ab})_{1\leq a,b\leq d}$, which is the inverse of the covariance matrix if the latter is regular, as they equal the negative scaled entries $g_{ab}(g_{aa}g_{bb})^{-1/2}$. For dynamic data, the partial spectral coherences, which are transforms of the partial correlations between the component processes at all time lags into the frequency domain, can analogously be derived from the inverse of the spectral density matrix.

Partial correlation graphs reveal the essential linear relations between multiple variables. There is some (linear) relation between two variables a and b (possibly mediated by other variables) if they are *connected* by a *path*, i.e. if vertices $a = a_0, \ldots, a_l = b$, $l \geq 1$, exist such that there is an edge between each pair of successive vertices. A *connectivity component* of an undirected graph is a maximal subset of pairwise connected variables.

Although graphical models are based on pairwise relations given all the remaining variables, the *global Markov property* allows statements for situations, where we do condition on a separating subset only. Hence, this property is interesting for dimension reduction. It is equivalent to the pairwise Markov property if the spectral density matrix is regular at all frequencies. More precisely, the global Markov property implies that all partial correlations between variables in a set A and variables in a set B are zero given the linear effects of the variables in a set S separating A and B in G. Moreover, if S separates A and B in G then the course of the variables in B does not provide information (in terms of linear regression) on the variables in A given the variables in S and vice versa. This means, for reducing the conditional variance of variables in A (B) we should consider linear combinations of variables in $A \cup S$ ($B \cup S$). Therefore, separations in the partial correlation graph provide information on suitable directions for rotations and on subsets of variables which may be analysed individually.

For exemplification we study a 11-variate time series simulated from a VAR(1)-model. The model parameters have been fitted to a physiologic time series of heart rate, blood pressures etc. as analysed in Keller (2000). Figure 1 shows the partial correlation graph derived by estimating the partial spectral coherences from 600 observations using the program Spectrum (Dahlhaus and Eichler, 2000). Following a suggestion by Brillinger (1996) we measure the strength of the partial correlations via the area under the partial spectral coherences and use gradually distinct edges for illustration (Gather, Imhoff and Fried, 2002). There are three groups of strongly related variables, while the other partial correlations are rather small. Furthermore, variables 10 and 11 do not have any strong partial correlations. Hence, there may be between 3 and 6 strong relationships and accordingly we need between 5 and 8 PC's to capture the systematic variability. As this is a rule of thumb only we should additionally consider the percentage of explained variability. Experimenting with other processes as well we found about 500 observations to be necessary to detect the important associations in 11-variate time series. The "true" partial correlation graph for the underlying stochastic process estimated from 5000 observations is also shown in Figure 1.

In order to get meaningful components we use the partition $\{X_1, X_2, X_3, X_4\}$,

Fig. 1. Partial correlation graph estimated from a simulated time series of length 600 (left) and true partial correlation graph (right).

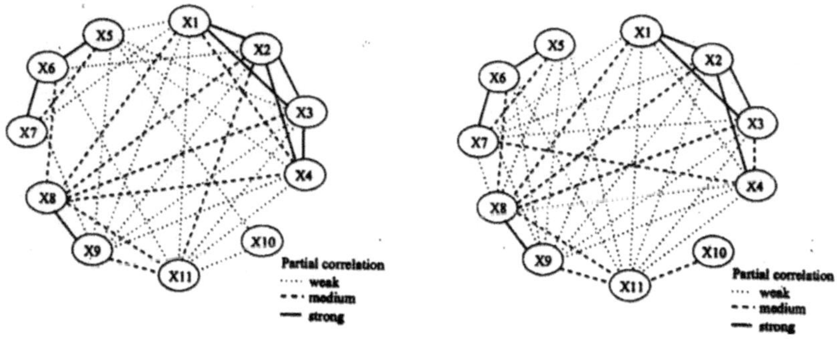

$\{X_5, X_6, X_7\}$, $\{X_8, X_9\}$, $\{X_{10}\}$, $\{X_{11}\}$ and then extract PC's for the former subsets individually treating X_{10} and X_{11} separately. This agrees with existing medical knowledge as these subsets represent intrathoracic pressures, arterial pressures, heart rate and pulse, temperature and pulsoxymetry respectively in the physiological time series. Figure 2 depicts the component processes corresponding to the subsets $\{X_1, X_2, X_3, X_4\}$ and $\{X_5, X_6, X_7\}$ for another time series simulated from the same model with several patches of five subsequent innovative outliers inserted. Innovative outliers are outlying "common" shocks in the VAR(1)-process which influence all variables according to the relationships within the process. Here, using one dynamic PC for each subset we find that these PC's represent the important patterns, but there are smoothing effects as we apply temporal filtering. These PC's are weakly correlated as this is true for the corresponding subsets. Alternatively, one could first calculate a PC for $\{X_8, X_9\}$ and then dynamically regress the other variables on it, before we extract further PC's from the residuals. In this way, we could further reduce the remaining variability and reduce the correlations among the components exploiting the moderate correlations between $\{X_8, X_9\}$ and the other variables. However, this is computationally more expensive and the results would be more difficult to interpret for an operator.

4 Conclusion

Partial correlation graphs with different edges representing different strengths of associations reveal the essential relationships within multivariate time series. Graph separations can be used to identify subsets of closely related variables as well as "isolated" variables. This provides information on the number of PC's needed to extract the systematic variability from the data. Moreover, we may find suitable partitions of the variables and apply methods like blockwise PCA to derive interpretable PC's. In our experience the percentage of explained variability is typically not much smaller than the percentage obtained using the same number of "ordinary" PC's. On the other hand, partitioning the variables reduces computational costs, affords fewer observations and allows more meaningful results. Dynamic PCA allows to incorporate temporal dependencies and captures the systematic variability within multivariate time series better than static PCA. However, for dy-

Fig. 2. Simulated time series (360 observations) partitioned into subsets of strongly correlated variables with inserted patterns. Left: Variables 1 to 4 (top, solid) and corresponding PC (bottom, bold solid). Right: Same for variables 5 to 7.

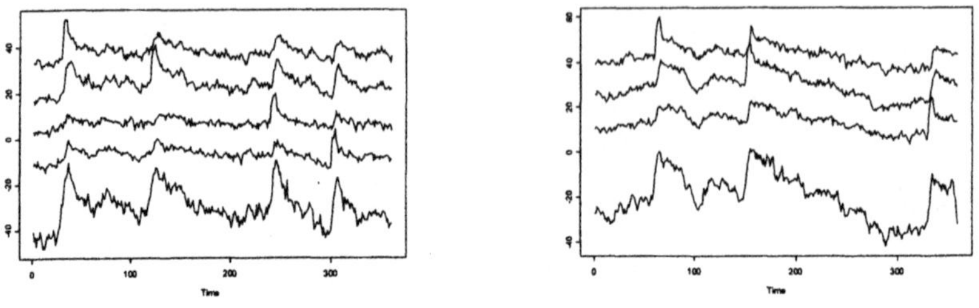

namic PCA more observations are needed to get reliable results than for static PCA. Therefore, dynamic PCA in its current form is mainly useful for retrospective pattern detection in multivariate time series, that is a difficult problem (Tsay, Peña and Pankratz, 2000). Often low order VAR-models are sufficient to capture the essential temporal dependencies between the variables. Dimension reduction techniques based on these model structures could be very useful for real-time process control.

References

Beale, E.M.L., Kendall, M.G. and Mann, D.W. (1967). The discarding of variables in multivariate analysis. *Biometrika*, **54**, 357-366.

Brillinger, D.R. (1981). *Time Series. Data Analysis and Theory.* San Francisco: Holden Day.

Brillinger, D.R. (1996). Remarks concerning graphical models for time series and point processes. *Revista de Econometria*, **16**, 1-23.

Casin, Ph. (2001). A generalization of principal component analysis to K sets of variables. *Computational Statistics & Data Analysis*, **35**, 417-428.

Cox, D.R. and Wermuth, N. (1996). *Multivariate Dependencies.* London: Chapman & Hall.

Dahlhaus, R. (2000). Graphical Interaction Models for Multivariate Time Series. *Metrika*, **51**, 157-172.

Dahlhaus, R. and Eichler, M. (2000). *Spectrum.* Program is available at http://www.statlab.uni-heidelberg.de/projects/graphical.models/

Gather, U., Imhoff, M. and Fried, R. (2002). Graphical models for multivariate time series from intensive care monitoring. *Statistics in Medicine*, to appear.

Keller, M. (2000). *Hauptkomponentenanalyse für intensivmedizinische Zeitreihen* (in German). Unpublished Diploma Thesis, Department of Statistics, University of Dortmund, Germany.

MacGregor, J.F., Jaeckle, C., Kiparissides, C. and Koutoudi, M. (1994). Process Monitoring and Diagnosis by Multiblock PLS Methods. *AIChE J.*, **40**, 826-838.

McCabe, G.P. (1984). Principal variables. *Technometrics*, **26**, 137-144.

Read, P.L. (1993). Phase portrait reconstruction using multivariate singular systems analysis. *Physica D*, **69**, 353-365.

Tsay, R.S., Peña, D., Pankratz, A.E. (2000). Outliers in multivariate time series. *Biometrika*, **87**, 789-804.

Comparing Two Partitions: Some Proposals and Experiments

Gilbert Saporta[1] and Genane Youness[2]

[1] Chaire de Statistique Appliquée-CEDRIC, CNAM, 292 rue Saint, 75003 Paris, France, saporta@cnam.fr
[2] CNAM-ISAE, BP 11 4661, Beirut, Lebanon, genaneb@terra.net.lb

Abstract. We propose a methodology for finding the empirical distribution of the Rand's measure of association when the two partitions only differ by chance. For that purpose we simulate data coming from a latent profile model and we partition them according to 2 groups of variables. We also study two other indices: the first is based on an adaptation of Mac Nemar's test, the second being Jaccard's index. Surprisingly, the distributions of the 3 indices are bimodal.

Keywords. Latent class, K-means, Rand index, Jaccard index, partitions

1 Introduction

When one observes two partitions of the same set of units, a natural question arises: do the partitions agree or disagree? One of the most popular measures of concordance is the Rand's measure of association (Hubert & Arabie 1985), which is based upon the number of pairs of units, which belong to the same clusters. A natural idea is to decide that the 2 partitions do not differ significantly if the index is larger than a critical value. We thus need to know, even approximately, the distribution of Rand's index under some null hypothesis. Few publications (Idrissi 2000) deal with that problem, and only under the hypothesis of independence. However this hypothesis is unrealistic, and departure from independence does not mean that there exists a strong enough agreement (Saporta 1997).

But the difficulty consists in conceptualising a null hypothesis of "identical" partitions and a procedure to check it.

In this communication we first remind the main properties of Rand's index and its distribution under independence. We propose an alternative to Rand's by using an idea derived from Mac Nemar's test for comparing proportions: here we compare proportions of discordant pairs. Finally we also use Jaccard's index. We simulate similar partitions coming from a common latent class model. Then we split arbitrarily the p variables into two groups and perform a partitioning algorithm on each set with the k-means's method.

2 Measures of agreement between partitions

2.1 Notations

Let V_1 and V_2 be two partitions (or two categorical variables) of n objects with the same number of classes k. If K_1 and K_2 are the disjunctives tables and N the corresponding contingency table with elements n_{ij}, we have: $N = K_1^{'}K_2$

Each partition V is also characterized by the $n \times n$ paired comparison table C with general term c_{ij}:

c_{ij}=1 if i and j are in the same class of V, c_{ij}=0 otherwise

We have $C_1 = K_1K_1^{'}$ and $C_2 = K_2K_2^{'}$

The four types of pairs of objects are:

Type 1: pairs belonging to the same class of V_1 and to the same class of V_2
Type 2: pairs belonging to different classes of V_1 but to the same class of V_2
Type 3: pairs belonging to the same class of V_1 but to different classes of V_2
Type 4: pairs belonging to different classes of V_1 and to different classes of V_2

If the respective frequencies of these four cases are : a, b, c, d, we have:

$$a + b + c + d = \frac{n(n-1)}{2}.$$

We note also $A = a + d$ (total number of agreements) and $D = b + c$ (total number of discordances)

2.2 Rand index

The Rand index (similar to Kendall's measure) is the proportion of agreements:

$$R = \frac{2A}{n(n-1)}$$

It may be proved that:

$$A = C_n^2 + \sum_{i=1}^{k}\sum_{j=1}^{k} n_{ij}^2 - \frac{1}{2}\left[\sum_{i=1}^{k} n_{i.}^2 + \sum_{j=1}^{k} n_{.j}^2\right]$$

We will not use the index corrected for chance by Hubert and Arabie (1985), but Marcotorchino's modified version (1991) for all n^2 pairs :

$$R = \frac{2\sum\sum n_{ij}^2 - \sum n_{i.}^2 - \sum n_{.j}^2 + n^2}{n^2}$$

which leads to a simple expression based in terms of paired comparisons:

$$R = \sum_{i=1}^{n}\sum_{j=1}^{n} \frac{(c_{ij}^1 c_{ij}^2 + \bar{c}_{ij}^1 \bar{c}_{ij}^2)}{n^2} \qquad \text{With } \bar{c} = 1 - c$$

Idrissi (2000) used this last formula to study the asymptotic normality of R under the hypothesis of independence. If the k classes are equiprobable, one finds that

$c^1_{ij}c^2_{ij} + \bar{c}^1_{ij}\bar{c}^2_{ij}$ has a Bernoulli distribution with parameter $1-\frac{2}{k}+\frac{2}{k^2}$. Hence:

$$E(R) = 1 - \frac{2}{k} + \frac{2}{k^2}$$

A. Idrissi claims that the Rand index between two categorical variables with k equiprobable modalities follows asymptotically a normal distribution with variance:

$$V(R) = \frac{1}{n^2}(1-\frac{1}{n})(1-\frac{2}{k}+\frac{2}{k^2})(\frac{2}{k}-\frac{2}{k^2})$$

This expression is not valid for small k (especially k=2) and only approximately true for large n since the c_{ij} are not independent due to the transitivity constraints

$$c_{ij} = c_{jk}c_{ik}$$

2.3 An adaptation of Mac Nemar's test

Mac Nemar's test is a well known non-parametric test used to check equality of proportions in matching samples:

a	b
c	d

For instance a represents the number of individuals who keep the same favorable opinion before and after a campaign, d the number of individuals who keep the same unfavorable opinion before and after, b and c are the frequency of those who are changing their opinion. The test statistic corresponding to the null hypothesis of equally changing opinions is: $Mc = \frac{b-c}{\sqrt{b+c}}$ and Mc has a normal distribution N(0,1) under H_0 for n large.

By using the test for the set of object pairs, we have a new way to measure the agreement between two partitions. It is easy to get :

$$Mc = \frac{\sum_i n_{i.}^2 - \sum_j n_{.j}^2}{2\sqrt{\frac{1}{2}(\sum_i n_{i.}^2 + \sum_j n_{.j}^2) - \sum_i \sum_j n_{ij}^2}}$$

In this case also, the transitivity relations between pairs goes against the assumption of independence between pairs.

2.4 The Jaccard's index

The Jaccard's index is a well-known measure of similarity between objects described by presence-absence attributes, used in cluster analysis. It counts the number of common attributes divided by the number of attributes possessed by at least one of the 2 objects.

Applied to the four types of pairs we have:

$$J = \frac{a}{a+b+c} = \frac{n - \sum_u \sum_v n_{ij}^2}{n + \sum_i \sum_j n_{ij}^2 - \sum_i n_{i.}^2 - \sum_j n_{.j}^2}$$

3 The latent class model

Now we have to define what we mean by "two partitions are close". Our approach consists in saying that the units come from the same common partition, the two observed partitions being noisy realisations. The latent class model is well adapted to this problem for getting partitions and have been used by Green and Krieger (1999) in their consensus partition research. More precisely, we use the latent profile model for numerical variables.

	Latent variables	
Observed Variables	*Qualitative*	*Quantitative*
Qualitative	Latent class	Latent traits
Quantitative	Latent profile	Factor analysis

Figure 1. Latent variables methods (Bartholomew & Knott 1999)

The basic hypothesis is the independency of observed variables conditional to the latent classes:

$$f(x) = \sum_k \pi_k \prod_j f_k(x_j / k)$$

The π_k are the proportion of classes and x is the random vector of observed variables, where the component x_j are independent in each class. Here we use the model only in order to generate data and not to estimate parameters.

For getting "near-identical partitions", we suppose the existence of such a partition for the population according to the latent profile model. Data are generated according to this model, with independent normal components in each class, in other words, a normal mixture model. Then we split arbitrarily the p variables into two sets and perform a partitioning algorithm on each set. The two partitions should differ only at random. We are thus enabled to get simulated sampling distributions of Rand, Mc Nemar or Jaccard's index. Our algorithm has four steps:

1. Generate the sizes n_1, $n_{2,}$ .., n_k of the clusters according to a multinomial distribution M(n; $\pi_1 \ldots \pi_k$)
2. For each cluster, generate n_i values from a random normal vector with p independent components
3. Get 2 partitions of the units according to the first p_1 variables and the last p-p_1 variables
4. Compute association measures

4 Empirical results

We applied the previous procedure with 4 equiprobable latent classes, 1000 units and 4 variables. The parameters of the normal distribution are chosen in such a way that $\left|m_{kj} - m_{k'j}\right| > 1.5\sigma_j$ for every j and k.

The number of iterations N is 1000. We present only one of our simulations (performed with S+ software).

Class n°1		*Class n°2*		*Class n°3*		*Class n°4*	
X1	N(1.2,1.5)	X1	N(-2,1.5)	X1	N(5,1.5)	X1	N(8,1.5)
X2	N(-10,2.5)	X2	N(0,2.5)	X2	N(-17,2.5)	X2	N(3.8,2.5)
X3	N(6,3.5)	X3	N(12,3.5)	X3	N(13,3.5)	X3	N(-5,3.5)
X4	N(-20,4.5)	X4	N(-12,4.5)	X4	N(0,4.5)	X4	N(7,4.5)

Table 1. The normal mixture model

The following figure shows the spatial repartition of one of the 1000 iterations.

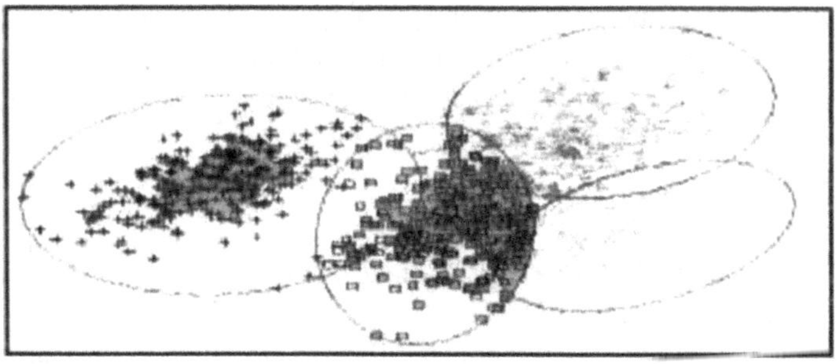

Figure 2. The first two principal components of one of the 1000 samples

Then, we compute 2 partitions with the k-means methods:the first one with X_1 and X_2, the other one with X_3 and X_4; we calculate the association indices 1000 times. Our results show that the distributions of these indices are far from a normal distribution, which is not surprising since the theoretical values should be high (close to 1 for Rand), but they are actually bimodal: this unexpected result has been observed systematically.

We noticed that all the observed Rand's values are over 0.72. Under the hypothesis of independence $E(R) = 0.625$, and with 1000 observations, independence should have been rejected for R>0.626 at 5% risk. The 5% critical value is much higher than the corresponding one in the independence case. It shows that departure from independence does not mean that the two partitions are close enough. However it is not possible to derive universal critical values since the distribution of R depends on the number of clusters, on their proportions and separability. An ad hoc bootstrap method may solve this problem.

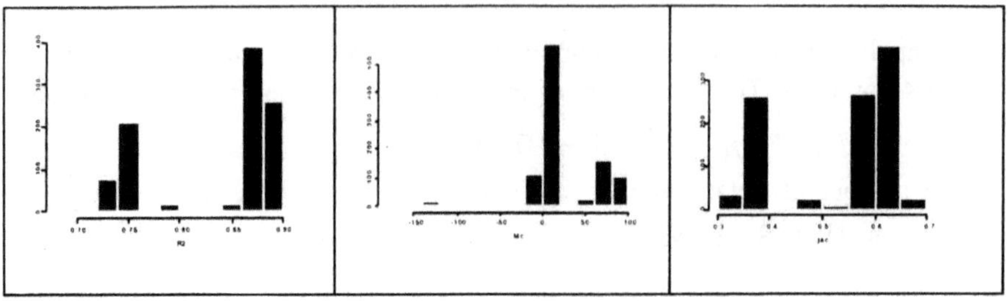

Figure 3. Distributions of Rand, Mac Nemar and Jaccard's indices.

5 Discussion

A latent class model has been used to deal with the problem of comparing close partitions and three agreement indices have been studied. The Rand index give the same importance to pairs in the same class, and to pairs in different classes of both partitions, which is arguable. Mac Nemar and Jaccard indices do not have this drawback. The distributions of the three proposed indices have been found very different from the case of independence and are bimodal. The bimodality might be explained by the presence of local optima in the k-means algorithm: we are studying this point. Finally, one has to add that agreement measures are only one of the many facets of comparing partitions.

References

Bartholomew, D.J. & Knott, M. (1999). *Latent Variable Models and Factor Analysis*, London: Arnold.

Green, P.& Kreiger, A. (1999). A Generalized Rand-Index Method for Consensus Clustering of Separate Partitions of the Same Data Base, *Journal of Classification*, **16**, 63-89.

Hubert, L. & Arabie, P.(1985). Comparing partitions, *Journal of Classification*, **2**, 193-198.

Idrissi, A. (2000). *Contribution à l'unification de Critères d'Association pour Variables Qualitatives*, Ph.D., Paris: Université Pierre et Marie Curie.

Marcotorchino, J.F.& El Ayoubi, N. (1991). Paradigme logique des écritures relationnelles de quelques critéres fondamentaux d'association, *Revue de Statistique Appliquée*, **39**, 2, 25-46.

Saporta, G. (1997). Problèmes posés par la comparaison de classifications dans des enquêtes différentes, in: *Proceedings of the 53rd Session of the International Statistical Institute*.

Comparison of Nested Simulated Annealing and Reactive Tabu Search for Efficient Experimental Designs with Correlated Data

Neil Coombes[1], Roger Payne[2] and Paulo Lisboa[3]

[1] Wagga Agricultural Institute, Pine Gully Rd, Wagga Wagga, NSW, 2650, Australia
[2] Statistics Department, IACR-Rothamsted, Harpenden, Herts, AL5 2JQ, UK.
[3] Department of Computing and Mathematical Sciences, Liverpool John Moores University, Liverpool, L3 3AF, UK.

Abstract. More complex analysis of correlated data has created a need for algorithms to find efficient designs for known correlation structures. Reactive Tabu Search (RTS) is adapted to the design of experiments with correlated error. It uses a scavenging descent local search and repeat returns to designs in the search path are used to invoke diversification strategies. Nested Simulated Annealing (NSA) is used in block design and adapted here for correlated data. NSA uses supercooling and reannealing to speed and continue the search. NSA is shown to be a less efficient algorithm than RTS as measured by interchanges tested and computing time.

Keywords. reactive tabu search, design optimisation, A-efficiency, correlated data

1 Introduction

There have been considerable advances recently in the methods for analysing correlated data. Gilmour *et al.* (1997) and Gilmour *et al.* (1998) give a thorough coverage in the context of agricultural field trials. These methods provide powerful extensions to conventional methods of analysis, catering for continuous trends of variability as well as the discrete blocking used in classical analysis of variance. The methods are applicable to trends in both space and time and may be useful in agricultural, environmental, ecological and medical contexts. To exploit the analysis techniques fully, new types of design are required to estimate treatment effects efficiently in the presence of complicated correlation structures. The search for A-efficient experimental designs is an example of an NP-hard discrete optimisation problem. The calculation of A-efficiency measures is itself computationally intensive so efficient search algorithms are vital.

Simulated annealing (SA) and Tabu search (TS) have been used successfully to construct small to moderately sized experiments, for example 5 by 5 row-column designs for 5 treatments (Eccleston and Chan, 1998). However, these techniques are inefficient for the larger designs used in many practical applications. For example, variety trials often have around 50 treatments replicated 3 or 4 times. Nested Simulated Annealing (NSA), (Whitaker, 1995) has been used in block design search and is an improvement on SA. Reactive Tabu Search (RTS), (Battiti and Tecchiolli, 1992) is a self-tuning heuristic from Operations Research which can be adapted well to the design situation.

Following a brief outline of the spatial model and design efficiency measures in Section 2, the two methods are described in Section 3 and their performance compared in Section 4.

2 The spatial model

The spatial mixed linear model for the data y presented in Gilmour *et al.* (1997) may be rewritten to emphasise the fixed effects of interest τ,

$$y = X_1\tau + X_2\beta + Zu + \xi + \eta \quad (1)$$

$\tau^{(t_1\times 1)}$ is the vector of treatment fixed effects of interest with corresponding design matrix $X_1^{(n\times t_1)}$. $\beta^{(t_2\times 1)}$ is a vector of fixed effects including an overall mean effect with design matrix $X_2^{(n\times t_2)}$. $u^{(b\times 1)}$ is a vector of random effects with design matrix $Z^{(n\times b)}$. $\xi^{(n\times 1)}$ is a random error vector exhibiting the spatial dependence present and $\eta^{(n\times 1)}$ is an independent random error vector. (u, ξ, η) are assumed to be pairwise independent.

From this model, let

$$y \sim N(X_1\tau + X_2\beta, \sigma^2 V)$$

where

$$V = Z_r Z_r'\gamma_r + Z_c Z_c'\gamma_c + \Sigma_{rc}\gamma_s + I\gamma_e \quad (2)$$

Here Z_r and Z_c are the row and column design matrices, and Σ_{rc} is the spatial process. Separable autoregressive processes for rows and columns, AR1 × AR1, are used in this paper. The current algorithm requires the user to specify a rectangular design layout, any fixed block dimensions, linear rows or columns, autoregressive parameters for rows and columns and the size of the spatial, row, column and error variance ratios, $(\gamma_s, \gamma_r, \gamma_c, \gamma_e)$.

The information matrix C of the elementary treatment contrasts $(\tau_i - \tau_j)$ is given by

$$C = X_1'V^*X_1$$

where

$$V^* = V^{-1} - V^{-1}X_2(X_2'V^{-1}X_2)^{-1}X_2'V^{-1} \quad (3)$$

In experimental design where all contrasts are of equal interest the usual measure of design optimality is the A-value, which is the trace of the generalised inverse of the information matrix. The aim of a design search algorithm is to find the permutation of the rows of X_1 which minimises the A-value for a given dependence structure V.

3 Search algorithms

New permutations of designs are obtained from an initial design by interchanging two rows of the treatment design matrix. The A-value is calculated for each interchange and search rules determine whether interchanges are accepted. The local search used in both algorithms is a scavenging descent which tests all valid positional interchanges for the design and accepts any interchange that improves the design as it is found.

3.1 Nested Simulated Annealing

SA uses a temperature parameter and a gradual cooling schedule to approximate the process of a cooling solid. Random changes to the design are accepted with probability $\min(1, \exp(-(A_{new} - A_{old})/T))$. NSA, (Whitaker, 1995), is an SA algorithm modified to reach better designs more quickly and to continue the search once an initial solution is found. NSA carries out SA at a number of pre-specified levels. The algorithm moves between the levels by supercooling, after improvement to the best A-value at the current level, and re-annealing after each local search is complete. The initial temperature is based on the maximum change in A-value for a sample of interchanges on the initial design. The rapid cooling divisor C is calculated to switch the algorithm to local search at the bottom level.

Nested Simulated Annealing schematic algorithm

```
set Anneal True, set Temperature T = T0, set bestA to Inf. at all levels
set Level to 0
set First_this_level True
repeat
   if T is Low then
      call Local_Search
      set Anneal False
   else
      if First_this_level then
         call Anneal_step for N random interchanges
      else
         call Anneal_step to first improvement or N interchanges
      endif
   endif
   if Anneal then
      if Improvement then
         store Best_this_level, T; invoke Supercool T = T/C
         Increase Level by 1
         set First_this_level True
      else
         Standard {Lundy or geometric} cool
         set First_this_level False
      endif
   else
      set Best_this_level to Inf. Decrease Level by 1
      if Level is not negative then
         restore Best_this_level, T; Reanneal
         set First_this_level True
         set Anneal True
      endif
   endif
until Level is less than 0
```

NSA sets the best A-value at each level to a high value, then searches for improvements using standard SA, testing a specified minimum number (N) of interchanges at each temperature until the latest design is an improvement on the initial design for that level. It stores the current best design and temperature at that level, carries forward the latest improved design and supercools to the next level. NSA quickly reaches a temperature where annealing ceases and a local search finds a local minimum. NSA then returns to the previous level and restores the temperature and design at that level. Annealing then continues until, finally, the local search takes place at Level 0.

3.2 Reactive Tabu Search

TS is a modified local search. The scavenging descent accepts any improvements to the best design as they are found and any improvement to the working design unless the interchange used returns two treatments to positions they occupied within a given tabu period. If no improvement is possible then the move increasing the A-value least is selected, again subject to the tabu rule. RTS (Battiti and Tecchiolli, 1992) uses the search history to modify the tabu period or to trigger a random shift in the search. Initially the tabu period is set to one, only preventing a return to the previously visited design. The tabu length is increased whenever a return to a repeat design is accepted. It is decreased when the time between repeat designs is greater than the moving average of times to repeats. Too many repeats affect a *chaos* parameter causing a number of random interchanges to be accepted.

Reactive Tabu Search schematic algorithm

```
Initialise: Working=Best={InitD,InitA}, history structures
repeat
   get a randomisation of positional swaps in List
   for each swap in List
      calculate Temp = update(Working,swap)
      if TempA improves BestA then
         update Best and Working designs, history
      else if (TempA < WorkingA) and (swap not tabu & TempD not in history) then
         update Working, history
      endif
   endfor
   if no improvement to Working then
      if all moves tabu then
         choose best move, update Working
      else
         choose best non-tabu move, update Working
      endif
      Tune search
      if Working in history then
         calculate time since last found, update movav
         increase tabu exclusion time
         if multi-repeat design then
            increment chaos counter
         endif
      else
         if tabu length stable then
            decrease tabu exclusion time
         endif
      endif
      if search chaotic then
         use n = fun(ran, movav) random interchanges to shift the design
         reinitialise memory structures
      endif
   endif
until Stop condition
```

Three history structures are used in RTS: the last time that each treatment occurred in each position in the design as measured by tested interchange number, the frequency of each positional swap used in the last h accepted interchanges, and the last h unique design permutations accepted. The first two are used to define tabu moves, while the latter is used to cause the search to diversify. RTS may be stopped by reaching a specified number of interchanges tested, by testing a specified number of randomisations of the interchanges, or by reaching a specified A-value.

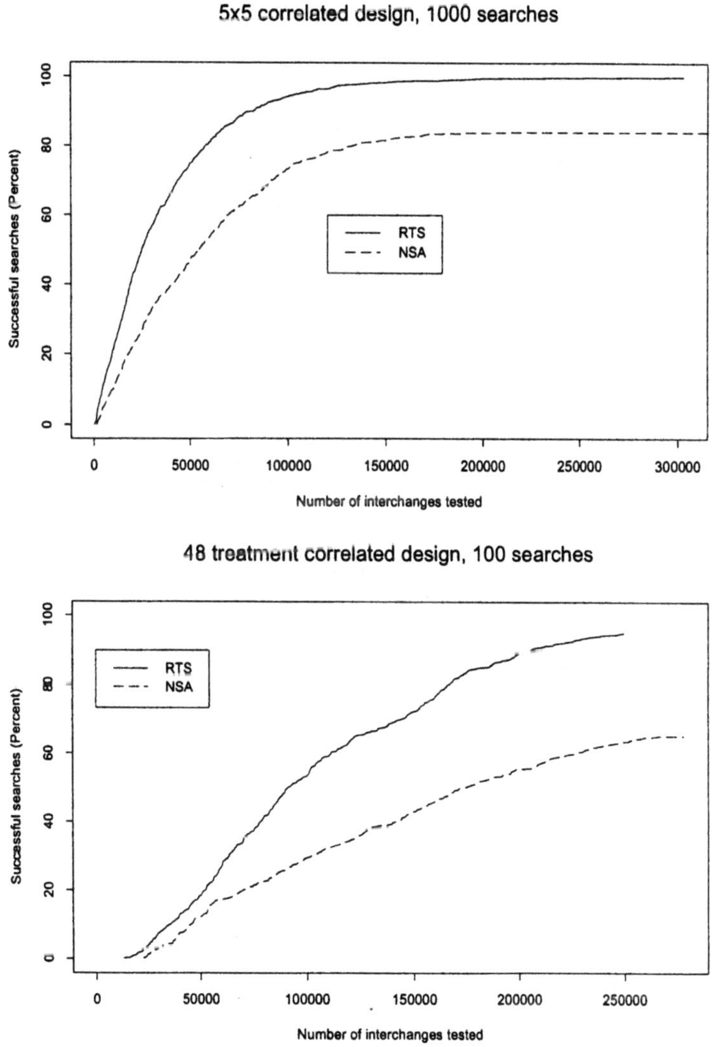

Fig. 1. Comparison of RTS and NSA reaching target A-value for 2 correlated designs.

4 Comparison of the algorithms

Algorithms must be tested by repeat searches for each design configuration because the random nature of the searches can produce quick solutions depending on a combination of algorithm, starting design and search path but not necessarily due to any advantage inherent in the algorithm.

NSA uses random interchanges at high temperatures to diversify the search and rapid cooling to intensify the search to a local minimum. RTS uses a scavenging local search to quickly find a local minimum and relies on repeat configurations and tabu length changes to diversify the search. The main disadvantage to NSA is its lack of intensive searching near local minima. The temperature parameter prevents the escape from the local minimum that would allow nearby improvements to be found. Re-annealing is effective at diversifying the search and may be advantageous in block designs where the design space has many equivalent designs near local minima. The main disadvantage with RTS is ensuring the search diversifies enough. The search can be held up close to minima in flat areas of the design space. A workaround may be used for block designs to encourage greater diversification by restricting the choice of interchanges that can be used for a non-improving move.

The examples used in Figure 1 are for AR1$\times$AR1 designs for 5 treatments completely randomised in 5 by 5 plots and 48 treatments in 16 by 12 plots with 8 by 6 corner replicates respectively. The small example has a known optimum which is found in a high proportion of searches by both methods. The larger example has no known optimum so a target value, approximately 99% of the best A-value found, was set for the search. For both examples RTS consistently finds the target A-value more often for the number of interchanges tested. For the small example the cost in computing time per thousand interchanges tested is approximately 0.062 seconds for RTS and 0.048 seconds for NSA. For the larger example the cost is 2.225 seconds for RTS and 2.215 seconds for NSA. Adjusting for these time differences RTS still finds the target A-values more quickly than NSA. Note that NSA has an overhead at the start of the algorithm in establishing one-interchange A-value changes. RTS has an initial cost in setting up memory structures then ongoing costs of maintaining them and checking for repeat designs. For larger problems, the time to calculate A-values dominates computing time.

RTS is a flexible algorithm and can be used in design search for multi-site, factorial, multiple objective and unreplicated designs. The results produced in these types of design are encouraging. RTS can be the basis of a general design package for a range of correlated designs.

References

Battiti, R. and Tecchiolli, G. (1992). The reactive tabu search. *ORSA Journal on Computing*, **6**, 126-140.

Eccleston, J. and Chan, B. (1998). Design algorithms for correlated data. In: *COMPSTAT 1998, Proceedings in Computational Statistics*, 41-52. Heidelberg:Physica-Verlag.

Gilmour, A.G., Cullis, B.R., Frensham, A.B. and Thompson, R. (1998). (Co)variance sytuctures for linear models in the analysis of plant improvement data. In: *COMPSTAT 1998, Proceedings in Computational Statistics*, 53-64. Heidelberg: Physica-Verlag.

Gilmour, A.G., Cullis, B.R. and Verbyla, A.P. (1997). Accounting for natural and extraneous variation in the analysis of field experiments. *Journal of Agricultural, Biological, and Environmental Statistics*, **2**, 269-293.

Whitaker, D. (1995). A nested simulated annealing algorithm. *Journal of Statistical Computation and Simulation*, **53**, 233-241.

Computational Connections between Robust Multivariate Analysis and Clustering

David M. Rocke[1] and David L. Woodruff[2]

[1] Department of Applied Science, University of California at Davis, Davis, CA 95616, USA
[2] Graduate School of Management, University of California at Davis, Davis, CA 95616, USA

1 Introduction

In this paper we examine some of the relationships between two important optimization problems that arise in statistics: robust estimation of multivariate location and shape parameters and maximum likelihood assignment of multivariate data to clusters. We offer a synthesis and generalization of computational methods reported in the literature. These connections are important because they can be exploited to support effective robust analysis of large data sets.

Recognition of the connections between estimators for clusters and outliers immediately yields one important result that is demonstrated by Rocke and Woodruff (2002); namely, the ability to detect outliers can be improved a great deal using a combined perspective from outlier detection and cluster identification. One can achieve practical breakdown values that approach the theoretical limits by using algorithms for both problems. It turns out that many configurations of outliers that are hard to detect using robust estimators are easily detected using clustering algorithms. Conversely, many configurations of small clusters that could be considered outliers are easily distinguished from the main population using robust estimators even though clustering algorithms fail.

There are assumed to be n data points in R^p and we may refer to them sometimes as a set of column vectors, $\{\boldsymbol{x}_i\} = \{\boldsymbol{x}_i | i = 1, 2, \ldots, n\}$. We are concerned here primarily with combinatorial estimators and restrict ourselves to those that are *affine equivariant.*

2 Robust Estimation and Clustering

2.1 Robust Estimation and Outlier Detection

The MCD was defined by Rousseeuw as that sample of size h that results in the lowest covariance determinant. Usually, h is chosen as the "half-sample size" $\lfloor (n + p + 1)/2 \rfloor$, which is the choice that maximizes the breakdown (Rousseeuw and Leroy 1987; Lopuhaä and Rousseeuw 1991). We define the MCD formally as the solution to the problem of selecting a set $H \subset N$ of size h so as to minimize $|W|$, where $N = \{1, 2, \ldots, n\}$ and where

$$W = \sum_{j \in H} (x_j - \bar{x}_H)(x_j - \bar{x}_H)^T,$$

and

$$\bar{x}_H = h^{-1} \sum_{j \in H} x_j.$$

The location and shape estimates are then x_H and $n^{-1}W$.

Rigorous analysis of the theoretical computational complexity of the MCD by Bernholt and Fischer (2001) implies that the problem must be addressed using heuristic algorithms that search for a good solution. The difficulty in constructing such algorithms is that if points that are "outliers" are included in H, they will distort the estimates of shape and location so as to make it difficult to detect that they are outlying.

An analysis of difficult forms is provided by Rocke and Woodruff (1996). An extremely plausible, yet still difficult, form of contamination is referred to as *shift outliers* (see Hawkins 1980 page 104). Shift outliers have the same shape and size as the main population, but a different location.

2.2 Maximum Likelihood Clusters

The problem of finding the maximum likelihood assignment of data points to clusters is similar, but the literature has developed separately for the most part. There is a very large literature devoted to clustering when there is a metric known in advance. However, in order to retain affine equivariance, we rely on the smaller but growing literature related to using metrics gleaned from the data itself.

A thorough examination of criteria based on the likelihood is given by Banfield and Raftery (1993). Their paper proposes a number of criteria that maximize the likelihood conditional on a clustering, under a number of assumptions about the relative sizes and shapes of the clusters. A popular method is to solve problem (MINW), (Friedman and Rubin 1967), which finds the clustering that minimizes the determinant of the pooled covariance $|W|$ where

$$W = \sum_{i=1}^{g} W_i,$$

$$W_i = \sum_{j \in H_i} (x_j - \bar{x}_H)(x_j - \bar{x}_H)^T,$$

and where $H_1, H_2, \ldots, H_g$ is a partition of N. This corresponds to maximum classification likelihood under the assumption that the data vectors are multivariate normal with cluster covariances such that $\Sigma_1 = \cdots = \Sigma_g$.

An objective that is similar from a computational standpoint is

$$\sum_{i=1}^{g} h_i \log \left| \frac{W_i}{h_i} \right|,$$

where $h_i = |H_i|$. The minimum corresponds to a maximum classification likelihood under the assumption of heterogeneous covariance matrices. It was first given by Scott and Symons (1971) and adjusted by Banfield and Raftery (1993). Call the problem with this objective function (MIND). In order to avoid singularities, as a practical matter a parameter $h_{\mathrm{min}} > p$ must be given for the minimum number of points assigned to each cluster. Difficult forms are discussed by Coleman and Woodruff (2000).

3 Neighborhoods

Although most were not written using the terminology of local search, the proposals in the literature for algorithms for robust estimation and cluster finding can be cast in that framework. This facilitates synthesis and some generalization. Local search is defined relative to an evaluation function for an optimization problem and a neighborhood structure.

3.1 Local Search

We define the generic hard problem to which local search algorithms are applied as

$$\min_{\tau} f(\tau) \quad \text{(P)}$$
$$\text{Subject to: } \tau \in \Xi$$

where the set Ξ is intended to summarize the constraints placed on the decision vector τ. Solution vectors that (do not) satisfy the constraints are said to be *(in)feasible.* The constrained optimization literature refers to all data for the problem—the data that specifies the *objective function* $f(\cdot)$ and Ξ—as (P). It is easy to see that the MCD, MINW and MIND estimators can all be stated in this form.

Neighborhoods are based on *moves* from one solution to another. All of the solutions that can be reached from a given solution in one move are said to be in the neighborhood of the solution. We use the notation $\mathcal{N}(\tau)$ to indicate the set of solutions that are neighbors of a solution τ.

Simplifying things somewhat to ease exposition, we can define *Steepest descent* as a general purpose procedure that begins with an initial solution, τ^0, and selects solutions at iteration $k > 0$ using the relation

$$\tau^k = \operatorname*{argmin}_{\tau \in \mathcal{N}(\tau^{k-1})} f(\tau)$$

(a tie breaking rule may be needed). The algorithm terminates when there are no lower objective function value solutions in the neighborhood of the current solution. Such a solution is referred to as a *local minimum.* A *first-improving* descent is similar but requires more notation and proceeds through an ordered neighborhood until an improving move is found which is then immediate made. After a move, the traversal of the neighborhood continues using the ordering (or some approximation to it). One possibility (for either steepest descent of first improving) is to repeat the descent many times, restarted at a random starting point each time the algorithm hits a local minimum.

Many general purpose optimization algorithms are based on combinatorial steepest descent (e.g., simulated annealing). An application of some of these methods to computation of the MVE is given by Woodruff and Rocke (1993).

3.2 Exchange Neighborhoods

For the MCD a sensible neighborhood is one where a point in H is exchanged with one not currently in H. We refer to this as an *exchange* or *swap* neighborhood. For (MINW) and (MIND) the corresponding neighborhood is one where a point is moved from one group to another. For solutions where the size constraints are not binding, the neighborhood has $(g-1)n$ solutions. There are fewer neighbors of solutions for which one or more of the size constraints is binding.

3.3 Constructive Neighborhoods

The swap neighborhoods can be classified as *transition* neighborhoods that move from one full solution to another. In contrast, *constructive* neighborhoods move from a partially specified solution to a more complete solution and *destructive* neighborhoods are the opposite. For the problems under consideration here, a constructive neighborhood would correspond to moves that transform a solution with some but not all data points assigned to one with one (or more) additional data point(s) assigned; a destructive neighborhood would correspond to moves that unassign one or more data points. So-called *greedy* algorithms can then be cast as steepest descent with a constructive neighborhood.

A constructive neighborhood for the MCD "surrounds" a set of points, H, that has between $p+1$ and h members. A subset of H (typically either empty or all of H) is required to be included in all neighbors; call this subset $\tilde{H}$. Finally, a subset of N is eligible for inclusion in any of the neighbors (typically all of N); call it $\tilde{N}$. Given a set H, moves to a new set H' must be such that all of the points in $\tilde{H}$ are in H' plus one or more points from $\tilde{N}$. This is summarized as follows:

p	Dimension of the data (given)
n	Number of data points (given)
h_{min}	Minimum cluster size (given)
N	Index set $1, \ldots, n$
H	Subset of N (currently) estimated to be in the majority population
$\tilde{N}$	Subset of N eligible for inclusion in the H during the next iteration
$\tilde{H}$	Subset of N required to be included in H during the next iteration

Algorithms based on steepest descent must specify the method of constructing an initial solution, an evaluation function, $\hat{f}(\cdot)$, and perhaps also a *refresh* period, ψ, that controls how many moves are allowed before corrections are made for the fact that $\hat{f}(\cdot)$ is based on an approximation to the current state of the neighborhood. Some of the algorithms in the literature have started with an initial set as large as a half-sample (e.g., Hawkins 1994), but many use a starting set of size $p+1$, and we have conducted simulation studies confirming this choice of size for computational reasons. There are three affine equivariant possibilities reported in the literature for picking a initial sets H to begin the descent process.

- Select $p+1$ points at random (RAND).
- Select $p+1$ points that are "good" based on a heuristic (HEUR).
- Select the $p+1$ points that have lowest Mahalanobis distance from the result of the last full solution constructed (WALK).

Clearly, use of the last choice results in an iterative algorithm that can be terminated either after some number of constructions, or when it reaches a fixed point. Refer to such an iterative algorithm as a *walking* algorithm. Note that K-means algorithms are generally (non-affine-equivariant) walking algorithms using this convention. Such algorithms are common in clustering, but apparently were first used in calculating the MCD by Hawkins (1999) and Rousseeuw and Van Driessen (1999) independently. A walking algorithm must be started using either RAND or HEUR.

For the estimators of interest to us, there are two move evaluation function commonly in use. One is based on Mahalanobis distances from the mean of points in H using the covariance matrix of points in H as the metric;

Algorithm	$\hat{f}(\cdot)$	$\tilde{H}$	ψ	Start/Restart	Walking?
Fast-MCD Rousseeuw 1999	MAHAL	$\emptyset$	∞	HEUR	Yes
Forward Atkinson 1994	UPDATE	H	1	HEUR	No
Improved FSA Hawkins 1999	UPDATE	$\emptyset$	∞	RAND	Yes
Hadi Hadi 1992	UPDATE	$\emptyset$	1	HEUR	No
Multout Rocke and Woodruff 1996	UPDATE	H	1	HEUR	No

Table 1. Summary of Affine Equivariant Constructive Neighborhoods for MVE/MCD Algorithms Reported in the Literature as cast in a Local Search Framework

i.e., select the point(s) $i \in \tilde{N}$ that minimize(s) $d^2_{S_H}(x_i, \bar{x}_H)$ where $d^2_{S_H}$ is the Mahalanobis distance under covariance matrix S_H, and $\bar{x}_H$ is the mean of the points in H. Call this evaluation method MAHAL. We indicate that multiple points might be selected, because if the refresh period is infinite, then one selects the h or $h-p-1$ (depending on the makeup of $\tilde{H}$) points that have lowest distance and all of the moves can be made at once. If the refresh period is one, then after each point was added, the values of $\bar{x}_H$ and S_H are updated. An alternative to MAHAL is to use update formulas to predict the effect of a move. The refresh period specifies how often the mean and covariance are recomputed from the current set H. Call this method UPDATE.

Table 1 gives a summary of constructive neighborhoods that have been reported in the literature for the MCD (and/or the MVE). In all cases, $\tilde{N} = N \setminus H$. Of course, this table provides only a summary of the constructive neighborhood used and not a complete description of the algorithms. The start-restart heuristic used by Fast-MCD is as follows: do a large number of random starts each followed by walking that is terminated after two constructive descents; the best ten results are then pursued with a convergent walking algorithm. The heuristic reported for use with Forward is to select the $p+1$ points closest to the mean of all of the data under the metric for the data. Hadi suggest the use of a non-affine equivariant starting heuristic, but his algorithm is otherwise affine equivariant. Multout uses the result of a lengthy search based on swap neighborhoods to find a starting point for a constructive descent that is very similar to Forward. Many of the methods are being updated so that this table represents only the state of affairs at the time of this writing. Our main goal is to demonstrate that the local search framework is very useful as a means of synthesizing the evolving methods. This notation generalizes to the clustering problems as shown in Rocke and Woodruff (2002).

4 Conclusions

In this paper we have drawn on concepts from local search to demonstrate strong connections between algorithms for two important problems in statistics: robust estimation of multivariate location and shape parameters and maximum likelihood assignment of multivariate data to clusters. We provided a synthesis and generalization of computational methods reported in the literature. These connections are important because they can be exploited to support effective robust analysis of large data sets.

References

Atkinson, A. C. (1994) "Fast Very Robust Methods for the Detection of Multiple Outliers," *Journal of the American Statistical Association*, **89**, 1329–1339.

Banfield, J.D. and A.E. Raftery (1993), "Model-Based Gaussian and Non-Gaussian Clustering," *Biometrics*, **49**, 803-821

Bernholt, T. and P. Fisher (2001), "The Complexity of Computing the MCD-Estimator," Technical Report, Lehrstuhl Informatik 2, Universität Dortmund, Germany.

Coleman, D.A. and D.L. Woodruff (2000), "Cluster Analysis for Large Data Sets: Efficient Algorithms for Maximizing the Mixture Likelihood," *Journal of Computational and Graphical Statistics*, to appear.

Friedman, H.P. and J. Rubin (1967), "On some Invariant Criteria for Grouping Data", *Journal of American Statistical Association,* **62**, *1159-1178.*

Hadi, A.S. (1992) "Identifying Multiple Outliers in Multivariate Data," *Journal of the Royal Statistical Society, Series B*, 54, 761-771.

Hawkins, D.M., (1980) *The Identification of Outliers*, Chapman and Hall, London.

Hawkins, D.M. and Olive, D.J. (1999), "Improved Feasible Solution Algorithms for High Breakdown Estimation", *Computational Statistics and Data Analysis*, **30**, 1–11.

Hennig, C. (1998) "Clustering and Outlier Identification: Fixed Point Cluster Analysis," Rizzi, A., Vichi, M. and Bock, H.H. (eds.): Advances in Data Science and Classification, Springer, Berlin, 37–42

Lopuhaä, H. P. and Rousseeuw, P. J. (1991) "Breakdown Points of Affine Equivariant Estimators of Multivariate Location and Covariance Matrices," *Annals of Statistics*, **19**, 229–248.

Rocke, D.M., and D.L. Woodruff (1996), "Identification of Outliers in Multivariate Data" *Journal of the American Statistical Association*, 91, 1047-1061.

Rocke, D.M., and D.L. Woodruff (2002), "Multivariate Outlier Detection and Cluster Identification," submitted for publication.

Rousseeuw, P. J. and Leroy, A. M. (1987) *Robust Regression and Outlier Detection,* John Wiley, New York.

Rousseeuw, P. J. and Van Driessen, K. (1999) "A Fast Algorithm for the Minimum Covariance Determinant Estimator," *Technometrics*, **41**, 212-223.

Scott A.J. and Symons M.J. (1971), "Clustering based on Likelihood Ratio Criteria," *Biometrics*, **27**, 387-97.

Woodruff, D. L., and Rocke D. M. (1993) "Heuristic Search Algorithms for the Minimum Volume Ellipsoid," *Journal of Computational and Graphical Statistics*, **2**, 69–95.

Woodruff, D. L. and Rocke, D. M. (1994) "Computable robust estimation of multivariate location and shape in high dimension using compound estimators," *Journal of the American Statistical Association*, **89**, 888-896.

Computer Intensive Methods for Mixed-effects Models

José A. Sanchez and Jordi Ocaña

Statistics Department. University of Barcelona, Spain

Abstract. This work focuses on the development of a framework that implements computer intensive methods for mixed-effects models in an easy and efficient way, and special attention is given to flexible bootstrap techniques. Here we developed an S-Plus library, *nlme.bootstrap*, as an extension of the standard *nlme* library in S. This new library is designed for Monte Carlo and bootstrap methods for mixed-effects models. Some design and implementation aspects are described. To illustrate the use of the software, a simulation study on the performance of parametric bootstrap in the presence of non-normality is discussed.

Keywords. Mixed-effects models, simulation, bootstrap, statistical software.

1 Introduction

The aim of this study is to present the functionalities and describe the implementation of the *nlme.bootstrap* library.

Major statistical packages, like SAS, see Verbeke and Molenberghs (1997), or S-Plus, have their own libraries and macros for analysing mixed-effects models. Pinheiro and Bates (2000) developed the *nlme* library, which is part of the S-Plus standard software. The class of mixed-effects models considered in the *nlme* library assumes that both the random effects and the errors follow a gaussian distribution. Classical inference methods rely on the exact distribution or on the asymptotic properties of the estimators. Extensions to standard techniques (inference on functions of parameters, robust approaches and effect of model misspecification) are not fully covered by the *nlme* library. Alternative methods, like Monte Carlo simulation and bootstrap, can handle these issues but are computationally intensive. We are interested in an efficient implementation of these methods to extend the capabilities of the *nlme* library.

2 Library Design

2.1 Implementation details

nlme.bootstrap is an extension of the *nlme* library, and is oriented to the implementation of Monte Carlo and bootstrap techniques for mixed-effects models. The code is written in S and is designed to be incorporated in a future version of the *nlme* library. All analyses in this study were produced using S-Plus 6 running on a 700MHz Pentium III under Windows 2000. A version for Unix/Linux has been also developed, and there is another version for R. The first deliverable version of the library is pending testing by Pinheiro and Bates.

2.2 Strategy

The central task of the library is to generate huge collections of resampling pseudo-data sets, to fit mixed-effects models for each set and to extract coefficients and other statistics from them. The efficiency of the process is improved in two ways. First, all structures shared by the repeated model fitting over each pseudo-data set are initialised in a first step. Second, the procedure is accelerated by taking advantage of vectorized arithmetic, by dividing the process into two blocks: generation of pseudo-data and model fitting. Vectorized arithmetic implies the use of large data structures. To adjust memory requirements and speed, the size of the process is specified by two parameters: an "inner" parameter (number of pseudo-data generated on each block) and an "outer" parameter (number of blocks to process). This configuration facilitates the future parallelization of the process.

The implemented algorithm is designed to avoid unnecessary calculations. The fitting process takes into account what elements need to be calculated, depending on the user call. If residuals, random-effects or standard errors of estimates will not be used, like in ordinary parametric bootstrap, they are not computed, saving a considerable amount of time.

Several studies on the performance of the *nlme.bootstrap* library have been done, comparing these methods with a direct approach using an S program consisting on a single loop of *nlme* calls. The improvement is significant: time is reduced about five times for each fit, using a dataset of 300 cases.

2.3 Object model and functionalities

The main class in the library is called *bootstrap.lme*. It has several standard methods (*print*, *plot*, *summary*, *coef*, *pairs*, *intervals*) to integrate its interface with the *nlme* object model. Outputs of the different methods are similar to those from the *boot* library for general bootstrapping in S-Plus, developed by Davison and Hinkley (1997).

The constructor of the *bootstrap.lme* class requires a parameter that corresponds to a model fitted using *nlme* (or more precisely, an object of *nlme* class). Pseudo-data are generated according to the fixed coefficients and model matrices of this object. Users can specify how variance components are generated by means of two parameters. These parameters may be string values that specify several pre-implemented options for resampling (like "nonparametric" or "semiparametric") or specifics functions. In this way, all possible resampling strategies can be covered: parametric (using the theoretical distribution of variance components), sampling from the empirical distribution of random effects and/or residuals, or sampling from a non-parametric density estimation (using a kernel function).

Another possibility is to sample from a distinct model specification than the one used in the fitting process. This could be of interest, for example, in simulation studies of robustness and to compare models using bootstrap-based tests.

Optionally, the final object, which contains the matrix of coefficients for each fit, can be saved in a text file. Coefficients of random effects are saved using unconstrained parameterization (logarithm of standard deviations and generalized logits of correlations). All the methods have a parameter that indicates whether the corresponding output refers to the unconstrained or constrained parameters.

2.4 Using the *nlme.bootstrap* library

To illustrate the use of the methods in the library, we have chosen the "Oxboys" data, described in Goldstein (1987). The data consist of the height of a group of boys from Oxford, England, versus a standardized age. The Oxboys data frame has 234 rows and 4 columns (subject, age, height and occasion). We considered a linear mixed model with random effects in the intercept and the slope parameters to illustrate the outputs of the distinct methods.

$$height_i = (\beta_0 + b_{i0}) + (\beta_1 + b_{i1})age_i + \varepsilon_i$$
$$b_i \sim N_2(0, \Psi), \varepsilon_i \sim N_{n_i}(0, \sigma^2 I_{n_i})$$

Output 1 illustrates the use of *nlme.bootstrap* in the preceding data set and model. The computations reported by the *print* method include the bias and the standard error for each parameter, both calculated from the resampling object *Oxboys.boot* (Output 1).

Bootstrap confidence intervals can be obtained with the *intervals* method, which gives the normal, basic and percentile intervals for each parameter (Output 2).

Plotting functions give the histogram and probability plot for coefficients or functions of the coefficients in object *Oxboys.boot* (Figure 1).

The *pairs* method gives the matrix of scatter plots of coefficients (Figure 2).

The coefficients of each fit, including the log-likelihood, can be extracted using the *coef* method.

Output 1. Constructor of a *bootstrap.lme* object based on the Oxboys model with 1000 replications and normal distribution of random effects and residuals.

```
> Oxboys.boot<-bootstrap.lme(object = Oxboys.lme1, B = 1000, distRef =
norm, distRes = norm)
Resampling Method:
   Random effects: norm
   Residuals:      norm

Bootstrap Statistics :
                           original     bias       mean   std. error
               (Intercept) 149.3718 -0.03916  149.3326      1.59257
                       age   6.5255 -0.00135    6.5241      0.33586
     Subject.sd((Intercept))   8.0811 -0.11387    7.9672      1.11479
            Subject.sd(age)   1.6807 -0.02288    1.6578      0.24243
Subject.cor((Intercept),age)   0.6413 -0.00754    0.6337      0.12842
                     sigma   0.6599 -0.00092    0.6590      0.03464
                    logLik -362.0455  2.20705 -359.8384     10.99576
```

Output 2. *intervals* method, giving the percentile confidence interval for each parameter

```
> intervals(Oxboys.boot)$perc
                              lower     upper
               (Intercept)  146.0238  152.4166
                       age    5.8635    7.1789
     Subject.sd((Intercept))    5.8382   10.2049
            Subject.sd(age)    1.1836    2.1192
Subject.cor((Intercept),age)    0.3344    0.8415
                     sigma    0.5931    0.7282
                    logLik -381.5626 -337.6284
```

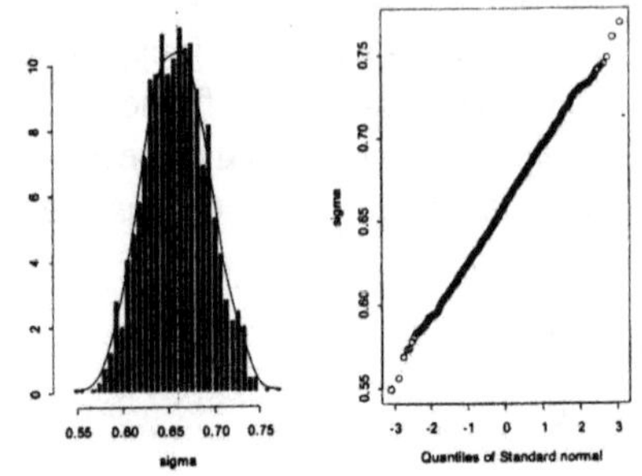

Fig. 1. *plot* method: empirical distribution of the sigma parameter

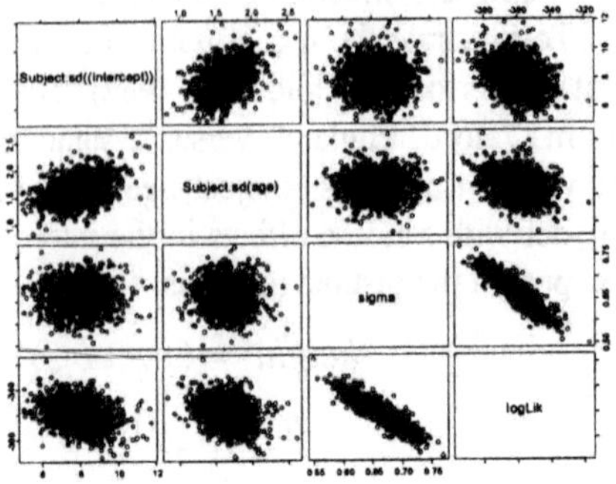

Fig. 2. *pairs* method: scatter plot matrix of random parameters and log-likelihood

3 Robustness of asymptotic inference for variance components: a simulation study

Estimation procedures based on Maximum Likelihood or Restricted Maximum Likelihood assume normal distribution for both, random effects and residuals. Sometimes, this assumption may be unrealistic. Asymptotic inference is based on properties of maximum likelihood estimators. The following example aims to determine the correctness of this inference in the presence of misspecification, and proposes an alternative based on parametric bootstrap, when the true, but not gaussian, distributions of random effects and residuals are known.

With the linear model associated with the Oxboys data, the library was used to construct data sets with the same design matrix but with the random effects and residuals generated according to distinct distributions (gaussian, uniform and exponential) sharing the same first and second moments. To reproduce the asymptotic confidence intervals for each parameter and fit, and to evaluate their empirical coverage and mean length (Table 1), the diagonal elements of the estimated variance-covariance matrices for fixed and random coefficients were saved.

Table 1. Mean length and empirical coverage for 95% confidence intervals based on asymptotic normality of estimators for unrestricted parameters. 1000 replications for each distinct distribution of random effects and residuals: Normal, Uniform and Exponential.

	Parameter	Mean	Normal	Uniform	Exponential
Fixed effects	(Intercept)	6.1684	0.930	0.946	0.905
	Age	1.3240	0.944	0.959	0.923
Random Effects	reStruct.Subject1	4.4807	0.942	0.992	0.687
	reStruct.Subject2	0.9823	0.941	0.965	0.796
	reStruct.Subject3	0.4777	0.944	0.965	0.816
	log.sigma	0.1352	0.948	0.993	0.720

The simulation study reveals that, in presence of skewness, the asymptotic confidence intervals tend to undercover the parameter value. Uniformity of random components has the opposite effect.

The library was also used to evaluate bias and standard error for the estimates of parameters by means of parametric bootstrap (Table 2).

We constructed confidence intervals based on bias correction and bootstrap standard errors and evaluated their coverage. 1000 replications were generated according to the model with exponential random effects and uniform residuals. We used parametric bootstrap (500 resamples) to estimate standard errors of estimates. Performances of confidence intervals based on asymptotic normality and on parametric bootstrap are shown in Figures 3 and 4. Undercoverage of random effects intervals and overcoverage of residual parameter intervals are clearly corrected using the parametric bootstrap.

Table 2. Bootstrap standard errors for estimates of parameters of fixed and random effects under distinct distribution assumptions on the basis of 1000 replications

Parameter	Estimate	Asymptotic SE	Normal	Uniform	Exponential
(Intercept)	149.3718	1.5854	1.6251	1.5069	1.6043
Age	6.5255	0.3363	0.3287	0.3189	0.3429
reStruct.Subject1	2.5052	0.1415	0.1532	0.1033	0.2780
ReStruct.Subject2	0.9349	0.1472	0.1634	0.1415	0.2454
ReStruct.Subject3	1.5207	0.4142	0.4349	0.3904	0.6327
Log.sigma	-0.4157	0.0524	0.0538	0.0381	0.0935

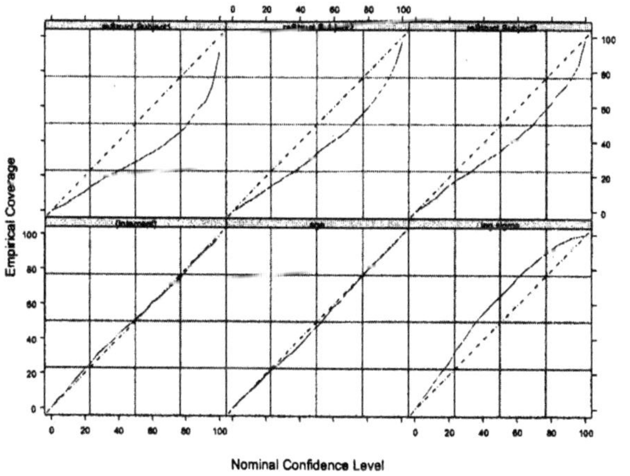

Fig. 3. Empirical coverage versus nominal level for asymptotic confidence intervals. Random effects: Exponential. Residuals: Uniform

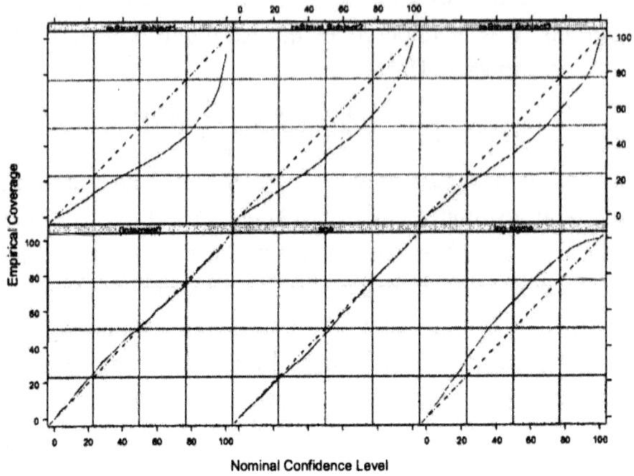

Fig. 4. Empirical coverage versus nominal level for confidence intervals based on Parametric Bootstrap. Random effects: Exponential. Residuals: Uniform

4 Final remarks

Implementation of computer intensive methods can achieve a higher degree of efficiency by taking profit of specificities of the algorithms to which they are applied. In this case, standard techniques for mixed-effects models are extended to provide a simple tool to evaluate techniques and improve inference; it also facilitates prospective studies. Development of the library has taken into account the integration with other existing libraries, like *boot* or *nlme*.

This approach will be extended to other statistical methodologies. We are working on the development of a similar library for Generalized Linear Mixed Models, taking as a basis the *glme* library recently developed by Pinheiro.

5 Acknowledgments

We thank Jose C. Pinheiro for his helpful comments and corrections of code. This work was supported by grants FIS 00/1130 and 2001SGR00067.

References

Davison A, Hinkley D (1997). *Bootstrap Methods and their application.* Cambridge: Cambridge University Press.

Goldstein H. (1987) *Multilevel Models in Educational and Social Research.* London: Griffin.

Pinheiro, J.C. Bates D. (2000). *Mixed-Effects Models in S and S-Plus.* New York: Springer-Verlag.

Verbeke G, Molenberghs G (1997) *Linear Mixed Models in Practice: A SAS oriented approach.* Lecture Notes in Statistics 126. New York: Springer-Verlag.

Construction of T-Optimum Designs for Multiresponse Dynamic Models

Dariusz Uciński[1] and Barbara Bogacka [2]

[1] Institute of Control and Computation Engineering, University of Zielona Góra, ul. Podgórna 50, 65–246 Zielona Góra, Poland

[2] School of Mathematical Sciences, Queen Mary, University of London, Mile End Road, London E1 4NS, U.K.

Abstract. The paper aims at developing the underlying theory and constructing an efficient procedure for determining optimal experimental conditions for discriminating between several rival multivariate statistical models where the expected response is given by ordinary differential equations. The method elaborated is validated on a simulation example.

Keywords. Nonlinear regression, dynamic systems, experimental design, process engineering

1 Introduction

Mechanisms of numerous processes encountered in engineering are not always fully known and experiments are done for their better understanding. Mechanistic models of these processes can often be obtained from physical reasoning, but there may be several plausible models. An experiment especially designed for discrimination between the competing models is a good source of information about the model fit using minimum experimental effort. Experimental design theory for precise estimation of the model parameters has been developed in recent years, c.f. Atkinson and Donev (1992), Pukelsheim (1993), Fedorov and Hackl (1997). However, the design problem for discrimination between models has drawn much less attention and has been developed for simple models only. Various criteria were considered in Atkinson and Fedorov (1975a,b), Burke *et al.*(1994), Stewart *et al.*(1998). The criterion, called T-optimality, introduced by Atkinson and Fedorov (1975a,b) in a single response case, has attracted our attention as, in a linear model, it has an interesting statistical interpretation as the power of a test for the fit of a second model when the first one is assumed true. In a nonlinear model this property holds asymptotically.

In the paper we consider T-optimum designs for discrimination between two rival multiresponse models in the form of ordinary differential equations with observations corrupted by white noise. The T-optimality criterion is generalised for this case: its use leads to solving a maximin problem. The design factors are the sampling strategy and experimental conditions such as temperature or initial process conditions. Both affect the accuracy of model discrimination. The inclusion of the experimental conditions, which are fixed for all measurements in a particular experimental run, complicate both the mathematical and computational problem of finding an optimum design. However, their inclusion is very important in applications.

2 Problem formulation

Consider a general non-linear model whose d responses can be observed up to additive random errors. The resulting observations $y_{ij} \in \mathbb{R}^d$ are given by

$$y_{ij} = \eta(x_i, \beta) + \varepsilon_{ij}, \quad i = 1, \ldots, n; \quad j = 1, \ldots, r_i, \quad \sum_{i=1}^{n} r_i = N, \tag{1}$$

where the nonlinear functional $\eta : \mathbb{R}^{q+s} \to \mathbb{R}^d$ is assumed to be the *true* model of the process. Here x_i stands for a setting of the vector $x \in X \subset \mathbb{R}^q$ of independent explanatory variables, (e.g. a time instant and/or spatial coordinates of a measurement), $x_i \neq x_\kappa$ whenever $i \neq \kappa$, and $\beta \in B \subset \mathbb{R}^s$ denotes a fixed vector of free parameters (e.g. temperature, pressure, initial concentrations, etc.), X and B are some known compact sets. The number of replications for a given setting x_i is denoted by r_i. The errors ε_{ij} are independent identically distributed normal random variables with mean zero and constant variance.

Control variables x_i's and β define process conditions, but the essential difference between the two is that x may vary from observation to observation (in (1) we thus have n different settings denoted by $x_1, \ldots, x_n$), whereas β is chosen only once and then retained fixed for all N observations.

Our basic assumption is that the model response $\eta(x, \beta)$ coincides with either $\eta_1(x, \beta, \tilde{p}_1)$ or $\eta_2(x, \beta, \tilde{p}_2)$, where functions $\eta_1 : \mathbb{R}^{q+s+m_1} \to \mathbb{R}^d$ and $\eta_2 : \mathbb{R}^{q+s+m_2} \to \mathbb{R}^d$ are given *a priori*, with $\tilde{p}_1 \in P_1 \subset \mathbb{R}^{m_1}$ and $\tilde{p}_2 \in P_2 \subset \mathbb{R}^{m_2}$ being fixed but unknown parameters (P_1 and P_2 denote some known compact sets). The purpose of the experiment is to determine which of the models η_1 or η_2 is true.

There is no loss of generality in assuming that the first model is true, i.e. $\eta(x, \beta) = \eta_1(x, \beta, \tilde{p}_1)$, where $\tilde{p}_1$ is regarded as known prior to the experiment (this value could be obtained based on some preliminary experiment). To discriminate between the models η_1 and η_2 as accurate as possible means to select x_i and β so as to maximize the non-centrality parameter defined as follows:

$$T_{12}^0(\xi_N, \beta) = \min_{p_2 \in P_2} \sum_{i=1}^{n} w_i \|\eta(x_i, \beta) - \eta_2(x_i, \beta, p_2)\|^2, \tag{2}$$

where $w_i = r_i/N$, $\|\cdot\|$ stands for the Euclidean norm, and the collection of variables

$$\xi_N = \begin{Bmatrix} x_1, \ldots, x_n \\ w_1, \ldots, w_n \end{Bmatrix} \tag{3}$$

is called the normalized N-observation exact design of the experiment. The x_i's are called the design or support points.

A good design for discriminating between the models will then provide a large lack of fit in terms of the sum of squares for the second model (Atkinson and Donev, 1992). What is more, for normally distributed errors ε_{ij} and large N, the noncentrality parameter $T_{12}^0(\xi_N, \beta)$ is proportional (in mean) to the logarithm of the ratio of the likelihoods associated with both the models (Ermakov and Zhigljavsky, 1983).

To make the problem more tractable, we extend the definition of the design to all probability measures ξ over X which are absolutely continuous with respect to the Lebesgue measure. This relaxation is generally accepted in optimum experimental design (Silvey, 1980; Cox and Reid, 2000; Atkinson and Donev, 1992; Walter and Pronzato, 1997; Pázman, 1986; Fedorov and Hackl, 1997). The corresponding generalization of the noncentrality parameter can be written as

$$T_{12}(\xi,\beta) = \min_{p_2 \in P_2} \int_X \|\eta(x,\beta) - \eta_2(x,\beta,p_2)\|^2 \, \xi(\mathrm{d}x). \tag{4}$$

A pair $(\xi^\star, \beta^\star)$ maximizing (4) is called the *local T_{12}-optimum design.*

3 T_{12}-optimality conditions

Let us make the following assumptions:

1. X, B and P_2 are compact sets,
2. η and $\partial\eta/\partial\beta$ are continuous functions on $X \times B$,
3. η_2 and $\partial\eta_2/\partial\beta$ are continuous functions on $X \times B \times P_2$.

Our main result characterizing T_{12}-optimum designs can be formulated as follows (its detailed highly technical proof will appear elsewhere):

Theorem 1. *Assume that the minimization problem defined in (4) possesses a unique solution $p_2^\star \in P_2$ for a pair $(\xi^\star, \beta^\star)$. Under Assumptions A1–A3 a necessary condition for $(\xi^\star, \beta^\star)$ to be T_{12}-optimal is that for each $x \in X$*

$$\|\eta(x,\beta^\star) - \eta_2(x,\beta^\star,p_2)\|^2 \le T_{12}(\xi^\star,\beta^\star) \tag{5}$$

and for each $\beta \in B$

$$\left\{ \int_X \frac{\partial}{\partial b} \Big[\|\eta(x,b) - \eta_2(x,b,p_2)\|^2 \Big] \Big|_{b=\beta^\star} \xi^\star(dx) \right\} (\beta - \beta^\star) \le 0. \tag{6}$$

The equality in (5) is attained at all support points of $\xi^\star$. Furthermore, for a fixed optimal $\beta^\star$, the set of all the corresponding optimal measures $\xi^\star$ is convex.
For problems without the free parameter β condition (5) becomes a sufficient one.

4 Numerical construction of T_{12}-optimum designs

Assume that an optimal design has the form (3) for a fixed n and that $w_i = \xi(x_i) \in [0,1]$. To shorten notation, let us introduce the vector $\gamma = (x_1,\dots,x_n,w_1,\dots,x_n,\beta_1,\dots,\beta_s) \in \Gamma$, where $\Gamma = \Big\{\gamma \in X^n \times [0,1]^n \times B : \sum_{i=1}^n w_i = 1\Big\}$, and write

$$J(\gamma,p_2) = \sum_{i=1}^n w_i \|\eta(x_i,\beta) - \eta_2(x_i,\beta,p_2)\|^2. \tag{7}$$

Thus our original optimum experimental design problem reduces to finding

$$\gamma^\star = \arg\max_{\gamma \in \Gamma} \Big\{ \min_{p_2 \in P_2} J(\gamma,p_2) \Big\} \tag{8}$$

This solution can be found using the following relaxation procedure outlined by Shimizu and Aiyoshi (1980):

Step 1. Choose an initial parameter vector $p_2^1 \in P_2$ and define the first set of representative values $Z^1 = \{p_2^1\}$. Set $k = 1$.

Step 2. Solve the current relaxed problem

$$\gamma^k = \arg\max_{\gamma \in \Gamma} \left\{ \min_{p_2 \in Z^k} J(\gamma, p_2) \right\}$$

Step 3. Solve the minimization problem

$$p_2^{k+1} = \arg \min_{p_2 \in P_2} J(\gamma^k, p_2)$$

Step 4. If $J(\gamma^k, p_2^{k+1}) \geq (1 - \epsilon) \min_{p_2 \in Z^k} J(\gamma^k, p_2)$, where ϵ is a small predetermined positive constant, then γ^k is a sought maximin solution, otherwise include p_2^{k+1} into Z^k, increment k, and go to Step 2.

It can be proved (Shimizu and Aiyoshi, 1980) that the above algorithm terminates in a finite number of iterations for any given $\epsilon > 0$. The technique can be regarded as a nonlinear cutting plane method.

5 Relevance of the approach to dynamic systems

The results shown in this section are in general form and they can be directly applied to deterministic dynamic systems discrimination of which is our primary motivation to study T-optimal designs. We shall concentrate on two competing models described by the following ordinary differential equations:

$$\mathcal{M}_\ell : \quad \frac{dz_\ell(t)}{dt} = f_\ell(t, z_\ell(t), \tilde{p}_\ell), \quad z_\ell(0) = z_0, \tag{9}$$

where t is time, z stands for a vector-valued function $z : T \to \mathbb{R}^\mu$ (also called the state), $T = [0, t_f]$ for a given t_f, $\tilde{p}_\ell$ signifies a vector of constant parameters being unknown to the experimenter, and $f_\ell : T \times \mathbb{R}^{\mu + m_\ell}$ is required to be continuous together with its Jacobian matrices $\partial f / \partial z$ and $\partial f / \partial p_\ell$, $\ell = 1, 2$.

Discrimination between models $\mathcal{M}_1$ and $\mathcal{M}_2$ is to be performed by making observations of the responses z_ℓ at discrete time instants $t_1, \dots, t_n$:

$$y_{ij} = h(z_\ell(t_i)) + \varepsilon_{ij}, \quad i = 1, \dots, n; \quad j = 1, \dots, r_i, \tag{10}$$

where $h : \mathbb{R}^\mu \to \mathbb{R}^d$ is continuously differentiable and defines the manner of observing the states. This leads to the question of how to select an appropriate sampling strategy and initial conditions for the state. This problem can be cast as the problem considered in Section 2 by making the following substitutions:

- define $X = T$, which entails only one explanatory variable corresponding to time, i.e. time instants $t_1, \dots, t_n$ are to be selected in place of $x_1, \dots, x_n$;
- fix $\beta = z_0$;

- replace $\eta_\ell(x, \beta, \tilde{p}_\ell)$ by $h(z_\ell(t))$ (note that z_ℓ depends implicitly on β and $\tilde{p}_\ell$).

As can be seen, the only complication here is that the function η is defined implicitly through the solution of the ordinary differential equation, but this obstacle is only of minor importance due to the availability of extremely efficient solvers performing numerical integration of vector differential equations.

6 Computational Example

Let us consider the T-optimum experimental design problem for discriminating between two models of a familiar simple pendulum. The true model has highly nonlinear form:

$$\ddot{z}_1(t) + 10\sin(z_1(t)) = 0, \quad z_1(0) = \beta, \quad \dot{z}_1(0) = 0, \tag{11}$$

where $t \in [0, 1.2]$. The rival model is the linearized version of (11):

$$\ddot{z}_2(t) + p\, z_2(t) = 0, \quad z_2(0) = \beta, \quad \dot{z}_2(0) = 0 \tag{12}$$

We assume that $p \in [5, 10]$ and $\beta \in [0, \pi/2]$.

In order to find an optimal design, $n = 2$ support points were assumed and the relaxation procedure outlined in Section 4 was implemented in SAS/IML software which gives access to a powerful and flexible programming language. The ordinary differential equations were solved using the built-in `ODE` subroutine. Maximization in *Step 2* was performed using the standard `NLPQN` (15 restarts for randomly generated initial conditions were performed at each iteration). In turn, for minimization in *Step 3*, an adaptive-random search algorithm was implemented and run (1500 calls were imposed as the limit for stopping its work). Switches to a local search were accomplished using the `NLPLM` subroutine.

After 5 iterations of the relaxation algorithm the following approximation to the optimal design was obtained:

$$\xi^\star = \begin{Bmatrix} 0.36 & 0.804 \\ 0.5 & 0.5 \end{Bmatrix}, \tag{13}$$

for the optimal initial condition $\beta^\star = \pi/2$.

Figure 1 shows the curve reflecting the behaviour of the RHS of (5) with respect to time. Its maximal value is attained at the two support points of the optimal design. This confirms the correctness of the results.

7 Concluding remarks

It has been shown how to generalize the approach based on the T-optimality criterion to multiresponse models with additional parameters which are to be fixed the same for all experimental runs.

Note that the outlined procedure remains valid if we allow for correlations between different responses, i.e. if the errors ε_{ij} are sampled from a distribution satisfying

$$\mathrm{E}[\varepsilon_{ij}] = 0, \quad \mathrm{E}[\varepsilon_{ij}\varepsilon_{\kappa\ell}^T] = C(x_i, \beta) \tag{14}$$

where $C(x_i, \beta)$ is symmetric and positive definite, and the mappings C and $\partial C/\partial \beta_r$, $r = 1, \dots, s$ are continuous.

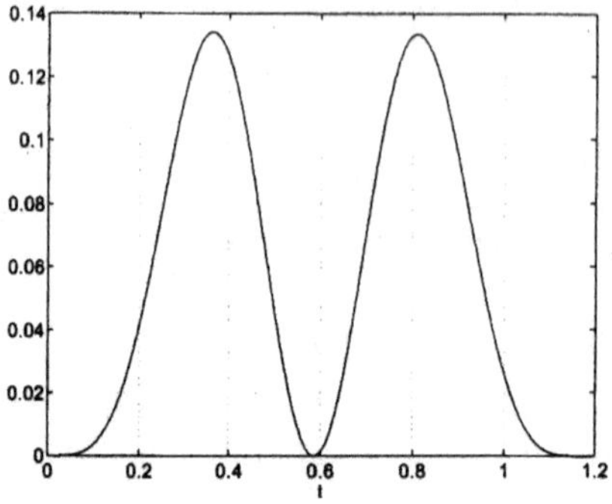

Fig. 1. The left-hand side of (5) vs. time for the T-optimal design.

Acknowledgments

This research of the first author was supported by the State Committee for Scientific Research under Grant No. 7 T11A 023 20. The University of Zielona Góra has been participating in the SAS educational programme since 2000.

References

Atkinson, A.C. and Donev, A.N. (1992). *Optimum Experimental Designs.* Ofxord: Clarendon Press.

Atkinson A.C. and Fedorov V.V. (1975a). The design of experiments for discriminating between two rival models. *Biometrika*, **62**, 57-70.

Atkinson A.C. and Fedorov V.V. (1975b). Optimal design: Experiments for discriminating between two several models. *Biometrika*, bf 62, 289-303.

Burke, A.L., Duever, T.A. and Pendilis, A. (1994). Model discrimination via designed experiments, Discriminating between the terminal and penultimate models on the basis of composition data. *Macromolecules*, **27**, 386-399.

Cox, D.R. and Reid, N. (2000). *The Theory of the Design of Experiments.* Boca Raton, Chapman and Hall.

Ermakov, S.M., Zhigljavsky, A. (1983). *Mathematical theory of experimental design.* Moscow: Nauka, (in Russian).

Fedorov, V.V. and Hackl, P. (1997). *Model-oriented design of experiments.* New York, Springer.

Pázman, A. (1986). *Foundations of Optimum Experimental Design.* Dordrecht, D. Reidel Publishing Company.

Pukelsheim, F. (1993). *Optimal design of experiments.* New York, Wiley.

Shimizu K. and Aiyoshi E. (1980). Necessary conditions for min-max problems and algorithms by a relaxation procedure. *IEEE Transactions on Automatic Control*, **AC-25**, 62-66.

Silvey, S.D. (1980). *Optimal design. An introduction to the theory for parameter estimation.* London, Chapman and Hall.

Stewart, W.E., Shon, Y. and Box, G.E.P. (1988). Discrimination and goodness of fit of multiresponse mechanistic models. *AIChE Journal*, **44(6)**, 1404-1412.

Walter, E. and Pronzato, L. (1997). *Identification of parametric models from experimental data.* Berlin: Springer.

Data Compression and Selection of Variables, with Respect to Exact Inference

Jürgen Läuter and Siegfried Kropf

Institute for Biometry and Medical Informatics, Otto von Guericke University Magdeburg, Leipziger Straße 44, 39120 Magdeburg, Germany

Keywords. Data compression, selection of variables, multivariate test

1 Introduction

In many applications of statistics, we are confronted with a large number p of different variables whereas the number n of independent individuals remains limited. The classical multivariate tests like Wilks' Λ or Hotelling's T^2 test do not attain sufficient power under these circumstances because they cannot take into account special parameter structures. Often overfitted parameter estimations and unstable behaviour arise, and confirmatory data analysis becomes difficult. We will nevertheless intend to investigate the multivariate data by exact statistical tests. Methods of dimension reduction and data compression are preferentially used.

We start from a sample of n independent p-dimensional observation vectors

$$(1) \qquad \boldsymbol{X} = \begin{pmatrix} \boldsymbol{x}'_{(1)} \\ \vdots \\ \boldsymbol{x}'_{(n)} \end{pmatrix} = \begin{pmatrix} x_{11} & \cdots & x_{1p} \\ & \vdots & \\ x_{n1} & \cdots & x_{np} \end{pmatrix}.$$

We will compress the data into so-called scores: $\boldsymbol{Z} = \boldsymbol{X}\boldsymbol{D}$. For this, a $p \times q$ coefficient matrix $\boldsymbol{D}$ is used ($q \leq p$). The original p-dimensional vectors $\boldsymbol{x}'_{(j)}$ are transformed into q-dimensional vectors $\boldsymbol{z}'_{(j)}$ $(j = 1, \ldots, n)$. The compression is based on principal components, that is, the matrix $\boldsymbol{D}$ is obtained as the solution of the eigenvalue problem

$$(2) \qquad \boldsymbol{G}\boldsymbol{D} = \boldsymbol{D}\boldsymbol{\Lambda} \quad \text{with} \quad \boldsymbol{G} = \sum_{j=1}^{n} (\boldsymbol{x}_{(j)} - \bar{\boldsymbol{x}})(\boldsymbol{x}_{(j)} - \bar{\boldsymbol{x}})' = (\boldsymbol{X} - \bar{\boldsymbol{X}})'(\boldsymbol{X} - \bar{\boldsymbol{X}}),$$

where $\bar{\boldsymbol{X}} = \boldsymbol{1}_n \bar{\boldsymbol{x}}'$ is the $n \times p$ mean matrix, $\boldsymbol{D} = (\boldsymbol{d}_1 \ \ldots \ \boldsymbol{d}_q)$ is the matrix of the eigenvectors $\boldsymbol{d}_h$ pertaining to the q largest eigenvalues λ_h $(\lambda_1 \geq \lambda_2 \geq \ldots,$ $\boldsymbol{D}'\boldsymbol{D} = \boldsymbol{I}_q$), and $\boldsymbol{\Lambda} = \boldsymbol{D}'\boldsymbol{G}\boldsymbol{D} = (\boldsymbol{Z} - \bar{\boldsymbol{Z}})'(\boldsymbol{Z} - \bar{\boldsymbol{Z}})$ is the diagonal matrix of these eigenvalues ($\boldsymbol{1}_n$ represents the vector consisting of n ones, $\boldsymbol{I}_q$ is the $q \times q$ identity matrix). In the following, we will treat different testing problems. We can show that it is sufficient to work exclusively with the compressed data $\boldsymbol{Z}$ of the diminished dimension q.

2 Tests and confidence regions for one group of individuals

Here we suppose that all vectors $x'_{(j)}$ have the same p-dimensional normal distribution $x'_{(j)} \sim N_p(\mu', \Sigma)$. The null hypothesis to be tested is $\mu' = \mu'_0$, where μ'_0 has a fixed given value. The covariance matrix Σ is unknown.

For this problem, the so-called spherical tests by Läuter (1996), Läuter, Glimm and Kropf (1998) are available. The tests use linear combinations of the p variables whose coefficients are functions of the "total sums of products matrix" $W = G + n(\bar{x} - \mu_0)(\bar{x} - \mu_0)'$. Hence, a data compression $Z_W = XD_W$ results that depends on the special null hypothesis $\mu' = \mu'_0$ considered.

However, we prefer the compression by the principal components from (2) which is invariant with respect to different null hypotheses. The following two theorems allow such a data compression and selection of variables with strictly keeping the level of significance α. The theorems are also consequences of our former results on spherical tests. They show that the sample-based principal components are an adequate tool to transform multivariate testing problems to a smaller dimension. Thus, principal components acquire a new important role in addition to their "naive" use for computational data compression [Reyment and Jöreskog (1993), Seber (1984)].

Theorem 1 on the principal components inference [Läuter and Glimm (2002)]. Consider a problem with p variables and q principal components $(1 \le q \le p)$. Let X have n independent rows, where $x'_{(j)} \sim N_p(\mu', \Sigma)$ for $j = 1, \dots, n$, $\operatorname{rank}(\Sigma) \ge q$, $n \ge q+1$. Let $D = (d_1 \ \dots \ d_q)$ be the weight matrix determined from (2), and $Z = XD$, $\bar{z}' = \bar{x}'D$. Then, if the null hypothesis $\mu' = \mu'_0$ is valid, the distribution of the statistic

$$(3) \quad F = \frac{(n-q)n}{q}(\bar{z}' - \mu'_0 D)\Lambda^{-1}(\bar{z} - D'\mu_0) = \frac{(n-q)n}{q}(\bar{x} - \mu_0)'D\Lambda^{-1}D'(\bar{x} - \mu_0)$$

is stochastically not larger than Fishers F distribution $F(q, n-q)$, that is,

$$(4) \qquad \Pr(F \ge F_{1-\alpha}(q, n-q)) \ \le \ \alpha \quad \text{for every } \alpha \in (0,1).$$

This theorem facilitates the application of the F test to the compressed data $Z = XD$. A significant value $F \ge F_{1-\alpha}(q, n-q)$ indicates that the p-dimensional null hypothesis $\mu' = \mu'_0$ can be rejected at the significance level α. However, this testing method is conservative, that is, its error of first kind is generally smaller than the given value α. Nevertheless, in many applications where a q-dimensional factor structure is actually present, the deviation from the nominal value α is slight. We should note that the result of Theorem 1 is not invariant against changes in the scales of the p variables.

Moreover, the procedure of Theorem 1 may be connected with selection of variables. If several models m are given, that is, several subsets of the complete set of p variables, then the score determination by (2) and the calculation of the statistic F by (3) can be carried out separately for each model m, where the variables not belonging to m are not utilized. We denote the number of variables of the model m by $p^{(m)}$. In the simplest special case, only the p models of single variables, with $p^{(m)} = 1$, are considered.

Theorem 2 on the principal components inference with model choice [Läuter and Glimm (2002)]. Assume several models $m = m_1, m_2, \ldots$ with the corresponding sizes $p^{(m)}$ and a number of principal components q with $1 \le q \le p^{(m)} \le p$ for all m. Let $\boldsymbol{X}$ have n independent rows, where $\boldsymbol{x}'_{(j)} \sim N_p(\boldsymbol{\mu}', \boldsymbol{\Sigma})$ for $j = 1, \ldots, n$, $\text{rank}(\boldsymbol{\Sigma}) \ge q$, $n \ge q+1$. Let, for any model m, $\boldsymbol{\Lambda}^{(m)}$ be the diagonal matrix of the q largest eigenvalues $\lambda_1^{(m)}, \ldots, \lambda_q^{(m)}$ of the corresponding restricted eigenvalue problem (2). Select the model $m = m_{\text{opt}}$ with the maximum value of $O^{(m)} = (p^{(m)})^{-q/2}\, \lambda_1^{(m)} \lambda_2^{(m)} \cdot \ldots \cdot \lambda_q^{(m)}$ as the "optimum model", and let $\boldsymbol{D}^{\text{opt}} = \boldsymbol{D}^{(m_{\text{opt}})}$, $F^{\text{opt}} = F^{(m_{\text{opt}})}$ be the corresponding weight matrix and the corresponding F statistic. Hence, if m_0 is the "true null set" of variables, that is, the set of variables i with $\mu_i = \mu_{0i}$, then the probability of both $m_{\text{opt}} \subseteq m_0$ and significance $F^{\text{opt}} \ge F_{1-\alpha}(q, n-q)$ is at most α.

As the consequence of Theorem 2, a resulting significance at the identified subset m_{opt} can be interpreted in such a way that deviations from the null hypothesis $\boldsymbol{\mu}' = \boldsymbol{\mu}'_0$ exist in this set and also, of course, in all greater sets of variables. Thus, "erroneous significances" occur with probability α, at most. To obtain significance of any subset, all models m contained in it can be utilized. Nevertheless, α adjustment like Bonferroni or Bonferroni-Holm is not necessary. Theorem 2 establishes a new method for the selection of variables which is based solely on the residual matrix $\boldsymbol{G}$. The knowledge of the mean values is not required.

Based on Theorem 1 or 2, confidence regions for the unknown mean vector $\boldsymbol{\mu}'$ can also be derived. Equation (3) yields the level $(1-\alpha)$ confidence region consisting of the values $\boldsymbol{\mu}'_0$ with

$$((n-q)n/q)\ (\bar{\boldsymbol{x}} - \boldsymbol{\mu}_0)'\boldsymbol{D}\boldsymbol{\Lambda}^{-1}\boldsymbol{D}'(\bar{\boldsymbol{x}} - \boldsymbol{\mu}_0)\ <\ F_{1-\alpha}(q, n-q). \tag{5}$$

In a geometric interpretation, this is an elliptic cylinder around $\bar{\boldsymbol{x}}'$ in the p-dimensional space. If we consider only one principal component, then (5) becomes

$$(n-1)\, n\ ((\bar{\boldsymbol{x}} - \boldsymbol{\mu}_0)'\boldsymbol{d}_1)^2/\lambda_1\ <\ F_{1-\alpha}(1, n-1), \tag{6}$$

which represents a "hyperdisk", the region between two hyperplanes, in the p-dimensional space. This is a basically new type of multivariate confidence regions. Actually, the regions are unbounded, but with respect to the most informative directions in the space, they often have very narrow limits. Theorem 2 provides the analogous relations with an additional model choice. Then, of course, the components of $\boldsymbol{\mu}'_0$ which correspond to variables outside the model m_{opt} remain completely unrestricted.

Example 1. We consider neurophysiological data by B. Bogerts (Magdeburg). The data refer to EEG measurements of 19 depressive patients. There are nine variables that correspond to different measuring sites on the patient's head (the channels 3 to 8 and 17 to 19). Each variable represents the absolute theta power difference between the beginning and the end of a six week treatment. The question to be answered is whether the mean vector $\boldsymbol{\mu}'$ deviates from zero and, if so, which are the most important channels to reflect this deviation. In this case, $p=9$, $n=19$, and we will use only one principal component $(q=1)$. Based on the estimates

$$\begin{matrix} & ch.3 & ch.4 & ch.5 & ch.6 & ch.7 & ch.8 & ch.17 & ch.18 & ch.19 \end{matrix}$$

$$\bar{\boldsymbol{x}}' = \begin{pmatrix} 0.871 & 1.589 & 1.037 & 1.146 & 0.851 & 0.853 & 1.422 & 0.751 & 0.995 \end{pmatrix},$$

$$\boldsymbol{G} = \begin{pmatrix} 157 & 172 & 102 & 95 & 70 & 53 & 162 & 69 & 66 \\ & 222 & 94 & 111 & 48 & 58 & 192 & 46 & 73 \\ & & 100 & 76 & 82 & 63 & 102 & 80 & 76 \\ & & & 91 & 55 & 67 & 108 & 68 & 75 \\ & & & & 93 & 52 & 62 & 74 & 58 \\ & & & & & 77 & 63 & 58 & 69 \\ & & & & & & 191 & 63 & 79 \\ & & & & & & & 125 & 49 \\ & & & & & & & & 100 \end{pmatrix},$$

Theorem 1 provides the weight vector

$$\boldsymbol{d}_1' = \begin{pmatrix} 0.414 & 0.459 & 0.313 & 0.309 & 0.228 & 0.217 & 0.451 & 0.239 & 0.255 \end{pmatrix}.$$

Thus, the compressed data are obtained by $\boldsymbol{Z} = \boldsymbol{X}\boldsymbol{d}_1$ which is only one column of 19 elements. If we apply these values for testing the null hypothesis $\boldsymbol{\mu}' = \boldsymbol{0}$, we obtain $F = 4.331$ which corresponds to the P value of $P = 0.0520$. A modified null hypothesis, for example, $\boldsymbol{\mu}' = -0.2\,(1 \;\; 1 \;\; \ldots \;\; 1)$, would lead to $F = 6.022$, $P = 0.0245$. Hence, this hypothesis is rejected at the level $\alpha = 0.05$. The confidence region (6) of the mean vector for the level $1-\alpha = 0.95$ is given by $-0.031 < \boldsymbol{\mu}_0'\boldsymbol{d}_1 < 6.475$. Applying Theorem 2 to all possible submodels, the optimum model of the three variables $ch.3$, $ch.4$ and $ch.17$ is found. The corresponding weight vector is

$$\boldsymbol{d}_1^{\text{opt}\,\prime} = \begin{pmatrix} 0.521 & 0.626 & 0 & 0 & 0 & 0 & 0.580 & 0 & 0 \end{pmatrix}.$$

The test of the null hypothesis $\boldsymbol{\mu}' = \boldsymbol{0}$ yields $F^{\text{opt}} = 3.253$, $P^{\text{opt}} = 0.0881$. The confidence region of the mean vector for $1-\alpha = 0.95$ is $-0.375 < \boldsymbol{\mu}_0'\boldsymbol{d}_1^{\text{opt}} < 4.921$.

3 Tests for several groups of individuals

The principal component scores $\boldsymbol{Z} = \boldsymbol{X}\boldsymbol{D}$ determined from (2) can also be used for comparisons of subgroups in the sample $\boldsymbol{X}$. We suppose that two normally distributed submatrices $\boldsymbol{X}^{(1)}$ and $\boldsymbol{X}^{(2)}$, consisting of $n^{(1)}$ and $n^{(2)}$ rows, respectively, are given which have to be tested for equality of the mean vectors. Additionally, the sample matrix $\boldsymbol{X}$ may contain other independent rows of an arbitrary distribution:

$$(7) \qquad X = \begin{pmatrix} X^{(1)} \\ X^{(2)} \\ \vdots \end{pmatrix}, \quad Z = \begin{pmatrix} Z^{(1)} \\ Z^{(2)} \\ \vdots \end{pmatrix} = \begin{pmatrix} X^{(1)}D \\ X^{(2)}D \\ \vdots \end{pmatrix}.$$

For such "internal" comparisons of means, the matrix G plays the role of the "total sums of products matrix". Therefore, the strategies of the spherical tests by Läuter (1996), Läuter et al. (1998) can be applied.

Theorem 3. Consider p variables and q principal components $(1 \le q \le p)$. Let X have n independent rows, where $x'_{(j)} \sim N_p(\mu^{(1)\prime}, \Sigma)$ for $j = 1, \ldots, n^{(1)}$, $x'_{(j)} \sim N_p(\mu^{(2)\prime}, \Sigma)$ for $j = n^{(1)} + 1, \ldots, n^{(12)}$, $\operatorname{rank}(\Sigma) \ge q$, $n^{(12)} = n^{(1)} + n^{(2)}$, $n^{(1)} \ge 1$, $n^{(2)} \ge 1$, $n \ge n^{(12)} \ge q + 2$. Let $D = (d_1 \; \ldots \; d_q)$ be the weight matrix determined from (2). The mean matrix of $Z^{(k)} = X^{(k)}D$ is denoted by $\bar{Z}^{(k)} = 1_{n^{(k)}} \bar{z}^{(k)\prime}$ $(k = 1, 2)$. Then, if the null hypothesis $\mu^{(1)\prime} = \mu^{(2)\prime}$ is valid,

$$(8) \quad F = \frac{n^{(12)} - 1 - q}{q} \frac{n^{(1)} n^{(2)}}{n^{(12)}} (\bar{z}^{(1)} - \bar{z}^{(2)})' \left(\sum_{k=1}^{2} (Z^{(k)} - \bar{Z}^{(k)})'(Z^{(k)} - \bar{Z}^{(k)}) \right)^{-1} (\bar{z}^{(1)} - \bar{z}^{(2)})$$

has exactly the F distribution $F(q,\ n^{(12)} - 1 - q)$. In the more general situation of Theorem 2, significance $F^{\mathrm{opt}} \ge F_{1-\alpha}(q,\ n^{(12)} - 1 - q)$ at a subset m_{opt} consisting only of variables i with $\mu_i^{(1)} = \mu_i^{(2)}$ can occur with the probability α, at most.

Addendum. Theorem 3 can be extended to arbitrary weights D and arbitrary model assessments $O^{(m)}$ (see Theorem 2) that are functions of the matrix G (with some regularity assumptions). More details are presented in Kropf and Läuter (2002).

Example 2. We consider gene expression data by M. Eszlinger, K. Krohn and R. Paschke (Leipzig) in patients with nodules in the thyroid gland. Tissue samples from the nodule and from the surrounding tissue have been taken from five patients with hot nodules and from five patients with cold nodules. The samples have been investigated with Affymetrix® gene chips measuring the expression values of 12 625 genes. Here in the example, we use the logarithmic expression values ("LogAvg") as delivered by the system, and we consider their differences from nodule and surrounding for each of the 12 625 genes in the ten patients. These differences are approximately normally distributed. We want to test the deviation between hot and cold nodules, and we are interested in finding the corresponding relevant genes.

Here, $p = 12\,625$, $n^{(1)} = 5$ (hot), $n^{(2)} = 5$ (cold), $n = 10$. We will apply Theorem 3 to compress the data X of dimension $p = 12\,625$ into scores $Z = XD$ of dimension $q = 3$. In this example, G is a huge $12\,625 \times 12\,625$ matrix. Therefore, special calculatory methods have to be used. The F test (8) yields $F = 26.20$ with

$P = 0.0008$. Therefore, the hypothesis $\boldsymbol{\mu}^{(1)\prime} = \boldsymbol{\mu}^{(2)\prime}$ is rejected if $\alpha = 0.05$. The analogous test for $q = 1$ leads to $F = 17.83$, $P = 0.0029$, the test for $q = 2$ results in $F = 7.94$, $P = 0.0158$.

Furthermore, we want to recognize the essential variables from the given 12 625 gene variables. Therefore, the models m_i each containing one of the variables $i = 1, \dots, p$ are investigated, where $q = 1$. In this case, the model assessment $O^{(m_i)} = g_{ii}$ holds. Maximization of the diagonal elements g_{ii} of $\boldsymbol{G}$ according to Theorem 3 and 2 leads to gene $i_{\text{opt}} = 10\,253$. However, the subsequent F test yields $F^{\text{opt}} = 1.65$, $P^{\text{opt}} = 0.2353$, which is not sufficient for a significance.

Based on the Addendum to Theorem 3, we can also apply other definitions of the weights $\boldsymbol{D}$ and the assessments $O^{(m)}$. As a trial for obtaining a relevant subset of variables from the complete set of 12 625 variables, we consider all subsets of 40 variables ($p^{(m)} = 40$) with the assessment $O^{(m)} = \text{tr}(\boldsymbol{G}^{(m)})$, where $\boldsymbol{G}^{(m)}$ is the submatrix of $\boldsymbol{G}$ corresponding to m. The maximization of $O^{(m)}$ yields a subset m_{opt} consisting of the 40 genes $i = 10\,253,\ 11\,859, \dots, 5785$. Then the principal component F test with these 40 variables and with the weight vector $\boldsymbol{d}_1^{\text{opt}}$ calculated from the eigenvalue problem $\boldsymbol{G}^{(m_{\text{opt}})}\boldsymbol{d}_1^{\text{opt}} = \text{Diag}(\boldsymbol{G}^{(m_{\text{opt}})})\boldsymbol{d}_1^{\text{opt}}\lambda_1^{\text{opt}}$ is attached. We get $F^{\text{opt}} = 9.48$, $P^{\text{opt}} = 0.0151$ and can conclude that the selected subset of 40 variables contains some relevant variables ($\alpha = 0.05$). The procedure keeps the multiple level 0.05 of significance. This means that such an optimum set m_{opt} belongs completely to the unknown null set of variables (fulfilling $\mu_i^{(1)} = \mu_i^{(2)}$) with probability 0.05, at the most.

This high-dimensional example demonstrates obviously the efficiency of our methods which are based on recent results in multivariate analysis.

References

Kropf, S. & Läuter, J. (2002). Multiple tests for different sets of variables using a data-driven ordering of hypotheses, with an application to gene expression data. *Biometrical Journal*, submitted.

Läuter, J. (1996). Exact *t* and *F* tests for analyzing studies with multiple endpoints. *Biometrics*, **52**, 964-970.

Läuter, J. & Glimm, E. (2002). A theorem on the principal components inference. *The Annals of Statistics*, submitted.

Läuter, J., Glimm, E. & Kropf, S. (1998). Multivariate tests based on left-spherically distributed linear scores. *The Annals of Statistics*, **26**, 1972-1988. Correction: 1999, **27**, 1441.

Reyment, R.A. & Jöreskog K.G. (1993). *Applied factor analysis in the natural sciences.* Cambridge: Cambridge University Press.

Seber, G.A.F. (1984). *Multivariate observations.* New York: Wiley.

Data Extraction from Dense 3-D Surface Models

Mitchum Bock[1], Adrian Bowman[1], Janet Bowman[2] and Paul Siebert[2]

[1] Department of Statistics, University of Glasgow, Glasgow, Scotland, UK
[2] 3D Matic Research Laboratory, Department of Computer Science, University of Glasgow, Glasgow, Scotland, UK

Keywords. Shape, curves, surfaces, procrustes, principal components, facial modelling, mesh conformation

1 Introduction and background

Stereo-photogrammetry involves the use of pairs of cameras to extend standard photographic methods by recovering depth information through triangulation (Ayoub et al. 1998). The resulting data consist of a point-cloud of three-dimensional observations, providing a digital representation of the surface of the object of interest. The number of observations can typically be many thousands, providing a rich source of information on object shape. This can be represented efficiently as an irregular triangular mesh, in Virtual Reality Modelling Language form.

This technology is currently being used in a study of the growth of children's faces. There have been very few quantitative studies of growth, especially in three dimensions. This study involves an additional longitudinal component in that repeated measurements are taken over time, from 3 months to 2 years for one cohort of children and from 3 years to 5 years for another. Stereo-photogrammetry provides a practical means of gathering these data. In particular, the very fast capture time involved is important in minimising the effects of movement in the subject. The clinical aims of the study are to contrast the facial shape and growth of healthy children with those from children who have been born with a cleft lip and/or palate and who have subsequently undergone surgical repair.

This short paper describes the interface between the computer science, which has produced informative and realistic 3-D models of real surfaces, and the statistical analysis, which attempts to answer the research questions. In particular, several types of data, each extracted from the 3-D stereophotographic models, are described and methods of analysis appropriate to each type are briefly outlined. The types of data follow a natural progression of increasing complexity and information, namely: 1. point data consisting of between 30-40 anatomically meaningful landmarks; 2. curve data obtained by interpolating anatomical points and 3. surface data represented by a dense regular mesh of points, anchored at the anatomical landmarks and conformed to the surface of the 3-D model.

2 A hierarchy of data extraction

The raw information collected from the stereo-camera system is both rich and complex. The aim of the following sections is to highlight the different levels of data which can be extracted and to indicate appropriate methods of analysis. Point-based methods have the advantage of simplicity and well established methods of analysis while curve and surface approaches may be more informative at the cost of more sophisticated analysis.

2.1 Point data

The analysis of landmark data, consisting of a configuration of anatomical points for each object, has a long history and a set of tools for analysis which are now well established. The recent text by Dryden & Mardia (1998) gives an authoritative description of this research area. Figure 1 shows a baby face with a number of landmarks identified using the computing tool mentioned in Section 1. For cleft cases, most interest centres on the nasal and upper lip regions. A helpful strategy is therefore to characterise the variation present in the facial shape of non-cleft children. Cleft repair cases can then subsequently be plotted on a corresponding scale to identify and quantify the differences in facial shape from non-cleft cases.

Since each stereo-photograph has its own frame of reference, it is necessary to standardise the collected landmark configurations for location and orientation. If we are interested simply in shape, and not in size, then we also need to standardise for scale. This can all be done through the technique of Generalised Procrustes Analysis (Dryden & Mardia page 87). This technique was applied to the landmarks from a small set of 44 three year old non-cleft cases, using information from landmarks around the nostrils and lower nasal region. The end result is sets of landmarks, which are all comparable. The top left panel of Figure 2 displays the variation remaining at each landmark position while the crosses show the mean configuration.

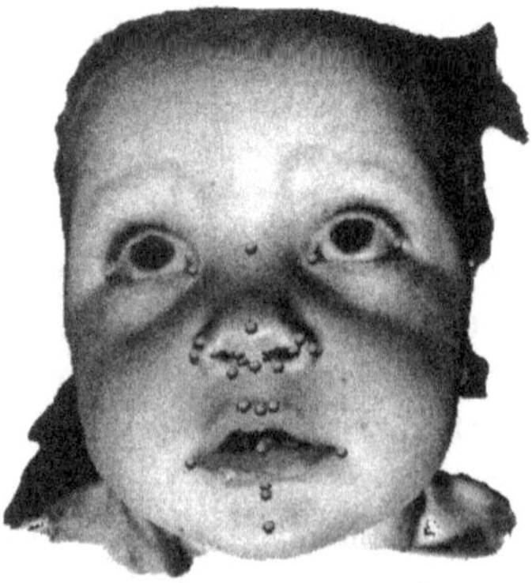

Figure 1: Snapshot of a 3-D photorealistic model with anatomical landmarks identified

A more informative description of shape variation is available through the application of principal components analysis (Dryden & Mardia page 96). The top right panel of Figure 2 represents the first principal component by displaying the movements of the landmarks from the mean as we travel along +/- 3 standard deviations. The amount of variation in the data represented by the first principal component is modest at 25%. This is also a small, initial dataset and so any interpretations must be cautious at this stage. However, there is evidence of a broad pattern of narrowing of the nostrils with widening of the nose width, plus a small amount of asymmetry in the relative positions of the nasal and the upper landmarks. The bottom right panel of Figure 2 shows the variation represented in the second principal component, representing a further 15% of the variation in the data. Here the dominant feature is asymmetry in the position of the bases of the two nostrils.

The bottom left panel of Figure 2 plots the first two principal component scores, evaluated for both cleft and non-cleft cases. There seems to be strong evidence of a difference between the groups in the second principal component but not in the first. This matches the general perception that residual asymmetry can often be a strong feature in repaired cleft cases.

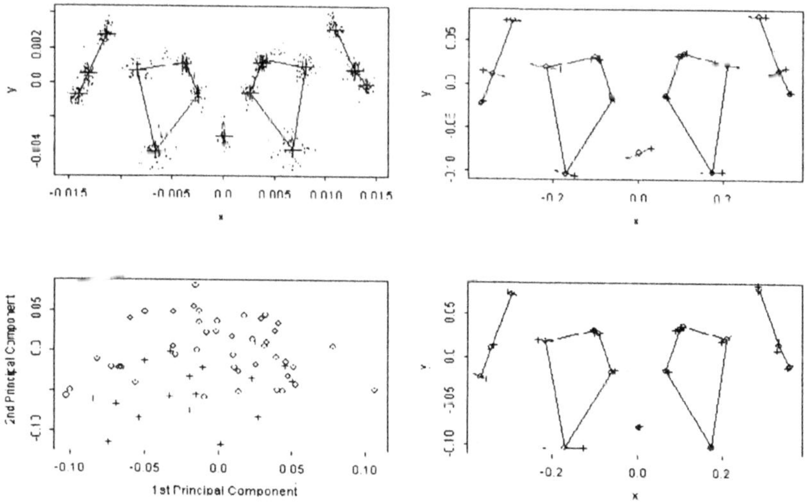

Figure 2: The top left panel shows the landmarks and means (crosses) after procrustes alignment. The right hand panels show the changes in shape associated with the first (top) and second (bottom) principal components of residual variation. The bottom left panels plot the first two PC scores for 14 cleft (crosses) and 44 non-cleft (circles) cases.

2.2 Curve data

Landmarks form an important starting point in the extraction of data from each stereo model. However, they clearly provide only a first approximation to the much more detailed shape information present in each model surface. A natural next step is to extract curves along each model surface, using anatomical landmarks as reference points. A computing tool has been developed to track curves along the facial surface between two identified landmarks. These points lie on the intersection of the model with a plane containing the landmarks and aligned by the average of the normal directions to the surface at the landmarks. The model is represented by an irregular, high density mesh as displayed in Figure 5 (top right panel). An extracted curve is then defined by a collection of points, which lie on this underlying mesh.

For the purposes of later analysis, it is important that there is a correspondence in the representations of these curves. This can be induced by the use of nonparametric smoothing. A functional data analysis formulation can be adopted, as in Ramsay & Silverman (1997), where a `parametric' representation is (x(t),y(t),z(t)), where t represents arc length along the curve. Nonparametric regression can be applied to the raw data (x_i, y_i, z_i, t_i) by separate smoothing of the pairs (t_i, x_i), (t_i, y_i) and (t_i, z_i). If it is necessary to hold the end-points fixed as anatomical landmarks, then additional weights can be incorporated to ensure this.

Figure 3 presents an initial analysis of curve data defining the ridge of the nose of 20 6-month-old babies. The variations represented by the first and second principal components are displayed. Variation in the prominence of the nose tip dominates the first principal component, whereas variation in the vertical location of the nose tip captured in the second principal component.

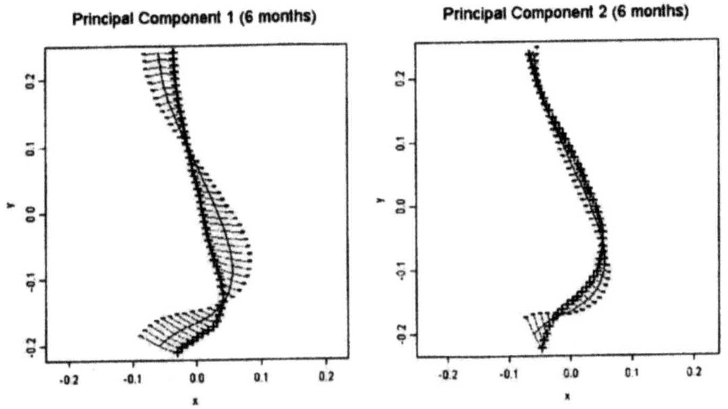

Figure 3: Variation in ridge curves, represented by the first two principal components patterns around the mean (solid line)

2.3 Surface data

The previous sub-sections describe the nature and analysis of point and curve data extracted from the 3-D models of babies faces. Ideally, however, we would like to use the data in its richest form, derived directly from the dense cloud of 3-D points, which define the model. The obstacle to using this raw data is the lack of correspondence between points in different models. All of the shape analysis techniques described above require this property of correspondence. Work is in progress to produce such a representation of the surface by conforming generic facial meshes to individual models. This section will briefly outline the process and present some of the initial results.

The process of representing surface data with correspondence properties involves three stages. Starting with a standard generic mesh representing the face we customise it to the needs of the study. In particular, the mesh was cut in half, edited to open the mouth slightly and move the vertices around the eyes, then reflected to produce a complete symmetric face. The second stage implements an algorithm (Ju and Siebert 2001) which takes the customised generic mesh and conforms it to a baby's model. Anatomical points (located on both the generic and the baby's model) are used to anchor the conformation and the algorithm warps the generic mesh for maximum agreement (see Figure 5 for results of this stage). The third stage involves the labelling and extraction of corresponding points for analysis.

For the data extracted from the conformed meshes to be of use, vertices that describe recognisable anatomical features must conform onto the corresponding features of models. The results of looking at the mid-line are promising (see Fig. 4) and suggest that further work looking at different features of interest is worthwhile.

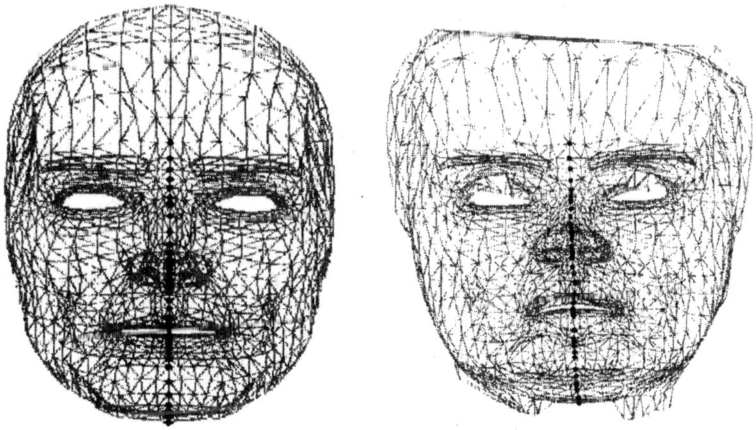

Figure 4: Customised generic mesh (left) with midplane points highlighted and a conformed mesh showing where the midplane points have been mapped to.

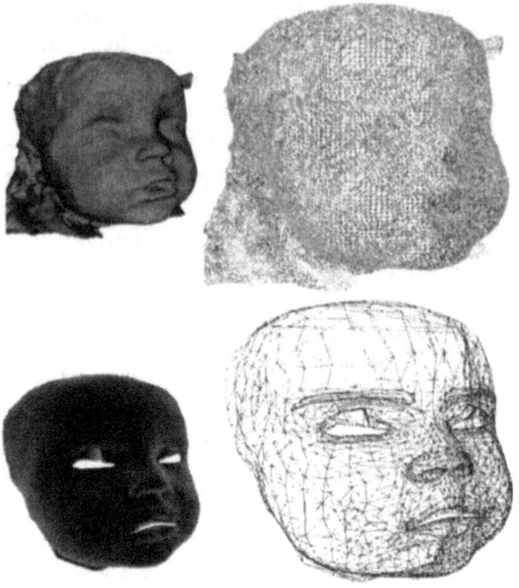

Figure 5: Original surface model (top left), underlying irregular mesh (top right), conformed generic mesh (bottom right) and conformed surface (bottom left)

3 Discussion

Although these methods have been discussed in the context of a specific application to the growth of children's faces, the generic nature of the tools makes them applicable to a wide range of other applications where the surface shape of objects is of prime interest. The focus of this paper has been on methods of data extraction, enabling standard methods of shape analysis to be applied. More advanced methods, allowing more detailed analysis, are under investigation.

Acknowledgements

This project is an interdisciplinary one also involving clinical colleagues, Ashraf Ayoub, Ann Garrahy, Kay Hood and Jill White, from the Glasgow Dental Hospital. We also thank Xiangyang Ju from the 3D Matic Research Laboratory for devising and programming the conformation algorithm. The financial support of the Chief Scientist's Office of the Scottish Executive, The Cleft Lip and Palate Association and the National Lotteries Board are gratefully acknowledged.

References

Ayoub, A.F., Siebert, P., Moos, K.F., Wray, D., Urquhart, C and Niblett, T. (1998). *A vision-based three-dimensional capture system for maxillofacial and surgical planning.* British Journal of Oral and Maxillofacial Surgery 36, 353-357.

Dryden. I.L. & Mardia. K.V. (1998). *Statistical Shape Analysis.* Wiley: New York.

Ju, X. and Seibert, J.P. (2001) Conforming generic animatable models to 3-D scanned data, *Proc.6th Numerisation 3-D/Scanning 2001 Congress*, Paris 2001

Ramsay. J.O. & Silverman. B.W. (1997). *Functional Data Analysis.* Springer-

Detection of Locally Stationary Segments in Time Series[1]

Uwe Ligges[1], Claus Weihs[1] and Petra Hasse-Becker[2]

[1] Fachbereich Statistik, Universität Dortmund, 44221 Dortmund, Germany

[2] Institut für Musik und ihre Didaktik, Universität Dortmund, 44221 Dortmund, Germany

Abstract . In many applications it is required to segment a time series into its locally stationary parts. Two applications are presented:

As a first example consider online monitoring of a BTA Deep-Hole-Drilling process. Here chattering and spiralling of the drilling tool should be avoided by process control. A second example is the analysis of vocal sound signals. It may be required to analyse only specific tones instead of a whole song.

Three new algorithms are introduced in this paper, all based on the theory of Dahlhaus (1997) and the analysis of the spectrum of time series, but with different methods to distinguish locally stationary parts of the signals.

Keywords. time series, segmentation, note classification, chattering prediction

1 Introduction

In many applications it is required to segment a time series into its locally stationary parts (cp. Dahlhaus (1997)), since on the one hand stationarity may be an assumption for further time series analysis, while on the other hand the transition areas may be of interest as in engineering or other technical applications.

Two quite different applications are presented:

As a first example consider online monitoring of a BTA (Boring and Trepanning Association) Deep-Hole-Drilling process (cp. Weinert et al. (2001)). Here chattering and spiralling of the drilling tool should be avoided by process control. This is possible by online monitoring of the process analyzing local parts of the signal of the drilling torque. The transition period to the start of chattering with an increased amplitude of the drilling torque is short, i.e. less than a second, so a fast algorithm has to be used.

A second example is the analysis of vocal sound signals. It may be required to analyse only specific tones instead of a whole song, e.g. for pitch tracking or conversion from wave-format into midi-format, so automatic detection of locally stationary parts of the wave is desirable.

Based on the algorithm by Adak (1998) for detection of locally stationary segments in time series, in this paper we develop new fast algorithms combining well known methods of time series analysis with fundamental facts of the area of the applications and measures which can be calculated quickly.

[1] This work has been supported by the Deutsche Forschungsgemeinschaft, Sonderforschungsbereich 475.

Correspondingly, three new algorithms are introduced in this paper, all based on the analysis of the spectrum of time series, but with different methods to distinguish locally stationary parts of the signals (cp. sections 2.2-2.5).

For the segmentation of vocal sound signals as a particular musical application, the algorithms are compared on artificial series of tones as well as on waves from real singing (cp. chapter 3). Notice that in this paper the term "note" is used for the graphical sign and the corresponding ideal musical sound planned by the composer, whereas the terms "sung note" or "tone" are used for the realized audio event corresponding to a note.

For the BTA Deep-Hole-Drilling process the Kolmogorov-Smirnov method yields satisfying results, i.e. "early" chattering and spiralling prediction. Other methods applicable for monitoring of the process, involving AR models or kernel densities, are described in Busse et al. (2001).

2 Segmentation algorithms

Since the segmentation of vocal sound signals and online monitoring of the drilling torque (for BTA Deep-Hole-Drilling) are obviously special cases of change point detection, at first a look at already existing methods was taken. It can be assumed, that the expectation of the time series is equal to zero.

2.1 Algorithm by Adak

Since change is not necessarily abrupt, but could also be somehow "smooth", as a first step the algorithm for "Time–Dependent Spectral Analysis of Nonstationary Time Series" (AA) introduced by Adak (1998) was implemented. This algorithm has been developed particularly for the segmentation of seismological data and speech, so it was hoped that it would also be useful for the prediction of chattering and the segmentation of vocal sound.

Unfortunately, with vocal sound data AA resulted into many errors of two kinds: falsely identifying change points in the middle of tones and not detecting change points between different tones. This leads to the conclusion that AA is not an appropriate algorithm for the segmentation of vocal sound signals.

In the case of online monitoring of a BTA Deep-Hole-Drilling process, using this algorithm online, real-time computation is not possible with recent computers.

In section 2.2 a new algorithm will be developed, nevertheless taking interesting ideas from Adak's algorithm.

2.2 Segmentation using KS–distance

Since there were many problems with the tree structure in the AA algorithm, we pass through the time series with a window of span n, for each window calculating the periodogram.

To compare two periodograms, the Kolmogorov–Smirnov–distance D^{KS} of their empirical spectral distributions is used. In principle, if this distance is larger than a threshold $w \in [0,1]$, a change point is detected. To fix w, the real number T_r of tones in a song is utilized. For different singers different values of D^{KS} indicate a change point, so it might be necessary to adjust the threshold w for each times series. For details see Ligges et al. (2002).

A change point at the beginning of the b-th part (interval) of the time series will be detected, if and only if the following three inequalities are valid:

$$D^{KS}(\hat{S}_{b-1}, \hat{S}_b) > w\,,\ D^{KS}(\hat{S}_{b-2}, \hat{S}_b) > w \text{ and } D^{KS}(\hat{S}_{b-1}, \hat{S}_{b+1}) > w\,, \quad (1)$$

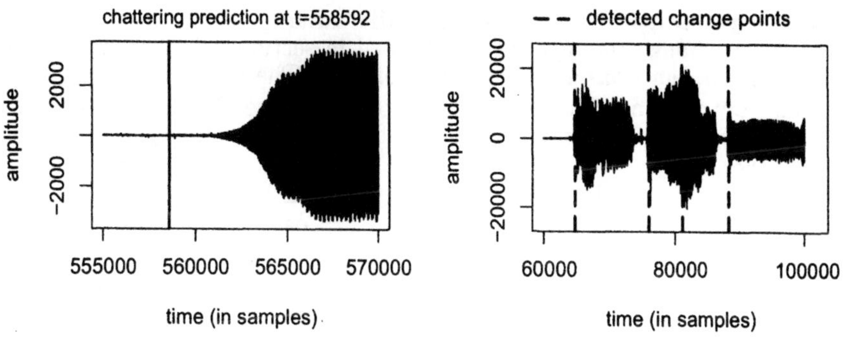

Fig. 1. (a) Chattering prediction on the drilling torque of a BTA-Deep-Hole-Drilling process. (b) Fragment of a segmented vocal sound signal.

where $D^{KS}(\hat{S}_{b-1}, \hat{S}_b)$ is the Kolmogorov–Smirnov–distance of the empirical spectral distributions $\hat{S}_{b-1}$ and $\hat{S}_b$. The three inequalities are required to make the algorithm more robust against noise like vibrato. For theory regarding vocal performances (e.g. definition of "vibrato") compare Seidner and Wendler (1997).

In the following discussion KS will be extended to allow for overlapping parts (e.g. by $\frac{n}{2}$ points) which are compared for change point detection.

KS is much better in segmentation of vocal sound signals than AA (a comparison in chapter 3). Nevertheless it is not really satisfying because of error rates above 20 percent on real singers data. In figure 1 (a) it is shown, however, that KS is already satisfactory for online monitoring of the drilling process, because prediction of chattering is very early. Thus, no further algorithm was developed for this example.

2.3 Exact determination of fundamental frequency

In the following sections an exact determination of the fundamental frequency is desirable. Assume the wave was sampled with 11kHz and the window for which the periodogram was calculated covers $n = 512$ observations. Then the values of the periodograms can be determined only for the following Fourier frequencies (in Hertz): $21.53, 43.07, \ldots, 5512.50$. Therefore, since the difference between the frequencies corresponding to the very low notes E and F, e.g. is only 5 Hertz, tones corresponding to these two notes cannot be expected to be well separable, so frequencies have to be well estimated from the periodogram in a more sophisticated way. Comparing mean, harmonical mean, geometrical mean, the straightforward "highest–peak" estimator and the following estimator (cp. Weihs et al. (2001))

$$\hat{\lambda} = \lambda_h + \frac{\lambda_s - \lambda_h}{2} \left(\frac{v_s}{v_h} \right)^{\frac{1}{e}}, \tag{2}$$

which averages the frequency λ_h of the highest peak with value v_h and the frequency λ_s of the highest peak of the direct neighbors of λ_h with value v_s of the periodogram, the last estimator is the best of these four.

Simulating wave files as described in Weihs et al. (2001) shows, that the maximal error of the approximation described in (2) was much better than any of the other compared estimators. Further on, with this method the true frequency can be estimated accurately enough to be able to separate even very low tones.

2.4 Segmentation using halftone distance

Let us now change the algorithm described in section 2.2 in an attempt to solve the problem from a musical point of view. The Kolmogorov–Smirnov–distance D^{KS} can be replaced by the halftone distance D^{HT}, which is defined as the difference in halftones between the frequencies λ_1 and λ_2 (cp. Berg and Stark (1982)):

$$D^{HT}(\lambda_1, \lambda_2) := 12 \cdot log_2 \left(\frac{\lambda_1}{\lambda_2} \right). \tag{3}$$

Since the halftone distance is an implicitly given musical measure, in principle no threshold has to be adjusted like in KS, since to detect a change point, the distance in halftones between the fundamental frequencies (cp. section 2.3) of two neighboring parts of the time series must be equal or larger than one.

This algorithm (HT) works analogously to the KS algorithm with the following distinctions. Instead of the empirical spectral distribution $\hat{S}_b$ the fundamental frequency λ_b has to be estimated and the formulas (1) have to be replaced by
$|D^{HT}(\lambda_{b-1}, \lambda_b)| > 0.9$, $|D^{HT}(\lambda_{b-2}, \lambda_b)| > 0.9$ and $|D^{HT}(\lambda_{b-1}, \lambda_{b+1})| > 0.9$.

On the one hand these three inequalities are required to make the algorithm more robust against vibrato, on the other hand many singers are sliding from one tone to the other, so the "absolute" difference of 1 between two halftone is being reduced heuristically to 0.9.

2.5 Segmentation using note classification by fundamental frequency

Another measure to distinguish tones, and so to segment the time series, is the note a tone corresponds to. Under the assumption that the frequency tuning of instruments is known, e.g. diapason $a' = 440$ Hertz, it is possible to perform a classification. For each part of the vocal time series the fundamental frequency can be estimated from the periodogram as already described. After that, this part can be classified as follows: given λ_0 is the (known) frequency of a reference note, this can be achieved by using the fundamental frequency note classifier $C_{\lambda_0}^{FF}(\lambda) := \left\lfloor 12 \cdot log_2 \left(\frac{\lambda}{\lambda_0} \right) + \frac{1}{2} \right\rfloor$, which is derived in a straightforward manner from formula (3), for a fundamental frequency λ.

All in all this algorithm (NC) works in an analogous manner to that described in section 2.4, except that the three equations formerly labelled with (1) have to be replaced by $|C_{\lambda_0}^{FF}(\lambda_{b-1}) - C_{\lambda_0}^{FF}(\lambda_b)| > 0$, $|C_{\lambda_0}^{FF}(\lambda_{b-2}) - C_{\lambda_0}^{FF}(\lambda_b)| > 0$ and $|C_{\lambda_0}^{FF}(\lambda_{b-1}) - C_{\lambda_0}^{FF}(\lambda_{b+1})| > 0$.

A typical segmentation of a fragment of the vocal sound signal from a semi-professional bass singer performing "Tochter Zion" is shown in figure 1 (b). In this particular fragment NC has segmented the time series very accurately.

3 Comparison of the algorithms

Three kinds of comparisons were done on vocal sound signals. As already mentioned in section 2.1, the algorithm by S. Adak (AA) is neither appropriate for the segmentation of vocal sound signals nor for the chattering prediction of a BTA Deep-Hole-Drilling process, because it has a large runtime and it produces inaccurate results, which is obvious when looking at the results in section 3.1. So in the last two comparisons this algorithm is not compared with the others.

3.1 Simulation: artificial series of tones

In a first test of the algorithms two waves (16 bit, 11kHz) of 25 tones each are artificially generated and the duration of the tones is randomly chosen between 0.05 and 1 second with abrupt change points. Also the pitch is randomly chosen (from D to f'''), except for 6 tones with fixed pitch in order to test the algorithms on neighboring halftones. The difference between the two generated waves is the weighting of fundamental frequencies and overtones (sine–waves), which is set as follows: fundamental frequency 70% (15%), 1. overtone 20% (70%) and 2. overtone 10% (15%), values for the second wave in brackets. The parameter n (span of the window) was set to 512 for all algorithms.

The following number of errors was discovered for the algorithms:

AA 19 (23), KS 1 (1), HT 4 (2), NC 2 (3),

where "error" means that the particular algorithm did not detect a change point or the algorithm detects a change point at a place without any real change points (i.e. more than 512 data points away from a real one). Since there are no large differences in the segmentation exactness of the latter three algorithms, more comparisons have to be done.

3.2 Simulation: An artificial performance of "Tochter Zion"

In a second step artificial waves of "Tochter Zion" were generated. The duration of a half note corresponds to 1 sec and between two notes there are breaks of $\frac{1}{100}$ seconds duration. Four versions are required for the different types of singing voices (soprano, alto, tenor, bass). For each of these versions an experimental design is used to simulate different types of voices related to the weighting of fundamental frequency and its first five corresponding overtones. All of these six frequencies are weighted with any of the following ratios $\frac{0}{6}, \frac{1}{6}, \ldots, \frac{6}{6}$ with the restrictions that the sum of the weights must be equal to one and that the fundamental frequency must have at least a weighting of $\frac{1}{6}$. This design results in 1008 waves totally.

	soprano			alto			tenor			bass		
	KS	HT	NC	KS	HT	NC	KS	HT	NC	KS	HT	NC
Min.	8.00	8.00	8.00	8.00	6.00	5.00	8.00	3.00	6.00	8.00	3.00	3.00
Mean	8.01	8.41	8.00	8.01	7.77	7.41	8.03	7.56	7.94	8.04	12.66	8.51
Max.	11.00	9.00	8.00	11.00	9.00	8.00	10.00	10.00	10.00	11.00	23.00	16.00

Table 1. Number of errors in a segmentation of artificially generated waves of "Tochter Zion"

In table 1 statistics about error rates corresponding to the four types of singing voices are shown for the algorithms KS, HT and NC. The number 8 appears in many places in that table, because 8 times there are two neighboring notes with the same pitch only interrupted by breaks of $\frac{1}{100}$ seconds duration. Obviously these tones are problematic to be distinguished. KS seems to be the most robust algorithm (particularly for bass singers), while the other two algorithms in some circumstances have very small minimal error rates.

3.3 Segmentation on real singers' performances

Since in the simulation described in section 3.2 no noise and no vibrato was added to the artificially generated sine waves, we finally use real data to find out, whether the algorithms work on real vocal sound signals appropriately.

The algorithms are compared by results of segmentations of 17 versions of the classical song "Tochter Zion" (Händel) performed by different real singers (from real amateurs to real professionals) as described in Weihs, et al. (2001). The number of errors in this experiment is shown in table 2, where each column (2-18) represents one particular singer.

Obviously the segmentation algorithm using note classification (NC) is much better in error rate than the other algorithms on real singers performances.

	soprano				alto						tenor			bass				sum
	1	2	3	4	1	2	3	4	5	6	1	2	3	1	2	3	4	
KS	35	20	31	38	25	18	31	22	26	28	27	40	35	31	24	31	34	496
HT	49	24	32	36	26	29	26	25	19	24	39	28	23	28	27	32	33	500
NC	27	19	20	25	22	18	12	9	6	12	12	17	16	27	22	21	23	308

Table 2. Segmentation errors on 17 real singers' performances of "Tochter Zion"

4 Conclusion

In section 2.2 it is shown that algorithm KS already satisfies for online monitoring of a BTA Deep-Hole-Drilling process, because prediction of chattering is possible early.

The algorithms KS and HT (sections 2.2 and 2.4) are comparably good w.r.t. error rates and computation time consumption. The algorithm NC (cp. section 2.5) is the best of the here described algorithms on segmentation of real singers' vocal performances.

References

Adak, S. (1998). Time-Dependent Spectral Analysis of Nonstationary Time Series. *Journal of the American Statistical Association*, **93**, 1488-1501.

Berg, R. E. and Stork, D. G. (1982). *The Physics of Sound.* New Jersey: Prentice-Hall.

Busse, A. M., Hüsken, M. and Stagge, P. (2001). Offline-Analyse eines BTA-Tiefbohrprozesses. *Technical Report 16/2001.* SFB 475, Department of Statistics, University of Dortmund, Germany.

Brockwell, P. J. and Davis, R. A. (1991). *Time Series: Theory and Methods.* New York: Springer.

Dahlhaus, R. (1997). Fitting Time Series Models to Nonstationary Processes. *The Annals of Statistics 1997*, **25**, 1-37.

Ligges, U., Weihs, C. and Hasse-Becker, P. (2002). Detection of locally stationary segments in time series – algorithms and applications. *Technical Report.* SFB 475, Department of Statistics, University of Dortmund, Germany.

Seidner, W. and Wendler, J. (1997). *Die Sängerstimme.* Berlin: Henschel.

Weihs, C., Berghoff, S., Hasse-Becker, P. and Ligges, U. (2001). Assessment of Purity of Intonation in Singing Presentations by Discriminant Analysis. In: Kunert, J. and Trenkler. G. (2001). *Mathematical Statistics and Biometrical Applications*, 395-410. Lohmar: Josef Eul Verlag.

Weinert, K., Webber, O., Busse, A., Hüsken, M., Mehnen, J. and Stagge P. (2001). In die Tiefe: Koordinierter Einsatz von Sensorik und Statistik zur Analyse und Modellierung von BTA-Tiefbohrprozessen. In: Spur, G. (2001) *ZWF, Zeitschrift für wirtschaftlichen Fabrikbetrieb*, **5**. München: Carl Hanser Verlag.

Detection of Outliers in Multivariate Data: A Method Based on Clustering and Robust Estimators

Carla M. Santos-Pereira[1] and Ana M. Pires[2]

[1] Universidade Portucalense Infante D. Henrique, Oporto, Portugal and Applied Mathematics Centre, IST, Technical University of Lisbon, Portugal.

[2] Department of Mathematics and Applied Mathematics Centre, IST, Technical University of Lisbon, Portugal.

Keywords. Multivariate analysis, Outlier detection, Robust estimation, Clustering, Supervised classification

1 Introduction

Outlier identification is important in many applications of multivariate analysis. Either because there is some specific interest in finding anomalous observations or as a pre-processing task before the application of some multivariate method, in order to preserve the results from possible harmful effects of those observations. It is also of great interest in supervised classification (or discriminant analysis) if, when predicting group membership, one wants to have the possibility of labelling an observation as "does not belong to any of the available groups". The identification of outliers in multivariate data is usually based on Mahalanobis distance. The use of robust estimates of the mean and the covariance matrix is advised in order to avoid the masking effect (Rousseeuw and Leroy, 1985; Rousseeuw and von Zomeren, 1990; Rocke and Woodruff, 1996; Becker and Gather, 1999). However, the performance of these rules is still highly dependent of multivariate normality of the bulk of the data. The aim of the method here described is to remove this dependence.

2 Description of the method

Consider a multivariate data set with n observations in p variables. The basic ideas of the method can be described in four steps:

1. Segment the n points cloud (of perhaps complicated shape) in k smaller subclouds using a partitioning clustering method with the hope that each subcloud (cluster) looks "more normal" than the original cloud.
2. Then apply a simultaneous multivariate outlier detection rule to each cluster by computing Mahalanobis-type distances from all the observations to all the clusters. An observation is considered an outlier if it is an outlier for every cluster. All the observations in a cluster may also be considered outliers if the relative size of that cluster is small (our proposal is less than $2p+2$, since for smaller number of observations the covariance matrix estimates are very unreliable).
3. Remove the observations detected in 2 and repeat 1 and 2 until no more observations are detected.
4. The final decision on whether all the observations belonging to a given cluster (not previously removed, that is with size greater than $2p+1$) are outliers is based on a table of between clusters Mahalanobis-type distances.

There is no need to fix k in advance, we suggest to apply several values of k and observe the results (there is of course a limit on k, which depends on the number of observations and the number of variables).

The choice of the partitioning clustering method is crucial and may depend on some previous exploratory analysis of the data: if a not too complicated shape is expected than the classical k-means method (which tends to produce hyperspherical clusters) is adequate; with more complicated shapes, methods more robust to the "spherical cluster" assumption (like for instance the model based clustering of Banfield and Raftery, 1992) are to be preferred. It is also necessary to take into account the size of the data set. For large data sets the method *clara* of Kaufman and Rousseeuw (1990) may be appropriate.

By a simultaneous multivariate outlier detection rule we mean a rule such that, for a multivariate normal sample of size n_j, no observation is identified as an outlier with probability $1 - \alpha_j$, $j = 1, \ldots, k$ (see Davies and Gather, 1993). If $\alpha_j = 1 - (1 - \alpha)^{1/k}$ an overall level α can be guaranteed for mixtures of k multivariate normal distributions. An observation, $\mathbf{x}$, with squared Mahalanobis distance $d^2 = (\mathbf{x} - \hat{\boldsymbol{\mu}}_j)^T \hat{\boldsymbol{\Sigma}}_j^{-1} (\mathbf{x} - \hat{\boldsymbol{\mu}}_j)$ greater than a detection limit $c(p, n_j, \alpha_j)$ is declared an outlier relatively to the jth cloud. The constant $c(p, n_j, \alpha_j)$ is asymptotically $\chi^2_{p;\beta}$, with $\beta = (1 - \alpha_j)^{1/n_j}$. However, as discussed in Becker and Gather (2001), for not very large samples and depending on the estimators $\hat{\boldsymbol{\mu}}_j$ and $\hat{\boldsymbol{\Sigma}}_j$, there may be large differences to the asymptotic value. Those authors suggest that more reliable constants can easily be determined by simulation. For the estimators $\hat{\boldsymbol{\mu}}_j$ and $\hat{\boldsymbol{\Sigma}}_j$ we may choose the classical sample mean vector and sample covariance matrix, or, for greater protection against masking, robust analogues of them.

In step 4, let the observations (not removed in the previous steps) be labelled as $\mathbf{x}_{ij}$, $i = 1, \ldots, n_j$, $j = 1, \ldots, k$, where n_j are the sizes of the final k clusters (note that $\sum n_j$ may be smaller than n). Then define

$$D_{lm} = \min_{i=1,\ldots,n_l} (\mathbf{x}_{il} - \hat{\boldsymbol{\mu}}_m)^T \hat{\boldsymbol{\Sigma}}_m^{-1} (\mathbf{x}_{il} - \hat{\boldsymbol{\mu}}_m) .$$

A given cluster, l, is said to be disconnected from the remaining data if $D_{lm} > c(p, n_m, \alpha_m)$, for all $m \neq l$, and connected otherwise. Disconnected clusters are suspicious of containing only outliers, however such a decision is application dependent and must be taken case by case.

The proposed procedure is affine invariant (that is, the observations identified as outliers do not change under affine transformations) if the location and scatter estimators are affine equivariant and if the clustering method is affine invariant. Note that the k-means method does not have this property.

3 Simulation study

In order to evaluate the performance of the above method and to compare it with the usual method of a single Mahalanobis distance we conducted a simulation study with:

- Three clustering methods: k-means, *pam* (partioning around medoids, from Kaufman and Rousseeuw, 1990) and *mclust* (model based clustering for gaussian distributions, from Banfield and Raftery, 1992), each of them with $k = 3, 4, 5$.

- Two pairs of location-scatter estimators: classical $(\bar{\mathbf{x}}, \mathbf{S})$ and Reweighted Minimum Covariance Determinant (Rousseeuw, 1985) with an approximate 25% breakdown point (denoted RMCD25), which has better efficiency than the one with (maximal) %50 breakdown point. (For the first estimators we used the asymptotic detection limits while for the RMCD25 the detection limits were determined previously by simulation with 10000 normal data sets.)
- Overall significance level: $\alpha = 0.1$.
- Classify disconnected clusters as outliers.
- Eight distributional situations:
  1. Normal ($p = 2$) without outliers, 150 observations from $N_2(\mathbf{0}, \mathbf{I})$.
  2. Normal ($p = 2$) with outliers, 150 observations from $N_2(\mathbf{0}, \mathbf{I})$ plus 10 outlying observations from $N_2(\mathbf{10.51}, 0.1\mathbf{I})$.
  3. Normal ($p = 4$) without outliers, 150 observations from $N_4(\mathbf{0}, \mathbf{I})$.
  4. Normal ($p = 4$) with outliers, 150 observations from $N_4(\mathbf{0}, \mathbf{I})$ plus 10 outlying observations from $N_4(\mathbf{8.59}, 0.1\mathbf{I})$.
  5. Non-normal ($p = 2$) without outliers, 50 observations from $N_2(\boldsymbol{\mu}_1, \boldsymbol{\Sigma}_1)$, 50 observations from $N_2(\boldsymbol{\mu}_2, \boldsymbol{\Sigma}_2)$ and 50 observations from $N_2(\mathbf{0}, \boldsymbol{\Sigma}_1)$, with $\boldsymbol{\mu}_1 = (0, 12)^T$, $\boldsymbol{\Sigma}_1 = \text{diag}(1, 0.3)$, $\boldsymbol{\mu}_2 = (1.5, 6)^T$ and $\boldsymbol{\Sigma}_2 = \text{diag}(0.2, 9)$.
  6. Non-normal ($p = 2$) with outliers, 150 observations as in the previous case plus 10 outlying observations from $N_2((-2, 6)^T, 0.01\mathbf{I})$
  7. Non-normal ($p = 4$) without outliers, 50 observations from $N_4(\boldsymbol{\mu}_3, \boldsymbol{\Sigma}_3)$, 50 observations from $N_4(\boldsymbol{\mu}_4, \boldsymbol{\Sigma}_4)$ and 50 observations from $N_4(\mathbf{0}, \boldsymbol{\Sigma}_3)$, with $\boldsymbol{\mu}_3 = (0, 12, 0, 0)^T$, $\boldsymbol{\Sigma}_3 = \text{diag}(1, 0.3, 1, 1)$, $\boldsymbol{\mu}_4 = (1.5, 6, 0, 0)^T$ and $\boldsymbol{\Sigma}_4 = \text{diag}(0.2, 9, 1, 1)$.
  8. Non-normal ($p = 4$) with outliers, 150 observations as in the previous case plus 10 outlying observations from $N_4((-2, 6, 0, 0)^T, 0.01\mathbf{I})$.

For each combination we run 100 simulations using the statistical package S-plus. Figure 1 shows one of the generated data sets for situation 6. The superimposed detection contours for two of the methods show that the outlier detection with previous clustering is much more appropriate to this situation.

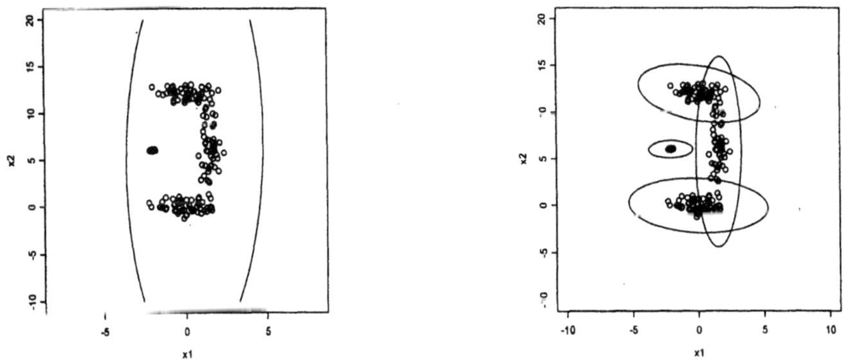

Fig. 1. Non-normal bivariate data with outliers and simultaneous detection contours ($\alpha = 0.1$) with: single robust Mahalanobis distance using RMCD25 (left panel); *mclust* with $k = 4$ and RMCD25 (right panel).

Tables 1 and 2 give the results of the simulation in terms of "proportion of runs with correct identification of all the outliers," p_1, "average proportion of outliers masked over the runs," p_2, "average proportion of swamping over the runs," p_3, and "proportion of runs without swamping" (see the simulation study of Kosinski, 1998).[1] A good method should have $p_1 = 1$, $p_2 = 0$, small p_3 and $p_4 = 1 - \alpha = 0.9$. For the normal data all the methods behave very well, except for some masking with the classical Mahalanobis distance, which is not surprising. For nonnormal data the best performance is achieved by *mclust* with $k = 4$ and $k = 5$, without significant differences between the classical and the robust estimators. The failure of the k-means method is justified by the remarkable non-sphericity of the distributions chosen.

4 Examples

We have used the proposed method (with the same variants chosen for the simulation study) on the HBK data set (also analysed by several authors, like Rousseeuw and Leroy, 1985, or Rocke and Woodruff, 1996). All the variants detected exactly the 14 known outliers.

A more complicated data set was also analysed: the "pen-based automatic character recognition data" (available from http://www.ics.uci.edu/~mlearn /MLSummary.html). We have selected one of the classes, corresponding to the digit "0" with 1143 cases on 16 variables. An interesting feature of this data is that the characters can be plotted making it easy to judge the adequacy of the outlier detection (which is very convenient for checking on competing methods but very tedious to do for all the observations, besides the aim is to perform automatic classification). The single Mahalanobis distance with classical estimators revealed 106 outliers. The single Mahalanobis distance with RMCD25 pointed 513 observations (!!!) most of them looking quite unsuspicious. For the other options with clustering (only k-means and *pam*, since *mclust* is not adequate for data of this size) those numbers varied from 95 to 140. Very aberrant observations were detected by all the methods including classical Mahalanobis distance. However, ten strange observations (looking more like a "6" than a "0") were detected by all the clustering methods but not by the classical Mahalanobis distance. In future work we intend to assess the impact of this on the accuracy of the discrimination process for the whole data set.

5 Conclusions

From the simulation study and from the examples we conclude that the proposed method is promising. However, some refinements may be necessary. We intend to try other (more efficient) robust estimators and to perform a more extensive simulation study, namely to investigate the performance on larger data sets, higher contamination, and the instability of the indicator p_4.

References

Banfield, J.D. and Raftery, A.E. (1992). Model-based Gaussian and non-Gaussian clustering. *Biometrics*, **49**, 803-822.

Becker, C. and Gather, U. (1999). The masking breakdown point of multivariate outlier identification rules. *Journal of the American Statistical Association*, **94**, 947-955.

[1] For the cases without outliers p_1 and p_2 are not defined (nd).

Situation	Clust. Method	k	MD with RMCD25 p_1	p_2	p_3	p_4	MD with $\bar{\mathbf{x}}$ and $\mathbf{S}$ p_1	p_2	p_3	p_4
Normal $p=2$ Without outliers	k-means	3	nd	nd	0.000	1.00	nd	nd	0.002	0.78
	"	4	nd	nd	0.005	0.53	nd	nd	0.001	0.87
	"	5	nd	nd	0.003	0.67	nd	nd	0.002	0.85
	pam	3	nd	nd	0.000	1.00	nd	nd	0.002	0.78
	"	4	nd	nd	0.005	0.58	nd	nd	0.002	0.78
	"	5	nd	nd	0.004	0.65	nd	nd	0.001	0.81
	mclust	3	nd	nd	0.000	1.00	nd	nd	0.001	0.93
	"	4	nd	nd	0.003	0.75	nd	nd	0.000	0.95
	"	5	nd	nd	0.002	0.78	nd	nd	0.001	0.92
	none	1	nd	nd	0.001	0.92	nd	nd	0.001	0.91
Normal $p=2$ With outliers	k-means	3	1.00	0.00	0.000	1.00	1.00	0.00	0.002	0.76
	"	4	1.00	0.00	0.006	0.51	1.00	0.00	0.002	0.69
	"	5	1.00	0.00	0.004	0.62	1.00	0.00	0.001	0.85
	pam	3	1.00	0.00	0.000	1.00	1.00	0.00	0.002	0.73
	"	4	1.00	0.00	0.005	0.54	1.00	0.00	0.002	0.75
	"	5	1.00	0.00	0.003	0.66	1.00	0.00	0.002	0.75
	mclust	3	1.00	0.00	0.000	1.00	1.00	0.00	0.001	0.85
	"	4	1.00	0.00	0.003	0.72	1.00	0.00	0.001	0.82
	"	5	1.00	0.00	0.003	0.72	1.00	0.00	0.001	0.93
	none	1	1.00	0.00	0.001	0.88	0.84	0.16	0.001	0.93
Normal $p=4$ Without outliers	k-means	3	nd	nd	0.000	1.00	nd	nd	0.001	0.91
	"	4	nd	nd	0.001	0.94	nd	nd	0.000	0.95
	"	5	nd	nd	0.000	0.99	nd	nd	0.000	0.99
	pam	3	nd	nd	0.000	1.00	nd	nd	0.001	0.91
	"	4	nd	nd	0.001	0.93	nd	nd	0.001	0.93
	"	5	nd	nd	0.000	0.95	nd	nd	0.000	0.99
	mclust	3	nd	nd	0.000	1.00	nd	nd	0.000	0.98
	"	4	nd	nd	0.001	0.91	nd	nd	0.000	1.00
	"	5	nd	nd	0.001	0.94	nd	nd	0.001	0.98
	none	1	nd	nd	0.000	0.95	nd	nd	0.001	0.91
Normal $p=4$ With outliers	k-means	3	1.00	0.00	0.000	1.00	1.00	0.00	0.003	0.67
	"	4	1.00	0.00	0.007	0.44	1.00	0.00	0.003	0.69
	"	5	1.00	0.00	0.005	0.57	1.00	0.00	0.001	0.82
	pam	3	1.00	0.00	0.000	1.00	1.00	0.00	0.002	0.76
	"	4	1.00	0.00	0.005	0.52	1.00	0.00	0.002	0.81
	"	5	1.00	0.00	0.004	0.65	1.00	0.00	0.001	0.87
	mclust	3	1.00	0.00	0.000	1.00	1.00	0.00	0.002	0.80
	"	4	1.00	0.00	0.005	0.64	1.00	0.00	0.001	0.87
	"	5	0.99	0.01	0.003	0.71	1.00	0.00	0.001	0.89
	none	1	1.00	0.00	0.001	0.84	0.60	0.40	0.004	0.94

Table 1. Results of the simulation study for normal data.

Becker, C. and Gather, U. (2001). The largest nonidentifiable outlier: a comparison of multivariate simultaneous outlier identification rules. *Computational Statistics and Data Analysis*, **36**, 119-127.

Davies, P.L. and Gather, U. (1993). The identification of multiple outliers. *Journal of the American Statistical Association*, **88**, 782-792.

Kaufman, L. and Rousseeuw, P.J. (1990). *Finding Groups in Data: An Introduction to Cluster Analysis.* New York: Wiley.

Kosinski, A.S. (1998). A procedure for the detection of multivariate outliers. *Computational Statistics and Data Analysis*, **29**, 145-161.

Rocke, D.M. and Woodruff, D.L. (1996). Identification of outliers in multi-

Situation	Clust. Method	k	MD with RMCD25				MD with $\bar{\mathbf{x}}$ and S			
			p_1	p_2	p_3	p_4	p_1	p_2	p_3	p_4
	k-means	3	nd	nd	0.199	0.14	nd	nd	0.063	0.60
	"	4	nd	nd	0.017	0.77	nd	nd	0.037	0.72
	"	5	nd	nd	0.005	0.82	nd	nd	0.022	0.77
Non-normal	pam	3	nd	nd	0.193	0.16	nd	nd	0.029	0.74
$p = 2$	"	4	nd	nd	0.012	0.73	nd	nd	0.030	0.63
Without	"	5	nd	nd	0.006	0.70	nd	nd	0.036	0.73
outliers	mclust	3	nd	nd	0.046	0.81	nd	nd	0.041	0.73
	"	4	nd	nd	0.018	0.71	nd	nd	0.044	0.75
	"	5	nd	nd	0.001	0.83	nd	nd	0.024	0.84
	none	1	nd	nd	0.001	0.90	nd	nd	0.000	0.99
	k-means	3	0.94	0.06	0.213	0.06	0.14	0.86	0.035	0.81
	"	4	0.38	0.62	0.104	0.52	0.00	1.00	0.067	0.61
	"	5	0.50	0.50	0.023	0.75	0.51	0.49	0.040	0.72
Non-normal	pam	3	0.89	0.11	0.207	0.11	0.18	0.82	0.046	0.73
$p = 2$	"	4	0.84	0.16	0.169	0.23	0.00	1.00	0.045	0.74
With	"	5	0.83	0.17	0.008	0.72	0.53	0.47	0.045	0.71
outliers	mclust	3	0.21	0.79	0.054	0.78	0.17	0.83	0.047	0.75
	"	4	0.99	0.01	0.026	0.80	0.99	0.01	0.043	0.73
	"	5	0.94	0.06	0.009	0.80	1.00	0.00	0.014	0.85
	none	1	0.00	1.00	0.000	0.96	0.00	1.00	0.000	1.00
	k-means	3	nd	nd	0.181	0.25	nd	nd	0.018	0.78
	"	4	nd	nd	0.028	0.71	nd	nd	0.026	0.82
	"	5	nd	nd	0.008	0.85	nd	nd	0.040	0.76
Non-normal	pam	3	nd	nd	0.124	0.47	nd	nd	0.016	0.80
$p = 4$	"	4	nd	nd	0.007	0.82	nd	nd	0.037	0.81
Without	"	5	nd	nd	0.005	0.89	nd	nd	0.033	0.76
outliers	mclust	3	nd	nd	0.034	0.86	nd	nd	0.006	0.90
	"	4	nd	nd	0.007	0.88	nd	nd	0.012	0.85
	"	5	nd	nd	0.002	0.91	nd	nd	0.013	0.89
	none	1	nd	nd	0.001	0.90	nd	nd	0.000	0.97
	k-means	3	0.76	0.24	0.205	0.21	0.07	0.93	0.019	0.82
	"	4	0.13	0.87	0.045	0.57	0.01	0.99	0.074	0.75
	"	5	0.16	0.84	0.015	0.69	0.26	0.74	0.040	0.79
Non-normal	pam	3	0.66	0.34	0.168	0.33	0.06	0.94	0.016	0.79
$p = 4$	"	4	0.88	0.12	0.131	0.36	0.02	0.98	0.043	0.8
With	"	5	0.74	0.26	0.013	0.79	0.21	0.79	0.042	0.78
outliers	mclust	3	0.23	0.77	0.291	0.53	0.18	0.82	0.171	0.64
	"	4	1.00	0.00	0.013	0.84	0.99	0.01	0.009	0.88
	"	5	0.94	0.06	0.007	0.92	0.98	0.02	0.015	0.90
	none	1	0.00	1.00	0.001	0.90	0.00	1.00	0.000	0.96

Table 2. Results of the simulation study for non-normal data.

variate data. *Journal of the American Statistical Association*, **91**, 1047-1061.

Rousseeuw, P.J. (1985). Multivariate estimation with high breakdown point. In: *Mathematical Statistics and Applications, Volume B, eds. W. Grossman, G. Pflug, I. Vincze and W. Werz* 283-297. Dordrecht: Reidel.

Rousseeuw, P.J. and Leroy, A.M. (1985). *Robust Regression and Outlier Detection.* New York: Wiley.

Rousseeuw, P.J. and von Zomeren, B.C. (1990). Unmasking multivariate outliers and leverage points. *Journal of the American Statistical Association*, **85**, 633-639.

Development of a Framework for Analyzing Process Monitoring Data with Applications to Semiconductor Manufacturing Process

Yeo-Hun Yoon[1], Young-Sang Kim[2], Sung-Jun Kim[3] and Bong-Jin Yum[2]

[1] EDS Team, Samsung Electronics, San ♯24, Nongseo-ri, Kiheung-eup, Yongin-city, Kyunggi-do, 449-711, Korea
[2] Department of Industrial Engineering, Korea Advanced Institute of Science and Technology, 373-1, Kusong-dong, Yusong-gu, Taejon, 305-701, Korea
[3] Deloitte Consulting, 19F, Seoul Finance Center Bldg., 63, Mukyo-dong, Chung-gu, Seoul, 100-170, Korea

Abstract. A semiconductor manufacturing process consists of hundreds of steps, and produces a large amount of data. These process monitoring data contain useful information on the behavior of a process or a product. After semiconductor fabrication is completed, dies on a wafer are classified into bins in the EDS (Electrical Die Sorting) process. Quality engineers in semiconductor industry are interested in relating these bin data to the historical monitoring data to identify those process variables that are critical to the quality of the final product. Data mining techniques can be effectively used for this purpose. In this article, a framework for analyzing semiconductor process monitoring and bin data is developed using the data mining and other statistical techniques.

Keywords. Semiconductor manufacturing, data mining, decision tree, artificial neural network, Taguchi parameter design

1 Introduction

A semiconductor manufacturing process consists of hundreds of steps, and produces a large amount of data. In each step, a number of process variables and performance characteristics are measured and stored in databases. These process monitoring data contain useful information on the behavior of a process or a product, and deserve serious attention from the viewpoint of process or product improvement. However, analyzing such data is difficult in that they usually consist of many variables correlated with each other, the signal-to-noise ratios are small, and the number of observations are large.

After semiconductor fabrication is completed, dies on a wafer are classified into bins in the EDS (Electrical Die Sorting) process. For instance, bin '*p*' corresponds to perfect dies, bin '*r*' to repairable dies, and the rest to various types of defective dies. Quality engineers in semiconductor industry are interested in relating these bin data to the historical monitoring data to identify those process variables that are critical to the quality of the final product. Data mining and statistical techniques can be effectively used for this purpose.

Han and Kamber (2001) define data mining as 'the task of discovering interesting patterns from large amounts of data where the data can be stored in databases, data warehouses, or other information repositories'. It is an interdisciplinary field combining work from database systems, statistics, and artificial intelligence (or machine learning). This article develops a framework for analyzing semiconductor process monitoring and bin data using the data mining and other statistical techniques.

2 Literature review

Data mining techniques have been used in semiconductor manufacturing for yield improvement. Sorell (2000) emphasized the importance of such tasks as 'projection definition and goal setting' and 'data preparation' before data mining is performed. Mieno *et al.* (1999) applied a regression tree method to failure analysis in LSI manufacturing. Irani *et al.* (1993) used generalized ID3 (GID3) algorithm for diagnosis and optimization of a reactive ion etching process, and for knowledge acquisition for an expert system. Kittler and Wang (2000) discussed the advantages of decision tree algorithms and their enhancements for yield improvement, and compared these methods to regression and Kruskal-Wallis analyses. However, these previous works presented only a part of the whole analysis process. In process monitoring data analysis, it is important to develop an overall analysis strategy. Otherwise, one might undergo the process of trial and error repeatedly. In this article, a framework for data preparation, rule extraction, and confirmation is developed for process monitoring data analysis and applied to a semiconductor manufacturing process.

3 Description of data

In a semiconductor manufacturing process, a manufacturing lot consists of a number of wafers. After a lot is processed, electrical testing is performed, and dies on a wafer are sorted and classified into bins depending on whether a die successfully passes required tests or fails.

From an actual semiconductor process, eight important production steps were selected. The data gathered from these steps consist of the following information.

1) Lot ID
2) EDS bin data
3) ADI (After Development Inspection) and ACI (After Cleaning Inspection) CD (Critical Dimension) in 8 steps ($ADI_1 \sim ADI_8$ and $ACI_1 \sim ACI_8$)
4) Overlay alignment variables in 8 steps (4 variables in each step, namely, α_i, β_i, γ_i, and δ_i, where $i = 1, \cdots, 8$)
5) Times of EDS and ADI/ACI/Alignment measurements

The ADI and ACI CD values were collected at several points on each wafer and their means are used in the present study. There are three data sheets, each stores values of a part of the above variables at the corresponding steps.

4 General framework and results

The analysis procedures consist of four stages : (1) data preparation (e.g., cleaning and merging); (2) basic statistics, plotting, and data partitioning; (3) decision tree analysis; and (4) training an artificial neural network (ANN) followed by Taguchi parameter design (Phadke, 1989).

Data preparation is one of the most important steps for analyzing process monitoring data. After the data set is arranged, various types of plots are generated to visualize and explore the properties of data. The initial exploration results are used to partition the whole data set into more homogeneous ones. Then, data mining techniques are applied to the individual data sets for extracting useful information. A decision tree analysis is conducted to identify critical process variables that influence the bin data, and subsequently their new specifications can be derived. Finally, the rules from the decision tree analysis are confirmed and their robustness are checked using a trained ANN and the Taguchi parameter design methodology (Phadke, 1989).

4.1 Data preparation

The first step of the analysis is data cleaning, i.e., treatment of the variables and observations that have missing values. Since the ADI_6 and ACI_6 variables have too many missing values, they are decided to be deleted from the analysis. In addition, the observations that have any missing value are also deleted. Finally, data sets are merged into a single set by matching the Lot ID in each data sheet. The number of observations in the final data set was 888.

4.2 Basic analysis

The summary statistics of all variables are calculated, namely, means, variances, signal-to-noise ratios, etc. Then, the data are plotted in various forms. For instance, plots between process variables, plots between process and product variables, time series plots, and histograms. Figures 1 and 2 respectively show the plots between the yield and an ACI variable and the time series plot of ACI_3, where yield is the sum of bins '*p*' and '*r*'. In Figure 1, two outliers that have very small yields (dotted circles) are deleted because they were considered to be due to some abnormalities that occurred during the process operation. However, the point represented as a solid circle is kept since this type of point can sometimes occur during the normal operating condition and may contain useful process information. From Figure 2, the data set is divided into two according to the ACI_3 variable since, at a certain point in time t_1, the ACI_3 values show a significant change.

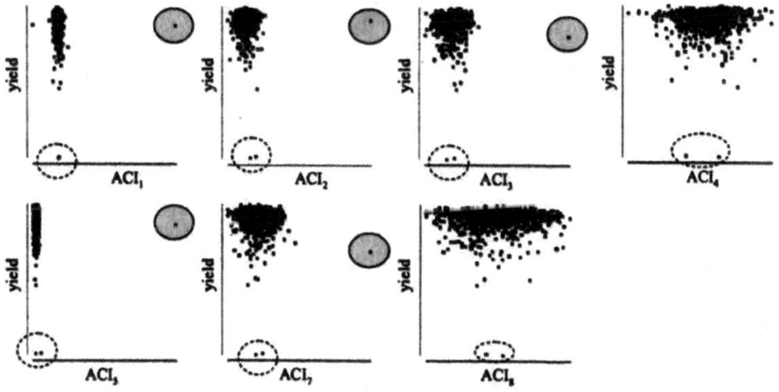

Fig. 1. Plots of Yield vs. ACI Variables

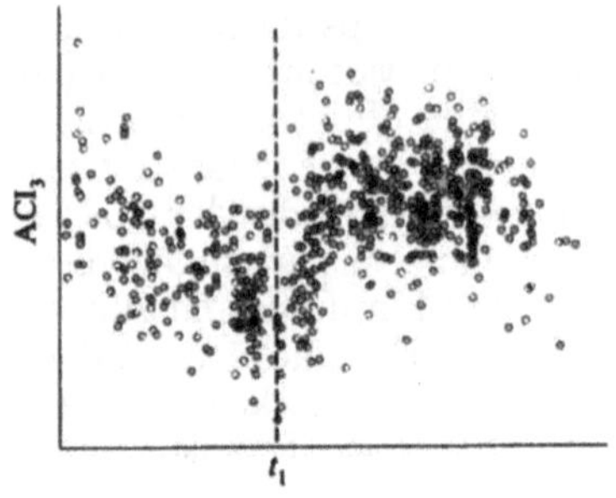

Fig. 2. Time Series Plot of ACI_3

4.3 Decision tree analysis

For each divided data set, a decision tree analysis is conducted for selected bins. From the decision tree analysis, critical process variables are identified and their desirable ranges are proposed to decrease the proportions of defective bins. Figure 3 is the final decision tree for the bin '*a*' and the results for other bins are shown in Table 1.

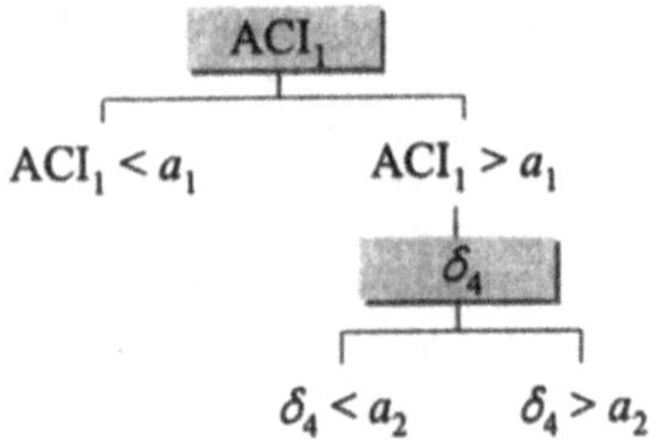

Fig. 3. Decision Tree for bin '*a*'

From Table 1, new process specifications are proposed to decrease the proportions of defective bins. For instance, if ACI_1 is less than a certain value a_1, then the bin '*a*' ratio decreases implying that the specification for ACI_1 needs to be changed below a_1. On the other hand, it is also found that even when ACI_1 is greater than a_1, the bin '*a*' ratio can be made small by changing the value of alignment δ_4 below a certain value a_2. Since an alignment variable is easier to adjust than an ACI variable in the actual production environment, the latter is a more desirable rule in terms of process operability.

4.4 Artificial neural network

An ANN is trained and the Taguchi parameter design is applied to the trained ANN to confirm the rules and check their robustness. An ANN is trained using the bin '*a*' data since it is the most important bin among the defective ones. The optimal configuration of the ANN is determined by an experimental

Table 1. Critical Process Variables and Their Desirable Ranges

Defective Bins	Critical Process Variables	Ranges of Process Variables
bin '*a*'	ACI_1, δ_4	$ACI_1 < a_1$ $ACI_1 > a_1$ and $\delta_4 < a_2$
bin '*b*'	ADI_3, ADI_5, ADI_7	$ADI_3 > b_1$ $ADI_3 < b_1$ and $ADI_7 < b_2$ $ADI_5 < b_3$
bin '*c*'	ACI_3, ACI_5	$ACI_3 > c_1$ and $ACI_5 < c_2$

method (see Kim *et al.*, 2001). The purpose of the Taguchi parameter design is to experimentally determine an optimal setting of design or controllable variables in a system such that the system performance is robust against various noise or uncontrollable variables (see Phadke, 1989 for more details).

In the present study, the design variable corresponds to the process condition which has two levels, namely, existing and proposed conditions in terms of ACI_1 and δ_4. At the existing condition, ACI_1 and δ_4 variables are set to their mean values and, at the proposed condition, they are set to $a_1 + \sigma_{ACI_1}$ and $a_2 - \sigma_{\delta_4}$, respectively, where σ_{ACI_1} and σ_{δ_4} are the standard deviations of ACI_1 and δ_4, respectively.

Twenty three variables whose variabilities are relatively large are selected as noise variables. Levels of a noise variable are determined as $\mu \pm \sigma$ where μ and σ are the mean and standard deviation of the noise variable, respectively. A Placket-Burman design with 24 runs (Montgomery, 1997) is adopted as the noise matrix. Then, the design and noise matrices are crossed to have 2×24 experimental runs and the ANN predicts the bin '*a*' values at the existing and proposed conditions as shown in Figure 4.

		1	2	···	24		
		1	1	···	2	A	
		1	1	···	2	B	
		⋮	⋮		⋮	⋮	
		2	1	···	2	W	SN
1	Existing Condition	bin '*a*' values					-19.02
2	Proposed Condition						-17.48

Fig. 4. Experimental Design, Predicted Bin '*a*'s and SN ratios

At each condition, a signal-to-noise (SN) ratio is calculated as a performance measure of robustness based on the bin '*a*' values obtained at the 24 noise conditions. Since the bin '*a*' is the-smaller the-better characteristic, the corresponding SN ratio is calculated as follows (Phadke, 1989) :

$$SN = -10 \log \left(\frac{1}{24} \sum_{j=1}^{24} y_j^2 \right) \tag{1}$$

where y_j is the bin '*a*' value at the jth noise condition. Since the SN ratio of the proposed condition (-17.48) is greater than that of the existing condition

(-19.02), the robustness of the rule generated by the decision tree analysis for bin '*a*' is confirmed.

5 Conclusion

In this article, a framework for analyzing process monitoring data is developed and applied to a semiconductor manufacturing process. Statistical and data mining techniques adopted include classification (i.e., decision tree analysis), data visualization and exploration, and Taguchi parameter design using a trained ANN. It is believed that the developed framework can be applied not only to semiconductor manufacturing processes, but also to other industrial production processes.

References

Han, J. and Kamber, M. (2001). *Data Mining : Concepts and Techniques*, Morgan Kaufmann Publishers.

Irani, K.B., Cheng, J., Fayyad, U.M. and Qian, Z. (1993). Applying Machine Learning to Semiconductor Manufacturing, *IEEE Expert*, **8(1)**, 41-47.

Kim, Y.S., Yum, B.J. and Kim, M. (2001). Robust Design of Artificial Neural Network for Roll Force Prediction in Hot Strip Mill, *Proceedings of International Joint Conference on Neural Networks*, 2800-2804, Washington, D.C.

Kittler, R. and Wang, W. (2000). Data Mining for Yield Improvements, *Proceedings of the International Conference on Modeling and Analysis of Semiconductor Manufacturing*, 270-277, Tempe, Arizona.

Mieno, F., Sato, T., Shibuya, Y., Odagiri, K., Tsuda, H. and Take, R. (1999). Yield Improvement Using Data Mining System, *IEEE International Symposium on Semiconductor Manufacturing Conference Proceedings*, 391-394, Santa Clara, CA.

Montgomery, D.C. (1997). *Design and Analysis of Experiments*, 4th Ed., John Wiley & Sons, Inc.

Phadke, M.S. (1989). *Quality Engineering Using Robust Design*, Prentice Hall, Inc.

Sorell, M.A. (2000). Data Mining, Strip Mining and Other Historical Data High Jinx, *Proceedings of the International Conference on Modeling and Analysis of Semiconductor Manufacturing*, 293-299, Tempe, Arizona.

Different Ways to See a Tree - KLIMT

Simon Urbanek

Department of Computer Oriented Statistics and Data Analysis,
University of Augsburg, Germany

Abstract. Recursive partitioning trees offer a valuable tool to analyze structure in datasets. There are many ways to display various structures contained in a tree. This paper describes different means of visualization of a tree with our prototype software, KLIMT (Klassification - Interactive Methods for Trees), for interactive graphical analysis of trees.

Keywords. classification and regression trees, visualization, interactive software, exploratory data analysis

1 Introduction

The value of trees is recognized not only in statistics, but also in many different fields such as machine learning, botany and medical decision-making. Trees provide valuable way of displaying structure in datasets. One of the major advantages of tree-based models is the fact, that their interpretation is quite intuitive and the structure often easy to explain. The models themselves impose hardly any restrictions on the underlying problem and dataset (Breiman [1984]).

Today it is fairly easy to grow many trees using various algorithms, such as the Bayesian approach (Chipman, George and McCulloch [1998]) or greedy algorithms (Ripley [1996]). In order to be able to choose a possibly best model we need to work with the tree, analyze it and explore it. Static tree plots display only one view of a tree, but dependencies or special cases can better be detected with multiple views.

KLIMT is an interactive software for exploratory data analysis of tree models. It is a stand-alone application with interfaces to R/S/S-plus software packages to allow seamless integration. The bidirectional interface is based on the RSJava package from Omegahat and provides similar features as the ggobi/R model described by Temple Lang and Swayne [2001]. For platforms unsupported by Omegahat a flat file interface can be used. Both interfaces allow to start KLIMT within R/S/S-plus by a simple `Klimt(tree,dataset)` command. Technical details about both interfaces and the application can be obtained from the documentation section of the KLIMT project homepage.

Many traditional plots are supported by KLIMT, such as histograms, barchars, scatterplots or boxplots. The interactive features include selection, queries, zooming, variation of displays, multiple views, pruning and linked highlighting for all plot types. The software meets most requirements for interactive software as described by Unwin [1999]. We have recently added new approaches to visualization of trees in KLIMT. The next section describes variations of the rather classical, hierarchical interpretation of a tree, whereas different alternative approaches of displaying a tree are presented in section 3.

2 Hierarchical views

Unlike many statistical plots trees don't have a fixed or exactly defined graphical representation. Even if we concentrate on the hierarchical structure the variety of different plotted trees is huge. Three different ways of drawing the same tree are shown in Fig. 1. We want to concentrate on the graphical properties and alternative visualization of trees here, therefore we have chosen an easily interpretable tree generated from the Fisher/Anderson Iris dataset with the *tree* library for R. The dataset is well known and the size of the tree allows us to illustrate the visualization aspects even in this static context without interactivity which is provided in KLIMT to allow analysis of complex tasks. Plot A was drawn with R's native *plot* function, all other plots were created by KLIMT.

Placement of the nodes is one of the most important factors. In some plots the distance between levels of the tree height is constant, in other plots it is proportional to some property of the tree such as the deviance gain in a split (see plot A). It is also possible to display all leaves on the same level to allow better comparison amongst them (plot C). Usually the child nodes are placed below their parents symmetrically, i.e. the centers of parent node and children nodes build an isosceles triangle, but various techniques, such as equidistant partitioning of each level can be used. KLIMT offers various placing algorithms, but also allows the user to freely modify the tree by dragging individual nodes or entire branches.

Not only the node placement varies, but also the means of displaying nodes and connecting them. It is possible to visualize additional information by using different symbols for nodes, e.g rectangles of various sizes, where the size is proportional to the population of a node (plots B and C).

Conventionally trees are plotted in top-bottom orientation, but for deep trees this may cause problems because the screen has usually more room in the horizontal than in the vertical direction. KLIMT alternatively allows the tree to be displayed in left-right orientation (see plot C) to avoid this shortcoming.

Overloading plots with information can offset the benefits of the plot, in particular its ability to provide information at a glance. If there is too much information attached to each particular node it is often not possible to display more than two levels of the tree on a screen or a page. Therefore additional tools are necessary to keep track of the overall structure in order not to get lost. Most of these tools, such as zoom, pan, overview window or toggling of labels are available in interactive context only.

Especially for analysis, visualization of additional information is required. There are basically two possibilities of providing the information: Integration of the information in the tree visualization or use of external linked graphics.

Direct integration is limited by the spatial constraints posed by the fixed dimension of a computer screen or other output medium. Its advantage is the immediate impact on the viewer and therefore easier usage. It is recommended to use this kind of visualization for properties that are directly tied to the tree, such as the node size or the criterion used for the growth of the tree.

External linked graphics are more flexible, because they are not displayed directly in the tree structure for each node separately, but are only logically linked to a specific node. Spatial constraints are less of a problem because one graphic is displayed instead of many for each node. The disadvantage of linked graphics is that they must be interpreted more carefully. The viewer has to bear in mind the logical link used to construct the graphics as it is not visually attached to its source (node in our case).

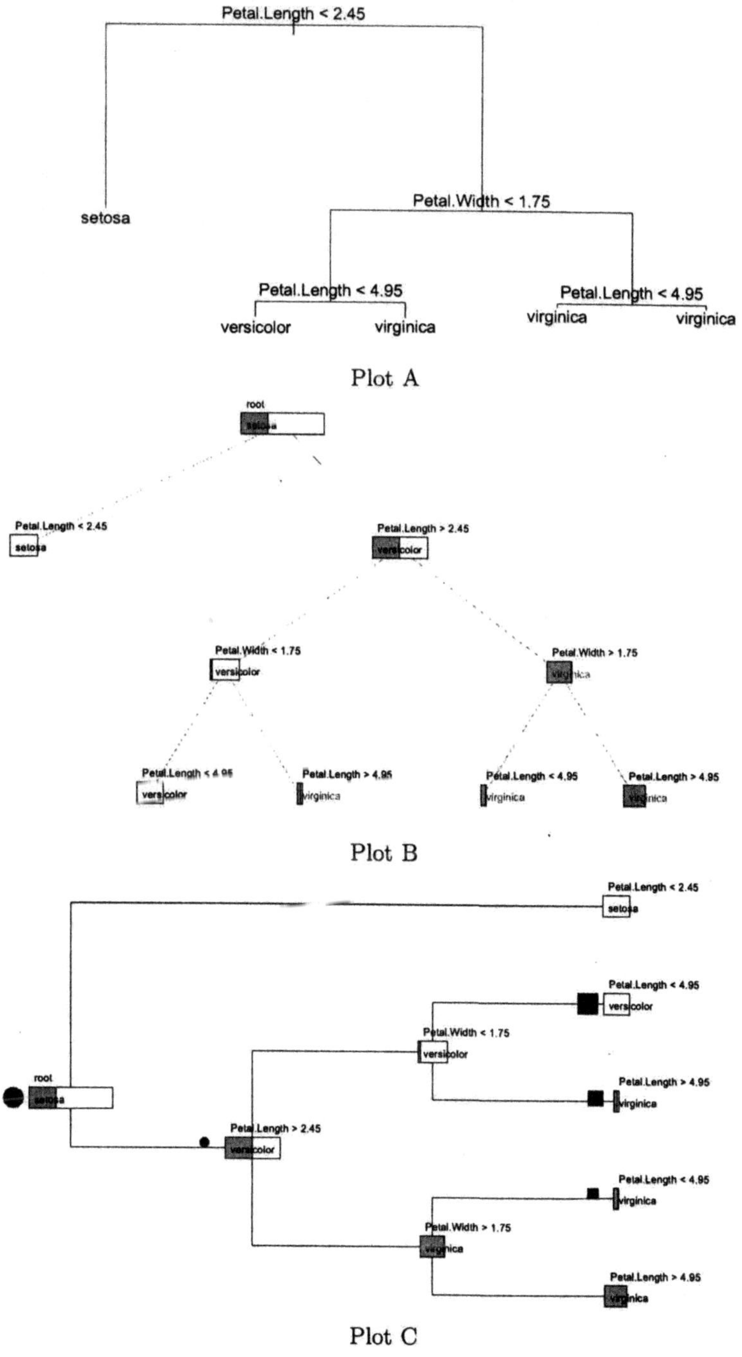

Fig. 1. Three different ways to visualize the same tree.

KLIMT supports linked highlighting both on case level as well as on node level for linking between trees and plots. Directly related information such as node size or deviance gain are visualized directly in the tree structure.

3 Alternative views

The plot of the hierarchical structure of a tree is not the only way to see a tree. If only two measurement variables are involved in the decisions, a scatterplot of these two along with partitioning lines describes the entire tree. Each rectangular partition corresponds to a leaf and bears either a class name (for classification trees) or the predicted value (for regression trees). An example of such an *enhanced scatterplot* is given in Fig. 2.

If more than two variables are used in the splitting rules of the tree, the enhanced scatterplot represents only a two-dimensional projection of the measurement space, orthogonal to the axes. The interpretation of such a plot must be done more carefully, because splits on variables other than the plotted ones cannot be visualized. Still the scatterplot is helpful when examining individual splits and can be valuable for detection of additional structures in the data.

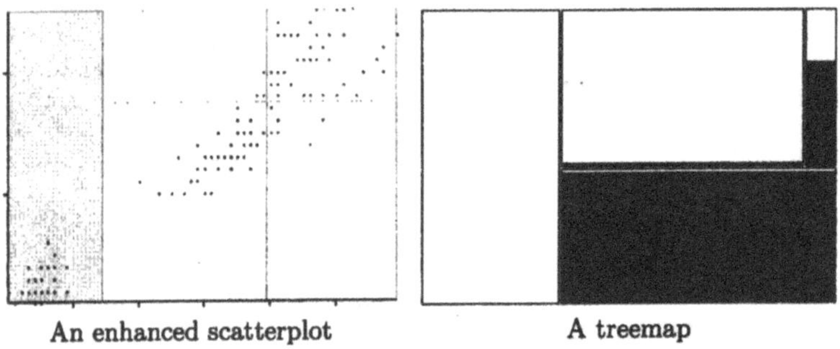

Fig. 2. Alternative ways to visualize a tree.

This approach is especially useful for continuous variables. Unfortunately the principle cannot be generalized for categorical variables, even in the two-dimensional case, because it is not always possible to construct a sequence of categories such that an n-way split of a node results in exactly n continuous partitions of the category space. This is true for any $n > 1$.

Instead of looking at the measurement space it is possible to consider the number of cases in each node. The corresponding graph shown in Fig. 2 is structurally similar to a mosaicplot and is often called a *treemap*. The plot is constructed as follows. The basis is a rectangular region representing all cases thus corresponding to the root. For each child of the root the region is partitioned horizontally into pieces proportional to the number of cases in each node. If the node is not a leaf, its space is now partitioned vertically according to the size of its children. This procedure is repeated recursively until a leaf is reached while the partitioning direction alternates between horizontal and vertical for each level as illustrated in Fig. 3.

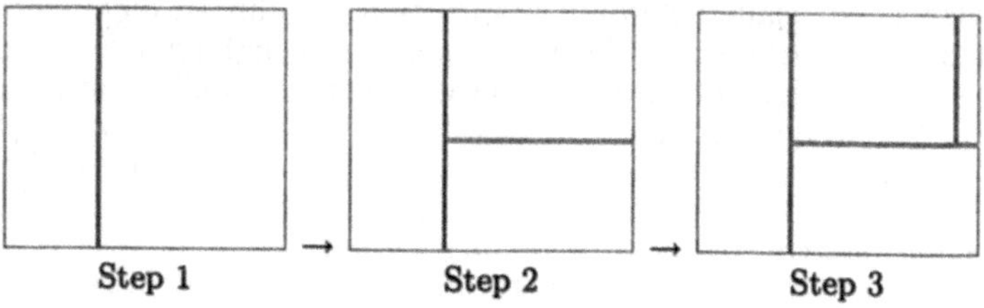

Fig. 3. Construction of a treemap.

The advantage of treemaps compared to scatterplots is that the limitation of two variables does not apply and even splits on categorical variables can be used. In a two-dimensional, continuous case the corresponding partitions in each plot can be mapped in a bijective fashion, but the area used by the same partition in each plot differs. In a scatter plot the size of a partition is given by the scale of the variables, whereas in a treemap the size is proportional to the number of cases in that partition.

When highlighting is applied, selected cases in the scatterplot are represented by points of a different color and/or size. In a treemap the number of selected cases is proportional to the volume of differently colored area within a partition, usually filled from bottom to top as if water was poured into the partitions. The proportion of the height of such highlighting to the total height of a partition is equal to the proportion of selected cases to the total number of cases in the partition. Therefore treemaps are useful for comparing proportions in the dataset, whereas enhanced scatterplots offer a way to recognize individual points, such as outliers or points at the edge of a split.

In order to directly compare leaves it is possible to use special plots that are a combination of treemaps and spineplots. The construction of the plot is done like a treemap where partitioning is not performed in alternate directions, but only in the horizontal direction. The resulting plot resembles a spineplot except that individual spines correspond to leaves of the tree and not classes of a variable. Therefore we refer to this plot as a *spineplot of leaves* as illustrated in Fig. 4.

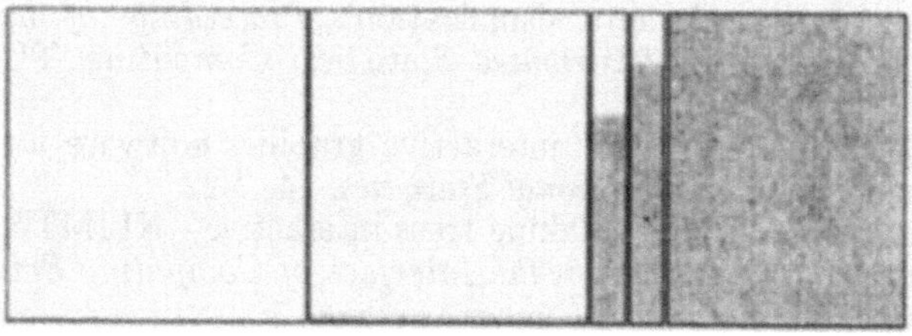

Fig. 4. A spineplot of leaves.

This view is especially helpful in conjunction with linked highlighting. The filled area is proportional to the number of cases highlighted in the corresponding leaf. In Fig. 4 all *virginica* species are selected allowing visual comparison of the absolute proportions in all nodes. Another property of spineplots is that relative proportions inside each spine correspond to the height of the filled area and hence are directly comparable. This means that both absolute and relative comparisons amongst leaves are possible at a glance.

The disadvantage of both spineplots of leaves and especially treemaps is the fact that identification of a certain node within the plot is somewhat difficult. Labeling as proposed for mosaicplots by Hoffman [2001] resulting in doubledecker-plots is not possible, because in general each level of the tree involves different variables in the splits. Direct interactive query remains the most appealing solution in this case.

KLIMT implements all three proposed plots. For spineplots of leaves the identification of individual spines is simplified, because the sequences of leaves in the hierarchical tree plot and the corresponding spineplots for leaves are identical.

3.1 Conclusion

A tree model can be observed from many different angles and each view displays various aspects of the model and underlying dataset. Beside the usual plots emphasizing the hierarchical aspect of a tree, our software KLIMT provides additional views such as enhanced scatterplots, treemaps and spineplots of leaves. All pictures in this paper were generated by KLIMT except for Fig. 1, Plot A. The combination of those plots and the interactive features of KLIMT such as hot linking of all views and plots, queries and immediate manipulations of plots, provides an analyst with a versatile tool for exploratory analysis of classification and regression trees.

References

Breiman, L., Friedman, J., Olshen, R. and Stone, C. (1984) *Classification and regression trees*, Wadsworth

Chipman, H., George, E. and McCulloch, R. (1998) "Bayesian CART model Search (with discussion)", *Journal of the American Statistical Association*, 93, 935-960

Hofmann, H. (2001) *Graphical tools for exploration of multivariate categorical data*, BoD, Norderstedt

Klimt project, `http://www.klimt-project.com`

Omegahat project, `http://www.omegahat.org`

Ripley, B. D. (1996) *Pattern Recognition and Neural Networks*, Cambridge University Press

Temple Lang D., Swayne D. F. (2001) "ggobi meets R: an extensible environment for interactive dynamic data visualization", *Proceedings of the 2nd International Workshop on Distributed Statistical Computing*, TU Vienna

Unwin, A. R. (1999) "Requirements for interactive graphics software for exploratory data analysis", *Computational Statistics*, 14, 7-22

Urbanek S. and Unwin, A. R. (2001) "Making trees interactive - KLIMT", *Proceedings of The 33rd Symposium on the Interface of Computing Science and Statistics*

e-stat: A Web-based Learning Environment in Applied Statistics

Erhard Cramer, Katharina Cramer, and Udo Kamps

University of Oldenburg, Department of Mathematics,
D-26111 Oldenburg, Germany,
e-stat@uni-oldenburg.de, www.emilea.de

Abstract. Within the "New Media in Education Funding Programme" the German Federal Ministry of Education and Research (bmb+f) supports the project "e-stat" to develop and to provide a multimedia, web-based, and interactive learning and teaching environment in applied statistics called EMILeA-stat. The structure of EMILeA-stat, its scope and objectives is sketched, and some screen-shots of interactive applets are shown. In the Compstat 2002 proceedings several other contributions on EMILeA-stat can be found.

Keywords. Applied statistics, web-based, interactive, teaching environment, learning environment

1 Introduction

Within the "New Media in Education Funding Programme" the German Federal Ministry of Education and Research (bmb+f) supports the project "e-stat" (project period April 2001 – March 2004) to develop and to provide a multimedia, web-based, and interactive learning and teaching environment in applied statistics called EMILeA-stat, which is a registered brand name. The funding programme aims to encourage structural changes in education and to stimulate the learning software market in Germany.

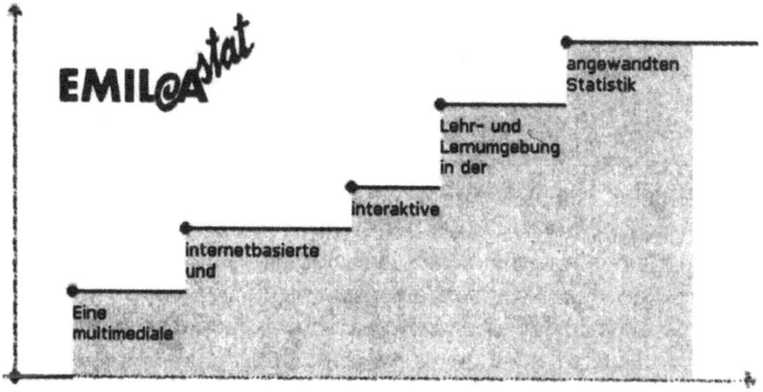

Fig 1.1. EMILeA-stat: An **E**lectronic, **M**ultimedia, **I**nteractive, and internet-based **L**earning and Teaching **e**nvironment in **A**pplied **stat**istics

The EMILeA-stat group is organized in several teams located at seven German universities, i.e., the universities of Bonn (Prof. Dr. Joachim Buhmann), Berlin (Humboldt-Universität, Prof. Dr. Wolfgang Härdle and Prof. Dr. Bernd Rönz), Dortmund (Prof. Dr. Ursula Gather and Prof. Dr. Claus Weihs), Karlsruhe (Prof. Dr. Karl-Heinz Waldmann), Münster (Prof. Dr. Heinz Lothar Grob), Oldenburg (leading university, Dr. Erhard Cramer, Prof. Dr. Udo Kamps (project leader), Prof. Dr. Claus Möbus, Prof. Dr. Dietmar Pfeifer and Prof. Dr. Kristina Reiss) and Potsdam (Prof. Dr. Hans Gerhard Strohe). Moreover, in test and evaluation phases of using EMILeA-stat other universities will be involved, too. At the present time about 70 people are co-working in developing und realizing the learning and teaching environment. It is supported by further partners in advice and in authorship of modules and it cooperates with economic partners such as SPSS Software, BertelsmannSpringer Science+Business Media (Springer Verlag), and MD*Tech Method & Data Technologies (XploRe-Software). For more details we refer to the project's web page www.emilea.de.

2 The aims of EMILeA-stat

In the following we sketch the structure of EMILeA-stat, its scope and objectives, and we show some screen shots of interactive applets. Statistical and quantitative thinking and acting have become fundamental skills in several branches of natural sciences, life sciences, social sciences, economics, and engineering. Models, tools, and methods, which have been developed in statistics, are applied in modelling and data analysis, e.g., in business and industry, in order to obtain decision criteria and to gain more insight into structural correlations.

Owing to these various applications and the necessity of using statistical methodology there have to be consequences with respect to learning and teaching. The organization of teaching processes has to be changed: For example, pupils should get to know elementary and application-oriented statistics which requires that statistics and data analysis, theoretically and practically, have to become part of teachers' studies at the universities and at in-service training courses. Moreover, students of many different disciplines with a statistics impact should be familiar with basic and advanced statistics. These goals may be achieved by working with EMILeA-stat. Being an essential part of statistical education it may be applied in schools, universities, and in further vocational training. Supervised and self-directed learning will also be possible because of a basic concept, which offers the opportunity to tailor individual courses covering specific learning needs.

EMILeA-stat supports teaching in different manners and can hence be seen as an innovative high-quality contribution to the learning facilities of the present and the future. Statistics-modules, views, and scenarios will be well suited to each other and complexly linked. The presentation of the material takes in account that different types of users have different needs, such that EMILeA-stat may be used in teaching, selective support of teaching as well as web-based learning. By this it is possible to address and to reach pupils, students, and working persons in the fields of bio and life sciences, computer science, engineering mathematics, psychology, statistics, different areas of economics, and social sciences with just

one system. EMILeA-stat will also serve as a kind of information-forum in "everyday-statistics" and it will offer the opportunity for the everyday use as an intelligent statistical encyclopedia.

3 The System EMILeA-stat

EMILeA-stat is developed as a user-oriented product with an open source concept. The chosen architecture is also open to guarantee the efficient applicability and practical adaptation to respective needs of teaching as well as learning. It will be possible - and the present teams particularly desire it - to incorporate user-specific views and scenarios. Hence, people teaching statistics are invited to contribute statistical content in a broader sense. It is planned that non-commercial education will be free of charge and freely accessible via Internet.

The following topics will be contained in EMILeA-stat: descriptive and inductive statistics, explorative data analysis, interactive statistics, graphical representations and methods, basic mathematics needed in statistics, probability theory, statistical methods in finance and insurance mathematics, modelling and prediction of data in financial markets, statistical methods in marketing, scenarios stock-market, productions and virtual company, experimental design, statistical quality management, business games, statistical methods in controlling, statistical ecology.

Due to the integration of user-oriented views and scenarios, which are near to real world applications, web-based learning and teaching with EMILeA-stat will be possible in different ways:

(1) Method-led: Learning of statistical methods along a prescribed way of modules including terms, methods, examples, exercises, etc.

(2) Problem-led: The user describes his/her problem in a simplified way and a consulting component proposes statistical theory and methods for the statistical analysis of the respective data.

(3) View-led: Examples from a specific area of statistical applications, e.g., economics, psychology, serve as a motivation to deal with statistical terms and methods. Views, understood as a specific subject environment, incorporate the abstract curriculum with user-oriented introductions, examples, and exercises.

(4) Scenario-led: Starting from a complex and maybe confusing scenario, e.g., a virtual company together with a huge amount of data from different subdivisions of an enterprise, the user should - on the job - learn to detect, to measure as well as to value structures by applying statistical tools.

(5) Data-led: The complexity of the learning and teaching environment EMILeA-stat enables a self-contained introduction to exploratory data analysis.

For the presentation of the basic statistical contents, the EMILeA-stat system offers three levels of abstraction in order to take into account different methodological backgrounds of future users. The modules are written in an elementary level (presentation in a popular scientific way by assuming no or only a low previous knowledge), a basic level (like base courses in applied statistics for students, e.g., of economics, psychology, and social sciences), and an advanced level (containing deeper material and special topics within the broad field of statistics and applied probability).

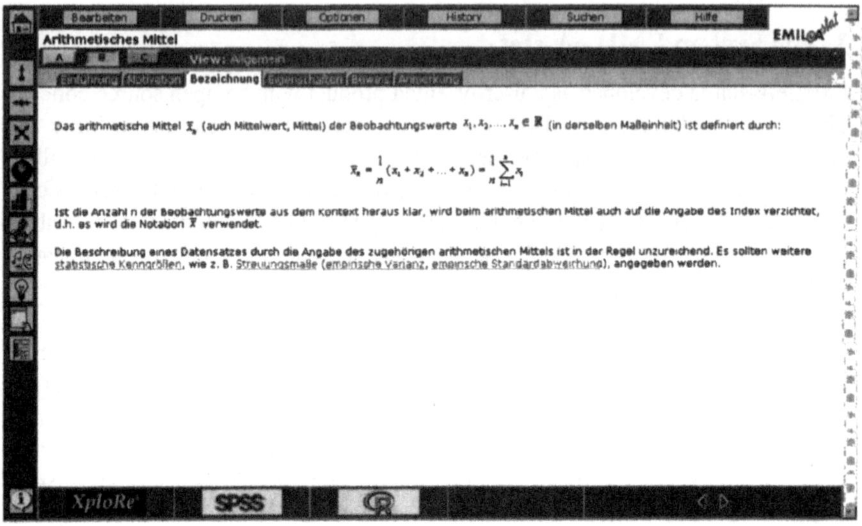

Fig 3.1. The graphical user interface of EMILeA-stat

4 Descriptive Statistics and Interactivity

EMILeA-stat covers a variety of topics of statistics and probability. One field of contents will be descriptive statistics including the main tools of presenting data. Starting with elementary visualizations like pie charts, bar and line graphs, Box plots, stem-and-leaf plots, histograms, and plots of the empirical distribution function, traditional measures like mean, standard deviation, quantiles, etc. are considered as well as measures of relationships between measurement or categorical variables. Methods to describe economic data like price index numbers and measures of concentration, e.g., Lorenz curves and Gini coefficients, are illustrated by a variety of examples and exercises. The set of univariate approaches is completed by regression and time series analysis. Finally, procedures of multivariate data analysis like cluster analysis, principal component analysis, factor analysis, and multidimensional scaling supplement the tool box of descriptive statistics and exploratory data analysis.

In the interactive learning and teaching environment EMILeA-stat the statistical content is supported by visualizations, such as interactive Java-Applets, throughout.
In the following some screen-shots of Java-Applets within EMILeA-stat are presented. The instructions accompanying the graphics indicate the concept of learning by discovery enabled via interactivity.

4.1 Histogram (see Fig 4.1.)

The histogram applet provides many interactive features. Already included examples may be modified by adding new data points. They can be given numerically or by clicking with the right mouse button under the x-axis. Moreover, existing data values (red points) can be moved to the right and left with the left mouse button. The axes are automatically rescaled. Each bar can be split of into two bars by clicking with the mouse into the respective bar. The endpoints of the bars can also be shifted. The number of bars or the length of the intervals can be specified. An optimal bandwith may be chosen according to methods given in the statistical literature. Finally, the numerical values for the intervals, frequencies, etc. are given in a table.

Das Histogramm

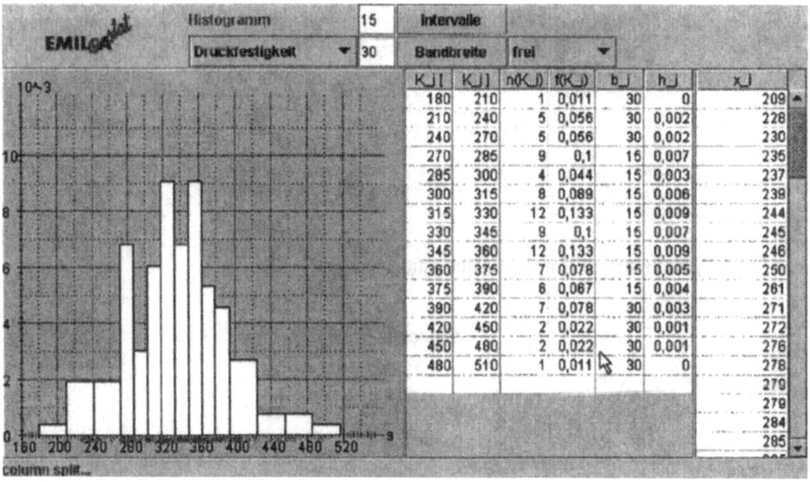

Fig 4.1. Interactive applet: Histogram

4.2 Mean and Median (see Fig 4.2.)

By using the left and right button of the mouse new data points can be added and given ones can be moved to the right or left, respectively. The user can study the effects of these procedures on the mean and median by observing the behavior of the red and green arrows. For instance, this applet is very useful to demonstrate the effect of outliers on the two location measures presented. The numerical values of the data are given in the window on the right.

4.3 Lorenz curve (see Fig 4.3.)

The 'Lorenz curve' applet is based on five market competitors each having a turnover between 0 and 100. At the beginning each competitor has the same turnover 50 which results in a straight line of the Lorenz curve. The user can modify it by moving the black points on the axes. This causes directly a reaction of the Lorenz curve which differs now from the start configuration. The deviation is colored in yellow. The table on the right contains the numerical values which are necessary to compute a measure of concentration, e.g., the Gini coefficient. It is automatically updated if the turnovers are changed.

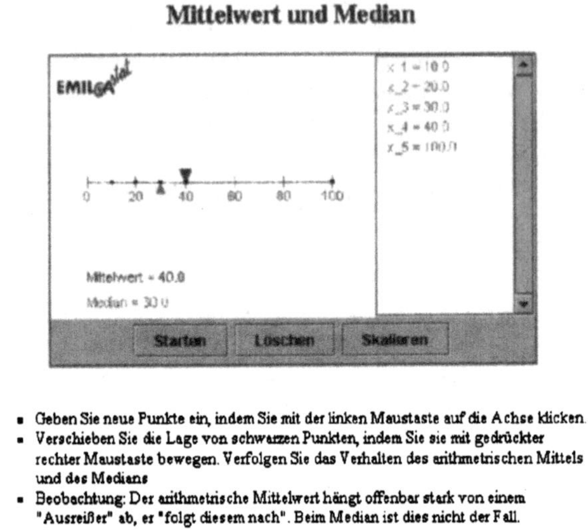

Fig 4.2. Interactive applet: Mean and median

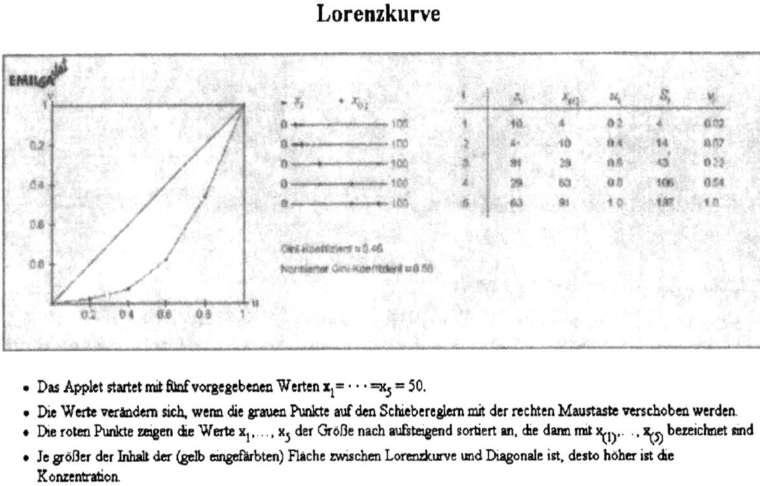

Fig 4.3. Interactive applet: Lorenz curve

5 Contact

Remarks, suggestions, and queries are welcome. Please contact Dipl.-Math. Katharina Cramer (project coordinator) or Prof. Dr. Udo Kamps (project leader) University of Oldenburg, Department of Mathematics, 26111 Oldenburg, Germany, e-stat@uni-oldenburg.de, www.emilea.de.

e-stat: Automatic Evaluation of Online Exercises

Knut Bartels[1]

[1] LS Statistik und Ökonometrie, Universität Potsdam, PF 900327, 14439 Potsdam, Germany

Abstract. The implementation of tools for online exercises into the e-stat environment is described here. Besides the usual types of online exercises special emphasis is put on the automatic evaluation of free text answers. This can be formulated as a problem of classification which is based on an analysis of the free text with statistical methods of computational linguistics.

Keywords. Online exercises, automatic evaluation, classification of free text, natural language processing.

1 Starting points

This is a report on a work in progress and most of the results are preliminary. Nevertheless there are some new methods in online exercises that will be implemented into e-stat which are worth a presentation in this early state already. In this short paper first an overview on the types and fields of usage of online exercises is given. The next section covers the evaluation of answers given as free text. Finally the implemetation into the e-stat environment with some technical aspects concludes this contribution. Since the first natural language of e-stat is German, the development is oriented to the processing of German texts. But it turns out that the choice of language is not restricting the general method of analyzing the input if the linguistic tools are available and therefore examples can be presented in English here.

2 Remarks on online exercises

Basically any classical exercise or problem as given in textbooks or exams can be published on some server and thus becomes an online exercise. But the availability over a net is just a formal characteristic. The special qualities of online exercises are the direct way of communication between author and users offered by the web, the possibility of interaction offered by the graphical user interfaces, and the connection to other services e.g. for statistical computing.

There is a multitude of exercises already available in the net, most written by (groups of) lecturers for their own courses and most in English language. Some commercial tools for authoring online exercises exist, for example *Questionmark, Quia, WebCT/Ciad(Triads), Lotus learning space*. These remarks base on a (wide but naturally incomplete) web search for online exercises with special attention paid to the systems just mentioned and to statistical content.

2.1 Purposes

Online exercises must be distinguished by the purpose they serve. The two main fields are assessments or examinations with results that are used for decisions (e.g. for grading, on employment or certification) on the one hand and practice exercises for learning and assessing progress on the other.

For the first field the absence of ambiguities and the exactness of the evaluation is crucial. The ubiquity of access is less important and the possibility of connecting to third parties is often not allowed for this purpose.

For the second the feedback on given answers is essential while the one or other rare mistake in an evaluation may be tolerated if there is an instance for users to complain to or ask for a second opinion. The ease of access to the exercises and supporting resources should be available without limitations.

2.2 Types of questions

The by far most used types of questions are *discrete choice* and *short entry* and, of course, combinations of these.

The most used type of *discrete choice* refers to all kinds of exercises that are to be solved by pointing, clicking and moving objects, e.g. multiple choice, matching pairs, sequencing, classification or setting sliders. The common characteristic of these exercises is the finite and usually small set of possible responses (or classes of responses when moving objects is involved). Thus authoring the exercise involves assigning a value to each possible response, from simple true or false, over scores out of a total to explanatory feedback texts or all of these.

The second largest class of exercises asks for a *short entry* that is not one out of a set of given ones. Typical settings are to fill in missings in a short text, to give the numeric result of a calculation, to fill a table or to state relevant keywords. The main difference to discrete choice is that the possible answers are not limited that strictly. Usually the input is restricted by the characters allowed and the length of the entry. Assigning values to possible answers now first involves creating classes of similar answers which then can be handled as in discrete choice. Some methods of constructing these classes are simply naming exactly one true and all others false, setting intervals for numeric solutions, and checking against regular expressions for short text strings.

2.3 Methods of evaluation

As described in the last subsection, the method of evaluation is bound to the type of question. If the exactness of the evaluation is the main issue, as in an examination, then only discrete choice and simple short entries are possible candidates. In online exercises for educational purposes giving feedback is more important than absolute exactness - note that even teachers make mistakes or do not understand a student's solution sometimes. Providing the user with feedback is generally possible in two ways. The first is the evaluation by a teacher where the resources of the net are only used for transmitting exercise, answer and response (principle: "email to tutor"). The second is an automatic evaluation without any human interaction where only types of questions can be posed that come with an implemented algorithm for evaluation.

While the first way is adequate for assignments of homework, it is clearly not an option for an online learning platform that should give instant response. Nevertheless it has the advantage that the teacher knows the user, can give individual advice and is able to compare answers from different users. The second way gives instant responses but has the disadvantage that it limits the type of questions to simpler structured ones. Usually the algorithm of evaluation is simple and is performed at the client, that is the users computer, without any notification to the author.

A combination of these methods is a *server side automatic evaluation* of the answers. This keeps most benefits of both ways: it allows for individual assistance, if the user wishes, and for an assessment and improvement of the exercises themselves, e.g. grade of difficulty, common errors, etc. But the possible types of questions are still limited by the available algorithms for evaluation. To push that limit we try to extend the type of possible questions to free text input.

3 Evaluation of free text

Why do we want answers as free text input? Well, the answer to a discrete choice question is strongly biased by the given set of alternatives and the usual strategy is to passively eliminate the given wrong answers without actively formulating a correct answer oneself. For assessing a student's ability to use a statistical method, we believe that questions have to be posed for the reason why a method of analysis was chosen and for the interpretation of the numeric result without giving any other hints. Short entry questions can support this - ask for the result and the name of the method - but the interpretation or reasoning must be given in free text then.

As a consequence of the considerations in section 2.2 one can conclude as follows: The more freedom the user has for answering the exercise the more difficult the evaluation becomes. Now, if we allow for free text input as answers, we should have the supremum of difficulties in evaluating them.

There certainly is some truth to it, but the real problem is mitigated by the following issues:

- We still pursue the method of constructing a limited number of classes of possible answers. Evaluating the free text then becomes a classification problem.
- We sacrifice a bit of the exactness and allow the system to result in an indecidable state where the system asks for more detailed input or refers to a human.
- The more answers to an exercise we have collected and supervisedly evaluated, the less likely it becomes that completely different new answers are given.

In short summary the given answer to be evaluated is compared with known answers. These are previously given by students or expected by the author, and are correct answers as well as faulty ones. To each answer a well defined response is given, so that answers map to responses. The inverse mapping classifies the answers according to the different responses. The classification then is based on scores from an analysis of the free text with simple string comparison, regular expression pattern matching and, what is the novelty here, statistical methods of computational linguistics.

In the field of *artificial intelligence* and *machine learning* a lot of research is done on automatic classification of (longer) texts. But this does not solve the problem of finely classifying a text of a few sentences that should contain a statement on a known topic. In *natural language processing* formal methods as first steps towards the ultimate goal of understanding language are developped. One of these methods, part-of-speech tagging, we employ for analyzing free text input.

3.1 Part-of-speech tagging

Part-of-speech tagging consists of assigning a tag to each unit of input. Here the units are words, which we define as strings beginning with nonwhites and ended by whitespace, and the tags are labeling the words according to their syntactical use, that is their part-of-speech. Additionally we extract the lemmas for all words, especially for those that underly grammatical flexion and are thus difficult to detect by simple string comparison. The following two (somewhat artificial) examples that may be interpretations of the numerical result from a calculation of a price index shortly illustrate how tagging works and how it helps in classifying free text.

WORD	TAG	LEMMA	(meaning of tag)
On	IN	in	(preposition)
average	JJ	average	(adjective)
prices	NNS	price	(normal noun, plural)
for	IN	for	(preposition)
roses	NNS	rose	(normal noun, plural)
rose	VBD	rise	(verb, past tense)
about	RB	about	(adverb)
2.3	CD		(cardinal number)
%	NN	%	(normal noun)
.	SENT	.	(sentence ender)

This tagging is based on the common *Penn Treebank tag set* for English and was performed with the tree-tagger of Schmid (1994). Note that the two occurrences of the string "rose" were correctly tagged and reduced to their lemmas that can be checked by further procedures.

A similar example in German reads:

WORD	TAG	LEMMA	(meaning of tag)
Durchschnittlich	ADJD	durchschnittlich	(adverbial adjective)
sind	VAFIN	sein	(verb, auxiliar, finite)
die	ART	d	(article)
Preise	NN	Preis	(normal noun)
für	APPR	für	(preposition)
Stiegen	NN	Stiege	(normal noun)
um	APPR	um	(preposition)
2,3	CARD	card	(cardinal number)
%	NN	%	(normal noun)
gestiegen	VVPP	steigen	(verb, full, participle perfect)
.	$.	.	(sentence ender)

For German a different tagset, namely STTS, had to be chosen, of course. Note the extraction of the lemma *"steigen"* from the participle *"gestiegen"*. Here string comparison would fail completely.

The method of tagging involves Hidden Markov Models. The tagged sequence of words is chosen with maximal probability

$$\mathsf{P}(w_1,\ldots,w_n,t_1,\ldots,t_n) = \mathsf{P}(w_n|w_1,\ldots,w_{n-1},t_1,\ldots,t_n)\cdot \mathsf{P}(t_n|w_1,\ldots,w_{n-1},t_1,\ldots,t_{n-1})\cdot \mathsf{P}(w_1,\ldots,w_{n-1},t_1,\ldots,t_{n-1}),$$

where w_i denotes the i-th word and t_i its tag. Under the simplifying assumptions that the probability of a word w_n depends only on its own tag t_n and that the probability of a tag depends only on the two preceding tags (trigram) one obtains after recursion

$$P(w_1,\ldots,w_n,t_1,\ldots,t_n) = \prod_{i=1}^{n} P(w_i|t_i)P(t_i|t_{i-2},t_{i-1}) \quad .$$

The conditional probabilities of word/tag-pairs $P(w_i|t_i)$ and the trigrams $P(t_i|t_{i-1},t_{i-2})$, are estimated from an initially given *corpus* of correctly tagged sentences. For further details on part-of-speech tagging see e.g. Manning & Schütze (1999).

3.2 Processing free text answers

The first step of processing the input is a normalization, that is a spell check and a transformation to a standardized form of special characters (umlauts, punctuation, etc.). Then the text is syntactically parsed with statistical handling of ambiguities on the basis of a corpus of statistical texts with a special weight on answers to statistics exercises. Especially the base of already known answers, which should be at least 30 different sentences for a simple question as *"Interprete your numerical result."*, are integrated.

Now the tagged input is compared with the base of all known answers. First two scores from string comparison are calculated for each answer in the base (maximal lenght of string in input contained in known answer, and vice versa). If there is an almost perfect agreement already here on string level, which happens quite often if the base is large, the sequences of tags are compared, too, and in case of agreement the result is the class, that is the response text, of this answer from the base.

In the more likely case that there is no perfect agreement, a scoring method is applied to classify the tagged input. This scoring is based on keywords, word/tag-pairs and their order. The lemmas of the words are checked for keywords. One set of keywords is specific to the question. From keywords rules are build that get scores if satisfied. If for example the tagged text of the first example in section 3.1 was a correct answer, an appropriate rule could look like

"increase <u>or</u> rise <u>or</u> (become <u>and</u> (great <u>or</u> expensive))".

Another set of keywords is a large list of words that have a meaning in statistics and still another contains words of offensive language. These are applied to each input to check whether it is some reasonable approach to answer the question at all.

By combining and weighting these single percentages and scores a most similar answer from the base is determined. If the degree of correspondence exceeds a certain level of confidence the class of this answer determines the response. If not, an excuse is sent back to the user for the system not being able to analyze the given input together with hints to change it appropriately. A first milestone is to classify at least 95% of the given input to an exercise correctly.

An alternative approach would be a completely different tagset, not referring to the syntax of a natural language as in our examples, but with a semantical annotation of indicating the possible meanings in statistics. But the development of an appropriate tagset is a very challenging task which we will not pursue in the near future.

4 Integration to e-stat

Interactive exercises are written as Java-applets for integration to the e-stat environment. Any input, not only those of free text type, will be sent to a server for processing and the result is sent back and shown in the applet. All input is saved on the server for analyzing it regularily. The default mode of solving the exercise is anonymously, but a login to the system is possible so that individual profiles for the users can be set up.

As support for authoring exercises for e-stat a basic Java-applet for the connection to the server will be published and authors will be provided manuals to write and maintain their exercises. Though the response is calculated automatically, the base of known answers and the fine settings for the classification must be supervised. The exercises can be edited over a web interface or as an XML document.

The results of this work in progress will be published and updated on the e-stat web pages at *www.emilea-stat.de* and the links to the Potsdam group or directly at *www.uni-potsdam.de/u/statistik* and the links to e-stat found there.

References

Abney, S. (1996). *Statistical Methods and Linguistics.* In Klavans, J. & Resnik, P. (eds) The Balancing Act. Cambridge, MA: MIT Press.

Brants, T. (1999). *TnT - A Statistical Part-of-Speech Tagger.* In Proceedings of the Sixth Applied Natural Language Processing Conference ANLP-2000, Seattle, WA.

Charniak, E. (1997). *Statistical Techniques for Natural Language Parsing.* AI Magazine 18(4): 33-44.

Manning, C.D. & Schütze, H. (1999) *Foundations of statistical natural language processing.* The MIT press.

Schmid, H. (1994). *Probabilistic part-of-speech taging using decision trees.* In: International Conference on New Methods in Language Processing, Manchester, UK. pp. 44-49.

Yarowsky, D. (1995). *Unsupervised Word Sense Disambiguation Rivaling Supervised Methods.* In: Proceedings of the 33rd Annual Meeting of the Association for Computational Linguistics, Cambridge, MA. pp. 189-196.

NEGRA Korpus, Sonderforschungsbereich 378, Universität des Saarlandes. http://www.coli.uni-sb.de/sfb378/projects/NEGRA-en.html

Commercial software: Questionmark: *www.questionmark.com*; WebCT/ciad/Triads: *www.webct.com*, *www.derby.ac.uk/ciad*; Quia: *www.quia.com*.

e-stat: Basic Stochastic Finance at School Level

Christian Mohn[1] and Dietmar Pfeifer[2]

[1,2] Fachbereich Mathematik, Universität Oldenburg, 26111 Oldenburg, Germany

Abstract. This text shows possibilities how basic facts of mathematics of finance can be introduced at school level by using e-stat. The basic facts ground on the discrete option pricing theory in the onc-period binomial-tree-model. A simple numerical example is used to explain the term of "arbitrage" and its importance to financial mathematics. The example leads into two interactive Java-applets which give an impression of the ability of computers to bring complex contents to a better understanding by interactivity.

This text will also give an idea of how to teach basic elements of mathematical stochastics by using mathematics of finance such as probability and expectation.

Keywords. Stochastic education at school, options, option pricing theory, discrete mathematics of finance, financial markets, binomial-tree-model, Java-applets, combination of options, arbitrage, e-stat, e-learning

1 Introduction

1.1 The importance of teaching financial mathematics at school

In general financial mathematics is supposed to be quite a difficult subject to examine, which is in general only dealt with at university level. At least since we have publications like EBERLEIN (1998) based on the Cox-Ross-Rubinstein-model this conjecture is no longer true. It is even possible to introduce basic aspects of probability theory by demonstrating simple ideas of financial mathematics.

In e-stat we combine this fact with the possibilities of interactivity using a computer. This does not only enable teaching financial mathematics at school level but gives also motivations and excites pupils in order to get better teaching results.

1.2 Teaching strategy

The course we produced for e-stat is called "Basic facts of financial mathematics". It starts with simple explanations of common assets used at the stock exchange and leads afterwards to an easy pricing problem in the binomial-tree-model. Basic facts of probability and expectation can be explained during this part. Essential is a look from the economic point of view with the terms of "arbitrage", "leverage effect" and "combination of options". The course follows in part the monographs of PLISKA (1997), UHLIR AND STEINER (2001) and BINGHAM AND KIESEL (1998).

2 Options and Arbitrage

An *option* is the right to trade a certain amount of an asset for a fixed price, called "*strike price*" or "*exercise price*", at a future date, called *"time of maturity*". If it is possible to use the right within a certain time period the option is called *American* option, if it can only be exercised at the time of maturity it is called *European* option. In the following text we will focus on European options. A *call* option gives the owner the right to buy an asset, a *put* option confers the right to sell.

Of course the exertion of an option depends on the asset price. In case of a call option it only makes sense to exercise if the price of the underlying asset is higher than the strike price. If the price is lower it is cheaper to buy the asset directly at the stock exchange than to purchase it using the option.

There are a few other types of options called "exotic options" which differ in some modalities of the owners right.

2.1 Notations

In this text we will not refer to any transaction costs resulting from the stock exchange. The following designations are significant: *T:* the **T**ime of maturity, *X:* e**X**ercise price, strike price, S_t: **S**tock price/asset price at time $t \in [0,T]$, *i:* riskless **I**nterest rate. As said above the exertion of an option depends on the ratio of the asset price at time *T* and the strike price *X*:

	call	put
exercise option	$S_T > X$	$S_T < X$
do not exercise option	$S_T \leq X$	$S_T \geq X$

The main goal is now to find the „correct" price C_0 of a call option (respectively P_0 of a put option) at time 0, which is the time of buying the option. We therefore look first at the price of an option at its time of maturity *T*, where we have:

$$C_T = \max\{S_T - X;0\} = (S_T - X)^+ \quad \text{(positive-part)}$$

$$P_T = \max\{X - S_T;0\} = (X - S_T)^+ .$$

2.2 Modelling a financial market

We will now consider the so-called *Two-Step-One-Period-Model*, i.e. we assume that the asset price in the beginning of the trading-period S_0 increases till the time of maturity *T* up to $S_T^+ > S_0$ or decreases down to $S_T^- < S_0$, and we require $S_T^- < X < S_T^+$. Let $p \in (0,1)$ the probability of increase; Thus $1-p$ is the probability of decrease. The following example can be found in an article of EBERLEIN (1998). Our asset is a stock with a start price $S_0 = 100€$; with probability $p = 0.4$ it increases to the value of $S_T^+ = 130€$, respectively decreases with probability $1-p = 0.6$ to the

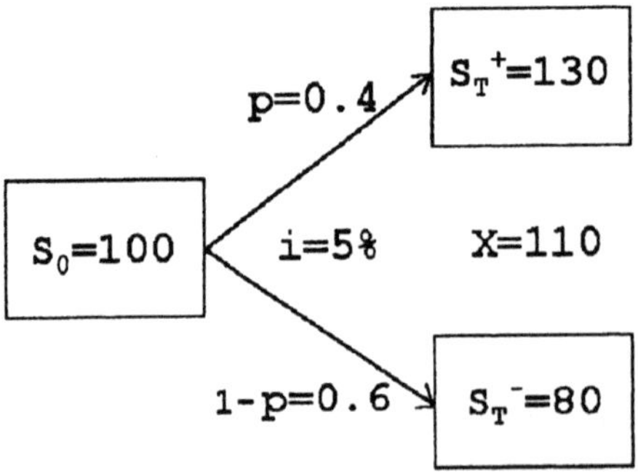

value of $S_T^- = 80€$. The strike price is set to $X = 110€$, the interest rate is $i = 5\%$. We call the interest rate factor $r = (1+i)$ and the discount factor $v = \frac{1}{r}$. Which price C_0 has the writer of a call option (i.e. the buyer of the option) on this stock to pay?

2.3 Finding correct option prices

To answer this question we will first look at the discounted option prices, giving $v \cdot C_T^+ = 19.05€$ and $v \cdot C_T^- = 0€$. We will now consider two possibilities of finding the option price C_0:

2.3.1 First possibility: Fair game "writer vs. market"

We interpret the model as a fair game. The writer buys an option at time 0 and gains a certain amount of money in case of an increasing stock price whereas he loses his money in case of a decreasing stock price. He expects an average gain of zero. Thus, a formula to calculate the "fair" option price uses the expectation:

$$C_0 = E(v \cdot C_T) = p \cdot v \cdot C_T^+ = \mathbf{7.62}$$

The problem of this strategy is the restriction to the writer to trade with options exclusively. If he uses both options and stocks for trading he is able to achieve a result shown in table 2.1.

On the left side we can see the action the writer does. He starts with zero money. At time 0 he sells short two stocks and buys five call options and lends the rest of his money for credit. At time T he has to pay for his stock, gets back his credit and exercises his options in case of an increasing stock price. At the end he earns 10€ independent of the behaviour of the stock price.

$t = 0$		$t = T$	
action	account movements	$S_T^+ > X$	$S_T^- \leq X$
short sell 2 stocks	+200.00	–260.00	–160.00
buy 5 call options	–38.10	+100.00	0.00
give credit	–161.90	+170.00	+170.00
balance / sum	0.00	+10.00	+10.00

Table 2.1. Example of arbitrage

Situations like the one in table 2.1 we call "arbitrage". Of course we want to prevent such situations. We will therefore look at a second possibility to find a correct option price.

2.3.2 Second possibility: The arbitrage-free market

We demand $S_T^+ > r\, S_0$ and take a look at the following table 2.2.

$t=0$		$t=T$	
action	account movements	$S_T^+ > X$	$S_T^- \leq X$
short sell n stocks	$+n\,S_0$	$-n\,S_T^+$	$-n\,S_T^-$
buy m call options	$-m\,C_0$	$m\left(S_T^+ - X\right)$	0.00
give credit	$-(n\,S_0 - m\,C_0)$	$r(n\,S_0 - m\,C_0)$	$r(n\,S_0 - m\,C_0)$
balance / sum	0.00	0.00	0.00

Table 2.2. Arbitrary formulas to calculate the option price

From the two columns below $t = T$ we obtain $-n\,S_T^+ + m\left(S_T^+ - X\right) = -n\,S_T^-$, leading to the ratio $h = n/m$ (so-called *hedge ratio*) given by

$$h = \frac{S_T^+ - X}{S_T^+ - S_T^-};$$

for the corresponding call price we obtain the unique solution $C_0 = h\left(S_0 - v\,S_T^-\right)$. In the example above we have $h = (130-110)/(130-80) = 0.4$ and $C_0 = \mathbf{9.52}$. The proof of C_0 being the arbitrage-free price is not very difficult.

2.4 Illustration by interactivity

The question is now which way we take to motivate these ideas at school. Within e-stat, a standard tool is the use of Java applets. In the following Java applet (fig. 2.1) an arbitrage-table as seen above is realized. In the lower left part of the applet the results of all the transfers of stocks and options can be found, exactly as in table 2.1. In the upper part it is possible to edit all the parameters of the *Two-Step-One-Period-Model.* That enables pupils to examine the idea of arbitrage with their own numbers, which is crucial to memorize problems and ways of acting in financial markets. For instance does the applet give the opportunity to make exercises where some parameters are given, e.g. stock prices and type of options, and the others are to find out by the pupils and to test their solution. The applet works as follows:

In the first line of the applet (starting with "Aktien:" [stocks]) can be chosen between buying or selling stocks. Then it is filled in, how many stocks are to buy or sell. On the right side of the first line the stock prices at time $t = 0$ and $t = T$ can be entered.

In the second line (starting with "Optionen:" [options]) almost same choices as in the first line are made but with the option instead of the stock. Additionally it can be chosen whether to trade with put- or call-options and of course there is no need to enter on option price at time $t = T$.

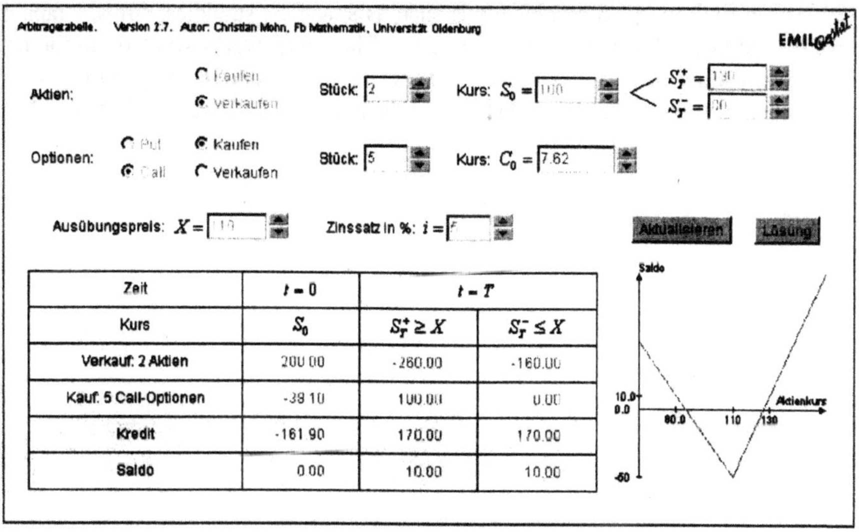

Fig 2.1. Java-applet of an Arbitrage-table with graphical output

In the third line (starting with "Ausübungspreis:" [strike price]) the strike price of the option and the interest is entered. On the right side there are two buttons. The left one, called "Aktualisieren" [actualize] calculates the new arbitrage table in the lower left and also the graphic in the lower right corner with all changes that have been made in the first three lines. The right button called "Lösung" [solution] calculates the correct hedge ratio and option price in an arbitrage-free market, dependent of the stock prices and the choice of buying or selling stocks and of trading with puts or calls. Therefore the upper part of the applet is different coloured: All data that are needed to calculate the correct hedge ratio and option price are orange and all data that are changed by this calculation are green, that is, number of stocks and options, option prices and buying or selling options.

The arbitrage table in the lower left corner is a fixed output, directly controlled by the values of the first three lines of the applet as described with table 2.1.

The graphical output visualizes the total gain or loss in dependence of the stock price. Thus the behaviour aside from the two stock prices S_T^+ and S_T^- is visible.

3 Combination of options

Another aspect of the concept of arbitrage is the investigation of combining different options. Actually this is only an enhancement to the concept above: there is no longer only one kind of option to trade with but several. Still we need to calculate the correct prices for all options we use. This is possible with a little bit more complicated theory of financial mathematics, which shall not be explained in this place (however it is done in another course of e-stat). What we want to show here is again the power of interactivity and the increase of motivation of pupils by enabling them to find out things by themselves. In the Java applet (fig. 3.1) created for this matter, it is possible to combine up to five European put or call options with parameters of one's own choice. The result is a graphical output of the payoff with all chosen options (thin line) and the combination of them which is the sum (thick line). Thus, pupils are enabled to find out good combinations such as strangles, straddles, strips, straps or even a butterfly-call-spread by trial and error. The applet works similarly to the one above. In the lower five lines you can edit for up to five options all parameters and calculate the graphical output by clicking on the button "Aktualisieren" [actualize].

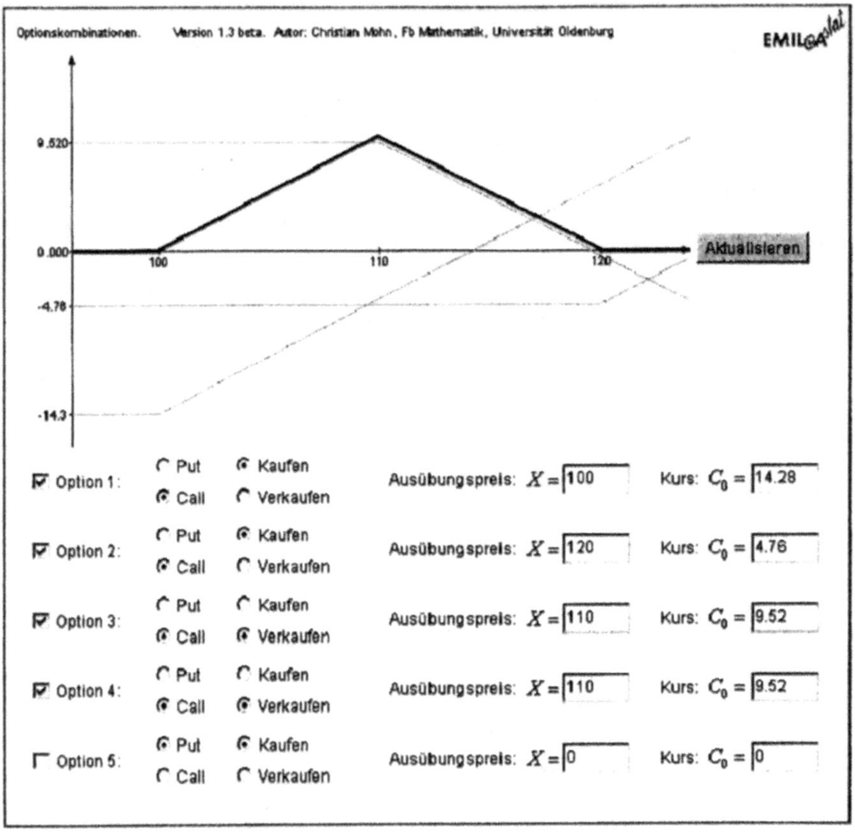

Fig 3.1. Beta-Version of a Java-applet that calculates the sum of a combination of options

References

N.H. Bingham, R. Kiesel (1998): *Risk-Neutral Valuation.* Pricing and Hedging of Financial Derivatives. Springer-Verlag, London.

E. Eberlein (1998): *Grundideen moderner Finanzmathematik.* Mitteilungen der Deutschen Mathematiker-Vereinigung, Heft 3, 10 - 20.

D. Lamberton, B. Lapeyre (1997): *Stochastic Calculus Applied to Finance.* Chapman & Hall, London.

S.R. Pliska (1997): *Introduction to Mathematical Finance.* Discrete Time Models. Blackwell Publishers, Massachusetts.

H. Uhlir, P. Steiner (2001): *Wertpapieranalyse.* 4. Aufl., Physica-Verlag, Heidelberg.

e-stat: Development of a Scenario for Statistics in Chemical Engineering

Claus Weihs, Martin Kappler

Statistic Department, University of Dortmund, 44221 Dortmund, Germany

Abstract. In the environment of e-stat, a learning software for statistical methods, we develop a scenario for statistics in chemical engineering. In this scenario, data analysis methods and design of experiments are applied to the optimisation of the styren process. According to Moore (1997), learning by means of individual action is the most important new model of pedagogy. In particular, the combination of statistical content, learning in realistic data situations, and the use of multimedia with various interaction facilities supports effective and modern learning best. All these concepts are realised in the presented scenario utilising modern internet technology.

Keywords. scenario, statistical education, e-learning, learning-software, multimedia, design of experiments, chemical engeneering

1 Introduction

At least since the big attention achieved by the scenario of the Club of Rome "The Limits to Growth" (Meadows et al., 1972), scenarios are a well accepted means for knowledge acquisition in research.

As part of the learning software e-stat we develop a scenario, which allows the user to apply capabilities learned in an e-stat course in a realistic environment. Also, the scenario can be used to acquire new knowledge in a field with little prior knowledge. This is achieved by 'transferring' the user into a chemical engineering environment letting him or her plan the production, evaluate the outcomes, and try to optimise the process by means of statistical data analysis and experimental design.

According to Moore (1997), learning by means of individual action is the most important new model of pedagogy. In particular, the combination of statistical content, learning in realistic data situations, and the use of multimedia with various interaction facilities supports effective and modern learning best, following Moore. Moreover, the facility to fit tempo and detail to individual needs makes an individual learning process possible. All these concepts are realised in the presented scenario of the learning software e-stat.

In this paper we will comment on scenarios in e-stat (section 2), and the styren process (section 3). We will indicate the structure of the styren scenario (section 4), and the statistical methods included (section 5). Finally, we will also discuss the technical realisation on the internet.

2 What is a scenario in e-stat?

Scenarios in e-stat are independent environments, in which learners can apply their already available statistical knowledge with consummate ease in realistic situations and can acquire new methods in order to achieve given goals. At this point the e-stat kernel only serves as a help system to solve problems that emerge during the scenario and to refer to descriptions of statistical methods that should be used.

By means of the presented scenario of the styren production process the users of e-stat have a facility to get acquainted with statistical data analysis of realistic data, as well as with design of experiments in a realistic environment.

By the intuitive structure of the scenario the user is enabled to manage the learning steps by him- or herself, to find structure in the data, to get an intuitive understanding of the process, and to design experiments to optimise the process in order to maximise yield.

3 Styren process

Styren is one of the most important intermediate products in chemical industry. It is used in many technical synthetic materials, e.g. in Styrofoam. Styren is produced on the basis of ethyl benzene and (water) steam in an adiabatic reactor (see figure 1).

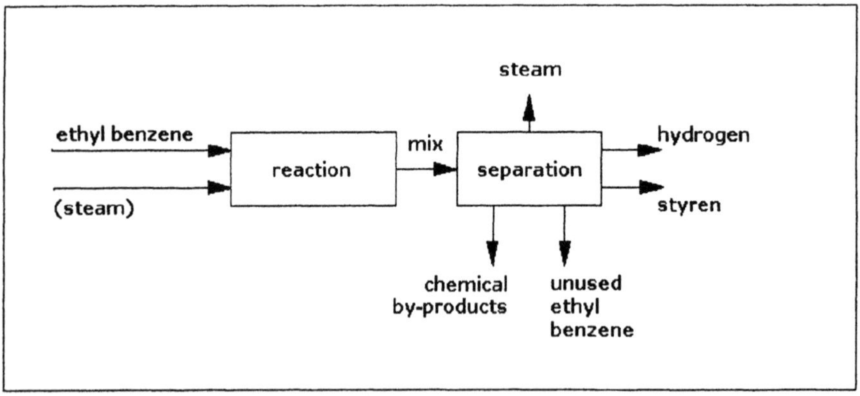

Figure 1. Diagram of the reactor

The main reactions taking place in the reactor are listed in table 1. The first reaction is shifted to the right by support of a catalyst. The other reactions, which can't be suppressed, result in the by-products shown in figure 1.

As a result the production process of styren depends mainly on the input (amount) of ethyl benzene. But there are also many other influences like pressure, temperature of the substances and the reactor respectively and properties of the catalyst. Some of these variables are controllable, others are only observable. All variables *may* have an influence on the styren yield but don't *have to*. It is the task of the user to identify those variables which are really relevant.

1.	$C_6H_5\text{-}C_2H_5$ (ethyl benzene)	$\underset{}{\overset{cat.}{\rightleftharpoons}}$	$C_6H_5\text{-}C_2H_3 + H_2$ (styren)
2.	$C_6H_5\text{-}C_2H_5$	$\rightleftharpoons$	$C_6H_6 + C_2H_4$
3.	$C_6H_5\text{-}C_2H_5 + H_2$	$\rightleftharpoons$	$C_6H_5\text{-}CH_3 + CH_4$
4.	$C_2H_4 + 2\ H_2O$	$\rightleftharpoons$	$2\ CO + 4\ H_2$
5.	$CH_4 + H_2O$	$\rightleftharpoons$	$CO + 3\ H_2$
6.	$CO + H_2O$	$\rightleftharpoons$	$CO_2 + H_2$

Table 1. Main reactions

Theoretically, the styren yield can be determined by means of a set of differential equations from the values of the influential factors. This can be used to simulate the production. Since the influential variables, however, are themselves influenced by random variation, 'noise' has to be added in order to obtain a realistic data set.

4 Structure of the scenario

By calling the scenario the user reaches its starting page, where by means of a user identification the status of the last use of the scenario is recaptured. This way, all former activities, results and notes are re-available, and, moreover, the user's individual state of knowledge may be taken into account for the selection of helping aids and exercises. A guest status is planned also.

After registration the user enters he main page of the scenario. Here, information about the design of the scenario, its aims and use is available in form of a guided tour. There, background information about the process, reaction equations, and influential factors are included also.

From the main page, the user can also re-activate former results, graphics and notes. Moreover, the e-stat kernel may be activated to learn about statistical methods and terms.

The main functionality of the main page is the start of the statistical tasks of the scenario. From there the user can analyse the process factors by means of statistical methods. The user can choose between different analyses, can start the corresponding programs and react on the presented results.

The user is supported by a context-sensitive helping system, which includes descriptions of possible analysis steps as well as of statistical methods. Most of the help pages are taken from the e-stat kernel including interactive examples. Moreover, the user is guided by means of questions leading to the next analysis steps or animating the user to 'play' with the data independently.

5 Statistical methods

5.1 Quality assessment

The scenario will be roughly divided into two parts. In the first part the user works with a fixed data set of 100 observations representing past runs of the production.

The statistical methods available in this part of the scenario are the means of descriptive and inductive statistic. The user can get familiar with the data by examining, e.g. mean, median, variance, histogram, boxplot etc. of the variables. To determine the dependence structure in the data, bivariate methods can be used, e.g. correlation, scatter plot, regression, time series analysis etc. Specifically, influences of the process factors on styren yield should be of interest to the user.

As this data set is fixed, it is possible to include some educational objectives like treatment of missing values, mistyping or outliers. That way, the usefulness of the various statistical methods is demonstrated.

5.2 Quality optimisation

In the second part of the scenario the user is asked to go through the most important steps of quality optimisation, i.e. screening, modelling, and optimisation (Weihs and Jessenberger, 1999). In each step experimental designs can be chosen. The styren process is simulated with the corresponding values of the influential factors, and the results can be evaluated afterwards.

5.3 Enhancements

At present no further steps of quality assurance like stabilisation, control, risk analysis and characterisation are planned because of staff and time limitations. But as the scenario, just as the whole environment e-stat, is designed in open source, it is possible to implement these steps later.

6 Technical realisation

As the scenario is provided with a high degree of interactivity and shall be general accessible via the internet, a method is required to generate dynamic webpages. Two such approaches will be presented.

6.1 CGI / Perl

The Common Gateway Interface (CGI) offers the facility to generate dynamic web pages on server side. For this, CGI only serves as a connection for the communication of the webserver and an external program that generates a website (or graphic, etc.) (see Conolly, 1999).

The only requirements for this external program, the CGI-processor, are having access to the environment variables as well as reading from standard input (STDIN) and writing to standard output (STDOUT). Almost every programming or script language allows to develop such programs. The use of Perl (Practical Extraction and Report Language) as CGI-processor is very common. Perl is an interpreter language that has its roots in Unix but is available today for (nearly) all operating systems.

Linking of other programs as third step applications is possible by invoking them from the CGI-processor, commiting data and/or orders for the third step application and generating a new website out of the results passed back (see figure 2).

However concerning performance and security the use of CGI and Perl is not without drawbacks (see Bergsten, 1999).

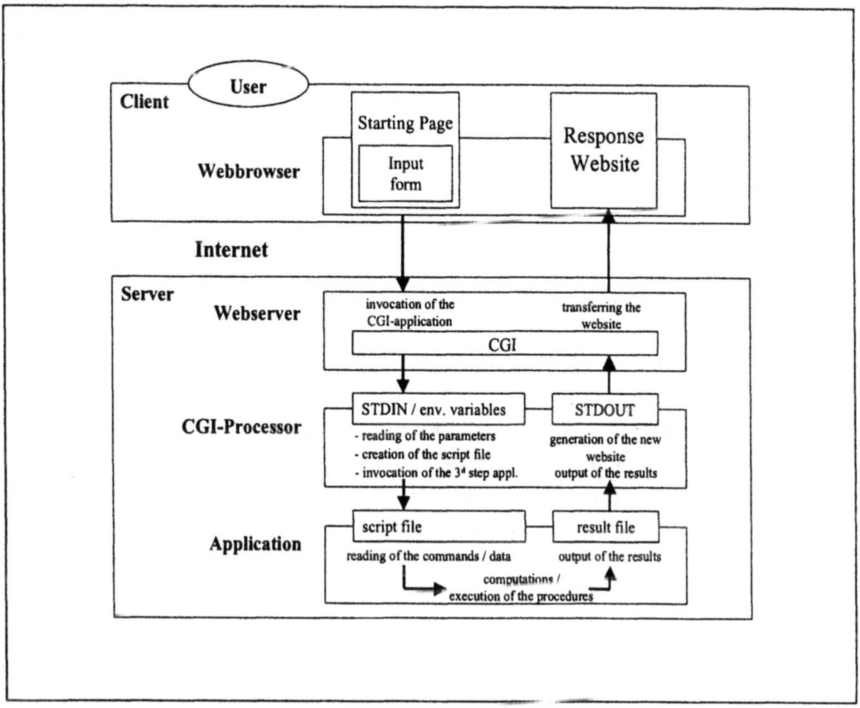

Figure 2. Diagram of CGI processing

6.2 Java

The Java technology provides developers with a platform-independent method for building web-based applications. Due to the strictly object oriented structure of Java the development as well as the maintenance of programs are frequently easier than they are with other programming languages.

In general the use of Java offers two approaches for interaction of the user with the program. These are applets and servlets.

Applets are programs that are transferred completely to the user's computer and are running in the browser integrated in HTML.

Servlets are developed to benefit of the advantages of the Java platform, as portability, performance, reusability and crash protection, and to solve the problems of using CGI (see Bergsten, 1999). A servlet can almost be thought of as an applet that runs on the server side (see Sun Microsystems, Inc., 2002). In contrast to CGI programs servlets execute all requests in one process leading to a higher performance.

The larger the programs are the more suitable is the use of servlets instead of applets because of the shorter downloading time.

Due to the mentioned advantages, easier development and maintenance as well as performance and download time, the use of Java and servlets is planned for the realisation of the scenario.

7 Conclusion

By means of modern internet technology we are developing an e-stat scenario for styren production. This scenario includes statistical methods for data analysis and design of experiments applied to optimise the chemical reaction process. This way, the user can learn to apply these methods with ease in a realistic environment.

References

Bergsten, H. (1999). *An Introduction to Java Servlets*. [Online] available URL: http://www.webdevelopersjournal.com/articles/intro_to_servlets.html (20.02.2002)

Conolly, D. (updated 1999). *Common Gateway interface*. [Online] available URL: http://www.w3.org/CGI/ (20.02.2002)

Kamps, U., Cramer, K., Ralf, C. & Kopp, T. (updated 2002). *e-stat*. [Online] available URL: http://www.emilea.de/ (20.02.2002)

Meadows, D. H., Meadows, D. L., Randers, J. & Behrens III., W. W. (1972). The Limits to Growth: A Report to The Club of Rome. *Club of Rome*.

Moore, D. S. (1997). New Pedagogy and New Content: The Case of Statistics. *Int. Stat. Rev.*, 65, 2, 123-165.

Sun Microsystems, Inc. (updated 2002). *Java Servlet Technology*. [Online] available URL: http://java.sun.com/products/servlet/index.html (20.02.2002)

Weihs, C., Jessenberger, J. (1999). *Statistische Methoden zur Qualitätssicherung und -optimierung*. Wiley-VCH, Weinheim

e-stat: Web-based Learning and Teaching of Statistics in Secondary Schools

Claudia Pahl, Petra Lipinski, and Kristina Reiss

Department of Mathematics, Carl von Ossietzky Universität Oldenburg, 26111 Oldenburg, Germany

Abstract. The student course of the e-stat project, which is currently under development, will use multimedia support in order to implement an interactive, internet-based learning and teaching environment. This educational environment for learning statistics will include a variety of tools in order to support self-initiated learning processes for approaching statistical problems and finding their solutions.

Keywords. Statistical education, web-based learning, multimedia, teaching environment, learning environment, interactive learning

1 Introduction

Teaching mathematics is a complex endeavor. There are no easy recipes for helping all students to learn and understand mathematics and to enable them to make use of mathematics as a problem-solving tool. Effective teaching requires pedagogical and psychological knowledge about teaching and learning, understanding what students know and need to learn and then challenging and supporting them to learn it well (NCTM 2000). To achieve these goals, challenging and supporting learning environments must be provided to initiate the student's learning processes.

In this context, the project "e-stat" supported by the German Federal Ministry of Education and Research (bmb+f) will develop a multimedia, interactive, computer- and web-based learning and teaching environment in applied statistics called EMILeA-stat. The student course is a part of EMILeA-stat which addresses pupils in secondary schools. The aim of the student course is to provide an interactive student-directed environment for statistical education.

In the following paragraphs, we want to explain the educational and psychological aspects which are essential for the current development of the course. Then we will demonstrate how educational principles may be used to define a suitable learning environment for statistics education. Finally, we will discuss aspects of a probable assessment of the EMILeA-stat student course.

2 Theoretical aspects of implementing a learning environment

2.1 Learning and teaching as process-oriented activities

Reinmann-Rothmeier and Mandl (2001) support an "integrated view" on teaching and learning. This view, which may be regarded as a moderate constructivist approach, is based on problem-oriented learning environments which foster self-regulated learning and cooperative learning as well. Moreover, problem-oriented learning environments should contribute to the students' motivation. The problems presented in such an environment should be challenging, but carefully adapted to the prior knowledge and the problem-solving abilities of the students. Finding a balance between "construction" (by the student) and "instruction" (by the teacher) plays a crucial role. Different empirical investigations have shown that a problem-oriented learning environment can contribute to successful learning processes (Krapp & Weidenmann 2001).

Implementing a problem-oriented learning environment will ask for some methodological decisions with respect to the role of the respective teacher, the program and the role of the students. In the following we will concentrate on the most important issues which will play a role in the EMILeA-stat student course.

The notion of a problem is used in a variety of meanings in different contexts. In educational research it was found out that authentic problems are best for initiating a student's learning process. Moreover, these learning environments should initiate actions performed by the students. With the help of these constructivist actions the pupils will be able to learn independently and self controlled. Research suggests that real world problems, which are relevant in the student's view, motivate to acquire new knowledge and new skills. In addition to that, situated learning and authentic problems ensure that students acquire applicable knowledge (Krapp & Weidenmann 2001).

Knowledge and skills should not remain restricted to specific situations. On the contrary, the same content should be learned in several different contexts with respect to various abstraction levels. Different ways of looking at the problems can support the process of learning and understanding.

Learning through different motivations, namely multimodality, should be taken into consideration. Texts and illustrations which are supplemented by video simulations and sound effects can simultaneously address several senses. Individual cognitive and non-cognitive conditions are regarded equally. A supply of knowledge which considers multi-codal and multi-modal presentation manners adds to a raised availability and particularly applicability of the knowledge.

It isn't possible to do without instructions, such as directing, presenting and explaining. Its renunciation would rather easily lead to an excessive challenge which would ask too much of the students (Krapp & Weidenmann 2001). The learning environment must provide simultaneously various possibilities of independent learning in complex situations and the knowledge necessary for the handling of problems.

A vast application of these constructivist approaches seems important. Especially, because surveys about the knowledge and skills of students, e.g. PISA (Programme for the International Student Assessments), discovered deficiencies in students' mathematical abilities. These deficiencies are caused among other things by a product-oriented atmosphere in the mathematics classroom. The surveys have also shown that product-oriented teaching in schools will probably not stimulate mathematical thinking, understanding, and application of this knowledge.

2.2 The role of statistics

The need to understand and be able to use mathematics, especially statistics, in everyday life, in professional education, and at the workplace in our fast-changing and increasingly technological world has never been greater and will continue to increase (NCTM 2000). Particularly statistics is a very application-oriented and close-to-reality field of mathematics. For instance, the availability of relevant empirical data material in the work place is ever growing in today's society. Furthermore, statistics are often "misused to sway public opinion on issues or to misrepresent the quality and effectiveness of commercial products" (NCTM 2000, 48). Only with sufficient knowledge about statistics and corresponding skills are the students able to critically reflect on statistical analysis and results, e.g. presented in a graphic format, and become informed citizens and intelligent consumers. In today's information-based society it has to be the school's task to prepare their students according to the new challenges which will await the students in the "real world" (NCTM 2000).

2.3 Signifiance of the computer

The multimedia computer which plays a more and more essential role in our highly technological society can produce a vivid and rich learning environment with challenging problems. Along with the world wide web the computer makes it possible to employ various important educational and psychological principles of the moderate constructivist use of learning (Krapp & Weidenmann 2001).

For example, programs can include possibilities of differentiation, variety, and alternation. Presentation manners vary by text, pictures, animation and narration. Further, the programs offer a multiplicity of types of tasks and give a differentiated feedback to the students (Krapp & Weidenmann 2001).

The self-control of the learners can be carried out to a considerable degree. With explorative teaching programs the learning ways are almost freely determinable by the user.

With the aid of simulated experiments, the reality can be modeled flexibly. An example of simulation experiments represents the "virtual urn".

If students acquire knowledge through the process of solving problems independently, they gain theoretical and practical knowledge much more efficiently. In the process of acting independently, the students can gain new insights, experience and at the same time can train cognitive skills. With every action an idea will be built up, which enables a mental action connected with the concrete level. And finally it's possible to transfer this mental action onto a symbolic level (Krapp & Weidenmann 2001).

2.4 The computer system

The e-stat system will combine content and technical possibilities with a high diversity of methodical and educational approaches. It is planned to implement an interactive, internet-based environment for learning and teaching statistics. A similar project is for example ActivStats (Vellemann 2000) which presents an introductory statistics course in a multimedia computer-based format. On the one hand it has all features of a traditional statistics textbook (text, graphics, exercises), and on the other hand this program puts emphasis on interactive lessons, videos, data visualization, and data analysis (Peschek 1998). However, ActivStats is limited to a certain group of people who buy the software. Being part of the world wide web EMILeA-stat has the advantage of providing many possibilities and is freely accessible to everybody who has internet.

There is also a feature which makes EMILeA-stat unusual. It offers statistical contents of different levels of complexity for people with various knowledge and needs (basic level A, medium level B, demanding level C). This feature is hardly fulfilled to a similar degree by any other computer-based learning programs, such as ActivStats. In addition to that, the modular structure of contents enables to use only certain parts of EMILeA-stat (modules). For example, when required, teachers may utilize modules of EMILeA-stat which are easily accessible via internet. Consequently, EMILeA-stat takes different types of users and several possibilities of use into consideration. However, this complexity can entail "cognitive overload" and "lost in hyperspace". Too many propositions and too many ways to act could cause the user to spend a lot of time to understand the handling of the user interface. Confusing possibilities of navigation are problematic for the orientation in the learning environments (Tergan 1997). To avoid these disadvantages it is necessary to create a simple and clear user interface.

Furthermore, EMILeA-stat contains methods for the integration of existing statistical engines (for example SPSS and XploRe), virtual teaching scenarios and an automated glossary. Contributions by people who are teaching statistics are welcomed.

3 Practical ways of including the educational aspects into the project - some examples

As a result of the educational aspects mentioned above, the student should be confronted with complex and productive tasks and situations. The problems should be oriented towards the student's environment, experiences and interests. Thus, students can take advantage of their world knowledge and everyday experiences to work on the mathematical problems and to solve them individually. Moreover, the subject matter of the e-stat project makes use of authentic material, such as data from daily life environments. For instance, we took data material out of a survey about organization of youth's leisure time to use it in our e-stat project.

In addition to that, many pages will have links to one or more data sets from real statistical studies. These data sources are permanently updated and offer current data material. The application-oriented procedures help students to organize and systemize their experiences, to learn from real situation-scenarios and to critically reflect on information material. Furthermore, students can use their new-gained knowledge in various ways and after all learn to cope with the expectations that their environment faces them with. In the area of "data and graphics" for example, students should be encouraged to critically reflect on the data and graphics which their environment offers. In this respect, not only perfect model graphics should be dealt with, but also graphics which convey a wrong impression on the facts.

While designing the EMILeA-stat course special attention is paid to the following aspects: working independently, effective communication and interaction among peer students, and the collecting of experiences on their own through self-active work. In order to support these abilities the student course will offer ideas how to create questions and to design appropriate statistical studies, furthermore ideas on how to design questionnaires and how to collect data to answer the questions. In this context, it seems very appropriate to encourage students to make their own surveys in the areas of their interest. Suggestions for the processing and evaluation of data material are offered. Possibilities exist to paste own data into Java-applets in order to process this data in a simple way and to end up with graphic evaluations with the help of the Java-applets (see Fig 1). These Java-applets are currently under development. Various ways and possibilities of data material evaluations will be offered. The student should learn by discovering and acquiring new knowledge. The opportunity of participating in collecting data enables the student to develop a substantial data competence by taking a critical look at various statistical problems. These examples show that we want to provide an environment which gives the students the opportunity to solve statistical problems in interactive processes. The interactive process should be characterized by student-directed actions.

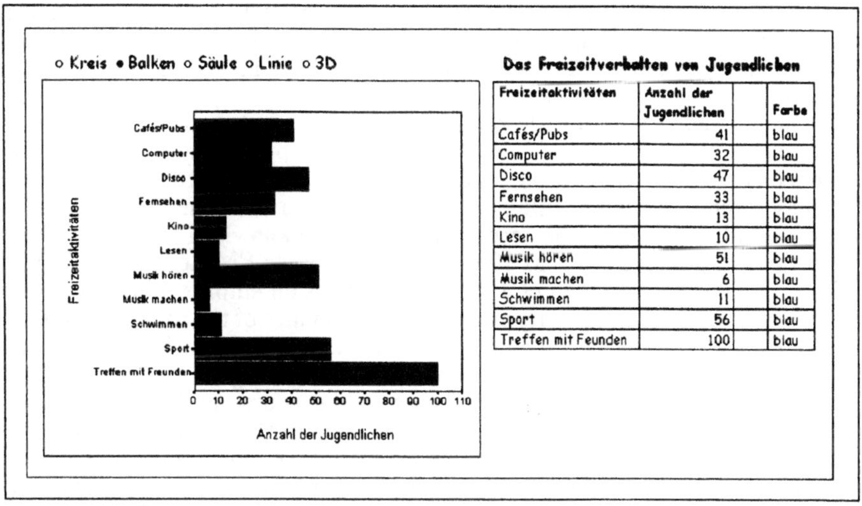

Fig 1 A Java-applet that offers a simple possibility to process data material

Another way to encourage interactive processes and independent student actions are the integration of simulations. They can be a very effective contribution to student's learning. Simulations show to be especially effective if students are asked to make hypotheses about possible outcomes before the actual start of the simulation. From a pedagogical-psychological point of view simulations, supported by tutoring work, are very valuable for the learning process, because they reduce the problem of transferring. In this context, the learning environment to a very large extent covers "real situations". At the same time simulations are more learner-friendly, because they make experiences possible which the real world does not always offer. The simulation-learning game helps to develop the student's systematical thinking. The person using the simulation for learning can build up a dynamic mental model of how the system functions (Krapp & Weidenmann 2001).

The EMILeA-stat student course offers many different types of tasks. Students will be able to examine their personal level of achievement. This will, for example, be possible by fill-in-the-blank texts and multiple choice tests. The student's input to these tasks will be followed by a student-oriented feedback. This feedback goes beyond "correct" or "incorrect", but rather responds to the individual mistake.

The EMILeA-stat student course will be implemented as a learning environment which provides challenging, non-routine problems and supports active, self-initiated learning processes, as it is suggested by recent research results in mathematics education (Baumert, Lehmann et al. 1997; Reiss, Klieme & Heinze 2001).

4 Conclusion

Interactive learning environments offer experiences which can hardly be achieved by conventional teaching and learning methods. "Virtually" based learning environments may to a high degree fulfill the claims for action-orientedness, problem-orientedness, authenticity, interactivity, learner-directed action, variety and choice. Consequently, the use of an interactive learning environment like e-stat, which probably will be finished by March 2004, can improve and support (not replace) the conventional teaching. Technology should only be one component along with teachers, books, and other media to provide a flexible learning environment, which meets the needs of the greatest number of learners.

In a nutshell, educational concepts have to be taken into consideration in a way that mathematical knowledge and appropriate logical structure are connected with a presentation which successfully addresses the students. Only if these prerequisites are fulfilled, the student can learn mathematics while understanding it, and while actively building new knowledge from experience and prior knowledge (NCTM 2000). To what degree EMILeA-stat will fulfill these educational claims in real use, still depends on empirical investigations.

In order to assess at an early stage the functionality of the user interface and the efficiency of different arrangements of learning and teaching, evaluation should take place soon. In general, we plan to investigate students' interest in interactive and multimodal elements of the course and to what extent these elements can support the learning process. In this way, early discovered problems and disadvantages could be avoided in the final version. This part of the evaluation will take place on the basis of theories and approaches which come from pedagogical psychology .

References

Baumert, J., Lehmann, R. et al. (1997). *TIMSS - Mathematisch-naturwissenschaftlicher Unterricht im internationalen Vergleich: Deskriptive Befunde.* Opladen: Leske + Budrich.

National Council of Teachers of Mathematics (NCTM) (2000). *Principles and Standards for School Mathematics.* United States of America.

Peschek, W. (1998). Didaktische Überlegungen zu ActivStats, einer interaktiven Hypermedia-Lernumgebung für die Stochastik. In: Winkelmann, B. (Ed.), *Mathematikdidaktische Untersuchungen anhand der Betrachtung des Lernens von Mathematik mit Hypermedia.* Occasional Paper Nr. 171, IDM, Universität Bielefeld , 20-32.

Krapp, A., Weidenmann, B. (2001). *Pädagogische Psychologie: Fourth Edition.* Weinheim: BeltzPVU.

Reinmann-Rothmeier, G. & Mandl, H. (2001). Unterrichten und Lernumgebungen gestalten. In: A. Krapp & B. Weidenmann, *Pädagogische Psychologie*, 601 - 646. Weinheim: Beltz.

Reiss, K., Klieme, E. & Heinze, A. (2001). Prerequisites for the understanding of proofs in the geometry classroom. In: M. van den Heuvel-Panhuizen (Ed.), *Proceedings of the 25th Conference of the International Group for the Psychology of Mathematics Education.* Vol. 4, Utrecht (The Netherlands): Utrecht University.

Tergan, S.-O. (1997). Hypertext und Hypermedia: Konzeption, Lernmöglichkeiten, Lernprobleme. In: Issing, L., Klime, P.(Ed.), *Information und Lernen mit Multimedia*, 123 - 137. Vol. 2. Weinheim, Basel: BeltzPVU.

Vellemann, P. (2000). *ActivStats.* CD-ROM for Windows and Mac. Data Description, Inc., Addison Wesley Longman: Ithaca, NY.

EMILeA-stat: Structural and Didactic Aspects of Teaching Statistics through an Internet-based, Multi-medial Environment

Ulrike Genschel[1], Ursula Gather[1], Anita Busch[1]

[1] Department of Statistics, University of Dortmund, 44221 Dortmund, Germany

Keywords. Statistical education, multimedia, internet-based learning environments, pedagogy, interactivity

1 Introduction

There have been initiatives to reform college level statistical education in the United States. In particular, the American Statistical Association has recommended that introductory statistics courses rely more on computers and automated computations (Cobb, 1992). Moore (1997) has advocated instruction through a constructive model of learning, in which students are encouraged to learn through their own activities and teachers are to de-emphasise lecture-based classroom education.

In contribution to the discussion on teaching techniques in statistics, we will illustrate the design and the written content of a new interactive software for statistical education, called EMILeA-stat (currently under development). We suggest a way of teaching statistical theory/methodology via an internet-based, multi-medial environment. Success in teaching statistics requires an educator to create a level of greater accessibility and relevance in instruction; a tenant of often overlooked "humanistic" educational psychology is that students learn best what they want and need to know (Gage and Berliner, 1991). EMILeA-stat can help to realize this educational environment supported by meaningful multi-medial instruction.

Through the structure of the presented educational material, EMILeA-stat aims at drawing out and maintaining a maximum level of student activity to foster a greater understanding and appreciation of statistics among students. That is, the software intends to stimulate student interest in material and student facilitate self-learning through a two-way flow of information (e.g. interactive applets or possibilities for self-evaluation). However, part of our philosophy is that the instructor still assumes a traditional role by setting learning objectives and guiding the students to achieve these goals with EMILeA-stat providing an important tool in the teaching process. The instructor encourages students to create a personal, optimal level of processing information and discover their most productive learning pace and activities. This approach has advantages for memory retention as well (Rumelhart and McClelland, 1986). Further potential benefits of multi-medial instruction are presented by Velleman and Moore (1996).

2 Project characteristics

2.1 Stratified levels of difficulty and "views"

A characteristic feature of EMILeA-stat is the structuring or presentation of nearly all material at three different degrees of difficulty, called levels A, B, and C, respectively. Level A is meant to impart the general idea of a statistical concept. Formulas and rigorous mathematical details are strictly omitted. Level B, which will be the assumed level of instruction in this talk, is a medium level of difficulty, to address college students not majoring in Statistics. Here, the practical use of a statistical procedure is presented along with a basic statistical justification for the procedure's formulation or origin. Level C is designed and intended for more mathematically orientated and advanced students. Content emphasizes statistical theory, adding mathematical derivations for the statistical tools described at levels A or B.

By establishing a hierarchy of educational levels in EMILeA-stat, we hope to provide instructors greater flexibility in customizing their instruction to match the needs and backgrounds of new statistics students. Teachers can potentially introduce material using more intuitive explanations and simpler language at one level (A or B), sacrificing some rigor and technicalities with the intention of making statistics more accessible to a larger student audience. More theoretical or technically involved discussions can be introduced at higher levels (B or C), possibly even after the instructor has received some measure of student feedback.

An advantage of the multi-medial environment is the capacity of providing significantly more examples than are possible in a classroom setting, where time is a practical constraint. This increases the opportunities for student exposure to statistical applications (both inside and outside the classroom). In EMILeA-stat, these applications will vary in subject and content in order to address audiences of different backgrounds and interests. To demonstrate the relevancy of statistical tools, EMILeA-stat creates separate so called "views", enabling students to explore statistics within an applied subject area of most practical importance and benefit to the student. Views may also differ in content, depending on the relevancy of a statistical concept for a specific field of application. For example, EMILeA-stat will offer a "view" for business science and economics students, as well as a view emphasizing psychological applications. A "general" overall view, without a specific target area of application, will also exist and incorporate examples gathered from a variety of problem scenarios. Furthermore, EMILeA-stat attempts to incorporate "real-world" related problems and associated data to make statistical tools generally more appealing to students.

2.2 Further structural features

Throughout the entire section of testing and estimation theory (our current project task), we provide a motivational example to begin or introduce each new chapter or topic. These examples will be used subsequently throughout the chapter for purposes of illustration and cohesion, helping the students to better understand the selection and implementation of statistical methods from the onset of data analysis. Aside from these guiding examples, EMILeA-stat seeks to provide supplementary examples and interactive exercises to increase student familiarity with the chosen learning material and exposure to further applications. EMILeA-stat also tries to implement graphical illustrations wherever possible, to promote the use and literacy of graphs in statistics and to visually support more theoretical statistical concepts.

In our internet-based teaching approach, presented material is intended and formulated to be as concise as possible. Conciseness is closely related to clarity of the written content. To avoid endless scrolling in an internet environment, we stress the need for clear organizational separations and arrangement to simplify the presentation for students and instructors, while permitting quick access to any information or content. (EMILeA-stat is also intended as a reference tool.) Moreover, the same structured format is maintained throughout the software so that potential users will recognize a consistent pattern of material arrangement, which should increase student comfort and ability in a repeated process of student-directed learning.

At the end of each section/chapter, EMILeA-stat provides to give the student a chance for self-assessment "quizzes" with context validity, to permit the student a chance for self-assessment of acquired knowledge and skills and also allow for the possibility of immediate, corrective feedback. Question types will vary in style and form to assess student understanding at various levels of learning. Possibilities range from standard multiple choice questions with one or more correct answers, or "fill-in-the-blank" responses, to short answer questions where students are required to complete more involved computational or graphical tasks for example.

In addition to general problems for the purpose of practice and self-evaluation, (resembling homework-type problems), we want to emphasize that an important goal of EMILeA-stat is to promote and allow collaborative problem solving by students in groups, motivated by the idea that students can help each other in learning. Creating opportunities for students to explain classroom material, using their own understanding and words, can be difficult for instructors (though arguably beneficial to the students). EMILeA-stat can allow for interactive group work and provide content and focus for student discussion in statistics. An ideal number of students in such "laboratory" groups could potentially range from 2-6, highly depending on the learning goals and strengths of the students, which have to be assessed by the instructor.

2.3 Interactive features

To illustrate some of the interactive features of EMILeA-stat, we will present an interactive Java applet to enhance student understanding of the Wilcoxon Rank Sum Test. The Java applet is constructed so that the student receives opportunities to practice the steps needed to perform the test. Potentially, students can practice and perform a statistical method using data sets which may be provided, randomly generated or even student-collected. We will also present a proposal for a multiple choice test, incorporating some of the features mentioned above, as well as illustrate the implementation of the motivational example.

References

Cobb, G. (1993). Teaching statistics. In: *Heeding the Calls for Change: Suggestions for Curricular Action, Ed. L. A. Steen, pp. 3-43.* Washington, D.C.: Mathematical Association of America.

Moore, David S. (1997) New Pedagogy and New Content: The Case of Statistics. *International Statistical Review*, **65**, 2, 123-137.

Rumelhart, D., Mc Clelland, J. (Eds.). (1986). *Parallel distributed processing: Explorations in the microstructure of cognition.* Cambridge, MA: MIT Press.

Tanur, Judith M. (1997). Discussion : Leading Horses to Water. *International Statistical Review*, **65**, 2, 159-161.

Velleman, P.F., Moore, D.S. (1996). Multimedia for teaching statistics : promises and pitfalls. *American Statistician*, **50**, 217-225.

Evaluating the GPH Estimator via Bootstrap Technique

Silvia Golia[1]

[1] Dipartimento di Metodi Quantitativi, Università di Brescia, C.da S.Chiara 50, 25422 Brescia, Italy

Abstract. In this work a bootstrap method for long memory processes is presented. The method is based on a combination of the discrete wavelet transform and the stationary bootstrap and is a development of the one presented by Percival et al. (2001). This bootstrap scheme is empirically tested calculating the GPH estimator for fractional gaussian noise time series. The set of bootstrap estimations is compared with the set of estimations obtained via MonteCarlo simulation procedure.

Keywords. Bootstrap, discrete wavelet transform, GPH estimator, long memory process

1 Introduction

In the last decades many researchers have found that the so-called long-memory processes (LMP) are effective models for a wide variety of time series. By definition, a LMP is a process which exhibits an hyperbolic rate of decay of the autocorrelation function or has a spectral density function with a pole at zero (Beran, 1994). This kind of process has been widely studied and methods for the estimation of the Hurst exponent H, which characterizes LMP, have been proposed. With regard to these estimators for H, in many cases analytical distibutional results are available asymptotically or under specific assumptions (i.e. gaussianity). It is interesting and useful to try to identify the properties of any specific estimator for H without any constraint on the distribution of the process or the size of the available time series. The natural way to do it is to apply a bootstrap method. The original jackknife and bootstrap were developed for independent and identical distributed (i.i.d.) data, so they cannot be roughly applied to non-i.i.d. cases like time series and other dependent data. For this purpose, a special class of bootstrap methods for stationary time series has been developed in the last decade. Two different approaches can be identified. The first is a parametric model-based approach, which is founded on the idea of fitting parametric models first and then resampling from the residuals (Efron and Tibshirani, 1993). On the contrary, the second approach is based on a nonparametric, purely model-free bootstrap scheme. The bootstrap technique presented in this paper belongs to the latest approach. The method is based on the use of the discrete wavelet transform (DWT) and is a development of the one presented by Percival et al. (2001).

The paper is organized as follows. In section 2 recalls on wavelet decomposition and theoretical properties of the wavelet coefficients are presented. A detailed description of the bootstrap procedure is given in section 3 and in section 4 the resampling algorithm is tested calculating the GPH estimator (Geweke and Porter-Hudak, 1983) for fractional gaussian noise (FGN) time series. Conclusions follow in section 5.

2 Wavelet Decomposition

Let $\mathbf{x} = \{x_1, x_2, \ldots, x_n\}$ be a single realization of a generic stochastic process and n an integer multiple of 2, i.e. $n = 2^J$ with J any positive integer. The DWT of $\mathbf{x}$ is an orthonormal transform $\mathcal{W}$ that yields a vector of coefficients $\mathbf{W} = \mathcal{W}\mathbf{x}$. The orthonormal condition on $\mathcal{W}$ implies that the time series can be reconstructed from its coefficients via $\mathbf{x} = \mathcal{W}'\mathbf{W}$ (Daubechies, 1992).

The rigorous definition of the wavelet coefficients is done in the framework of the multiresolution analysis (MRA). A MRA of $L^2(\mathbf{R})$ consists of a nested sequence of closed subspaces V_j, $j \in \mathbf{Z}$ of $L^2(\mathbf{R})$ with a certain number of properties. A multiresolution approximation is entirely caracterized by the scaling function ϕ, also referred to as the *father wavelet.* Dilates and translates of the scaling function, defined by $\phi_{j,k}(x) = 2^{j/2}\phi(2^j x - k)$, $k \in \mathbf{Z}$, constitute an orthogonal bases for each subspace V_j. For further details see Mallat (1998).

The DWT can be described in terms of wavelet and scaling filters. For a given wavelet basis, the filter $\mathcal{H}$ is represented by the sequence $\{h_k\}_{k\in\mathbf{Z}}$ given by $h_k = \langle \phi, \phi_{1,k} \rangle$, where again ϕ is the scaling function and $\langle \cdot, \cdot \rangle$ the inner product. The filter $\mathcal{H}$ is known as the *low pass filter.* The connection between the scaling function, the related low pass filter $\mathcal{H}$ and the *mother wavelet* ψ is given by $\psi(x) = \sum_{k\in\mathbf{Z}} (-1)^k h_{1-k} \phi_{1,k}(x)$. Scaling function coefficients at level $j-1$ are obtained applying the filter $\mathcal{H}$ to the sequence $c_{j,\cdot} = \{c_{j,k}\}_{k\in\mathbf{Z}}$:

$$c_{j-1,\cdot} = \mathcal{H} c_{j,\cdot}$$

and a more general rule gives $c_{j-m,\cdot} = \mathcal{H}^m c_{j,\cdot}$. The so called *high pass filter* $\mathcal{G}$ is define by: $g_k = (-1)^k h_{1-k}$, $k \in \mathbf{Z}$ and it is used to calculate the wavelet coefficients at level $j-1$ as follows:

$$d_{j-1,\cdot} = \mathcal{G} c_{j,\cdot}$$

A more general rule gives $d_{j-m,\cdot} = \mathcal{G}\mathcal{H}^{m-1} c_{j,\cdot}$. Filters $\mathcal{H}$ and $\mathcal{G}$ are called *quadrature mirror filters.*

The interesting aspect of the DWT is that it approximately decorrelates LMP. Wornell (1996) provided theoretical evidence that the wavelet coefficients are wide-sense stationary and rather weakly correlated at each scale and across scales for $1/f$ stochastic processes (theorems 3.5 and 3.6, pag 50-51). The $1/f$ family of statistically self-similar random processes are generally defined as processes having power spectra $f(\lambda)$ obeying a power law relationship of the form $f(\lambda) \sim \sigma^2/|\lambda|^\gamma$ with the spectral parameter γ related to the Hurst exponent H according to $\gamma = 2H + 1$. Theorem 3.3 in Wornell (1996, pag 42) states that a fractional Brownian motion corresponding to $0 < H < 1$ and the associated fractional Gaussian noise, which will be used in section 4, are $1/f$ processes. Moreover the correlation between scales can be made arbitrarily small by increasing the length of the wavelet filter (Craigmile et al. 2000).

3 The Algorithm

Many purely model-free bootstrap methods for stationary time series has been developed in the last decade. Some of them are based on "blocking" arguments, i.e. the data are divided into blocks of consecutive observations. The blocks of length l, which can basically be non-overlapping (Carlstein,

1986) or overlapping (blockwise or moving-block bootstrap, Künsch, 1989), are resampled with replacement and the selected blocks are concatenated in order to obtain resampled time series. The obtained pseudo time series is not stationary, even if the original series is stationary. For this reason Politis and Romano (1994) proposed the stationary bootstrap method.

The main difference lies in the fact that the block length is not fixed but random. In fact, let $L_1, L_2, \ldots$ be a sequence of i.i.d. random variables having geometric distribution with parameter p so that $P(L_i = k) = p(1-p)^{k-1}$ for $k = 1, 2, \ldots$ and independent of the L_i, let $I_1, I_2, \ldots$ be a sequence of i.i.d. random variables having uniform distribution on $\{1, 2, \ldots, n\}$. The blocks are of the type: $\mathcal{B}(I_i, L_i) = (x_{I_i}, x_{I_i+1}, \ldots x_{I_i+L_i-1})$. The parameter p is related to the blocks length l of the moving-block method by the relation: $p = l^{-1}$.

Nevertheless, in general these methods do not seem to be applicable to time series yielded by complicate data generating processes (DGP) such as, for example, a long memory or a chaotic DGP (Golia and Sandri, 2001).

Given the DWT of the time series under study, the main idea of the method proposed in this paper is to apply the stationary bootstrap method at the wavelet coefficients computed at each scale. Then the reversible property of the wavelet transform is invoked and a resampled time series is obtained.

The proposed resampling algorithm can be described in details as follows:

1. starting from the observed time series $\{x_t\}_{t=1}^n$, compute DWT to obtain the wavelet and scaling coefficients vectors;
2. resample the vectors of coefficients computed at each scale via the stationary bootstrap method using different parameters p varying on each scale;
3. apply the inverse of the DWT to the resampled vectors of coefficients in order to obtain a resampled time series.

The choice to apply the stationary bootstrap to the wavelet coefficients calculated at each scale relies on the fact that they are wide-sense stationary and weakly correlated, as stated in section 2.

The quantities involved in the algorithm are the parameter p, the coarsest level of approximation of the signal and the quadrature mirror filter. As stated at the beginning of this section, the parameter p in the stationary bootstrap method is connected with the block length l in the moving-block bootstrap algorithm. Moreover it is well known that the bias and variance of a block bootstrap estimator depends on the block length. The problem of the optimal block size choice is addressed in Hall et al. (1995). For the problem of the estimation of bias and variance of an estimator $l \sim n^{1/3}$. The block length is used in order to determine p which will be $p = n^{-1/3}$.

The orthonormal quadrature mirror filter used in the DWT can be generated using different types of mother wavelet such as, for example, the Haar, the Coiflet, the Daubechies or the Symmelet wavelet.

4 Empirical Evidences

In this section an application of the bootstrap method described in section 3 to a fractional gaussian noise process (Mandelbrot and Van Ness, 1968) with Hurst exponent $H = 0.7$ and $H = 0.9$ is presented. First of all, a MonteCarlo experiment is conducted in order to derive values for the mean and standard deviation of H estimated by the GPH estimator (Geweke and Porter-Hudak, 1983) and its empirical distribution. 3000 FGN series of length $n = 2^{10} =$

1024 are generated and recentered by subtracting off its sample mean. For each of these adjusted series the estimation of H via GPH estimator[1] is calculated. Table 1 reports the sample mean and sample standard deviation of the estimations which will be considered as the "true" mean and standard deviation of the estimator. Then, given a simulated series, 200 resampled

	Mean	Std. Dev.
H = 0.7	0.709	0.064
H = 0.9	0.916	0.065

Table 1. Mean and standard deviation for the 3000 estimations via Monte Carlo experiment

series of length $n = 1024$ are generated. As stated in section 3, the values of the parameter p in the stationary bootstrap method vary on each scale depending on the length of the coefficients vectors lw and are given by $p = lw^{-1/3}$. The coarsest level of approximation of the signal is fixed equal to 4 and the quadrature mirror filter used is the Daubechies filter with 4 vanishing moments. Daubechies filters are minimal phase filters that generate wavelets which have a minimal support for a given number of vanishing moments. For each resampled series the GHP method is applied in order to get an estimation of H.

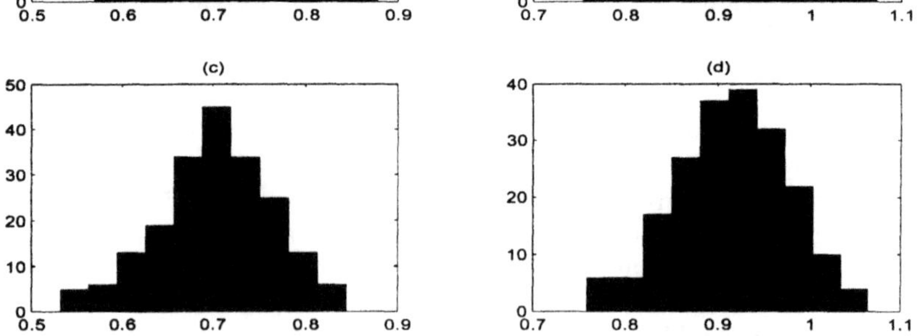

Fig. 1. Histograms of 200 estimations of the parameter H via GPH method for FGN with $H = 0.7$ computed on (a) the MonteCarlo and (c) the bootstrap procedures and FGN with $H = 0.9$ computed on (b) the MonteCarlo and (d) the bootstrap procedures.

[1] The GPH estimation is calculated on $n^{0.7}$ Fourier frequencies

Figure 1 shows the histograms of the 200 estimations of H coming from the application of the bootstrap method and 200 simulated series via MonteCarlo procedure.

Beside the visual evidence, in order to test if the values of the estimates of H computed using the resampled series are reproducing the real randomness of the chosen estimator, the following three nonparametric tests are applied: the t-test (equal mean), the test for equal variance and the Kolmogorov-Smirnov test (equal distribution). On the basis of the p-values reported in

	Equal Mean	Equal Variance	Equal Distribution
H = 0.7	0.186	0.324	0.568
H = 0.9	0.767	0.97	0.972

Table 2. p-values of the three nonparametric tests

table 2 we can accept the whole set of hypotheses for both the FGN processes with $H = 0.7$ and $H = 0.9$. The results give confirmation that in the present application the resampling method is working correctly.

Moreover for both the processes 30 different series of size 1024 are simulated. For each of them 100 bootstrap series are generated and the respective estimations of H are calculated. Then, from each of the 100 estimations the sample mean and standard deviation are computed. One set of 30 sample means and one set of 30 sample standard deviations are obtained.

A t test is conducted on the set of the sample means in order to verify if their mean is equal to 0.709 for $H = 0.7$ (0.916 for $H = 0.9$). The p-values reported in the first column of table 3 allow to say that the resampled series in mean give an average estimation of H equal to the "true"[2] value.

	Mean	Std. Dev.
H = 0.7	0.627	0.349
H = 0.9	0.874	0.121

Table 3. p-values of the t-tests on the set of 30 estimations of sample means and standard deviations.

Moreover a t test is performed on the set of the sample standard deviations in order to verify if their mean is equal to 0.064 for $H = 0.7$ (0.065 for $H = 0.9$). The p-values reported in the second column of table 3 allow to say that the resampled series give an average standard deviation for the estimation of H equal to the "true"[2] value.

5 Conclusions

Due to the particular structure of the autocorrelation function, standard methods such as moving-block or stationary bootstrap are not applicable in

[2] "True" in the sense of the MonteCarlo experiment

a LM context. In this paper a bootstrap method for LMP based on DWT, development of the one presented by Percival et al. (2001), is illustrated and tested on the FGN process with two different values of H. From the results reported in section 4 it seems clear that the resampled series are able to reproduce the real randomness of the estimator GPH extracted from a MonteCarlo procedure. A Daubechies filter with 4 vanishing moments and coarsest level of approximation equal to 4 in the DWT seem to give good results in this study, but extra work is needed in order to explore if this combination works for other LMP and if the performances can be improved.

The work has been performed using the Matlab 5.3 software and the WaveLab 802 library available at *http://www-stat.stanford.edu/~wavelab/*.

Acknowledgments. I wish to thank G. De Luca and M. Sandri.

References

Beran, J. (1994). *Statistics for long-memory processes.* New York: Chapman & Hall.

Carlstein, E. (1986). The use of subseries values for estimating the variance of a general statistic from a stationary sequence. *The Annals of Statistics*, **14**, 1171-1179.

Craigmile, P.F., Percival, D.B. and Guttorp, P. (2000). The impact of wavelet coefficient correlation on fractionally differenced process estimation. *NRCSE Technical Report No. 49.*

Daubechies, I. (1992). *Ten lectures on wavelets.* Philadelphia: SIAM CBMS-NSF series, vol. 61.

Efron, E. and Tibshirani, R. (1993). *An introduction to the bootstrap.* New York: Chapman & Hall.

Geweke, J. and Porter-Hudak, S. (1983). The estimation and application of long memory time series models. *Journal of Time Series Analysis*, **4**, 221-238.

Golia, S. and Sandri, M. (2001). A resampling algorithm for chaotic time series. *Statistics and Computing*, **11**, 241-255.

Hall, P., Horowitz J.L. and Jing B. (1995). On blocking rules for the bootstrap with dependent data. *Biometrika*, **82**, 561-574.

Künsch, H.R. (1989). The jackknife and the bootstrap for general stationary observations. *The Annals of Statistics*, **17**, 1217-1241.

Mallat, S. (1998) *A wavelet tour of signal processing.* San Diego: Academic Press.

Mandelbrot, B.B. and Van Ness, J.W. (1968). Fractional Brownian Motion, fractional noise and application. *SIAM Review*, **10**, 422-437.

Percival, D.B., Sardy, S. and Davison, A.C. (2001). Wavestrapping Time Series: Adaptive Wavelet-Based Bootstrapping. In: *Nonlinear and Nonstationary Signal Processing, edited by W.J. Fitzgerald, R.L. Smith, A.T. Walden and P.C. Young*, 442-470. Cambridge: Cambridge University Press.

Politis, D.N. and Romano, J.P. (1994). The stationary bootstrap. *Journal of the American Statistical Association*, **89**, 1303-1313.

Wornell, G.W. (1996). *Signal processing with fractals: a wavelet approach.* Upper Saddle River: Prentice-Hall PTR.

Evolutionary Algorithms with Competing Heuristics in Computational Statistics

Josef Tvrdík[1], Ivan Křivý[1] and Ladislav Mišík[2]

[1] Department of Computer Science, University of Ostrava, 701 03 Ostrava, Czech Republic

[2] Department of Mathematics, University of Ostrava, 701 03 Ostrava, Czech Republic

Abstract. The paper presents a new class of evolutionary algorithms based on the competition of different heuristics. The algorithm was applied to solving some optimization problems of computational statistics, namely to estimating the parameters of non-linear regression models, constrained M-estimates and optimizing the smoothing constants in the Winters exponential smoothing. The results showed that the evolutionary algorithm with competing heuristics can be successfully used in solving some global optimization problems of computational statistics.

Keywords. Evolutionary algorithms, heuristics, global optimization, nonlinear regression, robust estimates, time series analysis

1 Algorithm and its convergence

We consider evolutionary algorithms (EAs) intended for finding the global minimum of complicated objective functions. A more detailed formulation of the optimization problem is given in Mišík et al. (2001). The algorithm can be described as follows:

- I: Generate an initial population $\mathcal{P}_0 = \{\mathbf{x}_1, \mathbf{x}_2, \cdots, \mathbf{x}_N\}$ chosen as an independent identically distributed sample according to the probability measure p_0 and set $k = 0$.
- C_k: Copy a portion of M best points of $\mathcal{P}_k$ directly into the new population. Here the best points are those with the lowest values of the objective function f and M is an integer from $\{1, 2, \cdots, N-1\}$.
- G_k: Generate a new trial point according to the probability measure $p_{k+1} = \pi(\mathcal{P}_k)$ and include it to the new population when fulfilling the condition $\mathcal{C}$. Repeat this procedure until the new population is complete.
- M_k: With the probability m_{k+1} replace a randomly chosen point by its mutation.
- R: Set $k = k + 1$ and go back to C_k.

According to the results of Solis and Wets (1981) and Mišík (2000), the EAs are convergent provided that for every measurable subset S of the searching domain D we have $\prod_{k=1}^{\infty}(1 - p_k(S)) = 0$, where $p_k(S)$ denotes the probability of producing a new trial point in the set S just in the k-th iteration. It is not necessary to use the only heuristic for producing a new trial point in the step G_k. The use of more different competing heuristics was proposed, see e.g. Tvrdík et al. (2001). Considering h heuristics at disposal, the new trial points can be generated by any of the heuristics with the probability q_i, $i = 1, 2, \dots, h$, depending on the successfulness of the i-th heuristic during preceding part of the optimization process. Several strategies for evaluating the successfulness of the individual heuristics can be applied. One of them is based on simple counting the number of successful insertions of new trial points, n_i,

$$q_i = \frac{n_i + n_0}{\sum_{j=1}^{h}(n_j + n_0)}, \tag{1}$$

where $n_0 > 0$ is a constant. Setting $n_0 > 1$ prevents a dramatic change in q_i by one random successful use of the i-th heuristics. Another strategy takes into account the relative position w_i of the trial points when they are inserted into the new population of points ordered in a nondecreasing sequence with respect to their objective function values,

$$q_i = \frac{W_i + w_0}{\sum_{j=1}^{h}(W_j + w_0)}, \tag{2}$$

where $W_i = \sum w_i$ over a previous period of the process and $w_0 > 0$ is an input parameter, which can be set to the mean value of w_i.

The EA with competing heuristics and without explicit mutation (i.e. the probability of mutation $m_{k+1} = 0$ in the step M_k) is convergent if at least one of the heuristics in use is convergent when using alone. In practice this can be ensured by including the heuristics generating new trial points uniformly distributed over D. On the other hand, it was shown in Mišík et al. (2001) that in some practical tasks better results can be achieved using a nonconvergent heuristics instead of the theoretically convergent algorithm. That is why a natural question arises to estimate the probability that the random search produces a proper trial point. Therefore, we tried to estimate the probability p that just the uniform search produces a new trial point in a set $S \subset D$ with its Lebesgue measure $\lambda(S) > 0$ during all the process. Starting from the sufficient condition for the convergence of the EA (see Solis and Wets, 1981), the following approximate relation between p and $\lambda(S)$ was derived (Tvrdík et al., 2002)

$$\frac{n\ \lambda(S)}{2h\ \lambda(D)} < p < \frac{n\ \lambda(S)}{h\ \lambda(D)}, \tag{3}$$

where n is the total number of trial points generated during the optimization process. Because of using the linear approximation, the validity of this relation is restricted to such sets S that the ratio $\lambda(S)/\lambda(D)$ is sufficiently small.

2 Implementation of EA with competing heuristics

The structure of the implemented algorithm with competing heuristics can be briefly written as follows.

procedure EA
generate $\mathcal{P}$ (an old population of N points taken at random from D)
repeat
  find the worst point in $\mathcal{P}$, $\mathbf{x}_{worst}$, with the highest value of f
  copy M best points of $\mathcal{P}$ into new population $\mathcal{Q}$, $1 \le M < N$

repeat
  repeat
    generate a new trial point **y** by some heuristics applied to $\mathcal{P}$
  until $f(\mathbf{y}) < f(\mathbf{x}_{worst})$
  insert the next trial point into $\mathcal{Q}$
 until $\mathcal{Q}$ is completed to N points
 replace $\mathcal{P}$ by $\mathcal{Q}$
until stopping condition is true

The heuristics in this procedure is any non-deterministic rule which gives a new point $\mathbf{y} \in D$. We applied eleven heuristics in the implementation:

- four heuristics based on randomized reflection in the simplex (Křivý and Tvrdík, 1995),
- four heuristics originated by differential evolution (Storn and Price, 1997),
- two heuristics derived from evolutionary strategy (e.g. Bäck, 1996 or Kvasnička et al., 2000),
- uniform random search.

Two variants of the EA with competing heuristics were tested: *Competing1* using equation (1) and *Competing2* using equation (2) for the evaluation of probabilities q_i. The common input parameters were set as follows: $N = 10d$ (d is the dimension of searching domain D), $M = N - 1$ and stopping condition in the form $f_{(N/2)} - f_{(1)} \le \varepsilon$ provided that $f_{(1)} \le f_{(2)} \le \dots \le f_{(N)}$. The value of ε was adjusted to $1 \times 10^{-15} \times$ total sum of squares for nonlinear regression tasks and 1×10^{-7} for others. To avoid the degeneration of evolutionary process it is sometimes useful to reset the current values of q_i to their starting values. When any probability q_i decreases below a minority limit δ, i.e. when $q_i < \delta$, the values of n_i or w_i are reset to 0 what means that the probabilities are reset to their starting values, $q_i = 1/h$.

The percentage R of runs resulting in a point very near to the global minimum and the average number of objective function evaluations NE were the variables measured in the experiments.

3 Applications

Both implementations of the EA with competing heuristics were applied to solving the following optimization problems:

- estimating the parameters of some nonlinear regression models presented by the National Institute of Standards and Technology, U.S.A. (www.itl.nist.gov/div898/strd),
- making the constrained M-estimates in linear regression models (Arslan et al., 2000),
- optimizing the values of smoothing constants in the time series analysis when using the Winters algorithm with multiplicative seasonal adjustment (Winters, 1960).

3.1 Optimization of nonlinear regression models

Both *Competing1* and *Competing2* variants of the algorithm were tested in the estimation of parameters of non-linear regression models. The tasks for testing were chosen from the Nonlinear Least Squares Datasets of the National Institute of Standards and Technology, U.S.A. The testing experiments were carried out with all eight tasks of higher level of difficulty. The examples are not easy for standard algorithms used in statistical packages, see McCullough and Wilson (1999), McCullough (2000). Twenty runs for each task were performed. The results are given in Table 1. Identification of the tasks is the same as in NIST Datasets.

Table 1. Results of estimating parameters of nonlinear regression models

		Competing1			*Competing2*		
Task	d	R	$\overline{NE}$	sd(NE)	R	$\overline{NE}$	sd(NE)
Bennett5	3	100	77504	33757	100	71623	26559
BoxBOD	2	100	1086	78	100	1057	85
Eckerle4	3	100	1473	147	100	1858	122
MGH09	4	100	14090	1631	100	13305	1987
MGH10	3	100	39050	2887	100	39596	1123
Rat42	3	100	2834	129	100	2785	218
Rat43	4	100	3321	185	100	3868	204
Thurber	7	60	13121	969	85	12649	1229

3.2 Constrained M-estimates

Constrained M-estimates of the location parameters in a linear regression model are considered according to Arslan et al.(2000). The model can be written as $\mathbf{y} = \mathbf{X}\beta + \mathbf{e}$, where $\mathbf{y}$ is a n-dimensional response vector, $\mathbf{X}$ an $(n \times d)$ design matrix, β a d-dimensional vector of parameters and e the error vector. Defining the residuals as $r_i = y_i - \hat{y}_i$, $i = 1, 2, \ldots, n$, $\hat{\mathbf{y}} = \mathbf{X}\beta$, the CM-estimation problem is to find the global minimum of

$$L(\beta, \sigma) = \frac{1}{n} \sum_{i=1}^{n} \rho(\frac{r_i}{\sigma}) + \ln(\sigma)$$

over $\beta \in R^d$ and a scatter parameter $\sigma > 0$ subject to the constraint

$$\frac{1}{n} \sum_{i=1}^{n} \rho(\frac{r_i}{\sigma}) \leq \varepsilon \rho(\infty),$$

where $\varepsilon \in (0, 1)$. Using truncated Huber's function for $\rho(t)$

$$\rho(t) = \begin{cases} \frac{t^2}{2} \text{ for } |x| \leq G, \\ \frac{G^2}{2} \text{ for } |x| > G \end{cases}$$

with $G = 2.795$ and $\varepsilon = 0.5$, we could find the CM-estimates of both the location and scatter parameters for the following two tasks:

- Hertzsprung-Russel diagram for the star cluster CYG OB1 (Rousseeuw and Leroy, 1987),
- example from Robust Regression Tutorial of NCSS by Hintze (2001).

The values of R, $\overline{NE}$ and the standard deviation of NE are given in Table 2. As regards the CYG OB1 task, the CM-solution is $\hat{y}_i = -8.533 + 3.058x_i$.

Table 2. Reasults of making CM-estimates in linear regression models

		Competing1			*Competing2*		
Task	d	R	$\overline{NE}$	sd(NE)	R	$\overline{NE}$	sd(NE)
CYG OB1	3	90	3040	200	80	2893	166
NCSS	7	85	6420	678	95	6275	591

3.3 Optimization of smoothing constants

This section deals with an application of evolutionary algorithms to solving the problem of time series smoothing. The time series of some Czech labour market descriptors were treated using the Winters exponential smoothing algorithm with multiplicative seasonal adjustment.

The Winters multiplicative seasonality algorithm is based on the following formulas:

$$a(t) = \alpha \frac{X_t}{Sz_{t-L}(t-L)} + (1-\alpha)[a(t-1) + b(t-1)],$$

$$b(t) = \beta[a(t) - a(t-1)] + (1-\beta)b(t-1),$$

$$Sz_t(t) = \gamma \frac{X_t}{a(t)} + (1-\gamma)Sz_{t-L}(t-L),$$

where $a(t), b(t)$ and $Sz_t(t)$ are estimates of the time series level, its slope and its seasonal factor at time t, respectively, α, β, γ are smoothing constants and L represents the number of periods per year.

The smoothing constants α, β and γ were optimized with respect to the criterion SSE, the searching space being constrained to $0 < \alpha, \beta, \gamma \leq 1$. The empirical formulas proposed by Cipra (1986) were used for estimating the initial values $a(0), b(0)$ and $Sz_t(t)$ for $t = -(L-1), -(L-2), \ldots, 0$.

The original data (extended by the Czech Ministry of Labour and Social Affairs) covered the period from January 1995 to January 2001. The following four labour market descriptors were studied: labour force, number of job applicants, unemployment rate, and number of job vacancies. For each individual time series at least ten independent runs were performed. In all the cases the global minimum was found with the reliability $R = 100$.

The resulted optimum values of smoothing constants showed that some of them reach their upper limit value, which indicates that the mechanism generating them has recently gone through some fundamental changes. Using the optimum values of smoothing constants, all the time series were smoothed and the forecasts for the year 2001 were calculated and compared with the real data as well as with the forecasts resulted from the Box-Jenkins methodology when applicable. It was found that in some cases the forecasts resulted from the Winters smoothing algorithm are even better than those based on the Box-Jenkins approach.

4 Conclusions

The evolutionary algorithm with competing heuristics was applied to several problems of the searching for the global minimum in computational statistics. As regards nonlinear regression models, the algorithm was highly reliable but time-consuming in two tasks. The CM-estimates were a bit less reliable. In smoothing constants optimization the algorithm proved to be sufficiently fast and reliable in all runs. The self-adaptation feature of the algorithm with competing heuristics makes the setting of input parameters easier comparing with other evolutionary algorithms. Thus, the algorithm can be used with its default parameter settings in statistical programs.

References

Arslan, O., Edlund, O. and Ekblom, H. (2000). Algorithms for Constrained M-Estimates. In: *COMPSTAT 2000, Short Communications and Posters*, 143–144, Utrecht: Statistics Netherlands.

Bäck, T. (1996). *Evolutionary Algorithms in Theory and Practice.* New York: Oxford University Press.

Cipra, T. (1986). *Time Series Analysis with Applications in Economics* (in Czech). Prague: SNTL Press.

Hintze, J. (2001). *NCSS and PASS, Number Cruncher Statistical Systems.* Kaysville, Utah: WWW.NCSS.COM.

Křivý, I. and Tvrdík, J. (1995). The Controlled Random Search Algorithm in Optimizing Regression Models. *Comput. Statist. and Data Anal.*, **20**, 229–234.

Křivý, I. and Tvrdík, J. (2001). Stochastic optimization in smoothing time series of labour market descriptors. In: *Bulletin of ISI*, Tome LIX, Book 1 - Contributed Papers, 253-254. Seoul: ISI Press.

Kvasnička, V., Pospíchal, J. and Tiňo, P. (2000). *Evolutionary algorithms* (in Slovak). Bratislava: Slovak Technical University Press.

McCullough, B.D., Wilson, B. (1999). On the accuracy of statistical procedures in Microsoft Excel 97, *Comput. Statist. and Data Anal.*, **31**, 27–37.

McCullough, B. D. (2000). The Accuracy of Mathematica 4 as a Statistical Package. *Comp. Statistics*, **15**, 279–299.

Mišík, L. (2000). On Convergence of a Class of Evolutionary Algorithms. In: *Proceedings of MENDEL 2000, 6th International Conference on Soft Computing*, 97–100. Brno: Technical University Press.

Mišík, L., Tvrdík J. and Křivý, I. (2001). On Convergence of a Class of Stochastic Algorithms. In: *Proceedings of ROBUST 2000* (J. Antoch and G. Dohnal eds), 198 - 209, Prague: JČMF Press.

Solis, F. J. and Wets, R. J-B. (1981). Minimization by Random Search Techniques. *Mathematics of Operations Research*, **6**, 19–30.

Rousseeuw, P. J. and Leroy, A. M. (1987). *Robust Regression and Outlier Detection.* New York: John Wiley & Sons.

Storn, R. and Price, K. (1997). Differential Evolution – a Simple and Efficient Heuristic for Global Optimization. *J. Global Optimization*, **11**, 341–359.

Tvrdík, J., Křivý, I. and Mišík, L. (2001). Evolutionary Algorithm with Competing Heuristics. In:*Proceedings of MENDEL 2001, 7-th Int. Conference on Soft Computing*, 58-64. Brno: Technical University Press.

Tvrdík, J., Mišík, L. and Křivý, I. (2002). Competing Heuristics in Evolutionary Algorithms. Accepted to the 2nd Euro-ISCI (Košice, July 16-19, 2002) to appear in *Studies in Computational Intelligence series.* Berlin: Springer-Verlag.

Winters, P. R. (1960). Forecasting Sales by Exponentially Weighted Moving Averages. *Management Science*, **6**, 324–342.

Acknowledgement. This research was supported by the grant 402/00/1165 of the Czech Grant Agency.

Exact Nonparametric Inference in R

Torsten Hothorn[1,*] and Kurt Hornik[2]

[1] Institut für Medizininformatik, Biometrie und Epidemiologie
Friedrich-Alexander Universität Erlangen-Nürnberg,
Waldstraße 6, D-91054 Erlangen, Deutschland

[2] Institut für Statistik, Wirtschaftsuniversität Wien, Augasse 2-6,
A-1090 Wien, Österreich

Abstract. For many of the classical tests, including the Wilcoxon signed rank and rank sum and the Ansari-Bradley tests, exact unconditional distributions of the test statistics can be obtained using recursion formulae provided that the underlying distribution functions are continuous. For every score function special algorithms are needed. Moreover, they are not valid for tied scores. However, the classical tests can be viewed as special cases of permutation tests. We use the shift algorithm introduced by Streitberg & Röhmel (1986) for the computation of the conditional distribution of a permutation test for integer valued scores. Implementation details and generalizations to situations with rational or real scores are given.

Keywords. Permutation Test, Exact Conditional Inference, Confidence Interval

1 Introduction

Let $x = (x_1, \dots, x_m)$ and $y = (y_1, \dots, y_n)$ denote real valued observations (we restrict ourself to the 2-sample problem in the following). It is a common problem to establish a "difference" between the two groups, that is to reject the null hypothesis of "identity" of x and y. In the framework of permutation tests, the null hypothesis is formulated in terms of exchangeability: the joint distribution $F_{x,y}$ of x and y is invariant with respect to all permutations of the observations

$$H_0^P : F_{x,y} = F_{x,y} \circ \sigma, \text{ for all } \sigma \in \text{Symm}(N)$$

where Symm(N) is the group of all permutations σ of $1, \dots, N = n + m$.

The classical nonparametric procedures are designed for testing the usual null hypothesis that the observations are independent and identically distributed according to common continuous distribution functions F_x and F_y:

$$H_0^C : F_x = F_y.$$

* Financial support from Deutsche Forschungsgemeinschaft, grant SFB 539-A4/C1, is gratefully acknowledged.

Let $a = a((x, y))$ denote an N-vector of scores assigned to the vector of observations (x, y). In the framework of permutation tests, the data scores $a = (x, y)$ are used. A sensible test statistic is the sum of the x-values

$$T = \sum_{i=1}^{m} a_i \tag{1}$$

which we will refer to as "permutation statistic" in the following. With respect to the classical procedures, simple linear rank statistics (e.g. Hájek et al., 1999) can be formulated as special cases of the statistic (1). For example, the Wilcoxon rank sum statistic can be computed as

$$W = \sum_{i=1}^{m} a_i,$$

where $a = \text{rank}(x, y)$, i.e., the sum of the mid ranks of the x-sample. For many simple linear rank statistics, the exact unconditional null distributions can be obtained using recursion formulae provided that the underlying distribution functions are continuous. However, for tied observations the exact conditional null distribution is of special interest. The distribution of the permutation statistic (1) for tied observations can be computed using the shift algorithm by Streitberg & Röhmel (1986).

The `R` package `exactRankTests` uses the shift algorithm for the computation of the exact null distribution, conditional on the data, of the permutation statistic as well as simple linear rank statistics. We give a short description of the algorithm and illustrate the use of the package with special emphasize to confidence intervals for the parameters of interest as well as formula based interfaces.

2 Exact Conditional Null Distribution

Streitberg & Röhmel (1986) and Streitberg & Röhmel (1987) introduced the shift algorithm for the determination of the null distribution of T in (1) for possibly tied samples. In this section we review the basic idea and implementation of the shift algorithm and comment on the incorporation of real or rational scores.

2.1 Shift Algorithm

Let $h(m, t)$ denote the number of subsets $\{i_1, \dots, i_m\}$ of $\{1, \dots, N\}$, each subset having m elements and total sum of the corresponding scores of $t = a_{i_1} + \cdots + a_{i_m}$. Consequently,

$$P_{H_0}(T(\sigma a) = t) = \frac{m!n!}{N!} h(m, t).$$

The coefficients can be represented by their generating function

$$H_m^N(x) = \sum_{t=0}^{a_\bullet} h(m, t) x^t,$$

where $a_\bullet = \sum_{i=1}^{N} a_i$ is the total sum of the scores. Streitberg and Röhmel point out that there is no straightforward recursion formula for the determination of the one-dimensional generating function; however, one can simplify the problem by enlarging it. Consider the simultaneous computation of all $H_m^N, m = 0, \ldots, N$. The corresponding, now two-dimensional, generating function

$$H_N(x,y) = \sum_{m=0}^{N} H_m^N(x) y^m$$

can be formulated recursively

$$H_N(x,y) = (1 + x^{a_1}y)(1 + x^{a_2}y) \cdots (1 + x^{a_N}y). \tag{2}$$

The derivation of the recursion is given in Streitberg & Röhmel (1986), where the following algorithm is proposed. The generating function H is represented by a matrix $\mathbf{H}_N = [h(m,t)]$ with $N+1$ rows (numbered $m = 0, \ldots, N$) and $a_\bullet+1$ columns (numbered $t = 0, \ldots, a_\bullet$). This matrix is computed recursively, starting from $\mathbf{H}_0 = [1]$. The recursion formula

$$H_i(x,y) = (1 + x^{a_i}y) H_{i-1}(x,y)$$

follows directly form (2). For the matrix representation, this recursion is implemented by shifts both of the rows and columns:

$$\mathbf{H}_i = \left[\begin{array}{c|c} \mathbf{H}_{i-1} & \begin{array}{ccc} 0 & \ldots & 0 \\ \vdots & \ddots & \vdots \\ 0 & \ldots & 0 \end{array} \\ \hline 0 \;\ldots\; 0 & \underbrace{0 \;\ldots\; 0}_{a_i} \end{array} \right] + \left[\begin{array}{c|c} \begin{array}{ccc} 0 & \ldots & 0 \end{array} & 0 \;\ldots\; 0 \\ \hline \begin{array}{ccc} 0 & \ldots & 0 \\ \vdots & \ddots & \vdots \\ 0 & \ldots & 0 \end{array} & \mathbf{H}_{i-1} \end{array} \right].$$

where the zero column block in the second matrix also has width a_i.

This procedure is iterated for $i = 1, \ldots, N$; resulting in a matrix $\mathbf{H}_N = [h(m,t)]$ with rows $m = 0, \ldots, N$ and columns $t = 0, \ldots, a_\bullet$. Multiplying the mth row of $\mathbf{H}$ with the factor $m!(N-m)!$ leads to a matrix

$$\mathbf{H}^* = [h^*(m,t)] = [h(m,t) \cdot m!(N-m)!] .$$

The probability that the sum of the first m scores (out of N scores) is equal to t is then given by $h^*(m,t)/N!$.

2.2 The Incorporation of Real or Rational Scores

The algorithm of Streitberg and Röhmel is defined for integer valued scores only. To be able to deal with real or rational scores $a_i > 0$, $i = 1, \ldots, N$; we follow a suggestion by Hothorn & Lausen (2001): the cost of the algorithm increases as $a_\bullet$ increases. Therefore, we want to limit $a_\bullet$ and map the scores a_i into $\{1, \ldots, N\}$ by taking the integer part of the scores multiplied by $N/\max(a_i)$. This preserves the shape of the scores and results in a similar test.

3 Confidence Intervals

Bauer (1972) proposed a general method for the construction of confidence intervals and point estimators for special alternatives, e.g. shift in location or scale, using rank statistics. As an example, we treat the computation of confidence intervals for the shift in location by the Wilcoxon rank sum test. Assume a shift alternative, i.e.

$$H_1^C : F_y(z) = F_x(z - \Delta) \quad \forall z \in \mathbb{R}, \Delta \neq 0.$$

Consider the Wilcoxon statistic $W(\delta)$ as a function of $x_1, \dots, x_m, y_1 - \delta, \dots, y_n - \delta$. Bauer (1972), among others: see Hollander & Wolfe (1973), proved that the Wilcoxon statistic only changes in points $\delta_{ij} = y_j - x_i; i = 1, \dots, m; j = 1, \dots, n$. Let w_α denote the α-quantile of the null distribution of the Wilcoxon rank sum statistic. A $(1 - 2\alpha)$ confidence interval for Δ is given by the set $[\delta(w_\alpha), \delta(w_{1-\alpha} + 1))$, where $\delta(i)$ denotes the ith element of the increasing ordered δ_{ij}.

The computation and ranking of δ_{ij} for all $i = 1, \dots, m$ and $j = 1, \dots, n$ is feasible for $\max(m, n) \leq 50$. For large samples we therefore compute asymptotic confidence sets as follows. As long as both expectation and variance of a statistic are known (also for tied samples), we are able to standardize each statistic using

$$\mathrm{E}(W) = \frac{m}{N} \sum_{i=1}^{N} a_i \text{ and } \mathrm{Var}(W) = \frac{mn}{N^2(N-1)} \left\{ N \sum_{i=1}^{N} a_i^2 - \left(\sum_{i=1}^{N} a_i \right)^2 \right\}.$$

The limiting distribution of a standardized simple linear rank statistic is known to be standard normal (see Hájek et al., 1999). Let z_α denote the α quantile of the standard normal distribution. We derive a $(1 - \alpha)$ confidence set (l, u) with l and u being the solutions of the minimization problems

$$l = \mathrm{argmin}_{\delta \in \mathbb{R}} \mid (W(\delta) - \mathrm{E}W(\delta))/\sqrt{\mathrm{Var}W(\delta)} - z_\alpha \mid$$
$$u = \mathrm{argmin}_{\delta \in \mathbb{R}} \mid (W(\delta) - \mathrm{E}W(\delta))/\sqrt{\mathrm{Var}W(\delta)} - z_{1-\alpha} \mid$$

which can efficiently by solved using `uniroot`.

Currently, confidence intervals and point estimates for a shift in location are implemented in `wilcox.test` whereas `ansari.test` offers confidence intervals and point estimates for the ratio of scale parameters.

Confidence intervals which can be derived by inverting permutation tests are given in Röhmel (1996) and are computed in `perm.test`. For continuous data, the confidence intervals can be interpreted as "the largest shift in location ... consistent with the observations" and for binary data they cover the difference in success rates with the prespecified level (see Röhmel, 1996, for the details).

4 Using exactRankTests

exactRankTests implements the functions dperm, pperm and qperm. As it is standard in R/S they give the density, the probability function and the quantile function. We illustrate the use of the package by example data of Diastolic Blood Pressure (see Table 1) from Mehta & Patel (1998). Note that the sample sizes are very small (4 and 11 observations, respectively) and that ties occur.

Control	80 94 85 90 90 90 108 94 78 105 88
Active	94 108 110 90

Table 1. Diastolic Blood Pressure

We first assume that the two vectors Active and Control contain the data. The two sided exact conditional p-value of the permutation statistic (1) can be computed as follows:

```
R> Control <- c(80,94,85,90,90,90,108,94,78,105,88)
R> Active <- c(94, 108, 110,90)
R> pperm(sum(Control), c(Control, Active), length(Control),
         alternative="two.sided")
[1] 0.1040293
```

where the arguments to pperm are the statistic itself, the scores (here: the data scores) and the number of observations in the first sample. We mentioned that simple linear rank statistics are special cases of the permutation statistic (1). Therefore, the two sided exact conditional p-value of the Wilcoxon statistic can be computed by the following lines.

```
R> bpranks <- rank(c(Control, Active))
R> pperm(sum(bpranks[seq(along=Control)]), bpranks, length(Control),
         alternative="two.sided")
[1] 0.0989011
```

Of course, this kind of computation is very low level. Therefore, high level functions with appropriate user interface exist for the most important statistics: the permutation statistic as well as Wilcoxon statistic. Assume that the data is organized in a data frame bpdata with the variable bpres containing the blood pressure values and a factor group at two levels Active and Control. The Wilcoxon test can now be computed using a formula based interface.

```
R> wilcox.exact(bpres ~ group, data=bpdata, conf.int=TRUE)

        Exact Wilcoxon rank sum test

data:  bpres by group
```

```
W = 35, p-value = 0.0989
alternative hypothesis: true mu is not equal to 0
95 percent confidence interval:
 -4 22
sample estimates:
difference in location
                   9.5
```

The function perm.test offers a unified interface to permutation statistics as well as confidence intervals as given by Röhmel (1996). For the blood pressure example, the computations are done by:

```
R> perm.test(bpres ~ group, conf.int=TRUE)

        2-sample Permutation Test

data:  bpres by group
T = 402, p-value = 0.1040
alternative hypothesis: true mu is not equal to 0
95 percent confidence interval:
 -3.00000 20.66667
```

5 Conclusion

The package exactRankTests, distributed under GPL and available under cran.r-project.org, offers exact conditional nonparametric inference of permutation and rank tests including confidence intervals.

References

Bauer, D. F. (1972). Constructing confidence sets using rank statistics. *Journal of the American Statistical Association*, **67**(339), 687–690.

Hájek, J., Šidák, Z., & Sen, P. K. (1999). *Theory of Rank Tests.* London: Academic Press, 2nd edn.

Hollander, M. & Wolfe, D. A. (1973). *Nonparametric statistical inference.* New York: John Wiley & Sons.

Hothorn, T. & Lausen, B. (2001). On the exact distribution of maximally selected rank statistics. *Preprint, Universität Erlangen-Nürnberg, submitted.*

Mehta, C. R. & Patel, N. R. (1998). *StatXact-4 for Windows.* Cytel Software Cooperation, Cambridge, USA.

Röhmel, J. (1996). Precision intervals for estimates of the difference in success rates for binary random variables based on the permutation principle. *Biometrical Journal*, **38**(8), 977–993.

Streitberg, B. & Röhmel, J. (1986). Exact distributions for permutations and rank tests: An introduction to some recently published algorithms. *Statistical Software Newsletters*, **12**(1), 10–17.

Streitberg, B. & Röhmel, J. (1987). Exakte Verteilungen für Rang- und Randomisierungstests im allgemeinen *c*-Stichprobenfall. *EDV in Medizin und Biologie*, **18**(1), 12–19.

Exploring the Structure of Regression Surfaces by using SiZer Map for Additive Models

Raya Miranda, R.[1], Martínez Miranda, M.D.[1] and González Carmona, A.[1]

[1] Department of Statistics and O. R., University of Granada, Spain

Abstract. In this work we study the structure of regression surfaces for additive models by using the graphic tool SiZer Map. This kind of graph was introduced by Chaudhuri and Marron (1999) for density and regression estimation. To estimate the regression model we use the backfitting algorithm, Buja *et al.* (1989), with local linear smoothers, Opsomer and Ruppert (1997), implemented with binning methods, Fan and Marron (1994).

Keywords. Additive model, backfitting, SiZer Map, bandwidth, binning

1 Introduction

SiZer Map is a useful graphic tool for exploring features in data sets used to estimate the regression model. The graph uses different colours to show where the curve significantly increases and decreases, with different levels of smoothing, working with confidence bands for the derivatives. It could also be used to show the behaviour of the estimates on a wide range of values for the smoothing parameter and we could decide which is the optimum amount of smoothing from the data.

This graphic method for exploring features of data has only been developed for one dimensional smoothing regression. An important problem for extending SiZer to higher dimensions is how to do the map. The map would have shaded regions in three dimensions, which is rather difficult to visualize. Although, when the multidimensional model is considered as an additive model, this problem disappears. Additive models for fitting multidimensional regression surfaces permit to consider each variable separately, so it is possible to extend SiZer Map to additive models easily. We will have the same number of SiZer Map as the number of available variables.

We have carried out simulation experiments with several different bivariate regression models. We have made a computational implementation of this graphic method with Matlab 5.10.421.

2 Estimating by local linear regression

Consider the following problem of bivariate regression: Let $((X_1, Z_1, Y_1), \ldots, (X_n, Z_n, Y_n))$ be a set of independent observations of a tridimensional variable, (X, Z, Y), which satisfaces the model

$$Y_i = m(X_i, Z_i) + \varepsilon_i,$$

where m is the unknown regression function, that is,

$$m(x_0, z_0) = \mathrm{E}\left[Y | (X, Z) = (x_0, z_0)\right],$$

and the residuals, ε_i, are independent random variables with mean 0 and variance σ^2.

The additive model suposes that the regression function can be written as the sum of smooth univariate functions in the covariates, X and Z, i. e.,

$$m(x_0, z_0) = \alpha + m_1(x_0) + m_2(z_0)$$

for each point (x_0, z_0). To ensure identifiability of the model we include the intercept α, and assume that $\mathrm{E}[m_1(X_i)] = \mathrm{E}[m_2(Z_i)] = 0$.

There are several methods to estimate the additive components. The most important are the backfitting algorithm and the marginal integration method proposed by Linton and Nielsen (1995) for the bivariate additive model and generalized to the multi-dimensional setting by Hengartner (1996) and the backfitting-projection algorithm of Linton *et al.* (1997). In Opsomer and Ruppert (1997) we can find the solution of the backfitting algorithm with local linear smoothers in each variable for a bivariate additive model and it has been extended for a backfitting algorithm with local polynomial regression smoothers for a D-dimensional ($D > 2$) additive model in Opsomer (2000). This estimation requires the selection of one or more univariate smoothing methods.

Now, we give some notation. Let $\mathbf{y} = (Y_1, \ldots, Y_n)^T$, $\mathbf{x} = (X_1, \ldots, X_n)^T$ and $\mathbf{z} = (Z_1, \ldots, Z_n)^T$. The vectors $\mathbf{m_1} = (m_1(X_1), \ldots, m_1(X_n))^T$ and $\mathbf{m_2} = (m_2(Z_1), \ldots, m_2(Z_n))^T$. Let $s_{1,x}$, $s_{2,z}$ be the equivalent kernels for the local polinomial regression in x and z for each component. The expression for the first component is $s_{1,x} = e_1^T (X_x^T W_x X_x)^{-1} X_x^T W_x$, where $e_1 = (1,0)^T$,

$$W_x = \mathrm{diag}\left\{h_1^{-1} K\left((X_1 - x)/h_1\right), \ldots, h_1^{-1} K\left((X_n - x)/h_1\right)\right\},$$

for some kernel function K and bandwidth parameter h_1 and

$$X_x = \begin{bmatrix} 1 & (X_1 - x) \\ \vdots & \vdots \\ 1 & (X_n - x) \end{bmatrix}.$$

The expression for $s_{2,z}$ will be similar.

Let S_1 and S_2 represent the matrices whose rows are the smoothers at $\mathbf{X}$ and $\mathbf{Z}$

$$S_1 = \begin{pmatrix} s_{1,X_1}^T \\ \vdots \\ s_{1,X_n}^T \end{pmatrix}, \qquad S_2 = \begin{pmatrix} s_{2,Z_1}^T \\ \vdots \\ s_{2,Z_n}^T \end{pmatrix}.$$

Then, the vector of estimations is defined as, $\widehat{\mathbf{m}} = \widehat{\alpha} + \widehat{\mathbf{m}}_1 + \widehat{\mathbf{m}}_2$ and the component vectors, $\widehat{\mathbf{m}}_1 = (\widehat{m}_1(X_1), \ldots, \widehat{m}_1(X_n))$, $\widehat{\mathbf{m}}_2 = (\widehat{m}_2(Z_1), \ldots, \widehat{m}_2(Z_n))$, are the solutions to the set of estimating equations given by

$$\begin{bmatrix} I & S_1^* \\ S_2^* & I \end{bmatrix} \begin{bmatrix} \widehat{\mathbf{m}}_1 \\ \widehat{\mathbf{m}}_2 \end{bmatrix} = \begin{bmatrix} S_1^* \\ S_2^* \end{bmatrix} \mathbf{y},$$

where $S_1^* = (I - \mathbf{1}\mathbf{1}^T/n)S_1$, $S_2^* = (I - \mathbf{1}\mathbf{1}^T/n)S_2$. The backfitting algorithm converges to the solution:

$$\widehat{\mathbf{m}}_1 = \left\{I - (I - S_1^* S_2^*)^{-1}(I - S_1^*)\right\} \mathbf{y}$$

$$\widehat{\mathbf{m}}_2 = \left\{I - (I - S_2^* S_1^*)^{-1}(I - S_2^*)\right\} \mathbf{y}, \tag{1}$$

provided the inverse exist, with $\widehat{\alpha} = \overline{\mathbf{y}}$; and then the explicit estimator expression is:

$$\widehat{\mathbf{m}} = \left\{\mathbf{1}\mathbf{1}^T/n + 2I - (I - S_1^* S_2^*)^{-1}(I - S_1^*) - (I - S_2^* S_1^*)^{-1}(I - S_2^*))\right\} \mathbf{y}.$$

So the estimation of the regression function in any point will be $\widehat{m}(x_0, z_0) = \widehat{\alpha} + \widehat{m}_1(x_0) + \widehat{m}_2(z_0)$, where

$$\widehat{m}_1(x_0) = s_{1,x_0}(\mathbf{y} - \widehat{\mathbf{m}}_2), \quad \widehat{m}_2(z_0) = s_{2,z_0}(\mathbf{y} - \widehat{\mathbf{m}}_1).$$

We define now $\mathbf{y_1} = \mathbf{y} - \mathbf{m_2}$. Then, a fast binned implementation of the local linear smoother to estimate $\widehat{m}_1(x_0)$, Fan and Marron (1994), applied to $\mathbf{y_1}$ instead of $\mathbf{y}$, will be

$$\widehat{m}_1(x_0) = \frac{\overline{S}_2(x_0)\overline{T}_0(x_0) - \overline{S}_1(x_0)\overline{T}_1(x_0)}{\overline{S}_2(x_0)\overline{S}_0(x_0) - \overline{S}_1(x_0)^2}$$

where

$$\overline{S}_l(x_j) = \sum_{j'=1}^{g} k_{1,j-j'} c_{j'} x_{j'}^l, \quad \overline{T}_l(x_j) = \sum_{j'=1}^{g} k_{1,j-j'} y_{1,j'}^{\Sigma} x_{j'}^l,$$

$$k_{1,j-j'} = K_{h_1}\left(x_j - x_j'\right).$$

being $y_{1,j'}^{\Sigma}$, for the bins sums of the $\mathbf{y_1}$. Analogously for $\widehat{m}_2(x_0)$.

3 SiZer Map for additive models

SiZer Map will permit to verify the signification of features in the surface, such as peaks and valleys in a family of smoothers $\{\widehat{m}(x, z; h_1, h_2) : (h_1, h_2) \in \mathbf{H}\}$, based in confident bands for the derivative, $D\widehat{m}(x, z)$, in the scale space defined by $\mathbf{H}$.

In the univariate case, in the scale space defined by the variable, x, and the bandwidth, h, the SiZer color map will be black (blue in colours version) to indicate locations where the derivative is significantly positive, white (red) where the derivative is significantly negative, and gray (purple) where the derivative is not significantly different from 0. A darker shade of gray is used in areas where the data is too sparse for SiZer to be effective.

The confident limits for the derivative are expressed in this case:

$$\widehat{m}'(x) \pm q\widehat{SD}(\widehat{m}'(x)),$$

where the quantile, q, is calculated from normal approachs, or using bootstrap methods (more details about this topic can be found in the paper of Chaudhuri and Marron (1999)).

3.1 Development of SiZer Map with binning methods

Now, we develop SiZer Map for the first component, m_1; since for m_2 is similar and we have omitted it.

Consider the estimation of the first derivative of m_1 at the point, x_0,

$$\widehat{m}_1'(x_0) = s_{1,x_0}^{(1)}(\mathbf{y} - \widehat{\mathbf{m}}_2),$$

being $s_{1,x_0}^{(1)} = e_2^T(X_{x_0}^T W_{x_0} X_{x_0})^{-1} X_{x_0}^T W_{x_0}$, with $e_2 = (0,1)^T$. Let $\mathbf{y_1} = \mathbf{y} - \widehat{\mathbf{m}}_2$.

An efficient binned approximation of the local linear derivative estimate $\widehat{m}_1'(x_0)$ is

$$\widehat{m}_1'(x_0) \simeq \frac{\overline{T}_1(x_0) - \overline{T}_0(x_0)\overline{X}(x_0)}{\overline{S}_2(x_0) - \overline{S}_1(x_0)^2/\overline{S}_0(x_0)}.$$

and

$$\overline{X}(x_j) = \overline{S}_1(x_j)/\overline{S}_0(x_j).$$

Since the derivative estimator is a weighted sum of the $\mathbf{y_1}$, we can use

$$\operatorname{Var}(\widehat{m}_1'(x_0)|X,Z) = \operatorname{Var}\left(n^{-1}\sum_{i=1}^{n} W_1(x_0,X_i)\,\mathbf{y_1}|X,Z\right)$$

$$= \sum_{i=1}^{n} \sigma^2(\mathbf{y_1}|X,Z)(W_1(x_0,X_i))^2.$$

A binned approximation to $\sigma^2(\mathbf{y_1}|X,Z)$, based on calculations familiar from simple linear regression, is

$$\sigma^2(\mathbf{y_1}|(X,Z) = (x_j,z_j)) \approx \left(1 - \widehat{\rho}(x_j)^2\right)\widehat{\sigma}(x_j)^2,$$

where

$$\widehat{\sigma}(x_j)^2 = \frac{\overline{U}_0(x_j)}{\overline{S}_0(x_j)} - \left(\frac{\overline{T}_0(x_j)}{\overline{S}_0(x_j)}\right)^2$$

and

$$\widehat{\rho}(x_j)^2 - (\widehat{m}_1'(x_j))\left(\frac{\overline{S}_0(x_j)\overline{S}_2(x_j) - \overline{S}_1(x_j)^2}{\widehat{\sigma}(x_j)^2\,\overline{S}_0(x_j)}\right)$$

using the notation below and $\overline{U}_0(x_j) = \sum_{j'=1}^{g} k_{j-j'} y_{1,j'}^{2\Sigma}$ for $y_{1,j'}^{2\Sigma}$ denoting the bin sums of the $y_{1,j}^2$. Our binned approximation to the conditional variance is now

$$\operatorname{Var}(\widehat{m}_1'(x_j)|X_1,\ldots,X_n) \approx \frac{\overline{V}_2(x_j) - 2\overline{V}_1(x_j)\overline{X}(x_j) + \overline{V}_0(x_j)\overline{X}(x_j)^2}{\left(\overline{S}_2(x_0) - \overline{S}_1(x_0)^2/\overline{S}_0(x_0)\right)^2}$$

where $\overline{V}_l(x_j) = \sum_{j'=1}^{g} k_{j-j'} c_{j'} x_{j'}^l \widehat{\sigma}^2(\mathbf{y_1}|(X,Z) = (x_{j'},z_{j'}))$.

Then, the intervals represented in SiZer Map are built for the first component, as:

$$\widehat{m}_1'(x) \pm q\sqrt{\widehat{\operatorname{Var}}(\widehat{m}_1'(x))}.$$

According to the statement previously made, to calculate the quantile, q, it is possible several approximations. Chaudhuri and Marron (1999) suggested three estimates, one of them is based on bootstrap methods that add a big computational cost. In our job we use two gaussian approximations for these quantile.

The simplest approximation, that we denote by q_1 is $q_1 = \Phi^{-1}\left(1-\frac{\alpha}{2}\right)$. The second approximation, q_2, is based in the normal approximation, but it is defined as a variable amount in the bandwidth network considered, $q_2 = q_2(h_1)$. In its definition takes part an estimated effective sample size, ESS, given by:

$$\text{ESS}(x, h_1) = \frac{\sum_{i=1}^{n} K_{h_1}(x - X_i)}{K_{h_1}(0)}$$

so then,

$$q_2(h_1) = \Phi^{-1}\left(1 + (1-\alpha)^{\frac{1}{m(h_1)}}\right),$$

where $m(h_1) = \frac{n}{L^{-1}\sum_{l=1}^{L} \text{ESS}(x_l, h_1)}$, and the x_l, $(1 \leq l \leq L)$, are the value network of X that define the scale space together with the bandwidths.

4 Simulations

The aim of this job was, as well as ilustrating the working of SiZer Map for the additive models, to permit a valuation of the plug-in bandwidth selector proposed by Opsomer and Ruppert (1998) for additive models.

The functions that define the model are:

$$m_1(x) = 1 - 6x + 36x^2 - 53x^3 + 22x^5$$

$$m_2(z) = \text{sen}(5\pi z).$$

The residuals, ε, have been generated according to a $N(0,1)$, and the vector of covariables, (X, Z), according to a bivariate normal distribution,

$$(X, Z) \rightarrow N\left((0.5, 0.5)^T, \begin{pmatrix} 1/9 & \rho \\ \rho & 1/9 \end{pmatrix}\right),$$

for values of ρ: 0, -0.25, -0.5, 0.25 y 0.5. And two sizes of sample, $n = 200$ and 500.

For each function were made two kind of graphics, the map and a graphic which represent the estimations of the components for different values of the bandwidth. The Figure 1 shows these graphics for the case $\rho = 0$ and $n = 200$.

References

Buja, A., Hastie, T. J. and Tibshirani, R. (1989). Linear Smoothers and Additive Models (with discussion). *The Annals of Statistics*, **17 (2)**, 453-555.

Chaudhuri, P. and Marron, J. S. (1999) SiZer for Exploration of Structures in Curves. *Journal of the American Statistical Association*, **94 (447)**, *Theory and Methods*, 807-823.

Fan, J. and Marron, J. S. (1994) Fast Implementations of Nonparametric Curve Estimators. *Journal of Computational and Graphical Statistics*, **3**, 35-56.

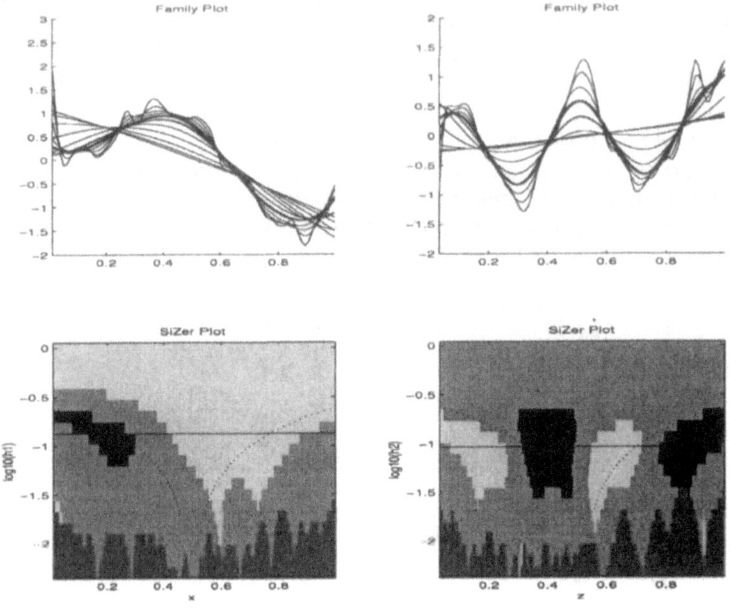

Fig. 1. In the two first family plots, the single lines show different regression estimates for the additive component and the thick line is the optimal bandwidth using the plug-in selection method given by Opsomer and Ruppert (1998). Below, the SiZer Map associated to each component; the thick line represents the optimal bandwidth of Opsomer and Ruppert.

Hastie, T. J. and Tibshirani, R. (1990) Generalized Additive Models. Chapman & Hall.

Hengartner, N. W. (1996) Rate Optimal Estimation of Additive Regression via the Integration Method in the Presence of Many Covariates. Electronic article, http.//www.stats.yale.edu.

Linton, O. B., Mammen, E. and Nielsen, J. P. (1999) The Existence and Asymptotic Properties Algorithm Under Weak Conditions. *The Annals of Statistics*, **27** **(5)**, 1443-1490.

Linton, O. B. and Nielsen, J. P. (1995) A Kernel Method of Estimating Structured Nonparametric Regression Based on Marginal Integration. *Biometrika*, **82**, 93-100.

Opsomer, J. D. (2000) Asymptotic Properties of Backfitting Estimators. *Jounal of Multivariate Analysis*, **73**, 166-179.

Opsomer, J. D. and Ruppert, D. (1997) Fitting a Bivariate Additive Model by Local Polynomial Regression. *The Annals of Statistics,* **25** **(1)**, 186-211.

Opsomer, J. D. and Ruppert, D. (1998) A Fully Automated Bandwidth Selection Method for Fitting Additive Models. *Journal of the American Statistical Association,* **93** **(442)**, *Theory and Methods*, 605-619.

Fast and Robust Filtering of Time Series with Trends

Roland Fried[1] and Ursula Gather[1]

[1] Department of Statistics, University of Dortmund, 44221 Dortmund, Germany

Abstract. Fast and robust methods are needed for denoising time series data measured with high sampling frequencies. In intensive care e.g. physiological variables like the heart rate are observed in short time intervals. Systematic changes have to be detected quickly and distinguished from clinically irrelevant short term fluctuations and artifacts. Median filtering works well if there is no substantial trend in the data but improvements are possible by approximating the data by a local linear trend.

Keywords. Signal extraction, linear regression, level shifts, outliers

1 Introduction

Tukey (1977) suggests median filtering as a simple approach to smooth noisy time series, clean the data from artifacts and extract the current level. Atypical observations (outliers) and sudden changes (level shifts) in the data can be identified in this way. However, median filtering does not consider trend periods although in view of high sampling frequencies most changes occur gradually. Fitting a local linear trend instead of a local level may improve the approximation and provide information on trends.

In a recent study, Davies, Fried and Gather (2002) compare the properties of L_1 regression, the repeated median and the least median of squares (LMS) for fitting a local linear trend to a time series. They find the LMS to resist a large percentage of outliers without showing a significant bias, while the repeated median performs somewhat better than L_1 regression. However, calculation of the LMS is time consuming as its computational complexity for fitting a line is $O(n^4)$ (Stromberg, 1993). Therefore, we cannot apply the LMS to time intervals. Instead we can use a faster method like the repeated median (complexity $O(n^2)$), or we can preprocess the data using e.g. the median (complexity $O(n)$) or the repeated median and apply the LMS to the more sparsely sampled outcome of this preprocessing.

In order to develop a procedure for non-parametric filtering of a time series with trends we compare the performance of methods for preprocessing. Thereafter we apply combinations of the LMS or the repeated median with such a preprocessing procedure to real and simulated time series and compare the results to the outcome of a "long" repeated median.

2 Preprocessing

Let $y_1, \ldots, y_N$, $N = n_1 n_2$, be a sequence of measurements, $n_1 = 2m + 1$. For preprocessing the data we first approximate the level μ_{jn_1+m+1} at the center of each time interval $jn_1 + 1, \ldots, (j+1)n_1$, $j = 0, \ldots, n_2 - 1$. Thereafter we can apply a method for robust linear regression to the approximates

$\tilde{\mu}_{jn_1+m+1}, j = 0, \ldots, n_2 - 1$. For ease of notation we restrict the discussion to the first time interval centered about $m+1$ in the following.

2.1 Methods for preprocessing

A simple possibility for approximation of μ_{m+1} is to use the median $\tilde{\mu}_{Me}$ of $y_1, \ldots, y_{n_1}$. Alternatively, one can take the slope within the current time window into account and apply the repeated median T_{RM} (Siegel, 1982)

$$\tilde{\mu}_{RM} = med_i(y_i - \tilde{\beta}_{RM}(i-m-1)),$$
$$\tilde{\beta}_{RM} = med_i\left(med_{j\neq i}\frac{y_i - y_j}{i-j}\right)$$

or L_1 regression T_{L1} defined by

$$(\tilde{\mu}_{L1}, \tilde{\beta}_{L1}) = \text{argmin}\left\{(a,b) : \sum_{i=1}^{n_1} |y_i - a - b(i-m-1)|\right\}. \tag{1}$$

Davies et al. (2002) find the small-sample performance of T_{RM} and T_{L1} to be rather similar, although T_{RM} has the larger breakdown point, that is $\lfloor n/2 \rfloor / n \approx 0.5$. The breakdown point of T_{L1} for design points on a lattice as in the situation considered here is approximately $1 - 1/\sqrt{2} \approx 0.293$.

Fast computation of T_{L1} is possible using a descent technique (Sposito, 1990). An one-step or a two-step approximation starting from the median for the level and zero slope might be interesting because of the robustness of the starting point and its computational speed. In the iterations, we alternate between calculation of the generalized median of the scaled deviations from the current level with weights $|j - m + 1|$, and calculation of the median of the trend-corrected observations. However, we find the breakdown point of the k-step estimator, $k \geq 2$, to be the same as the breakdown point of T_{L1}. Therefore we do not consider such k-step approximations in the following.

2.2 Efficiency

For a comparison of the small-sample properties of the distinct methods we simulate data from the model

$$Y_t = \mu + \beta(t - m - 1) + \epsilon_t, \quad t = 1, \ldots, 2m+1,$$

with $\mu = 0$ and several slopes β. The errors ϵ_t are always Gaussian white noise with mean zero and unit variance. The value of m is determined by requiring on the one hand a certain stability (m large) and on the other hand the demands made by the on-line nature of the application (m small). We consider small to moderate $m = 5, 10, 15, 30$ corresponding to $n_1 = 11, 21, 31, 61$.

As a first step we derive the relative efficiencies of $\tilde{\mu}_{Me}$, $\tilde{\mu}_{L1}$ and $\tilde{\mu}_{RM}$ with respect to the least squares estimator $\tilde{\mu}_{L2}$. Since for high-frequency measurements the standard deviation of the measurement error will typically be large in comparison to realistic slope sizes, we consider $\beta = 0, 0.01, 0.05, 0.1$ (Table 1). For the level, $\tilde{\mu}_{L1}$ is always more efficient than $\tilde{\mu}_{Me}$, which is slightly more efficient than $\tilde{\mu}_{RM}$ if $\beta = 0$. The relative efficiency of $\tilde{\mu}_{Me}$ decreases with increasing $|\beta|$ and increasing sample size. For $\beta = 0.1$ fixed, increasing the sample size further than $n_1 = 31$ does almost no longer decrease the MSE of $\tilde{\mu}_{Me}$. In general, the median is not even consistent if we extend the sample range, but it is consistent if we consider in-fill asymptotics corresponding to increasing the sampling frequency.

Table 1. Efficiencies relative to L_2 regression (in percent) measured by the simulated MSE for $\tilde{\mu}_{Me}$, $\tilde{\mu}_{L1}$ and $\tilde{\mu}_{RM}$ for $N(0,1)$ errors.

m	β	$\tilde{\mu}_{Me}$	$\tilde{\mu}_{L1}$	$\tilde{\mu}_{RM}$	m	β	$\tilde{\mu}_{Me}$	$\tilde{\mu}_{L1}$	$\tilde{\mu}_{RM}$
5	0.00	66.0	69.2	65.0	15	0.00	64.8	66.7	64.3
5	0.01	66.5	69.4	65.3	15	0.01	64.2	66.3	63.9
5	0.05	65.5	69.6	65.6	15	0.05	60.1	66.6	63.7
5	0.10	63.8	69.2	65.2	15	0.10	49.1	67.0	63.7
10	0.00	64.6	66.8	63.9	30	0.00	64.5	65.5	64.1
10	0.01	64.4	66.7	63.9	30	0.01	63.3	65.6	63.8
10	0.05	63.1	67.7	64.1	30	0.05	49.0	65.4	63.4
10	0.10	57.3	68.2	64.2	30	0.10	29.4	65.9	64.2

2.3 Computation times

The computation time needed is important if many variables are to be monitored simultaneously or other algorithms are run concurrently. Table 2 shows the mean times of applying the functionals to 1000 samples of sizes 11 to 31 using a self-written FORTRAN program on a Sun workstation ultra spark with 170 MHz and 320 MB Ram. While $\tilde{\mu}_{Me}$ needs considerably more time than the arithmetic mean $\tilde{\mu}_{AM}$, it is much cheaper than $\tilde{\mu}_{L1}$ and $\tilde{\mu}_{RM}$.

The computation time needed for the whole procedure can be calculated from these values as follows: If we apply method A in the first step to n_1 observations and then apply method B to n_2 subsequent outcomes of the first step the computation time will be approximately (with obvious notation) $n_2C_{n_1}^A + C_{n_2}^B$ for the first n_1n_2 observations but only $C_{n_1}^A + C_{n_2}^B$ thereafter.

Table 2. Mean time C_n^A of applying method A to 1000 samples of size n (in seconds).

	C_n^{AM}	C_n^{Me}	C_n^{L1}	C_n^{RM}
$n = 11$	0.0075	0.042	0.55	0.89
$n = 21$	0.0096	0.047	2.41	2.63
$n = 31$	0.0120	0.177	4.72	4.38

2.4 Outliers

In applications like intensive care many outliers occur. For the sake of brevity we restrict attention to a sample of size $n_1 = 21$ and replace an increasing number of observations 0(1)10 by additive outliers of increasing size 0(2)10 at random time points in the window. We concentrate on one-sided positive outliers as these constitute a more difficult challenge than outliers with random sign. Each of the 121 cases is simulated 2000 times and the squared bias, variance and mean square error are calculated. In view of the results obtained above and in Davies et al. (2002) we only consider $\tilde{\mu}_{Me}$ and $\tilde{\mu}_{RM}$ for preprocessing further on, cf. Figure 1. Both methods are able to resist up to six outliers. For more outliers, the comparison depends on the slope: $\tilde{\mu}_{Me}$ resists outliers better in case of a small slope ($\beta = 0$), while $\tilde{\mu}_{RM}$ is better in case of a rather large slope ($\beta = 0.2$), the differences being due to bias.

Fig. 1. Steady state ($\beta = 0$, left) and trend ($\beta = 0.2$, right): Simulated squared bias (top) and MSE (bottom) for the level. Repeated median o and median Δ.

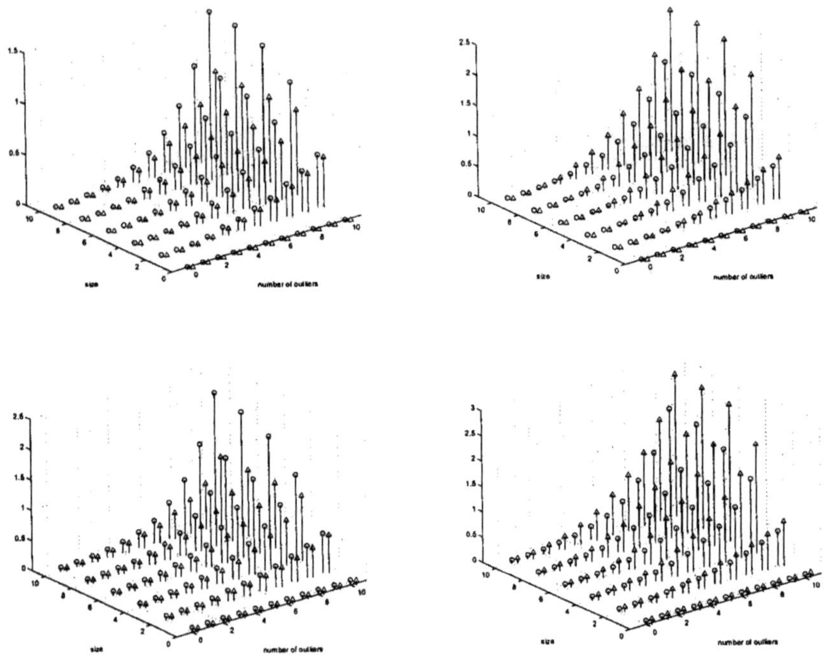

3 Combination of the procedures

Figure 2 shows a simulated time series with 5% of the observations replaced by isolated and 10% by patchy outliers (i.e. 15% outliers altogether) as well as some approximates of the level and the slope. The choice of the outliers is guided by our experience with real physiologic time series measured every second. We used $n_1 = 11$ observations for preprocessing and $n_2 = 31$ in the second step. This former is motivated by the previous findings that we should choose n_1 small when using the median, while the latter is due to the findings of Davies et al. (2002) who consider 31 observations to be sufficient to get a reliable slope approximate. The two-step approximate combining the repeated median and the median performs at least as good as the repeated median applied to the full time interval with $n_1 n_2$ observations and it is computationally much less expensive. This is also true w.r.t. the slope approximates not shown here. T_{RM} as well as the combination methods provide good approximates in the steady state as well as in the trend period.

Sudden changes in the slope or the level can better be detected using T_{LMS} instead of T_{RM}. However, T_{LMS} shows instabilities w.r.t. certain monotone (random) patterns in the data.

Similar results are obtained when applying the methods to the heart rate of a patient observed every second, cf. Figure 3. T_{LMS} bridges the decrease at the beginning but it still signals a trend when the level has already stabilized. T_{RM} and T_{RM} in combination with $\tilde{\mu}_{RM}$ perform almost identical, while T_{RM} in combination with $\tilde{\mu}_{AM}$ is strongly influenced by some negative outliers.

Fig. 2. Simulated time series (dotted), underlying level (bold solid) and level estimates: T_{RM} with $\tilde{\mu}_{Me}$ (dashed-dotted), T_{LMS} with $\tilde{\mu}_{Me}$ (dashed), T_{RM} (solid).

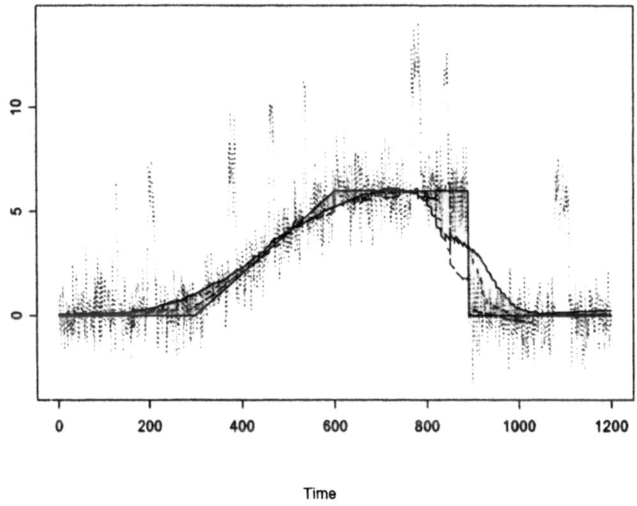

4 Conclusion

The suitable choice of a filtering procedure for fitting of local linear trends strongly depends on the application. The suitable length of the time windows should be guided by the expected length of outlier patches and the minimal length of relevant structural changes. If many variables are to be monitored simultaneously, application of T_{LMS} or T_{RM} to a very long time window may not be possible. E.g., in our simulation example calculation of the T_{RM} from the long time window takes more than 200 times longer than using a combination of two repeated medians. In theory, using such a combination reduces the breakdown point as it is asymptotically 1/4 instead of 1/2 for each of the approximates. However, in practice more outliers can be resisted as outliers will often occur in long patches if the sampling frequency is high. If the

Fig. 3. Heart rate (dotted) and level estimates: T_{RM} with $\tilde{\mu}_{Me}$ (dashed-dotted), T_{RM} with $\tilde{\mu}_{AM}$ (dashed-dotted-dotted), T_{LMS} with $\tilde{\mu}_{Me}$ (dashed), T_{RM} (solid). T_{RM} with $\tilde{\mu}_{Me}$ and T_{RM} are almost identical.

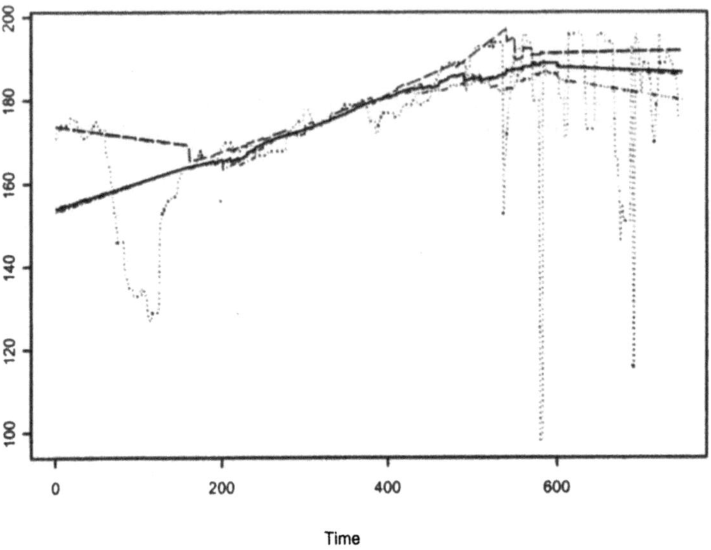

sampling frequency is high in comparison to realistic slope sizes, application of $\tilde{\mu}_{Me}$ with a small window width seems sufficient for preprocessing as it is rather efficient, fast and can withstand several outliers. Since its relative performance in comparison to $\tilde{\mu}_{RM}$ worsens with increasing window width and increasing slope sizes, $\tilde{\mu}_{RM}$ may sometimes be a better choice.

References

Davies, P.L., Fried, R. and Gather, U. (2002). Robust signal extraction for on-line monitoring data. Technical Report 1/2002, SFB 475, University of Dortmund, Germany.

Siegel, A.F. (1982). Robust regression using repeated medians. *Biometrika*, **68**, 242–244.

Sposito, V.A. (1990). Some properties of L_p-Estimators. In: K.D. Lawrence and J.L. Arthur (eds.). *Robust Regression. Analysis and Applications*, 23–58. New York: Marcel Dekker.

Stromberg, A.J. (1993). Computing the exact least median of squares estimate and stability diagnostics in multiple linear regression. *SIAM J. Sci. Comput.*, **14**, 1289–1299.

Tukey, J.W. (1977). *Exploratory Data Analysis*. Reading, Mass.: Addison-Wesley.

Functional Principal Component Modelling of the Intensity of a Doubly Stochastic Poisson Process

Ana M. Aguilera, Paula R. Bouzas[1] and Nuria Ruiz-Fuentes[2]

[1] Department of Statistics and Operation Research, University of Granada, 18071-Granada, SPAIN
[2] Department of Statistics and Operation Research, University of Jaén, 23071-Jaén, SPAIN

Abstract. An estimator for the intensity process of a doubly stochastic Poisson process is presented, having no statistical previous knowledge of it. In order to give a statistical structure of the intensity, a functional Principal Components Analysis is applied to k estimated sample paths of the intensity built from k observed sample paths of the point process.

Keywords. Doubly stochastic Poisson process, functional principal component analysis

1 Introduction

Estimation problems for observed doubly stochastic Poisson processes are of great theoretical and practical importance because they model wide range of point (or counting) phenomena. Many examples can be found in several fields of science as engineering and communications systems (Ochi, 1990), information theory (Boel and Beneš, 1980), engineering mechanics (Di Paola and Falsone, 1999, Grigoriu, 2000), reliability (Landers *et al.*, 2001), economics (see Bouzas *et al.* (2001)) and more examples of different kind are explained in Snyder and Miller (1991), Grigoriu (1995) and Last and Brandt (1995), among others.

As it is known, estimates of the intensity process appear as a natural component in expressions for the counting and time statistics. Estimating the intensity when it is known its first and second moments is complicated but widely studied (see Snyder and Miller (1991)).

This paper presents an estimator for the intensity process of a doubly stochastic Poisson process (DSPP) when k long enough trajectories of the process are observed but no other statistical knowledge of the intensity. Having observed the k sample paths of the DSPP at discrete times, other k shorter discrete sample paths of the intensity are estimated by means of the moving average estimator of each original trajectory. Then, Functional Principal Component Analysis (FPCA) is applied to the estimated sample paths of the intensity, providing an estimated orthogonal expansion of the intensity process in terms of uncorrelated random variables. Some properties of the point estimator are also pointed out.

An application of the estimation method to a simulated DSPP is given in Section 4.

---

[1] E-mail: paula@ugr.es
This research was partially supported by Dirección General de Investigación, Ministerio de Ciencia y Tecnología, under Project BMF 2000-1466.

2 Estimation of the intensity of a DSPP

The problem is described as follows. Let $\{N(t) : t \geq t_0\}$ be a DSPP with intensity $\{\lambda(t, x(t)) : t \geq t_0\}$ where $\{x(t) : t \geq t_0\}$ is the information process. It is, $\{N(t) : t \geq t_0\}$ is a Poisson process with intensity $\{\lambda(t, x(t)) : t \geq t_0\}$ conditioned to the information process. We will just suppose that $\lambda(t, x(t)) \equiv \lambda(t)$ is a second order and continuous in quadratic mean stochastic process whose sample paths are square integrable functions.

Our purpose is to estimate $\lambda(t, x(t))$, having no previous knowledge about its structure, from k independent trajectories of $N(t)$ observed on the interval $[t_0, t_0 + rT)$, and denoted by

$$\{N_\omega(t) : \omega = 1, \ldots, k\}.$$

In order to estimate a sample path for $\lambda(t)$ from each N_ω $(\omega = 1, \ldots, k)$, we will first cut each one into pieces to be able to use an estimator of the intensity. Choosing a partition in $[t_0, t_0 + rT)$, defined by the nodes

$$t_0 = T_0 < T_1 < \ldots < T_i < \ldots < T_r = t_0 + rT,$$

with $T_i - T_{i-1} = T$ for all $i = 1, \ldots, r$, we can cut the sample path N_ω into r shorter sample paths in $[T_{i-1}, T_i)$, denoted by $N_{\omega,i}(t)$. They are chosen of equal length to take advantage of the whole information of the sample path N_ω in next step. These new sample paths are independent due to the independence of the increments of a DSPP, so they can be seen as observed in a same interval $[t_0, t_0 + T)$.

We will define these r sample paths as

$$N_{\omega,i}(t) = N_\omega(t + (i-1)T) \quad t \in [t_0, t_0 + T)$$

with $\omega = 1, \ldots, k$ and $i = 1, \ldots, r$.

Next, for every $\omega = 1, \ldots, k$, we will define a point estimator of the intensity using the r new sample paths $\{N_{\omega,i} : i = 1, \ldots, r\}$. First of all, let us choose a partition in $[t_0, t_0 + T)$, defined by the nodes

$$t_0 = t_{\omega,0} < t_{\omega,1} < \ldots < t_{\omega,j} < \ldots < t_{\omega,p_\omega} = t_0 + T,$$

that could be different for every initial N_ω.

Let us propose

$$\hat{\lambda}(t_{\omega,j}) = \frac{1}{r} \sum_{i=1}^{r} \frac{N_{\omega,i}(t_{\omega,j}) - N_{\omega,i}(t_{\omega,j-1})}{t_{\omega,j} - t_{\omega,j-1}} = \frac{1}{r} \sum_{i=1}^{r} \frac{N_{\omega,i}(t_{\omega,j-1}, t_{\omega,j})}{t_{\omega,j} - t_{\omega,j-1}}$$

as the estimated value of the intensity in $t_{\omega,j}$, $j = 1, \ldots, p_\omega$.

Next, we study the main properties of this estimator. Let us notice that

$$E[\hat{\lambda}(t_{\omega,j})] = \frac{1}{r} \sum_{i=1}^{r} \frac{E[N_{\omega,i}(t_{\omega,j-1}, t_{\omega,j})]}{t_{\omega,,j} - t_{\omega,j-1}} = \frac{\displaystyle\int_{t_{\omega,j-1}}^{t_{\omega,j}} \lambda(\sigma)\, d\sigma}{t_{\omega,j} - t_{\omega,j-1}}$$

and if $\max_{j=1,\ldots,p}(t_{\omega,j} - t_{\omega,j-1}) \to 0$, when $p_\omega \to \infty$, then the estimator is unbiased, it is

$$E[\hat{\lambda}(t_{\omega,j})] \to \lambda(t_{\omega,j}).$$

Also, we can prove that

$$Var[\hat{\lambda}(t_{\omega,j})] = \frac{1}{r^2} \frac{\sum_{i=1}^{r} Var[N_{\omega,i}(t_{\omega,j-1}, t_{\omega,j})]}{(t_{\omega,,j} - t_{\omega,j-1})^2} = \frac{1}{r} \frac{\int_{t_{\omega,j-1}}^{t_{\omega,j}} \lambda(\sigma)\, d\sigma}{(t_{\omega,j} - t_{\omega,j-1})^2} \xrightarrow{r \to \infty} 0$$

so that the estimator is consistent.

3 Modelling the intensity by FPCA

We have now p_ω estimated values of the intensity in the interval $[t_0, t_0 + T)$, for every $\omega = 1, \ldots, k$, that can be seen as k estimated sample paths of the intensity. These estimated sample paths of the intensity have been derived from the initial k observed sample paths of the DSPP. We propose now to apply FPCA in order to model the structure of the intensity of the DSPP.

Then we will obtain the following orthogonal expansion of the intensity process in terms of the first q principal components (p.c.'s)

$$\lambda^q(t) \simeq \hat{\mu}[\lambda(t)] + \sum_{s=1}^{q} \hat{\xi}_s \hat{f}_s(t)$$

where $\hat{\mu}[\lambda(t)]$ is the estimation of the expectation of the intensity from the k estimated samples and $\hat{f}_s(t)$ are the eigenfunctions of the second order integral equation

$$\int_{t_0}^{t_0+T} \hat{C}_\lambda(t,u)\, \hat{f}_s(u)\, du = \hat{\lambda}_s \hat{f}_s(t), \qquad t \in [t_0, t_0 + T] \tag{1}$$

with $\hat{C}_\lambda$ the estimator of the covariance of the intensity from the estimated samples and $\hat{\xi}_s$ is the *sth* principal component defined by

$$\hat{\xi}_s = \int_{t_0}^{t_0+T} (\hat{\lambda}(t) - \hat{\mu}[\lambda(t)])\, \hat{f}_s(t)\, dt.$$

In order to solve the second order integral equation (1) and estimate the FPCA from discrete observations, different numerical methods have been developed in the last years by different authors. In any case the trajectories are approximated in a finite dimension space, generated by a base, so that the FPCA is reduced to multivariate PCA. A detailed study can be seen in Ramsay and Silverman (1997) and Valderrama *et al.* (2000).

4 Application

The estimator proposed in Section 2 has been applied to a simulated DSPP in which the random intensity is uniformly distributed in $[0,1]$ for each $t \in [t_0, t_0 + T)$.

In order to observe the behaviour of the estimator we define the following measurement of the error committed in each trajectory

$$\varepsilon_\omega = \frac{1}{r} \sum_{i=1}^{r} \varepsilon_{\omega,i}, \quad \omega = 1, \ldots, k$$

where

$$\varepsilon_{\omega,i} = \frac{1}{p_\omega} \sum_{j=1}^{p_\omega} [\lambda_{\omega,i}(t_{\omega,j}) - \hat{\lambda}_{\omega,i}(t_{\omega,j})]^2, \quad i = 1, \ldots, r.$$

There have been simulated $k = 150$ trajectories of the DSPP in $[t_0, t_0 + rT) \equiv [0, 100)$. We took $r = 10$ and $p_\omega = 20$, for every $\omega = 1, \ldots, 150$ so calculating the point estimator, there have been obtained 150 trajectories of the intensity in $[0, 10]$ estimated in 20 nodes of the interval. The mean of the errors became 0.175 and the standard deviation 0.035 so we can say the error is small and that it keeps stable.

Applying Section 3, we have estimated the FPCA using the 150 estimated paths of the intensity. It has been obtained that the first p.c. explains approximately 86% of the variance and the first five p.c.'s more than 99%, so they and their principal functions allow us to rebuild in a continuous way the intensity process.

References

Boel, R. and Beneš, V. (1980). Recursive Nonlinear Estimation of a Diffusion Acting as the Rate of an Observed Poisson Process. *IEEE Transactions on Information Theory*, **26 (5)**, 561-575.

Bouzas, P.R., Aguilera, A.M. and Valderrama, M.J. (2002). Forecasting a class of Doubly Stochastic Poisson Process. *Statistical Papers*, in press.

Di Paola, M. and Falsone, G (1999). Higher order statistics of the response of linear systems excited by polynomials of filtered Poisson pulses. *Probabilistic Engineering Mechanics*, **14**, 55-62.

Grigoriu, M. (1995). *Applied non-gaussian processes.* Prentice Hall: New Jersey.

Grigoriu, M. (2000). A spectral representation based model for Monte Carlo Simulation. *Probabilistic Engineering Mechanics*, **15**, 365-370.

Landers, T.L., Jiang, S.T. and Peek, J.R. (2001). Semi-parametric PWP model robustness for log-linear increasing rates of occurrence of failures. *Reliability Engineering and System Safety*, **73**, 145-153.

Last, G. and Brandt, A. (1995). *Marked Point processes on the real line. (The dynamic approach)*. Springer-Verlag: New York.

Ochi, M.K. (1990). *Applied Probability and Stochastic Processes (In Engineering and Physical Sciences)*. Wiley: New York.

Ramsay, J.O. and Silverman, B.M. (1997). *Functional data analysis.* Springer-Verlag: New York.

Snyder, D.L. and Miller, M.I. (1991). *Random point processes in time and space. (2nd Edition).* Springer-Verlag: New York.

Valderrama, M.J., Aguilera, A.M. and Ocaña, F.A. (2000). *Predicción dinámica mediante análisis de datos funcionales.* La Muralla-Hespérides: Madrid.

Growing and Visualizing Prediction Paths Trees in Market Basket Analysis

Massimo Aria[1], Francesco Mola [2] and Roberta Siciliano [1]

[1] Dipartimento di Matematica e Statistica,
Università di Napoli Federico II, Italia
via Cintia, Monte S. Angelo, I-80126, Napoli
{aria; roberta}@unina.it

[2] Dipartimento di Economia,
Università di Cagliari, Italia
Viale Fra' Ignazio, I-90123, Cagliari
mola@unica.it

Abstract. This paper provides a new approach to Market Basket Analysis taking as measure unit the monetary value of each choice in the transaction. Furthermore, instead of association rules suitable prediction rules are defined in order to consider the causal links between different items. Two methodologies will be presented in order to grow predictive paths through oriented graphs (either trees and neural networks) which facilitate the visual comparison among different basket typologies.

Keywords. Tree-based models, segmentation, factorial representation, assortment choices, neural network, latent basket, mixing parameter.

1 The framework

Market Basket Analysis is a methodology aimed at identifying associations between a large number of products bought by different consumers in a specific location, such as a hypermarket. The analyzed data usually consist of all buying transactions done by the clients in a certain period. Starting from the whole set of transactions, the main aim of the analyst is to understand the interdependence structure between the sales of the different products available. Association Rules (Agrawal et al. 1995; Hofmann et al., 2000) have been proposed in order to discover heavy associations and relevant patterns between different items in a set of transactions. A formal statement of the problem is the following: let $I \equiv \{I_1, \ldots, I_m, \ldots, I_M\}$ be a set of binary attributes, called items. Let $\mathbf{T}$ be a database of T transactions. Each transaction $\mathbf{t}$ is represented as a binary vector, which generic element is $t_m = 1$ if $\mathbf{t}$ includes the item I_m, and $t_m = 0$ otherwise. Let A be a set of some items in I. An association rule is an implication of the form $A \rightarrow B$, where A is a set of some items in I, and B is a single item in I that is not present in A. The rule $A \rightarrow B$ holds in T with confidence c if $c\%$ of transactions in T that contain A also contain B. The rule $A \rightarrow B$ has support s in the transaction set $\mathbf{T}$ if $s\%$ of transactions in $\mathbf{T}$ contain $A \cap B$. In statistical terms the support of rule $A \rightarrow B$ is the frequency with which two items occur together while the confidence is the conditional frequency of item B given item A. In classical theory support and confidence are maximum likelihood estimators of the probability of $A \cap B$ and the conditional probability of B given A, respectively. The typical procedure to discover interesting rules is then to generate all possible rules with minimum support and minimum confidence. In this way, it is relevant to count the number of transactions containing a given item as well as which items are included in a given basket.

2 A new approach to the problem

From the business management point of view the above mentioned approach is unsatisfying. Indeed, it only considers whether a certain item is chosen or not (0 or 1) and it does not take into account the quantity bought of each item. This, however, is the data in which the manager is interested. To overcome this drawback we like to consider quantities as well as prices and take as measure unit the monetary value of the transaction. In this way, we are also able to weight the items in the transaction and to order them.
Let $\mathbf{Q}$ be the $T \times M$ matrix of occurrences of the items in the T transactions, namely the t-th row records the quantities bought of each item in the t-th transaction. Let $\mathbf{D}_p = diag(p_m)$ denote the diagonal matrix including the prices of the M items. By definition, the matrix resulting as $\mathbf{V} = \mathbf{QD}_p$ includes the monetary values spent on each item in the T transactions; the T-dimensional vector resulting as $\mathbf{s} = \mathbf{Vu}$ gives the total amount spent in each transaction, whereas the M-dimensional vector $\mathbf{r} = \mathbf{V}^T\mathbf{u}$ the total amount spent on each item. Similar measures of confidence and support can be defined taking account of monetary values rather than frequencies.
We assume to classify the items into G groups as well as the transactions into H groups. The stratifying variable for the items could be defined considering the typology of the items, the typology of the market baskets, the quality class of the items. Instead, the transactions can be distinguished according a time variable (different day time periods), or a space variable (various cashes), etc. As a result, the $T \times M$ matrix $\mathbf{V}$ can be viewed as structured as $H \times G$ sub-matrices, where the generic sub-matrix $\mathbf{V}_{hg}$, for $h = 1, \dots, H$ and $g = 1, \dots, G$, includes t_h transactions for m_g items, with $\sum_h t_h = T$ and $\sum_g m_g = M$. In order to select the blocks of matrices which are of some interest for the analysis, and to omit the remaining ones, we can make use of a dummy variable, called *picking-up variable*, δ_{hg} which is equal to one if the matrix $\mathbf{V}_{hg}$ is selected and zero otherwise. Each sub-matrix is pre-multiplied by the corresponding picking-up variable, yielding to the final data matrix $\mathbf{X}$ to be analyzed.

3 Market Basket Mining Tool Box

The proposed tool box for Market Basket Mining (MBM), which makes use of MATLAB routines, is organized to include several methods depending on the task of the user and the data to be considered in the analysis. The system is interactive so that the user is called to choose among several objectives leading also to the choice among several alternative strategies of analysis. The main structure of the menu is the following:
Desk: to select the business target, which requires the input of the data matrix and the assumptions;
Explore: to explore the data matrix according to one of the implemented tools of analysis;
Visual: to visualize intermediate and final results for example the assortment choices, the prediction paths, the causal links among the items, as complementary to the output of the explorative analysis.
Model: for decision-making and forecasting purposes.
In the desk analysis, it is crucial to define the business target of the market basket analysis. Not necessarily the frequencies of the items are of main importance for defining the product allocation in the hypermarket. Often the hypermarket manager deals with problems like: *"On which products depends the maximization of the sales of my product with the highest added value?"*, or: *"What is the expected additional loss for the whole assortment eliminating the item with the lowest added value?"*
Two main strategies can be followed depending on the *business target* and *prior information*. We can perform methods belonging to **supervised learning** in presence of one or more response variables (i.e., segmentation, regression modelling, latent budget modelling), or methods of **unsupervised learning** in absence of prior criterion variables (i.e., cluster analysis, principal component analysis, log-linear modelling). A further distinction is between **global analysis** and **local analysis**: the former concerns the analysis of the *complete data set* and might require suitable dimensionality reduction methods in presence of very large data sets; the latter instead provides a *partial analysis* within the available data set in order to satisfy specific goals. In the following, we present two methodologies: first, a supervised learning method for global analysis, facing a typical problem of data mining in market basket analysis; second, a unsupervised model for local analysis to explain market basket behaviors. In the first case, we define a response or criterion variable which is the target product or output and the set of items which are the inputs. We introduce a new methodology, where it is also possible to fix a weighting system for the transactions (i.e., to discriminate a transaction where only one product in great quantity is bought with respect to another one where several products in small quantity have been chosen), as well as a cost function for the different items (i.e., to discriminate the different possible allocations of the products into the hypermarket). In the second case, we will use adapt latent budget model to market basket analysis.

3.1 Global Exploration by Prediction Path Trees

We believe that a suitable approach to data reduction in data mining is tree-based procedures. In the particular case of market basket data, the tree allows to identify prediction paths which might determine the choice of a target product (i.e., the output is a dummy variable), the degree of preference (i.e., the output is an ordinal variable), or the monetary value (or its increment) of a product (i.e., the output is numerical).

As exploratory tool we provide a new methodology that can be included into the framework of two-stage segmentation (Mola and Siciliano, 1997; Siciliano and Mola, 1998), emphasizing the role of the inputs to be globally considered before selecting the best split. Main idea in this paper is to provide a tree-growing via a factorial method considering *linear discriminant functions.* This approach seems particularly convenient to deal with large data sets structured according to a stratifying variable.

We assume that using the picking-up variable the matrix $\mathbf{X}$ to be analyzed is structured to include the monetary values of the M items bought in a set of transactions. For sake of brevity, we assume this set consists of just T transactions (or equivalently $H = 1$) while the items are classified into G groups. In this way, the matrix can be understood as partitioned into G sub-matrices $\mathbf{X}_g$, each consisting of m_g item values. Due to the large offer of items in an hypermarket, it is expected a very large number of items compared to the number of transactions. In this example, any standard tree procedure is not feasible for several reasons: computationally expansive, only few typological inputs can be identifies, unstable solutions. Instead, an alternative methodology is proposed. Our general methodology named TS-DIS (Two-Stage segmentation via DIScriminant analysis) can deal with both numerical and categorical output. In the following, we present the numerical case. As an example of numerical output, we might consider the monetary values spent on the most preferred item, in this way discovering the items links which determine an high expense of the target product.

In our approach, we distinguish between prospective and retrospective splits of the transactions at a given node: a *prospective split* is found considering all possible partitions of any input features into two sub-groups; a *retrospective split* is found considering all possible partitions of the output into two sub-groups. The best split at a given node is found among the prospective splits. Instead, we can define the optimal split of the transactions as the retrospective split which maximizes the between group deviation of the output without considering the inputs. This is a theoretical partition since it can be not necessarily produced by any observed split of the input features. Let $\tilde{Y}$ denote a dummy output which discriminates the two sub-groups of transactions according to the optimal solution provided by the best retrospective split. The recursive partitioning at each node applies the following two stages:

I. Factorial analysis: For each group of within group correlated inputs we find a factorial linear combination of inputs such to maximize the predictability

power to get the optimal split of the transactions;
II. Multiple splitting criterion: Among the possible splits that can be deduced from the linear combinations determined in stage one we find the best factorial (multiple) split of the transactions.
Stage I allows to reduce the dimensionality of the problem passing from G inputs $(X_{1g},\ldots,X_{m_g g})$ to G linear combinations of inputs $Z_g = \sum_{j=1}^{m_g} \alpha_j X_{jg}$, where α_j are the values of the eigenvector associated to the largest eigenvalue of the matrix $\mathbf{W}_g^{-1}\mathbf{B}_g$, being $\mathbf{W}_g$ the within group deviation matrix and $\mathbf{B}_g$ the between class deviation matrix. The new inputs Z_g are all normalized such to have mean equal to zero and variance equal to one. In this respect, they will play the same role in the greedy selection of the best split in stage II, thus producing unbiased splits in the procedure.
Stage II provides to define automatically the factorial multiple split of the transactions into two sub-groups. The Z_g are the candidate splitting variables which generate the set S of prospective splits. These can be interpreted as multiple splits being defined on the basis of a combination of inputs. The best multiple split is found maximizing the between class deviation of the output $s^* \equiv argmax_s\{Dev_s(B)\}$ for any split s in the set S.
We can also calculate the efficiency measure based on the ratio between the between class deviation due to the best observed split, i.e., $Dev_{s^*}(B)$, and the between class deviation due to the optimal split, i.e., $Dev_{k^*}(B)$. This measure could be also used as a stopping rule for the tree-growing.

3.2 Local Decisions by Latent Budget Model

The latent budget model is a reduced-rank probability model to decompose a table with constant row-sum data, namely compositional data such as for example time budgets and conditional proportions derived from a contingency table (de Leeuw and van der Heijden, 1991). In market basket analysis, any $t_h \times m_g$ matrix $\mathbf{V}_{hg}$ includes monetary values and it can be converted into a matrix of compositional data dividing each row element by its margin. In this way, we define a matrix of observed budgets, i.e., the baskets, that can be fitted by latent budget model. The decomposition is a mixture of K positive factorial terms (called latent budgets or *latent baskets*) satisfying restrictions likewise conditional probabilities. In matrix formulation, the latent budget model can be written as $\mathbf{P} = \mathbf{A}\mathbf{B}^T$, where $\mathbf{P}$ is the $t_h \times m_g$ matrix of theoretical baskets, $\mathbf{A}$ is the $t_h \times K$ matrix of mixing parameters $\pi_{k|i}$ (with $\sum_k \pi_{k|i} = 1$, for $i = 1,\ldots,t_h$ and $k = 1,\ldots,K$), $\mathbf{B}$ is the $K \times m_g$ matrix of latent baskets $\pi_{j|k}$ (with $\sum_j \pi_{j|k} = 1$ for $j = 1,\ldots,m_g$ and $k = 1,\ldots,K$). An high value of the mixing parameter $\pi_{k|i}$ means an high probability of the i-th transaction to fall into the k-th latent basket, in this way determining a classification of the t_h transactions into K groups or budgets. The latent baskets parameters $\pi_{j|k}$ can be compared with the mean budget, that is found as $\sum_k \pi_{j|k}\pi_{+k} = \pi_{j+}$ being π_{+k} the probability to fall into the k-th latent basket and π_{j+} the probability to prefer the j-th item. If the latent basket $\pi_{j|k}$ is higher than the mean budget π_{j+} means that given the k-latent basket the selection of the j-th item increases and vice versa. Thus, we characterize the clusters of the transactions identifying their typological products and in this way the market basket behavior. The model parameters can be estimated by weighted least-squares method minimizing the function $f = SSQ(\mathbf{W}(\mathbf{P} - \mathbf{AB}^T))$, where $\mathbf{W}$ is a diagonal matrix, assigning a weighting systems to the transactions. This approach is known as unconditional latent budget analysis (Siciliano and Mooijaart, 1999) and it can be also understood as a supervised neural network connecting sequentially transactions, latent baskets and items with causal links.

4 Conclusions

Main advantages of MBM tool box are related to the possibility of better planning the marketing policies of a hypermarket. As an example, from a promotional viewpoint, if two or more products results sequentially associated, it will be sufficient to promote only one of them to increase sales of both. From a merchandizing viewpoint, they should be allocated on the same shelf in the layout of a hypermarket. In the oral presentation we will show the results of applications of the proposed methods on original real data sets, being actually under experimentation at a hypermarket in the area of Naples.

References

Agrawal, R., Mannila, H., Srikant, R., Toivonen, H. and Verkamo, A.I. (1995). Fast discovery of association rules. *Advances in Knowledge Discovery and Data Mining*, 307–328, AAAI/MIT Press, Menlo Park, CA.

de Leeuw, J., van der Heijden, P.G.M. (1991). Reduced-rank models for contingency tables, *Biometrika*, 78, 229-232.

Hofmann, H., Siebes, A., Wilhelm A. (2000). Visualising Association Rules. In *KDD-2000 - Proceedings of the Sixth ACM SIGKDD International Conference on Knowledge Discovery and Data Mining*, pages 227-235, ACM, New York.

Kim, H., Loh, W.Y. (2001). Classification Trees with Unbiased Multiway Splits, *Journal of the American Statistical Association*, 96, 454, 589-604.

Mola, F., Siciliano, R. (1997). A Fast Splitting Procedure for Classification Trees, *Statistics and Computing*, 7, 208-216.

Siciliano, R., Mola, F. (1998). Ternary Classification Trees: a Factorial Approach, in Blasius, J. and Greenacre, M. (Eds.): *Visualization of categorical data*, New York: Academic Press.

Siciliano, R., Mooijaart, A. (1999). Unconditional Latent Budget Analysis: a Neural Network Approach, in S. Borra, R. Rocci, M. Vichi, M. Schader (Eds.): *Advances in Classification and Data Analysis*, Springer-Verlag, Berlin, 127-136.

Improved Fitting of Constrained Multivariate Regression Models using Automatic Differentiation

Trevor Ringrose and Shaun Forth

Applied Mathematics and Operational Research Group, Cranfield University, RMCS Shrivenham, Swindon SN6 8LA, Great Britain.

Abstract. Regression models for multivariate response surfaces are proposed, and an illustration is given of numerical maximisation of likelihoods using automatic differentiation being superior to that using finite differences.

Keywords. automatic differentiation, multivariate regression, response surface methods

1 Introduction

Response Surface Methodology (RSM) is a collection of techniques for designing and analysing experiments, with the response variable viewed as defining a surface over the explanatory variables. The fitted surface is often a second order polynomial regression model, and it is common to use 'Canonical Analysis' to express the surface in its canonical form, usually centred at the stationary point and with orthogonal axes defining the main directions of curvature. These orthogonal axes and their associated curvatures are the eigenvectors and eigenvalues of a symmetric matrix formed from the coefficients of the second-order terms in the model. This canonical form can aid interpretation of the fitted model, and can discover 'ridges', defined by zero or near-zero eigenvalues, where explanatory (input) variables can be changed in certain ratios without changing the response (output) variable.

In many cases there are several response variables, but approaches to the multivariate problem are less well developed, with response surfaces and canonical forms still being fitted separately, leading to difficulties in interpretation. However, the response surfaces may share similar features, e.g. the axes or the stationary points may be similar to each other, and we might then ask whether the observed differences could be assigned to random variation.

Hence we propose fitting constrained multivariate regression models directly in their canonical form, with constraints applied in various combinations on the eigenvectors, eigenvalues and stationary points. A particularly promising model is that where all sets of eigenvectors (orthogonal axes in the canonical form) are constrained to be equal, since this aids interpretability and seems quite likely in practice. This is strongly related to the Common Principal Component (CPC) method for grouped multivariate data (Flury, 1988), where each group is constrained to have the same principal component axes (eigenvectors of the covariance matrix) but with no constraints on the eigenvalues. Similarly the stationary points can be constrained to equality or the surfaces can be constrained to be scalar multiples of each other.

Assuming normal errors, these models can be fitted numerically with maximum likelihood and the appropriateness of the constraints assessed with the likelihood ratio test. The constraints are implemented either by re-expressing the problem or by constrained optimisation using Lagrange multipliers. All optimisation here was performed using the MATLAB command `fmincon`.

Numerical optimisation of any smooth objective function is more efficient and robust if the appropriate first (and possibly second) derivatives are known. If not then conventionally they must be approximated using finite-differences, with the ensuing problem of choice of step-size (a balance between truncation error accuracy and floating point round-off errors). Fitting the models proposed here using finite-difference gradients proved to be unreliable with problems of both accuracy and robust convergence. We therefore used the alternative of automatic differentiation to supply derivatives, accurate to machine round-off error, directly from the objective function.

2 Constrained multivariate regression models

Suppose we have n observations on p response and q explanatory variables. A second order polynomial model can conveniently be rewritten as

$$E(Y) = X\beta = X_0\beta_0 + X_1\beta_1 + X_2\beta_2 \tag{1}$$

where Y is the matrix of n observations by p responses, X_0, X_1, X_2 contain the columns of the design matrix X corresponding to block/constant, linear and quadratic terms respectively, and β_0, β_1 and β_2 similarly contain columns of the parameter matrix β. In addition, the superscript $^{(j)}$ will be used to denote the j-th response, so for example $\boldsymbol{\beta}_1^{(j)}$ is the (column) vector of linear-term parameters for response j, i.e. the j-th column of β_1. We assume that the p responses are normally distributed with covariance matrix Σ.

The canonical form of the j-th response surface can be derived from a symmetric $q \times q$ matrix $B^{(j)}$, whose diagonal elements are the coefficients of the square terms in $\boldsymbol{\beta}_2^{(j)}$ and whose off-diagonal elements are half the coefficients of the cross-product terms. Its eigendecomposition is $B^{(j)} = M^{(j)}\Lambda^{(j)}M^{(j)T}$ where the eigenvalues $\lambda_h^{(j)}$ are the diagonal elements of $\Lambda^{(j)}$ and the orthogonal eigenvectors $\boldsymbol{m}_h^{(j)}$ are the columns of $M^{(j)}$. This matrix is defined so that, if $\boldsymbol{x}_i^T$ is the i-th row of X_1, the i,j-th element of $X_2\beta_2$ in (1) can be rewritten as

$$\{X_2\beta_2\}_{ij} = \boldsymbol{x}_i^T B^{(j)} \boldsymbol{x}_i = \boldsymbol{x}_i^T M^{(j)} \Lambda^{(j)} M^{(j)T} \boldsymbol{x}_i = \sum_{h=1}^{q} \lambda_h^{(j)} \left(\boldsymbol{m}_h^{(j)T} \boldsymbol{x}_i\right)^2 \tag{2}$$

Hence all observations can be expressed with respect to these axes, which define the principal axes of curvature on the response surface, with the eigenvalues and eigenvectors giving the curvatures and directions respectively. The canonical form then has its stationary point at $\boldsymbol{x}_0^{(j)} = -0.5(B^{(j)})^{-1}\boldsymbol{\beta}_1^{(j)}$.

The usual unconstrained multivariate regression model can therefore be fit directly in its canonical form by rewriting the model as in (2) and maximising the log-likelihood numerically, using Lagrange multipliers to constrain the $M^{(j)}$ to be orthogonal. Constrained regression models in fact require fewer constraints when the likelihood is maximised in its canonical form, since usually (2) can be simplified. For example, to fit the CPC-type model we put $M^{(j)} = M \ \forall\, j$ and so have only one matrix to constrain to orthogonality.

3 Automatic Differentiation

Automatic Differentiation (AD) concerns the mathematical and software process of taking a (possibly) vector-valued function $\mathbf{y} = \mathbf{f}(\mathbf{x})$ defined by computer code and creating new code that calculates $\mathbf{Jf}(\mathbf{x}) = [\partial \mathbf{f_i}/\partial \mathbf{x_j}]$, the Jacobian. The underlying premise of AD is that computer codes define functions in terms of the few well defined mathematical constructs given by the programming language's arithmetic operations and intrinsic functions. Hence each statement in the code may be differentiated and, in *forward (or tangent) mode* AD, directional derivatives are propagated through the entire code using the chain rule of differentiation in time $O((1+n)C(\mathbf{f}))$ where n is the number of directional derivatives and $C(\mathbf{f})$ is the time required to calculate the function. Adjoint, or reverse mode, AD enables calculation of the gradient of a scalar valued function in time $O(C(\mathbf{f}))$ i.e. independent of the number of inputs (the so-called *cheap gradient result*). Adjoint mode AD requires propagating adjoint values backwards through the code and requires complicated compiler-like tools or storing and manipulating a record (or *tape*) of all the calculations performed. Well established AD tools are available for the differentiation of codes written in Fortran (ADIFOR, TAMC, Odyssèe, AD01) and C (ADIC, ADOL-C). See references in Griewank (2000) and Corliss et al (2002) for more details.

Since release 5.0 MATLAB has supported object-oriented (OO) programming. The ADMAT package of Verma (1998) clearly demonstrated the feasibility of using AD in MATLAB via these OO techniques. We found ADMAT to run slower than theory predicts, leading to development of the MAD package.

3.1 MAD - Matlab Automatic Differentiation

MATLAB utilises several intrinsic classes of variables and associated intrinsic functions. For example the `double` class consists of arrays of floating point numbers and functions such as `plus` to add two variables of class `double` together. The function `plus` is called by the MATLAB interpreter each time it encounters the plus sign `+` between two such variables and returns another appropriate value of type `double`. We say that the operator `+` is *overloaded* by the function `plus`. To facilitate forward mode AD in MATLAB we have designed and implemented two new classes of variable `fmad` and `derivvec`. The `fmad` class is used to store an array variable's value and a number of directional derivatives for that variable. Now if the MATLAB interpreter encounters such `fmad` variables it looks to the `fmad` class for an associated `plus` function.

3.1.1 The `fmad` Class

The `fmad` class is made up of functions written so as to propagate values and derivatives of variables. There is a class constructor necessarily called `fmad`. For two arrays `valx`, `derivx` of class `double` the statement `x=fmad(valx,derivx)` returns a variable `x` of `fmad` class. Within this variable `x` are stored its value `x.value = valx` and derivative `x.derivs = derivx`[1]. Assuming `valx` is an array of size (or dimensions) `[i1,i2,...,in]` then the supplied `derivx` must consist of an integer multiple nd of `nel`=`i1`×`i2`×...×`in` values. Each successive set of these `nel` values, taken in array element order, is interpreted as a directional derivative of `x`.

A number of functions associated with the `fmad` class are also supplied. For example, when MATLAB attempts to execute the statement `z = x + y`, for either or both `x` and `y` of `fmad` class, then MATLAB's interpreter detects that the

[1] the dot denotes components of a structure in MATLAB

intrinsic MATLAB plus function is inappropriate and invokes the overloaded plus function of figure 1. The MATLAB intrinsic function isa(x,'fmad') re-

```
function z=plus(x,y)
if isa(x,'fmad')&isa(y,'fmad')
    value=x.value+y.value;
    deriv=x.deriv+y.deriv;
elseif isa(x,'fmad')
    value=x.value+y;
    deriv=x.deriv;
else
    value=x+y.value;
    deriv=y.deriv;
end
z.value=value;
z.deriv=deriv;
z=class(z,'fmad');
```

Fig. 1. plus function for fmad class

turns the value true if x is of class fmad and false otherwise. The MATLAB class function takes a MATLAB structure and returns a variable of the class given by the second argument. With this knowledge we see that such code is simple and intuitive, requiring little MATLAB expertise to understand.

MATLAB is an interpreted language and variables are not declared. A variable's class is determined by the function (and its inputs) which create it. Since fmad functions return fmad variables, then once a first variable is defined to be of fmad class via the class constructor fmad, all other variables depending on the first will be of fmad class. In turn any other variable, depending via intermediates on the first will also be of fmad class. In this manner derivatives are propagated through a user's code and any modifications of MATLAB code to enable AD are minimal. MAD provides accessor[2] functions getvalue and getderivs to obtain the values and directional derivatives of fmad variables. So we may initialise one fmad array variable x, perform some numerical operations and then extract derivatives. For example, given the function $y = \sum_{i=1}^{5} x_i^2$ with $\mathbf{x}$ =[1 2 3 4 5], we may calculate the directional derivative in the x_1 direction [1 0 0 0 0], i.e. $\partial y/\partial x_1$ by,

```
x=fmad([1 2 3 4 5],[1 0 0 0 0]);
y=sum(x.^2);
dydx1=getderivs(y)
```

which correctly calculates the value 2. In the above we have utilised overloading of the sum and power intrinsics. With MAD all directional derivatives are calculated to machine precision so there are no approximation errors as is the case with finite-differencing. Unlike computer algebra packages, MAD readily handles arbitrarily complicated MATLAB code (loops, branches user-defined functions, etc.) and does not experience the exponential explosion in the complexity of terms common to such packages.

To deal with arbitrary numbers of directional derivatives we have designed and implemented the derivvec class.

[2] In object oriented programming an accessor function allows an application programmer to get a value from an object of a class without any knowledge of the internal structure of the object.

3.1.2 The derivvec Class

In the `fmad` constructor function described above, and for the case when the number of directional derivatives `nd` is greater than one, then the vector of the size `sx` of `valx` (`sx = size(valx) = [i1 i2 ... in]`) and the supplied derivatives `derivx` are passed to the `derivvec` class constructor function. This function stores the derivatives in the returned `derivvec` class variable as a (two-dimensional) matrix of size `[nel nd]` so that the jth column of the matrix stores the jth directional derivative, "unrolled" in array element order. Functions are provided to add and subtract variables of `derivvec` class, multiply them by matrices etc., all via operator overloading and in such a manner that when writing the `fmad` functions we may write MATLAB expressions correct for a single directional derivative, assured that the coding of the `derivvec` class will handle arbitrary numbers of directional derivatives.

We claim that our forward mode AD package of MAD is an improvement over that of ADMAT (Verma 1998) due to this separation of derivative manipulation from function value calculation because it: (a) improves clarity of the `fmad` class functions as seen in figure 1, allowing users to add their own functionality; (b) makes performance optimisation possible since, for a large number of directional derivatives, most of the floating point operations are performed in a few `derivvec` functions; and (c) for functions with sparse Jacobians and which use arrays of arbitrary rank (ADMAT restricted to vectors), allows use of variables of MATLAB's `sparse` matrix class within the `derivvec` class.

For our example $y = \sum_{i=1}^{6} x_i^2$ we may trivially calculate the gradient of $\mathbf{y}$,

```
x=fmad([1 2 3 4 5 6],eye(6));
y=sum(x.^2);
dydx=squeeze(getderivs(y))'
```

which returns `dydx= [2 4 6 8 10 12]`, and where `eye(6)`$= \mathbf{I}_6$, the identity matrix and `squeeze` is used to eliminate singleton dimensions.

4 Results

To demonstrate the superiority of optimisation using automatic differentiation rather than finite differences to calculate the gradients for `fmincon` we performed a small experiment on the unconstrained model fitted directly in its canonical form, since in this case we know the correct answer and can directly compare the two numerical solutions. Data were randomly generated from underlying unconstrained or CPC-type models, with p and q each 2 or 6 and n chosen to give a central composite design. For each combination, 500 data sets were generated and the unconstrained model fitted using finite differences and MAD. For regression parameters, eigenvalues and eigenvectors the differences between the directly-fitted results and usual least-squares results were calculated and summarised. For all the above parameters, the mean absolute difference for finite differences was consistently of the order 10^{-3} whereas for automatic differentiation it was 10^{-6}. This suggests that in other cases optimisation using automatic differentiation is likely to be more accurate than that using finite differences. Finite differences also took 4–5 times longer.

We also fit the CPC-type model to the three responses measuring dough expansion in the dough mixing data from Gilmour & Ringrose (1999), where $n = 28, p = 3, q = 3$. Here finite differences took around 380 seconds, MAD around 54 seconds (on a SUN Ultra 10 workstation). The results were equivalent to about 3 decimal places for all parameters, with the MAD likelihood larger in the 12th decimal place. The likelihood ratio test suggests that the model fits adequately.

Unconstrained								
Y_1			Y_2			Y_3		
1st	2nd	3rd	1st	2nd	3rd	1st	2nd	3rd
1.25	0.81	0.28	0.53	0.17	0.05	0.85	0.51	0.29
-0.22	0.31	0.92	-0.16	-0.85	-0.51	-0.14	0.61	0.78
-0.92	0.26	-0.31	-0.97	0.04	0.23	-0.88	0.29	-0.38
0.33	0.91	-0.23	0.17	-0.53	0.83	0.46	0.74	-0.49

	Constrained		
Y_1	0.26	1.17	0.92
Y_2	0.07	0.40	0.27
Y_3	0.24	0.87	0.55
	0.91	0.01	0.41
	-0.36	0.51	0.78
	-0.20	-0.86	0.47

Table 1 : Eigenvalues (above) and eigenvectors (below).

From table 1, e.g. the first unconstrained eigenvalue-eigenvector pair for Y_3 is 0.85 and $(-0.14, -0.88, 0.46)$, while in the CPC-type model it is 0.87 and $(0.01, 0.51, -0.86)$. In general the eigenvalues are all fairly close to the unconstrained ones, while some of the eigenvectors look quite different. However, given that one expects eigenvectors to have high variances this is not too surprising, and the constrained eigenvectors still show the same patterns as the unconstrained ones.

5 Conclusions

The proposed models offer the potential of more interpretable multivariate response surfaces, at the expense of more difficult model-fitting. However, automatic differentiation offers a straightforward way of calculating derivatives of complicated functions directly from the function-evaluation routine, hence increasing the speed and accuracy of numerical optimisation.

Automatic differentiation is therefore widely applicable, in particular for improved maximisation of complicated likelihood functions, and the package MAD implements this in the MATLAB environment.

References

(ed.) Corliss, G., Faure, C., Griewank, A., Hascoet, L. and Naumann, U. (2002). *Automatic Differentiation: From Simulation to Optimization.* New York: Springer.

Flury, B. (1988). *Common Principal Components and Related Multivariate Models.* New York: Wiley.

Forth, S.A. (2001). *MAD - A Matlab Automatic Differentiation Toolbox, Version 1, beta Release, the Forward Mode.* AMOR Report 2001/5, Cranfield University (RMCS Shrivenham).

Gilmour, S.G. & Ringrose, T.J. (1999). Controlling processes in food technology by simplifying the canonical form of fitted response surfaces. *Applied Statistics*, **48**, 91–101.

Griewank, A (2000). *Evaluating Derivatives: Principles and Techniques of Algorithmic Differentiation.* Philadelphia: SIAM.

Verma, A. (1998). ADMAT: Automatic Differentiation in MATLAB Using Object Oriented Methods. In: *Proceedings of SIAM Interdisciplinary Workshop on Object Oriented Methods for Interoperability*, p174-183. Philadelphia: SIAM.

Imputation of Continuous Variables Missing at Random using the Method of Simulated Scores[1]

Giorgio Calzolari [2] and Laura Neri [3]

[2] Dipartimento di Statistica "G.Parenti", Università di Firenze
Viale Morgagni 59, I-50134 Firenze, Italy
[3] Dipartimento di Metodi Quantitativi, Università di Siena
Piazza San Francesco 8, I-53100 Siena, Italy

Abstract. For multivariate datasets with missing values, we present a procedure of statistical inference and state its "optimal" properties. Two main assumptions are needed: (1) data are missing at random (MAR); (2) the data generating process is a multivariate normal linear regression. Disentangling the problem of convergence of the iterative estimation/imputation procedure, we show that the estimator is a "method of simulated scores" (a particular case of McFadden's "method of simulated moments"); thus the estimator is equivalent to maximum likelihood if the number of replications is conveniently large, and the whole procedure can be considered an optimal parametric technique for imputation of missing data.

Keywords. Simulated scores, missing data, estimation/imputation, structural form, reduced form

1 Introduction

Empirical research in economic and social science often suffer from missing data. There are three major problems created by missing data. First, if the nonrespondents are systematically different from the respondents and we do not take into account the difference, analysis may be biased. Second, missing data imply loss of information, so estimates will be less efficient than planned. Third, besides theoretical problems like bias and efficiency, a big practical problem is that tools for effectively treating datasets affected by missing values are not readily available. Several methods have been (and continue to be) developed to draw inference from data sets with missing data (Little and Rubin, 1987). The most attractive for applied researcher are those forcing the incomplete dataset into a rectangular complete-data format, where missing values are filled by imputation of suitable estimates.

The main applied approach to the general problem of obtaining valid inferences when facing missing data is the *Multiple Imputation* technique. This idea was explicitly proposed in Rubin (1978) and a decade later the basic reference textbook was published (Rubin, 1987). The currently available solution to this problem is to create multiple imputations specifying one "encompassing multivariate model" for the entire data set (at least conditional on completely observed variables), and then using fully principled likelihood/bayesian techniques for analysis under that

[1] We are grateful to Monica Billio, Fabio Corradi, Emanuela Dreassi, Giampiero M. Gallo, James Lepkowsky, Fabrizia Mealli, Donald B. Rubin and Federico M. Stefanini for suggestions and comments, but retain full responsibility for the contents of this paper.
We gratefully acknowledge financial support from CNR and MURST-MIUR through projects "Stochastic models and simulation methods for dependent data" and "Employment and unemployment in Italy: measurement issues and behavioural analyses".

	X	Y_1	Y_2	...	Y_p
i=1					?
2				?	
.					
.					?
.		?	?	?	?
.					
n			?	?	

Table 1. Dataset with missing values.

model. This generates a posterior distribution for the parameters of the model and a posterior predictive distribution for the missing values (given the model specifications and the observed data). The primary example of such approach is Schafer's freeware, based on Schafer (1997), which involves iterative Markov Chain Monte Carlo (MCMC) computations; Rubin (2000) explains the advantage and disadvantage of such a method. Among the disadvantages, we point out on the fact that iterative versions of software for creating multiple imputations are not always immediately usable for real applications by the typical analyst dealing with missing data; it often needs experts to face with potentially misleading "non convergent" MCMC and other possible difficulties.

In this paper we introduce a method, feasible for data analysts, for creating imputations when dealing with a general missing data *pattern* of continuous variables assuming that missing data are MAR (Missing At Random, Rubin, 1976), and the data generating process is a multivariate normal linear regression. We can obtain the multiple imputed datasets by repeating several times (each time till convergence) the iterative "least-squares estimation/multivariate normal imputation" procedure. We show that, at convergence of the iterative procedure, the estimator is a "method of simulated scores" (MSS, see Hajivassiliou and McFadden, 1990, a particular case of McFadden's (1989) method of simulated moments MSM).

For precision's sake, we must say that through the paper we always refer to a "single" imputation; this makes explanations easier, without loss of generality.

Analytical details on the iterative estimation/imputation process, as well as detailed proofs of the propositions, can be found in Calzolari and Neri (2002).

2 Estimation/imputation based on structural form

A schematic representation of an incomplete dataset is shown in Table 1 where the n rows represent the observational units, X and Y represent variables recorded for those units; X denotes the column(s) of complete variables; in the p columns of Y, question marks identify missing values (they can occur anywhere, in any *pattern*). Our hypothesis through this paper is that the missing data are MAR.

Formally, Y denotes an $n \times p$ matrix of data, Y_{obs} denotes the observed portion of Y, Y_{mis} denotes the missing portion (so that $Y = [Y_{obs}, Y_{mis}]$), X is an $n \times k$ matrix without missing values. When considering a particular column of Y, the portion (rows) of X, corresponding to its missing values, is indicated as X_{mis}.

The task of generating imputations is often a hard task, except in some simple cases such as datasets with only one variable affected by missing values or very special patterns of missingness. The main difficulty is to find a solution for imputing a general pattern of missing data preserving the original association structure of the data. In this paper we deal with this problem when the Y matrix is composed of variables defined on a continuous scale, and when it may be reasonable to use the multivariate normal model to create imputations (Rubin, 1987).

We consider a set of normal linear regression models. We solve the technical problem introducing some convenient modifications into the "sequential regression multivariate imputation" (SRMI) method by Raghunathan, Lepkowsky, Van Hoewyk, and Solenberger (1997), which is adopted by the imputation software (*IVE-ware*). They build the imputed values by fitting a sequence of regression models and drawing values from the corresponding predictive distribution, under the hypothesis of MAR mechanism, infinite sample size and simple random sampling. The method follows a bayesian paradigm. Each imputation consists of c "rounds". Round 1 starts regressing the variable with the fewest number of missing values, say Y_1, on X, and imputing the missing values with the appropriate regression model. Assuming a flat prior for the regression coefficient, the imputations for the missing values in Y_1 are drawn from the corresponding posterior predictive distribution. After Y_1 has been completed, the next variable with the fewest number of missing values is considered, say Y_2; observed Y_2 values are regressed on (X, Y_1) and the missing values are imputed, and so on. The imputation process is then repeated in rounds 2 through c, modifying the predictor set to include all the Y variables except the one used as the dependent variable. Repeated cycles continue for a pre-specified number of rounds, or until stable imputed values occur (convergence in distribution).

The method we propose follows the SRMI method, but introduces a convenient modification of the variance covariance matrix estimator. Practically, the procedure starts exactly as the SRMI (round 1): we estimate the coefficients of the linear regression model related to the variable with fewest missing values (let be Y_1), by OLS, using the Y_1 observed part $(Y_{obs,1})$. Supposing that $\widehat{\Pi}_1$ is the estimated regression coefficient and $\widehat{\sigma}_{11}$ the residual variance, then the imputed value set is

$$\widetilde{Y_1} = X_{mis,1}\widehat{\Pi}_1 + \sqrt{\widehat{\sigma}_{11}}\ \widetilde{u}_1$$

where $\widetilde{u}_1$ is a vector of independent pseudo-random standard normal deviates.

So we have a first set of completed values for Y_1 and we attach it as an additional column to X. We then regress the next variable with fewest missing values (say $Y_{obs,2}$) against X and the completed Y_1 and use the OLS estimated coefficients and variance for an imputation step that completes Y_2. Going on, the first round ends when all the missing values are completed. As the SRMI's authors put in evidence, the updating of the right hand side variables after imputing the missing values depends on the order in which we select the variables for imputation. Thus, the imputed values for Y_j involve only $(X, Y_1, ..., Y_{j-1})$, but not $Y_{j+1} ... Y_p$. For this reason the procedure continues to overwrite the imputations for the missing values iteratively. In any iteration after the first round, we always have complete data for all variables, part of which are observed, part have been imputed in the previous iteration. The system of regression equations has, as dependent variable for each equation, the variable to be "imputed if missing" and has on the right hand side all the others variables

$$\begin{aligned}
Y_1 &= X\gamma_{11} + Y_2\gamma_{12} + Y_3\gamma_{13} + ... + Y_p\gamma_{1p} + \varepsilon_1 \\
Y_2 &= X\gamma_{21} + Y_1\gamma_{22} + Y_3\gamma_{23} + ... + Y_p\gamma_{2p} + \varepsilon_2 \\
&\dots \\
Y_p &= X\gamma_{p1} + Y_1\gamma_{p2} + Y_2\gamma_{p3} + ... + Y_{p-1}\gamma_{pp} + \varepsilon_p
\end{aligned} \tag{1}$$

The $\gamma_{11}, \gamma_{21}, ..., \gamma_{p1}$ are scalars or $(k \times 1)$ vectors depending on X being a single column or a $(n \times k)$ matrix, while all the other γ-s are scalars and the ε-s have a cross-equations multivariate normal distribution.

Equations (1) represent a system of simultaneous equations in structural form. The jointly dependent variables Y appear also on the right hand side of the equations, while the variables X play the role of "exogenous" variables. Such a system is obviously underidentified, as it violates the order condition for identification (eg. Greene, 2000, sec. 16.3.1): infinite sets of γ-values would be observationally equivalent. It is therefore useless (or impossible) to apply estimation techniques suitable for simultaneous equation systems, like two or three stage least squares, full information maximum likelihood, etc. Nevertheless we can estimate each equation separatly by OLS as in SRMI approach. After coefficients have been estimated by OLS, we compute from residuals the estimate of the $(p \times p)$ variance covariance matrix, say $\widehat{\Psi}$. Differently from the SRMI method, we use the Cholesky decomposition of the matrix $\widehat{\Psi}$ to produce vectors of pseudo-random numbers for imputation, thus considering also covariances besides variances.

When a value of Y_1 is missing, we impute the value obtained from the right hand side of the first equation in (1) where: the γ -s are at the previous iteration estimated value; the value(s) of X is (are) observed; the values of the Y on the right hand side are in any case complete (some of them are observed, the others have been imputed in the previous iteration); the value of ε_1 is "jointly" produced with ε_2, ..., with ε_p by the pseudo-random generator with a cross-equations variance-covariance matrix equal to the last estimated $\widehat{\Psi}$. The same is done for the second equation in (1), filling missing values of Y_2, and so on.

Repeated cycles continue until convergence on the estimated parameters has been achieved.

A question naturally arises: why and when does the the iterative estimation/imputation procedure converge? (Even if the MCMC context is different from ours, still recently Horton and Lipsitz, 2001, p. 246 point out that convergence "remains more of an art form than a science"). Considering the procedure as it has just been described, the answer does not result obvious. In order to answer this question, it is convenient to think that the structural form system can be easily transformed in a reduced form (Greene, 2000, sec. 16.2.3) and to think at the sequence of iterations in a different order, as if iterations were "grouped". Let's first see what happens if we keep parameter values fixed (the γ -s and the Cholesky decomposition of the matrix $\widehat{\Psi}$), and we only iterate substitutions of imputed values on the right hand side of equations (1). These iterated substitutions (e.g. Thisted, sec. 3.11.2) are "exactly" the steps of the well known Gauss-Seidel method for the "simultaneous solution" of the system of equations (also called by econometricians "stochastic simulation", because of the presence of the ε terms). The simultaneous solution, using econometric terminology, is the well known derivation of the "reduced form" (or "restricted" reduced form) from the "structural form". The reduced form system has variables Y only on the left hand side, while the right hand side includes only the variable(s) X and the error terms. Thus, till we hold parametes fixed at some values, the iterated substitution of imputed values will converge to the reduced form derived from the structural form (or restricted reduced form). Now we can re-estimate parameters (with OLS on the structural form) and start again a new cycle of iterated substitutions in (1), and so on.

The strictly thighted sequence of estimations and imputations, for each structural equation, of the SRMI method has thus been reordered, disentangled and converted into a sequence of iterations that are conceptually much more manageble. In each iteration, an OLS estimation of "all" the structural form equations (1), using observed and previously imputed values, is followed by the "simultaneous solution", i.e. transformation into reduced form, that produces "all" the values of the variables Y which are then imputed. Studying the convergence of this new sequence of estimation and imputation phases becomes more manageable, as it will be clear in Section 3.

The SRMI method and the one just proposed follow different paradigms. The former is based on the bayesian paradigm and the latter on the frequentist paradigm. Beyond this difference, it is important to put in evidence the main technical difference. The SRMI method draws the random normal deviates of the imputation step for each equation "independently"; the method we propose considers stochastic terms drawn from a multivariate normal distribution with a cross-equations variance-covariance matrix (Ψ) estimated from residuals.

3 Properties of the estimator

The "good" properties of the estimator discussed above are ensured by the following propositions:

Proposition 1 *For a complete data set, the reduced form parameters estimator, derived from the the* OLS *estimator of the strucural form parameters (γ-s and Ψ), is equal to the* OLS *direct estimator of the reduced form parameters.*

Proposition 2 *The* OLS *estimator of the reduced form parameters, at convergence of the estimation/imputation procedure, is a* MSS *(Method of Simulated Scores) estimator with "one" replication (Hajivassiliou and McFadden, 1990).*

We briefly discuss the implications and consequences of the two propositions; proofs and a more detailed discussion can be found in Calzolari and Neri (2002).

Working on the structural form system (1) has the advantage of being computationally simple as well as rather intuitive. The discussion on convergence of the iterated imputations with fixed parameters (section 2) and *Proposition 1* ensure that we can get exactly the same results if we work directly on the reduced form, estimating its parameters directly by OLS, and using such an estimated reduced form for imputation. However, even if the estimation phase would be simple (even simpler than for (1)), the imputation phase would be much more complex. For each pattern of missing data we should, in fact, specify the appropriate imputation function, with pseudo-random errors that should be conditional on the Y-s observed in that pattern. Since there are 2^p possibly different patterns of missingness, the technical solution would be very hard. Also, there would be no substantial simplifications over the exact maximum likelihood approach, where up to 2^p different conditional densities should be specified, according to which Y-s are observed in each pattern. That's why it is preferable to work, in practice, with the structural form (1).

Nevertheless, passing to the reduced form is necessary for our proof, because the reduced form is much more manageable from the analytical point of view. The reduced form is a simple system of multivariate normal linear regression equations, without endogenous variables on the right hand side. With a complete data set, the OLS normal equations are exactly the same as: *score=0* (this holds both for coefficients and covariance matrix of the reduced form). But since data have been completed by imputation (simulation), the score is, in fact, a "simulated score". As a consequence, we have *Proposition 2*. "One" replication means that only "one" set of pseudo-random error terms are generated and used for imputation. As a consequence, the (asymptotic) variance of the estimated parameters is larger than for maximum likelihood; however, if we perform the same procedure with more replications, the variance decreases, and the simulated scores estimator can reach the maximum likelihood efficiency if the number of replications is conveniently large (in principle, infinitely large).

4 Conclusion

In this paper we have introduced a method for imputation of missing data, assuming data generated by a multivariate normal linear regression model and a MAR missing data mechanism. The method is based on an iterated estimation/imputation procedure. Besides its technical simplicity and feasibility, the peculiarity of the method is in the properties of the estimator. First of all the parameters estimator is consistent and asymptotically normal. Moreover, being a simulated scores estimator, its efficiency can be improved by increasing the number of replications. Finally, the estimator becomes as efficient as maximum likelihood if the iterative procedure is replicated a sufficiently large number of times, each time iterating to convergence.

We dealt with the missing data problem in the context of a linear normal model in which some observations of some variables (treated as "endogenous" variables) were missing, while other variables (treated as "exogenous" variables) were completely observed.

The imputation approach described can be used to create a single imputation with a variance estimation procedure taking into account the uncertainty due to missing data or can be part of a framework of multiple imputation.

Of course, in many real cases missing data do not affect only continuous variables. The problem exists also for categorical data, count data, or censored variables: generalization of the method here proposed is left to future research.

References

Calzolari G., Neri L. (2002): "A Method of Simulated Scores for Imputation of Continuous Variables Missing At Random", *Quaderni del Dipartimento di Statistica "G. Parenti"* No. 49, Università degli Studi di Firenze.

Gourieroux C., Monfort A. (1996): *Simulation-Based Econometric Methods.* Oxford University Press.

Greene W. H. (2000): *Econometric Analysis* (fourth edition). Upper Saddle River, NJ: Prentice-Hall, Inc.

Hajivassiliou V., McFadden D. (1990): "The Method of Simulated Scores, with Application to Models of External Debt Crises", Cowles Foundation Discussion Paper No. 967, Yale University.

Horton N. J. and Lipsitz S. R. (2001): "Multiple Imputation in Practice: Comparison of Software Packages for Regression Models with Missing Variables", *The American Statistician*, 55, 244-255.

Little R. J. A., Rubin D. B. (1987): *Statistical Analysis with Missing Data.* New York: Wiley.

Mc Fadden D. (1989): "A Method of Simulated Moments for Estimation of Discrete Response Model without Numerical Integration", *Econometrica*, 57, 995-1026.

Raghunathan T. E.: www.isr.umich.edu/src/smp/ive.

Raghunathan T. E., Lepkowski J., Van Voewyk J., Solenberger P. (1997): "A Multivariate Technique for Imputing Missing Values Using a Sequence of Regression Models", Technical Report, Survey Methodology Program, Survey Research Center, ISR, University of Michigan.

Rubin D. B. (1976): "Inference with Missing Data", *Biometrika*, 63, 581-592.

Rubin D. B. (1978): "Multiple Imputations in Sample Surveys-A Phenomeno-Logical Bayesian Approach to Nonresponse", *The Proceeding of the Survey Research Methods Section of the American Statistical Association*, 20-34, with discussion and reply.

Rubin D. B. (1987): *Multiple Imputation for Nonresponse in Survey.* New York: Wiley.

Rubin D. B. (2000): "The Broad Role of Multiple Imputations in Statistical Science", in *Proceeding in Computational Statistics, 14th Symposium, Utrecht-The Netherlands, 2000*, ed. by J. G. Bethlehem and P. G. M. van der Heijden. Vienna: Physica-Verlag, 3-14.

Schafer J. L. (1997): *Analysis of Incomplete Multivariate Data.* London: Chapman & Hall.

Thisted, R. A. (1988): *Elements of Statistical Computing.* New York: Chapman and Hall.

Induction of Association Rules: Apriori Implementation

Christian Borgelt and Rudolf Kruse

Department of Knowledge Processing and Language Engineering
School of Computer Science
Otto-von-Guericke-University of Magdeburg
Universitätsplatz 2, D-39106 Magdeburg, Germany

Abstract. We describe an implementation of the well-known *apriori algorithm* for the induction of association rules [Agrawal *et al.* (1993), Agrawal *et al.* (1996)] that is based on the concept of a *prefix tree*. While the idea to use this type of data structure is not new, there are several ways to organize the nodes of such a tree, to encode the items, and to organize the transactions, which may be used in order to minimize the time needed to find the frequent itemsets as well as to reduce the amount of memory needed to store the counters. Consequently, our emphasis is less on concepts, but on implementation issues, which, however, can make a considerable difference in applications.

Keywords. association rule, apriori algorithm, prefix tree, item coding

1 Introduction

Association rule induction is a powerful method for so-called *market basket analysis*, which aims at finding regularities in the shopping behaviour of customers of supermarkets, mail-order companies, on-line shops and the like. With the induction of association rules one tries to find sets of products that are frequently bought together, so that from the presence of certain products in a shopping cart one can infer (with a high probability) that certain other products are present. Such information, expressed in the form of association rules, can often be used to increase the number of items sold, for instance, by appropriately arranging the products in the shelves of a supermarket (they may, for example, be placed adjacent to each other in order to invite even more customers to buy them together).

The main problem of association rule induction is that there are so many possible rules. For the product range of a supermarket, for example, which may consist of several thousand different products, there are billions or even trillions of possible association rules. It is obvious that such a vast number of rules cannot be processed by inspecting each one in turn. Efficient algorithms are needed that restrict the search space and check only a subset of all rules, but, if possible, without missing important rules.

The importance of a rule is usually measured by two numbers: Its *support*, which is the percentage of transactions that the rule can be applied to (or, alternatively, the percentage of transactions, in which it is correct), and its *confidence*, which is the number of cases in which the rule is correct relative to the number of cases in which it is applicable (and thus is equivalent to an estimate of the conditional probability of the consequent of the rule given its antecedent). To select interesting rules from the set of all possible rule, a *minimum support* and a *minimum confidence* are fixed.

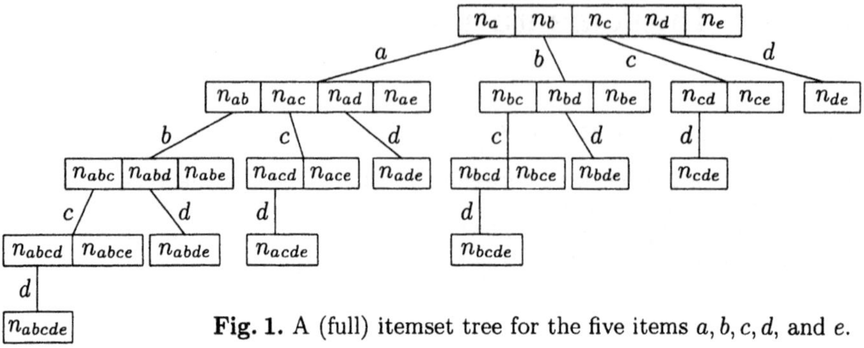

Fig. 1. A (full) itemset tree for the five items a, b, c, d, and e.

Among the best known algorithms for association rule induction is the *apriori algorithm* [Agrawal *et al.* (1993), Agrawal *et al.* (1996)]. This algorithm works in two steps: In a first step the *frequent itemsets* (often misleadingly called *large itemsets*) are determined. These are sets of items that have at least the given minimum support (i.e., occur at least in a given percentage of all transactions). In the second step association rules are generated from the frequent itemsets found in the first step. Usually the first step is the more important, because it accounts for the greater part of the processing time. In order to make it efficient, the apriori algorithm exploits the simple observation that no superset of an infrequent itemset (i.e., an itemset not having minimum support) can be frequent (can have enough support).

In this paper we describe an implementation of the apriori algorithm that has gained some popularity in the research community. This program was written by the first author of this paper and can be retrieved free of charge at `http://fuzzy.cs.uni-magdeburg.de/~borgelt/software.html`.

2 Itemset Trees

In order to find the frequent itemsets, we have to count the transactions they are contained in. Our implementation is based on the idea to organize the counters for the itemsets in a special kind of *prefix tree*, which not only allows us to store them efficiently (using little memory), but also supports processing the transactions as well as generating the rules. The structure of a (full) itemset tree as our implementation uses it is shown in Figure 1. Each n_S denotes a counter for an itemset S. The edge labels on a path from the root to some node specify the common part of the itemsets for which there are counters in that node. Since the common part would be a prefix if we were dealing with sequences instead of sets, such a data structure is commonly called a *prefix tree*. That we are dealing with sets, not sequences, is the reason, why this tree structure is unbalanced: n_{abc}, for instance, is the same as n_{bca} and therefore only one of these counters is needed.

During the first step of the apriori algorithm this tree is created level by level. That is, in a first traversal of the transactions the frequencies of the one element sets are determined (the root node is created), in a second traversal the two element sets are checked (the second tree level is created—the children of the root) and so on. Of course, in doing so, some branches of the tree can be pruned, because by exploiting the simple observation mentioned above we can find out whether a branch can contain frequent itemsets or not.

3 Node Organization

There are different data structures that may be used for the nodes of this tree. In the first place, we may use simple vectors of integer numbers to represent the counters. The items (note that we only need one item to distinguish between the counters of a node, see above) are not explicitly stored in this case, but are implicit in the vector index. Alternatively, we may use vectors, each element of which consists of an item identifier (an integer number) and a counter, with the vector elements sorted by the item identifier.

The first structure has the advantage that we do not need any memory to store the item identifiers and that we can very quickly find the counter corresponding to a given item (we simply use the item identifier as an index), but it has the disadvantage that we may have to add "unneccessary" counters (i.e., counters for itemsets, of which we know from the information gathered in previous steps that they must be infrequent), because the vector may not have "gaps". This problem can only partially be mitigated by enhancing the vector with an offset for the first element and a size, so that unneccessary counters at the margins of the vector can be discarded. The second structure has the advantage that we only represent the counters we actually need, but it has the disadvantage that we need extra memory to store the item identifiers and that we have to carry out a binary search in order to find the counter corresponding to a given item.

A third alternative would be to use a hash table per node. However, although this reduces the access time, it increases the amount of memory needed, because for optimal performance a hash table must not be too full. In addition, it does not allow us to exploit easily the order of the items (see below). Therefore we do not consider this alternative here.

Obviously, if we want to optimize speed, we should choose simple counter vectors, despite the gap problem. If we want to optimize memory usage, we can decide dynamically, which data structure is more efficient in terms of memory, accepting the costs for the binary search if necessary.

It should also be noted that we need a set of child pointers per node, at least for all levels above the currently added one (in order to save memory, we should not create child pointers before we are sure that we need them). For organizing these pointers we have basically the same options as for organizing the counters. However, if the counters have item identifiers attached, we have an additional possibility. We may draw on the organization of the counters, using the same order of the items and leaving child pointers nil if we do not need them. This can save memory, even though we may have unneccessary nil pointers, because we need not store item identifiers a second time.

4 Item Coding

It is clear that the way in which the items are coded (i.e., are assigned integer numbers as identifiers) can have a significant influence on the gap problem for counter vectors mentioned above. Depending on the coding we may need large vectors with a lot of gaps or we may need only short vectors with few gaps. A good heuristic approach to minimize the number and size of gaps seems to be the following: It is clear that frequent itemsets contain items that are frequent individually. Therefore it is plausible that we have only few gaps if we sort the items w.r.t. their frequency, so that the individually frequent items receive similar identifiers (and, of course, infrequent items are discarded entirely). In this case it can be hoped that the offset/size representation of a vector can discard the greater part of the uneccessary counters, because these can be expected to cluster at the vector margins.

Extending this scheme, we may also consider to code the items w.r.t. the number of frequent pairs (or even triples etc.) they are part of, thus using additional information from the second (or third etc.) level to improve the coding. This extension, however, is not yet incorporated in the publicly available version of our program (version 4.0 at the time of this writing).

5 Recursive Counting

The itemset tree is not only an efficient way to store the counters, it also makes processing the transactions very simple, especially if we sort the items in a transaction ascendingly w.r.t. their identifiers. Then processing a transaction is a simple doubly recursive procedure: To process a transaction for a node of the tree, (1) go to the child corresponding to the first item in the transaction and process the remainder of the transaction recursively for that child and (2) discard the first item of the transaction and process it recursively for the node itself (of course, the second recursion is more easily implemented as a simple loop through the transaction). In a node on the currently added level, however, we increment a counter instead of proceeding to a child node. In this way on the current level all counters for itemsets that are part of a transaction are properly incremented.

By sorting the items in a transaction, we can also apply the following optimizations (this is a bit more difficult—or needs additional memory—if hash tables are used to organize the counters and thus explains why we are not considering hash tables): (1) We can directly skip all items before the first item for which there is a counter in the node, and (2) we can abort the recursion if the first item of (the remainder of) a transaction is beyond the last one represented in the node. Since we grow the tree level by level, we can even go a step further: We can terminate the recursion once (the remainder of) a transaction is too short to reach the level currently added to the tree.

6 Transaction Representation

The most simple way of processing the transactions is to handle them individually and to apply to each of them the recursive counting procedure described in the previous section. However, the recursion is a very expensive procedure and therefore it worth considering how it can be improved. One approach is based on the fact that often there are several similar transactions, which lead to a similar program flow when they are processed. By organizing the transactions into a prefix tree (an idea that has also been used by Han *et al.* (2000) in a different approach) transactions with the same prefix can be processed together. In this way the procedure for the prefix is carried out only once and thus considerable performance gains can result. Of course, the gains have to outweigh the additional cost of constructing such a transaction tree. This extension, however, is also not yet incorporated in the in the publicly available version of our program (version 4.0 at the time of this writing).

7 Experimental Results

We conducted several experiments in order to study the effects of the different optimizations we discussed above. The results we present here were selected, because they demonstrate these effects most clearly.

item sorting	node structure	transaction representation	transaction prep. (s)	subset check (s)	memory used (Mb)
none	vector	individually	—	30.5	152.1
indiv., descending	vector	individually	—	36.1	171.2
indiv., ascending	vector	individually	—	26.2	103.2
pairs, descending	vector	individually	—	23.9	83.5
pairs, ascending	vector	individually	—	18.9	63.3
none	dynamic	individually	—	37.6	51.2
indiv., descending	dynamic	individually	—	44.3	50.8
indiv., ascending	dynamic	individually	—	33.8	52.6
none	vector	prefix tree	0.1	24.1	152.1
indiv., descending	vector	prefix tree	0.1	31.5	171.2
indiv., ascending	vector	prefix tree	0.1	16.2	103.2
none	dynamic	prefix tree	0.1	28.6	51.2
indiv., descending	dynamic	prefix tree	0.1	38.0	50.8
indiv., ascending	dynamic	prefix tree	0.1	20.4	52.6

Table 1. Results on the webview dataset (max. 8 items, min. support 0.056%).

item sorting	node structure	transaction representation	transaction prep. (s)	subset check (s)	memory used (Mb)
none	vector	individually	—	187.9	167.4
indiv., descending	vector	individually	—	173.4	66.3
indiv., ascending	vector	individually	—	80.3	67.5
pairs, descending	vector	individually	—	176.4	73.5
pairs, ascending	vector	individually	—	89.0	69.6
none	dynamic	individually	—	219.1	59.8
indiv., descending	dynamic	individually	—	205.2	52.6
indiv., ascending	dynamic	individually	—	80.8	67.4
none	vector	prefix tree	0.2	30.3	167.4
indiv., descending	vector	prefix tree	0.2	22.9	66.3
indiv., ascending	vector	prefix tree	0.2	27.6	67.5
none	dynamic	prefix tree	0.2	42.7	59.8
indiv., descending	dynamic	prefix tree	0.2	32.6	52.6
indiv., ascending	dynamic	prefix tree	0.2	28.2	67.4

Table 2. Results on the census dataset (max. 8 items, min. support 0.05%).

We used two datasets: The BMS-webview-1 dataset, which was also used as a benchmark by Zheng *et al.* (2001) and which is derived from the KDD-Cup 2001 data [Kohavi *et al.* (2000)]. In contains clickstream data from Gazelle.com, a small legcare company, which no longer exists. Each transaction describes a web session consisting of all the product detail pages viewed. The second dataset is the well-known census dataset (also known as adult dataset) from the UCI machine learning repository [Blake and Merz (1998)], which contains data from the US Census Bureau. Each record of this dataset describes one tax payer. We prepocessed this dataset by discretizing numeric attributes and replacing the values of a record by expressions of the form "*attribute=value*". These two dataset exhibit very different characteristics and are thus very well suited to show the differences between the optimizations studied above.

The results of our experiments are shown in Table 1 for the webview dataset and in Table 2 for the census dataset. The fixed parameters are a maximum number of eight items per frequent itemset for both datasets and a minimal support value of 0.056% for the webview dataset and a minimal support value of 0.05% for the census dataset. The other parameters used are indicated in the first columns of the tables. Item sorting refers to whether items were sorted based on individual frequencies or on number of frequent pairs they are part of. The node structure is either a pure counter vector or a dynamic, memory minimizing choice between the options discussed above. The third column states whether transactions were processed individually or organized into a prefix tree. All times were measured on an AMD Athlon Thunderbird 1.33GHz system running S.u.S.E. Linux 7.3. The apriori program was compiled with gcc version 2.95.3. The memory used for the itemset tree was computed using an estimate for the overhead of the memory management system. It is consistent with output of the `top` system utility.

From these tables it can be seen that sorting the items can considerably reduce memory usage as well as processing time, with ascending sorting faring best (conjectured reason: it leads to more nodes having shorter vectors thus mitigating the gap problem). Considering frequent pairs is not necessarily better than sorting w.r.t. individual frequencies. The dynamic node structure leads to minimal memory usage, but often at a considerable loss in performance. The gains from organizing the transactions in a prefix tree clearly outweigh the construction costs. In addition, with prefix trees, the performance loss of the memory minimization is less severe.

8 Conclusions

In this paper we discussed several ways to optimize the performance of the classic apriori algorithm for inducing association rules. As our experimental results confirm, considerable performance improvements can be achieved with relatively simple modifications. However, which optimizations are best for a given problem can depend heavily on the dataset used.

References

Agrawal, R., Imielienski, T., and Swami, A. (1993). Mining Association Rules between Sets of Items in Large Databases. In: *Proc. Conf. on Management of Data*, 207–216. New York: ACM Press

Agrawal, A., Mannila, H., Srikant, R., Toivonen, H., and Verkamo, A. (1996). Fast Discovery of Association Rules. In: Fayyad *et al.* (1996), 307–328

Blake, C.L. and Merz, C.J. (1998). *UCI Repository of Machine Learning Databases.* Dept. of Information and Computer Science, University of California at Irvine
`http://www.ics.uci.edu/ mlearn/MLRepository.html`

Fayyad, U.M., Piatetsky-Shapiro, G., Smyth, P., and Uthurusamy, R., eds. (1996). *Advances in Knowledge Discovery and Data Mining.* Cambridge: AAAI Press / MIT Press

Han, J., Pei, J., and Yin, Y. (2000). Mining Frequent Patterns without Candidate Generation. In: *Proc. Conf. on the Management of Data (SIGMOD'00, Dallas, TX).* New York: ACM Press

Kohavi, R., Bradley, C.E., Frasca, B., Mason, L., and Zheng, Z. (2000). KDD-Cup 2000 Organizers' Report: Peeling the Onion. SIGKDD Exploration 2(2):86–93.

Zheng, Z., Kohavi, R., and Mason, L. (2001). Real World Performance of Association Rule Algorithms. In: *Proc. 7th Int. Conf. on Knowledge Discovery and Data Mining (SIGKDD'01).* New York: ACM Press

Intelligent WBT: Specification and Architecture of the Distributed, Multimedia e-Learning System e-stat

Claus Möbus, Bernd Albers, Stefan Hartmann, Jochen Zurborg

Innovative Learning and Teaching Systems,
Department of Computing Systems, University of Oldenburg

Keywords. Architectures, Authoring Systems, Cognitive Approaches, Distributed Learning Environments, Intelligent Distance Learning, Internet Environments, Learning Environments, Content-Engineering, Intelligent Web Based Training (I-WBT), Specification of e-Learning Systems, Unified Modeling Language (UML), Extensible Markup Language (XML),

1. Introduction

Modern e-Learning Systems (eLS) are expected to be innovative not only concerning comprehensive representation of content enriched by multimedia, but also in the integration of learning situations in contexts suitable for students. Suitable, motivating contexts can be "fun" as found in strategic games or business simulations or of a more "serious" variety in the form of virtual data labs. In the new BMBF Project EMILeAstat (e-stat) 13 partners from different organisations are cooperating to construct such an innovative intelligent web based training (I-WBT) system for applied statistics. The German Federal Ministry of Education and Research finances e-stat by means of the NMB funding program "Neue Medien in der Bildung" (New Media in Education).

Special emphasis is placed on reuse and sharing of contents, clean separation of factual contents and its didactical motivated presentation, as well as the avoidance of proprietary solutions. E-stat is an attempt to go beyond the scope of existing WBT systems by using a strong integrating concept in combining content with a high diversity of methodical and didactical approaches. This ambitious approach creates the need for new research and evaluation. For example, a method for the presentation of coherent and user-adaptive content (learning objects) supplied by a variety of sources has to be found. The development of an eLS can be approached from different angles. This paper describes the architecture of e-stat from a knowledge and content engineering point of view, while applying cognitive-science criteria where necessary. All components were developed in a close cooperation with the statistical content providers.

2. The I-WBT-System "e-stat"

Since e-stat integrates different learning-methods, scenarios, and a consulting component into a knowledge landscape, existing methods of specification[123] have to be modified accordingly, to ensure a systematic method for production of content. For development and analysis purposes we make use of concepts and notations supplied by object orientated analysis (OOA) and object orientated design (OOD)[4]. The specification will be realized on demand during the course of the project.

2.1 Specification of e-stat

In the beginning the wind rose[5] (Fig.1) was used as a metaphor for the e-stat idea. It was meant to express e-stats ambition to supply applicable solutions with changing didactical demands[67] (e.g. instructional, cognitive and constructive): courses of differing levels of complexity for mathematicians, managers, psychologists and engineers but also for people with a special need of practical experience such as industrial technicians.

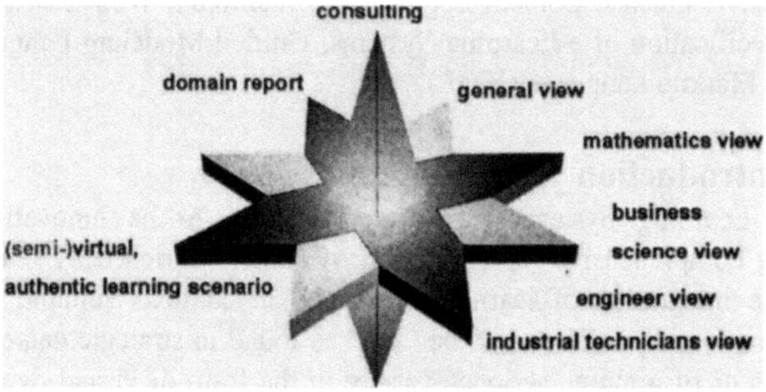

Figure 1: The Wind Rose

e-stat furthermore contains methods to integrate existing statistical engines, (semi-) virtual learning scenarios, an automated glossary, and the case based consulting component for the "hasty user". The wind rose metaphor is a visualisation of the e-stat project suitable for people without expertise in computer science.

Next the wind rose was transferred into use-cases of the semiformal UML-Notation[8] (Fig. 2). A use-case is the typical application of e-stat in a user-specific context. Due to the open nature of the e-stat system, the process of defining new use-cases has not been finalized.

The next step in the OOA constitutes the construction of the static system structure using a class diagram.

A class diagram represents classes with their attributes and operations, inheritance and associations between them. A class defines structure (attribute), behaviour (operations), and relations (associations and inheritance structures) for a collection of certain objects[9]. As can be seen in our class diagram, e-stat is a composite aggregate of views (Fig. 3). Views are shared aggregates of scenarios, courses, course units and concepts. Concepts have recursive structure. They can be built up by text blocks (text leafs) module frames, and/or concepts. This architecture ensures the representation of hierarchically organized lessons. Module frames are again composite aggregates of modules, which are the smallest building blocks of the e-stat landscape.

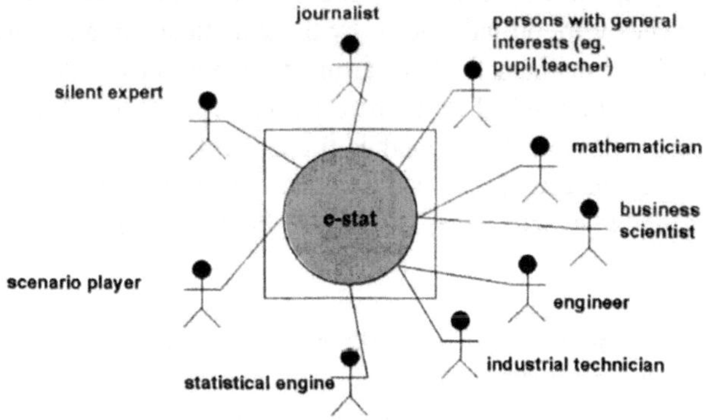

Figure 2: Use-Cases of e-stat

Scenarios are shared aggregates of situations, simulations, courses, and course units. Aggregation (shared aggregate) exists if whole-part relationship holds good between objects belonging to the corresponding classes. Objects can be part of several aggregate objects. Composition (composite aggregate) defines a stronger form of aggregation. In addition to the whole-part relationship there is the following constraint: each object belongs to only one aggregation object at one point in time[9].

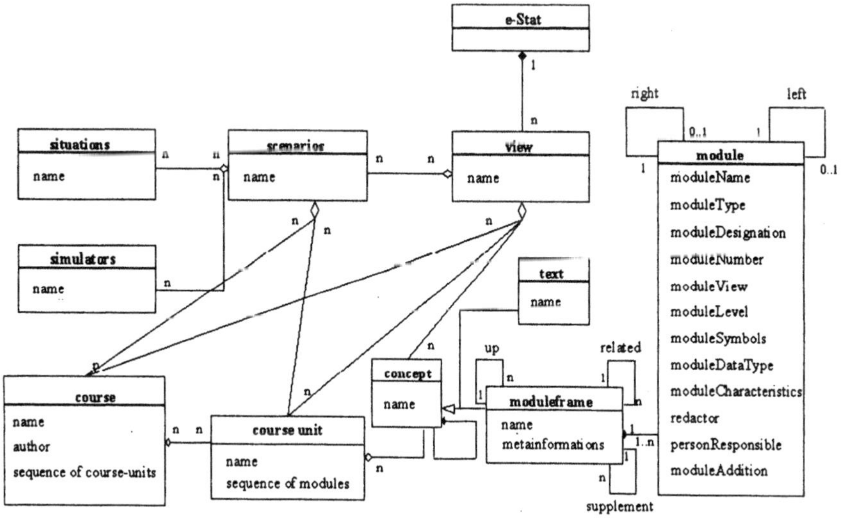

Figure 3: Class Diagram of the Content Aspect

Module frames (e.g. median) can be considered to be a single concept. Thus they can be part of any number of different views (e.g. psychological, mathematical or general view). Therefore, module frames can contain a number of different modules, which each have to be associated with a single view. To reuse modules (e.g. definitions) in a certain view they have to be referenced explicitly by that view. Modules constitute the smallest knowledge-units supplied by content providers. Modules can be of the following types: definition, expression, proof, procedure, example, exercise, solution, comment, text, data, audio, video, XploRe, SPSS, R, etc. Types were provided to us by the statistical content providers, which are members of the mathematical department of our university. Inside module frames, modules are interlinked to define a partial order (e.g. „X depends on Y"). Modules can also point to external module frames or modules contained in external module frames. A conceptual map can be created automatically using this pointer structure. This ontology will be used for case based reasoning in the consulting component of e-stat.

2.2 Implementing e-stat using a 4-Tier Architecture

Most classes of the class diagram are implemented using the standardised XML language. XML allows to create semantic tags additional to syntactic tags (HTML). Authors receive the respective document type definitions (DTDs) to generate valid class objects (Fig. 4). We use DTDs to define elements that can exist in all documents, the attributes those elements can have, the hierarchical structure of elements within elements, and the order in which the elements will appear throughout the documents. The module structure is the result of an interactive process between domain experts, content provider, and developers.

```
<?xml version="1.0" encoding="UTF-8"?>
<!--module: root element of module document -->
<!ELEMENT module (moduleKeywords?, moduleContent, moduleAuthors, moduleSymbols, moduleBibliography?, moduleEnc
<!--module attributes:identifying module-->
<!ATTLIST module
    moduleName CDATA #REQUIRED
    moduleType (definition | theorem | proof | method | example | exercise | solution | comment | text | audio | applet | video | X
    moduleDesignation CDATA #REQUIRED
    moduleAddition CDATA #REQUIRED
    moduleNumber CDATA #REQUIRED
    moduleView CDATA #REQUIRED
    moduleLevel (A | B | C) #REQUIRED
    moduleDataType (ordinal | metric | nominal | none) #REQUIRED
    moduleCharacteristics CDATA #REQUIRED
>
<!--moduleKeywords: used for searching-->
<!ELEMENT moduleKeywords (#PCDATA)>
<!--modulePosition: specifies position of module by right and left-->
<!ELEMENT modulePosition (right?, left?)>
<!ELEMENT right (targetModule)>
<!ELEMENT left (targetModule)>
<!--moduleContent: content (texts, images, etc) of module-->
<!ELEMENT moduleContent (heading?, p*)>
<!-- - - - - - - - - - - - - - - - - - - - - - - - - - - - - - - - - -->
<!--p: paragraph-->
<!ELEMENT p (#PCDATA | b | i | u | br | center | cite | code | targetModule | a | p | table | ul | ol | formula | img | javaApplet | sa
<!ATTLIST p
    id ID #IMPLIED
>
<!--formula: only contains name of formula without '<format>'-->
<!ELEMENT formula EMPTY>
<!ATTLIST formula
    formulaName CDATA #REQUIRED
    id ID #IMPLIED
>
<!--img: like html img element-->
<!ELEMENT img (caption?)>
<!ATTLIST img
    src CDATA #REQUIRED
    align (top | middle | bottom) "middle"
```

Figure 4: The e-stat-DTD

DTDs for modules and module frames have been developed. Independent DTDs for courses (-units) and scenarios will be supplied. Modules are specified by following attributes taken form the class-diagram (Fig. 3): moduleName, moduleType, moduleDesignation, moduleNumber, moduleView, moduleLevel, modulSymbols, moduleDataType, moduleCharacteristics, redactor, personResponsible, and moduleAddition. They are marked as required, which means their values are necessary to identify one module. Additionally, ontological links serve to construct the conceptual map of the modules, which is essential for the consulting-component.

Other optional attributes are moduleKeywords, moduleBibliography (corresponding to the BibTex-Format), moduleConnection, and modulePosition. The last two elements can both be used to define a partial order on modules presented to the students in the graphical user interface (GUI).

Figure 5 shows a screenshot of the student GUI, which is based on a sketch provided us by our statistical partners at our university. The example module presents the concept mean in German. The first row of the GUI shows the administration operations like edit (Bearbeiten), print (Drucken), options (Optionen), history (History), search (Suchen), and help functions (Hilfe). Below the administrating row is the name of the module frame: mean (Arithmetisches Mittel). The information about level, view and the current course follows. Then the partial order is presented in the form of tabs. All these rows show information about the current module. In fig.5 the tab description (Bezeichnung) is selected and the content part of the GUI contains the description of the concept mean. At the bottom of the window some external statistical engines can be choosen. The column on the left-hand side shows navigation capabilities like home, up, related and supplement. Visualizations, interactive exercises, other multimedia enrichments, and information about the authors of the module are further parts of the panel on the left hand side of the screen.

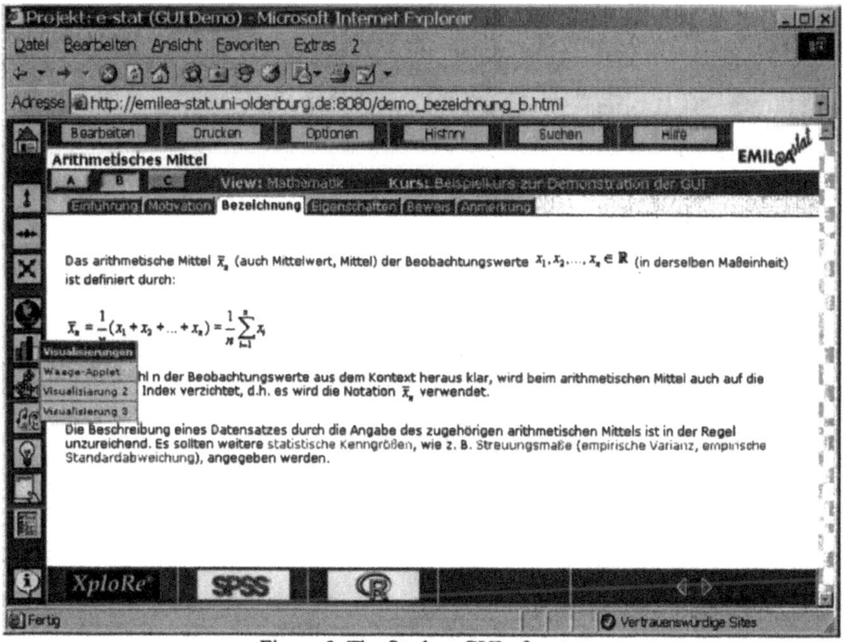

Figure 5: The Student GUI of e-stat

On the bottom right of the window the navigation for the course is located.
Depending on the separation of content and layout the authors will also get a XSL-file, which is responsible for the layout. While developing content, authors have preview permanently.
e-stat is implemented using a 4-tier architecture. The presentation tier supplies content providers and students with suitable graphical user interfaces (GUIs). The view-author GUI will be powerful enough to enable authors to construct a course for his particular target group from the certified e-stat modules by means of a system similar to „shopping cart" systems used by many e-shops. It should only be necessary to construct new modules in very special cases.
Authors can use current XML-editors (as seen in Fig. 4) instead of a special e-stat content-authoring GUI.
In the logic-tier we use an Apache server, which is installed in Oldenburg. On this server our e-stat-control-system is implemented. Interactions with our database and other statistical-engines will be managed in this tier, as well as the handling of the user-administration.
The native XML-database TAMINO is represented in the data-tier. Statistical engines (Xplore, SPSS, qs-stat) and scenario engines for simulation (e.g. handling business planning, production or stock exchange) are part of the application tier.

3. Contact

Prof. Dr. Claus Möbus, Innovative Learning and Teaching Systems, Department of Computing Systems, University of Oldenburg, D-26111 Oldenburg, Germany mailto:moebus@informatik.uni-oldenburg.de

[1] Aviation Industry CBT Committee (AICC)
[2] Learning Technology Standards Committee (LTSC) mit den Work Groups IEEE P1484.1 Architecture and Reference Model WG: Learning Technology Systems Architecture (LTSA) und IEEE P1484.12 Learning Object Metadata WG: LOM Standards
[3] Alliance of Remote Instructional Authoring and Distribution Networks for Europe (ARIADNE)
[4] BALZERT, H., Lehrbuch der Objektmodellierung, Heidelberg: Spektrum Akademischer Verlag, 1999, ISBN 3-8274-0285-9
[5] Förderantrag an das BMBF, Förderkennzeichen 08NM058A, Dezember 2000, Projektantrag 229
[6] JANK, W. & MEYER, H., Didaktische Modelle, Frankfurt a. M.: Cornelsen Scriptor, 1994, ISBN 3-589-21012-5
[7] BRUNS, B. & GAJEWSKI, P., Multimediales Lernen im Netz: Leitfaden für Entscheider und Planer, Berlin: Springer Verlag, 1999, ISBN 3-540-65428-3
[8] BOOCH, G. & RUMBAUGH, J. & JACOBSON, I., The unified Modeling Language User Guide, Addison-Wesley, 1999, ISBN 0-201-57168-4
[9] BALZERT, H., UML kompakt, Heidelberg: Spektrum Akademischer Verlag, 2001, ISBN 3-8274-1054-1

Interactive Exploratory Analysis of Spatio-Temporal Data

Johannes M. Dreesman

Governmental Institute of Public Health Service Lower Saxony, Roesebeckstr. 4-6, D-30449 Hannover, Germany

Abstract. An approach for the exploratory analysis of the space-time interaction in spatio-temporal-data is presented. The approach uses the interactive brushing technique, known from exploratory data analysis (EDA), in order to connect time series with maps. By applying this approach to infectious disease surveillance data, we gain further insight into the space-time-relationship of these data and obtain a tool for detecting outbreaks of infectious diseases.

Keywords. Graphical Data Analysis, Spatial Modelling, Space-Time Interaction, Brushing, Surveillance Data

1 Analysis of spatio-temporal data

Currently there is a strong interest in models for the analysis of spatio-temporal data, e.g. in epidemiology, where data on the spatial and temporal occurrence of cancer or infectious diseases are routinely collected. A major goal of these models is the identification of clusters in space and/or time where diseases occur more often than expected. Further objectives include the prediction of the spatial and temporal spread of infectious diseases like influenza (Cressie & Mugglin, 2000). Spatio-temporal models may include a temporal, a spatial and a space-time interaction component. Because of the complexity of these models, inference is usually carried out in a Bayesian framework using Markov chain Monte Carlo (Knorr-Held & Besag, 1998). To support the modelling process, methods for exploring such data are needed. Time series graphs and maps are established tools for exploring the time and space dimension respectively, but there is a lack of tools for exploring the interaction component.

2 Analysis of infectious disease surveillance data

Monitoring and control of infectious diseases is one of the main tasks of public health authorities. The most important data source is the reporting system for notifiable diseases. In Germany a new infectious disease law was empowered at the beginning of 2001. One major innovation was the reorganisation of the transmission process of the reports from the local health authorities to the states' and the national authorities. One of the objectives of the analysis at the state and the national level is the detection of non-localised "outbreaks", i.e. an increase of the number of cases beyond expected levels. There are many possible artefacts which could cause such an increase. Thus, if an increase is detected, it may be seen as a hypothesis and it has to be checked first, whether the outbreak is real. If an outbreak was verified, immediate control measures might be established, e.g. the attempt to identify and remove the source of infection or the vaccination of persons at risk. Besides outbreak detection, another important objective of the analysis is the observation and prediction of the spread of diseases with an epidemic behaviour, like influenza or other viral diseases, in order to obtain an "early warning system".

These tasks have to be done routinely, i.e. every week, for nearly 50 notifiable diseases. Therefore automatic procedures are preferable, but they are difficult to develop and apply. Especially the detection of outbreaks requires the estimation of the expected "endemic" level from the past. For this purpose robust procedures are needed. First, because seasonality, trends and outbreaks in the past have to be considered. Second, because the diseases differ strongly regarding their frequency, and the procedures should apply satisfactorily to very frequent as well as to rare diseases (Farrington & Beale, 1997). In addition, outbreaks may vary strongly in their spatial and time extension. Concerning these problems, exploratory data analysis (EDA) is best suited for this task, in particular if the hypothesis generating character of the analysis is taken into account.

Figure 1. Map of incidence rates for reported Rotavirus infections from week 5 to week 8 of the year 2001 in Lower Saxony

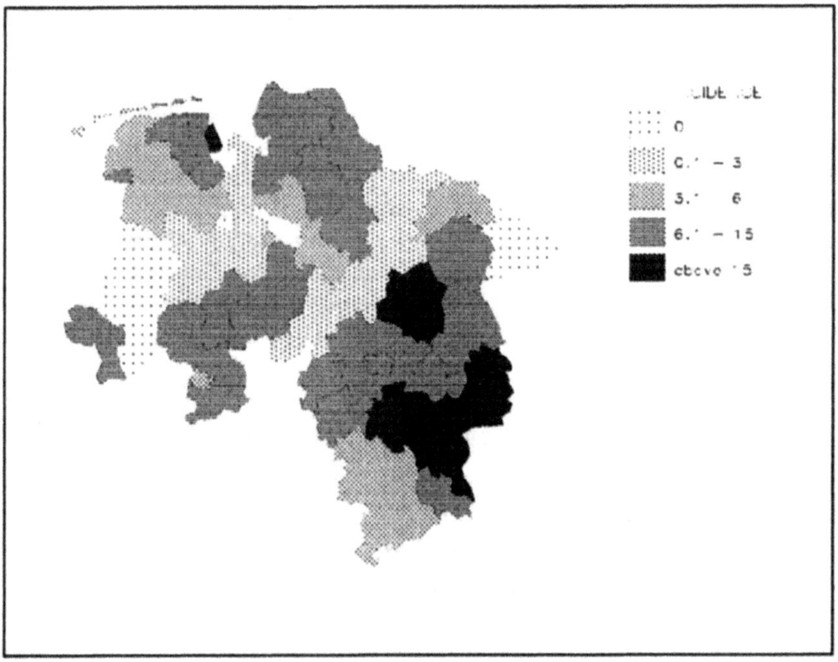

Since an important aspect of the analysis of infectious diseases is their spatial distribution, the representation by maps is an established method for this analysis. Figure 1 shows the incidence rates for Rotavirus in the German federal state Lower

Saxony for the weeks 5 to 8 of 2001. Rotavirus, which causes a rather harmless enteritic disease mainly in children, was the most frequently reported pathogen during the first weeks after the new infectious disease law came into force.

3 Interactive exploratory data analysis

From a statistical point of view, maps of spatial data are an exploratory tool. The goal is to visualise data and to formulate hypotheses. Maps are not very well suited as a confirmatory tool. Even maps of incidence rates hardly allow the identification of significantly increased incidences, since the variance and thus the significance of an incidence rate depends on the size of the underlying population, which usually is quite heterogeneous. Thus, in incidence maps significantly elevated incidences in districts with high population density (towns) often attract less attention than non significant incidences in large, but less densely populated districts (Lawson et al., 1999). Therefore mapping of case counts or incidence rates often is the first step of spatial data analysis and may be followed by other methods for assessing the significance of the observed effects.

This exploratory character of the visualisation of spatial data comes to the foreground in exploratory spatial data analysis (ESDA). Haining et al. (1998) defined ESDA to be an extension of EDA to the problem of detecting spatial properties of data sets where, for each attribute value, there is a locational datum. The major goals are the detection of spatial patterns in data, the formulation of hypotheses which are based on or are about the geography of the data and the assessment of spatial models. For these purposes, as in EDA, visual and robust techniques are used. The special feature is the link between numerical and graphical methods on the one hand and the map on the other hand. The map is the essential additional tool for ESDA.

An important approach to achieve the above goals are interactive graphical methods where objects in different graphs are linked to each other. This linking technique is strongly related to the brushing technique. Brushing was first used for scatterplot matrices showing all pairwise scatterplots of a multivariate data set. The brush denotes a rectangle that is superimposed on one of the scatterplots and can be moved and resized by the data analyst using the mouse. All points from the current scatterplot that are inside the brush are highlighted, as are the points from other scatterplots that correspond to these points (Becker & Cleveland, 1987).

In interactive spatial graphs this approach is used to link objects in statistical graphs like points in scatterplots or bars in histograms to the corresponding spots or regions in a map (Wilhelm & Steck 1998). If a data object is selected in the statistical graph by the mouse, the corresponding region in the map will be highlighted. There are a few programs or program libraries for spatial data analysis where such a linking strategy is incorporated, e.g. REGARD (Haslett et al., 1991), SPLANCS (Rowlingson & Diggle, 1991) or more recently LiveMap (Brunsdon, 1998). LiveMap is an add-on package for XLISP-STAT. The LISP-code adds the mapping functions to the kernel of XLISP-STAT, which provides the linking strategy. Thus, the maps may be linked to other XLISP-STAT graphs like scatterplots or histograms.

4 A tool for interactive exploratory analysis of surveillance data

Figure 2. Interactive analysis of incidence rates of reported Rotavirus infections for the weeks 5 to 8 of the year 2001 in Lower Saxony

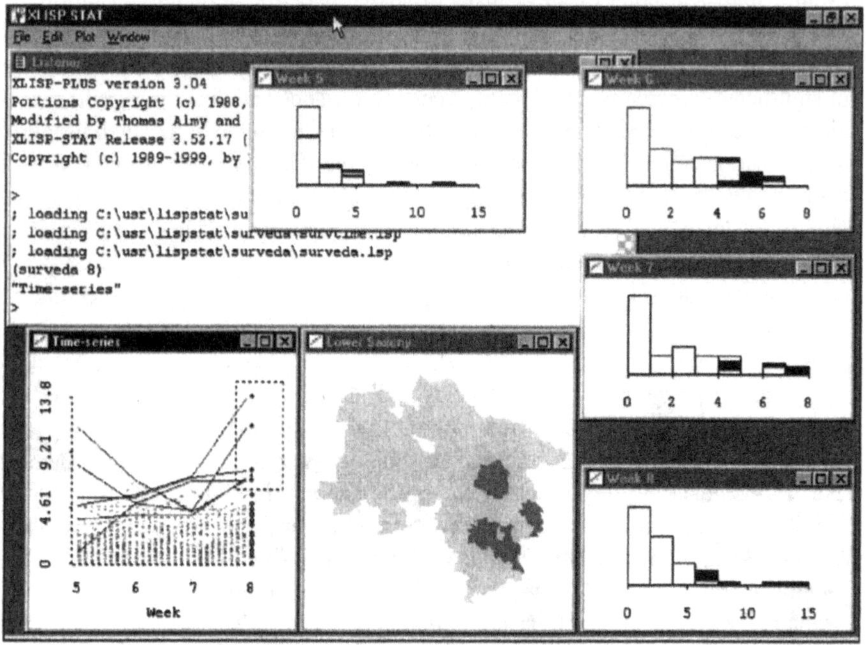

Based on LiveMap and XLISP-STAT we developed a system for analysing infectious disease surveillance data from Lower Saxony. Figure 2 shows the system at work by example of the Rotavirus incidence rates of the 47 districts and self-administered cities of Lower Saxony for the weeks 5 to 8 of 2001. The hardcopy of the XLISP-STAT-screen contains a command window and six graphics windows. The window titled "Lower Saxony", where the map of the state is shown, is generated by the LiveMap-package. The bottom-left window, entitled "Time-series", shows 47 time series of incidence rates over 4 weeks, one for each district. These time series, apart from those selected by the rectangle, are presented by dashed lines and therefore difficult to distinguish. In the remaining 4 windows, for each of the 4 weeks the 47 incidence rates are presented by a histogram. The ordering of the weeks is from the top left graph to the bottom right graph. The six graphs are connected by links, giving a joint tool for interactive analysis. Each graph contains 47 graphical objects, corresponding to the 47 districts. There are 47 time-series, 47 regions in the map or 47 histogram pieces. The links connect the graphical objects corresponding to the same district. Highlighting an object in one of the windows will highlight the corresponding graphical object in every other windows.

In the time series window of Figure 2, the 6 time series with highest incidence rates in week 8 are selected by a rectangle. The selected time series are highlighted by changing their colour from grey to black and the line type from dashed to solid. The final points of the time series change from circles to filled dots. In the map, the districts to which the selected time series refer are highlighted by a darker colour and can immediately be identified. It becomes obvious, that the selected districts are close to each other in the south east of Lower Saxony. In the bottom right histogram, ("week 8"), the districts with the highest incidence rates are also highlighted. From the remaining histograms ("week 5", "week 6", "week 7"), where the distributions of incidence rates in the preceding weeks are given, it can be observed that the selected districts also had high incidence rates in the weeks before. This also could have been seen from the time-series graph. Thus, the histograms don't give much additional information, but they serve as a an example, how further graphs may be connected. While the time-series graph and the histograms can be used to select the districts according to the incidence rates, the map allows to select the districts to be highlighted directly by using the mouse.
The system enables us to analyse the interaction of the spatial and time dimension in an interactive way. It is easily seen whether districts with higher incindence rates cluster in space and how the spatial pattern changes over time. For the Rotavirus infections, taken as example here, the observed spatial cluster doesn't indicate that there is a joint source of infection, but it tells us something about the spatial heterogeneity of the activity of this pathogen.

5 Conclusions

Spatio-temporal modelling is an important issue in the analysis of data that are routinely collected in epidemiology, ecology, or census statistics. The development of computing facilities and algorithms, especially in spatial statistics and Bayesian analysis, enables us to use more complex and realistic models. In addition, there was a strong development in the field of geographic information systems, in order to administer and visualise such data, but these tools are rather static. Therefore, we should also use the computing facilities to explore the data in a more dynamic way by using interactive methods.
As noted by Bailey and Gatrell (1995), the goal of such interactive systems is the presentation of the data in a dynamic and imaginative way by using maps. The purpose is not primarily to draw elegant maps, but to provoke questions and to gain immediate feedback.
The detection of outbreaks in infectious disease surveillance is a rather exploratory task. Since the extension in space and time is not known in advance, one needs special exploratory tools incorporating these two dimensions. Interactive spatial data analysis can be very helpful for this task since it allows a joint visualisation of data in several dimensions, e.g. the space and time dimension. We presented a prototype system for analysing infectious disease data from Lower Saxony in this way and found that a new perspective on this data was opened, which could be rather helpful for detecting outbreaks and for analysing spatial heterogeneity.

References

Bailey, T.C. & Gatrell, A.C. (1995). *Interactive spatial data analysis*. Harlow: Longman Scientific & Technical.

Becker, R.A. & Cleveland W.S. (1987). Brushing scatterplots. *Technometrics,* **29**, 127-142.

Brunsdon, C. (1998). Exploratory spatial data analysis and local indicators of spatial association with XLISP-STAT. *The Statistician*, **47**, 471–484.

Cressie, N. & Mugglin A.S. (2000): Spatio-temporal hierarchical modeling of an infectious disease from (simulated) count data. In: *COMPSTAT2000 Proceedings in computational Statistics*, 41–52, Heidelberg: Physica-Verlag.

Farrington, C.P. & Beale, A.D. (1997). The detection of outbreaks of infectious disease. In: *Geomed97*, 97 – 117. Heidelberg: Springer.

Haining, R., Wise, S. & Ma J. (1998). Exploratory spatial data analysis in a geographic information system environment. *The Statistician,* **47**, 457–469.

Haslett, J., Bradley, R., Craig, P., Unwin, A. & Wills G. (1991) Dynamic graphics for exploring spatial data with application to locating global and local anomalies. *The American Statistician,* **45**, 234-242.

Knorr Held, L., & Besag, J.E. (1998): Modelling risk from a disease in time and space. *Statistics in Medicine*, **17**, 2045 – 2060.

Lawson, A. B., Böhning, D., Biggeri, A., Lesaffre, E. & Viel, J.F. (1999). Disease mapping and its uses. In: *Disease Mapping and Risk Assessment for Public Health*, 3 – 13. New-York: Wiley.

Rowlingson, B.S. & Diggle, P.J. (1991). SPLANCS: spatial point pattern analysis code in S-Plus. *Computers and Geoscience,* **19**, 627 – 655.

Wilhelm, A. & Steck, R. (1998). Exploring spatial data by using interactive graphics and local statistics. *The Statistician, 47*, 423–430.

Interactive Graphics for Data Mining

Daniela Di Benedetto
[1] Department of Computeroriented Statistics and Data Analysis,
Augsburg University, D-86135 Germany
daniela.dibenedetto@math.uni-augsburg.de

Abstract. Today information analyses require investigation of very large raw data sets and the study of relationships between numerous and different kinds of variables. Data Mining techniques have been developed for large data sets, but they produce very many results and these can be difficult to interpret. Adding interactive controls to the standard graphics can help a lot to sort through the results and understand them. This paper reviews also some tools which could give new and interesting ideas on how to develop new interactive representations, and investigates how they scale up for large data sets, with particular regard to multivariate continuous data and tools such as Trellis Displays and Parallel Coordinates, especially as presented in Manet and Cassatt interactive software.

Keyword. Interactive graphics, Data Mining, Parallel Co-ordinates, Trellis.

1 Introduction

In 1637, when René Descartes presented first Cartesian Coordinate plot, data collection and representation required just human effort, hand writing and paper: things are completely different today. Towards the end of the last century, progress in computer science made several major contributions in implementing direct manipulation graphics, making Cartesian Coordinates appear on computer screens as a first raw version of modern visualisation tools (Becker, Cleveland,1996)

Enabling and empowering software in such a way as to make realistic, reliable and understandable visualisations, working with ever larger data sets is, for Data Mining, essential, but as most statistical graphics have primarily been designed for smaller data sets than arise in Data Mining, it is necessary to see whether they can be amended to cope with these new tasks.

While on the one hand Data Analysis must face the problem of manipulating huge amounts of raw data, on the other hand it produces an equally large number of results whose relationship with raw data, must be effectively managed (Unwin, 2001). Another challenge is representing multidimensional data when the number of variables enormously increases. If looking at a data set with more then three or four variables is difficult for our imagination, it is even more complicated to display them successfully in a scientific way (Keim, Kriegel 1993).

Data sets may be classified according to their size (little data set, large data set), kind (categorical, continuous), and the number of variables to be considered (one or two, a few, many, infinite): different graphical tools are better for different set types.

Trellis displays easily represent continuous data sets with a large number of observations, whereas Mosaic plots visualise very well categorical data without being limited by the data set size and, also taking account of missing values (a very common feature of large statistical surveys).

These graphical tools are affected by the number of variables, as no more than five or six can be handled without very much loss of quality of visualisation: Parallel Coordinate displays are the answer for the need to represent within two dimensions, a data set, where the number of variables can be also very large. Cassatt (Winkler, 2000) provides features and tools which enable the investigation of data sets by using interactive Parallel Coordinate with very good results.

We also investigate the structural relationship between Trellis and Mosaic Plot displays, which found a brilliant interactive implementation in Manet (Hoffmann, 2000).

2 Trellis Displays and interactive tools

When attempting to display all data in one plot is unsatisfactory, Trellis graphics break up the data into smaller chunks, based on conditioning on other variables, and plots each chunk separately. The most common version displays a scatterplot in each panel, but it is possible to choose other graphical representations

Displaying many variable combinations in only one display, as Trellis (Becker et al 1996) do is more restrictive than what an interactive control permits: displaying one or many linked views of the total information at a time, making the information itself more interpretable and enabling users interaction. For this purpose, adding an interactive control, could be interesting especially for variable selection and ordering.(Theus, 1995)

For illustration we consider here the Quakes data set, giving the locations of 1000 seismic events of MB > 4.0 observed in a cube near Fiji since 1964. The original data set is a data frame with 1000 observations on 5 numerical and continuous variables (Harvard PRIM-H project).

The subject of this section is the representation of a subset of its obtained selecting long, lat and depth variables, by Trellis display as implemented in R1.4.1 (library base).

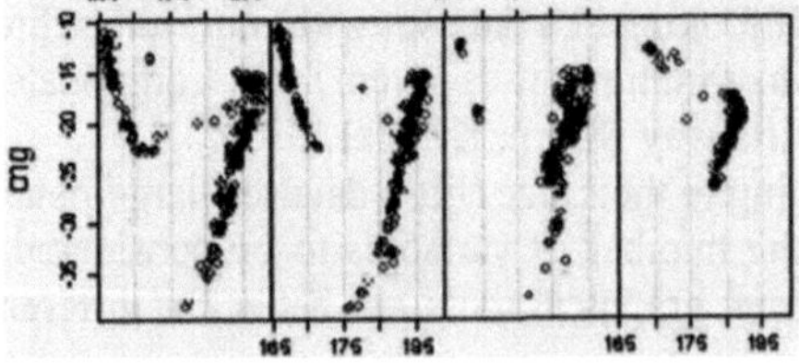

Fig.1 Trellis from R of long, lat and depth variables, extracted from Quakes dataset. We can see how the number of observations could badly affect visualisation quality

In fig.1 we display a Trellis of Long vs. Lat, given Depth, split into four categories. Each panel represents the relationship between the two variables for a certain depth range: it results as a snapshot of a partial view of the total phenomenon, and other views may be necessary to gain a complete picture.

Also, in this example, we display just 1000 observations, but it is immediate how confused the display would appear, if the number of observations was larger (a very common aspect of today's research).

If we look at the structure of the variable lat, in fig.2, separating also this variable in two categories could seem opportune: we will see later which role these two categories of the same variable play according with the others.

A Mosaic Plot is a display, which manipulates categorical variables without any regard of how big categories are. Manet offers its fully interactive version

There is a strong relationship between Trellis displays and Mosaic Plots, and we want to discuss it graphically. The main objective of Mosaic Plots is to uncover relationships between all variables keeping the original data frame structure, whereas, other displays for categorical data, visualise just points or areas proportional to the counts, or can display (as a Trellis Display can) relationships between variables just in two-dimensional "sub-data-frames". Mosaic Plots are not affected by the size of the data set, as they are an area-based display and not a point-based display.

Graphically, the representation order of the single variables is very important in Mosaic Plots: to arrange continuous variables displayed in Trellis to be plotted in a Mosaic plot, following the typical ordering procedure, they must be discretised as the conditional variable whose segments are separately represented in each chunk; discrete variables, can be plotted according to our logical intention and not just as the consequence of a formal constraint (Hoffmann 2000).

Mosaic Plots are plotted in a big rectangle which represents the total number of observations, each intersection dimension between two categorical variables is then represented by a smaller rectangle whose area is proportional to the intensity of intersections themselves. In Trellis displays intersections are represented in equally dimensioned cells, using scatterplots, barcharts or boxplots. In the version with bar chart the display seem to be very close to the Mosaic plot one; also the case with scatterplot give an idea of how the two displays could be intuitively related.

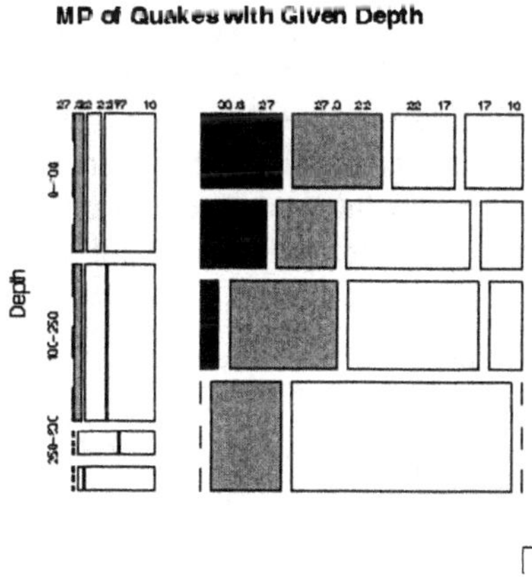

Fig.2 R Mosaic Plot It comes directly from transforming the corresponding Trellis.

The idea to transform directly a Trellis Display into a Mosaic Plot, is to draw a scatterplot, dividing both axes in appropriate intervals, according to segmentation rules, then choosing which set of segments fit most the data set; by obvious consequence, also the Cartesian quadrant will be divided into two-dimensional intervals. The results will be a "mosaic" with all different tiles. Then, each category combination can be represented by a rectangle with area proportional to the number of observations falling inside each related two-dimensional interval and then plotted according to Mosaic Plot rules.

Leaving the Trellis continuous structure, for the Mosaic plot categorical one, could seem to involve an information loss: this is partially true, but what we gain is a more informative combination of the total of the category belonging to all of the investigated variables. Furthermore, Mosaic Plot quality display and informative power are not influenced by the number of observations and treatment of missing values makes an informative contribution. At this stage, all the interactive tools would be imported, and, also there could be interactive variables selection with the aim of deciding whether to display one plot at time or a whole panel as in the original. Variables can also be differently ordered, and recursive intersections selected.

In fig.3 a different parametrisation of the Mosaic Plot is shown: the "Bar Chart Mosaic Plot", illustrates the strong relationship between Trellis and Mosaic Plot. We plot the two Lat segments, in two lines, and then four Long levels, and finally, within each Long level four Depth levels. By using interactive tools, we selected bins related to the second Lat variable category in quakes data set: now its behaviour within the whole data set it is clear and comparable, category by category.

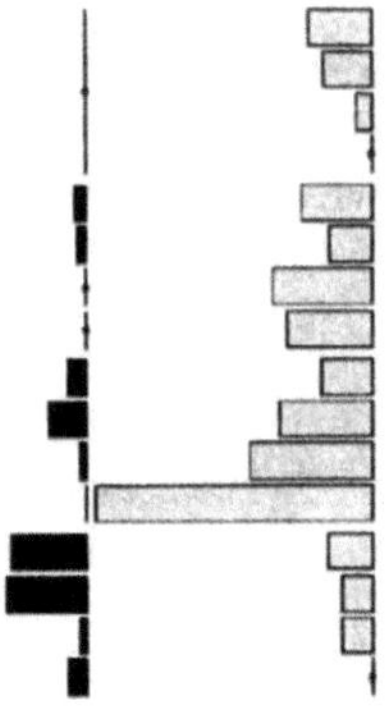

Fig.3 Barchart Mosaic Plot, Manet Manet provides a set of interchangeable options for drawing Mosaicplots: one of them is the Barchart option, from which it is easy to see which role the depth variable(lower values plotted on the left side) plays: maybe the segmentation which could appear to be obvious from the Trellis display, could be improved by choosing a partition algorithm producing a smallest number of totally empty classes but optimising within-homogeneity and between-heterogeneity.

While transforming Trellis in Fig.1 to the corresponding Mosaic Plot in Fig.3 we used only static tools offered by R1.4.1. This transformation consists of many elementary steps, which R can very easily handle, but which can be enormously reduced if interactive software is available. Here we have, for instance, two different examples of a Mosaic Plot representing the same data, which, at a first sight, suggests some changes in the segmentation. An interactive tool ensures sufficient flexibility to make changes without starting again from the initial point of a display building process.

3 Interactive Parallel Coordinates

Parallel Coordinates (Inselberg, 1998) contribute to Exploratory Data Analysis by being able to display a large number of variables in a 2-dimensional space. They offer a very good display of many continuous variable but with restricted drawing of lines and are mostly useful when implemented in interactive software such as Cassat. The first advantage of Parallel Coordinates is its absolute simplicity of representation of very complex data sets. It must be interpreted differently than a standard 2-dimensional display and by using rules determined by the fundamental Point-Line duality of Parallel Coordinates: from this duality all the relationships representations follow.

Parallel Coordinate can displays Time Series (Fig.4) giving a different view of the data. We can connect by lines each individual on each variable axis (years in this case) or display boxplots and related statistics by year. Interactive tools enable sorting according to selected statistics, which allow comparative performance assessment.

The figure shows the "Yields of Wheat on the Broadbalk Fields at Rothamsted" Time series (Fisher, 1924) in such a way to have each year's observations on a different axis. There are 18 time series observed annually over a 76 year period.

In Fig.4 we selected by zooming the last part of the Time Series as we noticed that was a very interesting one because of some unusual behaviour. For instance, we highlighted a Time Series which showed a high value in most of these years, except in 1917, which falls right in the middle of the First World War: this is the only year in all 76 axis where this series had an average value. The Zoom interactive tool plays an important role when the number of variables increases.

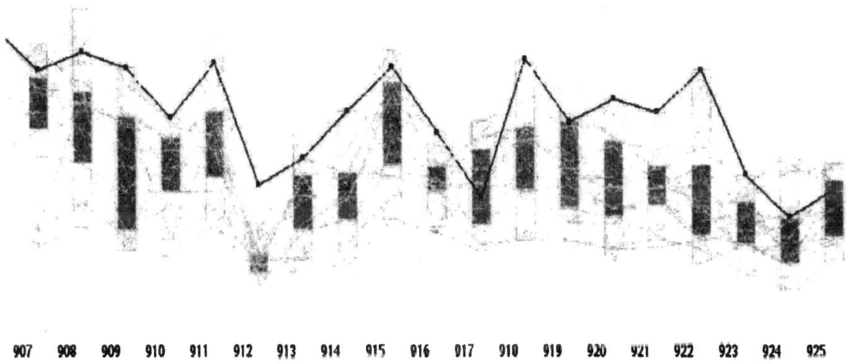

Fig.4 Parallel Coordinate plot of Fisher Grain time series.

We could also have sorted according to one of many relevant statistics to further investigate these behaviours, or rescaled variables stardandising them, or grouping, selecting and querying those Time series which show a particular path. Interactive query and selection tools enable the user to query the Data Base by visually selecting the desired individuals, graphically isolating (highlighting) them, and then obtaining from the system related information and characteristics. This is a query system inverted with respect to the common one, referred to as a star system Data Base interrogated by SQL select commands.

Cassatt can manipulate a reasonably large number of observations (though with 40.000 cases can appear messy), but its value is greatest when there are many variables (it has been tested that it is not a problem displaying up to 300 variables) and not a huge number of observations, as is true for any Parallel Coordinates system.

4 Conclusions

Visualising Data is always difficult, but is more difficult for large data sets because of: the number of data, the number of variables and the different kinds of data. An additional problem arises when exploring data in that new information gained progressively influences the analysis.

Trellis Display and Mosaic Plot offer different and interesting data visualisations, but the interactive options and the possibility to visualise the structure between all the variables with only one display, make it interesting to study the relation between Trellis and Mosaic Plot, and transform the first in the second one: here we tried this transformation and we realised that while information may be lost because of the discretising process, more information may be gained about the total data and variables structure.

Parallel Coordinates, on the other hand, is a very flexible tool whose approach can handle high dimensional data set with very good and interesting performances, because of the special geometry of its display and also because of the interactive implementation: for instance, zooming tools and sorting empower Parallel Coordinate′ ability to work with a very large number of variables.

Parallel Coordinates can display many different kinds of data, and in this case was shown to be a good tool for displaying Time Series: to manipulate them by using interactive Parallel Coordinate, gives a different perspective and can add some value to the visual approach.

From the computer science point of view, interactive visualisation in general, considerably reduces computational flows and the number of operations to carry out: in fact interactivity consists in re-using results just obtained by previous elaboration, and in automatically translating user action (through computer device usage) in automatic data elaboration and also with the aim of getting the best and most informative results. Empowering interactive data mining is an important way to optimise the use of existing software and hardware and, at the same time, to stimulate improvements in software and hardware technology. Statistically, this approach provides greater flexibility and facilitates a more exploratory approach making real the opportunity to display huge data sets.

References

Becker, R., Cleveland, W.S., Shyu, M-J. (1996). *The Visual Design and Control of Trellis Display*. JCGS, 5, 123-155.

Fisher, R.A.,(1924), *The influence of rainfallon the yield of wheat at Rothamsted*, Phil. Trans.Roy.Soc. London B213, 89-142

Hofmann, H. (2000). *Exploring categorical data: interactive mosaic plots*. Metrika 51(1), 11-26.

Inselberg, A. (1998). *Visual Data Mining with Parallel Coordinates*. Computational Statistics, 13(1), 47-63.

Keim, D.A., Kriegel,H.P.(1993) *Possibilities and Limits in Visualising Large Amounts of Multidimensional Data*, Proc. Int. Workshop on Perceptual Issues in Visualization, San Jose, CA

Theus M.(1995) *Trellis Display vs.Interactive Graphics , Computational Statistics,* Computational Statistics

Unwin, A. R. (1999). *Requirements for interactive graphics software for exploratory data analysis.* Computational Statistics, 14, 7-22.

Unwin, A. R., Hofmann, H., Bernt, K. (2001). *The TwoKey Plot for Multiple Association Rules Control. In L. De Raedt Siebes*, A. (Ed.), PKDD, (pp. 472-483). Freiburg: Springer.

Winkler, S. (2000). *CASSATT* www1.math.uni-augsburg.de/Cassatt/

Least Squares Reconstruction of Binary Images using Eigenvalue Optimization

Stéphane Chrétien[1] and Franck Corset[2]

[1] Département de mathématiques, Université de Franche-Comté, 16 route de Gray, 25030 Besançon Cedex, France

[2] INRIA Rhône-Alpes, 655 avenue de l'Europe, Montbonnot, 38334 Saint Ismier Cedex - France

Abstract. The goal of this paper is to present an application to binary image least-squares estimation of some recent results in the semi-definite programming approximation theory of some combinatorial problems due to Goemans and Williamson, Yu. Nesterov and others. In particular, we show in a very simple fashion that a good suboptimal solution may be obtained via eigenvalue optimization.

Keywords. Least-squares estimation, Semi-Definite programming, eigenvalue optimization, Combinatorial optimization.

1 Introduction

A standard (penalized) linear least-squares estimation problem is an optimization problem of the form

$$\min_{x\in\mathbb{R}^n} \|y - Ax\|^2 + x^t P x$$

where $y \in \mathbb{R}^m$ is an observation vector, $A \in \mathbb{R}^{m\times n}$ is a linear operator, $x \in \mathbb{R}^n$ is the original vector parameter to be recovered and $P \in \mathbb{R}^{n\times n}$ is a penalization matrix. In practice, the penalization quadratic term $x^t P x$ is introduced to take into account the correlations between the componants of x. Here, we focus on more special problems where x is required to be a binary vector. The binary linear least-squares estimation problem

$$\min_{x\in\mathbb{R}^n} \|y - Ax\|^2 + x^t P x \text{ subject to } x \in \{-1, 1\}^n$$

is therefore a combinatorial optimization problem.

Standard methods for solving this complex estimation problem are Simulated Annealing or Genetic algorithms. Notice that the methods usually used either do not necessarily provide binary outputs and thus "blind" truncation has to be performed in order to retrieve 0-1 solution, or do not garante any performance bound in polynomial time.

Our approach is quite different. We propose an eigenvalue relaxation which provides an upper bound to the objective function of an equivalent problem. Using Lagrangian duality theory (cf. [2, Chap. XII]) and a randomized method invented by Goemans and Williamson [1], we show how a good (and sometimes optimal) binary solution can be recovered from the presented scheme. Some simulations results are proposed in the final section.

The main advantage of our eigenvalue relaxation over the standard Semi-Definite Programming (SDP) relaxation of [1] or [3] is that the variables to be optimized lie in $\mathbb{R}^{n+1}$ instead of $\mathbb{R}^{\frac{n(n+1)}{2}}$ for SDP, resulting in high reduction of the computational effort.

Notations. Transposition is denoted by $\cdot^t$. The set of real symmetric matrices of order n is denoted by $\mathbb{S}_n$. The maximum eigenvalue of $X \in \mathbb{S}_n$ is denoted by $\lambda_{\max}(X)$. For any vector u in, say $\mathbb{R}^d$, $D(u)$ is the diagonal matrix whose diagonal elements are the components of u in the same order. The vector e is the vector of ones. The diagonal vector of a matrix X in $\mathbb{S}_n$ is denoted by $\text{diag}(X)$. The identity matrix is denoted by I.

2 Eigenvalue relaxation

The first step towards an efficient approximation of the solutions to our problem is a formulation as a polynomial programming problem. Indeed, is easily seen that the constraint $x \in \{-1,1\}^n$ is equivalent to $x_i^2 - 1 = 0$, $i = 1,\ldots,n$. Further notice that our solution will be the same if instead of minimizing the quadratic objective function, we maximize its opposite. Finally, we obtain the following equivalent problem

$$\max_{x\in\mathbb{R}^n} -x^t(A^tA+P)x + 2y^tAx \text{ subject to } x_i^2 - 1 = 0, \quad i = 1,\ldots,n \tag{3}$$

the optimal value of which being the opposite of the original least-squares objective fonction. This last problem can again be converted into the more convenient form.

$$\max_{x\in\mathbb{R}^{n+1}} x^tBx \text{ subject to } x_i^2 - 1 = 0, \quad i = 1,\ldots,n+1 \tag{4}$$

with

$$B = \begin{bmatrix} -(A^tA+P) & A^ty \\ y^tA & 0 \end{bmatrix} + \mu I$$

where μ can be chosen such that $B \succcurlyeq 0$. To see this we use a well known trick. Introduce an additional variable x_{n+1}. Then the objective $-x^tAx + 2y^tAx$ is equal to $-x^tAx + 2x_{n+1}y^tAx$ with the constraint $x_{n+1} = 1$. Moreover, it is equivalent from the viewpoint of solution vector x to add to the objective the constant $(n+1)\mu$ which itself is equal to $\mu(x^tx + x_{n+1}^2)$ for all binary vector x. Problem (3) has therefore the same solutions as

$$\max_{x\in\mathbb{R}^{n+1}} x^tBx \text{ subject to } x_i^2 - 1 = 0, \quad i = 1,\ldots,n \text{ and } x_{n+1} = 1.$$

Now, notice that a solution of this last problem can be easily recovered from (4) as follows : assume that a solution x of (4) is given. If $x_{n+1} = 1$ we are done. Otherwise $-x$ is also solution. Thus, problem (3) and problem (4) have the same solutions.

We then obtain the following result

Theorem 1 *The optimum value of the (convex) maximum eigenvalue minimization problem*

$$\min_{u\in\mathbb{R}^{n+1}} \lambda_{\max}\left(B + D(u) - \frac{1}{n+1}e^t u\right) \tag{7}$$

is an upper bound to the optimum value of (3).

Proof We use Lagrangian duality theory as presented in [2]. The Lagrangian function is given by

$$\begin{aligned} L(x,u) &= x^t B x + \textstyle\sum_{i=1}^{n+1} u_i(x_i^2 - 1) \\ &= x^t(B + D(u))x - e^t u \end{aligned}$$

since $\sum_{i=1}^{n+1} u_i(x_i^2 - 1) = x^t D(u)x - e^t u$. An important remark is that we can restrict the space of possible solutions to the sphere $\mathcal{S}_n = \{x \in \mathbb{R}^{n+1} \mid \|x\|^2 = n+1\}$ which obviously includes the set $\{-1,1\}^{n+1}$. The dual function $\theta(u)$ is defined by

$$\theta(u) = \sup_{x\in\mathcal{S}_n} L(x,u)$$

where the supremum is attained for all $u \in \mathbb{R}^{n+1}$ since $L(\cdot,u)$ is continuous and $\mathcal{S}_n$ is compact. Therefore

$$\begin{aligned} \theta(u) &= \max_{x\in\mathcal{S}_n} x^t(B + D(u))x - e^t u \\ &= \max_{x\in\mathcal{S}_n} x^t(B + D(u))x - \tfrac{1}{n+1} e^t u x^t x \text{ since } x^t x = n+1 \\ &= \max_{x\in\mathcal{S}_n} x^t(B + D(u) - \tfrac{1}{n+1} e^t u)x \end{aligned}$$

and thus, we finally obtain $\theta(u) = \lambda_{\max}\left(B + D(u) - \frac{1}{n+1}e^t u\right)$. The fact that $\theta(u)$ is an upper bound to the optimal value of problem (4) is a standard result from duality theory. See [6] for details. □

The following proposition proves the existence of a minimizer u^* of θ.

Proposition 1 *The eigenvalue dual function θ admits at least one minimizer.*

Proof. See [6] □

It is important to realize what we have done : Theorem 1 allows to replace the original NP-hard problem (3) by the (simpler) convex problem (7) which is thus solvable in polynomial time. It should be clear that the convex dual problem may not give the optimal value of problem (3). Therefore problem (7) is called a convex relaxation of problem (3). The difference between the respective optimal value of (7) and (3) is called the duality gap.

A well know result in duality theory gives a typical situation where the duality gap is zero and an optimal solution can be recovered.

Proposition 2 *If u^* is a minimizer of θ and θ is differentiable at u^*, then any x^* maximizing $L(x, u^*)$ over S_n is a binary solution to (3).*

Proof. See [6] □

The maximum eigenvalue is not always differentiable. In convex analysis, subdifferentiability is the adequate generalization of differentiability. See for instance [2] for an exhaustive treatment of the subject and [4] for first and second order analysis of the maximum eigenvalue function. Moreover, the eigenvalue function $\min_{u \in \mathbb{R}^{n+1}} \lambda_{\max}\left(B + D(u) - \frac{1}{n+1} e^t u\right)$ can be minimized very efficiently using bundle methods, e.g. the $\mathcal{U}$-Newton algorithm of Oustry [4].

3 Worst case bound and randomized approximation

The previous section provided an upper bound to the transformed problem (3). Our next question is : how good is this approximation ? The answer is given by a theorem of Yu. Nesterov [3]. Nesterov's bound is obtained as follows. Problem (4) is in fact equivalent to the matrix optimization problem

$$\max_{X \in \mathbb{S}_n} \text{trace}(BX) \text{ such that } X \succcurlyeq 0, \text{ diag}X = I, \text{ rank}X = 1 \tag{11}$$

To see this, realize that the rank condition implies that $X = xx^t$ for some $x \in \mathbb{R}^{n+1}$, which implies that $\text{diag}X = (x_1^2, \ldots, x_n^2)^t$. Thus the constraint diagX $= I$ implies that $x \in \{-1, 1\}^{n+1}$. Finally, $\text{trace}(BX) = x^t Bx$ and we are done. Nesterov's relaxation consists in removing the rank condition which is the only nonconvex constraint of problem (11). The resulting problem given by

$$\max_{X \in \mathbb{S}_n} \text{trace}(BX) \text{ such that } X \succcurlyeq 0 \text{ and diag}X = I \tag{12}$$

is convex and is called a Semi-Definite Programming (SDP) relaxation of our combinatorial problem.

Nesterov proved the following result.

Theorem 2 (3) *Let $val(4)$ and $val(12)$ denote respectively the optimum value of (4) and (12). Then, the SDP relaxation problem (12) satisfies $val(12) \geq val(4) \geq \frac{2}{\pi} val(12)$.*

The fact that the optimum value of our eigenvalue relaxation equals the optimal value of SDP relaxation results is in fact a well known result. The following proposition shows how to recover a solution of the SDP relaxation from a solution of our eigenvalue minimization problem.

Proposition 3 *The eigenvalue relaxation and the SDP relaxation provide the same bound. Moreover, let (e_j^*), $j = 1, \ldots, r$ be a family of eigenvectors associated to $\lambda_{\max}(B + D(u) - \frac{1}{n+1} e^t u)$ whose multiplicity is denoted by r. Let (α_j),*

$j = 1,\dots,r$ be a family of nonnegative real numbers with $\sum_{j=1}^{r} \alpha_j = 1$ such that $\sum_j \alpha_j (e_j^{*2} - 1) = 0$. Then, the matrix X^* defined by $X^* = \sum_j \alpha_j e_j^* e_j^{*\mathsf{T}}$ is a solution to the Semi-Definite problem (12). Moreover, the optimal values of (12) and (7) are equal.

Proof. See [6]. □

Using these results, we are now in a position to present the randomized algorithm for reconstruction of a binary solution. This randomized method was discussed extensively in [1] and [3], and we refer the reader to these references for further details. The algorithm works as follows :

- compute $V = X^{1/2}$
- fix a positive integer k and pick k vectors z^i, $i = 1,\dots,k$ at random with uniform distribution on the sphere $\mathcal{S}_n$.
- for all i set $x^i = sign(V^t z^i)$ where sign is taken component-wise.
- take $i_0 = \max_{i=1,\dots,k} {x^i}^t B x^i$.

Then, with very high probability, ${x^{i_0}}^t B x^{i_0}$ is very close to $val(12)$. See [3] for details.

4 Simulation results

We applied the eigenvalue optimization technique to the problem of binary image denoising. In this problem, y is the observed image which is just the original corrupted by additive Gaussian white noise with standard deviation equal to .5, the operator A is the identity matrix and the penalization matrix P is the matrix associated with the quadratic form

$$\sum_i \sum_{j \in \mathcal{V}(i)} (x_i - x_j)^2$$

where $\mathcal{V}(i)$ is the simplest neighborhood of point i on the image consisting of first order neighbors except the diagonal ones. In this simulations, we programmed a standard bundle method with Wolfe's line-search to minimize the eigenvalue problem. The original image is a small portion of a printed letter. Application to the full character denoising problem will be performed in the near future. We obtained the results given in figure 1 below.

5 Conclusion

In this paper, we presented an approach to binary linear least-squares estimation based on Lagrangian duality and eigenvalue optimization. As opposed to previous strategies, binary sub-optimal solutions can be obtained via a randomized algorithm, producing an objective value within a garanteed interval near the optimal value. The eigenvalue approach has the advantage of reducing the dimension of the solution space in comparison to Semi-Definite Programming strategy. Recent methods such as second order bundle methods allow to expect efficient resolution of large scale problems.

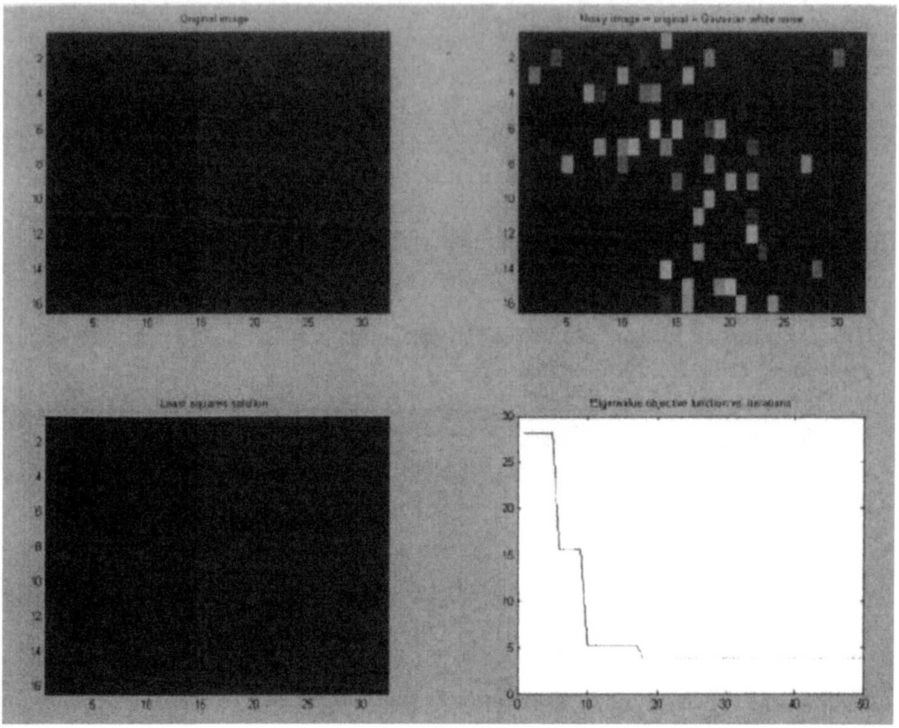

Figure 1: Simulations results

References

[1] Goemans M. and Williamson D., Improved approximation algorithms for maximum cut and satisfiability problems using semi-definite programming. *Journal of the ACM* (1996).

[2] Hiriart-Urruty J.-B. and Lemaréchal C., *Convex analysis and optimization algorithms*, Springer (1993).

[3] Nesterov Yu., Quality of semi-definite relaxation for nonconvex quadratic optimization, *Core discussion paper 9719* Center for Operations Research & Econometrics, Louvain-la-Neuve (1997)

[4] Oustry F., A second order bundle method to minimize the maximum eigenvalue function, *Math. Programming*, **89** no. 1 Ser. A, pp. 1–30 (2000).

[6] Chrétien S. and Corset F., Eigenvalue relaxation of some binary linear least-squares estimation problems, Technical Report, Mathematics department, Université de Franche-Comté, *in preparation.*

Locally Adaptive Function Estimation for Categorical Regression Models

Alexander Jerak and Stefan Lang

Department of Statistics, University of Munich, Ludwigstr. 33,
80539 Munich, Germany.

Abstract
In this paper we present a nonparametric Bayesian approach for fitting unsmooth or highly oscillating functions in regression models with categorical responses. The approach extends previous work by Lang et al. (2002) for Gaussian responses. Nonlinear functions are modelled by first or second order random walk priors with locally varying variances or smoothing parameters. Estimation is fully Bayesian and uses latent utility representations of categorical regression models for efficient block sampling from the full conditionals of nonlinear functions.

Keywords adaptive smoothing, MCMC, nonparametric Bayesian regression, random walk priors, unsmooth functions, variable smoothing parameter.

1 Introduction

Nonparametric methods for fitting smooth curves, such as kernel, local or spline regression, are now widely available and accepted. However, these methods can have bad performance when estimating unsmooth functions which have jumps, edges, or are highly oscillating. Two prominent approaches in nonparametric regression with Gaussian responses that adapt to such spatial heterogeneity are local regression with variable bandwidth (Fan and Gijbels, 1995) or wavelet shrinkage regression (Donoho and Johnstone, 1994). Currently, these methods are restricted to metrical responses and there is a clear lack of methodology for non-Gaussian responses.

In this paper we present a nonparametric fully Bayesian method for fitting unsmooth and highly oscillating functions in regression models with *categorical responses*. The approach extends recent work by Lang et al. (2002) for Gaussian responses. Our approach uses a two-stage prior for the unknown regression function. The first stage are first or second order random walk models as a discretized Bayesian version of the common roughness penalty for smoothing splines. The second stage consists of analogous smoothness priors for varying variances of the random walk model errors used in the first stage. The varying variances in our method correspond to variable smoothness parameters and make the prior more flexible for modelling functions with differing curvature.

Bayesian inference is based on latent utility representations of categorical regression models, see e.g. Albert and Chib (1993). The approach is particularly useful for probit models, because in this case the full conditionals of unknown parameters are Gaussian and efficient MCMC sampling schemes developed for Gaussian responses can be exploited.

The performance of our approach is illustrated by selected results from an extensive simulation study.

2 Bayesian locally adaptive probit models

2.1 Categorical response models

Consider regression situations, where observations (y_t, x_t), $t = 1, \ldots, T$, on a categorical response y with categories $r \in \{1, \ldots, k\}$ and covariates x_t are given. Categorical response models may be motivated from the consideration of latent utilities. For the case of a *nominal response* y with unordered categories $1, \ldots, k$, let U_r be a latent utility associated with the r th category. Assume that U_r is given by

$$U_r = \eta_r + \varepsilon_r, \tag{1}$$

where η_r is a predictor depending on covariates and parameters, and $\epsilon_1, \ldots, \epsilon_k$ are random errors. For simplicity, we assume for the rest of the paper that η_r depends only on a function $f_r(x)$ of a single (metrical) covariate x, i.e. $\eta_r = f_r(x)$. The primary focus is thereby on modelling highly oscillating functions or functions with discontinuities. Extensions to situations with more than one covariate or with different covariates for each predictor η_r are straightforward and pose no further difficulties.

Following the principle of random utility the observable response y is determined by

$$y = r \quad \Leftrightarrow \quad U_r = \max_{j=1,\ldots,k} U_j, \tag{2}$$

i.e. in choice situations the alternative is chosen which has maximal utility. Since only differences of utilities are identifiable we may set $U_k = 0$ for the reference category k. If the ϵ's are i.i.d. normal, one gets the independent probit model. Assuming that i.i.d. error variables follow the extreme value distribution yields the multinomial logit model.

For the case of an *ordered response* y, the widely used cumulative threshold models can be constructed in a similar way, see e.g. Albert and Chib (1993).

2.2 Prior models

For Bayesian inference the unknown functions f_r, or more exactly the corresponding vectors of function evaluations, are considered as random and must be supplemented by appropriate prior distributions. For simplicity, we assume that the covariate values x_t are all distinct, observed on an equidistant grid and ordered. In the following let f denote one of the unknown functions

f_r. Define $f_t := f(x_t)$ and let $f = (f_1, \ldots, f_T)'$ be the vector of function evaluations. Then, a common prior for a smooth function f is a first or second order random walk model

$$f_t = f_{t-1} + u_t \quad (RW(1)) \quad \text{or} \quad f_t = 2f_{t-1} - f_{t-2} + u_t \quad (RW(2))$$

with Gaussian errors $u_t \sim N(0, \tau^2)$ and diffuse priors $f_1 \propto constant$, or f_1 and $f_2 \propto constant$, for initial values, respectively. The amount of smoothness is controlled by the variance parameter τ^2 which corresponds to the smoothing parameter in a classical approach. To estimate the amount of smoothness simultaneously with the unknown function a highly dispersed inverse gamma prior $\tau^2 \sim IG(a, b)$ is assigned in a further stage of the hierarchy.

For unsmooth or highly oscillating functions, as primarily considered in this paper, the assumption of a global variance or smoothing parameter is however not appropriate. Following Lang et al. (2002) we therefore replace the constant variance τ^2 by *locally adaptive* variances $\tau_t^2 = exp(h_t)$. For the parameters h_t we add a second smoothness prior in the form of first or second order random walks for $h = (h_q, \ldots, h_T)$, i.e.

$$h_t = h_{t-1} + v_t \quad \text{or} \quad h_t = 2h_{t-1} - h_{t-2} + v_t, \quad v_t \sim N(0, \sigma^2).$$

The index q depends on the choice of the prior for f, for a RW(1) $q = 2$ and for a RW(2) $q = 3$. Once again, a highly dispersed inverse gamma prior is assigned to the variance parameter σ^2.

Alternatively we may assume *independent* local variances τ_t^2 instead of dependent variances as in the formulation above. This may be particularly useful for functions with discontinuities. Assuming i.i.d. inverse gamma priors

$$\tau_t^2 \sim IG(\nu/2, \nu/2)$$

the marginal distribution of the errors is a Student distribution with ν degrees of freedom. The case $\nu = 1$ of a Cauchy distribution is of special interest as a robust prior and is used for the rest of this paper. In the following, the approach with independent local variances is refered to as *hierarchical t-formulation.*

2.3 Bayesian inference via MCMC

For categorical probit models, as considered in this paper, a useful and very efficient sampling scheme can be developed on the basis of the latent variables with Gaussian errors defined in (1). Bayesian inference is based on the posterior augmented by the latent variables $U_t = (U_{t1}, \ldots, U_{tk})$:

$$p(f, h, \tau^2, U|y) \propto p(y|U)p(U|f) \prod_{r=1}^{k-1} \{p(f_r|h_r)p(h_r|\sigma_r^2)p(\sigma_r^2)\},$$

with $p(y|U) = \prod_t p(y_t|U_t)$, $f = (f_1', \ldots, f_{k-1}')'$, $h = (h_1', \ldots, h_{k-1}')'$ and $\tau^2 = (\tau_1^2, \ldots, \tau_{k-1}^2)'$. The conditional likelihood $p(y_t|U_t)$ is determined by the mechanism (2) and is given by

$$p(y_t|U_t) = \sum_{r=1}^{k} I(max(U_{t1},\ldots,U_{tk}) = U_{tr})I(y_t = r). \tag{3}$$

The full conditionals for the U_t's are truncated normals with truncation points determined by the restrictions (3). The advantage of augmenting the posterior by the latent variables is that the full conditionals for functions f_r become Gaussian, allowing the usage of the sampling schemes developed for Gaussian responses in Lang et al. (2002) with only minor changes. More specifically, additional drawings from the full conditionals for the U_t's are necessary. Updating of f, h and τ^2 can be done exactly as described in Lang et al. (2002) using the current values of the latent utilities as the response variables.

3 Simulation study

A simulation study was carried out to illustrate the performance of our locally adaptive approach. Due to page limitations the presenation is restricted to binary probit models and a rescaled version of the Doppler function for f considered by Donoho and Johnstone (1994). Binary responses $y = (y_1,\ldots,y_T)$ were generated by setting $\eta_t = f(x_t)$ and drawing $B(1,\Phi(\eta_t))$ distributed random variables y_t, $t = 1,\ldots,400$. The function f is the rescaled Doppler function shown in Figure 2 (solid lines).

We generated 250 replications of the model and applied our approach to each replication. Usually, the best results were obtained by using a RW(2) prior for f and a RW(1) prior for varying variances (in the following abbreviated by RW2VRW1). The presentation is therefore restricted to this case. For comparison we also applied the hierarchical t-formulation (TRW2) and a simple RW(2) for f with a global variance (RW2).

Figure 1 displays boxplots of $log(MSE) = log(1/T\sum_{t=1}^{T}(f(x_t) - \hat{f}_t(x_t))^2)$ for the various estimators and Figure 2 shows function estimates averaged over the 250 replications together with the true function. Finally, Figure 3 shows the average estimated local variances and weights for RW2VRW1 and TRW2, respectively. From Figures 1 - 3 we can draw the following conclusions:

- In terms of MSE and bias the approach RW2VRW1 with dependent local variances clearly outperforms the competing approaches RW2 with a global variance and the hierarchical t-formulation TRW2 with independent local weights.

- The estimated variance function for RW2VRW1 accurately reflects the spatial variability of the Doppler function, while the estimated weights of TRW2 do not.

We also analyzed the coverage of pointwise credible intervals, which are estimated by computing the respective quantiles of the sampled function evaluations. For a nominal level of 80 % we obtained an average coverage of 70 % for the approach with a global variance RW2, 73 % for the hierarchical t-formulation TRW2 and 77 % for the model RW2VRW1 with dependent local variances. Thus, the average coverage for the approach RW2VRW1 is at least close to the nominal level whereas for TRW2 and RW2 the average coverages are considerably less than the nominal level.

4 Conclusions

This paper presents a practical approach for fitting highly oscillating or unsmooth functions in categorical regression models. The simulation study in Section 3 suggests that for highly oscillating functions the approach with locally dependent variances performs superior to the hierarchical t-formulation and simple random walk models with a global variance. Currently the methodology is restricted to probit models. An extension to other link functions, particularly to logit models, is subject to future research.

References

Albert, J. and Chib, S. (1993). Bayesian analysis of binary and polychotomous response data. *Journal of the American Statistical Association*, **88**, 669-679.

Donoho, D. L. and Johnstone, I. M. (1994). Ideal spatial adaption by wavelet shrinkage. *Biometrika*, **81**, 425-455.

Fan, J. and Gijbels, I. (1995). Data-driven bandwidth selection in local polynomial fitting: variable bandwidth and spatial adaption. *Journal of the Royal Statistical Society B*, **57**, 371-394.

Lang, S., Fronk, E.-M., Fahrmeir, L. (2002). Function estimation with locally adaptive dynamic models. To appear in *Computational Statistics*.

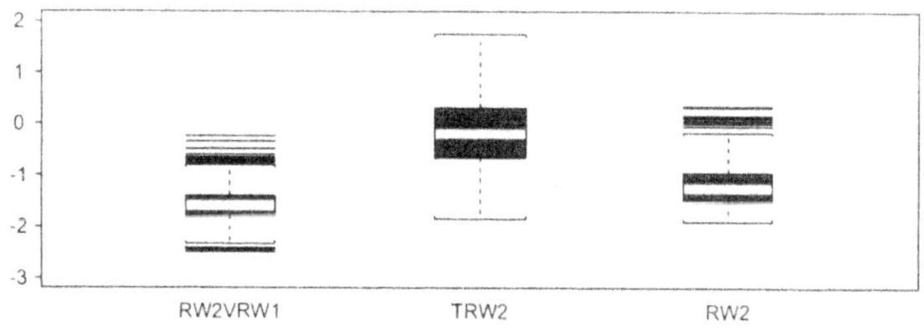

Fig. 1. Doppler function: boxplots of $log(MSE)$ for the three estimators.

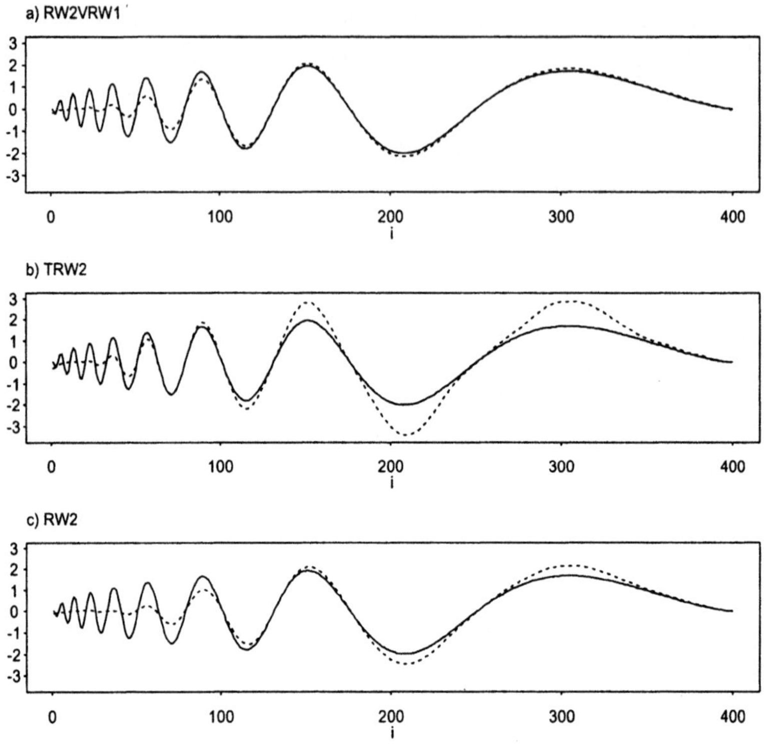

Fig. 2. Doppler function: average estimates (dashed lines) and true function (solid lines).

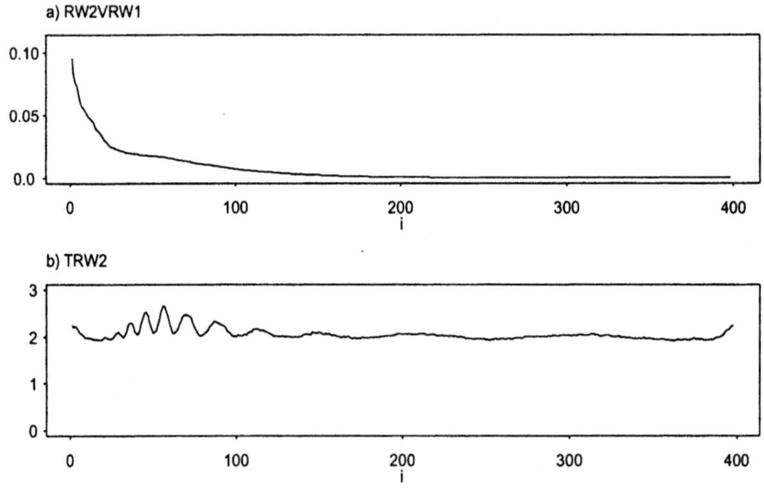

Fig. 3. Doppler function: average estimates of variance or weights for RW2VRW1 and TRW2.

Maneuvering Target Tracking by using Particle Filter Method with Model Switching Structure[1]

Norikazu IKOMA[1], Tomoyuki HIGUCHI[2], and Hiroshi MAEDA[1]

[1] Dept. of Computer Engineering, Faculty of Engineering, Kyushu Institute of Technology, Fukuoka 804-8550, JAPAN
[2] The Institute of Statistical Mathematics, Tokyo 106-8569, JAPAN

Abstract. Tracking problem of maneuvering target is treated with assumption that the maneuver is unknown and its acceleration has abrupt changes sometimes. To cope with unknown maneuver, Bayesian switching structure model, which includes a set of possible models and switches among them, is used. It can be formalized into general (nonlinear, non-Gaussian) state space model where system model describes the target dynamics and observation model represents a process to observe the target position. Heavy-tailed uni-modal distribution, e.g. Cauchy distribution, is used for the system noise to accomplish good performance of tracking both for constant period and abrupt changing time point of acceleration. Monte Carlo filter, which is a kind of particle filter that approximates state distribution by many particles in state space, is used for the state estimation of the model. A simulation study shows the efficiency of the proposed model by comparing with Gaussian case of Bayesian switching structure model.

Keywords. Bayesian modeling, target tracking, non-Gaussian distribution, multiple model, switching structure, particle filter.

1 Introduction

Problem of target tracking has been investigated actively after Kalman filter algorithm had been proposed, e.g.[Singer 1970]. Since the middle of 1980s, solution to this problem have been applied to, e.g., beam pointing control of a phased array radar, where benchmark problem is presented by [Blair and Watson 1994]. In this application, interacting multiple model that includes constant velocity model, constant thrust model and constant speed turn model is used with Kalman filter for state estimation [Blom and Bar-Shalom 1988].

Recently, a state estimation methods for nonlinear non-Gaussian state space model, which are called particle filter in general, are proposed: [Gordon et al. 1993], [Kitagawa 1996], and [Isard and Blake 1998]. These particle filters use many number of particles in state space to approximate non-Gaussian distribution of state estimate. Their ideas are considered as the special realization of sequential Monte Carlo method[Liu and Chen 1998]. For nonlinear or non-Gaussian model, particle filter can achieve more precise estimation of the state than the one of Kalman filter since Kalman filter only approximates

[1] This study was carried out under the ISM Cooperative Research Program (2001-ISM·CPR-2024).

the state distribution by Gaussian(uni-modal) while the actual one might be multi-modal.

The particle filters allows us to use nonlinear structure for the target tracking problem. Bayesian switching structure model that includes a set of possible models is applied to the problem and good performance is reported in [McGinnity and Irwin 2001]. Bayesian switching is also related to self organizing model[Kitagawa 1998] that automatically tune the hyper-parameters of the model by augmenting the state vector with hyper-parameters. This idea is generalized to switching the model structure by adding indicator vector of the model to the state vector [Higuchi 2001].

While simultaneous consideration of the multiple models is effective for dealing with the target tracking problem, non-Gaussian distribution for describing the system noise is also worthy of being considered. A use of heavy-tailed uni-modal distribution to follow abrupt changes of target's acceleration is proposed with particle filter[Ikoma et al. 2001]. A representative one of such distribution is Cauchy distribution. By assuming Cauchy distribution to the system noise, which corresponds to increments of acceleration, good performance of tracking is accomplished both for constant period and abrupt changing time point of acceleration.

In this paper, we propose the use both of Bayesian switching structure and heavy-tailed uni-modal distribution simultaneously in a tracking problem. A simulation study shows the efficiency of the proposed model by comparing with Gaussian system noise case of Bayesian switching structure model.

2 Model

Basic model for target tracking is firstly introduced, where, acceleration is assumed to be constant in continuous model, i.e., first derivative of the constant assumed element is according to a uni-modal distribution of 0 mode. The basic model is extended to Bayesian switching structure model that includes a set of possible models(candidate models), e.g., constant velocity model, constant acceleration model, and so on, with state vector that consists of position, velocity, acceleration, and jerk (and higher derivatives if needed) of the target.

To cope with the unknown maneuver, the state vector is extended to include indicator variable to select one model among the possible models. Markov switching is used to allow the indicator variable to evolve in the system model. To follow abrupt change of acceleration, we assume heavy-tailed uni-modal distribution for system noise. It will achieve good performance both for constant acceleration period and abrupt changing time point due to the heavy-tail property. The model is formalized as a nonlinear non-Gaussian state space model with system model described above and observation model that represents position observation process of the target.

2.1 Basic model

Position of the target in one-dimensional space is treated here, and is represented by $r(t)$ where t stands for continuous time index. Let acceleration of the target, $a(t)$, be a maneuver and it is assumed to be unknown. System model, which describes dynamics of the target, can be written in stochastic differential equation

$$\begin{bmatrix} \dot{r}(t) \\ \dot{s}(t) \\ \dot{a}(t) \end{bmatrix} = \begin{bmatrix} 0 & 1 & 0 \\ 0 & 0 & 1 \\ 0 & 0 & 0 \end{bmatrix} \begin{bmatrix} r(t) \\ s(t) \\ a(t) \end{bmatrix} + \begin{bmatrix} 0 \\ 0 \\ 1 \end{bmatrix} v_a(t) \tag{1}$$

where $s(t)$ is velocity of the target, and $v_a(t)$ is white Gaussian noise with 0 mean and variance ${\tau_a}^2$.

By discretizing the continuous system model eq.(1) with sampling time T (i.e. sampling points becomes $t = T_0 + kT$ with discrete time index k), with 0-th order hold assumption to the system noise such that $v_k^{(a)} = v_a(kT)$, and by denoting $r_k = r(kT)$, $s_k = s(kT)$, $a_k = a(kT)$, we have a discrete time system model

$$\begin{bmatrix} r_k \\ s_k \\ a_k \end{bmatrix} = \begin{bmatrix} 1 & T & T^2/2 \\ 0 & 1 & T \\ 0 & 0 & 1 \end{bmatrix} \begin{bmatrix} r_{k-1} \\ s_{k-1} \\ a_{k-1} \end{bmatrix} + \begin{bmatrix} T^3/3! \\ T^2/2 \\ T \end{bmatrix} v_k^{(a)}. \tag{2}$$

Note that acceleration of eq.(2) is modeled by random walk, and effects of the random walk model's terms($a(t)$ and $v_a(t)$) during the sampling time T appear in transition matrix and the vector apearing in the second term of right hand side in eq.(2).

From the state of the system, observation y_k is target position corrupted with observation noise w_k,

$$y_k = \begin{bmatrix} 1 & 0 & 0 \end{bmatrix} \begin{bmatrix} r_k \\ s_k \\ a_k \end{bmatrix} + w_k \tag{3}$$

where w_k is assumed to be Gaussian white with 0 mean and variance σ^2.

2.2 Switching

Due to the assumption that maneuver of the target is unknown, acceleration of the target, which is element of random walk in system model eq.(2), may be 0 for some time, may have some certain non-zero value for the another time, and may have changes between these values. This causes model mismatching to apply eq.(2) while acceleration constant period. To cope with this, we prepare candidate models of different element of random walk, i.e., position, velocity, acceleration, jerk(difference of acceleration) and so on, and switch system model among them.

Candidate models are as follows. Firstly, position constant(random walk) model is

$$r_k = r_{k-1} + Tv_k^{(r)} \tag{4}$$

where $v_k^{(r)} = v_r(kT)$ with white Gaussian system noise $v_r(t)$ with 0 mean and variance ${\tau_r}^2$. Next, velocity constant(random walk) model is

$$\begin{bmatrix} r_k \\ s_k \end{bmatrix} = \begin{bmatrix} 1 & T \\ 0 & 1 \end{bmatrix} \begin{bmatrix} r_{k-1} \\ s_{k-1} \end{bmatrix} + \begin{bmatrix} T^2/2 \\ T \end{bmatrix} v_k^{(s)} \tag{5}$$

where $v_k^{(s)} = v_s(kT)$ with white Gaussian system noise $v_s(t)$ with 0 mean and variance ${\tau_s}^2$. Finally, jerk constant(random walk) model is

$$\begin{bmatrix} r_k \\ s_k \\ a_k \\ c_k \end{bmatrix} = \begin{bmatrix} 1 & T & T^2/2 & T^3/3! \\ 0 & 1 & T & T^2/2 \\ 0 & 0 & 1 & T \\ 0 & 0 & 0 & 1 \end{bmatrix} \begin{bmatrix} r_{k-1} \\ s_{k-1} \\ a_{k-1} \\ c_{k-1} \end{bmatrix} + \begin{bmatrix} T^4/4! \\ T^3/3! \\ T^2/2 \\ T \end{bmatrix} v_k^{(c)} \tag{6}$$

where, $c_k = c(kT)$ stands for jerk, and $v_k^{(c)} = v_c(kT)$ with white Gaussian system noise $v_c(t)$ with 0 mean and variance ${\tau_c}^2$.

Among these candidate models(eq. (2), (4), (5), and (6)), one is selected and used as a system model. To denoted the selection, an indicator that specifies the selected model is introduced into the state of the model. The indicator is denoted by i_k, and its value is equal to the highest order of the element, i.e., 1 means constant position, 2 velocity, 3 acceleration, and 4 jerk. It is switched according to Markov process with transition matrix (which consists of transition probability), for example(in case of four candidate models),

$$\Pr\{i_k = i | i_{k-1} = j\} = \begin{bmatrix} 0.950 & 0.025 & 0.000 & 0.000 \\ 0.050 & 0.950 & 0.025 & 0.000 \\ 0.000 & 0.025 & 0.950 & 0.050 \\ 0.000 & 0.000 & 0.025 & 0.950 \end{bmatrix} \tag{7}$$

where column correspond to the indicator value before transit, and row corresponds to the after.

2.3 Heavy-tailed system noise

Since the property of unknown maneuver, acceleration of the target may have abrupt change. It is represented by system noise term $v_k^{(a)}$ in acceleration constant model eq.(2). With Gaussian system noise, its variance must be increased to follow this abrupt change, however, on the other hand, stability for constant acceleration period will be lost. To satisfy both properties of following and stability simultaneously, uni-modal heavy-tailed distribution is employed for system noise. Where, uni-mode represents the small fluctuation with high probability for stable period and heavy-tail bears abrupt change with low probability. Cauchy distribution is typical for such distribution, and we use it in simulation study.

3 Simulation study

One-dimensional trajectory shown in Fig.1(a) is used, which has small observation noise with $N(0, \sigma^2)$, $\sigma^2 = 10^{-4}$. Acceleration of the trajectory is shown in Fig.1(b). In all figure, horizontal axis shows discrete time index k.

By applying Bayesian switching structure model for Gaussian case(ordinary) and Cauchy(our proposal), we have obtained the estimate of position, velocity, acceleration, and jerk. As for the condition of Monte Carlo filter, number of particles is set to 50,000. system noise variances ${\tau_r}^2$, ${\tau_s}^2$, ${\tau_a}^2$, and ${\tau_c}^2$ are determined by grid search for maximizing the likelihood of the data.

Acceleration result(median) for both model are shown in Fig.2 by solid line, with actual one by dashed line. It can be seen that Cauchy estimates keeps stable at constant acceleration period without loss of following property of abrupt changing points.

Evolution of model indicator variable, which is involved in state vector together, for both model are shown in Fig.3. Indicator value 1(or 1 and 2) are majority for beginning part(i.e., constant position), 3 is major in the middle of the series, 2 is major at the ending part(constant velocity). By looking the result, the most appropriate model is majority almost all the period of the series.

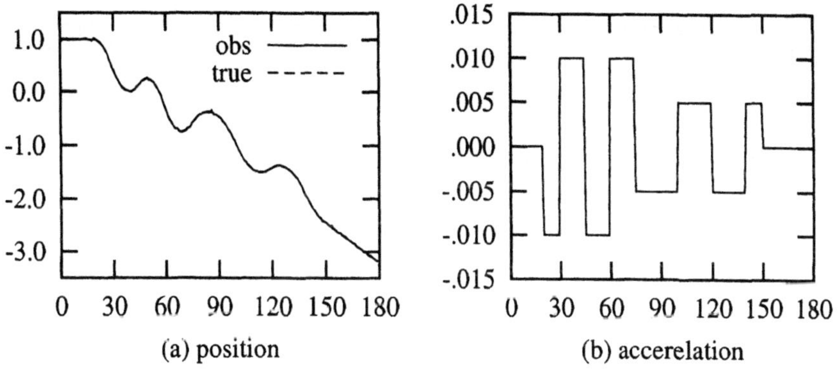

Fig. 1. Trajectory of target, (a)position(observation and true) and (b)acceleration.

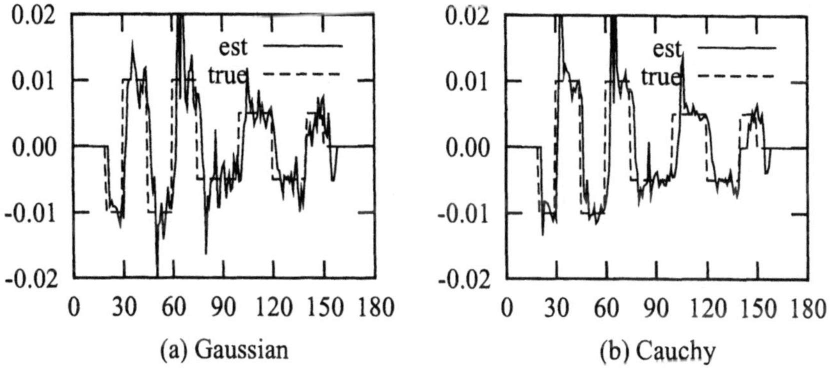

Fig. 2. Acceleration result: median (a)Gaussian and (b)Cauchy.

4 Conclusion

By exploiting an advantage of particle filter that allows us to treat nonlinear and non-Gaussian time series model tractably, we propose the use both of Bayesian switching structure(nonliner) and heavy-tailed uni-modal distribution(non-Gaussian) simultaneously in a target tracking problem. A simulation study that treats simple one-dimensional space tracking problem shows the efficiency of the proposed model by comparing with Gaussian system noise case of Bayesian switching structure model.

The model can easily be extended to multi-dimensional position with nonlinear observation equation by radar(polar coordinate). It will be an interesting future work.

References

Bar-Shalom, Y., Chang, K.C., and Blom, H.A.P. (1985). Tracking a maneuvering target using input estimation versus the interacting multiple model algorithm. *IEEE Trans. on Aerospace and Electronic Systems*, **25**, 2, 296-300.

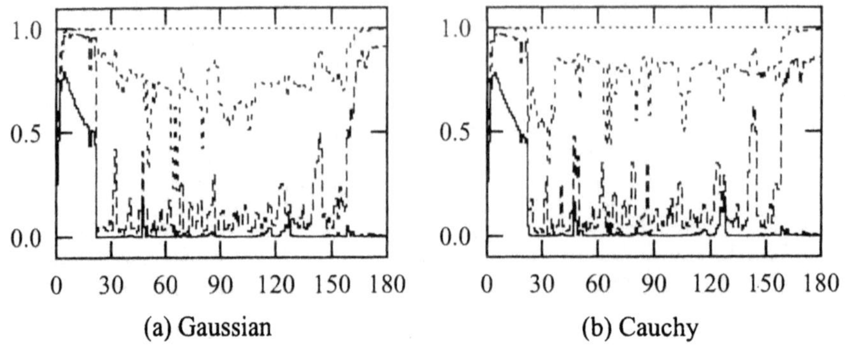

Fig. 3. Switching result: accumulated ratio of possible models (a)Gaussian and (b)Cauchy.

Blair, W.D., and Watson,G.A. (1994). IMM algorithm for solution to benchmark problem for tracking maneuvering targets. *Proceedings of the SPIE*, **2221**, 476-488.

Blom, H.A.P., and Bar-Shalom,Y. (1988). The interacting multiple model algorithm for systems with Markovian switching coefficients, *IEEE Trans. on Automatic Control*, **33**, 8, 780-783.

Gordon, N.J., Salmond, D.J., and Smith, A.F.M. (1993). Novel approach to nonlinear/non-Gaussian Bayesian State Estimation. *IEE Proceedings-F*, **140**, No.2, 107-113.

Higuchi, T. (2001). Evolutionary time series model with parallel computing. *Proc. of The 3rd JAPAN-US Joint Seminar on Statistical Time Series Analysis*, 183-190.

Ikoma, N., Ichimura, N., Higuchi, T, and Maeda, H. (2001). Particle filter based method for maneuvering target tracking. *Proc. of IEEE International Workshop on Intelligent Signal Processing*, 3-8.

Isard, M., and Blake, A. (1998). CONDENSATION – Conditional density propagation for visual tracking. *Journal of Computer Vision*, **29**, No.1, 5-28.

Kitagawa, G. (1998). A self-organizing state-space model. *Journal of the American Statistical Association*, **93**, 443, 1203-1215.

Kitagawa, G. (1996). Monte Carlo filter and smoother for non-Gaussian nonlinear state space models. *Journal of Computational and Graphical Statistics* **5**,No.1,1-25.

Liu, J.S., and Chen, R. (1998). Sequential Monte Carlo methods for dynamic systems. *Journal of the American Statistical Association*, **93**, 1032-1044.

McGinnity, S., and Irwin, G.W. (2001). Manoeuvring target tracking using a multiple-model bootstrap filter. *Sequential Monte Carlo Methods in Practice*, Doucet, A, ed., Springer, 479-497.

Singer, R.A. (1970). Estimating optimal tracking filter performance for manned maneuvering targets. *IEEE Trans. Aerospace and Electronics Systems*, **Vol.AES-6**, No.4, pp.473-483.

mathStatica: Mathematical Statistics with *Mathematica*

Colin Rose[1] and Murray D. Smith[2]

[1] Theoretical Research Institute, Bellevue Hill, Sydney, NSW 2023, Australia.
[2] Department of Econometrics and Business Statistics, University of Sydney, NSW 2006, Australia.

Abstract. This paper presents **mathStatica** (2002), a completely general toolset for doing mathematical statistics with *Mathematica* (Version 4). **mathStatica** defines statistical operators for taking expectations, finding probabilities, deriving transformations of random variables and so on. Importantly, **mathStatica** is not tied to a set of pre-specified statistical distributions. Rather, it is designed to derive statistics such as moments, cumulative distribution functions, characteristic functions, and other generating functions for user-defined distributions. **mathStatica** supports discrete and continuous distributions — univariate and multivariate. Applications to inference include: estimation (moment unbiased, minimum variance unbiased, best unbiased, maximum likelihood: symbolic and numeric), curve-fitting (Pearson and Johnson systems, non-parametric kernels), asymptotics, decision theory, and moment conversion formulae (for conversion between cumulants, raw moments, and central moments: univariate and multivariate). **mathStatica** accompanies the book: Rose and Smith (2002), Mathematical Statistics with *Mathematica* (Springer Texts in Statistics).

Keywords. Computational statistics, *Mathematica*, **mathStatica**

1 Introduction

The use of computer software in statistics is far from new. Indeed, literally hundreds of statistical computer programs exist. Yet, underlying existing programs is almost always a numerical/graphical view of the world. *Mathematica* can easily handle the numerical and graphical sides, but it offers in addition an extremely powerful and flexible symbolic computer algebra system. The **mathStatica** software package builds upon that symbolic engine to create a sophisticated toolset specifically designed for doing mathematical statistics. This forms the basis of Rose and Smith (2002), a Springer-Verlag text entitled *Mathematical Statistics with Mathematica* that ships with CD-ROM, custom palettes, online help, and live interactive chapters. Section 2 focuses on **mathStatica**'s basic functions: plotting, expectations, probability, transformations. Section 3 provides a quick look at some more specialised functions, such as the calculation of Fisher Information, the construction of densities of order statistics, finding unbiased estimators of population moments, and moments of moments.

2 Core Functions

2.1 Getting Started

mathStatica adds over a hundred new functions to *Mathematica*. But most of the time, we can get by with just a few of them. To illustrate, let us suppose that X and Y have a Gumbel bivariate Exponential distribution with joint density $f(x, y)$:

$$f(x, y) \;=\; e^{-2(x+y)}\,(e^{x+y} + \alpha(e^x - 2)(e^y - 2))$$

with domain of support $x > 0$, $y > 0$, where parameter α is such that $-1 < \alpha < 1$. In *Mathematica*, we enter this as:

```
f  =  e^(-2 (x+y)) (e^(x+y) + α (e^x - 2) (e^y - 2) );
domain[f]  =  {{x, 0, ∞}, {y, 0, ∞}} && {-1 < α < 1};
```

Here is a plot of $f(x, y)$ when $\alpha = -0.8$:

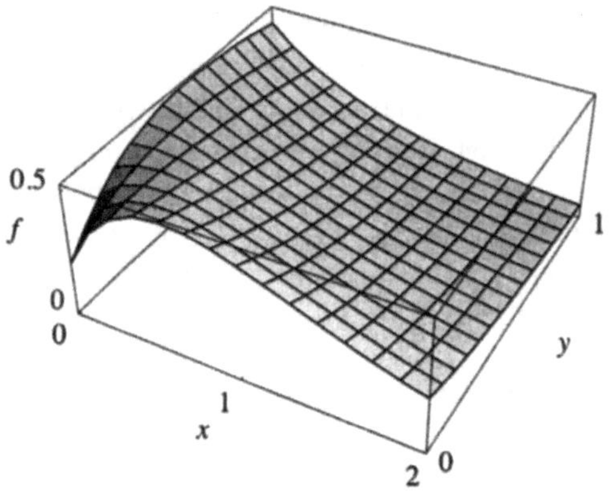

Here is the joint distribution function, namely $P(X \le x, Y \le y)$:

```
Prob[{x, y}, f]
```

$$e^{-2\,(x+y)}\;(-1 + e^x)\;(-1 + e^y)\;(e^{x+y} + \alpha)$$

Here is Cov(X, Y), the covariance between X and Y:

```
Cov[{x, y}, f]
```

$$\frac{\alpha}{4}$$

More generally, here is the variance-covariance matrix:

```
Varcov[f]
```

$$\begin{pmatrix} 1 & \frac{\alpha}{4} \\ \frac{\alpha}{4} & 1 \end{pmatrix}$$

Here is the marginal density of X:

```
Marginal[x, f]
```

e^{-x}

Here is the conditional density of Y, given $X = x$:

```
Conditional[y, f]
```

— Here is the conditional pdf $f(y \mid x)$:

$e^{x-2(x+y)} (e^{x+y} + (-2 + e^{x}) (-2 + e^{y}) \alpha)$

Here is the bivariate moment generating function $E[e^{t_1 X + t_2 Y}]$:

```
mgf = Expect[e^(t1 x + t2 y), f]
```

— This further assumes that: $\{t_1 < 1, t_2 < 1\}$

$$\frac{4 - 2 t_2 + t_1 (-2 + (1 + \alpha) t_2)}{(-2 + t_1) (-1 + t_1) (-2 + t_2) (-1 + t_2)}$$

Differentiating the moment generating function is one way to derive moments. Here, for example, is the product moment $E[X^2 Y^2]$:

```
D[mgf, {t1, 2}, {t2, 2}] /. t_ → 0 // Simplify
```

$$4 + \frac{9\alpha}{4}$$

which we could otherwise have found directly with:

```
Expect[x^2 y^2, f]
```

$$4 + \frac{9\alpha}{4}$$

Multivariate transformations pose no problem to **mathStatica** either. For instance, let $U = \frac{Y}{1+X}$ and $V = \frac{1}{1+X}$ denote transformations of X and Y, to U and V. Then, using **mathStatica**'s `Transform` function, we can find the joint pdf of random variables U and V, as follows:

```
Transform[{u == y/(1 + x), v == 1/(1 + x)}, f]
```

$$\frac{e^{\frac{-2-2u+v}{v}} \left(4 e \alpha - 2 e^{\frac{1}{v}} \alpha - 2 e^{\frac{u+v}{v}} \alpha + e^{\frac{1+u}{v}} (1 + \alpha)\right)}{v^3}$$

2.2 Discrete Random Variables

mathStatica handles discrete random variables in the same way it handles continuous random variables. The only difference is that, when we define a discrete density, we add the 'flag' `{Discrete}` to tell the software that the random variable is discrete. To illustrate, let the discrete random variable $X \sim \text{NegativeBinomial}(r, p)$ with pmf $f(x)$:

```
        f  =  Binomial[r + x - 1, x] p^r (1 - p)^x;
domain[f]  =  {x, 0, ∞} && {Discrete} && {0 < p < 1, r > 0} ;
```

Here is a plot of $f(x)$:

```
PlotDensity[f /. {p → 1/2, r → 10}];
```

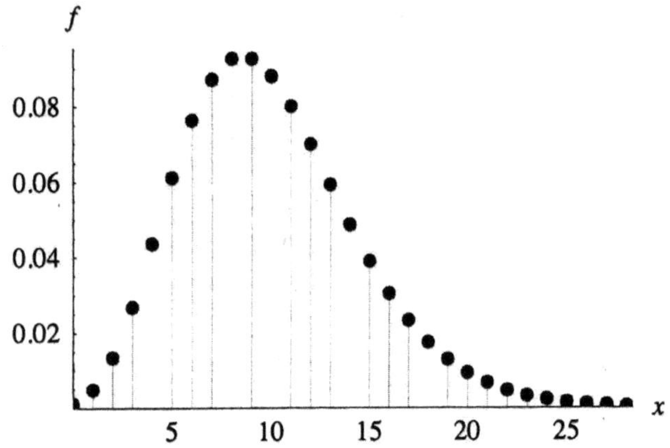

The mean $E[X]$ and variance of X are given by:

```
Expect[x, f]
```

$$\left(-1+\frac{1}{p}\right) r$$

```
Var[x, f]
```

$$\frac{r-p\,r}{p^2}$$

The probability generating function is the solution to $E[t^X]$:

```
Expect[t^x, f]
```

$$p^r\,(1+(-1+p)\,t)^{-r}$$

3 Specialised Functions

3.1 Fisher Information

The Fisher Information on a parameter can be constructed from first principles using the `Expect` function. Alternatively, we can use **mathStatica**'s `FisherInformation` function, which automates this calculation. To illustrate, let us suppose that $X \sim \text{InverseGaussian}(\mu, \lambda)$ with pdf $f(x)$:

$$\mathbf{f} = \sqrt{\frac{\lambda}{2\,\pi\,x^3}}\;\mathrm{Exp}\left[-\lambda\,\frac{(x-\mu)^2}{2\,\mu^2\,x}\right];$$

$$\mathbf{domain[f]} = \{x, 0, \infty\}\ \&\&\ \{\mu > 0, \lambda > 0\};$$

Then, Fisher's Information on (μ, λ) is the (2×2) matrix:

```
FisherInformation[{μ, λ}, f]
```

$$\begin{pmatrix} \frac{\lambda}{\mu^3} & 0 \\ 0 & \frac{1}{2\,\lambda^2} \end{pmatrix}$$

3.2 Order Statistics

Let random variable X have a Logistic distribution with pdf $f(x)$:

$$\mathbf{f} = \frac{e^{-x}}{(1+e^{-x})^2}; \qquad \mathbf{domain[f]} = \{x, -\infty, \infty\};$$

Let $(X_1, X_2, \ldots, X_n)$ denote a sample of size n drawn on X, and let $(X_{(1)}, X_{(2)}, \ldots, X_{(n)})$ denote the ordered sample, so that $X_{(1)} < X_{(2)} < \cdots < X_{(n)}$. The pdf of the rth order statistic, $X_{(r)}$, is given by the **mathStatica** function:

```
OrderStat[r, f]
```

$$\frac{(1+e^{-x})^{-r}\,(1+e^{x})^{-1-n+r}\,n!}{(n-r)!\,(-1+r)!}$$

The joint pdf of $X_{(r)}$ and $X_{(s)}$, for $r < s$, is given by:

```
OrderStat[{r, s}, f]
```

$$\frac{e^{x_s}\,(1+e^{-x_r})^{-r}\,(1+e^{x_s})^{-1-n+s}\,\left(\frac{1}{1+e^{x_r}}-\frac{1}{1+e^{x_s}}\right)^{-r+s}\,\Gamma[1+n]}{(-e^{x_r}+e^{x_s})\,\Gamma[r]\,\Gamma[1+n-s]\,\Gamma[-r+s]}$$

3.3 Unbiased Estimation of Population Moments; Moments of Moments

mathStatica can find unbiased estimators of population moments. For instance, it offers h-statistics (unbiased estimators of population central moments), k-statistics (unbiased estimators of population cumulants), multivariate varieties of the same, polykays (unbiased estimators of products of cumulants) and more. Consider the k-statistic k_r which is an unbiased estimator of the rth cumulant κ_r; that is, $E[k_r] = \kappa_r$, for $r = 1, 2, \ldots$. Here are the 2nd are 3rd k-statistics:

```
k2 = KStatistic[2]
k3 = KStatistic[3]
```

$$k_2 \to \frac{-s_1^2 + n\, s_2}{(-1+n)\, n}$$

$$k_3 \to \frac{2\, s_1^3 - 3\, n\, s_1\, s_2 + n^2\, s_3}{(-2+n)\, (-1+n)\, n}$$

As per convention, the solution is expressed in terms of power sums $s_r = \sum_{i=1}^{n} X_i^r$.

Moments of moments: In 1928, Fisher published the product cumulants of the k-statistics, which are now listed in reference bibles such as Stuart and Ord (1994). Surprisingly, the 'known' solutions are not always correct. Here is the correct solution to $\kappa_{2,2}(k_3, k_2)$:

```
CumulantMomentToCumulant[{2, 2}, {k3[[2]], k2[[2]]}]
```

$$\frac{288\, n\, \kappa_2^5}{(-2+n)\,(-1+n)^3} + \frac{288\,(-23+10\,n)\,\kappa_2^2\,\kappa_3^2}{(-2+n)\,(-1+n)^3} + \frac{360\,(-7+4\,n)\,\kappa_2^3\,\kappa_4}{(-2+n)\,(-1+n)^3} + \frac{36\,(160-155\,n+38\,n^2)\,\kappa_3^2\,\kappa_4}{(-2+n)\,(-1+n)^3\,n} +$$
$$\frac{36\,(93-103\,n+29\,n^2)\,\kappa_2\,\kappa_4^2}{(-2+n)\,(-1+n)^3\,n} + \frac{24\,(202-246\,n+71\,n^2)\,\kappa_2\,\kappa_3\,\kappa_5}{(-2+n)\,(-1+n)^3\,n} + \frac{2\,(113-154\,n+59\,n^2)\,\kappa_5^2}{(-1+n)^3\,n^2} +$$
$$\frac{6\,(-131+67\,n)\,\kappa_2^2\,\kappa_6}{(-2+n)\,(-1+n)^2\,n} + \frac{3\,(117-166\,n+61\,n^2)\,\kappa_4\,\kappa_6}{(-1+n)^3\,n^2} + \frac{6\,(-27+17\,n)\,\kappa_3\,\kappa_7}{(-1+n)^2\,n^2} + \frac{37\,\kappa_2\,\kappa_8}{(-1+n)\,n^2} + \frac{\kappa_{10}}{n^3}$$

4 References

Fisher, R.A. (1928), Moments and product moments of sampling distributions, *Proceedings of the London Mathematical Society*, series 2, **30**, 199–238.

mathStatica (2002), `http://www.mathStatica.com`

Rose, C. and Smith, M.D. (2002), *Mathematical Statistics with Mathematica*, Springer-Verlag: New York.

Stuart, A. and Ord, J.K. (1994), *Kendall's Advanced Theory of Statistics*, volume 1, 6th edition, Edward Arnold: London (also Wiley: New York).

5 Acknowledgements

The **mathStatica** package was partly developed while Rose was visiting Wolfram Research, and while Smith was visiting the Sonderforschungsbereich 386, Institut für Statistik, Ludwig-Maximilians-Universität, Munich. Rose gratefully acknowledges a Visiting Scholar Grant from Wolfram Research, while Smith acknowledges gratefully the support of the Alexander von Humboldt-Stiftung. ■

MCMC Model for Estimation Poverty Risk Factors using Household Budget Data

Ene Käärik[1], Ene-Margit Tiit[1], Mare Vähi[1]

[1] Institute of Mathematical Statistics, University of Tartu, Estonia

Abstract. An important task for analysts of official statistical data is the determination of social risk groups, especially in poverty risk. As the data used in such analyses (Household Income and Expenditures surveys) have rather high non-response rate and may be affected by different sources of errors, the problem of model stability is important. One possibility to check the stability of a model is to use different generation procedures.

1 Equivalence scale and absolute poverty line for Estonia

In 1999, with the support of UNDP, the equivalence scale was calculated for Estonia, using the empirical data of household consumption. It was found that the scale 1:0,8:0,8 occurred to be the most suitable one for Estonia (taking into account several factors, including considerably high percentage of individual consumption in households total consumption). Using that scale the absolute poverty line was calculated integrating different components of poverty, such as food poverty, life-style poverty, living-room poverty, etc. Inflation coefficients have been used to correct the poverty line yearly.

The number of households living under the poverty line has dropped in Estonia from 37,6% in 1997 to 24,7% in 2000. However the problem still exists: Which are the most important risk factors of poverty? Is it possible to model the poverty?

2 Data

To answer the question, the data of national HIES were used. In this survey about 6000 households (almost 1% from the total number of Estonian households) were interviewed every year. In the course of the interview people fill in diaries about their consumption (during a month) and indicate the total income for the given month. Also, the household structure and detailed description of household members (their social and employment status, education, etc) were recorded.

Literature presents a long list of possible poverty risk factors, such as unemployment, large number of dependent household members, lack of adult male household members, low level of education, living in rural area, too young or too old household members, illness, etc. All possible factors were checked and the most influential ones were chosen as potential risk factors. These factors were:

- Number of dependent household members;

- Number of working household members;
- Education level of the household, defined as highest education level of the head or his/ her spouse;
- Living place (urban/ rural);
- Ethnicity (Estonian speaking and non-Estonian households, defined by the head of household).

Surprisingly, the age was not an important poverty factor. In Estonia, the poverty of young people, that is common in several European countries, is not a problem. Also, the poverty of elderly people is not a severe problem, as the retired persons usually belong to the third/fifth income decile.

All models used the Household Budget Data of 2000 collected by Estonian Statistical Office. The sample size was about 6000 households.

3 Classical models

In the first step, the poverty risk was estimated using two different models for grouped data, using *logit* as the link function:

$$Y_i = \ln\left(\frac{p_i}{1-p_i}\right), \quad p_i = \frac{v_i}{n_i}, \quad i = 1, 2, \ldots, G,$$

where n_i is the number of households in the group i, and v_i – the number of poor households among them. The groups were formed using four factors: number of dependents (5 levels), living place (2 levels), ethnicity (2 levels) and education (4 levels). Four of the groups ($G = 80$) were empty. All the remaining groups were used in the following analyses.

As the first step, the model including main effects of all listed factors was constructed. Two covariates were also included – these were variable X measuring the average number of employed members per household in the given group, and its square X^2. The model's parameters were calculated using the following two assumptions about the distribution of Y:

1. Classical assumptions: Y and the residuals have normal distribution with equal variance in all groups; the parameters of the model have been estimated using LS method (equal to the MLE);
2. Generalized assumptions: the response r_i (poverty risk) is binomially distributed $r_i \sim B(p_i, n_i)$, for grouped data (i = 1,2, ..., G), the link function is *logit*, the parameters of the model are estimated using MLE method (the optimization technique for estimating the regression parameters is the Fisher-scoring algorithm).

3.1 Results

The parameters of the model were estimated using different methodology, and the results are quite similar (see Table 3.1). In most cases the differences is about 5–10% from the value of the parameter. The model is quite well interpretable.

Table 3.1 Estimates of parameters

	Model Type / Factors	Generalized		Classical	
		Parameter	p	Parameter	p
a_0	Intercept	-2,360	<0,001	-2,166	<0,001
a_1	Number dependents	0,583	<0,001	0,612	<0,001
a_2	Ethnicity	0,487	<0,001	0,455	<0,001
a_3	Living place	-0,443	<0,001	-0,476	<0,001
a_4	Higher Education	-1,185	<0,001	-1,178	<0,001
a_5	Secondary Vocational Educ.	-0,450	<0,001	-0,482	<0,001
a_6	Secondary Education	-0,336	0,002	-0,346	<0,001
a_7	X (average number of working members)	1,376	0,0023	0,965	<0,001
a_8	X^2	-0,743	0,0025	-0,537	<0,001

- The most influential factor to reduce the poverty risk is education, especially the higher education of the head/ spouse of head of the household.
- Living in rural area increases the poverty risk of a household while living in urban area reduces the risk.
- Not being able to speak Estonian increases the poverty risk.
- Having large number of dependents household members increases the poverty risk.
- The average number of working people X has somewhat contradictory character. That seems to be caused by the fact that the number of dependents in household is positively correlated with the number of working people (see also Figure 3.1). Using this information, the quadratic term was added to the model. Although it caused some multicollinearity, the result is well interpretable – the households with bigger number of working members have less poverty risk.
- The difference in values of R^2 is caused by the fact that in the classical model the inter-class variability was not accounted.

Figure 3.1 Dependence of poverty risk on the average number of working household members

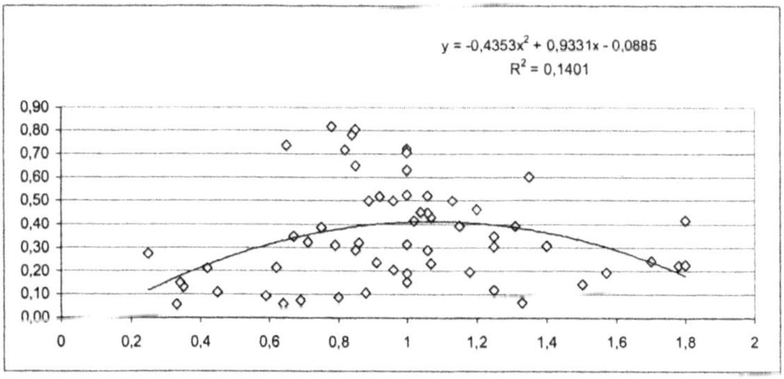

4 MCMC models

During the second stage the model was constructed using Markov Chain Monte Carlo (MCMC) method to illustrate how to use this tool for building models by Bayesian approach, and large sets of data from official statistics were used

In the more detailed MCMC model we used a logistic regression and assumed that the poverty risk has a binomial distribution.

Let p_i be the probability of poverty on the i-th level (group), and let us assume that poverty risk is binomially distributed $r_i \sim B(p_i, n_i)$, $i = 1,2, \ldots, G$.

4.1 Logistic models

We consider two logistic models, the first one with random effects (for accounting the over-dispersion) and the second one without random effect.

The logistic model with random effects is the following:

$$logit(p_i)=\alpha_0 + \alpha_1 x_{1i} + \alpha_2 x_{2i} + \alpha_3 x_{3i} + \alpha_4 x_{4i} + \alpha_5 x_{5i} + \alpha_6 x_{6i} + \alpha_7 x_{7i} + \alpha_8 x_{8i}\ b_i,$$

with arguments x_{ji} ($j = 1, \ldots, 8$ indicate the factors from Table 4.11 and $i = 1, \ldots, 76$ the groups).

The parameters α_j ($j = 1, \ldots, 8$) of the model were assumed to be random (Bayesian approach) with noninformative priors. Due to large samples we had basis to assume that the regression coefficients are normally distributed with unknown parameters and the random effect has normal distribution $b_i \sim Normal(\mu, \tau)$. The hyperparameters of this term were assumed to be random: mean having noninformative normal distribution and variance – noninformative gamma distribution. The dataset was grouped, and 76 nonempty groups were used as empirical data for the model.

4.2 Results

To get suitable models in both cases, 10,000 iterations were made using WinBUGS1.3. After studying the simulation history and quantiles' graphs, the length of burn-in period was chosen equal to 5,000. The simulation statistics of both models is presented in Table 4.1

Table 4.1 Esimates of parameters

	Mean		SD		2.5%		97.5%	
Model	1	2	1	2	1	2	1	2
a0	-3.878	-2.351	0.542	0.181	-4.812	-2.706	-2.768	-1.984
a1	0.583	0.584	0.042	0.035	0.501	0.516	0.665	0.654
a2	-0.349	-0.339	0.112	0.089	-0.566	-0.517	-0.131	-0.167
a3	-0.506	-0.449	0.125	0.095	-0.751	-0.635	-0.269	-0.259
a4	-1.259	-1.187	0.173	0.139	-1.608	-1.465	-0.937	-0.916
a5	0.518	0.472	0.095	0.081	0.330	0.317	0.703	0.630
a6	-0.432	-0.440	0.098	0.075	-0.629	-0.587	-0.241	-0.293
a7	1.388	1.409	0.561	0.363	0.341	0.650	2.635	2.113
a8	-0.679	-0.765	0.293	0.204	-1.329	-1.155	-0.146	-0.328
μ	1.422	–	0.459	–	0.613	–	2.179	–

The resulting histograms for parameters are smoothed and the resulting marginals posteriors show a normal-like form (for example see Figure 4.1).

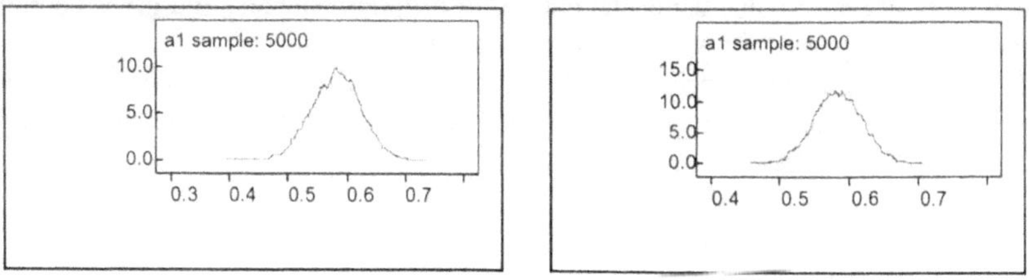

Figure 4.1 Kernel density plots for parameters a1 (number of dependents) model 1 and model2

To analyze the convergence of the process, two parallel chains were used. For the model with random effects the most problematic nodes (a7 and μ) were monitored in detail. After 650 iterations convergence diagnostics of Gelman and Rubin, based on normal approximations, gives the value of convergence statistics less than 1.2 (see Figure 4.2). Markov chain convergence conditions are satisfied for *a7*, convergence of random term μ is slower (see Figure 4.2).

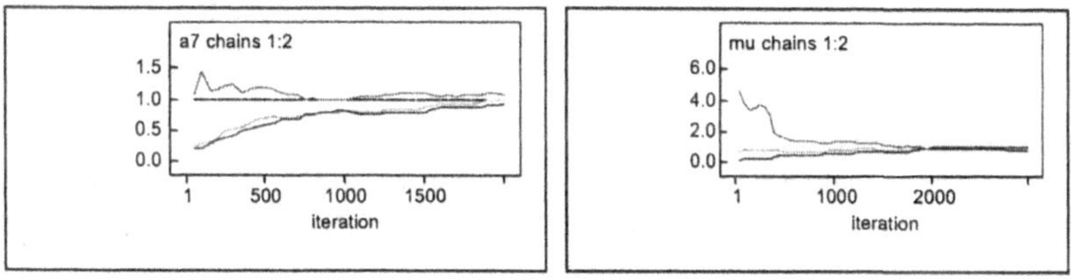

Figure 4.2 Convergence analysis for node a7 and μ (mu).
Gelman-Rubin statistics (upper line)

The convergence analysis of the second model (without random effects) gives almost similar results, so we follow the node of the quadratic term *a8* (Fig 4.3).

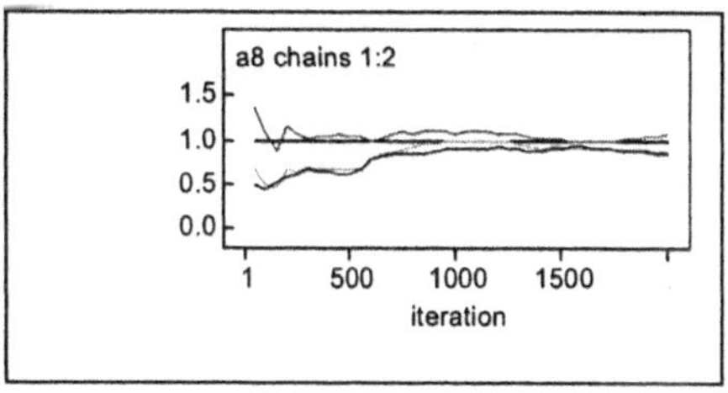

Figure 4.3 Convergence analysis for node a8: Gelman-Rubin statistics (upper line)

5 Discussion

This work shows that the Bayesian approach provides us good perspective to build rather complicated models using the data from different sources, and gives additional information compared with usual regression models.

Two of the used four models – usual logit and MCMC without random term – fitted very well (the difference in parameters less than 1%). From this fact the following conclusions can be drawn:

- *Logit* was successful choice as link function.
- Overdispersion was not a problem in the given model.
- The model coefficients estimated by different methods were similar to each other; hence the stability of the model has been proved.
- The graphs demonstrate that the distribution of coefficients in MCMC-models are quite close to normal.

From the shape of distribution curves of OR's and regression's parameters we drew a hypothesis that some of the explanatory variables do not have monotonic influence on the poverty risk, e.g. the risk functions surface must have several modes. More detailed analysis proved it to be true: the professional secondary education increased the poverty risk (compared with the general secondary education) for households living in countryside and for households from ethnic minorities.

The comparison demonstrated that the coefficients calculated using different procedures were quite similar, hence the stability of the model was proved.

Acknowledgements

The work was supported by the Office of the Minister of Population and Ethnic Affair of Estonia. The authors are grateful to Kaja Sõstra for helpful discussions during the research.

References

Cowles, M.K. & Carlin, B.P (1996). Markov Chain Monte Carlo convergence diagnostics: a comparative review. In: *Journal of the American Statistical Association,* vol **91**, 883-904.

Gilks, W.R., Richardson, S. & Spiegelhalter, D.J. (1997). *Markov Chain Monte Carlo in practice.* Champman&Hall.

Household Living Niveau 2000, Statistical Office of Estonia, Tallinn, 2001.

MD*Book online & e-stat: Generating e-stat Modules from LaTeX

Rodrigo Witzel and Sigbert Klinke
Humboldt-Universität zu Berlin, School of Business and Economics, Institute of Statistics and Econometrics, Spandauer Str. 1, 10178 Berlin, Germany

Abstract. We describe the basics of MD*Book and MD*Book online which we use together with XploRe to create interactive electronic books. We generate different output formats (Postscript, PDF, HTML and MM*Stat-like formats) for the internet as well as for CD and/or local installation.

The e-stat project for creating a knowledge base in statistics needs statistical input from various authors to cover a larger area in statistics. The teaching material, which we want to incorporate in the e-stat project, is already available in LaTeX and we have used MD*Book to convert it from LaTeX to HTML. Interactivity in our electronical books is incorporated via a applet which connects to XploRe - a statistical software.

Thus we extended our tool MD*Book, and therefore also MD*Book online, such that we can generate the XML file (and further files) which are the base for an e-stat module.

We first describe MD*Book and MD*Book online. Finally we describe the conversion process and show an reused example from MM*Stat as e-stat module.

Keywords. MD*Book, MD*Book online, MM*Stat, E-Books, internet, conversion from LaTeX to HTML, conversion from LaTeX to XML.

1 The basics of MD*Book

The MD*Book project has it roots in three projects: 1) the translation of LaTeX tutorials for the XploRe helpsystem; 2) the "Golden Solution" which integrates XploRe, a statistical software, as an client applet into HTML and 3) the MM*Stat project (see Rönz 2001) for teaching basic statistics.

We developed two programs: `tex2sk` and `mdbook2`. The program `texsk` decomposes a set of LaTeX source files and generates a control file. This control file is used in `mdbook2` to generate all output formats (see Figure 3) and it contains other informations, e.g., if we generate a CD or an internet version etc.

`mdbook2` has five main options: `-ps`, `-pdf`, `-html`, `-java` and `-xpl`. `-ps` and `-pdf` generate a Postscript document or a PDF document, respectively. Since the translation from LaTeX to HTML takes time, these options were planned to determine, if everything compiles under LaTeX without problems.

The `-html` and `-java` options both generate HTML versions: `-html` without using JavaScript in the HTML pages and `-java` generates a MM*Stat-like version. A difference between the two versions is that the first one runs under Netscape and Internet Explorer whereas the intensive use of JScript in MM*Stat leads to the current situation that Internet Explorer 5.x or higher is necessary to run the MM*Stat versions.

Fig. 1. The "Golden Solution" for the quantlet `MMStatLinReg`. In the upper file cards we see *execute* for executing the XploRe code and *edit* to edit the XploRe code.

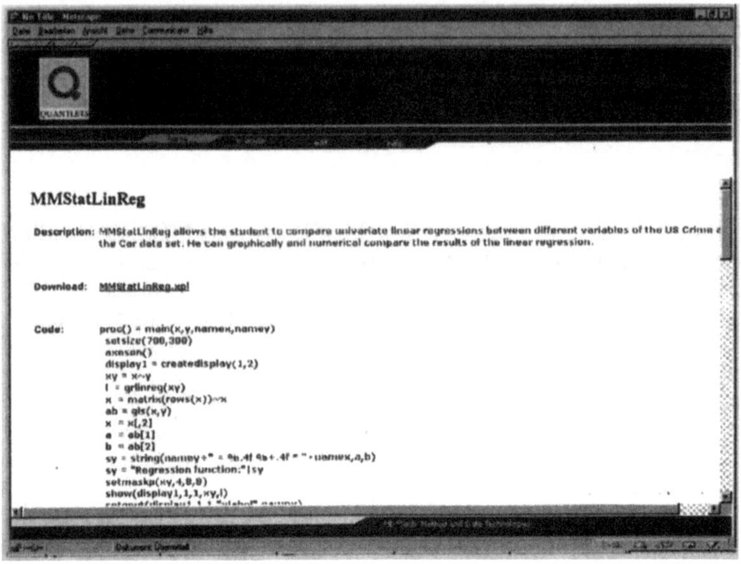

Fig. 2. *Enhanced* example in section 11.2 in the english MM*Stat version generated with MD*Book. The LaTeX code was reused for an example e-stat module generated with MD*Book.

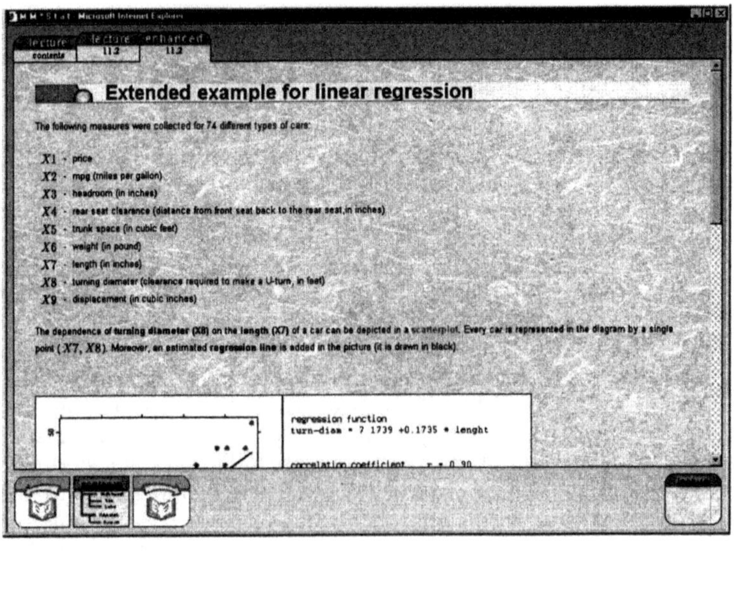

Fig. 3. Generation of an *e-book* from LaTeX sources and XploRe quantlets with MD*Book.

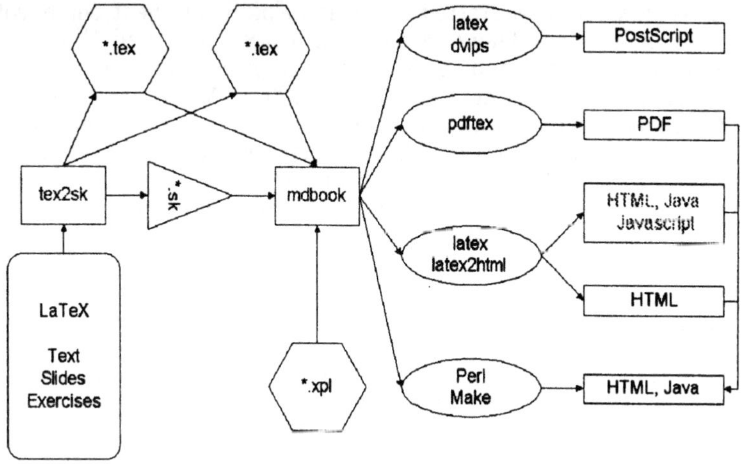

Fig. 4. Architecture of MD*Book online.

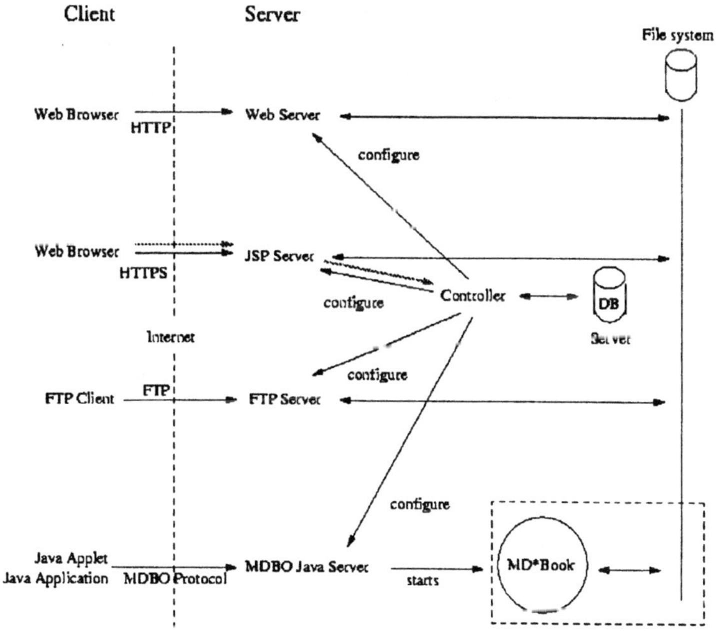

The last option, `-xpl`, generates the "Golden Solution" for XploRe quantlets (see Figure 1).

In the last few years, we started to develop XploRe as client/server software in addition to the standalone versions for Unix/Linux and Windows. The client is written in Java and can run as an application as well as an applet in a HTML page (Lehmann and Kleinow 2001). A link in the text points to a HTML page which allows us to embed the XploRe client applet (the "Golden solution" in Figure 1). This technique works for HTML documents (internet, CD) as well as for PDF documents (internet). In the upper file cards of Figure 1, we see two different interaction types: 1) *execute*, execute the XploRe code immediately; 2.) *edit*, show the XploRe code in the client editor and execute it later.

2 MD*Book online

MD*Book Online wraps MD*Book so that it can be used in internet browsers and via FTP clients. MD*Book itself can only be used with a valid UNIX account in our intranet and it is started by entering command lines in a UNIX shell. The desire to use MD*Book's sophisticated abilities to create interactive books outside of our intranet leads to the development of such a wrapper.

The MD*Book system is composed of a server and communicating clients for usage in the internet (see Figure 4). The server's environment on which MD*Book runs does not change. The client running on an user's computer tunnels only input and output to the MD*Book server. Usually such client algorithms are simple and programmed quickly. Therefore they are already available for a large variety of hardware and operating systems.

But client/server systems are always distributed algorithms which means that you have to care about synchronizing clients and distributing resources among clients. Synchronizing mechanisms have to work when MD*Book is started repeatedly in the same directory. The confidential nature of e-book sources requires authentic access to the server's resources. To solve those two problems we introduce work spaces and locks that prevent running more than one MD*Book at the same time.

Because we use only free widely available client/server systems in our implementation of MD*Book Online we also developed a general model and a general implementation of wrappers for command line software.

The workflow of an e-book author consists of storing texts, diagrams and interactive items via FTP in a MD*Book Online work space first, initializing the production of a desired e-book format by Java applet and finally downloading produced books via FTP or internet browser. For all other tasks, like viewing documentation, templates for e-books and e-book previews, it is sufficient to use an internet browser.

3 From LaTeX to e-stat modules

The aim of the e-stat project is to build a knowledge database for statistics, which should be used to improve teaching.

The statistical knowledge will be represented by modules in XML files and algorithm working with them. These files have to follow syntactical rules which are described in several XML-DTDs.

The manual generation of the XML files requires some experience, but can be done quickly with comfortable XML editors. Unfortunately this does not

Fig. 5. Oldenburg preview of the example module for linear regression. Note the link MMStatLinReg.xpl to the "Golden solution" page beyond the graphics such that the user can reproduce the graphic itself.

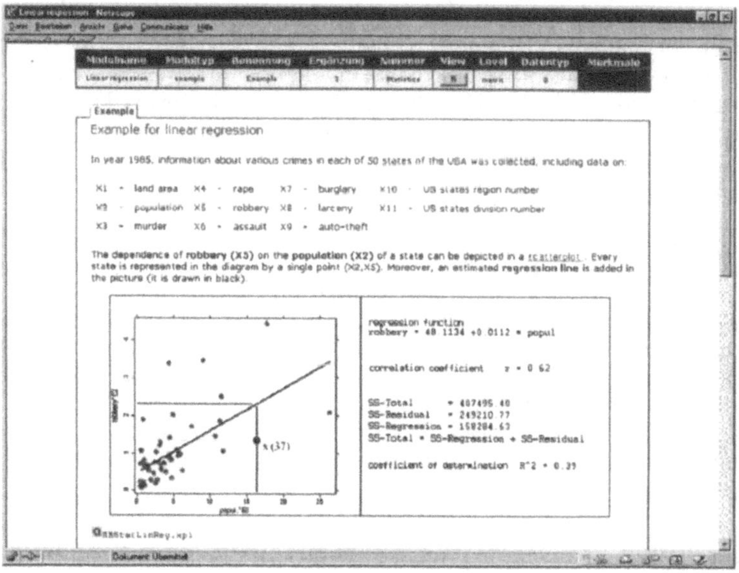

hold for the generation of mathematical formulas. The time consuming effort for the generation of the formulas as well as for the XML files in general, will reduce the growth of the e-stat knowledge database. This will become even more serious after the support of the BMB+F has ended and the authors of e-stat modules will have to finance the generation of new knowledge module by other sources.

The conversion from LATEX to XML, embedded in MD*Book, allows authors to generate their knowledge modules from LATEX sources.

A mapping from LATEX to e-stat knowledge modules can only be incomplete, since the modules contain information which are not part of LATEX sources.

Mainly LATEX consists of text, which can be structured by chapters, paragraphs, etc. But in LATEX does not exist any required semantic structure. In contrast, the modules have attributes, e.g., unique identification or meta information like keywords or relation to other modules.

Thus an author has to insert all necessary information in his/her LATEX source file. We do not require the authors to learn new editors or special file formats. The information will be integrated via special LATEX file, which can be downloaded from the MD*Book online pages. The main step of the conversion is again the use of `latex2html` for converting LATEX text and formulas to HTML. e-stat macros will be converted in HTML4 `<object>` tags. In these tags we can store nearly any information without losing the compability with HTML4 standard. Since a part of the e-stat macros can not evaluated correctly, we use `text4ht` as a another conversion tool. Since `tex4ht` work on DVI files, we can use any LATEX macro. The third step combines both conversion to one HTML file. It contains all necessary module information. To ensure that a standard XML parser can decompose the HTML file into e-stat modules, `tidy` is used. The algorithm of decomposition is written in XSLT. Additionally it generates with `ttM` the MathML representation of the formulas. In the final step, again written in XSLT, a HTML representation of the XML module is generated. The author can control the conversion in a browser without analyzing a XML file.

Our conversion script as well as all used tools are free and can be downloaded from our Web site. But the simplest way to convert LaTeX sources to e-stat modules is the use of our MD*book online interface.

4 An example

Finally, we present an example module which is taken from the english version of MM*Stat (chapter 11.2, first "enhanced" example). We reused the already existing LaTeX code. This example is an example for simple linear regression.

We integrated a link to the "Golden Solution" of the XploRe quantlet **`MMStatLinReg`** which allows the user to analyze the variables of a data set about crimes with simple linear regression. The quantlet was taken from the *interactive* example for chapter 11.2 in MM*Stat.

We have developed our own view (see internet address below) of the modules since the original one (see Figure 5) can not handle several modules which might be in one LaTeX file.

References

Lehmann, H. and Kleinow, T. (2001): *Client/Server based Statistical Computing.* Proc. of the ISM symposium "Statistical software in the Internet age". ISM, Tokyo, 2001.

Rönz, B. (2001): *The Multimedia-Project MM*Stat for Teaching Statistics.* Proc. of the ISM symposium "Statistical software in the Internet age". ISM, Tokyo, 2001.

Internet addresses

Example (base URL):
`http://apus.wiwi.hu-berlin.de/ sigbert/cs2002/`

— LaTeX file	`csmodul.tex`
— Oldenburg preview	`estat/opreview/001_csmodul.html`
— Berlin preview	`estat/bpreview/_index.html`
— Golden solution	`mmeng/MMStatLinReg.html`

MD*Book	
— online:	`http://www.md-book.com`
— Mailinglist:	`mailto:mdbook@rz.hu-berlin.de`
E-Books:	`http://www.xplore-stat.de/~ebooks`
MM*Stat/	
MD*Booklet:	`http://www.md-stat.com`
XploRe 4:	`http://www.xplore-stat.de`
Rodrigo Witzel	`http://ise.wiwi.hu-berlin.de/~witzel`
Sigbert Klinke	`http://ise.wiwi.hu-berlin.de/ sigbert`

Missing Data Incremental Imputation through Tree Based Methods

Claudio Conversano, Carmela Cappelli

Department of Mathematics and Statistics, University of Naples Federico II
e-mail: {conversa, carmela.cappelli }@unina.it.

Abstract. Conditional mean imputation is a common way to deal with missing data. Although very simple to implement, the method might suffer from model misspecification and it results unsatisfactory for non linear data. We propose the iterative use of tree based models for missing data imputation in large data bases. The proposed procedure uses *lexicographic order* to rank missing values that occur in different variables and deals with these incrementally, i.e, augmenting the data by the previously filled in records according to the defined order.

Keywords. Missing Data, Data Mining, Lexicographic Order, Nonparametric Imputation, Tree-Based Models.

1 Introduction

Missing or incomplete data are a serious problem in many fields of research because they can lead to bias and inefficiency in estimating the quantities of interest. The relevance of this problem is strictly proportional to the dimensionality of the collected data. As an example, in data mining applications a substantial proportion of the data may be missing and predictions might be made for instances with missing inputs.

A common way for dealing with missing data is to discard records with missing values and restrict the attention to those record for which all data are observed. This approach relies on the restrictive assumption that missing data are Missing Completely At Random (MCAR), i.e., that the missingness mechanism does not depend on the value of observed or missing attributes. This assumption rarely holds and, when the emphasis is on prediction rather than on estimation, discarding the records with missing data is not an option (Sarle, 1998). An alternative and weaker version of the MCAR assumption is the Missing at Random (MAR) condition. Under a MAR process the fact that data are missing depends on observed data but not on missing data themselves. While the MCAR condition means that the distributions of observed and missing data are indistinguishable, the MAR condition states that the distributions differ but missing data points can be explained (and therefore predicted) by other observed variables. MAR and MCAR conditions are discussed in Little and Rubin (1987). In principle, a way to meet the MAR condition is to include completely observed covariates that are highly related to the missing ones. Actually, most of the existing methods for missing data imputation (discussed in Shafer, 1997) just assume the MAR condition and, in these settings, a common way to deal with missing data are conditional mean methods or model imputation, i.e., to fit a model on the complete cases for a given variable treating it as the outcome and then, for cases where this variable is missing, plug-in the available data in the model to get an imputed value. The most popular conditional mean method employs least squares regression but, in general, it can be unsatisfactory for nonlinear data and biased if model misspecification occurs.

2 Why using tree based models?

To overcome the shortcomings of the conditional mean imputation method, we consider a nonparametric approach which employs tree based models (Breiman *et* al, 1984) for missing data imputation.
Suppose a response variable y (which can take value either in the real line or in a set of previously defined classes) and a set of p covariates are observed on n objects and denoted by $\{(x_{1i}, \ldots, x_{ji}, \ldots, x_{pi}, y_i)\}_{i=1}^{n}$. A standard tree construction arises from a *divide and conquer* algorithmic strategy which recursively divides the data space into two subregions (subgroups of observations) according to a splitting criterion which aims to optimize classification or prediction for the cases to be split.
In the context of missing data imputation tree based models appear particularly promising because they enjoys two important properties:
a) they do not require the specification of a model structure;
b) they can deal with different type of predictors (numerical and categorical).
The basic idea is as follows: given a variable for which data are missing, and set of other d $(d < p)$ observed variables, the method works by using the former as response variable y and the latter as covariates $x_1, x_2, \ldots, x_d$. The resulting tree explains the distribution of the response variable in terms of the values of the covariates. Since the terminal nodes of the tree are homogeneous with respect to the response, they provide candidate imputation values for this variable.

3 The Method

To deal with the data presenting missing values in many covariates, we introduce an incremental approach based on a suitably defined data preprocessing. The underlying assumption is the MAR process.

Missing Data Ranking. Let $\mathbf{X}$ be the $n \times p$ original data matrix, with d completely observed variables. For any record, if q is the number of covariates with missing data, which maximum value is $Q = (p - d + 1)$, there might be up to $\sum_k \binom{Q}{q}$ type of missingness, ranging from the simplest case where only one covariate is missing to the hardest condition to deal with, where all the Q covariates are missing. We perform a two-way rearrangement of X, one with respect to the columns $(X_1, X_2, .., X_d, ., X_p)$ and one with respect to the rows $(1, 2, \ldots, m, \ldots, n)$. We propose to use a lexicographic ordering

(Cover and Thomas, 1991; Keller and Ullman, 2001) that matches the ordering by value, corresponding to the number of missing values occurring in each record. Practically, we form a string vector of length n that indicates the occurrence and the number of missing values for each row of $\mathbf{X}$. This allows to order $\mathbf{X}$ in a way that the first m rows ($m \ll n$) contain instances with no missing values and the remaining $(n - m)$ rows present missing values. Furthermore, columns also are ordered in the way that the first incomplete column X_d ($d \ll p$) presents the lowest number of missing values and it follows the complete observed ones. As a result, $\mathbf{X}$ is partitioned into four disjoint matrices as follows:

$$\mathbf{X}_{n,p} = \begin{bmatrix} \mathbf{A}_{m,d} & \mathbf{C}_{m,p-d} \\ \mathbf{B}_{n-m,d} & \mathbf{D}_{n-m,p-d} \end{bmatrix}$$

Note that, as a consequence of the ordering schema, only $\mathbf{D}$ contains missing values while the other three blocks are completely observed with respect to their rows and columns.

Incremental Imputation. Once that the different types of missingness have been ranked and coded, the missing data imputation is iteratively made using tree based models.

With respect to the records presenting only one missing attribute, a simple tree is used. Here, the variable with missing values is the response and the other observed variables are the covariates. The tree is built on the current complete data cases in $\mathbf{A}$ and the its results are used to impute the cases in $\mathbf{D}$. In fact, terminal nodes of the tree represent candidate "imputed values". Actual imputed values are get by dropping down the tree the cases of $\mathbf{B}$ corresponding to the missing values in $\mathbf{D}$ (for the variable under imputation), till a terminal node is reached. The conjunction of the filled-in cells of $\mathbf{D}$ with the corresponding observed rows in $\mathbf{B}$ generates new records which are appended to $\mathbf{A}$, that gains the rows whose missing values have been just imputed and a "new" column corresponding to the variable under imputation.

For records presenting multiple missing values, trees are used iteratively. In this case, according to the previously defined lexicographic ordering, the tree is first used to fill in the missing values of the covariate presenting the smallest number of incomplete records. The procedure is then repeated for the remaining covariates under imputation. In this way, we form as many trees as the number of covariates with missing values in the given missingness. In the end, the imputation of joint missingness derives from subsequent trees. This imputation process is incremental because, as it goes on, more and more information is added to the data matrix, both respect the rows and the columns. In other words, $\mathbf{A}$ is updated in each iteration, and the additional information is used for imputing the remaining missing inputs. Equivalently, the matrix $\mathbf{D}$ containing missing values shrinks after each set of records with missing inputs has been filled-in.

4 Empirical Evidence

We summarize the main results form several simulation settings and two real data applications. Concerning simulations, Table 1 summarizes the main features of these settings. In each setting, covariates are uniform in the interval $[0, 10]$ and values are missing with conditional probability $\pi = (1+e^{a+\mathbf{X}\mathbf{b}})^{-1}$, where $\mathbf{X}$ is the covariates matrix, a is a constant and $\mathbf{b}$ is a vector of coefficients (in either case some can be zero). Note that missing values generation depends only on some of the covariates.

Numerical response

	n	p	*missing variables*
sim1.n	500	3	$Y \sim N(-3 + 0.7X_1^2 - 0.3X_2^2, e^{0.3X_1+0.1X_2})$
sim2.n	1000	7	$Y_1 \sim N(X_1 - X_2^2, e^{0.2X_1-0.3X_2})$ $Y_2 \sim N(X_3 - X_4^2, e^{0.2X_3-0.3X_4})$
sim3.n	1000	7	$Y_1 \sim N(X_1 + e^{X_2}, 0.5X_1 - 0.2X_2)$ $Y_2 \sim N(X_3 - cos(X_4), 0.7X_3 - 0.4X_4)$

Binary response

	n	p	*missing variables*
sim1.c	500	3	$Y \sim Bin(n, \frac{e^{(X_1-X_2)}}{1+e(X_1-X_2)})$
sim2.c	1000	7	$Y_1 \sim Bin(n, (1 + e^{X_1+X_2})^{-1})$ $Y_2 \sim Bin(n, \frac{e^{sin(X_3)}+X_4}{1+e^{sin(X_3)}+X_4})$
sim3.c	1000	7	$Y_1 \sim Bin(n, (1 + e^{cos(X_1-X_2)})^{-1})$ $Y_2 \sim Bin(n, \frac{e^{sin(X_3)}+X_4}{1+e^{sin(X_3)}+X_4})$

Table 1. Simulation settings

For the numerical response case, we estimate the mean and the variance of the variable under imputation whereas, for the binary response case we focus on the estimation of the expected value π. Table 2 and 3 report the results of the estimated parameters from 100 independent samples randomly drawn from the original distribution function. For comparison purposes, we estimate the quantities of interest for the Unconditional Mean Imputation (UMI), Parametric Imputation (PI), Non Parametric Imputation (NPI) and Incremental Non Parametric Imputation (INPI). PI is based on linear least squares with constant variance while NPI is based on tree based models[1].
For the numerical case, we see that all the methods do pretty well concerning the mean but INPI outperforms the others in terms of variance. The same results were obtained for the estimation of the π parameter in the binary case, suggesting that the incremental use of trees overcomes variance underestimation and model misspecification.

[1] The terminal nodes are defined by means of the CART pruning method (Breiman *et* al, 1984))in the cross validation version.

	sim1.n		*sim2.n*				*sim3.n*			
	$\hat{\mu}$	$\hat{\sigma}$	$\hat{\mu}_1$	$\hat{\sigma}_1$	$\hat{\mu}_2$	$\hat{\sigma}_2$	$\hat{\mu}_1$	$\hat{\sigma}_1$	$\hat{\mu}_2$	$\hat{\sigma}_2$
All data	-639.2	916.5	-28.2	30.4	38.5	31.8	38.3	30.2	-27.8	29.9
UMI	-760.7	833.5	-33.5	27.2	26.9	29.6	45.2	26.1	-33.6	26.6
PI	-618.0	934.2	-27.4	30.8	37.7	30.8	37.5	31.0	-27.0	30.9
NPI	-612.0	904.3	-27.6	30.1	39.4	29.5	38.3	29.2	-27.1	29.2
INPI	-622.0	908.5	-27.7	30.4	37.3	31.5	38.3	30.3	-27.1	30.1

Table 2. Estimated means and variances for the numerical response case.

	sim1.n	*sim2.n*		*sim3.n*	
	$\hat{\pi}_1$	$\hat{\pi}_1$	$\hat{\pi}_2$	$\hat{\pi}_1$	$\hat{\pi}_2$
All Data	0.510	0.610	0.775	0.616	0.775
UMI	0.610	0.884	0.923	0.883	0.924
PI	0.551	0.699	0.851	0.700	0.859
NPI	0.539	0.667	0.897	0.740	0.849
NPII	0.514	0.633	0.845	0.676	0.813

Table 3. Estimated expected values for the categorical response case

In addition to the simulation study, we have tested the proposed incremental imputation method on two real data sets from the UCI Repository of Machine Learning databases (*www.ics.uci.edu/~mlearn/MLRepository*), namely the *Mushroom* dataset (8124 instances, 22 nominally valued attributes), and the *Boston Housing* dataset (506 instances, 13 real valued and 1 binary attributes). For each data set, we have generated missing values in 4 covariates, according to the MAR condition.

Table 4 and 5 summarize the results of the estimated parameters for the covariates under imputation. In the first example, the percentage of missing inputs for each covariate is: cap-surface(3%), gill-size (6%), stalk-shape(12%) and ring-number (19%). In three cases (gill-size, stalk-shape and ring-number) INPI reproduces the same marginal distributions of the clean data. Note that, in this case, the results provided by different approaches appear quite similar but, since the dimensionality of the original dataset is quite large ($n = 8124$), small differences in the estimation of π have to be considered important.

	cap-surface				*gill-size*		*stalk-shape*		*ring-number*		
	$\hat{\pi}_1$	$\hat{\pi}_2$	$\hat{\pi}_3$	$\hat{\pi}_4$	$\hat{\pi}_1$	$\hat{\pi}_2$	$\hat{\pi}_1$	$\hat{\pi}_2$	$\hat{\pi}_1$	$\hat{\pi}_2$	$\hat{\pi}_3$
All data	0.286	0.000	0.315	0.399	0.691	0.309	0.433	0.567	0.004	0.922	0.074
UMI	0.277	0.000	0.306	0.396	0.710	0.289	0.382	0.618	0.003	0.938	0.059
PI	0.277	0.005	0.324	0.393	0.680	0.320	0.433	0.567	0.003	0.915	0.081
NPI	0.271	0.000	0.316	0.412	0.689	0.311	0.438	0.562	0.004	0.920	0.076
INPI	0.277	0.000	0.319	0.403	0.690	0.310	0.433	0.567	0.004	0.922	0.073

Table 4. Estimated marginal distributions for the Mushroom dataset.

For the second example, the variable under imputation (with the percentage of missing inputs) are: weighted distances to 5 Boston employment centers (*dist*, 28%), nitric oxides concentration (*nox*, 39%), proportion of non-retail business acres per town (*indus*, 33%), number of rooms per dwelling (*rm*, 24%). The results, reported in Table 5, confirm the conclusion drawn from the simulation experiments.

	dist		*nox*		*indus*		*rm*	
	$\hat{\mu}$	$\hat{\sigma}$	$\hat{\mu}$	$\hat{\sigma}$	$\hat{\mu}$	$\hat{\sigma}$	$\hat{\mu}$	$\hat{\sigma}$
All data	3.795	4.431	0.555	0.013	11.136	47.064	6.285	0.478
UMI	3.993	3.703	0.579	0.009	11.659	31.374	6.276	0.389
PI	3.823	4.250	0.559	0.012	11.228	41.439	6.243	0.470
NPI	3.810	4.263	0.557	0.126	10.919	45.416	6.263	0.468
INPI	3.893	4.436	0.555	0.013	11.051	45.634	6.279	0.478

Table 5. Estimated mean and variance for the Boston Housing dataset.

5 Concluding remarks

We propose an automatic procedure that takes advantage of modern computing and allows to handle nonresponses in an easy and fast way, by defining a lexicographic order of the data. Numerical evidences showed that the use of tree based models in missing data incremental imputation is a promising approach, that can be fruitfully employed to impute missing values in both numerical and categorical covariates. Its effectiveness is particularly relevant for nonlinear data with noncostant variance.

Acknowledgements. Authors wish to thank Prof. J. Antoch (Charles University of Prague) and Prof. R. Siciliano (University of Naples Federico II) for their helpful comments.

References

Breiman, L., Friedman, J.H., Olshen, R.A., and Stone, C.J. (1984). *Classification and Regression Trees.* Monterey (CA): Wadsworth & Brooks.

Cover, T.M., Thomas, J.A., (1991). *Elements of Information Theory.* New York: John Wiley and Sons.

Keller, A.M., Ullman, J.D., (2001). *A Version Numbering Scheme with a Useful Lexicographical Order*, Technical Report, Stanford University.

Little, J.R.A, Rubin, D.B. (1987). *Statistical Analysis with Missing Data*, John Wiley and Sons, New York.

Sarle, W.S. (1998), Prediction with Missing Imputs, Technical Report, SAS Institute.

Schafer, J. L., (1997). *Analysis of Incomplete Multivariate Data*, Chapman & Hall.

Missing Values Resampling for Time Series

Andrés M. Alonso [1], Daniel Peña [2] and Juan J. Romo [2]

[1] Departamento de Matemáticas, Universidad Autónoma de Madrid, 28049 Madrid, Spain

[2] Departamento de Estadística y Econometría, Universidad Carlos III de Madrid, 28903 Getafe, Madrid, Spain

Keywords. Blockwise jackknife, blockwise bootstrap, missing values, time series

1 Introduction

In this work we are interested in the moving block jackknife (MBJ) and bootstrap (MBB) introduced in Künsch (1989) and independently in Liu and Singh (1992). These methods are valid to estimate the variance and the distribution of statistics defined by functionals of finite dimensional distributions, which includes robust estimators of location and scale, least squares estimators of autoregressive models and certain version of autocorrelations.

As it is usual in the jackknife methods, the variance estimator is obtained by a weighted sample variance of the statistic evaluated in a sample where some observations are deleted or downweighted. For independent data deleting observations is equivalent to assume that these observations are missing, but for autocorrelated data, as shown in Peña (1990), both procedures are very different. Here, we propose to substitute the observations in the deleted block by some missing value estimates which takes into account the data dependence structure. Then, the variance estimator is a weighted sample variance of the statistics evaluated in a "complete" series. This procedure could be interpreted as a smooth transition between the two parts with full weight in the MBJ.

We also extend this missing values replacement to the moving blocks bootstrap (MBB), in such a case the procedure resemble to a block joining engine similar to the matched-block bootstrap of Carlstein *et al.* (1998). Here, we propose to assume that there are missing observations between two consecutive blocks in MBB resamples and to substitute them by missing value estimates. In this way, the dependence structure is better preserved.

In Alonso *et al.* (2000), we establish the consistency of the proposed methods as variance and distribution estimators of the sample mean assuming the following conditions: (*i*) $\sum_{k=1}^{+\infty} k^2|\gamma_k| < \infty$, (*ii*) $\sum_{k=1}^{+\infty} k^2\alpha_k^{\delta/(6+\delta)} < \infty$, where γ_k is the k-th autocovariance, and α_k is the k-th α-mixing coefficient, with $\delta > 0$, and (*iii*) $\mathrm{E}[|X_t|^{6+\delta}] < \infty$.

The paper is organized as follows. Section 2 presents the MBJ with missing values techniques (M^2BJ) and the MBB with missing values techniques (M^2BB). The results of a simulation study comparing the MBJ and the M^2BJ, and the MBB and the M^2BB are presented in Section 3.

2 Resampling procedures

2.1 Moving missing block jackknife

Let $X_1, \ldots, X_N$ be observations from a stationary process $\{X_t\}_{t\in Z}$. Let us suppose that the statistic T_N, whose variance we want to estimate, can be written as $T_N = T(\rho_N^m)$, where $\rho_N^m = n^{-1}\sum_{t=1}^{n}\delta_{Y_t}$ is an empirical m-dimensional marginal measure, $n = N - m + 1$, the $Y_t = (X_t, \ldots, X_{t+m-1})$ are blocks of m consecutive observations and δ_y denotes the point mass at $y \in R^m$. The MBJ deletes or downweights blocks of m-tuples in the calculation of ρ_N^m:

$$\rho_N^{m,(j)} = (n - \|w_n\|_1)^{-1}\sum\nolimits_{t=1}^{n}(1 - w_n(t-j))\,\delta_{Y_t}, \text{ with } j = 0, \ldots, n-l, \quad (1)$$

where $\|w_n\|_1 = \sum_{i=1}^{l} w_n(i)$, $0 \leq w_n(i) \leq 1$ for $i \in Z$, $w_n(i) > 0$ if an only if $1 \leq i \leq l$, and l is the length of the downweighted block.

The MBJ variance estimator of T_N is defined as

$$\widehat{\sigma}^2_{Jack} = (n - \|w_n\|_1)^2 n^{-1}(n-l+1)^{-1}\|w_n\|_2^{-2}\sum\nolimits_{j=0}^{n-l}\left(T_N^{(j)} - T_N^{(\cdot)}\right)^2, \quad (2)$$

where $T_N^{(j)} = T_N(\rho_N^{m,(j)})$ is the j-th jackknife pseudo-value, $T_N^{(\cdot)} = (n - l + 1)^{-1}\sum_{j=0}^{n-l} T_N^{(j)}$ and $\|w_n\|_2^2 = \sum_{i=1}^{l} w_n(i)^2$.

In our approach we will use the following expression to calculate ρ_N^m:

$$\tilde{\rho}_N^{m,(j)} = n^{-1}\left(\sum\nolimits_{t=1}^{n}(1 - w_n(t-j))\,\delta_{Y_t} + \sum\nolimits_{t=1}^{n} w_n(t-j)\delta_{\widehat{Y}_{t,j}}\right), \quad (3)$$

where $\widehat{Y}_{t,j}$ is an estimate of Y_t supposing that it is a missing value in the j-th sample, and then calculate $\widetilde{T}_N^{(j)} = T_N(\tilde{\rho}_N^{m,(j)})$, for $j = 0, 1, \ldots, n-l$. In Figure 1, we illustrate the construction of a MBJ replica and a M^2BJ replica using the Wölfer annual sunspot index in the period 1770 - 1889 studied by Künsch (1989).

The M^2BJ variance estimator is defined by

$$\widetilde{\sigma}^2_{Jack} = n(n-l+1)^{-1}\|w_n\|_2^{-2}\sum\nolimits_{j=0}^{n-l}\left(\widetilde{T}_N^{(j)} - \widetilde{T}_N^{(\cdot)}\right)^2. \quad (4)$$

2.2 Moving missing block bootstrap

In the case of blockwise bootstrap, we define blocks of l consecutive observations $Y_{i,n} = (X_{i,n}, \ldots, X_{i+l-1,n})$, where $X_{i,n} = X_i$ if $i \in \{1, \ldots, n\}$ and

Fig. 1. (a) Original data, (b) MBJ with ten deleted observations and (c) M^2BJ with ten interpolated observations.

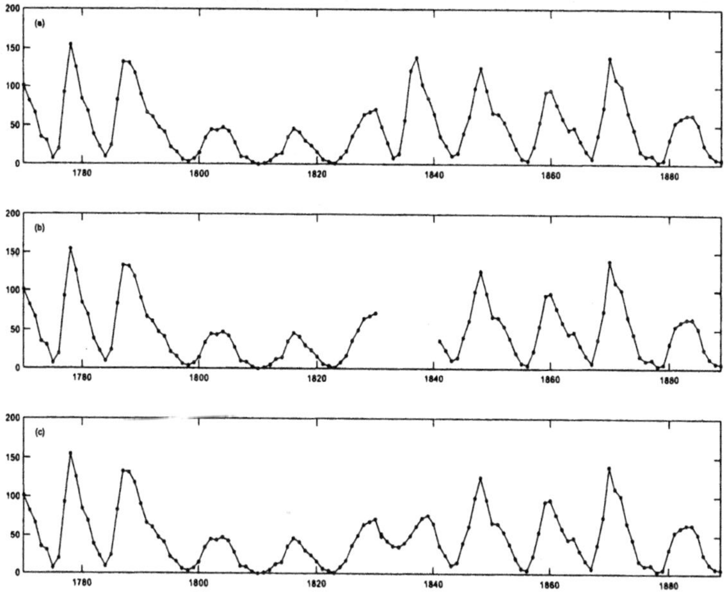

$X_{i,n} = X_{i-n}$ if $i \in \{n+1, \ldots, n+l-1\}$. Then $\{Y_{i,n}\}_{i=1}^{n}$ is used to obtain resamples $(Y_1^*, \ldots, Y_s^*)$ such that $\Pr^*\{Y_j^* = Y_{i,n}\} = 1/n$. The bootstrap statistic is $T_N^* = T_N(\rho_N^*)$, where $\rho_N^* = n^{-1}\sum_{t=1}^{n} \delta_{Y_t^*}$ and the bootstrap variance and distribution of T_N^*,

$$\mathrm{Var}^*\left(T_N^*\right) = \mathrm{E}^*\left[(T_N^* - \mathrm{E}^*\left[T_N^*\right])^2\right] \tag{5}$$

and

$$\Pr^*\left\{\tau_N(T_N^* - \mathrm{E}^*\left[T_N^*\right]) \leq x\right\} \tag{6}$$

are used as variance and distribution estimators of T_N. In (6), the normalizing constant τ_N should take into account the number of blocks s in the bootstrap resample; usually s is selected such that $n = sl$.

The method that we propose can be described as follows: given a MBB resample $(Y_1^*, \ldots, Y_s^*)$ where $l+k$ is the block size, the idea of moving missing blocks bootstrap (M^2BB) is to consider the last k observations in each block as missing values and substitute them by missing value estimates. In Figure 2, we illustrate the construction of MBB and M^2BB replicas considering one $(l+k = 9+1)$ or two $(l+k = 8+2)$ missing observations in each block.

The M^2BB statistic is $\widetilde{T}_N^* = T_N(\widetilde{\rho}_N^*)$, where $\widetilde{\rho}_N^* = n^{-1}\sum_{t=1}^{n} \delta_{\widehat{Y}_t^*}$, and $\widehat{Y}_t^* = Y_t^*$ if $t \in \{1, \ldots, l, l+k+1, \ldots, 2l+k, \ldots, (s-1)(l+k)+1, \ldots, (s-1)(l+k)+l\}$ and $\widehat{Y}_t^*$ is properly an estimate, otherwise. Then the bootstrap variance and

Fig. 2. (a) MBB replica, (b) $\mathrm{M^2BB}$ replica with one missing observation and (c) $\mathrm{M^2BB}$ replica with two missing observations.

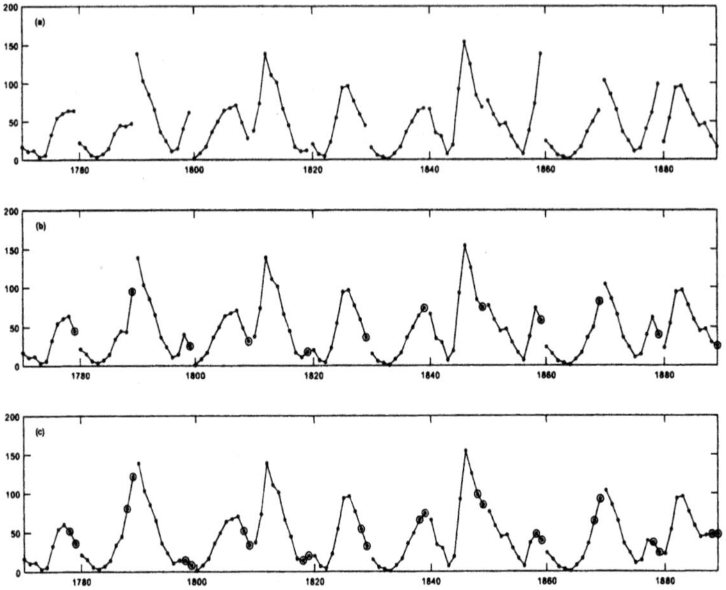

distribution of $\widetilde{T}_N^*$,

$$\mathrm{Var}^*(\widetilde{T}_N^*) = \mathrm{E}^*[(\widetilde{T}_N^* - \mathrm{E}^*[\widetilde{T}_N^*])^2] \tag{7}$$

and

$$\Pr{}^*\{\tau_N(\widetilde{T}_N^* - \mathrm{E}^*[\widetilde{T}_N^*]) \le x\}, \tag{8}$$

are used as variance and distribution estimators of T_N. In (8), τ_N is a function of $s(l+k)$.

3 Simulations results

In this section, we compare the performance of MBJ and $\mathrm{M^2BJ}$, and MBB and $\mathrm{M^2BB}$. We use the following autoregressive models $X_t = \sum_{i=1}^{p} \phi_i X_{t-i} + e_t$ considered by Bühlmann and Künsch (1994).

(M1) AR(1) $\phi_1 = 0.8$, e_t i.i.d. $\mathcal{N}(0,1)$.

(M2) AR(2) $\phi_1 = 1.372$, $\phi_2 = -0.677$, e_t i.i.d. $\mathcal{N}(0, 0.4982)$.

(M3) AR(5) $\phi_1 = 0.9$, $\phi_2 = -0.4$, $\phi_3 = 0.3$, $\phi_4 = -0.5$, $\phi_5 = 0.3$, e_t i.i.d. $\mathcal{N}(0,1)$.

As in Bühlmann and Künsch (1994) we choose the sample size $N = 480$ and $N = 120$. Our results are based on 1000 simulations, and block size range from $l = 1$ to $l = 95$ for $N = 480$, and from $l = 1$ to $l = 30$ for $N = 120$. The statistics T_N included in the simulation study are the sample mean and sample median (additional models and statistics could be found in Alonso *et al.* (2000)). We report the estimates for the following quantities: E_1, SD_1 are the average and the standard deviation of the statistic $\log \widehat{\sigma}_N^2$, and $MSE_1 = (\log \sigma_N^2 - E(\log \widehat{\sigma}_N^2))^2 + SD(\log \widehat{\sigma}_N^2)^2$ is the mean square error of the logarithm of the variance (as recommended in Carlstein *et al.* (1998)). The value of l given in L_1 is the block size producing the minimum MSE. The values E_2, SD_2, MSE_2, L_2 have the same interpretation and are computed for the proposed methods.

Table 1. MBJ y M^{2}BJ para estimar σ_N^2 de $\bar{X}_n$. (+) denotes cases where the MBJ outperforms the M^{2}BJ. (*) denotes cases where the M^{2}BJ outperforms the MBJ.

Model	N	$\log \sigma_N^2$	E_1	SD_1	L_1	MSE_1	E_2	SD_2	L_2	MSE_2
M1	480	3.21	2.95	0.25	35	0.130	2.90	0.24	45	0.152
M1	120	3.18	2.57	0.36	15	0.502	2.51	0.34	20	0.568
M2	480	1.68	1.61	0.09	2	0.012	1.65	0.09	7	0.009 *
M2	120	1.70	1.57	0.18	2	0.050	1.55	0.18	7	0.056
M3	480	1.82	1.67	0.14	15	0.045	1.64	0.16	20	0.056
M3	120	1.83	1.52	0.17	4	0.119	1.38	0.23	10	0.251 +

Table 2. MBJ y M^{2}BJ para estimar σ_N^2 de M_0. (*) denotes cases where the M^{2}BJ outperforms the MBJ.

Model	N	$\log \sigma_N^2$	E_1	SD_1	L_1	MSE_1	E_2	SD_2	L_2	MSE_2
M1	480	3.31	3.04	0.36	35	0.202	3.37	0.31	15	0.101 *
M1	120	3.29	2.67	0.52	20	0.654	2.81	0.48	15	0.458 *
M2	480	2.04	2.02	0.43	35	0.183	1.93	0.42	20	0.187
M2	120	2.05	1.97	0.56	15	0.315	1.98	0.52	10	0.273
M3	480	2.05	1.91	0.42	20	0.194	1.90	0.39	15	0.177
M3	120	2.03	1.71	0.50	10	0.356	1.77	0.52	6	0.337

Our main conclusions for jackknife methods are as follows: (a) In the cases where there exists a substantial difference between the two methods, the missing values replacement generally gives the least MSE. In particular, only in the case of the sample mean and models M3, has the MBJ a better performance; (b) for the median, and model M1 the M^{2}BJ outperforms the MBJ. Other simulation studies (not shown here) have confirmed this advantage of the proposed method in estimating the variance of autocovariance. As for the optimal value of l, it is larger in MBJ than in M^{2}BJ.

Table 3. MBB and M^{2}BB to estimate σ_N^2 of $\bar{X}_n$. (*) denotes cases where the M^{2}BB outperforms the MBB.

Model	N	$\log \sigma_N^2$	E_1	SD_1	L_1	MSE_1	E_2	SD_2	L_2		MSE_2
M1	480	3.21	2.93	0.24	30	0.135	3.00	0.19	5	15	0.079 *
M1	120	3.18	2.56	0.36	15	0.521	2.76	0.31	5	9	0.275 *
M2	480	1.68	1.60	0.11	2	0.018	1.70	0.10	1	1	0.011 *
M2	120	1.70	1.83	0.20	3	0.055	1.65	0.19	1	1	0.040 *
M3	480	1.82	1.66	0.16	15	0.052	1.87	0.11	3	5	0.014 *
M3	120	1.83	1.51	0.18	4	0.130	1.86	0.17	2	3	0.028 *

Table 4. MBB and M^2BB to estimate σ_N^2 of M_0. (*) denotes cases where the M^2BB outperforms the MBB.

Model	N	$\log \sigma_N^2$	E_1	SD_1	L_1	MSE_1	E_2	SD_2	L_2		MSE_2
M1	480	3.31	3.06	0.29	30	0.147	3.10	0.19	5	9	0.084 *
M1	120	3.29	2.76	0.44	15	0.471	2.90	0.36	5	10	0.277 *
M2	480	2.04	2.08	0.24	15	0.058	1.98	0.14	1	1	0.023 *
M2	120	2.05	2.07	0.33	9	0.111	1.94	0.23	1	1	0.066 *
M3	480	2.05	1.89	0.23	15	0.075	2.06	0.14	3	5	0.021 *
M3	120	2.03	1.80	0.29	4	0.137	2.06	0.22	2	3	0.051 *

In the comparison of bootstrap methods, we observe that: (a) In the cases where there exists a substantial difference between the two methods, the missing values replacement always gives least MSE; (b) for the mean and for the median, in all considered models, the M^2BB outperforms the MBB.

Acknowledgments
We would like to thank Mike Wiper for carefully reading that greatly improved the paper. This research was partially supported by the Plan Nacional I+D+I project BEC 2000-0167 and by Cátedra de Calidad BBVA.

References

Alonso, A. M., Peña, D. and Romo, J. (2000) Resampling time series by missing values techniques, *Working Paper*, **00-42**, Madrid: Universidad Carlos III de Madrid.

Bühlmann, P. and Künsch, H.R. (1994) Block length selection in the bootstrap for time series, *Research Report*, **72**, Zürich: Eidgenössische Technische Hochschule.

Carlstein, E., Do, K., Hall, P., Hesterberg, T. and Künsch, H. R. (1998) Matched-block bootstrap for dependent data, *Bernoulli*, **4**, 305-328.

Künsch, H. R. (1989) The jackknife and the bootstrap for general stationary observations, *The Annals of Statistics*, **17**, 1217-1241.

Liu, R. Y. and Singh, K. (1992) Moving blocks jackknife and bootstrap capture week dependence, In: *Exploring the Limits of Bootstrap*, 225-248. New York: Wiley.

Peña, D. (1990) Influential observations in time series, *Journal of Business & Economics Statistics*, **8**, 235-241.

ModelBuilder – an Automated General-to-specific Modelling Tool

Michał Kurcewicz
Department of Economics, Warsaw University, 00-241 Warsaw, Poland

Abstract. The general-to-specific methodology (also known as the London School of Economics, LSE, methodology) is one of the most widely used methods of econometric model construction. Recently, specification search algorithms that automate some stages of the general-to-specific methodology have been proposed. In this paper we present an enhanced specification search algorithm. Extensions include cointegration analysis and automated construction of Error Correction Models. Although, specification search algorithms were originally developed to analyse how well does the LSE modelling approach work in controlled conditions, they are also a valuable tool for empirical research. As an empirical example, an automated construction of a demand function for narrow money in Austria is presented. Results are similar to those obtained in a more traditional way - a stable money demand function is found.

Keywords. General-to-specific modelling, money demand

1 Introduction

Recent advances in automatic specification search (Hoover and Perez 1999, Hendry and Krolzig 1999) have shown that specification search algorithms are a valuable tool for the empirical researcher. However existing algorithms do not account for the non-stationarity of the analysed time series, which is a common feature of macroeconomic data. This article presents an extended specification search algorithm that explicitly allows for non-stationary data and attempts to construct error-correction models (ECMs).

The rest of the article is organised as follows. An extended specification search algorithm, allowing for non-stationary data is discussed in section 2. Section 3 contains an empirical example - an automated construction of a demand function for narrow money in Austria. Section 4 features a discussion of the current implementation. Finally, section 5 concludes.

2 An extended specification search algorithm

2.1 Cointegration analysis

The first stage of the extended specification search algorithm involves determining the order of integration of all time series present in the database. A time series X_t is called $I(1)$ (integrated of order one) if the difference $\Delta X = X_t - X_{t-1}$ is stationary while X_t is non-stationary. Similarly a non-stationary time series is $I(2)$ when twofold differencing is required to render it stationary. To determine the order of integration of individual variables the top-down Dickey-Pantula (Dickey and Pantula 1987) procedure was used. This procedure first tests weather a series is $I(2)$ versus the alternative of first-order integration (using an augmented Dickey-Fuller regression (Dickey and Said 1985)). If the null hypothesis is rejected a test of first order integration is performed. Since $I(2)$ analysis is currently not implemented, all series found to be $I(2)$ are differenced before further processing.

If the dependent variable is integrated the algorithm proceeds with cointegration analysis. A set of $I(1)$ variables is cointegrated if there exists a linear combination of these variables that is itself stationary (the coefficients of that linear combination form a cointegrating vector). To estimate the number of cointegrating vectors the Johansen test (Johansen and Jueslius 1990) is used. The algorithm stops if more than one cointegrating vector is found - in such a case system methods should be used to model the $I(1)$ variables (the current version of ModelBuilder support only single-equation methods). To verify cointegration between the dependent variable and some exogenous variables the ECM test (Ericsson and MacKinnon 1999) is used.

2.2 Formulation of the general model

If the dependent variable is stationary, the general model includes all variables present in the database ($I(1)$ variables enter the model as first differences). If the dependent variable is nonstationary and cointegrated with some exogenous variables, the algorithm attempts to construct an error correction model using integrated variables only. If no cointegration between the dependent variable and exogenous variables can be found, a model in first differences of integrated variables is built. In every case the general specification includes lagged dependent and exogenous variables - for yearly data 2 lags are included, whereas models for quarterly data contain 5 lags of all variables.

2.3 Presearch reductions

Because usually many variables in the general model are highly insignificant, a presearch reduction stage, as in Krolzig and Hendry (2000), was implemented. Presearch reductions aim (via F-tests) to impose zero restrictions on the least significant variables in the general specification. Since variables found jointly insignificant are eliminated from further analysis, only variables with p-values greater than 0.8 are considered for reduction at this stage. Monte-Carlo experiments have shown that presearch reductions can greatly improve performance of the specification search algorithm - without presearch reductions most initial specification (see below) are reduced to the same terminal specification, so most advantages of a multiple search path strategy are lost.

2.4 Specification search

Following Hansen (1999) we define the specification search algorithm in terms of initial specifications, search paths, a test battery and a measure of fit:

- **Initial specifications**. These are models that form the starting points of the algorithmic reduction. As in Hoover and Perez's algorithm they are obtained by deleting from the general model one of the 10 least significant non-significant variables (so at most 10 initial specifications are formed). An initial specification (if not rejected by the test battery - see below) becomes the first element in a search path. A default significance level of 5 % is used at this stage.
- **Search paths**. A search path is a sequence of model specifications, where the successor is a statistically valid restriction of its ancestor. Currently only zero restrictions are imposed - a new specification is obtained by deleting from the current specification the least significant variable among non-significant variables that the algorithm has not attempted to reduce at a earlier stage of the specification search. For the t-tests a default significance level of 5 % was adopted. For a reduction to be valid the reduced model must pass the test battery (see below). A model which

can not be further reduced (all variables are significant or reductions are leading to specifications that do not pass the test battery) becomes the terminal specification of a given search path. The algorithm keeps track of specifications that already have been visited so shared subpaths can be eliminated.

- **Test battery**. After every model reduction a battery of diagnostic tests is applied. These tests verify stability of the estimated parameters and check for potential problems like residual autocorrelation. Currently the battery consists of the following tests (which were also used in the original Hoover-Perez paper): Chow's forecasting and structural stability tests, Jarque-Bera's residual normality test, tests for residual autocorrelation and heteroscedasticity, and a test for ARCH effects. Since rejection by any test contained in the battery invalidates the current model reduction, the default significance level of each individual test is set to 1%.
- **Measure of fit**. Given that search paths can end in different terminal specifications, a method to select the "best" among the terminal models has to be developed. In their original algorithm Hoover and Perez used the standard error of regression as a measure of fit. However, other authors (Hansen 1999, Hendry and Krolzig 1999) found that this criterion favours over-fitted models and as an alternative the use of information criteria (which penalise models with a large number of regressors) or encompassing tests has been suggested. The present implementation of ModelBuilder uses Schwartz's Bayesian Information Criterion (BIC) (Schwartz 1978) to select the final model. That criterion has also the advantage of being a consistent one (Hansen 1999).

3 Modelling the demand for narrow money in Austria

Standard theories of money demand predict that money can be demanded for at least two reasons: as an inventory to smooth differences between income and expenditure streams and as a portfolio asset (Ericsson 1998). The theoretical long-run relationship is usually specified as:

$$M^d/P = L(Y, R),$$

where M^d is nominal money demand, P is the price level, Y is a scaling variable representing economic activity and R is the opportunity cost of holding money. The function L is increasing in Y and decreasing in R. In empirical work Y is typically proxied by GDP or industrial output, whereas for R the rate of return on assets not included in the analysed monetary aggregate is used.

3.1 Data

In our analysis seasonally unadjusted quarterly data for Austria for the period 1965:2 - 1994:4 were used. The series used are $m1$ - the log of narrow money supply (in real values), gdp - log of real GDP and int - the interest rate on long term government bonds (with a maturity of 10 years). Also, following recent research on money demand in Austria (Hayo 2000) a step dummy $str79$, was added to the database. This dummy variable represents the introduction (in 1979) of a new banking law.

3.2 Cointegration analysis of Austrian data

Dickey-Pantula unit root tests show that the three analysed series are $I(1)$ (see Table 1). Since the Johansen (1990) cointegration test (not reported) supports the existence of a single cointegrating vector, the algorithm proceeds

Table 1. Dickey-Pantula tests for Austria: 1965:2 - 1994:4. A constant, trend and seasonals were included in all regressions. Asterisks denote rejection of the null hypothesis at the 1% (**) and 5% (*) significance level (using asymptotic critical values). Lag length was chosen to minimise BIC.

Variable	*I(2) vs. I(1)*		*I(1) vs. I(0)*	
	lags	*t*-stat.	lags	*t*-stat.
gdp	5	-3.95*	5	-2.00
*m*1	1	-6.61**	1	-2.19
int	0	-6.49**	2	-3.55

Table 2. An automatically constructed error correction model for $\Delta m1$. Sample used: 1965:2 - 1989:4.

Variable	Coefficient	Std. Error	*t*-stat.	*p*-value
const	0.04298	0.00906	4.745	0.0000
ecm_{-1}	-0.3931	0.0692	5.683	0.0000
$q1$	-0.08042	0.0343	2.344	0.0213
$q2$	-0.1441	0.0153	9.411	0.0000
$q3$	0.04212	0.00968	4.349	0.0000
$\Delta m1_{-2}$	0.2703	0.0842	3.209	0.0019
Δgdp	0.5614	0.153	3.668	0.0004
$R^2 = 0.71 \quad \hat{\sigma}^2 = 0.023 \quad BIC = -4.44$				

with the construction an error correction model. The first step involves the estimation of an autoregressive distributed lags (ADL) model with current values and 5 lags of the variables $m1$ (dependent variable), *gdp* and *int*. The ADL also includes a constant, seasonals and the *str*79 dummy. The sample used in this and all subsequent regressions is 1965:2 - 1989:4, the remaining 20 observations are reserved for out-of-sample forecast evaluation. Solving the estimated equation for a static long run solution gives (standard errors in parentheses):

$$\begin{aligned} m1 = & \underset{(0.56)}{-5.56} - \underset{(0.013)}{0.014\, int} + \underset{(0.08)}{0.97\, gdp} - \underset{(0.036)}{0.17\, str79} \\ & - \underset{(0.21)}{0.05\, q1} - \underset{(0.24)}{0.38\, q2} - \underset{(0.21)}{0.31\, q3} \end{aligned}$$

We notice that coefficients on *gdp* and *int* have both the expected signs, however *int* is not significant in the long run. Residuals from the above equation (i.e. deviations from the long run equilibrium) form the error correction term. It is stored in the database as an additional variable (*ecm*).

In the next step the following general error correction model is estimated (seasonals omitted):

$$\Delta m1_t = \alpha_0 + \sum_{i=1}^{5} \alpha_{1_i} \Delta m1_{t-i} + \sum_{i=0}^{5} \alpha_{2_i} \Delta gdp_{t-i} + \sum_{i=0}^{5} \alpha_{3_i} \Delta int_{t-i} + \gamma ecm_{t-1} + \epsilon_t$$

Verifying whether $\gamma < 0$ is a test for cointegration of ml, gdp and int. The test statistic is the usual t-ratio, critical values can be found in Ericsson and MacKinnon (1999). In our case of a long run equation with a constant and 2 exogenous $I(1)$ variables (gdp, int) the asymptotic critical value is -3.50 (on a 5 % significance level). Since the t-ratio for $\hat{\gamma}$ is -4.10 we can reject the null hypothesis of no cointegration. However, it should be stated that critical values are affected by the presence of a step dummy in the long run equation.

3.3 Algorithmic reduction

The algorithmic reduction procedure takes as input the general error correction model. From that model 10 starting specification are formed by deleting the least significant variable among non-significant variables. Specification search resulted in two final models, the remaining 8 search paths were terminated due to reaching a already visited specification. The final model as selected by BIC is presented in Table 2. All variables of the selected model are highly significant, the omission of Δint in the final specification suggest that changes in the long run interest rate do not enter short dynamics. Δint was present in the second final model obtained. That model showed a better fit, as measured by the standard error of regression, but relative to the selected model, its out-of-sample forecasting ability was worser. This suggests that the significance of Δint in the second model was spurious.

To assess the stability of the automatically constructed money demand equation we performed in-sample stability tests (using the recursive Chow test) and out-of-sample forecasts (not reported). As no significant stability problems were detected, evidence for the existence of a stable money demand function for Austria was found.

4 Implementation

While developing ModelBuilder we decided against using an existing matrix-oriented language like GAUSS or MATLAB, but aimed at extending an existing very-high level language with the required functionality. ModelBuilder is written mainly in Objective Caml (Ocaml), a functional programming language descending from the Standard ML family. Ocaml offers many valuable features missing from common matrix-oriented languages like a strong type system catching many programming errors during compilation, high order functions, an elaborate module system and flexible error handling with exceptions.

The adoption of an open-source license for ModelBuilder allowed for significant code reuse - the implementation of special functions is based on the R statistical software (Ihaka and Gentlemen 1996), while code for random number generators was taken from the Gnu Scientific Library (Gallasi et al.). For matrix operations wrappers for the ATLAS (Whaley et al.) and LAPACK (Anderson et al.) libraries are provided.

5 Conclusions

In this paper an extended specification search algorithm that allows for non-stationary variables was proposed. It was illustrated by the construction of a demand function for narrow money in Austria. Results show that automatic specification search leads to similar outcomes as more traditional methods do.

However, the current implementation of ModelBuilder is somewhat limited. Future directions include mainly the automatic analysis of systems of equations - some preliminary results can be found in Krolzig (2001). Also implementing tests for weak exogeneity within the Johansen framework would be useful, as lack of weak exogeneity can negatively influent statistical inference (Hendry 1995). Many possibilities of improvement exists in the user interface - the current batch operation mode is suited only for expert users. To ease the less experienced users a graphical interface is under development.

References

Anderson, E. et al. (1999), Lapack Users' Guide, Society for Industrial and Applied Mathematics.

Dickey, D.A. and Pantula, S.G. (1987), Determining the Order of Differencing in Autoregressive Processes. *Journal of Business and Economic Statistics*, **5**, 455-461.

Ericsson, N.R. (1998), Empirical modeling of money demand, *Empirical Economics*, **23**, 295-315.

Ericsson, N.R. and MacKinnon, J.G. (1999), Distribution of Error Correction Tests for Cointegration, International Finance Discussion Paper 655, Board of Governors of the Federal Reserve System.

Gallasi, M. et al. (forthcoming), GNU Scientific Library Reference Manual. Network Theory Ltd.

Hansen, B.E. (1999), Discussion of "Data mining reconsidered". *Econometrics Journal*, **2**, 26-40.

Hayo, B. (2000), The Demand for Money in Austria. *Empirical Economics*, **25**, 581-603.

Hendry, D.F. (1995), On the Interactions of Unit Roots and Exogeneity, *Econometric Reviews*, **4**, 383-419.

Hendry, D.F. and Krolzig, H.-M. (1999), Improving on "Data mining reconsidered" by K.D. Hoover and S.J. Perez, *Econometric Journal*, **2**, 202-219.

Hoover, K.D. and Perez, S.J. (1999), Data mining reconsidered: Encompassing and the general-to-specific approach to specification search. *Econometrics Journal*, **2**, 167-191.

Ihaka, R. and Gentleman, R. (1996), R: A language for data analysis and graphics. *Journal of Computational and Graphical Statistics*, **5**, 299-314.

Johansen, S. and Juselius, K. (1990), Maximum Likelihood Estimation and Inference on Cointegration - With Applications to the Demand for Money, *Oxford Bulletin of Economics and Statistics*, **52**, 169-210.

Krolzig, H.-M. and Hendry, D.F. (2000), Computer automation of general-to-specific model selection procedures. *Journal of Economic Dynamics and Control*, **25**, 831-866.

Krolzig, H.-M. (2001), General-to-Specific Reductions of Vector Autoregressive Processes, *Friedmann, R., L. Knueppel and H. Luetkepohl (eds.), Econometric Studies - A Festschrift in Honour of Joachim Frohn, Muenster: LIT Verlag*, 129-157.

Ocaml. The Objective Caml Language. http://www.ocaml.org/.

Said, S.E and Dickey, D.A. (1985), Testing for unit root in autoregressive-moving average models of unknown order. *Biometrika*, **71**, 599-608.

Schwartz, G. (1978), Estimating the dimension of a model, *Annals of Statistics*, **6**, 461-464.

Whaley, C.R., Petitet A. and Dongarra, J.J., Automated Empirical Optimization of Software and the ATLAS Project, available at http://math-atlas.sourceforge.net/.

On the Use of Particle Filters for Bayesian Image Restoration

Ken Nittono[1] and Toshinari Kamakura[2]

[1] Graduate School of Science and Engineering, Chuo University, 1-13-27, Kasuga, Bunkyo-ku, Tokyo 112-8551, Japan
[2] Department of Industrial and Systems Engineering, Chuo University, 1-13-27, Kasuga, Bunkyo-ku, Tokyo 112-8551, Japan

Keywords. Particle filter, Sequential Monte Carlo method, Image restoration, Bayesian framework

1 Introduction

Image restoration is one of the wide variety of branches in image analysis. In recent work, the Bayesian approach to the restoration has attracted interest and much of this work involves the use of statistical modeling for images assuming Markov random fields (MRF), the stochastic technique based on Monte Carlo methods and maximum a posteriori (MAP) estimation.

The Bayesian approach ordinarily needs relatively large amount of computation time, especially for the type of problems of image processing because of high dimensionality and its massive configuration space. Geman and Geman (1984) proposed an approach to find MAP estimate for any given images. Due to the feature of Gibbs sampler along with simulated annealing, the method enables to find global maxima as MAP estimate, however, in practical situations the optimal annealing schedule for its convergence is not definite and in general it needs much iterative update calculation. Besag (1986) proposed an opposite scheme of iterated conditional modes (ICM). The scheme can considerably reduce computation time with the simple criterion for updating calculation, however, the restored images tend to retain relatively higher misclassification depending on the initial configuration or given images.

In this paper, we aim to study the application of particle filters to Bayesian image restoration. The particle filter is a Monte Carlo method for computing complex posterior distributions, and it provides simple and flexible algorithms for the computing (Doucent et al., 2001). Involving the virtue of the convenient algorithms of the particle filters, we attempt to compose a new method for Bayesian image restoration while using tractable algorighm.

2 Image model

We suppose that an image is composed of M pixels arrayed into rectangular lattice $S = \{1,\ldots,M\}$, and each pixel has a grey-level x_i ($i \in S$) which takes one of l values on $\Lambda = \{1,\ldots,l\}$. Let X_i be a random variable assigned to a pixel i

then a coloring $x=\{x_1,\ldots,x_M\}$ for an image can be regarded as a realization of a random field $X=\{X_i;\, i\in S\}$. Let y denote the given observed image as a realization of Y and x the unknown original image as a realization of X, respectively. According to Bayesian approach, MAP estimation is formalized to find a maximum of posterior distribution as follows,

$$p(x\mid y)=\frac{p(y\mid x)p(x)}{p(y)}. \tag{2.1}$$

We assume that given x, the random field Y is conditionally independent and its marginal distribution is denoted by $p(y_i\mid x_i)$, and observed image is degraded with an additive Gaussian noise, that is of the form $y_i=x_i+\eta_i$, where η_i is identically and independently distributed from a Gaussian distribution with mean zero and known variance σ^2. Note that the additivity and the Gaussianity are not essential matter for the formula of the sequential Monte Carlo method but a practical matter on which image restoration problem standing. We also assume that the original image x is a realization of a MRF described in terms of local characteristics,

$$P\{X_i=x_i\mid X_j=x_j, j\in S\setminus i\}=P\{X_i=x_i\mid X_j=x_j, j\in R_i\}\equiv p(x_i\mid x_{R_i}), \tag{2.2}$$

where $S\setminus i$ denotes any pixels in S except for i and R_i denotes the neighborhood composed of neighboring pixels of i. A set of adjacent four pixels is called the first-order neighborhood, a set of eight pixels containing diagonal four pixels additionally to the first-order one is called the second-order neighborhood and so on.

Following Geman and Geman (1984), we adopt another equivalent representation for the MRF, that is Gibbs distribution which is of the form,

$$p(x)=\frac{1}{Z}\exp\left\{-\frac{1}{T}U(x)\right\}\equiv\phi_T(x), \tag{2.3}$$

where T is the control parameter called temperature, Z is the normalising constant and $U(x)$ is the energy function as follows,

$$U(x)=\sum_{c\in C}V_c(x), \tag{2.4}$$

where c is a clique, C is the set of all cliques in S and $V_c(x)$ denotes an arbitrary function called a potential function. In our succeeding treatment we fix $T=1$, and in such case it is denoted by $\phi(x)$. In addition, we assume that pixels tend to take locally the same value of grey-levels.

3 Application of particle filters

On the computation of posterior distributions of Bayesian models such as dealing with observed data denoted by system and observation functions, much research has been attempted to make it tractable. The difficulty of the problem is the complexity of the posterior distributions such that involving non-Gaussianity, high dimensionality and nonlinearity. In recent several years, methodologies that utilize particle filters, called sequential Monte Carlo methods (Doucent et al.,

2001), have been proposed to overcome this problem. Kitagawa (1996) showed a filtering method called Monte Carlo filter using particle filters for a non-Gaussian nonlinear state space model. In the method, the filtering distribution is represented by a set of particles which are weighted proportional to the likelihood. The sequential Monte Carlo method provides recursive and also computationally tractable formulas for high dimensional posterior distributions which are based on such dynamic models in time series. With reference to this context, we attempt to apply the particle filters of the sequential Monte Carlo methods to the Bayesian image restoration.

For convenience, let us introduce x_0 which denotes the grey-level at the fictitious pixel as an initial state, and let $x_{0:i} = \{x_0, \ldots, x_i\}$ and $y_{1:i} = \{y_1, \ldots, y_i\}$ denote a partial coloring until pixel i for the original image and the observed image, respectively. Then we can define the representation of the recursive formula for the posterior distribution in (2.1), that is based on our image restoration problem, according to the sequential Monte Carlo approaches as follows,

$$p(x_{0:i} \mid y_{1:i}) = p(x_{0:i-1} \mid y_{1:i-1}) \frac{p(y_i \mid x_i) p(x_i \mid x_{R_i})}{p(y_i \mid y_{1:i-1})}. \tag{3.2}$$

Note that on our image model using Gibbs distribution, local characteristics in (2.2) are written as follows,

$$P\{X_i = x_i \mid X_j = x_j, j \in S \setminus i\} = \frac{\phi(x_{i=x_i})}{\sum_{l \in \Lambda} \phi(x_{i=l})} \equiv \Phi_{S\setminus i}(x), \tag{3.3}$$

where $x_{i=k}$ denotes the coloring such that the grey-level at pixel i is k, and furthermore, the energy function is decomposed by $U(x) = \sum_{c \in R_i \cup i} V_c(x) + \sum_{c \notin R_i \cup i} V_c(x)$, thus using allowable cancellations we obtain,

$$\Phi_{S\setminus i}(x) = \frac{\exp\{-\sum_{c \in R_i \cup i} V_c(x_{i=x_i})\}}{\sum_{l \in \Lambda} \exp\{-\sum_{c \in R_i \cup i} V_c(x_{i=l})\}} = \Phi_{R_i}(x) = p(x_i \mid x_{R_i}). \tag{3.4}$$

By the approach of sequential Monte Carlo method, the importance sampling weights for each particle in $\{x_{0:i}^{(j)}; j = 1, \ldots, N\}$ also can be defined by the recursive formula as follows,

$$\tilde{w}_i^{(j)} \propto \tilde{w}_{i-1}^{(j)} \frac{p(y_i \mid x_i^{(j)}) p(x_i^{(j)} \mid x_{R_i}^{(j)})}{\pi(x_i^{(j)} \mid x_{0:i}^{(j)}, y_{1:i})}, \tag{3.5}$$

where π is an arbitrary function called importance distribution, and $x_{R_i}^{(j)}$ denotes a local coloring in the neighborhood R_i based on the j-th sample. For our image restoration problem starting from the sampling from initial distribution $p(x_0)$, succeeding the calculation for particles once for each pixel i, and when i reaches M the final particles represent the posterior distribution $p(x_{0:M} \mid y_{1:M})$. Note that in the case that the importance distribution is the prior distribution, the importance weights are simply written by $\tilde{w}_i^{(j)} \propto \tilde{w}_{i-1}^{(j)} p(y_i \mid x_i^{(j)})$, and we adopt this case for our situation.

1. Generate N particles $\{x_0^{(j)}; j=1,...,N\}$ by the sampling from initial distribution $p(x_0)$.
2. At pixel i
   - 2-1. For each particle in $\{x_{0:i-1}^{(j)}; j=1,...,N\}$, compose the neighborhood R_i and generate a sample $\tilde{x}_i^{(j)}$ according to $p(x_i \mid x_{R_i}^{(j)})$.
   - 2-2. For each sample $\tilde{x}_i^{(j)}$ $(j=1,...,N)$, evaluate the importance weight $\tilde{w}_i^{(j)} = p(y_i \mid \tilde{x}_i^{(j)})$, and normalizing weights.
   - 2-3. Set $\tilde{x}_{0:i}^{(j)} = (x_{0:i-1}^{(j)}, \tilde{x}_i^{(j)})$.
   - 2-4. Generate N particles $\{x_{0:i}^{(j)}, j=1,...,N\}$ from the set $\{\tilde{x}_{0:i}^{(j)}; j=1,...,N\}$ by the resampling with replacement according to the weights $\tilde{w}_i^{(j)}$.
3. Set $i \leftarrow i+1$ and go to step 2.

Fig 1. The algorithm of particle filtering for image restoration

Fig. 1 shows the algorithm for the image restoration using the particle filter. The composition of R_i in step 2-1 is originally added by our image restoration problem and other modifications for the general algorithm of particle filtering is almost none. And thus the application method obtained by adapting the particle filters to the usual image restoration methods based on MRF, inherits the convenience of the filtering algorithm, and the restoration calculation should result in being tractable for high dimensional, that is large-sized, images.

4 Simulation

We conduct some simulations for an artificial image. The original image is composed of 64×64 pixels with $l=4$, and the observed image is degraded from the original image with Gaussian noise $\sigma^2 = 1$. For the comparison of the restoration, basically we adopt misclassification rate $\frac{1}{M}\sum_{i \in S} I(x_i, x_i^*)$ where x_i and x_i^* are grey-levels of the original and the restored image, and $I(a,b)$ is an indicator function. Difference $\frac{1}{M}\sum_{i \in S} |x_i - x_i^*|$ and mean square error $\frac{1}{M}\sum_{i \in S} (x_i - x_i^*)^2$ are additionally evaluated. For the simulation settings, we use uniform distribution as $p(x_0)$, and adopt the second-order neighborhood for R_i, and take account of only pair wise interaction for cliques as follows,

$$V_c(x) = \begin{cases} \psi_0(x_i, x_j) & \text{if } x_i = x_j \\ \psi_1(x_i, x_j) & \text{otherwise} \end{cases}, \tag{4.1}$$

and simply we set $\psi_0(x_i, x_j) = -1$ and $\psi_1(x_i, x_j) = 1$.

On the visiting rule for pixels we normally use a raster scan scheme, and in this case if one strictly use the "past" pixels in the sample path $x_{0:i}^{(j)}$ as the component of R_i, the pixels in R_i is constrained to the half of them, that is $\{(m-1, n-1), (m-1, n), (m-1, n+1), (m, n-1)\}$ for a pixel at row m column n. In order to obtain effective restoration, some appropriate utilization for the composition of R_i is required (Nittono and Kamakura, 2001), and thus we include observed data as the rest half of R_i for the accuracy of the estimation. As the restored image of the result of the calculation, we select the mode of the final particles.

Table 1. The effect of the restoration using the particle filter

	Misclassification	Difference	MSE
$N = 10$	0.1670	0.1895	0.2373
$N = 100$	0.1277	0.1406	0.1680
$N = 500$	0.1387	0.1565	0.1946
$N = 1000$	0.1052	0.1201	0.1519

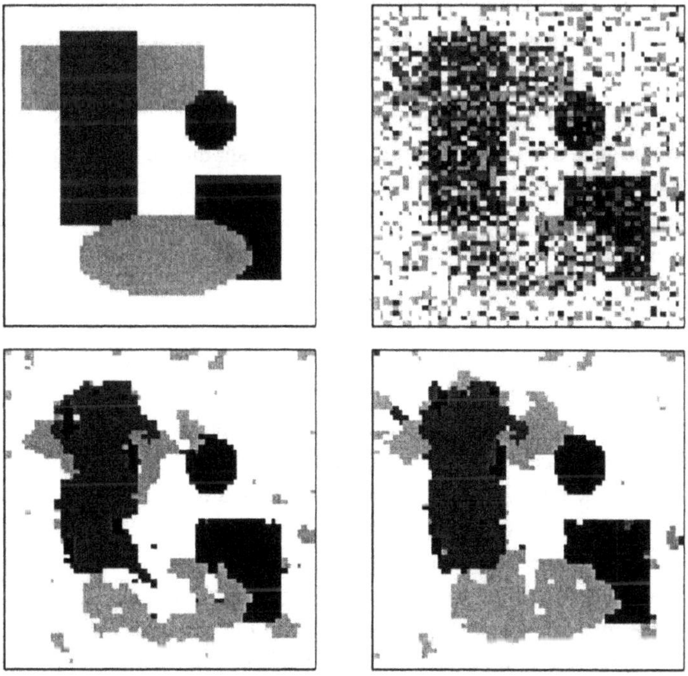

Fig 2. Original image (upper left), given observed image $\sigma^2 = 1$ (upper right), restored image $N = 10$ (lower left) and restored image $N = 1000$ (lower right)

Table 1 shows the effect of the restoration. In general, every rate tends to decrease as increasing the number of particles N except for $N = 500$ and an explanation for this is that there exist cases that some generation of outlying samples $\tilde{x}_i^{(j)}$ or $x_{0:i}^{(j)}$ in the process have influence on the estimation of joint distribution $p(x_{0:M}\ y_{1:M})$. Fig 2 shows the original, degraded and restored images. The restored image at $N = 1000$ seems to show the improvement for the inner regions of figures in comparison with the case at $N = 10$.

5 Discussion

As many authors have pointed out, the modelling of the image also has much influence on the effect of restoration. In our application, we adopt the Gibbs distribution for MRF and furthermore include some observed data for the composition of R_i in order to obtain the effective restoration, however, this should be noted and further studies for the modelling of the MRF and the structures of R_i, which expresses local configurations on images, should be also experimented as well as the usual image restoration when applying the particle filters.

We illustrated a construction of the algorithm for the Bayesian image restoration using particle filters. The general algorithm of the sequential Monte Carlo method is well applied without drastic modification and this fact results in again showing the wide range of its applicability. On the other hand, since a path of a particle $x_{0:i}^{(j)}$ corresponds to a (partial) coloring in our situation, the algorithm needs massive memory to store the paths of samples. And in addition, at the resampling step, simply for the exchange and copy of N particles for M pixels, it demands $O(NM^2)$ for computation time, thus elaborate improvement for the algorithm and data structure are required. And it should be noted that this aspect is also applied to general case of sequential Monte Carlo methods such that the dimension of posterior distribution is enormously high and if one needs joint distribution $p(x_{0:i}\ y_{1:i})$ rather than filtering distribution $p(x_i\ y_{1:i})$.

References

Besag, J. E. (1986). On the statistical analysis of dirty pictures (with discussion). *Journal of the Royal Statistical Society, Series B*, **48**, 259-302.

(ed.) Doucent, A., Freitas, N. D. and Gordon, N. (2001). *Sequential Monte Carlo Methods in Practice*. New York: Springer-Verlag.

Geman, S. and Geman, D. (1984). Stochastic relaxation, Gibbs distributions and Bayesian restoration of images. *IEEE Transactions on Pattern Analysis and Machine Intelligence*, **6**, 721-741.

Kitagawa, G. (1996). Monte Carlo filter and smoother for non-Gaussian nonlinear state space models. *Journal of Computational and Graphical Statistics*, **5**, 1-25.

Nittono, K. and Kamakura, T. (2001). Bayesian image restoration via varying neighborhood structure. *Journal of the Japanese Society of Computational Statistics*, **14**, to appear.

Optimally Trained Regression Trees and Occam's Razor

Petr Savický and Jan Klaschka

Institute of Computer Science, Academy of Sciences of the Czech Republic, Pod Vodárenskou věží 2, 182 07, Prague, Czech Republic, {savicky,klaschka}@cs.cas.cz

Abstract. Two bottom-up algorithms growing regression trees with the minimum mean squared error on the training data given the number of leaves are described. As demonstrated by the results of experiments with simulated data, the trees resulting from the optimization algorithms may have not only better, but also worse generalization properties than the trees grown by traditional methods. This phenomenon is discussed from the point of view of the Occam's razor principle.

Keywords. Regression trees, recursive partitioning, optimization, dynamic programming, bottom-up algorithms, generalization, Occam's razor

1 Introduction

Both classification and regression trees have traditionally been grown by the recursive partitioning, i.e. in a top-down manner, by a "local" optimization of individual splits. Savický *et al.* (2000) proposed two bottom-up algorithms aimed at the full optimization of *classification* trees. In Section 3 of the current paper we present analogous algorithms growing *regression* trees. As in the classification case, the algorithms are of exponential complexity, which restricts the applicability to tasks with a limited number of binary predictors.

Together with the new algorithms, we demonstrate in Section 4 selected results of numerical experiments with simulated data, focused on the generalization properties of the regression trees grown, i.e. on their accuracy on independent data drawn from the same distribution, not used for training.

Our experimental results show that, depending on the distribution and data set size, optimization may yield both better, and worse trees (with respect to generalization properties) than the recursive partitioning. Examples of similar phenomena in the classification context are described in Savický and Klaschka (2001).

In Section 5, the examples of situations where the trees resulting from optimization are inferior to those grown classically are discussed from the point of view of the limits of applicability of the Occam's razor principle.

2 Basics and notation

We consider regression problems with $P \geq 1$ binary $\{0,1\}$-valued predictors $X_1, \ldots, X_P$, and a real valued dependent variable Y. Thus, the data of a case consist of a real dependent variable value y, and a *predictor vector* $\mathbf{x} = (x_1, \ldots, x_P) \in \{0,1\}^P$.

A regression tree T tests one of the predictors in each of the nonterminal nodes, sending the cases with values 0 and 1 of the predictor into different branches. To each terminal node (leaf), T assigns a constant, defining thus a piecewise constant regression function f_T on $\{0,1\}^P$. By the size of the tree we mean the number $|T|$ of its terminal nodes.

By a *cube* we mean a subset of $\{0,1\}^P$ characterized by restricting some coordinates to constants (0 or 1). The dimension of the cube is the number of unrestricted coordinates.

Our algorithms are based on assigning trees to all cubes with at least one training case. By a *tree T for a cube B* we understand a tree that contains tests of only those predictors whose values are unrestricted in B, and which assigns estimates $f_T(\mathbf{x})$ only to $\mathbf{x} \in B$.

When T is a tree for a cube B and $\mathcal{L}$ a data set of N observations $(\mathbf{x}_n, y_n)$, $n = 1, \ldots, N$, the mean squared error of T is given by

$$R(T|\mathcal{L}) = \frac{1}{N} \sum_{\mathbf{x}_n \in B} \left(y_n - f_T(\mathbf{x}_n)\right)^2.$$

For $B = \{0,1\}^P$, $R(T|\mathcal{L})$ is the mean squared error in the standard sense.

Assuming that Q is the "true" distribution of the data, by the true error of a tree T for $\{0,1\}^P$ we mean the expected value $E_Q\left(Y - f_T(X_1, \ldots, X_P)\right)^2$ with respect to Q.

3 Tree optimization algorithms

In this section, we outline two algorithms growing regression trees, analogous to those for classification trees from Savický *et al.* (2000).

A tree T for a cube B is said to be *optimally trained* on data set $\mathcal{L}$, if $R(T|\mathcal{L}) \leq R(T'|\mathcal{L})$ holds, whenever T' is a tree for B and $|T'| = |T|$.

Let T be a tree for a cube B. The *error-complexity measure* of T with complexity parameter $\alpha \geq 0$ on data set $\mathcal{L}$ is defined as $R_\alpha(T|\mathcal{L}) = R(T|\mathcal{L}) + \alpha|T|$. Tree T for a cube B is *α-error-complexity optimal* on $\mathcal{L}$, if $R_\alpha(T|\mathcal{L}) \leq R_\alpha(T'|\mathcal{L})$, whenever T' is a tree for B.

Algorithm R1 grows, for a given $\alpha \geq 0$ and a training data set $\mathcal{L}$, an α-error-complexity optimal tree for $\{0,1\}^P$.

The structure of Algorithm R1 is similar to that of Algorithm I described in more detail in Savický *et al.* (2000). The algorithm assigns to every cube B that contains at least one observation, an α-error-complexity optimal tree. Using a dynamic programming approach, the algorithm starts from the cubes of zero dimension and proceeds in a bottom-up manner up to the whole set $\{0,1\}^P$. For each cube, an α-error-complexity optimal tree is constructed using the optimal trees for smaller cubes. The tree assigned to $\{0,1\}^P$ is the output of the algorithm.

Algorithm R2 grows trees of sizes from 1 to a given limit M, each of them being optimally trained on a training data set $\mathcal{L}$.

Algorithm R2 proceeds in steps similar to those of Algorithm II (for classification trees) described in Savický *et al.* (2000). A dynamic programming approach similar to that of R1 is applied, except that for each cube, trees of all considered sizes are constructed. The sequence of trees of sizes $1, \ldots, M$ assigned to $\{0,1\}^P$ is the output of the algorithm.

Clearly, the α-error-complexity optimal trees on $\mathcal{L}$ are optimally trained on $\mathcal{L}$, too. Thus, optimally trained trees are grown by both algorithms. While

Algorithm R2 grows the trees of all sizes up to M simultaneously, Algorithm R1 must be applied several times with different values of α to obtain trees of different sizes. Due to the simultaneous construction of M trees, Algorithm R2 needs M times larger memory than R1. On the other hand, R2 allows to obtain optimal trees of all sizes, while the set of sizes that can be obtained by R1 may not be contiguous. Proposition 2 in Savický *et al.* (2000) gives a characterization of sizes of the classification trees output by Algorithm I. A similar characterization holds for regression trees. According to our experience, Algorithm R2 yields slightly better results than Algorithm R1, as regards generalization properties of trees, when both methods use the same procedure of the selection of tree size.

Both Algorithms R1, R2 have been implemented in C++. The software, including the sources and other information sufficient to reproduce the experiments in the current paper, may be found at `http://www.cs.cas.cz/~savicky/trees`.

4 Numerical experiments

We performed numerical experiments with simulated data, focused on generalization properties of optimally trained trees. We addressed, on one hand, tasks where optimization was likely to give better results than classical tree-based methods. On the other hand, we have found distributions where optimally trained trees appear to be, as regards generalization properties, inferior to the trees grown by the recursive partitioning.

In this section we present two examples of the the results of such experiments – one example of each of these two types.

4.1 Distributions for generating the experimental data

The experiments dealt with in this paper used random samples from distributions described as follows: There are 10 predictors $X_1, \ldots, X_{10}$ whose joint distribution is uniform on $\{0,1\}^{10}$. The dependent variable Y is given by

$$Y = g(X_1, \ldots, X_6) + e,$$

where g is an appropriate function, and e is independent noise distributed as $N(0, \sigma^2)$.

For the distribution called *square of difference* (SQD), function g is given by

$$g(x_1, \ldots, x_6) = (x_1 + x_2 + x_3 - x_4 - x_5 - x_6)^2.$$

The results shown are for the level of noise $\sigma^2 = 1$.

For the problem referred to as *sum of squares* (SSQ). function g is

$$g(x_1, \ldots, x_6) = (x_1 + x_2 + x_3)^2 + (x_4 + x_5 + x_6)^2.$$

The noise level used in the experiment reported below was $\sigma^2 = 4$.

4.2 Experimental setting

For each of the two distributions mentioned above, we generated 100 training data sets of sizes $N = 10^2, 10^{2.25}, 10^{2.5}, \ldots, 10^{3.5}$, and 100 validation data sets of half size (both sizes being rounded to the nearest integers). For each training data set, trees of sizes from 1 up to 120 were grown using Algorithm R2. Of these, a single tree T_{opt} was selected as the tree attaining the least error on validation data.

In parallel, we constructed, using the same training data set for growing, and the validation data set for pruning, a tree T_{cart} with univariate splits by the "classical" program CART (see Breiman *et al.*, 1984). The $0SE$ rule was used for the selection of the "best" tree. Thus, the size selection methods used for Algorithm R2 and CART were similar.

Then, making use of the fact, that the true distribution of the data is known, we calculated the true errors of T_{opt} and T_{cart}. Means and SE of the 100 true errors were calculated for both T_{opt} and T_{cart} (see Tabs. 1 and 2). The 100 true errors obtained by Algorithm R2 were compared with those from CART by the matched pairs t-test.

4.3 Results

The results for the distribution SQD with $\sigma^2 = 1$ are summarized in Tab. 1. Optimally trained trees are markedly better than those from CART for all the considered data set sizes.

Data set size	True error mean±SE of 100 trees: Optimization	True error mean±SE of 100 trees: CART	Difference mean±SE	t	p-value of t $df=99$
100	3.747± 0.071	4.597±0.079	-0.850±0.105	-8.08	0.000000
178	3.132± 0.054	3.631±0.101	-0.499±0.117	-4.26	0.000046
316	1.883± 0.022	2.878±0.103	-0.995±0.107	-9.28	0.000000
562	1.335± 0.010	2.000±0.066	-0.665±0.065	-10.18	0.000000
1000	1.196± 0.003	1.571±0.037	-0.376±0.038	-9.99	0.000000
1778	1.109± 0.002	1.290±0.016	-0.181±0.016	-11.35	0.000000
3162	1.037± 0.001	1.175±0.009	-0.137±0.009	-15.92	0.000000

Table 1. Results of experiment with distribution SQD, noise variance $\sigma^2 = 1$. For each of the data set sizes, true errors of 100 trees T_{opt} and 100 trees T_{cart} are compared by the matched pairs t-test.

Tab. 2 shows the results for the SSQ distribution with noise variance $\sigma^2 = 4$. The situation is completely different from that in Tab. 1: The mean true errors of the optimally trained trees were larger than those of the trees constructed by CART for all the considered data set sizes. With the exception of the smallest and largest sizes, the t-tests of the differences result in $p < 10^{-6}$.

Data set size	True error mean±SE of 100 trees: Optimization	True error mean±SE of 100 trees: CART	Difference mean±SE	t	p-value of t $df=99$
100	11.840±0.154	11.275±0.190	0.565±0.201	2.82	0.005872
178	9.364±0.123	8.565±0.111	0.800±0.141	5.67	0.000000
316	7.255±0.065	6.728±0.061	0.526±0.072	7.36	0.000000
562	5.705±0.031	5.362±0.027	0.343±0.028	12.07	0.000000
1000	4.736±0.017	4.569±0.014	0.167±0.015	11.32	0.000000
1778	4.283±0.007	4.257±0.006	0.025±0.004	6.42	0.000000
3162	4.174±0.002	4.171±0.002	0.003±0.002	1.42	0.159440

Table 2. Results of experiment with distribution SSQ, noise variance $\sigma^2 = 4$. For each of the data set sizes, true errors of 100 trees T_{opt} and 100 trees T_{cart} are compared by the matched pairs t-test.

5 Evidence against the Occam's razor principle

Throughout this paper, we mean by the *Occam's razor principle* the following methodological suggestion: *Given two models with the same training-set error, the simpler one should be preferred because it is likely to have lower generalization error.* This is presented as the second form of Occam's razor in Domingos (1999), where an overview of the subject (including other interpretations of the principle) and a strong criticism of the above assumption may be found.

The preference of small classification models is supported, in some sense, theoretically by so called PAC learning theory (see, e.g., Blumer *et al.*, 1987): For smaller models, a higher *lower bound* of the probability of the correct classification is attained. It does not mean, nevertheless, that the actual probability of the correct classification by smaller models is higher, too.

There are various experimental results, see e.g. Murphy and Pazzani (1994), Webb (1996), Murthy and Salzberg (1995), demonstrating classification problems, where the Occam's razor principle is not applicable.

Our algorithms minimize the error on the training data given the model size. This is a similar, though not identical task to the minimization of the size given the error. Besides this, our experiment with the SSQ distribution, as reported in the Section 4, lacks a direct relationship with the Occam's razor principle, since the sizes of the trees T_{opt} and T_{cart} are unrelated. However, by a slight modification of the experiment, an example of a regression problem, where Occam's razor is not applicable, can be obtained.

The modified experimental setup is as follows. CART was used to construct a tree T'_{cart} using crossvalidation on the basis of training data only. Let ε be the error of T'_{cart} on the training data. Using Algorithm R2, we constructed a sequence of trees of sizes 1,...,120. Among those trees from this sequence, whose error was at most ε, we selected the tree T'_{opt} of the smallest size.

The experimental setup guarantees that both the size and error of T'_{opt} never exceed the same parameters of the corresponding tree T'_{cart}. By the Occam's razor principle, trees T'_{opt} could be expected to have true errors not exceeding those of trees T'_{cart}.

The results of the modified experiment, summarized in Tab. 3, show the opposite. The trees T'_{opt} have, on average, larger true errors than T'_{cart} for all the seven sizes considered, and all the differences are highly statistically significant.

Data set size	True error mean±SE of 100 trees: Optimization	True error mean±SE of 100 trees: CART	Difference mean±SE	t	p-value of t $df=99$
100	12.762±0.200	11.839±0.181	0.923±0.185	4.99	0.000003
178	9.685±0.124	8.872±0.101	0.814±0.114	7.13	0.000000
316	7.696±0.089	6.920±0.060	0.775±0.076	10.14	0.000000
562	5.813±0.039	5.472±0.030	0.340±0.037	9.12	0.000000
1000	4.816±0.021	4.649±0.018	0.168±0.014	12.06	0.000000
1778	4.321±0.010	4.290±0.008	0.031±0.005	6.85	0.000000
3162	4.189±0.004	4.179±0.004	0.010±0.002	5.62	0.000000

Table 3. Results of the modified experiment with distribution SSQ, noise variance $\sigma^2 = 4$. For each of the data set sizes, true errors of 100 trees T'_{opt} and 100 trees T'_{cart} are compared by the matched pairs t-test.

6 Discussion and conclusions

Depending on the problem, optimally trained regression trees may give better or worse generalization error than trees from the recursive partitioning. Further investigation of this phenomenon is needed in order to understand better the conditions, under which the optimization approach is useful. This can lead to improvements of the known tree construction techniques by using elements of optimization, where appropriate. At the same time, limits of the applicability of the Occam's razor principle could be better recognized.

A possible explanation of the failure of the optimization on the SSQ distribution, introduced in Section 4, lies in that the functions piecewise constant on rectangular regions have poor approximation properties for the underlying function. More generally, the optimized fitting to the training data may be expected to produce no benefit when the class of models is inadequate for a given type of problem.

For the SQD distribution (see Section 4), optimization was markedly better than the recursive partitioning. Note that the distribution is based on a simple nonmonotonous unimodal function. The results of our experiment suggest that even such a simple nonmonotonicity can make the problem hard for the recursive partitioning and, hence, make it easier for the optimization to win. The problems of this kind represent a reason for a further research of the optimization techniques.

Acknowledgement
This research was supported by grant 201/00/1482 from the Grant Agency of the Czech Republic.

References

Blumer, A., Ehrenfeucht, A., Haussler, D. and Warmuth, M. (1987). Occam's Razor. *Information Processing Letters* **24**, 377-380.

Breiman L., Friedman J.H., Olshen R.A. and Stone C.J. (1984). *Classification and Regression Trees*. Belmont CA: Wadsworth.

Domingos, P. (1999). The Role of Occam's Razor in Knowledge Discovery. *Data Mining and Knowledge Discovery* **3** (4), 409-425.

Murphy, P.M. and Pazzani, M.J. (1994). Exploring the Decision Forest: An Empirical Investigation of Occam's Razor in Decision Tree Induction. *J. of Art. Int. Res.* **1**, 257-275.

Murthy, S. and Salzberg, S. (1995). Lookahead and Pathology in Decision Tree Induction. In: *Proceedings of The Fourteenth International Joined Conference on Artificial Intelligence* (ed. C. Mellish), 1025-1031. San Mateo: Morgan Kaufmann.

Quinlan, J.R. (1993). *C4.5: Programs for Machine Learning.* San Mateo: Morgan Kaufmann.

Savický, P., Klaschka, J. and Antoch, J. (2000). Optimal Classification Trees. In: *COMPSTAT 2000, Proceedings in Computational Statistics* (eds. J.G. Bethlehem and P.G.M. van der Heiden), 427-432. Heidelberg: Physica-Verlag.

Savický, P. and Klaschka, J. (2001). Optimally Trained Classification Trees and Occam's Razor. *Technical Report 855*, Inst. of Computer Science, Academy of Sciences of the Czech Republic, Prague.
`http://www.cs.cas.cz/research/library/reports_800.shtml`

Webb, I.G. (1996). Further Experimental Evidence against the Utility of Occam's Razor. *J. of Art. Int. Res.* **4**, 397-417.

Parallel Algorithms for Bayesian Inference in Spatial Gaussian Models

Matt Whiley and Simon P. Wilson

Department of Statistics, Trinity College, Dublin 2, Ireland.

Abstract. Markov chain Monte Carlo (MCMC) implementations of Bayesian inference for latent spatial Gaussian models are very computationally intensive, and restrictions on storage and computation time are limiting their application to large problems. Here we propose three parallel algorithms for linear algebra calculations in these implementations. The algorithms' performance is discussed with respect to a simulation study, which demonstrates the increase in speed and feasible problem size as a function of the number of processors. We discuss how parallel algorithms may be useful more generally in MCMC schemes for these models.

Keywords. Bayesian inference, Markov chain Monte Carlo, Matrix inversion, Parallel algorithms, Spatial modelling

1 Introduction

Parallel processing offers the possibility to consider larger problems than would be possible with a serial machine. One common numerically intensive task is MCMC, and in this article we report on the effectiveness of using parallel algorithms in one particular such application that has considerable storage requirements and is computationally intensive, that of fitting latent Gaussian spatial models using the Bayesian approach. In this paper we propose three possible parallelisations of the MCMC algorithm and study and compare their performance in terms of speed-up and mixing properties of the resulting chains. Our conclusions are that parallel methods allow one to consider larger problems and do offer appreciable speed-up, but that certain parallelisations do reduce the mixing performance of the MCMC algorithm.

1.1 Markov Chain Monte Carlo and Parallel Computing

Parallel computation for MCMC has two possible benefits: it may reduce the time per iteration in any given problem and, by dividing storage requirements between processors, it allows a given MCMC algorithm to be run on larger data sets. There are two broad strategies which may be used, either seperately or together, to parallelise each MCMC iteration: 1. parallelise the calculations in the serial MCMC algorithm, 2. parallelise by partitioning θ into subsets that can be generated simultaneously.

The suitability of strategy 1 depends clearly on how the serial algorithm calculations can be parallelised, whereas the suitability of strategy 2 is a function of the dependencies between parameters; in the ideal case, all parameters are independent and can be updated simultaneously, and in the worst case each parameter is dependent on all others, and so simultaneous updating is impossible. In this paper we consider both strategies.

The issues of load balancing and communication overhead must also be considered. Load balancing tries to ensure that each processor is given the same amount of work to do, so that processors are not idle while others complete work. For strategy 2, one might assign subsets of components of θ to each processor so that the computation time to generate each subset is about equal. Communication overhead is the amount of data that must be transferred between processors. A large overhead will severely limit the increase in speed of the algorithm. In MCMC it is again a function of the amount of dependency between components of θ, since the current values of components of θ must be communicated between processors that are updating components with which they are dependent.

1.2 Spatial Gaussian Models, Inference and Matrix Calculations

Spatial Gaussian processes have found applications to many fields where spatial data exist. At a location s, such a process $Y(s)$ is Gaussian distributed, we assume with zero mean, and the covariance between the process at locations s_i and s_j is parameterised as a function of the distance $d(s_i, s_j)$ and other parameters θ, which include a variance σ^2 . The covariance function used in this paper is $\mathrm{Cov}(Y(s_i), Y(s_j)) = \sigma^2 e^{-\alpha d(s_i,s_j)}$, although many others have been proposed. In this paper we consider the common generalisation, as described in Diggle et. al. (1998), where $Y(s)$ is latent with data X conditionally independent given $Y(s)$ and additional parameters β. The dependence on $Y(s)$ is through the conditional mean $E(X_i \,|\, Y(s_i), \beta)$. Specifically, we shall consider the case in which we have Poisson random variables $X_1, \ldots, X_N$ defined at locations $s_1, \ldots, s_N$ with $\log(E(X_i \,|\, Y(s_i))) = Y(s_i) + \beta$. Data $x = (x_1, \ldots, x_N)$ are observed and the immediate goal of Bayesian inference is to compute $p(\theta, \beta, y \,|\, x)$.

Although it is natural to parameterise the correlation structure, this introduces a computational impediment to calculating both the likelihood and the proposal distribution for the y, which involve functions of the inverse of the covariance matrix, $\Sigma(\theta)$. This impediment is both in terms of computing time and storage space. This is particularly relevant to Bayesian and likelihood methods, since they must evaluate the likelihood and proposal distribution at very many different parameter values, and so several functions of the covariance matrix must be computed at each iteration.

As the sample size increases, the proportion of time spent computing the functions of $\Sigma(\theta)$ also increases. They are also the critical procedures in terms of memory. All our parallel approaches therefore concentrate on speeding up the calculation of these quantities.

2 Parallel Algorithms

2.1 Method 1: Parallelising Matrix Calculations

Our first parallelisation is to use strategy 1 alone. We mention here that we block update y from $p(y \,|\, x, \theta, \beta)$ using the ideas of Rue and Tjelmeland (2002), which improves the mixing properties of the chain.

Suppose the nth iteration of the MCMC algorithm has produced (θ_n, β_n, y_n). In addition, at the nth iteration, we would have computed and stored $\Sigma(\theta_n)^{-1}$ and $|\Sigma(\theta_n)|$. The $(n+1)$th iteration is then:

1. Propose $\tilde{y}$, which makes use of $\Sigma(\theta_n)^{-1}$, and compute the accept probability. Let $y_{n+1} = \tilde{y}$ or y_n according to acceptance or rejection of proposal.
2. Propose $\tilde{\beta}$. Let $\beta_{n+1} = \tilde{\beta}$ or β_n according to acceptance or rejection of proposal.

3. Propose $\tilde{\theta}$. Compute $\Sigma(\tilde{\theta})^{-1}$, $y_{n+1}^T \Sigma(\tilde{\theta})^{-1} y_{n+1}$ and $|\Sigma(\tilde{\theta})|$. Use these to compute the accept probability. Let $\theta_{n+1} = \tilde{\theta}$ or θ_n according to acceptance or rejection of proposal.

We merely comment here that all linear algebra calculations are carried out in parallel, for further details see Whiley and Wilson (2001).

There is an extensive literature on parallel algorithms for matrix operations; see Duff (1999) for a recent review of developments. The numerical linear algebra routines used here come from a library known as ScaLAPACK, a library of numerical algebra routines for parallel computers. The algorithms work by partitioning the matrix into roughly equally sized sub-matrices, assigning sub-matrices to different processors for parallel calculations, and transferring and co-ordinating the necessary information flow between the processors.

There are many issues of optimality in this procedure, to do with minimising information flow between processors, the size of sub-matrices and ensuring load balancing. We do not elaborate on these here, and only mention that ScaLAPACK arranges the processors in a virtual two dimensional $(r \times c)$ lattice called a process grid. This lattice of processors should be as close to square as possible, and in fact it can be beneficial to discard one or more of the processors if it allows a 'squarer' arrangement to be used.

2.2 Method 2: Parallelising by Partitioning Locations

This parallelisation uses both parallelising strategies, and samples y in subsets that are far enough apart to be deemed independent of each other. We begin by partitioning the spatial dataset S into a number of subsets such that $S = \cup_{k=1}^{K} S_k$. These subsets are chosen such that the correlation between non-neighbouring subsets falls below some pre-determined threshold. We note that in order to attain good load balancing it is desirable for the number of points in each subset to be roughly equivalent; see Sánchez and Wilson (2002) for further details. We can then update several non-neighbouring subsets simultaneously, and do so by making a Metropolis proposal for the distribution of y_k conditional on those subsets immediately surrounding it. Each proposal is accepted or rejected independently, and we continue until a proposal has been made for all K subsets. The result is a new value of the process on $\cup_{k=1}^{K} S_k$, composed of new and existing values of $y_1, \ldots, y_K$ according to whether the proposal was accepted or not at each subset.

As long as D is sufficiently large, this will be a good approximation to Gibbs sampling of each S_k in turn. The proposal for each y_k is done using the same method as in the first approach, but now we just require the much smaller inverse of $\Sigma_k(\theta)$. Thus, in this approach one never needs to calculate $\Sigma(\theta)^{-1}$, just K inverses of considerably smaller matrices.

Our MCMC algorithm is then that described in the previous section, but with the new update procedure for y described above.

There are many issues to address, in addition to how the algorithm speeds up calculation. First, there is an important distinction in the memory requirements of Method 1 and Method 2. In Method 1, the covariance matrix was divided between processors, whereas for Method 2 *each processor* requires a copy of the whole matrix. Thus, for very large problems, Method 2 may not be practicable. The method introduces an approximation to the sampling of the Gaussian process. Whether this approximation is reasonable must be investigated. By replacing a block update of y by updates of subsets, we may have altered convergence and mixing properties of the chain. We investigate these in Section 3.1.

2.3 Method 3: Use Full Conditional Distributions to Sample Partition

Method 2 may not perform well when the correlation length in the data is large, since the distribution of a subset given its neighbouring subsets will not be a good approximation to its full conditional. We therefore propose Method 3, which is then Method 2 but where we sample from the full conditional of each subset in the partition. Inherent in this method is inefficiency in the usage of the processors, since we have lost the ability to simultaneously update subsets, but we can still achieve some speed up by computing accept probabilities and proposals in parallel, and when updating θ and β we still parallelise the computation involved as detailed earlier.

3 Comparison of Performance

3.1 Performance per Iteration

We now compare the three parallelisation approaches. The model used is that described in section 1.2. Data are generated from this model with $\beta = 3$, $\sigma^2 = 0.5$ and $\alpha = 10$.

The algorithms were used on an artificially generated dataset on a 30×30 regular lattice. Process grids raging in size from 1×1 to 4×4 were used. For the two methods based on partitioning the dataset the data was divided up into 64 groups of roughly equal size.

In each case the algorithm was run for 100 iterations and the amount of CPU time needed was recorded. This was repeated 10 times for each process grid. Results are shown in Figure 1. The left hand figure shows the mean CPU time for each process grid divided by the number of processors in that grid, this is a crude measurement of the amount of work each processor in the grid must perform. The figure on the right shows the speedup achieved by each processor grid. Speedup is the mean CPU time for the single process case divided by the mean CPU time for each of the process grids considered. This figure also shows the performance ceiling, which is an K-fold speedup for an K processor grid and a common measure of the effectiveness of the parallelisation of an algorithm; the better the parallelisation the closer the algorithm will come to this ideal performance.

We observe that for all three algorithms increasing the number of processors does lead to performance gains. Method 2 is both computationally faster and gives better scaleability, i.e. it comes closest to the performance ceiling. On the whole, method 1 is the slowest, although for larger process grids there is little to choose between it and method 3. We also observe that method 1 does not scale well, and that method 3 only scales well for small numbers of processes. The likely explanation for method 3's poor scaleability for larger processes grids is the large communication overheads of this method, as the number of processes increases we eventually reach a point at which the amount of time spent on communication is outweighing the increase in processing power available to us.

3.2 Convergence

Although we have demonstrated that the second of our approaches to the parallelisation of this algorithm is more efficient than the others from a computational point of view we have yet to address the issue of convergence.

In order to address this issue we consider an example in which artificial data was generated on a 12×12 grid, with $\beta = 3$, and $\sigma^2 = 0.5$ and $\alpha = 10$. Each method was then run using the simulated data for a total of $1,000,000$

489

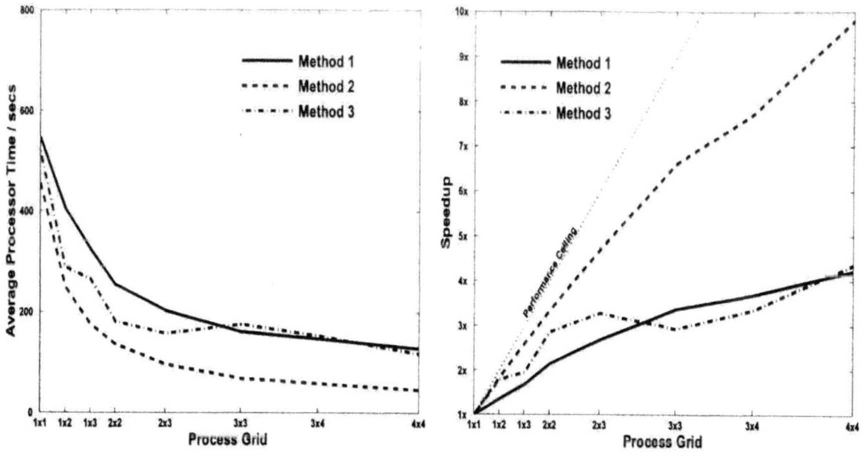

Fig. 1. Performance of the three parallelisation approaches as a function of number of processors. The left graph shows mean CPU time per processor, the right graph shows speedup.

iterations, with every 100th iteration being stored. For methods 2 and 3 the dataset was broken up into 9 regions of equal size, each being a 4×4 square. The three methods had approximate acceptance rates for the update of Y of 2%, 45% and 14% respectively; acceptance rates for β and θ were approximately 65% and 26% for all three methods.

Figure 2 shows the kernel density estimates for σ^2, α and β produced by the three methods. The density estimates are roughly equivalent. The main difference between the three runs was the degree of mixing in the chains. An examination of autocorrelation plots indicated poor mixing of β for method 1 and, to a lesser extent, method 3. Autocorrelation plots for α and σ^2 indicated reasonably good mixing for both parameters for all three methods. The poor mixing of β in method 1 is due to the low acceptance rates for the proposed updates of y. The effect of this is that the chain fails to explore the tails of the posterior marginal for β, in fact experimentation showed that very much longer runs than those used here were necessary in order for this method to fully explore the parameter space of the joint posterior distribution.

There is a high correlation between the value of β and $\bar{Y}$, this would indicate that ideally β and Y should be updated as a block. While extending the update of Y in method 1 to include a simultaneous update of β is feasible, such block updates are not possible when using methods 2 or 3 to update Y. The fact that mixing problems with β were observed for method 1 indicates that such an approach might be desirable.

3.3 Discussion

We conclude that all three methods presented here offer an improvement in terms of computational speed. Also, for the example presented here, there is little to choose between the three algorithms in terms of the resulting posterior density estimates, although there is evidence of poor mixing in

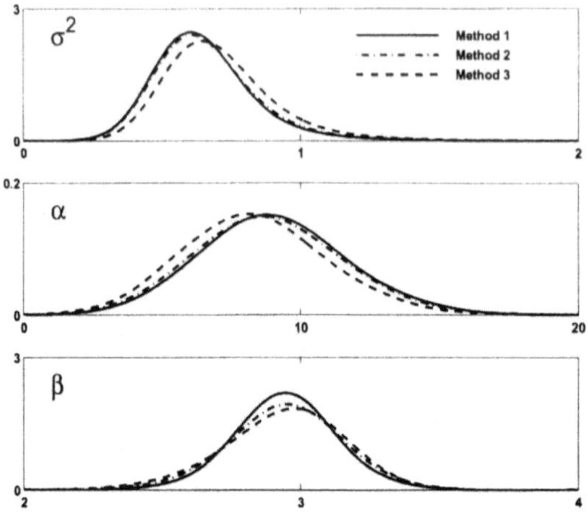

Fig. 2. Comparison of kernel density estimates for the posterior marginals of β, α and σ^2.

the chains produced by methods 1 and 3. The implication is that method 2 provides both good scaleability and, although it only approximates the full conditional distribution of y when updating, it also gives accurate posterior density estimates.

Acknowledgements
This research was part funded by the HEA through the IITAC project at TCD.

References

Diggle, P.J., Tawn, J.A. and Moyeed, R.A. (1998). Model-based geostatistics (with discussion). *Applied Statistics*, **47**, 299-350.

Duff, I.S. and van der Vorst, H.A. (1999). Developments and trends in the parallel solution of linear systems. *Parallel Computing*, **25**, 1931-1970.

Robles Sánchez, O.D. and Wilson, S.P. (2002). A Set Partitioning Algorithm for Use with a Parallel Approach to MCMC for Latent Gaussian Spatial Models. *Technical Report* 02/01, Department of Statistics, Trinity College Dublin.

Rue, H. and Tjelmeland, H. (2002). Fitting Gaussian Markov random fields to Gaussian fields. To appear, *Scandanavian Journal of Statistics.*

Whiley, M. and Wilson, S.P. (2001). A note on using parallel algorithms for implementing Bayesian inference in spatial Gaussian models. *Technical Report* 01/01, Department of Statistics, Trinity College Dublin.

Parameters Estimation of Block Mixture Models

Mohamed Nadif[1] and Gérard Govaert[2]

[1] LITA EA3097, Université de Metz, Ile du Saulcy, 57045 Metz, France

[2] HEUDIASYC, UMR CNRS 6599, Université de Technologie de Compiègne, BP 20529, 60205 Compiègne Cedex, France

Abstract. While most of the clustering procedures are aimed to construct an optimal partition of objects or, sometimes, of variables, there are other methods, called block clustering methods, which consider simultaneously the two sets and organize the data into homogeneous blocks. Recently, we have proposed a new mixture model called *block mixture model* which takes into account this situation. This model allows to embed simultaneous clustering of objects and variables in a mixture approach. We have considered this probabilistic model under the classification likelihood approach and we have developed a new algorithm for simultaneous partitioning based on the Classification EM algorithm. In this paper, we consider the block clustering problem under the maximum likelihood approach and the goal of our contribution is to estimate the parameters of this model. To reach this aim, we use an approximation of the likelihood and we propose an alternated-optimization algorithm to estimate the parameters, and to illustrate our approach, we study the case of binary data.

Keywords. Block mixture model, EM algorithm, fuzzy block clustering

1 Block mixture model

1.1 Notations

Firstly, we need some notation. Let $\mathbf{x}$ denote a $n \times r$ data matrix defined by $\mathbf{x} = \{(x_i^j); i \in I \text{ and } \in J\}$, where I is a set of n objects (rows, observations) and J is a set of r variables (columns, attributes). Even if our principal objective is to estimate the parameters of a block model which we will describe thereafter, we also use partitions of rows and columns. Then, for convenience, we represent the partition into g clusters of the sample either by $\mathbf{z} = (z_1, \ldots, z_g)$, where z_k is the kth cluster or by $\mathbf{z} = (\mathbf{z}_1, \ldots, \mathbf{z}_n)$, where $\mathbf{z}_i \in \{1, \ldots, g\}$ indicates the component of the observation i. Also, $\mathbf{z}_i = k$ can be represented by $\mathbf{z}_i = (z_{i1}, \ldots, z_{ig})$ with $z_{ik} = 1$ and $z_{ik} = 0$ otherwise. Then, the cluster z_k corresponds to the set of observations i such that $\mathbf{z}_i = k$ or $z_{ik} = 1$. The same notations are adopted for the partition $\mathbf{w}$, into m clusters of variables. In the following, the parameters g and m are assumed to be known.

1.2 Definition of the model

Recently, we have proposed a mixture approach to study the problem of block clustering (Govaert and Nadif, 2001). In this approach, the distribution of the data can be represented by the following probability density function

$$f(\mathbf{x};\theta) = \sum_{\mathbf{u}\in U} p(\mathbf{u};\theta)f(\mathbf{x}|\mathbf{u};\theta), \tag{1}$$

where U denotes the set of all possible partitions of $I \times J$ and θ contains all the unknown parameters of this mixture model.

In restricting this model to a set of partitions of $I \times J$ defined by a product of partitions of I and J, which will be supposed to be independent, we obtain the following decomposition

$$f(\mathbf{x};\theta) = \sum_{(\mathbf{z},\mathbf{w})\in \mathcal{Z}\times\mathcal{W}} f(\mathbf{x},\mathbf{z},\mathbf{w};\theta) = \sum_{(\mathbf{z},\mathbf{w})\in \mathcal{Z}\times\mathcal{W}} p(\mathbf{z};\theta)p(\mathbf{w};\theta)f(\mathbf{x}|\mathbf{z},\mathbf{w};\theta), \tag{2}$$

where $\mathcal{Z}$ and $\mathcal{W}$ denote the sets of all possible partitions $\mathbf{z}$ of I and $\mathbf{w}$ of J.

Let $\mathbf{p} = (p_1,\ldots,p_g)$ and $\mathbf{q} = (q_1,\ldots,q_m)$ where p_k and q_ℓ are the probabilities that a row and a column belong to the kth component and to the ℓth component respectively. Supposing that the x_i^j will be independent once $\mathbf{z}_i$ and $\mathbf{w}_j$ are fixed and denoting $\theta = (\mathbf{p},\mathbf{q},\alpha)$, we have $p(\mathbf{z};\theta) = \prod_{i=1}^n p_{\mathbf{z}_i}$, $p(\mathbf{w};\theta) = \prod_{j=1}^d q_{\mathbf{w}_j}$ and

$$f(\mathbf{x}|\mathbf{z},\mathbf{w};\theta) = \prod_{i,j} \varphi_{\mathbf{z}_i\mathbf{w}_j}(x_i^j;\alpha),$$

where $\varphi_{k\ell}(x,\alpha)$ is a probability density function defined on the real set $\mathbf{R}$ and parameterized by α. Finally, we obtain the following block mixture model

$$f(\mathbf{x};\theta) = \sum_{(\mathbf{z},\mathbf{w})\in \mathcal{Z}\times\mathcal{W}} \prod_i p_{\mathbf{z}_i} \prod_j q_{\mathbf{w}_j} \prod_{i,j} \varphi_{\mathbf{z}_i\mathbf{w}_j}(x_i^j;\alpha). \tag{3}$$

To tackle the problem of block clustering, we have used this model under the classification likelihood approach (Symons 1981) and we have successfully applied it to binary and continuous data using an algorithm based on the CEM algorithm (Celeux and Govaert, 1992). This approach has given an interpretation of some criteria proposed by Hartigan (1975), Bock (1979) and Govaert (1983, 1995) and has allowed us to propose new criteria. Now, we consider this problem under the maximum likelihood approach and we focus on the estimation of the parameters of the model.

1.3 Fuzzy block clustering

To estimate the parameters of the block mixture model under the maximum likelihood approach, one of the most widely used procedures is the EM algorithm (Dempster et al., 1977). But, in its application for block mixture models, difficulties arise due to the dependence structure in the model and approximations are required to make the algorithm tractable. A first approximation can be obtained by using the CEM algorithm (Govaert and Nadif, 2001) which replaces the E-step by a stochastic restoration step but it leads to biased parameter estimates. An improved estimator can be obtained using the approach of Hathaway (1986). As pointed out by the author, with the classic mixture model the EM algorithm can be viewed as an alternated maximization of the criterion $L_c(\mathbf{c},\theta)+E(\mathbf{c})$ where $\mathbf{c}$ is a $n\times g$ classification matrix which expresses a fuzzy partition of I, $L_c = \log f(\mathbf{x},\mathbf{c};\theta) = \sum_{i=1}^{n}\sum_{k=1}^{g} c_{ik}\log(p_k f(\mathbf{x}_i;\alpha))$ is the fuzzy completed log-likelihood (c_{ik} is the posterior probability of kth component membership of $\mathbf{x}_i$) and $E(\mathbf{c}) = -\sum_{i=1}^{n}\sum_{k=1}^{g} c_{ik}\log c_{ik}$ is the entropy function. Extending this approach to the block mixture model, we propose the following fuzzy clustering criterion

$$FC(\mathbf{c},\mathbf{d};\theta) = L_c(\mathbf{c},\mathbf{d};\theta) + E(\mathbf{c},\mathbf{d}) \tag{4}$$

where $\mathbf{c}$ denotes a rectangular $n\times g$ data matrix which expresses a fuzzy partition of rows and $\mathbf{d}$ denotes a rectangular $r\times m$ data matrix which expresses a fuzzy partition of columns.

Unfortunately and contrary to the classical mixture model, the optimisation of $FC(\mathbf{c},\mathbf{d};\theta)$ is not equivalent to the EM algorithm. But, using the interpretation of the EM algorithm of Neal and Hinton (1998), which extends the approach of Hathaway, we can show that our algorithm maximizes a pseudo likelihood (Besag, 1975) obtained by using the hypothesis

$$p(\mathbf{z},\mathbf{w}|\mathbf{x};\theta) = p(\mathbf{z}|\mathbf{x};\theta)p(\mathbf{w}|\mathbf{x};\theta).$$

Next, we seek to maximize the fuzzy clustering criterion $FC(\mathbf{c},\mathbf{d};\theta)$. To this end, we propose an iterative algorithm whose steps are carried out by the application of the EM algorithm on intermediate mixture models.

1.4 Algorithm

To maximize $FC(\mathbf{c},\mathbf{d},\theta) = FC(\mathbf{c},\mathbf{d},(\mathbf{p},\mathbf{q},\alpha))$, we propose to maximize the fuzzy clustering criterion with $\mathbf{d}$ and $\mathbf{q}$ fixed and then with $\mathbf{c}$ and $\mathbf{p}$ fixed. Indeed, if we add the hypothesis that the probability density function $\varphi_{k\ell}$ has a sufficient statistic u_i^ℓ and if we note $\mathbf{u}_i = (u_i^1,\ldots,u_i^m)$ and $\#\mathbf{d} = (\#d_1,\ldots,\#d_m)$ ($\#$ denotes the cardinality), we can write

$$FC(\mathbf{c},\mathbf{d},\theta) = FC(\mathbf{c},\theta/\mathbf{d}) + g(\mathbf{x},\mathbf{d},\mathbf{q}),$$

where

$$FC(\mathbf{c},\theta/\mathbf{d}) = \sum_{i=1}^{n}\sum_{k=1}^{g} c_{ik}\log\left(p_k\psi_k(\mathbf{u}_i;\alpha,\#\mathbf{d})\right) - \sum_{i=1}^{n}\sum_{k=1}^{g} c_{ik}\log c_{ik},$$

$\psi_k(\mathbf{u}_i;\alpha,\#\mathbf{d})$ is the probability density function of $\mathbf{u}_i$ and $g(\mathbf{x},\mathbf{d},\mathbf{q})$ does not depend on $\mathbf{c}$. Thus, the conditional fuzzy clustering criterion $FC(\mathbf{c},\theta/\mathbf{d})$ corresponds to the criterion associated to a classical mixture model, which can be written

$$\sum_{k=1}^{g} p_k \psi_k(\mathbf{u}; \alpha, \#\mathbf{d}).$$

Then, maximizing $FC(\mathbf{c}, \mathbf{d}, \theta)$ for $\mathbf{d}$ fixed is equivalent to maximize the conditional fuzzy clustering $FC(\mathbf{c}, \theta/\mathbf{d})$, which can be done by the EM algorithm applied to the mixture model $\sum_{k=1}^{g} p_k \psi_k(\mathbf{u}; \alpha, \#\mathbf{d})$.

As previously, the fuzzy clustering criterion can also be written as

$$FC(\mathbf{c}, \mathbf{d}, \theta) = FC(\mathbf{d}, \theta/\mathbf{c}) + g(\mathbf{x}, \mathbf{c}, \mathbf{d}),$$

where

$$FC(\mathbf{d}, \theta/\mathbf{c}) = \sum_{j=1}^{r} \sum_{\ell=1}^{m} d_{j\ell} \log(q_\ell \psi_\ell(\mathbf{v}^j; \alpha, \#\mathbf{c})) - \sum_{j=1}^{r} \sum_{\ell=1}^{m} d_{j\ell} \log d_{j\ell}.$$

Here $\mathbf{v}^j = (v_1^j, \ldots, v_g^j)$, where v_k^j is a sufficient statistic function of the $\{x_i^j; i \in z_k\}$, $\#\mathbf{c} = (\#c_1, \ldots, \#c_g)$ and $g(\mathbf{x}, \mathbf{z}, \mathbf{p})$ does not depend on $\mathbf{w}$. $FC(\mathbf{d}, \theta/\mathbf{c})$, which corresponds to the fuzzy clustering criterion is associated to the classical mixture model

$$\sum_{\ell=1}^{m} q_\ell \psi_\ell(\mathbf{v}; \alpha, \#\mathbf{c}).$$

Then, it can also be maximized by the EM algorithm applied to this mixture model.

To conclude this section, we briefly describe the different steps of the proposed algorithm.

1. Start from $\mathbf{c}^0$, $\mathbf{d}^0$ and θ^0, initial values of $\mathbf{c}$, $\mathbf{d}$ and θ.
2. Compute $(\mathbf{c}^{t+1}, \mathbf{d}^{t+1}, \theta^{t+1})$ starting from $(\mathbf{c}^t, \mathbf{d}^t, \theta^t)$:
   (a) Compute $\mathbf{c}^{t+1}, \mathbf{p}^{t+1}, \alpha'$ using on the data $(\mathbf{u}_1, \ldots, \mathbf{u}_n)$ the EM algorithm starting from $\mathbf{c}^t, \mathbf{p}^t, \alpha^t$.
   (b) Compute $\mathbf{d}^{t+1}, \mathbf{q}^{t+1}, \alpha^{t+1}$ using on the data $(\mathbf{v}^1, \ldots, \mathbf{v}^r)$ the EM algorithm starting from $\mathbf{d}^t, \mathbf{q}^t, \alpha'$.
3. Iterate the steps 2 until the convergence.

2 Application to binary data

In this section, we illustrate our block mixture model with its application to binary data. In this situation, the probability distribution $\varphi_{k\ell}(x_i^j; \alpha_k^\ell)$ is the Bernoulli distribution, which can be written as

$$\varphi_{k\ell}(x_i^j; \alpha_k^\ell) = (\alpha_k^\ell)^{x_i^j} (1 - \alpha_k^\ell)^{(1 - x_i^j)}.$$

The fuzzy complete log-likelihood can be written

$$FC(\mathbf{c},\mathbf{d},\theta) = FC(\mathbf{c},\theta/\mathbf{d}) + g(\mathbf{x},\mathbf{d},\mathbf{q}) \tag{5}$$

where

$$FC(\mathbf{c},\theta/\mathbf{d}) = \sum_{i=1}^{n}\sum_{k=1}^{g} c_{ik}\log\left(p_k\prod_{\ell=1}^{m}(\alpha_k^\ell)^{u_i^\ell}(1-\alpha_k^\ell)^{(d_\ell-u_i^\ell)}\right) - \sum_{k=1}^{g}\sum_{i=1}^{n} c_{ik}\log(c_{ik})$$

with $u_i^\ell = \sum_{j=1}^{r} d_{j\ell}x_i^j$ and $d_\ell = \sum_{j=1}^{r} d_{j\ell}$. The function $g(\mathbf{x},\mathbf{d},\mathbf{q})$ can be expressed as

$$g(\mathbf{x},\mathbf{d},\mathbf{q}) = -\sum_{j=1}^{r}\sum_{\ell=1}^{m} d_{j\ell}\log\frac{d_{j\ell}}{q_\ell}$$

which is a negative Kullback-Leibler divergence.

From (5), it is straightforward that for $\mathbf{d}$ fixed, $FC(\mathbf{c},\theta/\mathbf{d})$ corresponds to the decomposition of Hathaway associated to a classical mixture model with $\mathbf{u} = \{u_i^\ell; 1 \leq i \leq n, 1 \leq \ell \leq m\}$ data matrix. This justifies the use of the EM algorithm to estimate the parameters. In the same way, we can break the fuzzy complete log-likelihood $FC(\mathbf{c},\mathbf{d},\theta)$ into two terms $FC(\mathbf{d},\theta/\mathbf{c})$ associated to $\mathbf{v} = \{v_k^j; 1 \leq k \leq g, 1 \leq j \leq r\}$ data matrix and $g(\mathbf{x},\mathbf{c},\mathbf{p})$ which does not depend on $\mathbf{q}$.

3 Conclusions

In this work, we used a mixture model to take into account block clustering algorithms. Setting the problem of parameters estimation under the maximum likelihod approach, we have followed the Hathaway approach and we have proposed a fuzzy clustering criterion to estimate the parameters of the block mixture model. To maximize this criterion, we have used an alternated method based on the EM algorithm applied respectively on data matrices with reduced sizes. Unfortunately, the algorithm does not maximize the likelihood, as in the classical mixture model situation but maximizes only an approximation of the likelihood of the block mixture model. Nevertheless, our algorithm gives encouraging results on simulated binary data.

References

Bock, H. (1979) *Simultaneous Clustering of Objects and Variables, in Analyse des données et Informatique*. E. Diday et al. eds, Le Chesnay, INRIA, 187-203.

Besag, J.E. (1975) Statistical analysis of lattice systems. *The Statistician* , **24**, 179-195.

Celeux, G., Govaert, G. (1992) A classification EM algorithm for clustering and two stochastic versions. *Computational Statistics and Data Analysis*, **14**, 315-332.

Dempster, A., Laird, N. and Rubin, D. (1977) Mixture densities, maximum likelihood from incomplete data via the EM algorithm. *Journal of the Royal Statistical Society*, **39**, 1, 1-38.

Govaert, G. (1983) *Classification croisée.* Thèse d'Etat, Université Paris 6, France.

Govaert, G. (1995) Simultaneous clustering of rows and columns. *Control and Cybernetics*, **24**, 437-458.

Govaert G., Nadif M. (2001) Clustering with block mixture models. *submitted.*

Hartigan, J. (1975) *Clustering Algorithms.* New York, John Wiley.

Hathaway, R.J (1986) Another interepretation of the EM algorithm for mixture distributions. *Statistics & Probability Letters*, **4**, 53-56.

Neal, R.M., Hinton, G.E. (1998) *A view of the EM algorithm that justifies incremental, sparse, and other variants.* M.I jordan, editor, *Learning in Graphical Models*, Kluwer Academic Press, 355-358.

Symons, M.J. (1981) Clustering criteria and multivariate normal mixtures. *Biometrics*, **37**, 35-43.

Pattern Recognition of Time Series using Wavelets

Elizabeth Ann Maharaj

Department of Econometrics and Business Statistics, Monash University, Australia

Abstract: In this paper a pattern recognition procedure for time series using wavelets is developed. This is done by means of a randomization test based on the ratio of the sum of squared wavelet coefficients of pairs of time series at different scales. A simulation study using pairs of stationary and non-stationary time series and using the *Haar* and *Daubechies* wavelets reveals that the test performs fairly well at scales where there are a sufficient number of wavelet coefficients. The test is applied to a set of financial time series.

Keywords: Wavelet coefficients, Randomization test, Stationary and Non-Stationary Time Series

1 Introduction

Pattern recognition of time series is useful in many different situations. A financial application would be the comparison of the performance of different stocks and shares, or the comparison of interest rates between various countries. A geological application would be the detection of differences between the waveforms of an earthquake and of a nuclear explosion of similar strength. A medical application would be the comparison of different sections of a particular brain wave recording. Identification of similarities or differences in such time series is useful for decision making and forecasting.

Most existing time series comparison techniques are applicable to time series that are stationary, or to non-stationary time series that can be transformed to stationary time series by some simple transformation such as differencing. Basawa *et al.* (1984), Diggle and Fisher (1991), Guo (1999), Timmer *et al.* (1999), and Maharaj (2000) are amongst the authors who have put forward some of these comparison techniques. However some types of non-stationary time series, such as those that are non-stationary in the variance, cannot be transformed to stationary series by differencing or by using variance reduction techniques. Maharaj (2002) has considered the comparison of non-stationary time series using estimated evolutionary spectra (successive conventional estimated spectra of overlapping portions of the time series), which take into account structural changes over time. Using evolutionary spectra is likened to viewing the time series through a moving window of fixed length. However many time series require a more flexible approach where one can vary the window size to identify more accurately either the time or frequency at which particular events occur.

Wavelet analysis, which is a windowing technique with variable size regions, would be appropriate to analyze such time series (see Percival and Walden, 2000). Wavelet analysis allows the use of long time intervals (higher scales) where more precise low frequency information is available, and short time intervals (lower scales) where more precise high frequency information is available. In this paper we use wavelets to develop a procedure to identify similarities or differences in the patterns of pairs of time series.

In Section 2, using the notation in Percival and Walden (2000), a brief description of wavelets is given. In Section 3, the randomisation test using the wavelet coefficients is described. Section 4 describes the simulation study and its results while the application is discussed in Section 5.

2 Wavelets

Given a signal $\{x(.), -\infty < t < \infty\}$, the collection of coefficients $\{W(\lambda,t): \lambda > 0, -\infty < t < \infty\}$ is known as the continuous wavelet transform (CWT) of $x(.)$, where $W(\lambda,t) = \int_{-\infty}^{\infty} \psi_{\lambda,t}(u)\, x(u)\, du$ and $\psi_{\lambda,t}(u) \equiv \frac{1}{\sqrt{\lambda}} \psi\ ((u-t)/\lambda)$. λ is the scale associated with the transformation and t is its location. The function $\psi(.)$ that satisfies the properties $\int_{-\infty}^{\infty} \psi(u) du = 0$ and $\int_{-\infty}^{\infty} \psi^2(u) du = 1$, is a wavelet filter and is admissible if its Fourier transform $\Psi(\omega) \equiv \int_{-\infty}^{\infty} \psi(u) e^{-i\omega u} d\omega$ is such that $0 < \int_{-\infty}^{\infty} (|\Psi(\omega)|/\omega)^2 d\omega < \infty$. There are a large number of wavelet filters available, amongst them being the Haar and *Daubechies* wavelets.

In applications to time series observed over a discrete set of times, say, $t = 0, 1, 2, \ldots T\text{-}1$, we would be interested is the discrete wavelet transform (DWT) rather than the CWT. The DWT can be thought of as a judicious sub-sampling of W(λ, t) in which we deal with just 'dyadic' scales, that is, we need to pick λ to be of the form 2^{j-1}, $j = 1, 2. .$., and then within a given dyadic scale 2^{j-1}, we pick times t that are separated by multiples of 2^j. The DWT re-expresses a time series in terms of coefficients that are associated with a particular time and particular dyadic scale as well as one or more scaling coefficients. The wavelet coefficients are proportional to differences of averages of the time series observations at each scale, whereas the scaling coefficients are proportional to averages of the original series over the largest scale. The scaling coefficients reflect long-term variations and they would exhibit a similar trend to the original series. For the scale $\lambda = 2^{j-1}$, there are $T_j = T/2^j$ wavelets coefficients and the times associated with these coefficients are taken to be $(2n + 1)2^{j-1} - (1/2)$, $n = 0, 1, \ldots T_j - 1$. It is possible to recover the time series $x(t)$ from its DWT. Hence a time series and its DWT are actually two representations of the same mathematical entity. The DWT decomposes the energy in a time series across scales. If the time series $x(t)$ has 2^m observations and X_t, $W_{j,n}$ and V_n represent, the tth data value, the nth wavelet coefficient for scale 2^{j-1} and the nth scaling coefficient respectively, then the energy decomposition for the DWT is given by

$$\sum_{t=o}^{T} X_t^2 = \sum_{j=1}^{m} \sum_{n=0}^{T_j-1} W_{j,n}^2 + \sum_{n=o}^{m} V_n^2 \ .$$

$\sum_{n=0}^{T-1} W_{j,n}^2$ can be regarded as the contribution to the energy in the time series due to changes at scale 2^{j-1} whereas V_n^2 is that contribution to the energy in the time series due to changes at the last scale which consists of scaling coefficients An energy decomposition is closely related to the concept of wavelet variance, which partitions the variance of a process across dyadic scales. An important aspect of the DWT is that it de-correlates even highly correlated series, that is, the wavelet coefficients at each scale are approximately uncorrelated.

3 Pattern Recognition Procedure

Given two series $x(t)$ and $y(t)$ that we wish to identify as being similar or different, each of the series is decomposed into series of wavelet coefficients at different scales, using the *Daubechies* wavelets at levels 1 to 4. The *Daubechies* wavelets at level 1 is the *Haar* wavelet whereas at levels 2 and beyond, they are smoothed versions of the wavelets at previous levels.

A randomization test, which is a non-parametric test with which one can obtain the exact probability under the null hypotheses without making any assumptions about normality or homogeneity of variance, is applied at each scale. The null hypothesis at each scale is that there is no difference between the wavelets coefficients of the two series. The test statistic at each scale is the ratio of the sum of squared wavelet coefficients of the two series. That is

$$R(\lambda)=\sum_{j=0}^{T_j-1} W_y^2(\lambda,t)\Big/\sum_{j=0}^{T_j-1} W_y^2(\lambda,t)$$

where $W_x^2(\lambda,t)$ and $W_y^2(\lambda,t)$ are the squared wavelet coefficients of the series $x(t)$ and $y(t)$ respectively at scale $\lambda = 2^{j-1}$. Under the null hypothesis the distribution of $R(\lambda)$ will be invariant under 2^{T_j} possible interchanges between $W_x(\lambda,t)$ and $W_y(\lambda,t)$. In practice it will not be feasible to determine the distribution of $R(\lambda)$ but it can be approximated adequately by calculating $R_1(\lambda), R_2(\lambda), \ldots, R_s(\lambda)$ for some large number, say s-1 interchanges between $W_x(\lambda,t)$ and $W_y(\lambda,t)$. Then the significance probability of the observed $R(\lambda)$ value , say $R_1(\lambda)$ can be calculated as the proportion of values $R_1(\lambda)$, $R_2(\lambda), \ldots, R_s(\lambda)$ that is at least is as large as the values $R_1(\lambda)$. Since the wavelet coefficients are uncorrelated, the use of the randomisation test is appropriate.

4 Simulation Study[1]

4.1 Design

A simulation study to assess the size and power of the test, was carried out using pairs of series generated from various processes that are stationary in the mean and variance, non-stationary in the mean but stationary in the variance, non-stationary in the variance but stationary in the mean, non-stationary in both the mean and variance. Stationary series in both mean and variance, and series that were non-stationary in the mean were generated from the ARIMA process

$$(1-\phi B)(1-B)^d X(t)=(1-\theta B)\varepsilon(t) \qquad (1)$$

where ϕ and θ are the autoregressive and moving average parameters respectively, and $\varepsilon(t)$ a white noise process. For simulation purposes $\varepsilon(t)$ was assumed to be a zero mean Gaussian process with variance 1. To generate stationary series, $d = 0$ whereas to

[1] All programming was done in *MATLAB* with the associated wavelet toolbox being used to generate the wavelet coefficients. Programs are available from the author on request.

generate series non-stationary in the mean, $d=1$. Series that were non-stationary in the variance were generated from processes of the form:

$$X(t)=C(t)Z_t \text{ with } C(t)=exp\left(-(t-500)^2/2(200)^2\right) \quad (2)$$

and Z_t being the ARIMA process defined in Equation (1) with $d = 0$. Series that were non-stationary in the mean and variance were generated from a combination of processes defined in Equations (1) and (2) with $d = 1$ in Equation (1). In all cases, size was assessed using series that were generated from the following processes: ARIMA(1,*d*,0) with $\phi = 0$, and 0.5, ARIMA(0,*d*,1) with $\theta = 0.5$, and ARIMA(1,*d*,1) with $\phi = -0.6$ and $\theta = 0.3$. Power was assessed by taking series generated from the ARIMA(0,*d*,0) process against series generated from the ARIMA(1,*d*,0) process, with $\phi = 0.2, 0.4, 0.6$ and 0.8.

Series of lengths $T = 256$ and $T = 1024$ were considered using the *Daubechies* wavelets at levels 1 to 4. In what follows, we will refer to these wavelets as *db*1, *db*2, *db*3 and *db*4. For $T = 256$, the series can be decomposed into eight scales with one scaling coefficient whereas for $T = 1024$ the series can be decomposed into ten scales with one scaling coefficient.. To ensure that there were sufficient wavelet coefficients to carry out the tests, it was decided to decompose series of length $T = 256$ into 5 scales, 2^{j-1} ($j = 1, 2, 3, 4, 5$), so that each consisted of $T_j = T/2^j$ ($j = 1, 2, 3, 4, 5$) coefficients respectively. For $T = 1024$, the series were decomposed into 7 scales, 2^{j-1} ($j = 1, 2, 3, 4, 5, 6, 7$), so that each consisted of $T_j = T/2^j$ (j = 1, 2, 3, 4, 5, 6, 7) coefficients respectively.

One thousand interchanges were used in the randomisation test.

4.2 Discussion and Results

Size and power estimates of the test, some of which are given in Tables 1 and 2, were obtained at the 5% level of significance. For both $T = 256$ and $T = 1024$, for all time series types, the test had reasonably good size estimates on most scales for the *db*1 and *db*2 wavelets. However size tended to be overestimated for the *db*3 and *db*4 wavelets on some scales.

Table 1 Size and power estimates for the for series non-stationary in the mean and variance for $T = 1024$: *db*1 wavelet

	Size				Power			
Scale	AR(1) 0	AR(1) 0.5	MA(1) 0.5	ARMA(1,1) -0.6 0.3	AR(1) 0.2	AR(1) 0.4	AR(1) 0.6	AR(1) 0.8
1	0.02	0.10	0.03	0.07	0.02	0.23	0.93	1.00
2	0.06	0.05	0.03	0.09	0.98	0.90	0.99	1.00
4	0.08	0.08	0.03	0.08	0.41	0.89	1.00	1.00
8	0.10	0.03	0.05	0.03	0.23	0.73	1.00	1.00
16	0.08	0.04	0.04	0.07	0.22	0.43	0.85	1.00
32	0.06	0.04	0.06	0.02	0.12	0.23	0.54	0.99
64	0.08	0.08	0.12	0.06	0.09	0.12	0.30	0.62

Table 2 Size and power estimates for the for series non-stationary in the variance for $T = 1024$: *db*2 wavelet

	Size				Power			
Scale	AR(1) 0	AR(1) 0.5	MA(1) 0.5	ARMA(1,1) -0.6 0.3	AR(1) 0.2	AR(1) 0.4	AR(1) 0.6	AR(1) 0.8
1	0.04	0.05	0.07	0.09	0.35	0.92	0.99	1.00
2	0.01	0.06	0.06	0.03	0.12	0.15	0.27	0.43
4	0.06	0.05	0.05	0.07	0.30	0.70	0.98	1.00
8	0.02	0.03	0.03	0.05	0.15	0.65	0.98	1.00
16	0.07	0.03	0.05	0.04	0.09	0.42	0.81	0.94
32	0.10	0.03	0.01	0.10	0.12	0.28	0.54	0.98
64	0.04	0.02	0.04	0.04	0.12	0.21	0.36	0.67

For both $T = 256$ and $T = 1024$, for stationary series and series non-stationary in the variance, power tended to be lower at the Scale 2 compared to that at other scales. In general for $T = 256$, the power of the test for series non-stationary in the variance tended to be lower that for other types of series. However in general for $T = 1024$, the power estimates were quite good for all the times series types and wavelet types. For each time series type, there were very small differences in the power estimates between the four wavelet types.

5 Application

Time series plots of interest rates from July 1980 to June 2000, for four OECD countries: Australia, the United States, the United Kingdom and France that are non-stationary in both the mean and variance are given below. It can be seen that Australia's interest rate appears to follow a different pattern from the other countries for a long period of time, that is between March 1982 and March 1992, whereas with the exception of the short time period between March 1982 and November 1983, the interest rates for the UK, USA and France appear to move in tandem.

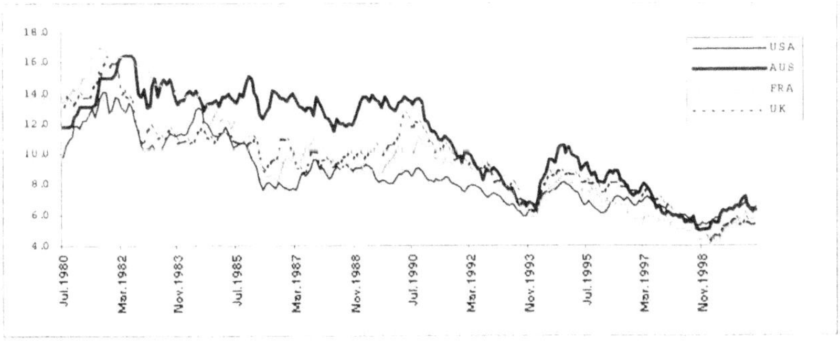

Fig 1 Interest rates for Australia, the United States, France and the United Kingdom: July 1980 - June 2000

Using the *db*1, and *db*2 wavelets, *p*-values of the test were obtained at Scales 1, 2, 4, 8 and 16. The *p*-values of the test for each pair of countries for the *db*1 wavelet are given in Table 3 from where it can be seen, there was no significant difference between the wavelet coefficients of the pairs [USA, UK], [USA, FRA] and [UK, FRA], at the 5% level, whereas for all the other pairs which include Australia, there was a significant difference between the wavelet coefficients at the 5% level on at least one scale, with the difference being highly significant at Scale 1 for the pairs [AUS, USA] and [AUS, FRA]. Similar observations were made for the *db*2 wavelet. These observations would appear to be consistent with the observations made from the graphs in Figures 1.

Table 3 *p*-value of the test to compare interest rates of the four OECD countries

	AUS/USA	AUS/FRA	AUS/UK	USA/UK	USA/FRA	UK/FRA
Scale			*db*1 wavelet			
1	0.004	0.004	0.042	0.096	0.689	0.392
2	0.061	0.661	0.170	0.369	0.530	0.683
4	0.115	0.054	0.348	0.488	0.657	0.295
8	0.422	0.599	0.937	0.707	0.677	0.959
16	0.083	0.442	0.413	0.087	0.136	0.639

6 Concluding Remarks

In summary then it would appear from the simulation study that the test performs reasonably well especially when the *db*1 and *db*2 wavelets are used. From the application, it can be seen that to some extent, the test can be quite successfully applied. Hence it can be concluded that this wavelet-based randomization test could be quite useful in comparing patterns of pairs of any type of time series, thus making a useful contribution to the area of time series classification.

References

Basawa, I.V., Billard, L. & Srinivasan, R. (1984). Large-sample tests of homogeneity for time series models. *Biometrika* **71**, 203-206.

Diggle, P.J. and Fisher, N.I. (1991). Nonparametric comparison of cumulative periodograms. *Appl. Statist.* **40**, 423- 434.

Guo, J. H. (1999). A non-parametric test for the parallelism of two first-order autoregressive processes. *Aust. N. Z. J. Stat.* **41**, 59-65.

Maharaj, E.A. (2000). Clusters of time series. *J. Classification* **17**, 297-314.

Maharaj, E.A. (2002). Comparison of non-stationary time series in the frequency Domain. To appear in *Comput. Statist. Data Anal.*

Percival, D.B. & Walden, A.T. (1999). *Wavelet methods for time series analysis.* Cambridge: Cambridge University Press.

Timmer, J., Lauk, M. Vach, W. & Lucking, C.H. (1999). A test for the difference between spectral peak frequencies. *Comput. Statist. Data Anal.* 30, 45-55.

Representing Knowledge in the Statistical System Jasp

Ikunori Kobayashi[1], Yoshikazu Yamamoto[1] and Takeshi Fujiwara[2]

[1] Tokushima Bunri University, 1314-1 Sanuki, Kagawa 769-2193, JAPAN
[2] The Graduate University for Advanced Studies, 4-6-7 Minami-Azabu, Tokyo 106-8569, JAPAN

Abstract. We describe a framework to assist processes of analyzing data in our statistical system Jasp. Recent statistical systems provide so many statistical methods that most users have difficulties to master how to use them properly. In addition, users are sometimes in danger of swallowing results from systems without thinking deeply. In order to prevent these problems, we have implemented rules of "condition – action" forms in Jasp classes to express heuristic knowledge for statistical analysis. By using these rules, Jasp can give advice to users about possible statistical analysis procedures and notify problems when they appear. This ability is useful for users, especially for students and novices in statistics or the Jasp system.

Keywords. Jasp, Class based object, Rule, Statistical knowledge

1 Introduction

We are developing a statistical system named Jasp (JAva based Statistical Processor) (Nakano, Fujiwara, Yamamoto & Kobayashi 2000) to provide a new data analysis environment to statisticians. Although we have several good statistical systems, such as SAS (SAS Institute Inc. 2002), S (Chambers 1998) and XploRe (Härdle, Klinke & Müller 1999), we designed Jasp to be able to use the latest computer technologies effectively by using features of the Java language.

As statistical systems provide us abundant procedures for solving diverse objectives of data analysis, it is indisputable that they are useful to help our statistical works and to contribute to the dissemination of statistical techniques. However, they often cause careless misuses of statistical techniques, because we can simply execute most of statistical procedures, for example, by just a click of mouse. Unfortunately, the easier to use a statistical system becomes, the more chances of misuses increase.

A user has to learn proper knowledge for using many built-in procedures to prevent such misuses, and must be able to judge whether results from the system are appropriate or not for the particular data. A statistical system usually is not able to evaluate the calculation results, such as statistics, figures and tables. This task should be performed by the user considering the object of the statistical analysis heuristically. Furthermore, a recent statistical system has so many functions that it is difficult to select suitable procedures among them for people who has less experience of statistical analysis, even if he or she studied it by reading some textbooks.

Statistical systems are required to solve these problems. As evaluation of the obtained results depends on the objective of the data analysis, it is very difficult to realize functions for it in a general purpose statistical system, even if we use techniques of AI research results (Russell & Norvig 1995, and Gale, Hand & Kelly 1993). However, it is possible to implement a function to inform the user of basic mistakes of using statistical procedures and suggestions about next possible procedures to investigate the data. It is also preferable to show a list of suitable instructions or commands and to provide a help to select appropriate ones depending on the state and the history of the analysis, because the list of all applicable functions is usually too long. For example, when a user obtains a basic regression model, a list of instructions to calculate statistics for diagnosing the model is preferable than a longer list for all the available regression procedures. We implemented such functions in a regression analysis supporing system (Kobayashi, Nakano & Watanabe 2002).

In this paper, we have implemented a general ability to assist users by using heuristic statistical knowledge stored in the Jasp classes. We note that the knowledge representation format in the system should be simple, because users wish to understand easily what kind of knowledge has been built in the system and how they works in statistical analysis. Thus, we decided to declare statistical knowledge in the similar style as methods in classes, i.e., knowledge is expressed as a rule which is embedded in a class definition and composed of a condition part and an action part. The scope of a rule is limited in the class. A rule also have a priority index to avoid conflict when several conditions of rules in the same class become true simultaneously.

2 Representing knowledge in Jasp

Most statistical systems have their programming languages as interfaces between systems and users to express various computational procedures easily and clearly, to draw graphics flexibly, and to customize functions for routine tasks. Jasp also has a programming language (Kobayashi, Fujiwara, Nakano & Yamamoto 2002). The Jasp language is a script language specially designed for statistical computations, and has two forms to express procedures: Jasp functions and Jasp classes. Jasp functions are used to define procedures of simple calculations, while Jasp classes provide a class based object-oriented programming framework for bundling several related Jasp functions. Furthermore, the Jasp language can handle classes of Java directly.

We decided to define rules in Jasp classes, because they are related to statistical objects which are defined by classes. Rules express heuristic knowledge about statistical notions. For example, conditions for data range are simple but useful rules: values of human height must be between 0 and 3 meters. We also have rules about orders of using statistical methods: diagnostic procedures must follow model building procedures. Levels of significance are important rules for evaluating hypothesis testing statistics: 5% is widely used, for example, in biological sciences, but may be too large in production engineering and quality control fields. We wish to define these rules in a simple general form, unless users have difficulties to formulate their knowledge in that forms.

We note that the original Jasp class is defined by the form

```
jaspclass CLASSNAME( SUPER ) BODY
```

where `CLASSNAME` is a name of the class, `SUPER` is the name of the super class. If there is no super class, `(SUPER)` can be omitted. `BODY` consists of constructors, methods, private functions. Private functions are Jasp functions

```
jaspclass MyLR(LinearRegression) {
  method MyLR( file ){
    this.LinearRegression( file )
    this.checked_vif = false                    // flag
  }
  rule check_VIF(0.5) {
    this.checked_vif == false
  }{
    flag = false
    for(i=1; i<=this.vif.size; i++){
      if(this.vif[i] >= 10){
        flag = true
      }
    }
    this.checked_vif = true
    if(flag){
      str = "One or more VIF values exceed 10.\n" +
            "This indicates the multicolinearity!\n"
    else {
      str = "VIF indicates no multicolinearity.\n"
    }
    str // to display the message in the CUI window.
  }
}
```

Fig. 1. A class with a rule to check VIF

which are effective only in the class, and methods are used as interfaces of private functions to the outside of the class. Constructors are methods that have the same name as their classes, and are called when an instance object is generated by users. Variables prefixed by "`this.`" are treated as fields (internal data) automatically in the constructors and the methods.

We extended this Jasp class to include rules at the similar position as methods and private functions. Rules are defined by the form

```
rule RULENAME( PRIORITY ) { CONDITION }{ ACTION }
```

where `RULENAME` is a name of the rule, `PRIORITY` is a real number which lies between 0 and 1. "1" indicates the highest priority of the rule in the class, and "0" indicates the lowest. This value expresses the relative importance of rules based on the programmer's knowledge. `CONDITION` shows the condition for the rule, and `ACTION` is the procedure which should be executed when the condition becomes true.

The condition part and the action part are implemented by Jasp functions. In the condition part, a Jasp function with no argument is available, returns a logical value (true or false) and is not allowed to change any field value. The action part can contain Jasp functions with no argument, which returns no value and can change some values of fields.

3 Examples of Rule definition

To illustrate the rule definition in Jasp classes, we show two simple examples in this section: first one is the rule to evaluate statistics, and the second one is the rule to navigate analysis.

Fig. 1 shows a Jasp class definition named MyLR, which inherits the built-in class LinearRegression that estimates a linear regression model by the least squares method. The class MyLR focuses on checking the Variance Inflation Factor (VIF) calculated in the super class LinearRegression. We define one method named MyLR and one rule named check_VIF in this class. The MyLR method is a constructor, because the name is as same as the class name. In this constructor, the constructor of the super class LinearRegression is called in the first line, and a field checked_vif is assigned to the logical value false in the next line. The condition of the check_VIF rule means "this rule has not checked yet". Therefore, the condition is satisfied just after the instance object of this class is created. In the action part, the vif field , which is defined in the super class and is filled, is checked to examine whether the VIF of each independent variable is greater than 10 or not. If at least one VIF values exceeds 10, a temporal variable flag is assigned to the logical value true. As the checked_vif field is assigned to true in the next line, the condition of this rule does not become true again, therefore, this rule is not used twice. At last, a message is shown in the Character User Interface (CUI) window according to the flag. Like this example, users can easily express a rule by using the inheritance mechanism in the object-oriented programming.

As rules are encapsulated in a class, rules become active when an instance object of the class is created in the Jasp interpreter. We show operations for it in the CUI window. We assume that the program has been loaded into the Jasp interpreter. To create an instance object of this class, we use the next command:

```
> my1 = MyLR("test_data.dat")
```

where ">" is a prompt in the CUI window. For checking rules, we send checkRules() message to an instance object by the command

```
> my1.checkRules()
  [check_vif(0.5)]
```

then the instance object searches rules whose conditions are true, and returns their names with local priorities as a list sorted by priorities. In this case, we are ready to see the VIF diagnostic result in the CUI window.

To accept the rule of the first element in the list, invokeFirst() message can be used.

```
> my1.invokeFirst()
  VIF indicates no multicolinearity.
```

Then, the instance object executes the action part of the activated rule.

Fig. 2 shows a part of a primitive program to define the class that represents test scores in a school. This class inherits NumericalMatrix class that expresses a matrix with numerical elements. It has one constructor with no argument and three rules. These rules show the order of analysis procedures about the test score.

In the CUI window, a command

```
> sr1 = TestScores()
```

generates an instance object named sr1. Like the former example, we send checkRules() message to an instance object for checking rules.

```
> sr1.checkRules()
  [read_data(1.0)]
```

```
jaspclass TestScores(NumericalMatrix) {
  method TestScores(){
    this.checked_range = false
    this.checked_graphs = false
  }
  rule read_data(1.0){
    this.data = null       // This field is defined in the super
  }{
    this.read_data()       // This method is defined in the super
  }
  rule view_graphs(0.8) {
    this.checked_graphs = false
    this.data != null
  }{
    this.view_graphs()     // This method is defined in the super
    this.checked_graphs = true
  }
  rule check_range(0.9) {
    this.checked_range == false
    this.data != null
  }{
    for(i=1; i<=this.row; i++){
      for(j=1; j<=this.col; j++){
      if(this.data[i,j] > 100)
        this.data[i,j] = 100
      if(this.data[i,j] <0)
        this.data[i,j] =0
      }
    }
    this.checked_range = true
  }
}
```

Fig. 2. A class with rules to indicate orders of procedures

Sending the invokeFirst() message to sr1, the method to read a data is called.

```
> sr1.invokeFirst()
  Please input a file name or an object :
```

Any rule can be invoked by using invokeRule() message with the name of a rule:

```
> sr1.checkRules()
  [check_range(0.9), view_graphs(0.8)]
> sr1.invokeRule("view_graphs")
```

In this case, the view_graphs() method defined in the super class is called and shows histograms and box plots of each column (subject) in the Graphical User Interface (GUI) window. The invokeRule() message also can accept integer which expresses the order of the rule as the argument:

```
> sr1.invokeRule(2)
```

To execute all rules in the list, `invokeAll()` message is available.

If users know the above commands about rules, they can proceed analysis automatically by following rules encapsulated in the particular class.

Rules in a class are inherited to the derived class by the inheritance mechanism of object-oriented programming. Then, a derived class can utilize the rules in the upper class as well as methods and fields of them.

4 Conclusion

If knowledge of statisticians is stored in statistical systems and can be used effectively, they become much more convenient especially for naive users. To approach this goal, we propose to represent and use statistical knowledge in the general purpose statistical system Jasp.

Knowledge in Jasp is described by a simple "condition – action" form as a rule stored in a Jasp class. By using these rules, Jasp can assist users by indicating its own evaluation of calculated statistics. Jasp can also indicate applicable procedures when users are at a loss what to do next in their data analysis processes and require some assistance.

Note that Jasp can not have any responsibility for the analysis results, in case only general knowledge about statistical techniques stored in Jasp is used. As objectives and characteristics of analyses vary substantially depending on data and users, users are responsible for considering them in their analysis processes. We think, however, that it is not difficult to implement rules for their particular purposes and domain knowledge about the data and to supply them for the later use in the similar analysis.

At present, rules are available in the CUI window. We plan to improve Jasp to use rules also in the GUI windows.

References

Chambers, J. M. (1998). *Programming with data: a guide to the S language*, Berlin:Springer.

Gale, W.A., Hand, D.J. & Kelly, A.E. (1993). Statistical applications of artificial intelligence. In: *Hand-book of Statistics 9: Computational Statistics*, 535–576. Amsterdam:Elsevier North-Holland.

Härdle, W., Klinke, S. & Müller, M. (1999). *XploRe – Learning Guide*, Berlin:Springer.

Kobayashi, I., Nakano, J. & Watanabe, K. (2002). A Regression Analysis Supporting System based on Extended Objects, *Bulletin of The Computational Statistics of Japan*, **14** (in Japanese) (to appear).

Kobayashi, I., Fujiwara, T., Nakano, J. & Yamamoto, Y. (2002). A Procedural and Object-Oriented Statistical Scripting Language, *Computational Statistics*, **17** (to appear).

Nakano, J., Fujiwara, T., Yamamoto, Y. & Kobayashi, I. (2000). A statistical package based on Pnuts. In: *COMPSTAT 2000 Proceedings in Computational Statistics* (Bethlehem, J.G. and van der Heijden, P.G.M. eds.), 361–366. Heidelberg:Physica-Verlag.

Russell, S. & Norvig, P. (1995). *Artificial Intelligence*, New Jersey:Prentice Hall.

SAS Institute Inc. (2002). Statistical Analysis System, (http://www.sas.com/).

Robust Estimation with Discrete Explanatory Variables

Pavel Čížek[1]

[1] Institut für Statistik und Ökonometrie, CASE, Humboldt-Universität zu Berlin

Abstract. The least squares estimator is quite sensitive to data contamination and model misspecification. This sensitivity is addressed by the theory of robust statistics which builds upon parametric specification, but provides methodology for designing misspecification-proof estimators by allowing for various "departures" of subsets of the data. Unfortunately, most of highly robust estimators developed within robust statistics cannot be easily applied to models containing binary and categorical explanatory variables. Therefore, we design a robust estimator based on least trimmed squares that can be used for any linear regression model no matter what kind of explanatory variables the model contains. Additionally, we propose an adaptive procedure that maximizes the efficiency of the proposed estimator for a given data set while preserving its robustness.

Keywords. Discrete explanatory variables, linear regression, robust statistics, least trimmed squares

1 Introduction

Regression estimation that involves discrete explanatory variables is quite natural in econometrics, for example, when we deal with dummy and categorical variables that are used to represent non-continuous characteristics such as an individual's gender or education. Such linear regression models are typically estimated using techniques based on the least squares (LS) principle. Although the least squares method is frequently used in regression analysis, mainly because of its simplicity and ease of use, it is quite sensitive to data contamination and model misspecification. This sensitivity is addressed by the theory of robust statistics which builds upon parametric specification, but provides methodology for designing misspecification-proof estimators by allowing for various "departures" of subsets of data. Advantages of one such robust estimator, least trimmed squares (LTS) estimator (Rousseeuw (1985)), over classical least squares based methods were demonstrated, for example, by Orhan, Rousseeuw, and Zaman (2001).

On the other hand, whereas discrete regressors do not cause any particular problems to standard estimation procedures such as LS, the situation is completely different in the case of many robust regression methods such as LTS. The main reason is that they completely reject a subset of observations and they can consequently exclude a group of observations defined by categorical variables from regression estimation. Hence, some discrete variables do not have to be identified. This is a serious shortcoming that was already addressed by Hubert and Rousseeuw (1997), for instance. Nevertheless, the existing remedies are limited only to a certain class of models (see Section 3) and that is why we propose here a general solution for robust estimation in the presence of discrete explanatory variables.

Motivated by the LTS estimator, we design a smoothed version of this estimator, removing thus the complete rejection of observations, the main cause of the problem. This solution can also add some further improvements to the LTS estimator, such as an increase of efficiency while preserving the robustness of LTS. The extent to which efficiency is improved and robustness is preserved depends heavily on the smoothing scheme used. Thus, we define first the smoothed LTS estimator in a general way and study its properties for a general smoothing scheme. Later, we propose a class of smoothing schemes and a rule that allows us, for a given data set, to adaptively find a smoothing scheme that maximizes the efficiency of the estimator while preserving its robust properties.

In the rest of this paper, we first give a brief overview of robust statistics (Section 2) and review existing attempts at robust estimation in the presence of discrete variables (Section 3). Further, we propose a smoothed version of LTS (Section 4) and discuss its asymptotic properties (Section 5). Finally, we describe the adaptive choice of smoothing (Section 6) and present some Monte Carlo simulations (Section 7).

2 Robust statistics

Robust statistics aims to study the behavior of parametric techniques under deviations from the standard assumptions of parametric models and to develop estimators that behave well not only under correct parametric specification, but also in the presence of "small" deviations from the parametric assumptions. Robust statistics offers two main approaches for developing new estimators (see Hampel et al. (1986)). First, estimators can be designed primarily to achieve maximum robustness possible. Second, the theory of robustness also supplies some "optimality" criteria that allow the construction of efficient estimators given certain robust qualities. we concentrate here on the first strategy, that is, on robust estimators with a very high degree of robustness.

Currently, there are several highly robust procedures for the linear regression model. Examples of these so-called high breakdown-point estimators are the *least median of squares* (Rousseeuw (1984)) and the *least trimmed squares* (Rousseeuw (1985)). Recently, the latter became more preferred to the former because it features better asymptotic performance and a fast and reliable approximation algorithm (Rousseeuw and Van Driessen (1999)). Both these estimators can withstand a high amount of contamination, but at the same time, they have inherent problems with estimation which includes discrete variables.

3 Existing approaches to robust estimation with discrete explanatory variables

There are estimators that are highly robust, and at the same time, can cope with discrete and categorical variables. The most general one, RDL_1 (Robust Distance and L_1 regression), was proposed by Hubert and Rousseeuw (1997). This estimator can handle only regression models, where binary variables enter the regression equation additively. Although the RDL_1 estimator is highly robust, it has several disadvantages. First, this estimator can hardly be generalized to more complicated models: for example, to general regression models with discrete variables including cross-effects or to nonlinear models (see Čížek (2001) for more details). Second, RDL_1 can be relatively easily influenced by misspecification occurring in dummy variables simply because it does not treat dummy variables in any special way (this is documented in Section 7).

4 Smoothed least trimmed squares

In this section, we introduce first the least trimmed squares estimator and then define a smoothed version of LTS that should allow the estimation of general linear regression models with discrete explanatory variables and obtain better properties than the existing robust estimators in the area of efficiency.

4.1 Linear regression model and least trimmed squares

Let us consider a linear regression model

$$y_i = x_i^T \beta + \varepsilon_i, \qquad i = 1, \ldots, n. \tag{1}$$

The least trimmed squares estimator $\hat{\beta}_n^{(LTS)}$ is defined as

$$\hat{\beta}_n^{(LTS)} = \arg\min_{\beta \in R^p} \sum_{i=1}^{h} r_{[i]}^2(\beta), \tag{2}$$

where $r_{[i]}^2(\beta)$ represents the ith order statistics of squared residuals $r_1^2(\beta), \ldots, r_n^2(\beta)$; $r_i(\beta) = y_i - x_i^T\beta$ and $\beta \in R^p$ (p denotes the number of estimated parameters). The *trimming constant* h has to satisfy $\frac{n}{2} < h \leq n$. This constant determines the robustness of LTS since definition (2) implies that $n - h$ observations with the largest residuals do not affect the estimator.

4.2 Definition of smoothed least trimmed squares

Consider a linear regression model (1). Moreover, let $w = (w_1, \ldots, w_n)$ be a vector of weights such that $w_1 \geq w_2 \geq \ldots \geq w_n \geq 0$. Then the *smoothed least trimmed squares estimator* (SLTS) is defined by

$$\hat{\beta}_n^{(SLTS,w)} = \arg\min_{\beta \in R^p} \sum_{i=1}^{n} w_i r_{[i]}^2(\beta), \tag{3}$$

where $\beta \in R^p$ is a p-dimensional vector of unknown parameters and $r_{[i]}^2(\beta)$, $i = 1, \ldots, n$, represent again the ordered sample of squared residuals $r_i^2(\beta)$. The estimator is quite similar to the weighted least squares estimator with one important difference: weights are assigned to the order statistics of squared residuals instead directly to the residual.

Clearly, the behavior and properties of the SLTS estimator are given entirely by the choice of weights. Let us provide several examples:

1. $w_1 = \ldots = w_n = 1$: SLTS is equivalent to the least squares;
2. $w_1 = \ldots = w_h = n/h$ for $\frac{n}{2} < h \leq n$ and $w_{h+1} = \ldots = w_n = 0$: SLTS is equivalent to the least trimmed squares;

3. $w_i = f(\frac{i}{n}; \omega_1, \ldots, \omega_m)$ for all $i = 1, \ldots, n$, where $f(x; \omega_1, \ldots, \omega_m)$ is a positive function on $\langle 0, 1\rangle$ parameterized by $\omega_1, \ldots, \omega_m \in R^m$. For example, such a function can be defined for two parameters $\omega \in \langle 0, \infty)$ and $\lambda \in (1/2, 1\rangle$ as

$$f(x; \omega, \lambda) = \frac{1}{1 + e^{\omega(x-\lambda)}} \tag{4}$$

Then we get a smoothing scheme $w_i = f(\frac{i}{n}; \omega, \lambda)$ for any given values of ω and λ, and moreover, we can choose among such smoothing schemes by selecting suitable values of these parameters.

Apparently, this estimator can share its robustness properties with the already reviewed LTS, at least for weights with very small $w_{h+1}, \ldots, w_n$ (e.g., (4) with $\omega \gg 1$). Additionally, once we restrict our attention only to strictly positive weights, we obtain an estimator that does not reject any observation completely, and therefore, discrete variables do not cause problems anymore.

Moreover, there are certainly many possibilities how smoothing can be chosen. The strategy that we discuss in this paper is the adaptive choice of weighting scheme. Consider, for example, a weighting scheme defined by (4). Then the corresponding SLTS estimate converges to LS for $\omega \to 0$ and to LTS with $h = [\lambda n]$ for $\omega \to \infty$. Thus, we can by means of the single parameter ω choose how far or close the corresponding SLTS estimator is to LS and LTS. In other words, we control the balance between the robustness of the estimator and the amount of information it employs from data.

5 Properties of smoothed least trimmed squares

Let us now discuss important properties of the SLTS estimator proposed in Section 4.2. We will first summarize the assumptions need for basic asymptotic results and then briefly review these results.

Assumption A. *Let weights $(w_{in})_{i=1}^n$ be generated by a stepwise everywhere positive function. Next, assume that the explanatory variables x_i have finite fourth moments and $n^{-1/4} \max_i \|x_i\| = O_p(1)$. Additionally, let the distribution function of the error term ε_i be absolutely continuous and symmetric around zero. The corresponding probability density is assumed to be bounded. Finally, the identification condition is to be satisfied, that is, for any $\varepsilon > 0$ and $U(\beta^0, \varepsilon)$ such that $B - U(\beta^0, \varepsilon)$ is compact, there exists $\alpha(\varepsilon) > 0$ such that it holds $\min_{\beta \in B - U(\beta^0, \varepsilon)} \sum_{i=1}^n w_i \,\mathsf{E}\, r_{[i]}^2(\beta) - \sum_{i=1}^n w_i \,\mathsf{E}\, r_{[i]}^2(\beta^0) > \alpha(\varepsilon)$.*

We present a weaker set of assumptions together with a set of remarks and propositions explaining that these assumptions are not restrictive in real applications in Čížek (2001). Notice especially the identification condition: although it is satisfied almost trivially in linear regression model, its presence allows for a straightforward generalization of the results to nonlinear regression. Now, we state the main asymptotic results, namely the consistency and asymptotic normality of SLTS.

Theorem 1. *Let Assumption A hold. Then the SLTS estimator $\hat{\beta}_n^{(SLTS, w_n)}$ is $\sqrt{n}$-consistent, $\sqrt{n}\left(\hat{\beta}_n^{(SLTS, w_n)} - \beta^0\right) = O_p(1)$, and asymptotically normal, $\sqrt{n}\left(\hat{\beta}_n^{(SLTS, w_n)} - \beta^0\right) \xrightarrow{F} N(0, V(w_n))$, as $n \to +\infty$.*

Proof: See Čížek (2001, Theorem 1). □

6 Adaptive choice of smoothing

Now we discuss the adaptive choice of weights for SLTS in more details. First, we describe the adaptive choice of weights theoretically based on an abstract decision rule. Second, we propose a specific decision rule for SLTS preserving robustness of LTS. We describe the procedure using smoothing in (4).

For weighting schemes generated by a function $f(x;\omega,\lambda)$ defined in (4), it holds that the corresponding SLTS estimator is more robust for $\omega \to \infty, \lambda \to 1/2$ because the largest residuals are assigned very small weights, and it is more efficient for $\omega \to 0, \lambda \to 1$ because all observations enter fully the objective function. Altogether, decreasing ω and increasing λ improve efficiency and decrease robustness of SLTS and vice versa.

Therefore, an adaptive choice of weights can work as follows. Start with the highest possible ω and $\lambda = 1/2$ to obtain the most robust estimate. The next step is to decrease ω and increase λ. This improves the efficiency of the estimator, but because it also decreases the robustness of SLTS, it is possible that the estimate is already adversely affected by contamination or other data problems. Hence, we need a decision criterion that indicates to which extent we can modify ω and λ without threatening the robustness of the estimator. Having a decision rule, we increase λ toward 1 and decrease ω toward 0 until the decision rule indicates that the estimate is not sufficiently robust. Thus, we obtain as low ω and as high λ as possible, which means as efficient an estimator as possible without threatening robust qualities of SLTS.

Such a decision rule can be based on regression residuals: we assume that the initial (most robust) estimate corresponding to some ω_0, λ_0 is consistent and we know that the principle of LTS is "to constrain the influence of observations with extremely large residuals on the estimate." Hence, we can construct a confidence interval for the $[\lambda_0 n]$ smallest residuals computed at ω_0, λ_0 (most robust choice) and require all *weighted* residuals for current values ω, λ to be inside this interval. Otherwise, some residuals are considered too large and we stop the adaptive choice process.

7 Simulations

In this section, we provide some simulation results documenting the features of SLTS with the adaptive choice of smoothing (for more detailed results, see Čížek (2001)). More precisely, we examine the effect of misspecification of categorical variables on the LS, RDL_1, and SLTS estimators. As mentioned in Section 3, RDL_1 is designed for a simple additive model, and moreover, it only takes care of continuous variables. On the other hand, SLTS should handle any type of regression model and any type of contamination.

Coefficient	Intercept	Slope	Dummy
LS	2.162 (0.041)	-1.001 (0.037)	2.836 (0.081)
SLTS	1.026 (0.060)	-1.000 (0.017)	3.978 (0.106)
RDL_1	1.268 (0.070)	-1.000 (0.023)	3.733 (0.116)

Table 1. Simulations for $y_i = 1 - x_i + 4d_i + \varepsilon_i$ with misspecified dummy variable for data sets of size $n = 200$. The standard deviations of estimates are in brackets.

The misspecification sensitivity is exemplified using a Monte Carlo simulation. we consider the model $y_i = 1 - x_i + 4d_i + \varepsilon_i$, where $\varepsilon_i \sim N(0,1)$ and

$d_i \in \{0, 1\}$. Further, assume that 20% percent of the observations have a misspecified binary variable d_i (it can correspond, for example, to wrong entries about the sex of individuals in a sample). The results obtained for sample size $n = 200$ and 1000 simulations are summarized in Table 1. Apparently, both LS and RDL_1 estimates are biased. Notice that the slope coefficient is estimated correctly, but the intercept and the effect of the dummy variable are wrong. On the contrary, the SLTS estimate provides unbiased results, which are not affected by the misspecification of the dummy variable.

8 Conclusion

In this paper, we introduced the smoothed least trimmed squares estimator and derived its asymptotic properties. Thus, we extended applicability of the LTS procedure to regression models that involve discrete explanatory variables. The resulting estimator is currently the only robust estimator with a high breakdown point that can be applied in these regression models.

We also constructed a procedure that adaptively chooses smoothing for SLTS and thus controls the balance between the robustness and the efficiency of the estimator. The adaptive procedure actually starts from an estimate close to LTS (most robust) and decides how far it can go towards least squares (improvement in efficiency) without endangering the robustness of SLTS.

On the other hand, we studied behavior of the adaptive choice of smoothing only for one class of generating functions (4). Hence, finding an optimal smoothing class with respect to the asymptotic variance of SLTS would be a very valuable improvement of SLTS and it is one of the main issues for further research.

References

Čížek, P. (2001): Robust estimation with discrete explanatory variables. *CERGE-EI Working paper*, **183**.

Hampel, F. R., Ronchetti, E. M., Rousseeuw, P. J., and Stahel, W. A. (1986): *Robust statistics, The approach based on influence function*. New York: Wiley.

Hubert, M., Rousseeuw, P. J. (1997): Robust regression with both continuous and binary regressors. *Journal of Statistical Planning and Inference*, **57**, 153–163.

Orhan, M., Rousseeuw, P. J., and Zaman, A. (2001): Econometric applications of high-breakdown robust regression techniques. *Economics Letters*, **71**, 1–8.

Rousseeuw, P. J. (1984): Least median of squares regression. *Journal of American Statistical Association*, **79**, 871–880.

Rousseeuw, P. J. (1985): Multivariate estimation with high breakdown point. In (ed.) W. Grossman, G. Pflug, I. Vincze, and W. Wertz, *Mathematical statistics and applications, Vol. B*, 283–297. Dordrecht, Netherlands: Reidel.

Rousseeuw, P. J., and Van Driessen, K. (1999): Computing LTS regression for large data sets. *Technical report, University of Antwerp*, submitted.

Robust Principal Components Regression

Sabine Verboven[1] and Mia Hubert[2]

[1] Department of Mathematics and Computer Science, University of Antwerp, Universiteitsplein 1, 2160 Wilrijk, Belgium

[2] Department of Mathematics, Katholieke Universiteit Leuven, Celestijnenlaan 200B, B-3001 Heverlee

Abstract. We consider the multivariate linear regression model with p explanatory variables X and $q \geq 1$ response variables Y. Moreover we assume that the regressors are multicollinear. This situation often occurs in the calibration of chemometrical data, where the X-variables correspond with spectra that are measured at many frequencies. It is well known that the classical least squares estimator has a large variance in the presence of multicollinearity. Moreover it can not be computed when $p > n$ since the X^tX matrix then becomes singular. Therefore many biased estimators have been proposed. A very appealing method is principal components regression (PCR) since it is easy to understand and to compute.

PCR first constructs a new set of uncorrelated explanatory variables, which are called the principal components. They correspond to the eigenvectors of the sample covariance matrix of the X-variables. The response variables are then regressed on these components using the (multivariate) least squares estimator.

Both stages of this procedure are however very sensitive to the presence of outliers in the data. We present a robust principal components regression method which also consists of two steps. First a robust principal components analysis is applied to the X-variables (Hubert and Rousseeuw, 2002), yielding a smaller set of k orthogonal regressors. We then regress the response variables Y on these regressors using the multivariate MCD-regression method (Rousseeuw et al. 2000).

To select the number of regressors, we propose a robust R^2. It is roughly defined as the proportion of the robust variance of Y which is explained by the robust fit. Moreover we propose several diagnostic plots, which allow us to visualize and to distinguish the possible outliers in the data.

Keywords. Principal component analysis, Principal component regression, Robust methods, High-dimensional data.

1 Principal component regression

We consider the multiple linear regression model with p explanatory variables and $q \geq 1$ response variables, measured at n observations. Using subscripts to denote the dimension of the matrices, and $\mathbf{1}_n$ for the column vector with all n components equal to 1, this model can be written as:

$$Y_{n,q} = \mathbf{1}_n \beta_0^t + X_{n,p} \mathcal{B}_{p,q} + E_{n,q} \tag{1}$$

with errors $E_{n,q}$ that are i.i.d. with zero mean and covariance matrix Σ_E.

To estimate the regression slopes $\mathcal{B}_{p,q}$ and the q-dimensional intercept β_0, we will typically apply the multivariate least squares regression (MLR) method (Johnson and Wichern, 1998). However, in the presence of multicollinearity, the classical MLR method has a very large variance. Moreover, if there are perfect linear dependencies among the regressors (e.g. when $p > n$) the least squares solution is not unique. High-dimensional data are often encountered in chemometrics, for instance for calibration where the predictors correspond with a complete spectrum measured at some hundreds of wavelengths.

To analyse such data sets, many biased estimators have been proposed, among which principal component regression (PCR). This is a very popular method since it allows for an easy interpretation and computation. PCR first applies PCA to the X-variables. This means that the dimension of the X-space is reduced by constructing a new set of $k < p$ orthogonal variables. They are called the principal components, the loadings or the factors and are obtained as the eigenvectors that belong to the k largest eigenvalues of the empirical covariance matrix of $X_{n,p}$. The projections of the data points on these loadings are called the scores $T_{n,k}$. In the second stage of PCR, the responses $Y_{n,q}$ are regressed on $T_{n,k}$ using the MLR method.

Although PCR is a very valuable tool to analyse multicollinear data, the method is very sensitive to outliers. Real data are often faced with outlying observations e.g. due to measurement errors. In the first stage of PCR the estimated covariance matrix of the regressors will be heavily affected by outliers in the X-space. Whereas outliers in X- and/or Y-space can cause the least squares fit to break down. That is why we propose to use robust estimators instead.

2 Robust Principal Component Regression

Analogously to PCR, we distinguish two stages in our Robust Principal Component Regression (RPCR).

2.1 Stage 1: Robust Principal Component Analysis

In the first step we apply a robust PCA method on the X variables. When p is small we can use the (reweighted) MCD estimator (Rousseeuw, 1984) or S-estimators of location and scatter (Rousseeuw and Leroy 1987, Davies 1987). For high-dimensional data we apply the ROBPCA method (Hubert and Rousseeuw, 2002). This algorithm itself consists of two main steps. In the first one the original p-dimensional X-space is reduced to a k-dimensional subspace that well fits the data. This dimension reduction step is followed by the robust estimation of the center $\hat{\mu}$ and the scatter matrix S of the projected data points. For this the FAST-MCD algorithm is used (Rousseeuw and Van Driessen, 1999). The robust principal components $P_{p,k}$ then correspond to the k eigenvectors of S, whereas the scores T satisfy:

$$T_{n,k} = (X_{n,p} - \mathbf{1}_n \hat{\mu}^t) P_{p,k}. \tag{2}$$

2.2 Stage 2: Robust multivariate regression

In the second step of RPCR, a robust regression of $Y_{n,q}$ on $T_{n,k}$ is performed. If we have several responses ($q > 1$), we apply the multivariate MCD-regression method (Rousseeuw et al. 2000). This estimator is, as its name suggests, based on the MCD estimator of location and scatter, and is here applied to the joint (X, Y) variables. It has the same breakdown value as the MCD estimator and can be easily computed with the FAST-MCD algorithm.

Note that if we only have one response variable we prefer the LTS regression estimator (Rousseeuw, 1984) since it has a higher finite-sample efficiency.

Also remark that we need to backtransform the estimated regression coefficients if we want to describe our model with respect to the original X-variables.

3 Selecting the optimal number of components

An important issue in PCR is the selection of the number of principal components. In the classical approach this is often obtained by choosing that number k which yields the minimum PRESS (predicted residual error sum of squares) or minimal RMSEP (root mean squared error of prediction). Both criteria evaluate the regression model by leaving out one observation at a time and computing its residual based on the regression of the remaining data points.

These criteria are however not useful when we apply robust methods. Then, the removal of one observation will not have a large influence on the estimated regression coefficients, and consequently neither on the residuals.

Therefore we propose to choose the number of factors based on a robust R^2. Assume that we have performed the RPCR method with j components. The MCD-regression not only yields regression coefficients, but also an estimate of the covariance matrix of the errors, denoted by $\hat{\Sigma}_{E,j}$, and an estimate of the covariance matrix of the responses $\hat{\Sigma}_Y$. The determinant of these matrices is used to define the robust R_j^2 based on j loadings:

$$R_j^2 = 1 - \frac{|\hat{\Sigma}_{E,j}|}{|\hat{\Sigma}_Y|}. \tag{3}$$

It gives the fraction of the total (robust) variance of Y which is explained by the j scores T, and thus is preferably close to 1.

We now select the number of loadings as the smallest j which gives a satisfactory value of R_j^2. We also make a plot of the R_j^2 versus j, with j ranging from 1 to a user-chosen maximal value. On such a figure we can quickly see when the increase of the R_j^2 stabilizes. Consider e.g. Figure 1, which is the R^2-plot of the Biscuit Dough data set that we will analyse in the next section. Here, we clearly see that the linear relation between the X and the Y variables is already well explained when we retain two components in the regression model.

4 Example

We will analyse the Biscuit Dough calibration data set described in Osborne et al. (1984). It contains 40 NIR spectra of biscuit dough with measurements every 2 nanometers, from 1200nm up to 2400nm. We first preprocessed the original data as in Marx and Eilers (1999) and we already omitted observation 23 because it is well-known to be an extreme outlier. So finally, we ended up with $n = 39$ spectra each of them representing $p = 601$ regressors. The responses are the percentages of 4 constituents in the biscuit dough: fat, sucrose, water and flour.

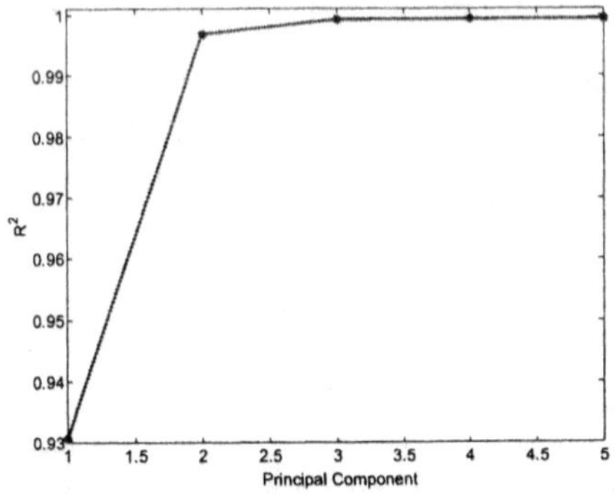

Fig. 1. The robust R^2-plot of the Biscuit NIR data set.

Based on the R^2-plot (Figure 1) we decided to retain $k = 2$ components. Remark that ROBPCA (which is solely applied on the X-variables) makes a screeplot from which we would decide to keep 6 components. But when we add the Y variables to perform the regression, we see that two scores already explain most of the variability of Y.

Now we would like to identify the outliers in this data set. For this we propose several diagnostic plots.

First we can visualise the outliers from the PCA step. For this we plot for each observation in the data set its orthogonal distance to the robust PCA-space versus its robust distance in X-space. Let l_j denote the eigenvalues of the scatter matrix S obtained with ROBPCA. Then, for each data point x_i its robust X-distance XD_i and its orthogonal distance OD_i are given by

$$XD_i = \sqrt{\sum_{j=1}^{k} \frac{t_{ij}^2}{l_j}} \quad \text{and} \quad OD_i = ||x_i - \hat{\mu} - P_{p,k} T_{k,i}^t|| \tag{4}$$

where the scores t_{ij} are obtained from (2). Observations whose X-distance exceeds the cut-off value $\sqrt{\chi_{k,0.975}^2}$ are outlying in the PCA-subspace. Data points with a large orthogonal distance are lying far from this k-dimensional subspace. The overview in Table 1 allows us to distinguish four types of observations.

Distances	small XD	large XD
small OD	regular observation	good PCA-leverage point
large OD	orthogonal outlier	bad PCA-leverage point

Table 1. Overview of the different types of observations based on their X-distance and their orthogonal distance.

Figure 2(b) shows this PCA-diagnostic plot for the Biscuit Doug data. We see that observations 7, 20, 21 and 24 are orthogonal outliers. They thus have a large orthogonal distance to the PCA plane but when we project them onto this PCA subspace they are indistinguishable from the regular observations.

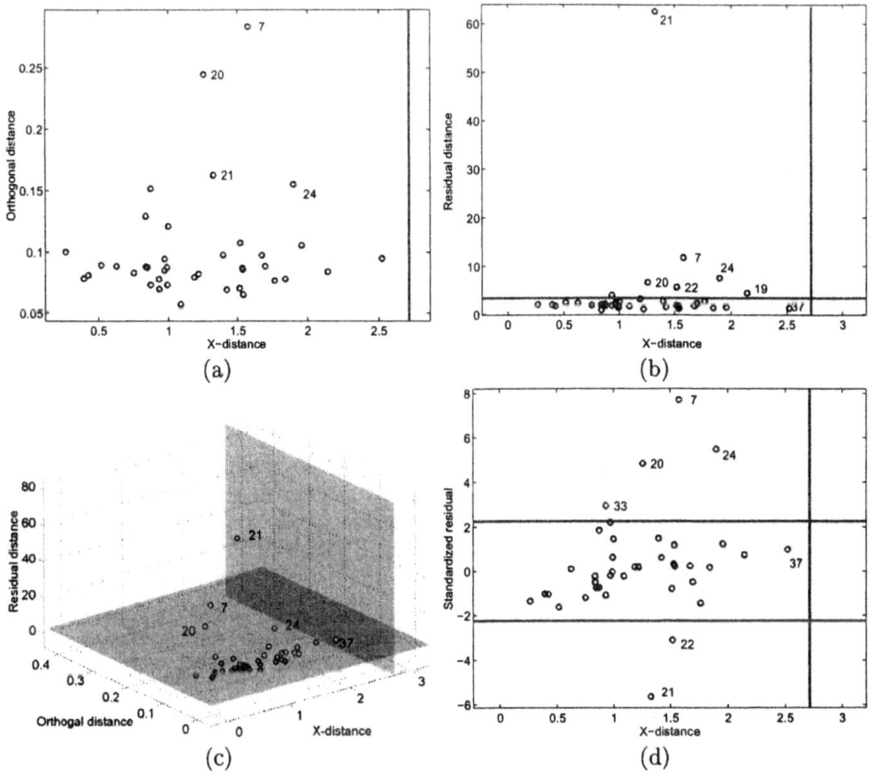

Fig. 2. Diagnostic plots of the Biscuit NIR data set: (a) PCA-diagnostic plot; (b) regression diagnostic plot; (c) 3D diagnostic plot; (d) diagnostic plot of univariate flour residuals.

Next, we can make diagnostic plots to classify the regression outliers. Similar to the diagnostic plots in Rousseeuw et al. (2000) we drawn on the horizontal axis the X-distance XD_i of each observation, and on the vertical axis the robust distance RD_i of its q-dimensional residual r_i, defined by:

$$RD_i = \sqrt{r_i^t \hat{\Sigma}_E^{-1} r_i} \tag{5}$$

Now, we can classify the data points according to Table 2. Observations whose residual distance exceeds $\sqrt{\chi^2_{q,0.975}}$ are called vertical outliers or bad leverage points, depending on their X-distance.

Distances	small XD	large XD
small RD	regular observation	good leverage point
large RD	vertical outlier	bad leverage point

Table 2. Overview of the different types of observations based on their X-distance and their residual distance.

Figure 2(a) shows this regression diagnostic plot for the Biscuit Dough data. Here, observation 21 appears as a very extreme vertical outlier. Also observations 7, 24, 22 and 20 have a large residual distance, whereas the biscuit dough with label 19 is rather a boundary case. If we combine Figures 2(a) and 2(b) we obtain the threedimensional diagnostic plot as shown in Figure 2(c).

Finally we have plotted in Figure 2(d) the standardized flour residuals, which are defined for each observation as

$$sr_{i,2} = (y_{i,2} - \hat{y}_{i,2})/\hat{\sigma}_2,$$

with $\hat{\sigma}_2^2$ the element on the second row and column of $\hat{\Sigma}_E$. On this plot we see that observation 21 is outlying but now so extreme as in Figure 2(b). Similar plots were obtained for the standardized residuals of fat, sucrose and water. We also computed the RPCR method on the four response variables separately. But then again biscuit dough 21 did not come out as such an extreme outlier. This illustrates that we have to take into account the covariance structure of the errors both for estimating the regression coefficients and for detecting the multivariate regression outliers.

References

Davies, L. (1987). Asymptotic behavior of S-estimators of multivariate location and dispersion matrices, *Annals of Statistics,* 15, 1269–1292.

Hubert, M. and Rousseeuw, P.J. (2002). ROBPCA: a new approach to robust principal component analysis, submitted.

Johnson, R.A. and Wichern, D.W. (1998). *Applied Multivariate Statistical Analysis, 4th edition.* Upper Saddle River: Pretence Hall.

Marx, B.D. and Eilers, P.H.C. (1999). Generalized linear regression on sampled signals and curves: a P-spline approach, *Technometrics,* **41,** 1-13.

Osborne, B.G., Fearn, T., Miller, A.R. and Douglas, S. (1984). Application of near infrared reglectance spectroscopy to the compositional analysis of biscuits and biscuit dough, *Journal of Scientific Food Agriculture,* **35**, 99-105.

Rousseeuw, P.J. (1984). Least median of squares regression, *Journal of the American Statistical Association*, **79**, 871-880.

Rousseeuw, P.J. and Leroy, A. (1987). *Robust Regression and Outlier Detection,* New York: John Wiley.

Rousseeuw, P.J., Van Aelst, S., Van Driessen, K. and Agulló, A. (2000). Robust multivariate regression, submitted.

Rousseeuw, P.J. and Van Driessen, K. (1999). A fast algorithm for the minimum covariance determinant estimator, *Technometrics*, **41**, 212-223.

Robust Time Series Analysis through the Forward Search

Luigi Grossi and Marco Riani

Dipartimento di Economia, University of Parma, Via Kennedy 6, 43100 Parma, ITALY
luigi.grossi@unipr.it, mriani@unipr.it

1 Introduction

The forward search (FS) is a powerful general method for detecting masked multiple outliers and for determining their effect on models fitted to the data (Atkinson and Riani, 2000). This method was originally introduced for models which assumed independent observations: linear and non linear regression, generalized linear models and multivariate analysis. In this paper we extend the forward search technique to the analysis of time series data. The basic ingredients of the FS are a robust start from an outlier-free subset of observations, a criterion for progressing in the search, which allows the subset to increase by one or more observations at each step, and a set of diagnostic tools that are monitored along the search. The robustness of the FS stems from the very definition of its algorithm, starting from "good" data points and including outliers at the end of the procedure. Computation of high-breakdown estimators is not required, except possibly at the starting stage. Indeed, the application of efficient likelihood or moment based methods at subsequent steps of the FS provides the analyst with more powerful tools than those obtained via traditional high-breakdown estimation.

The flexibility of the FS makes this procedure suited for extensions to areas other than multiple regression and multivariate analysis. This is especially true in time series, where it is often necessary to detect and model sudden or unexpected events. It is well known that outliers or structural changes in the observed time series may seriously damage identification and estimation of the suggested ARIMA or structural model (e.g. Koopman and Harvey, 1992; Chen and Liu, 1993), because they can introduce serious bias in the sample autocorrelation function. In time series the standard procedures for automatic outlier detection and correction consider four types of outliers, namely: additive (AO), innovational (IO), level shift (LS), and transitory change (TC) (Tsay, 1986; Chen and Liu, 1993). The AO represents a single spurious observation, the IO a pulse shock to the noise sequence which propagates to the observed time series, the LS a step function and the TC a spike that takes a few periods to disappear.

The main drawback of the procedures for detecting outliers in time series is that they start with the specification of a model for the observed series as if there were no outliers. In the ARIMA approach these procedures repeatedly use $T \times 4$ times the filter $\Pi = \phi(B)/\theta(B)$ based on the parameters estimated using all the observations, to obtain a statistic λ for each observation y_t, $t = 1, \dots, T$ and each type of outlier AO, IO, LS and TC. The maximum value of λ is compared with a predetermined critical value C to decide whether a potential outlier is present in the time series. Once an outlier is found, the filter Π is revised and reapplied for each observation and for each type of outlier in an iterative way. These procedures do not detect the k outliers at once, but proceed in several iterations detecting them one by one. Given that these approaches start with estimated parameters based on all T observations, as if no outlier were present in the data, it is clear that they may suffer from the well known masking and swamping effects. In order to avoid these problems and to be able to detect stretches of over influential observations, Bruce and Martin (1989) suggested leave-k-out diagnostics. The parameters of the model fitted to the full data set are compared with those generated by fitting the model to the data when a stretch of k points are taken to be missing. This method however, becomes computationally infeasible when k is large and may still suffer from masking and swamping effects if the number of outliers is greater than k.

In this paper we show how the forward search, free from masking and swamping problems, can detect masked multiple outliers in time series and their effect on the model fitted to the data.

2 The forward search in time series

The forward search is made up of three steps: choice of the initial subset, progressing in the search and diagnostic monitoring. In the following subsections we will examine in detail these three steps.

2.1 Choice of the initial subset

To initialize the FS, we start from a subset of observations which is intended to be outlier free. The initial subset is chosen among k blocks of contiguous observations and fixed dimension b. To find this subset we perform an exhaustive search of all possible blocks and choose the one which satisfies a least median of squares criterion. Specifically, if the suggested time series model is non stationary of order d, we divide the last $(T-d)$ observations of the time series into k subsamples, each made up of the first d initial data $y_1, \dots, y_d$ plus a set of $[(T-d)/k]$ contiguous observations where $[\cdot]$ is the integer part of a number. If $(T-d)/k$ is an integer we have exactly k subsamples. In the general case we have $k+1$ subsamples, the first k of size $d+[(T-d)/k]$, and the last of size $T-[(T-d)/k]k$. Without loss of generality in this paper we assume that $(T-d)/k = b$ (say) is an integer. The units forming subsample $r = 1, 2, \dots, k$, say $S_r^{(b+d)}$, are $y_1, \dots, y_d, y_{d+1+(r-1)b}, \dots, y_{d+br}$.

Now, let $v_{i,S_r^{(b+d)}}$ and $f_{i,S_r^{(b+d)}}$ be respectively the one step ahead prediction error and variance for unit i given observations in $S_r^{(b+d)}$. For a non-stationary time series of order d, minus twice the log of the normal likelihood based on the observations forming $S_r^{(b+d)}$ can be written as:

$$l(\theta_{S_r^{(b+d)}}) = b \ln(\hat{\sigma}_*^2) + \sum_{t=d+1+(r-1)b}^{d+rb} \ln f_t \tag{1}$$

where $\hat{\sigma}_*^2 = \sum_{t=d+1+(r-1)b}^{d+rb} v_t^2/f_t$ is the parameter which is concentrated out of the likelihood. Symbol $\hat{\theta}_{S_r^{(b+d)}}$ denotes the MLE of the hyperparameters found using only observations belonging to $S_r^{(b+d)}$. Now let $\tilde{v}_{t,S_r^{(b+d)}} = v_{t,S_r^{(b+d)}} / \sqrt{\hat{\sigma}_*^2 f_{t,S_r^{(b+d)}}}$, $t = d+1, \dots, T$ be the one step ahead standardized prediction error for each unit based on the hyperparameters estimated using observations belonging to $S_r^{(b+d)}$. We take as our initial subset of observations, the $(b+d)$-tuple which satisfies

$$\min_r [\tilde{v}^2_{[med],S_r^{(b+d)}}] \qquad r = 1,\ldots,k \tag{2}$$

where $\tilde{v}^2_{[l],S_r^{(b+d)}}$ is the lth ordered residual, among $t = d+1,\ldots,T$ and $med = [(T-d)/2]$.

When we evaluate the likelihood (1), we can use two alternatives: to consider the observations not belonging to $S_r^{(b+d)}$ as missing or not missing. In this paper we use this latter alternative. We will explore the other option in a subsequent paper.

Criterion (2) is similar to the least median of squares method for regression models with independent errors (Rousseeuw, 1984), except that here standardized residuals instead of raw residuals are considered. In conclusion, we take as our initial subset the stretch of data which minimizes the median of the squared one step ahead standardized prediction residuals. As the number of k subsamples considered increases, we increase the probability of having at least an initial subset of data which does not contain outliers. On the other hand, if the size of the initial subsets is too small, the estimates of the hyperparameters can be unstable. We suggest to choose $b \approx \sqrt{T}$ to ensure a balance between the statistical properties of the estimated hyperparameters and the robustness of the method.

2.2 Progressing in the search

Given a subset of dimension $m \geq b+d$, $m = b+d,\ldots,T$, the FS moves to dimension $m+1$ by selecting the $m+1$ observations with the smallest squared standardized prediction residuals, the observations being chosen by ordering all squared residuals $\tilde{v}^2_{t,S^{(m)}}$, $t = d+1,\ldots,T$. The first d observations always remain in the subset because they are necessary to initialize the Kalman filter. In order to avoid numerical problems with the initialization we have implemented the augmented Kalman filter collapsing it at step d as described by De Jong and Chu Chun Lin (1994). The squared standardized residuals $\tilde{v}^2_{t,S^{(m)}}$ at each step m are obtained minimizing the negative of the loglikelihood based on the observations forming the subset $S^{(m)}$, namely:

$$m \ln(\hat{\sigma}^2) + \sum_{t \in S^{(m)}} \ln f_{t,S^{(m)}}, \tag{3}$$

$\hat{\sigma}^2 = \sum_{t\in S^{(m)}} \tilde{v}^2_{t,S_{(m)}}$ is the parameter concentrated out of the likelihood.

2.3 Diagnostic monitoring

A major advantage of the FS approach is to provide the user with a number of informative pictures displaying all the diagnostics computed along the search. For instance, the anomalous behaviour of outliers or level shift is clearly revealed by the individual trajectories in the forward plot of standardized one step prediction errors, even when standard deletion diagnostics suffer from masking. An additional bonus of the forward search algorithm is that it is often possible to rank the observations following their entrance step into $S^{(m)}$. This gives an ordering of the data according to their degree of agreement with the null model, with observations furthest from it joining

$S^{(m)}$ in the last steps of the procedure. Finally, analyzing the individual trajectories of the units along the FS, we can detect the relationships among the observations: this helps to distinguish among the different types of outliers and interventions.

3 Some numerical work

As an illustration of the suggested approach, in this section we apply the forward search to the monthly series of the Italian Industrial Production Index from January 1981 to December 1996. We have used this series because these data have already been analyzed with different purposes by various authors (e.g. Kaiser and Maravall, 1999). The plot of the series, not given here for lack of space, shows a strong seasonal pattern with sharp peaks during the month of August and a slight increasing trend. To this series we have applied a basic structural model (Harvey, 1989) with trigonometric seasonality.

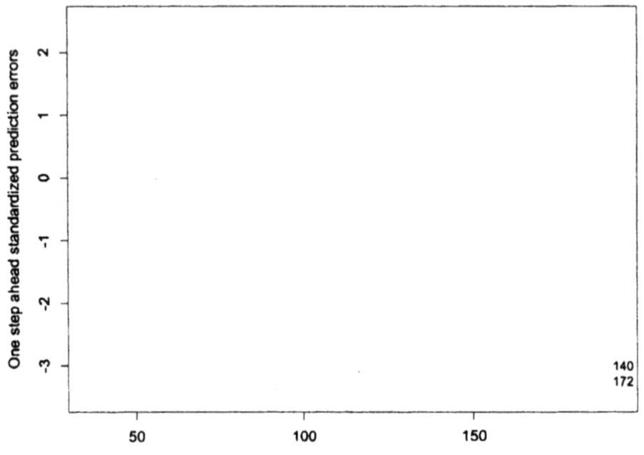

Fig. 1. One step ahead standardized forecast errors in each step of the FS: original data

Figure 1 is a classical representation used in forward search analysis to monitor residuals. It shows the one step ahead standardized prediction errors for each unit in each step of the FS. Each line refers to an observation. The plot shows that all cases have small standardized prediction errors apart from observations 140 (August 1992) and 172 (April 1995). Finally, it is interesting to notice that all lines remain stable throughout the search. This immediately and clearly shows that in these data there are no masked outliers.

We have then contaminated the data, introducing single and multiple outliers inflating original values by 20%. Figure 2 shows the one step ahead standardized prediction errors when there are two consecutive contaminated observations, 168 and 169 (December 1994 and January 1995). In this case the trajectory of observation 168 is very far from the bulk of the remaining observations and can be easily detected. Observation 169 can be easily detected too, but only when we observe the corresponding trajectory during the forward search. In the final step the prediction error of case 169 is not so large (2.41) and is close to the error of observation 64 which has not been contaminated. Note that the largest negative errors correspond to observations 170 (located immediately after the last contaminated observation) and 172. The trajectory of case 171 is inside the main bulk of the observations

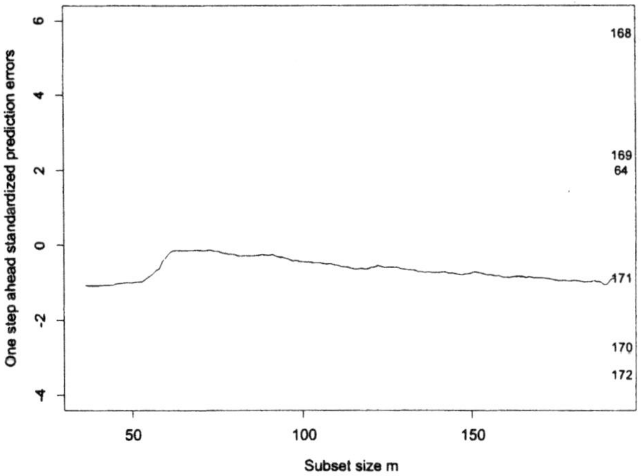

Fig. 2. One step ahead standardized forecast errors in each step of the FS: two consecutive artificial outliers (168 and 169). The curves which show a particular trajectory have been highlighted

but has a shape very similar to that of 169, 170 and 172 and very different from the remaining observations. In the case of a temporary level shift the FS enables easy detection of the length of the patch of outliers. In fact, at the end of a temporary positive level shift, there will be a strong negative residual which follows an opposite trajectory to the curve of the observation associated to the final level shift, as it happens in this example for unit 170.

In presence of a permanent level shift all the trajectories will remain constant throughout, but the observation which shows the change in level will always have a large positive or negative residual (Figure 3).

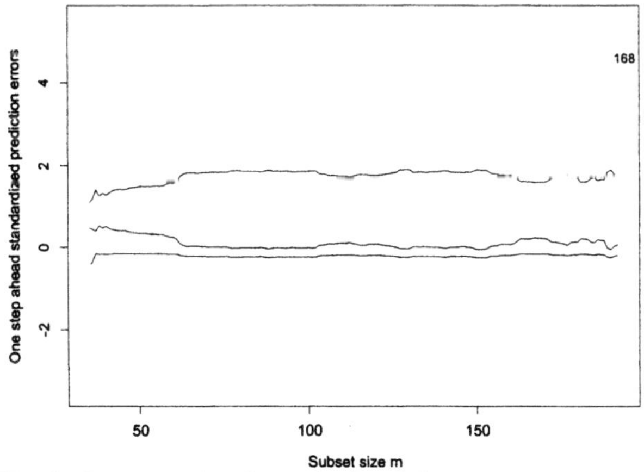

Fig. 3. One step ahead standardized forecast errors in each step of the FS: permanent level shift from y_{168}

Figure 4 shows the one step ahead standardized prediction errors when a temporary level shift is introduced from observation 160 (April 1994) to 169 (January 1995). As in the case of two additive outliers (Figure 2), the initial observations (160-162) of the temporary level shift are clearly evident. They show the three largest positive residuals in the central part of the forward search. Note at the end the sudden decrease of the residual for unit 162. At the final step this unit has a residual which is completely mixed with the "good" observations. The other observations belonging to the contaminated

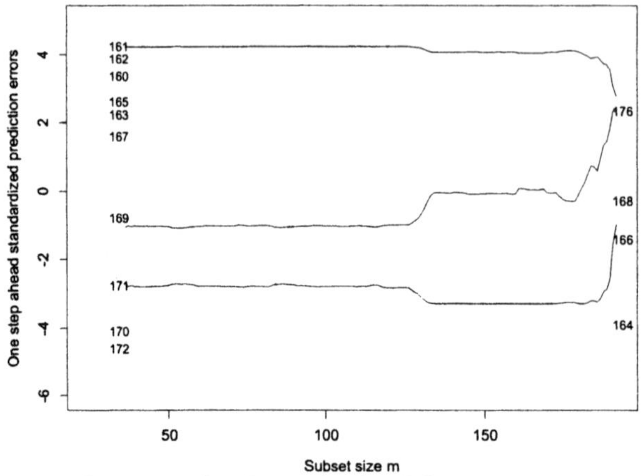

Fig. 4. One step ahead standardized forecast errors in each step of the FS: temporary level shift from y_{160} to y_{169}. The curves which show a particular trajectory have been highlighted

patch have a similar pattern with two sudden decreases. The first decrease when $135 \leq m \leq 137$ is due to the introduction in the subset of two contaminated observations. Note that observations 170, 171 and 172 (the first 3 observations after the contaminated patch) always show a large negative residual in the central part of the FS and a sudden increase in the final steps. Their behaviour is opposite to that of the contaminated units. Also very interesting is the specular behaviour of observations 164 (contaminated) and 176 (uncontaminated). The trajectory of observation 164 seems to agree with that of the other contaminated units. The curve associated with observation 176 is always inside the bulk of observations and goes out only in the last steps: this is a classical example of the swamping effect. Note that observations 164 and 176 correspond to August 1994 and August 1995, respectively, when production reaches the annual minimum and the analogy stresses the connection between one outlier and the value for the same month one year later.

References

Atkinson, A.C., Riani, M. (2000). *Robust Diagnostic Regression Analysis.* New York: Springer-Verlag.

Bruce, A.M., Martin, R. D. (1989). Leave-k-out diagnostic for time series, *Journal of the Royal Statistical Society*, **51**, 363-424.

Chen, C., Liu, L.M. (1993). Joint estimation of model parameters and outlier effects in time series. *Journal of the American Statistical Association*, **88**, 284-297.

De Jong, P., Chu Chun Lin, S. (1994). Stationary and non-stationary state space models. *Journal of Time Series Analysis*, **15**, 151-166.

Harvey, A.C. (1989), *Forecasting, Structural Time Series Models and the Kalman Filter.* Cambridge: Cambridge University Press.

Kaiser, R., Maravall, A. (1999), Seasonal outliers in time series. *Servicio de Estudios from Banco de Espana.*

Koopman, S.J. , Harvey, A. C. (1992). Diagnostic checking of unobserved components time series models. *Journal of Business and Economic Statistics*, **10**, 377-389.

Rousseeuw, P.J. (1984). Least median of squares regression. *Journal of the American Statistical Association*, **79**, 871-880.

Tsay, R.S. (1986). Time series model specification in the presence of outliers. *Journal of the American Statistical Association*, **81**, 132-141.

Rough Sets and Association Rules – Which is Efficient?

Daniel Delic, Hans-J. Lenz, Mattis Neiling

Free University of Berlin, Institute of Applied Computer Science,
Garystr. 21, D-14195 Berlin, Germany
daniel.delic@gmx.de, {hjlenz,mneiling}@wiwiss.fu-berlin.de

Abstract. We evaluate the rough set and the association rule method with respect to their performance and the quality of the produced rules. It is shown that despite their different approaches, both methods are based on the same principle and, consequently, must generate identical rules. However, they differ strongly with respect to performance. Subsequently an optimized association rule procedure is presented which unifies the advantages of both methods.

Keywords. Association rules, rough sets, bitmaps, hybrid association rules induction scheme, Apriori-algorithm

1 Introduction

Data mining methods are part of knowledge discovery from databases. Its objective is to extract nontrivial, beforehand unknown and potentially useful information. There exist several data mining procedures which differ in their methodology as well as in their data types.

In our work we focus on a comparison of the *association rules* with the *rough sets*. The association rule method was developed particularly for the analysis of databases, whose attributes possess only Boolean values. The rough set method, however, investigates databases. Its attributes can possess several values with the constraint, that a predefined set of attributes must exist on which the generated rules are based on. In order to be able to make a fair comparison, both procedures must be adapted in such a way, that they operate on the same data type. Our aim is to find out, which method is more efficient, and whether a combination of both methods improves the overall quality of unscrambled rules.

2 Association Rules

The association rule method was developed originally for the analysis of large databases of customer transactions. Each transaction consists of items purchased in a supermarket. In order to increase the analyzing capabilities of this method applied to data with more than Boolean attribute values, we defined the term *item* in a new way.

2.1 Attributes and Ranges

An item combines the name (label) of an attribute with one of its possible values, except *null* (see example 1). Each attribute has a finite number of values (range). Within a database each combination of attribute name and attribute value (except *null*) is distinguishable. Consequently there exists a set $A = \{a_1, a_2, a_3, ..., a_p\}$ with $p \in \mathbb{N}$ items in a database. An association rule then can be defined as follows: Given the subsets $X \subset A$ and $Y \subset A$ with $X \cap Y = \emptyset$, an association rule is an implication in the form $X \rightarrow Y$, with a confidence c $(0 \leq c \leq 1)$ and a support s $(0 \leq s \leq 1)$. The confidence is a measure for how much of the tuples that contain X, also contain Y. The support is the relative frequency of the set $X \cup Y$.

Example 1. For the attributes 'spaghetti' with domain(spaghetti) = {1, 0} = $\{a_1, a_2\}$ and 'tomato_ sauce' with domain(tomato_ sauce) = {1, 0} = $\{a_3, a_4\}$ we get the items a_1 = "spaghetti" and a_3 = "tomato_ sauce". The combinations a_2 = *null* and a_4 = *null* do not represent items of the database, because they are not distinguishable. For the attribute 'blood_ pressure' with domain(blood_ pressure) = $\{a_1, a_2, a_3\}$ = {*high*, *low*, *normal*} the following items result: a_1 = "blood_ pressure_ high", a_2 = "blood_ pressure_ low" and a_3 = "blood_ pressure_ normal".

2.2 Data Structure 'Bitmap'

It is known from Database Theory that bitmaps are efficient data structures (Jürgens/Lenz, 2001). Therefore the original relational schema has to be transformed into a bitmap table. The first column holds the tuple-addresses *TID*, each of the remaining columns contains a binary bitmap-attribute i.e. an item from the original data set (see Fig. 1). This structure is efficient, because the column-address of a bitmap-attribute doesn't change. Therefore it is possible to check directly very fast each tuple, by direct access for the appropriate column, whether the searched item is or is not available in the corresponding tuple of the original data set. If its present there is a '1', otherwise a '0' in the bitmap.

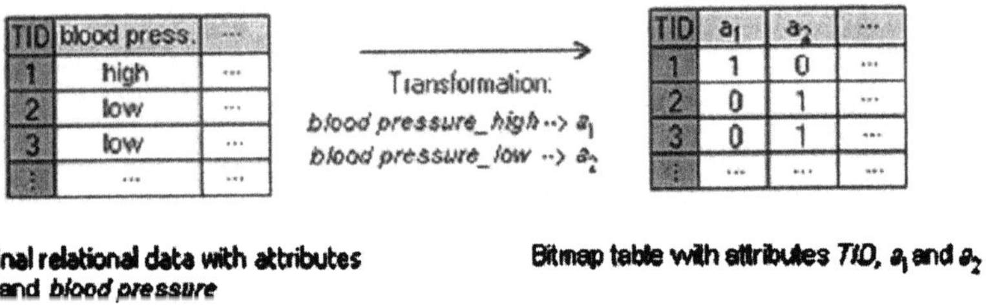

Fig. 1. Transformation of relational data into an efficient bitmap representation.

2.3 Rules Derivation

The procedure to generate association rules is based on the multi-pass-algorithm *Apriori*[1]. The principle idea of this algorithm is to iterate several times trough a database while searching for sets of items that occur in a sufficient large number. After each pass the number of items in those "large sets" is increased by one until all existing sets in the database are found. Subsequently the association rules are derived from these item sets.

[1] See *Agrawal/Srikant* (1994).

3 Rough Sets

The rough set theory was introduced by Zdislaw Pawlak[2]. It is a method for uncovering dependencies in data, which are recorded by relations. A detailed introduction to rough set theory can be found in *Munakata* (1998).

3.1 Model

The rough set method operates on data matrices, so called "information tables" (see Table 1). It contains data about the *universe* U of interest, *condition attributes* and *decision attributes*. The goal is to derive rules that give information how the decision attributes depend on the condition attributes.

Table 1. Information table.

universe	condition attributes		decision attribute
person	*temperature*	*blood pressure*	*heart problem*
Adams	normal	low	no
Brown	normal	low	no
Carter	normal	medium	jes
Ford	high	high	jes
Gill	high	high	no
Bellows	medium	medium	jes

A prerequisite for rule generation is a partitioning of U in a finite number of blocks, so called equivalence classes, of same attribute values by applying an equivalence relation. For example, the equivalence relation $R_1 = \{(u,v)|u(blood\,pressure, temperature) = v(blood\,pressure, temperature)\}$ would partition U(see Table 1) in three equivalence classes $U_1 = \{Adams, Brown\}, U_2 = \{Carter\}$ and $U_3 = \{Ford, Gill\}$. Depending on this classes, rules like e.g. "If temperature *normal* and blood pressure *low* then heart problem *no*." can be derived . Generally, no unique rule derivation is possible. For example, *Ford* and *Gill* have identical values of the condition attributes, but differ in their values of the decision attribute. In order to analyze such inconsistent tables, the concept of approximation spaces is used to determine the confidence of the derived rules.

A special feature of the rough set approach is the extraction of reducts. Due to reducts the creation of superfluous rules can be avoided. For example: Let there exist two rules with the same confidence value and the same decision attribute, whereby the set of condition attributes of the first rule is a superset of the set of condition attributes of the second rule. That means both rules describe the same fact and therefore the first rule is unnecessary, since its set of condition attributes is not minimal.

3.2 Rough Set Based Rule-Generation with *RS-Rules+*

Besides the "classic" version of rule generation with a fixed decision attribute (cf. Table 1), our algorithm *RS-Rules+* offers the possibility of varying the

[2] See Pawłak (1982).

selected decision attributes of a table. Each attribute of the table will be included either as decision attribute or as condition attribute. *RS-Rules+* is explained exemplarily on the basis of Table 2: The attributes of the original data set are $\{A\}, \{B\}$ and $\{C\}$. Each attribute has two non *null* values $A = \{a_1, a_2\}, B = \{a_3, a_4\}$ and $C = \{a_5, a_6\}$, so that there are six items (bitmap-attributes) for the resulting bitmap-table: $\{a_1\}, \{a_2\}$, $\{a_3\}, \{a_4\}, \{a_5\}, \{a_6\}$. In the first step all possible rules are constructed from all bitmap-attributes of the table. For example, for $A \rightarrow B$ four item-related rules can be set up: $a_1 \rightarrow a_3, a_1 \rightarrow a_4, a_2 \rightarrow a_3$ and $a_2 \rightarrow a_4$. All unnecessary rules and those which do not fulfil the minimum support are deleted. The condition attributes from the not deleted rules ($\{a_1\}$ and $\{a_4\}$), in pairs, build the condition attribute-sets for the second run[3] (i. e. $\{a_1, a_4\}$). Again all possible rules are produced and their support is measured. The rules with minimum support are stored ($\{a_1, a_4\} \rightarrow \{a_5\}$). Subsequently, the algorithm stops, because there is only one new condition attribute-set left.

Table 2. Rule-Generation with *RS-Rules+*.

	1st run	2nd run
New condition attribute-sets	$A = \{a_1, a_2\}$ $B = \{a_3, a_4\}$ $C = \{a_5, a_6\}$	$\{a_1, a_4\}$
Possible Rules	$\{a_1\} \rightarrow \{a_3\}$ $\{a_1\} \rightarrow \{a_4\}$ $\{a_1\} \rightarrow \{a_5\}$ $\{a_1\} \rightarrow \{a_6\}$ $\{a_2\} \rightarrow \{a_3\}$ $\{a_2\} \rightarrow \{a_4\}$ $\{a_2\} \rightarrow \{a_5\}$ $\{a_2\} \rightarrow \{a_6\}$ $\{a_3\} \rightarrow \{a_5\}$ $\{a_3\} \rightarrow \{a_6\}$ $\{a_4\} \rightarrow \{a_5\}$ $\{a_4\} \rightarrow \{a_6\}$	$\{a_1, a_4\} \rightarrow \{a_2\}$ $\{a_1, a_4\} \rightarrow \{a_3\}$ $\{a_1, a_4\} \rightarrow \{a_5\}$ $\{a_1, a_4\} \rightarrow \{a_6\}$
Rules with minsupport (valid rules)	$\{a_1\} \rightarrow \{a_3\}$ $\{a_4\} \rightarrow \{a_5\}$	$\{a_1, a_4\} \rightarrow \{a_5\}$
Condition attribute-sets with minsupport	$\{a_1\}, \{a_4\}$	$\{a_1, a_4\}$

4 Comparison of the Procedures

The transformation of the underlying data set into a bitmap-matrix and the modification of the rough set rules induction scheme give way for an

[3] The formation of condition attribute-sets is based on the same principle as the k-itemset construction in 2.3.

objective comparison of the both methods. Three benchmark data sets have been selected, which differ in size and number of attributes[4]:

1. *Car Evaluation Database*: 1728 tuples, 25 Boolean attributes in the bitmap-table.
2. *Mushroom Database*: 8416 tuples, 12 original attributes selected, 68 boolean attributes in the bitmap-table.
3. *Adult*: 32561 tuples, 12 original attributes selected, 61 boolean attributes in the resulting bitmap-table.

The needed computing times of the rough set algorithm were, depending on the benchmark data between 2 minutes and 4 hours, those of the association rule algorithm between 40 seconds and 45 minutes[5] (c.f. Table 3). Both procedures supplied same results for all examined tables, except for few cases due to reducts of the rough set algorithm[6]. However, not creating superfluous rules does not represent any difference in the principle of the rule production and can therefore not be seen as such. Hence we conclude that both procedures are equivalent with respect to the produced rules. This outcome can be justified also analytically: The necessary prerequisite for the generation of a rule is – for the association rule procedure – the occurrence of a certain k-itemset in a sufficient number of tuples, – for the rough set procedure – the occurrence of a set of condition attributes together with a decision attribute in a sufficient number of tuples. Hence in both approaches a certain quantity of items must occur in a sufficient number of records, in order to be able to induce a rule. Since the task of both methods is the extraction of all quantities which fulfill the given minimum support threshold – and since the presence of these quantities in a database is independent of the selected mining procedure – the same minimum support fulfilling itemsets are formed. Consequently this leads to the same set of rules.

5 Hybrid Association Rule Procedure *Apriori+*

Since the generated sets of rules of both procedures are almost equivalent, the selection of the procedure depends on the role of reducts. The best solution would be an additional extension of the faster association rule method with functions of the rough set procedure, such as formation of reducts and the ability to refer to a fixed decision attribute if needed. This hybrid method was realized with our algorithm "Apriori+": The derived rules are examined on reducts, unnecessary rules are deleted. If there is a given fixed decision attribute, all itemsets without this attribute can be ignored for rule generation.

A further benchmark confirmed the advantages of *Apriori+*: With all examined data sets the same rules (including reducts) as with *RS-Rules+* have been extracted, whereby the necessary computing times corresponded to almost 100% to the Apriori association rule procedure (see Table 3).

---

[4] The benchmark data can be found in *UCI Repository of Machine Learning Databases and Domain Theories* (URL: ftp.ics.uci.edu/pub/machine learning-databases/Adult, /car, /mushroom).

[5] All tests were executed on a PC with an AMD K6-2/400 processor.

[6] A more detailed description can be found in Delic (2001).

Table 3. Complexity (CPU time) grouped by *database*, *method* and *the use of the decision attributes*.

Database	Car Evaluation					Mushroom					Adult				
Minsupport	10%					35%					17%				
Minconfidence	75%					90%					94%				
Fixed Decision Attribute	Yes		No			Yes		No			Yes		No		
Method	RS+	Apr +	Apr	RS+	Apr +	RS+	Apr +	Apr	RS+	Apr +	RS+	Apr +	Apr	RS+	Apr +
CPU Time [min]	1,15	1,12	1,10	3,15	1,12	3,32	2,02	2	15	2,02	64	44	44	233	44

6 Summary

The association rule procedure, which was originally developed for processing attributes with Boolean domains, was enabled by the implementation of bitmap tables as efficient data structure to analyze also attributes with multivalue ranges. The Rough set procedure aims originally for generating rules with a fixed decision attribute. Evidently, any need of a a-priori-definition was removed.

On the basis of different data sets we compared the quality of produced rules and the necessary computing times of the algorithms. It turned out, that the rules of *RS-Rules+* and *Apriori+* do not differ. The necessary computing times were in favor of the association rule procedures.

Even so, any final judgement should be stated carefully, since many factors have a great impact. The needed time of computation e.g. depends also strongly on the implemented algorithms. Improvements with the rough set algorithm could quite lead to reduction of computing time. Another factor is the methodology of the rule generation. It cannot be excluded that there exists another rough set based procedure which is more efficient for rule generation than the one introduced here. The investigation on these and other related questions will be subject of further work.

References

Agrawal, R. and Srikant, S. (1994). Fast Algorithms for Mining Association Rules in Large Databases. In: *VLDB'94*, 487–499. Morgan Kaufmann.

Delic, D. (2001). Data Mining - Abhängigkeitsanalyse von Attributen mit Assoziationsregeln und Rough Sets. *MS thesis*. Free University of Berlin, Institute of Applied Computer Science, Berlin, Germany.

Jürgens, M. and Lenz, H.-J. (2001). Tree Based Indexes Versus Bitmap Indexes: A Performance Study. *International Journal of Cooperative Information Systems*, **10**, 355–376.

Munakata, T. (1998). Rough Sets. In: *Fundamentals of the New Artificial Intelligence*, 140–182. New York: Springer-Verlag.

Pawłak, Z. (1982). Rough Sets. *Int. J. Computer and Information Sci*, **11**, 341–356.

Skewness and Fat Tails in Discrete Choice Models

Rosa Capobianco[1]

[1] Department PE.ME.IS., University of Sannio, Benevento, 82100, ITALY

Abstract. In discrete choice models, the probability that the dependent variable will assume value 0 or 1 depends on a set of explanatory variables through a function F. In this paper we propose the class of skew Student t distribution as a function F in order to have a more flexible model that can simultaneously account for asymmetry and thick tails and such that probit and logit models can be considered as special cases. Two examples illustrate the performance of the proposed model.

Keywords. Dichotomous models, fat tails, skewness, Student t distribution.

1 Introduction

Discrete choice models, also known as qualitative or categorical response models, are very popular in the statistical literature. They are applied in several fields of research, such as economics and biometrics, in order to study the choice behaviour for decision makers or, more generally, "either-or" situations in which one alternative or another must be chosen (Griffiths et al., 1993). In particular, the dependent variable Y can be classified into one of two mutually exclusive categories, such as "success" or "failure", while the p independent variables can be both discrete or continuous.

Given a data set of size n, the outcome Y_i, for i=1,...,n, takes value 1 if a success is observed and 0 otherwise, with associated probability π_i and (1-π_i), respectively. Let $x=(x_1,\ldots,x_p)$ be the vector of explanatory variables, the probability for Y_i to assume 0 or 1 depends on a linear combination of the explanatory variables such as $x'\beta$, through a function F, where $\beta=(\beta_0,\beta_1,\ldots,\beta_p)$ is a vector of unknown parameters, so that $\pi_i=F(x'\beta)$. The choice of the F function determines different models. When F is the cumulative distribution function (c.d.f.) of the standardized normal random variable (r.v.), we have a probit model, while if F is the c.d.f. of the logistic r.v., we have the logit model (Griffiths et al., 1993). Both distributions are symmetric around 0 but because of their close similarity, it is difficult to distinguish statistically between them unless the sample size is very large or in cases where data are heavily concentrated in the tails (Amemiya, 1981). Alternatively we can consider as F the c.d.f. of the Student t distribution, strongly related with the logit link (Albert and Chib, 1993).

Discrete choice models with symmetric F, however, do not always provide the best fit to data when the probability of the outcome approaches 0 or 1 at different rates. In order to cope with this problem, other models have been proposed which use a different c.d.f. for F. For example, log-log models are defined with $F(x'\beta)$=1-exp[-exp($x'\beta$)], so that F is asymmetric and departs from 1 more sharply than it departs from 0 (Agresti, 1990). Chen et al. (1999) propose an alternative to probit models by considering as F the c.d.f. of Azzalini's skew-normal distribution (Azzalini and Della Valle, 1996). For a detailed review on asymmetric discrete choice models see Stukel (1988).

In this paper we propose a discrete choice model where F belongs to a class of distributions, proposed by Fernández and Steel (1998), that can simultaneously account for both skewness and fat tails. In particular we consider for F the c.d.f. of a skew Student t distribution which displays flexible tails and skewness controlled, respectively, by the degrees of freedom ν and by a skew parameter γ. Furthermore, it is easy to verify that probit and logit models can be obtained as special cases of this wide class of generalized linear models, for specific choices of ν and γ.

The outline of the paper is the following. In section 2 we introduce the class of distributions proposed by Fernández and Steel (1998), in section 3 we illustrate our proposal, while in section 4 we consider two applications of our model to real data sets.

2 The skew Student t distribution

Fernández et al. (1995) proposed a general method to introduce skewness in any unimodal and symmetric distribution. By applying this criterion to the Student t distribution, it is possible to get a class of distributions that can cope, at the same time, with skewness and fat tails, two features often observed in statistical applications.

Let f(x) be the density function (d.f) of a Student t distribution, the class of skewed distributions is obtained by inverse scaling f(x) on both side of the mode, i.e.

$$f(x\mid\lambda,\nu)=\frac{2}{\gamma+\frac{1}{\gamma}}\frac{\Gamma\left(\frac{\nu+1}{2}\right)}{\Gamma\left(\frac{\nu}{2}\right)\sqrt{\pi\nu}}\left\{1+\frac{x^2}{\nu}\left[\frac{1}{\gamma^2}I_{[0,+\infty)}(x)+\gamma^2 I_{(-\infty,0)}(x)\right]\right\}^{-(\nu+1)/2}$$

(Fernández and Steel, 1998). In what follows, we write X~Skew-t(γ,ν) to indicate that the r.v. X has a Skew t distribution. The parameter $\gamma\in(0,+\infty)$ controls the skewness while ν, the degrees of freedom, regulates the tail thickness. When γ varies between (0,1), the distribution has negative skewness, while for γ greater than 1 the skewness is positive. When γ equals 1 we get the symmetric Student t distribution. This approach does not affect the unimodality of the distribution and it is possible to control the amount of probability mass on both side of the model. Furthermore, the tail behaviour is not affected by the additional parameter γ.

This class of models is also appealing because it includes the asymmetric normal r.v. as limiting distribution for $\nu\to\infty$ while for $\nu\to\infty$ and γ=1 we get the normal r.v.. See Fernández and Steel (1998) for further details on the properties of $f(x\mid\gamma,\nu)$.

3 The statistical model

Let $(x_1,...,x_p)$ be p independent variables and define the unobservable r.v.

$$Y_i^* = \beta_0 + \beta_1 x_{i1} + ... + \beta_p x_{ip} + \varepsilon_i = x_i'\beta + \varepsilon_i,$$

for i=1,…,n. The dependent variable takes value 1 when $Y_i^* > 0$ and 0 otherwise.

In order to make a weak assumption on the distribution of Y_i^*, we assume for ε_i a flexible model, i.e. ε_i~Skew-t(γ, ν), so that the distribution of Y_i^* can be either symmetric or skewed, and can have either thick tails or not. Both skewness and fat tails can be verified through a test of hypothesis on the parameters. In particular, symmetry is tested by considering the hypothesis $H_0 : \gamma = 1$. Since a normal distribution is obtained when $\nu \to \infty$, a test on normal tails can be carried out in the following way. We construct a confidence interval for the degrees of freedom by collecting all the values of ν such that

$$\left\{-2\log\lambda_p(\nu) \leq \chi^2_{1-\alpha,1}\right\},$$

where $\lambda_p(\nu)$ is the profile likelihood ratio

$$\lambda(\nu_0) = \frac{\sup\limits_{\nu=\nu_0} \mathrm{L}(\theta)}{\mathrm{L}(\hat{\theta})},$$

with $\theta = (\beta, \gamma, \nu)$, while $\chi^2_{1-\alpha,1}$ is the (1-α)-th percentile of the chi-square r.v. with 1 degree of freedom. If the confidence interval is unbounded upwards, the normality hypothesis is not rejected and we get a skewed normal r.v. (Azzalini, personal communication).

In order to define π_i, we observe $Y_i = 1$ when $Y_i^* > 0$ or, equivalently, $\varepsilon_i > -x_i'\beta$. Thus

$$\begin{aligned}\pi_i &= P(y_i = 1) = P(\varepsilon_i > -x_i'\beta) = \\ &= 1 - P(\varepsilon_i < -x_i'\beta) = \\ &= 1 - F_\varepsilon(-x_i'\beta)\end{aligned} \tag{1}$$

where F_ε is the c.d.f. of the Skew-t distribution.

In an intuitive way, the larger the value of Y_i^*, the larger is the probability that $Y_i = 1$. However the increases in the probability rate are not constant for equal increases in Y_i^*. The rate of change in π_i, given a change in the j-th explanatory variable, is equal to

$$\frac{\partial \pi_i}{\partial x_{ij}} = f_\varepsilon(-x_i'\beta)\beta_j,$$

for j=1,…,p, where $f_\varepsilon(-x_i'\beta)$ is the d.f. of the Skew-t distribution, therefore the contribution of the j-th variable depends on the sign of the corresponding parameter.

The likelihood function is defined as

$$L(y_1,\dots,y_n)=\prod_{i=1}^{n}\pi_i^{y_i}(1-\pi_i)^{1-y_i}=$$

$$=\prod_{i=1}^{n}\left[1-F_\varepsilon(-x_i'\beta)\right]^{y_i}\left[F_\varepsilon(-x_i'\beta)\right]^{1-y_i},$$

thus maximum likelihood estimates of $(\beta_0,\dots,\beta_p)$ are calculated by applying a nonlinear optimisation procedure. It is convenient to adopt as initial values the estimates obtained by applying the probit model.

Finally it has to be remarked that the proposed model is also appealing because some generalized linear models can be considered as particular cases of the class (1):

1) For γ=1 we get the t-link;
2) For γ=1 and $\nu\to\infty$, F tends to the c.d.f. of the normal r.v., as a consequence, (1) converges to the probit model.
3) For γ=1 and ν=8, (1) approximates the logit model. In fact the logistic distribution is approximately equivalent to a Student t r.v. with ν=8 (Albert and Chib, 1993).

4 Examples

In this section we consider applications of the proposed model to two data sets frequently used in the literature to verify the performance of discrete choice models and show how, in both cases, it outperforms the probit model.

The first data set concerns the voting outcome, by state, in the 1976 U.S. presidential election (Greene, 1990). The outcome variable takes the value of 1 if the popular vote favored the Democratic candidate (Carter) and 0 if the vote favored the Republican candidate (Ford). The explanatory variables are the median income (x_1) and the median number of years of school completed by persons of 18 years of age or older (x_2), so that the model is

$$\pi_i=1-F\left\{-\left(\beta_0+\beta_1x_{i1}+\beta_2x_{i2}\right)\right\}.$$

Here Y_i^* can be interpreted as the propensity to vote for Jimmy Carter, which is assumed to be a linear function of the income and of the level of education. The estimated model is the following

$$\hat{\pi}_i=1-F\left\{-\left(\hat{\beta}_0+\hat{\beta}_1x_1+\hat{\beta}_2x_2\right)\right\}=1-F\left\{-\left(79.139-6.495x_2+0.037x_3\right)\right\},$$

i.e. an increase in the median income has a negative effect on the probability to vote for the Democratic candidate while the level of education has a positive effect. While the parameter estimates are quite similar to those obtained using a probit model ($\hat{\beta}_0=85.725$, $\hat{\beta}_1=-7.031$, $\hat{\beta}_2=0.039$), it is interesting to note that the estimated F has negative asymmetry and fat tails, the estimates of the skewness parameter and of the degrees of freedom are in fact, respectively, $\hat{\gamma}$=0.72 and $\hat{\nu}$=3.79. This last result is consistent with Albert and Chib (1993) analysis of the data set where they choose a t-link with 4 degrees of freedom to fit the data.

In order to make a comparison between the two approaches, we consider two indexes :

a) $R_1 = \dfrac{\text{number of wrong predictions}}{\text{total number of observations}}$,

where the predictions are $\hat{Y}_i = 1$ if $\hat{\pi}_i > 0.5$ and $\hat{Y}_i = 0$ otherwise,

b) $R_2 = \sum_{i=1}^{n} \dfrac{(y_i - \hat{\pi}_i)^2}{\hat{\pi}_i(1-\hat{\pi}_i)}$,

i.e. R_2 is the weighted residual sum of squares with the squared error $(y_i - \hat{\pi}_i)^2$ weighted by the inverse of its estimated variance (Maddala, 1988). Both indexes are to be minimized, so that we prefere the model such that R_1 and R_2 are smaller.

Regarding Greene data, we obtain the following results:

Model	R_1	R_2
Probit	0.24	41.03
Skew-t	0.22	33.51

Both R_1 and R_2 are minimum for the Skewt-t model, confirming the opportunity to adopt the proposed model.

The second example comes from Finney (1947). The explanatory variables are the volume of air inspired (x_1) and the rate of air inspired (x_2), while the response variable is the occurrence or nonoccurrence of a transient vaso-constriction on the skin of the digits. The estimates of the parameters $(\beta_0, \beta_1, \beta_2)$ are $(\hat{\beta}_0 = -8.323, \hat{\beta}_1 = 3.131, \hat{\beta}_2 = 2.054)$ whereas, using a probit model it yields $(\hat{\beta}_0 = -5.191, \hat{\beta}_1 = 2.116, \hat{\beta}_2 = 1.476)$. As in the previous example, the estimates of the parameter are quite similar in both cases but the estimated model for F is a Skew-t(1.788,19.012). In other words the distribution is positively skewed with high value of ν. In this case it makes sense to verify the normal tails of the distribution by using the procedure proposed in the previous section. Given that the confidence interval is unbounded upwards, we do not reject the hypothesis of normal tails so that the actual distribution is a skew normal r.v..

The indexes R_1 and R_2 are equal to

Model	R_1	R_2
Probit	0.13	31.39
Skew t	0.10	34.49

Once again it seems reasonable to use the Skew normal distribution to model Y_i^*.

References

Agresti, A. (1990). *Categorical Data Analysis*, New York: Wiley.

Albert, J. H. & Chib, S. (1993). Bayesian analysis of binary and polychotomous response data. *JASA*, **88**, 669-679.

Amemiya, T. (1981). Qualitative response models: a survey. *J. Economic Literature*, **19**, 1483-1536.

Azzalini, A. & Della Valle, A. (1996). The multivariate skew-normal distribution. *Biometrika*, **83**, 715-726.

Chen, M., Dey, D. K. & Shao, Q. (1999). A new skewed link model for dichotomous quantal response data, *JASA*, **94**, 1172-1186.

Fernández, C., Osiewalski, J. & Steel, M. F. J. (1995). Modelling and inference with v-spherical distributions, *JASA*, **90**, 1331-1340.

Fernández, C. & Steel, M. F. J. (1998). On bayesian modelling of fat tails and skewness, *JASA*, **93**, 359-371.

Finney, D. J. (1947). The estimation from individual records of the relationship between dose and quantal response, *Biometrika*, **34**, 320-334.

Greene, W. H. (1990). *Econometric Analysis*, New York: Macmillan Publishing Co.

Griffiths, W. E., Hill, R. C. & Judge, G. G. (1993). *Learning and Practising Econometrics*, New York: Wiley.

Maddala, G. S. (1988). *Introduction to Econometrics,* New York: Macmillan Publishing Co.

Stukel, T. (1988). Generalized logistic models, *JASA*, **83**, 426-431.

Aknowledgements

I wish to thanks prof. Mark Steel (University of Kent at Canterbury – U.K.), for his precious advice about the generation of the Skew-t distribution.

Standardized Partition Spaces[1]

Ursula Sondhauß and Claus Weihs

Department of Statistics, University of Dortmund, 44221 Dortmund, Germany

Abstract. We propose a standardized partition space (SPS) that offers a unifying framework for the comparison of a wide variety of classification rules. Using SPS, one can define measures for the performance of classifiers w.r.t. goodness concepts beyond the expected rate of correct classifications of the objects of interest. These measures are comparable for rules from so different techniques as support vector machines, neural networks, discriminant analysis, and many more. In particular, we are interested in assessing the reliability of classification rules when used for interpretation of the relationship between the values of predictors and the membership in classes.

We will demonstrate the high potential of SPS for the comparison of classification methods in a simulation study to analyse the following problem:

Given a medium number of predictors, (10-20), and a potentially complex relation between classes and predictors, one would expect flexible classification methods like support vector machines or neural networks to do better than simple methods like e.g. the linear discriminant analysis or classification and regression trees. Nevertheless, one often observes on real data sets, that the simple procedures do pretty well. Our assumption is, that simple methods are more robust against instability, and that the effect of instability superposes the effect of complexity of the relation. By instability we mean the deviation from the assumption that the collected data is some independent and identically distributed sample from some joint distribution of predictors and classes.

We analyse this problem with a simulation study using experimental design.

Keywords. Classification, Comparative Studies, Experimental Design

1 Standardized Partition Spaces

1.1 Argmax rules

The method of SPS is applicable to all classification methods that finally decide for a certain class $c, c = 1, ..., G$ following some argmax rule. Argmax rules are based on a transformation $\mathbf{m}(x) := (\mathbf{m}(x, 1), ..., \mathbf{m}(x, G))' \in \mathbf{M}$ of predictor values $x \in \mathbf{X}$ from predictor space $\mathbf{X}$ into some G-dimensional space of real numbers $\mathbf{M} \subseteq \mathbb{R}^G$. The vector $\mathbf{m}(x)$ is interpreted as a vector of membership values in the classes. Argmax rules assign an object with predictor values $x \in \mathbf{X}$ into the class with highest membership:

$$\mathbf{cl}(x, \mathbf{m}) = \arg \max_{c=1,...,G} \mathbf{m}(x, c).$$

[1] This work has been supported by the Deutsche Forschungsgemeinschaft, Sonderforschungsbereich 475.

We call $\max_{c=1,...,G} \mathbf{m}(x,c)$ 'assigment values' and the space of all observations that get assigned to the same class 'assigment area' of that class. Together, the assigment areas for all classes form the classifier's partition of the predictor space.

For example, all Bayes optimal classifiers are argmax classifiers.

1.2 Optimal Features

Our idea is motivated by the attempt to make any argmax rule comparable to the 'true' or 'best' Bayes optimal classifier that has complete knowledge (=: τ) about the distribution of predictors and classes on the population of objects. The Bayes optimal classifier assigns objects into the class with highest conditional probability:

$$\mathbf{cl}(x, \mathbf{p}_\tau) = \arg \max_{c=1,...,G} \mathbf{p}(c \mid x, \tau), \quad x \in \mathbf{X}.$$

This strategy minimizes the true expected error probability. Membership values of Bayes optimal classifiers all lie in the interval $[0,1]$ and sum up to 1. We denote this space of membership vectors by $\mathbf{M}^s \subset [0,1]^G$.

All available information in the predictors about the membership of objects in classes is coded in the membership values of the Bayes optimal classifier. Thus, Fukunaga (1990) calls them "optimal features".

If we had these optimal features, we could use them to answer questions of interest about the interplay of predictors and class membership. In real situations, they are unknown, and estimated membership functions of argmax classifiers may be used as surrogates. We are interested in the reliability of classification rules in this respect.

1.3 Scaling

In a first step, we want membership values to be directly comparable in size. We can not use their raw membership values. One obvious reason is that they may lie on various scales. They neither have to be non-negative nor add up to 1, as the optimal features do. Of course, we can standardize them ad-hoc into $\mathbf{M}^s$ by some monotone transformation without changing the final assignment into classes. But this might lead to patterns more influenced by the standardization procedure than by the rule's classification behaviour.

Less obvious, but equally important, even maximum membership values of argmax rules based on learnt conditional class probabilities are not a reliable measure for the membership of objects in the assigned classes, and thus not a reliable measure for the correctness of the rule's decision. They give information about the rule's performance from its own perspective only, whereas for comparisons, we would prefer a more objective view.

The details of the process of scaling are published in Sondhauß and Weihs (2001) . It is based on approximations of the empirical distribution of assignment values estimated on some test data within assignment areas with the beta-distribution.

All scaled membership vectors $\mathbf{m}^s(x) := (m^s(x,1), ..., m^s(x,G))'$ of any argmax rule lie in the same space $\mathbf{M}^s$, therefore we call this method the method of standardized partition spaces. Scaled membership vectors have the following properties:

- Scaled membership values are directly comparable in size.
- The average assignment value into each class approximates the correctness rate of that assignment on the test set.

- Scaled membership vectors of observations reflect as much as possible the position of the original membership vectors among each other within assignment areas.

1.4 Measures for the quality of scaled membership values

Following Hand's (1997) specifications for quality characteristics of conditional class probabilities, high quality membership estimations are characterized as follows: A high assessment (relative to the assessed membership in other classes) in the assigned class should be justified (accuracy), the relative sizes of membership in classes should reflect 'true' conditional class probabilities (precision), and membership values of objects in the different classes should be well-separated (non-resemblance).

Average precision on the test set is used for the scaling of membership values, and thus precision is no longer a quantity for comparison.

The measure of accuracy (Ac) is based on the Euclidean distances between scaled membership vectors $\mathbf{m}^s(x)$ and the vector representing the corresponding *true* class corners $\mathbf{e}(c(x))$ for the examples (x, c_x) in the test set $\mathbf{T}$. We standardize the mean of these distances such that a measure of 1 is achieved if all vectors lie in the correct corners, and zero if for all observations the membership in all classes is the same.

Analogously, the measure of the ability to separate (AS) is based on the Euclidean distances between scaled membership vectors $\mathbf{m}^s(x)$ and the vector representing the corresponding *assigned* class corner $\mathbf{e}(\mathbf{cl}(x, \mathbf{m}(x)))$. We standardize the mean of these distances in the same way as above.

1.5 Potential use of scaled membership values

Once we can trust the quantitative assessment of an object's membership in classes according to some rule, we can use them as surrogates for the optimal features. This makes all kinds of analyses possible for interpretation. E.g. for up to four classes, one can **visualize scaled membership vectors** of various classifiers on the same data in a barycentric coordinate system (a regular simplex) to explore differences in the classifiers' patterns. Also, observations with highest membership can be used as **prototypes** for objects in certain classes. To explore the connection of predictor variables with classes in a scale-independent manner, one can **plot membership values** versus the rank of the **predictor** variable for some data.

2 Experiment

We demonstrate the use of standardized partitions spaces in analyzing some astonishing phenomenon: simple classification methods often do pretty well on real data sets though their underlying premises about the true relation between predictors and classes can not reasonably be assumed to hold. We will implement a screening experiment to detect the main influencing factors for the performance of various classifiers.

In general, influencing factors for the goodness of any classification methods are data characteristics like the number of classes (we fix as three), the number of predictor variables (we fix as 12), the number of training objects (we vary), the number of training objects in classes (we use balanced design only), the number of missing values (we ignore this factor at this stage), and the form of the joint distribution of classes and predictors on objects.

Our main interest is on the influence of the form of the joint distribution of classes and predictors on objects.

One aspect of the joint distribution is the dependency structure between predictors. Often this makes the learning results less stable. Some classification methods, like e.g. the naïve Bayes, even assume independence of variables. In all cases, the analysis of the relevance of predictors for the detection of classes is obscured by this inner dependency.

Concerning the shape of the bivariate distributions between one predictor and the class, we define simplicity from the perspective of the classifier from a linear discriminant analysis: multivariate normal distributions with equal covariance matrices that lead to partitions with separating hyperplanes.

To see the effect of instability on the different types of classification methods, we model instability by deflected observations accounting for three factors: the percentage of deflected observations, the percentage of relevant variables in which the deflection takes place and the direction of the deflection.

Deflection only takes place on the test data. Thus the "true" generating process for the test data differs from that of the training data.

2.1 Classification Methods

We compared a set of classification methods, that are quite different in the assumptions they make about any underlying generating process of the data, and that are famous in different communities and for different tasks: the classifiers from linear and quadratic discriminant analysis 'LDA' and 'QDA' from statistics, the naïve Bayes 'NB' (famous for good performance in text classification), a Neural Network 'NN' (a tool for non-linear function approximation), the strikingly simple k-Nearest neighbor classifier 'k-NN', and some decision tree classifier 'rPart' (mainly famous in the machine learning community). In Sondhauß and Weihs (2001) you find a more detailed description of the implementation of these classification methods.

2.2 Quality Characteristics

The target values in our experiment are correctness rate (CR), accuracy (Ac) and ability to separate (AS). The more overlapping the true distributions are the more difficult is the problem as such. Thus we use as target values not the goodness criteria as such but their relation ratios (rCR, rAc, rAS) to the best that can be achieved: the values of these criteria for the optimal Bayes classifier.

2.3 Predictors

We also want to demonstrate the use of SPS for analysing the relevance of variables for the different classes. For that purpose we generate our predictor variables such that we have a clear concept for the relevance at least in the non-dependent case.

All univariate distributions have variance 1. They only differ in expectation, kurtosis and skewness. Table 1 gives the defined expectations, which are either zero, one, or the upper or lower 5%-quantile of the standard normal distribution, $u_{.05} \doteq -1.64$ and $u_{.95} \doteq 1.64$.

Table1.: Expectation of Predictors in Groups

V	G1	G2	G3	relevant for	V	G1	G2	G3	relevant for
V1	0	-1.64	1.64	G2 vs G3	V7	0	-1	1	G2 vs G3
V2	1.64	0	-1.64	G1 vs G3	V8	1	0	-1	G1 vs G3
V3	-1.64	1.64	0	G1 vs G2	V9	-1	1	0	G1 vs G2
V4	0	1.64	1.64	G1 vs rest	V10	0	0	0	Annoyance
V5	1.64	0	1.64	G2 vs rest	V11	0	0	0	Annoyance
V6	1.64	1.64	0	G3 vs rest	V12	0	0	0	Annoyance

We fix these expectations for the normal and the non-normal case. The resulting expectations, variances and covariances in the dependent case can be calculated.

2.4 Factors

We define the low (level=-1) and the high level (level=1) for our experimental design by looking at the simplicity of the learning task.

The more training data we have, the more information we have to learn classification rules, thus we consider the number of examples in the training set as influencing factor **TO**. We set $N = 1000$ as low (difficulty) level and $N = 100$ as high (difficulty) level.

To model the deviance of the true distribution from the normal distribution, we use the Johnson System (Johnson, 1949), where random variables can be generated such that they have pre-defined first four central moments. Low and high skewness values (0.1^2 and 1.15^2), and low and high kurtosis values (2.7 and 5.0) are selected such that all combinations are valid, and that a wide range of distributions in the skewness-kurtosis-plane is spanned.

On the low level of dependency **Dp**, we generate independent predictor variables $X_k, k = 1, ..., K$. The high level of dependence is constructed by calculating "inverted" variables: $\tilde{X}_k := \sum_{i=1}^{K} X_i - X_k$, $k = 1, ..., K$.

We do want deflection to be the deviance from the ordinary, thus the chosen high level of 40% for the percentage of deflected observations **DO** assures that more than a half of the observations is "ordinary". We define the low level as 10%. For the low level of the percentage of deflected variables **DV** only one of the nine relevant predictor variables is deflected, on the high level, all but one.

The direction of deflection **DD** of an observation is either determined by a shift of the value in each affected predictor variable towards its mean in the true group of the observation, or away from it towards the nearest true mean of this variable in another group.

2.5 Experimental design

To do the screening to detect the relevant factors for the relative performance of the analyzed set of classifiers, we use the standard Plackett-Burman design in Table 2 for seven factors. In this design the main effect of each factor is confounded with three interactions.

Table2.: Plackett-Burman-Plan with 8 Experiments

	Factor						
	TO	K	S	Dp	DO	DV	DD
	Confounded Interactions						
	-K*Dp -S*DD -DO*DV	-TO*Dp -S*DO -DV*DD	-TO*DD -K*DO -Dp*DV	-TO*K -S*DV -DO*DD	-TO*DV -K*S -Dp*DD	-TO*DO -S*Dp -K*DD	-TO*S -K*DV -Dp*DO
No.	Experimental level of factor in plan						
1	-1	-1	-1	-1	-1	-1	-1
2	-1	-1	1	-1	1	1	1
3	1	-1	-1	1	-1	1	1
4	1	1	-1	-1	1	-1	1
5	1	1	1	-1	-1	1	-1
6	-1	1	1	1	-1	-1	1
7	1	-1	1	1	1	-1	-1
8	-1	1	-1	1	1	1	-1

3 Results

In Table 3 we present the values of those coefficients of the approximated linear response functions that are significant to a five percent level, the measure of the fit of the linear model based on all factors R^2, and based only on the significant factors $R^2_{5\%}$.

The classifiers differ strongly in the significant main effects: NB, k-NN, and rPart react strongly negative on the dependency, whereas QDA, and LDA react both negative on kurtosis, most others react positively, only NN does not react at all on kurtosis. The reaction on the deflection is all in all not very strong (relatively to the best that could have been learnt), only LDA is reacting positively!

As mentioned above, main effects in the Plackett-Burman design are confounded with interactions. Thus interpretations have to be done carefully, especially for LDA, QDA, and NN, where low R^2-values indicate only mediocre model fit. E.g. the astonishing result that LDA reacts positively on skewness, may turn out to be the result of interactions. In future work, we will run additional simulations for these classifiers with experimental designs that account for potential interactions.

Table3.: Descriptions of the Approximated Response Functions

Meth.	Crit.	Intc.	TO	K	S	DP	DO	DV	DD	R^2	$R^2_{5\%}$
LDA	rCR	.9997	−.0189	−.0071	.0062				.0072	.89	.83
	rAc	.9973	−.0503	−.0158	.0173				.0175	.87	.81
	rAS	.999	−.0347	−.013	.0117				.0142	.89	.82
QDA	rCR	.9963	−.0583							.90	.82
	rAc	.9917	−.1334	−.0238						.91	.86
	rAS	.9933	−.0965	−.0192			.0165			.91	.88
rPart	rCR	.929	−.0966	.0479		−.1171				.91	.90
	rAc	.779	−.166	.1005		−.237				.91	.90
	rAS	.8873	−.1342	.0688		−.1806				.92	.91
NB	rCR	.999	−.1498	.1241		−.2816				.97	.96
	rAc	.9973	−.2308	.1665		−.5473	.0733			.97	.97
	rAS	.9983	−.2406	.1926		−.4381				.97	.96
k-NN	rCR	.9973	−.0586	.0286	.0167	−.0528	.0129			.98	.98
	rAc	.9927	−.1345	.0933		−.1213				.93	.91
	rAS	.9967	−.0983	.0426	.0319	−.0868	.0251			.98	.98
NN	rCR	.976	−.0384				.0139			.73	.65
	rAc	.9247	−.0894	.0396				−.0401	−.0586	.74	.72
	rAS	.9657	−.06				.0212			.74	.67

References

[Fuk90] Keinosuke Fukunaga. *Introduction to Statistical Pattern Recognitionen.* Academic Press, New York, 2nd edition, 1990.

[Han97] David J. Hand. *Construction and Assessment of Classification Rules.* John Wiley & Sons, Chichester, 1997.

[Joh49] N. L. Johnson. Bivariate distributions based on simple translation systems. *Biometrika*, 36:149–176, 1949.

[SW01] Ursula M. Sondhauss and Claus Weihs. Standardizing the comparison of partitions. Technical report, SFB 475, University of Dortmund, 2001. 31/01.

StatDataML: An XML Format for Statistical Data*

David Meyer[1], Friedrich Leisch[1], Torsten Hothorn[2] and Kurt Hornik[1,3]

[1] Institut für Statistik und Wahrscheinlichkeitstheorie, Technische Universität Wien, Wiedner Hauptstraße 8-10/1071, A-1040 Wien, Austria

[2] Institut für Medizininformatik, Biometrie und Epidemiologie, Friedrich-Alexander-Universität Erlangen-Nürnberg, Waldstraße 6, D-91054 Erlangen, Germany

[3] Institut für Statistik, Wirtschaftsuniversität Wien, Augasse 2-6, A-1090 Wien, Austria

Abstract. In order to circumvent common difficulties in exchanging statistical data between heterogeneous applications (format incompatibilities, technocentric data representation), we introduce an XML-based markup language for statistical data, called *StatDataML*. After comparing StatDataML to other data concepts, we detail the design which borrows from the language S, such that data objects are basically organized as recursive and non-recursive structures, and may also be supplemented with meta-information.

Keywords. Data Exchange, Data Design, XML

1 Introduction

Data exchange between different tools for data analysis and data manipulation is a common problem: different applications use different and often proprictary and undocumented formats for data storage. Import/export filters are often missing or insufficient, and if ever, focus on technical aspects (like storage modes and floating point specifications) in spite of supporting conceptional representation issues (like scales or representation of categorical data). The currently high costs for data exchange hence could be significantly reduced by the use of a well-defined common data exchange standard, if only because software packages would just need to provide one single mechanism.

The aim of this paper is to introduce such a data exchange standard for statistical data: the XML-based markup language *StatDataML*. The design borrows from the language S (see e.g., Chambers, 1998), such that data objects are basically organized as recursive structures (lists) and non-recursive structures (arrays), respectively. [1] Additionally, each object can have an attached list of properties (corresponding to S attributes), providing storage of meta-information.

* This research was supported by the Austrian Science Foundation (FWF) under grant SFB#010 ('Adaptive Information Systems and Modeling in Economics and Management Science').

[1] See Temple Lang & Gentleman (2001) for a more specific approach representing S objects in XML.

2 Requirements on Statistical Data

Statisticians need a data format which is both flexible enough to handle all different kinds of statistical data (from time series to decision trees), and specialized enough to incorporate statistical notions like scales and factors. Such a data format should feature:

- special symbols for infinities and undefined values
- special symbols for missingness ("not available")
- logical data
- categorical data (unordered, ordered or cyclic[2])
- numeric data (in the storage modes integer, real and complex)
- character data (strings)
- date/time information
- vectors (objects with elements of the same type)
- lists (objects with—possibly different—elements of any type)

Vectors should be indexable arbitrarily—in order to build matrices or multidimensional arrays. Lists allow for complex and even recursive structures (for they can contain lists again).

Table 1 compares some software products regarding these criteria: two families of mathematical programming languages (Splus[3]/R[4] and Matlab[5]/Octave[6]), statistical software (SPSS[7], SAS[8], Minitab[9]) and spreadsheets (Excel[10], StarCalc[11], Gnumeric[12]). In spreadsheets and Matlab/Octave, nominal data can only be represented by strings. Arrays of arbitrary dimension are supported by Splus/R and Matlab only. Complex numbers are only supported by Splus/R and Matlab/Octave. The latter cannot handle missingness. IEEE special values are not supported by Excel, StarCalc, SPSS, SAS and Minitab.

3 StatDataML

3.1 StatDataML is XML

For "statistical data" one would usually think of such things as tabular data, time series objects, responses and regressors or contingency tables. Programs

[2] like days of a week
[3] For Splus see: `http://www.insightful.com`
[4] For R see: Ihaka & Gentleman (1996) and `http://www.R-project.org`
[5] For Matlab see: `http://www.mathworks.com`
[6] For Octave see: `http://www.octave.org`
[7] For SPSS see: `http://www.spss.com`
[8] For SAS see: `http://www.sas.com`
[9] For Minitab see: `http://www.minitab.com`
[10] For Excel see: `http://www.microsoft.com`
[11] For StarCalc see: `http://www.staroffice.com` or `http://www.openoffice.com`
[12] For Gnumeric see: `http://www.gnumeric.org`

	R/Splus	Matlab	Octave	Excel/ StarCalc	Gnumeric	SPSS	SAS	Minitab
IEEE	yes	yes	yes	no	yes	no	no	no
NA	yes	no	no	yes	yes	yes	yes	yes
logical	yes	(yes)	no	yes	yes	yes	yes	yes
nominal		strings	strings	strings	strings	coding	yes	strings
unordered	yes					yes	yes	yes
ordered	yes					yes	yes	yes
cyclic	no	no	no	no	no	no	no	no
numeric	yes	yes			yes			
integer	yes	yes	no	no	no	(yes)	yes	no
real	yes	yes	yes	yes	yes	yes	yes	yes
complex	yes	yes	yes	no	no	no	no	no
character	yes	yes	yes	yes	yes	yes	yes	yes
date/time	yes	yes	yes	yes	yes	yes	yes	yes
arrays	yes	yes	no	no	no	no	no	no
matrix	yes	yes	yes	yes	yes	yes	yes	yes
lists	yes	yes	yes	no	no	no	no	no

Table 1. Data representation capabilites of different software packages

that produce such data store it on disk, using either a binary format or a text format. `StatDataML` files are XML files, thus ordinary text files, with extension `.sdml`, containing several XML elements (so called *tags*), which can be formally described with a special data definition language (DTD)—see the World Wide Web Consortium (2000) recommendation. In the following, we will go through the rules in the `StatDataML.dtd` file.

3.2 The File Header

```
<!ELEMENT StatDataML (description?, dataset?)>
```

The top level `StatDataML` element contains one `description` and one `dataset` element, each optional. It should contain the `StatDataML` namespace:

```
<StatDataML xmlns="http://www.ci.tuwien.ac.at/StatDataML">
   ...
</StatDataML>
```

3.3 The `description` element

```
<!ELEMENT description (title?, source?, date?, version?,
                       comment?, creator?, class?, properties?)>

<!ELEMENT title (#PCDATA)>
<!ELEMENT source (#PCDATA)>
<!ELEMENT date (#PCDATA)>
```

```
<!ELEMENT version (#PCDATA)>
<!ELEMENT comment (#PCDATA)>
<!ELEMENT creator (#PCDATA)>
<!ELEMENT class (#PCDATA)>
<!ELEMENT properties (list)>
```

The `description` element is used to provide meta-information about a dataset that is typically not needed for computations on the data itself. It consists of eight elements: `title, source, date, comment, version, creator` and `class` are simple strings (`PCDATA`), whereas `properties` is a `list` element (see next section). `date` should follow the ISO 8601 format (see below). The `creator` element should contain knowledge about the creating application and the `StatDataML` implementation, `properties` offers a well-defined structure to save application-based meta-information, and, finally, the `class` element will contain the class name, if any. There are some discussions about meta data in statistics[13]: one could think of extending the `description` element in such that extended information with logical markup can be stored.

3.4 The dataset element

```
<!ELEMENT dataset (list | array)>
```

We define a `dataset` element either as an array or as a list. We use arrays and lists as basic "data types" in StatDataML because every data object in statistics can be decomposed into a set of arrays and lists (as in Splus/R, or like the corresponding arrays and cell-arrays in Matlab). The basic property of a list is its recursive structure, in contrast to arrays which are always non-recursive. If one thinks about data as a tree, lists would be the branches and arrays the leaves.

3.4.1 Lists

A list contains three elements: `dimension, properties` and `listdata`:

```
<!ELEMENT list (dimension, properties?, listdata)>
<!ELEMENT listdata (list | array)*>

<!ELEMENT dimension (dim*)>
<!ELEMENT dim (e*)>
<!ATTLIST dim size CDATA #REQUIRED>
```

The `dimension` element contains one or more `dim` tags, depending on the number of dimensions. Each of them has `size` as a required attribute, and may optionally contain up to `size` names, specified with `<e>...</e>` tags. Note that arrays, like the whole dataset, can also have additional `properties` attached, corresponding, e.g., to attributes in S. In fact, the name "properties" was chosen because the name "attribute" is already used in XML itself. The `listdata` element may either contain arrays (with the actual data), or again lists, which allows for complex and even recursive structures.

[13] e.g., `http://www.gla.ac.uk/External/RSS/RSScomp/metamtg.html`

3.4.2 Arrays

```
<!ELEMENT array (dimension, properties?, (data | textdata))>
```

Arrays are blocks of data objects of the same elementary type with dimension information used for memory allocation and data access (indexing). The first two elements, `dimension` and `properties`, are identical to lists, only the `listdata` block is replaced by the `data` (or `textdata`) element which contains the data itself.

The data tag

```
<!ELEMENT data (e|ce|na)*>
<!ATTLIST data
  true CDATA "1"
  false CDATA "0"
  type (logical|nominal|numeric|character|datetime) "character"
  mode (unordered|ordered|cyclic|integer|real|complex) #IMPLIED>

<!ELEMENT na EMPTY>

<!ELEMENT e (#PCDATA|posinf|neginf|nan)* >
<!ELEMENT posinf EMPTY>
<!ELEMENT neginf EMPTY>
<!ELEMENT nan EMPTY>

<!ELEMENT ce (r,i) >
<!ELEMENT r (#PCDATA|posinf|neginf|nan)* >
<!ELEMENT i (#PCDATA|posinf|neginf|nan)* >
```

If `data` is used (especially recommended for character data), then each element of the array representing an existing value is encapsulated in `<e>...</e>` (or `<ce>...</ce>` for complex numbers). For missing values, `<na/>` has to be used, empty values are just represented by `<e></e>`.
The `type` attribute specifies the statistical data type, as logical, nominal, numeric, character and date/time. The optional `mode` attribute allows for further specification: `nominal` data could be `unordered` ("factors"), `ordered` and `cyclic` (e.g., days of week), whereas `numeric` data could be `integer`, `real` or `complex`.
As an example, consider a character dataset formed by color names, with one value missing (after `green`), and one being empty (after `blue`). The corresponding `data` section would appear as follows:

```
<data type="nominal" mode="unordered">
  <e>red</e> <e>green</e> <na/> <e>blue</e> <e></e> <e>yellow<e>
</data>
```

IEEE Number Format

The implementation is responsible for the correct casts. The number format has to follow the IEEE Standard for Binary Floating Point Arithmetic (see Institute of Electrical and Electronics Engineers, 1985), which is implemented by most programming languages and system libraries. However, the IEEE special values `+Inf`, `-Inf` and `NaN` must explicitly be specified by `<posinf/>`, `<neginf/>` and `<nan/>`, respectively, to facilitate the parsing process in case the IEEE standard were not implemented. These special values could appear as follows:

```
<data type="numeric" mode="real">
  <e>1.23</e> <e><posinf/></e> <e><nan/></e> <e>2.43</e>
</data>
```

Complex Numbers

Complex numbers are enclosed in `<ce>...</ce>` tags which contain exactly one `<r>...</r>` tag (for the real part) and one `<i>...</i>` tag (for the imaginary part). Apart from that, the same rules as for `<e>...</e>` apply:

```
<data type="numeric" mode"=complex">
  <ce> <r>12.4</r> <i>1</i> </ce>
</data>
```

Date and Time Information

Data of type `datetime` has to follow the ISO 8601 specification (see International Organization for Standardization, 1997). StatDataML should only make use of the complete representation in extended format:

CCYY-MM-DDThh:mm:ss±hh:mm

For example, the 12th of March 2001 at 12 hours and 53 minutes, UTC+1, would be represented as: 2001-03-12T12:53:00+01:00 .

3.5 Reference Implementations

Sofware is currently available for Matlab, Octave, Splus, R, and Gnumeric at `http://www.omegahat.org/` and under development for SPSS.

4 Conclusion

StatDataML seems general and flexible enough to cover most of statisticians' data representation needs, although one might miss features like data inheritance and support for distributed data (e.g. via URLs or SQL queries on databases). The representation of high level objects (like time series) and statistical models are currently under investigation.

References

Chambers, J. M. (1998). *Programming with Data: a guide to the S Language.* New York: Springer-Verlag.

Ihaka, R. & Gentleman, R. (1996). R: A language for data analysis and graphics. *Journal of Computational and Graphical Statistics*, **5**(3), 299–314.

Institute of Electrical and Electronics Engineers (1985). *IEEE Standard 754-1985 (R 1990), Standard for Binary Floating-Point Arithmetic.*

International Organization for Standardization (1997). *ISO 8601:1997, Data elements and interchange formats - Information Interchange - Representation of dates and times.*

Temple Lang, D. & Gentleman, R. (2001). RSXMLObjects: Reading and writing S objects in XML. S Package. http://www.omegahat.org/RSXMLObjects.

World Wide Web Consortium (2000). *Extensible Markup Language (XML), 1.0 (2nd Edition).* http://www.w3.org.

Statistical Computing on Web Browsers with the Dynamic Link Library

Akinobu Takeuchi[1], Hiroshi Yadohisa[2], Kazunori Yamaguchi[1], Michiko Watanabe[3], and Chooichiro Asano[4]

[1] College of Social Relations, Rikkyo University
[2] Department of Mathematics and Computer Science, Kagoshima University
[3] Department of Economics, Toyo University
[4] Department of Information Science, Soka University

Abstract. A statistical computing system using a user-friendly graphical user interface (GUI) to a pre-defined statistical engine is proposed. The GUI, which is constructed using a Web browser interface, provides access to a statistical engine program that has been implemented as a dynamic link library. Since a browser is used, the GUI can be constructed using popular and free tools such as HTML. Using HTML, we can describe the various information appropriate for the user's level and purpose, together with analysis tools.

Keywords. DLLSA/QC, DLL, Statistical software, World Wide Web

1 Introduction

Today, there is a developing trend in the computer industry towards making software more user-friendly. Recent software programs usually have a graphical user interface (GUI), but a common GUI that is suitable for a wide range of programs is very difficult to design and implement. Liu et al. (1995) and Yamaguchi et al. (1998) independently proposed solutions to this problem. Their solutions create an interface that is suited to users needs, which can then provide access to commercially available software for the statistical engine. Using the World Wide Web (WWW) interface as a model, Nakano (1998) tried to develop software with a user interface that was as independent of the computational engine as possible. The reader is also referred to the discussions on interface development in Thioulouse and Chevent (1996) and Sato (1997).

We previously described a project called DLLSA/QC, the Dynamic Link Library (DLL) for Statistical Analysis/Quality Control (Takeuchi et al., 2000, 2001), the purpose of which was to produce DLL versions of statistical analysis programs written in various programming languages. The DLLs were then made available in a library for potential users. This meant that individual users were able to customize their input/output interface according to their needs and working environment, while using the DLL as a statistical engine. Many Windows applications are capable of calling such DLLs. Currently, the Fortran sources in the NISAN system (Asano and Kimura, 1990) are the sources most often used for DLLSA/QC.

The purpose of the DLLSA/QC project was to enable statistical analysis that could not be done with commercially available software, by using the DLLSA/QC DLL as an statistical engine to other computational software.

This allows users to perform statistical analyses in their everyday environment. This has various advantages. For example, the latest computational methods can be used rather than those that are included in commercial software. However, appropriate software must be purchased for use as an interface. There are also some difficulties to overcome, such as the need for the user to have a basic programming knowledge in order to design, implement, and use the interface program.

In this paper, we propose a new statistical computing system called SCSWD (Statistical Computing System with Web browser interface and Dynamic link library) that allows users to employ popular tools such as HTML, forms, and CGI (Common Gateway Interface), which are usually available at little or no cost. HTML, especially, can be used to describe information in a manner that is appropriate to a user's needs and his level of expertise. The fact that several of the tools are free is a definite advantage.

2 Web browser interface for the DLL

In this section, we show how effective it is to use SCSWD. In addition, we describe how to make an interface program written in HTML and how to use the resulting product.

Our project team has previously developed a statistical DLL library (Takeuchi, et al., 2000). By using a DLL file as a statistical engine, developers can design and construct the interface with only the users needs in mind. However, developers often need detailed and extensive knowledge of particular programming languages to design and write the code for the interface. The extent to which the designer must know a programming language makes it difficult for a general user to construct the required interface.

There are several advantages in using a Web browser as an interface. They are:

1. The ability to add documents that correspond to analysis and/or data, to the interface,
2. The ability to use several pages as an interface by using hyperlinks,
3. The ability to use graphical objects (for example JPEG, GIF, etc.),
4. The ability to build an interface using active components such as animation GIF and Dynamic HTML.

These capabilities make it easy to develop an interface based on users' needs. In particular, our system assumes that users will want to construct and modify the interface to suit their needs. Therefore, it is important that users should be able to design and implement the interface.

Another important advantage of our system is its ability to handle data in a more secure manner. Several statistical analysis systems are currently available for Web browsers; however, they must send the data to a server via the Internet, which places the security of the data in jeopardy, whereas in our system the calculation is done on the local computer and the data is therefore more secure.

Although a JAVA-based system could be created with similar features, the interface and statistical engine components would need to be completely rewritten for each different application. Since our system is able to reuse the statistical engine DLL, the DLL has to be developed only once, which results in a considerable saving in programming development costs. The creation of a typical interface and the use of our system are described below.

2.1 Interface development

Figure 1 represents the SCSWD methodology. SCSWD has an interface component that is written in HTML and a program component for input and output of the data. We will now describe how to use a Web browser for the interface, which can, in turn, make use of a DLL file.

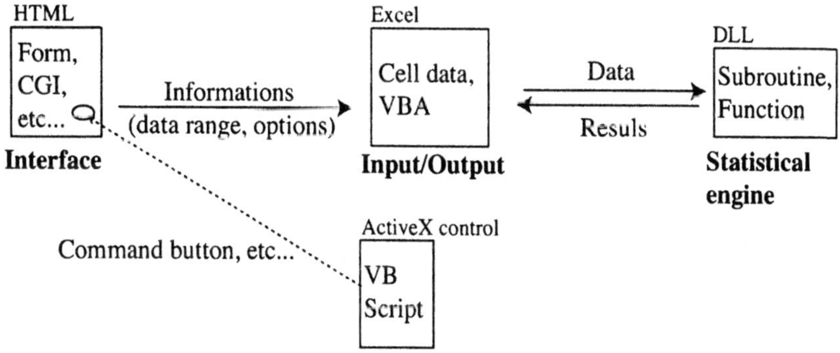

Figure 1. SCSWD system

2.1.1 An interface written in HTML

Many excellent books about HTML have been published. The reader is referred to them for further details on HTML. Our system makes extensive use of the form tag in specifying the choice of the data file and the options available to the user. The use of the form tag for the SCSWD system is summarized in Table 1. Figure 2 is a sample interface source file.

Table 1. Form tag list

Tag name	Feature
`<input type="file">`	Selecting the data file
`<input type="text">`	Inputting the text
`<input type="checkbox">`	Radio box

2.1.2 Input/Output and the ActiveX control program

The VBScript and Excel VBA macro programs are used to send the input data to the DLL and to receive the statistical engine results from the DLL. The user inputs the name of the data file and his selected options to the interface. Then, SCSWD copies this information to Excel cells by using VBScript and ActiveX. A basic knowledge of VBScript and ActiveX are needed to construct and use the interface. However, using the Microsoft ActiveX Control Pad allows a user to develop the interface relatively easily; the software for these programs is available on the Microsoft web page (http:// msdn.microsoft.com/ library/ default.asp?url=/ library/ enus/ dnaxctrl/ html/ cpad.asp). Once in Excel, the Excel VBA macro sends the information received from the VBScript to the DLL and displays the results received from the DLL. If necessary, the Excel VBA macro represents graphs and/or charts by using appropriate Excel functions.

```
<HTML>
<HEAD>
<TITLE>Hierarchical Cluster Analysis</TITLE>
</HEAD>

<BODY bgcolor="white">
<h2>Hierarchical cluster analysis</h2>

<form method="POST" name="open">
<b>I. Select a Excel file including the data in your
pc</b><br>
<input size=70 type="file" name="fdata">
</form>

...

<OBJECT CLASSID="CLSID:812AE312-8B8E-11CF-93C8-00AA00C08FDF"
ID="clus08_alx" STYLE="LEFT:0;TOP:0">
<PARAM NAME="ALXPATH" REF VALUE="clus08.alx">
</OBJECT>

</BODY>
</HTML>
```

Figure 2. HTML source

It should be noted that these two programs are used in SCSWD only for the input and output of the data and the results. A sample interface file and data files are available on our web site. A developer or a user can use SCSWD if they change the corresponding part of the sample files that are included with the interface example. These files are described in the next section.

2.2 How to use the interface

This section describes how to use our system using the files found on our web site. We explain how to use the sample **Interface** file (clus.html) and the **Input/Output** file (clus.xls) to call the **DLL** of the agglomerative hierarchical clustering algorithms (hcluster.dll). This DLL requires Windows95 and Excel97, or more current versions of these programs. On our web site, the files needed in the analysis are packed in one self-expanding file. Note that since hcluster.dll is created with Visual Fortran, users need to install Visual Fortran or the runtime version of Visual Fortran when they use this DLL. The Visual Fortran runtime files are available at no cost as a runtime distribution kit on the COMPAQ Computer Corporation web site (http:// www.compaq.com/ fortran/ visual/ redist.html). The user should open and read the clus.html and the clus.xls files first, in order to gain a basic understanding of how to use hcluster.dll. Since a Macro in Excel is used, click on [Enable Macros] when the choice is displayed. Each Excel sheet shows sample data and the results of an analysis. The following section describes how to input and analyze a new data set.

2.2.1 Preparing a Macro and Specifying the DLL Path

Steps i) to iv) need to be performed only once. Subsequent analyses can begin with step I) in Analysis subsection 2.2.2.

i) Select [Macro] from [Tools] in the Excel menu bar and click on the [Macros...] button.
ii) Choose "`Sheet1.run_mac`" from [Macro name] in the dialog box that appeared after the previous step, and click [Edit].
iii) The window containing the file path information of the `Sheet1` (Figure 3) then appears. Type in the local path for the location of the hcluster.dll file (see the marked area in Figure 3).
iv) Select [Save clus.xls] from [File] in the Visual Basic menu bar and close the Visual Basic window.

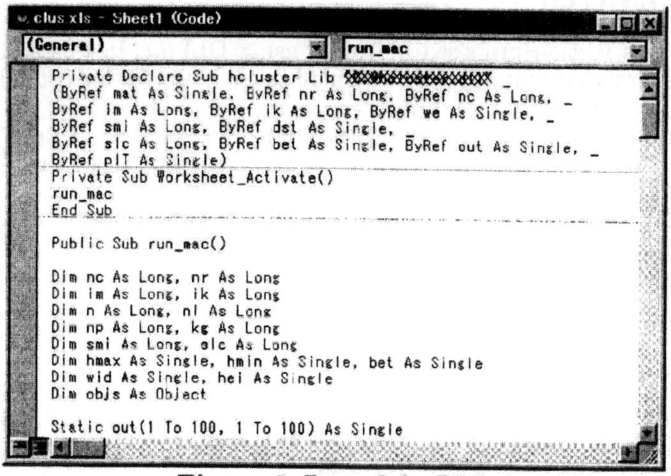

Figure 3. Pass of the DLL

2.2.2 Analysis

I) Input the data in the appropriate cells on the Excel sheet.
II) Open the clus.html file. When asked if using an ActiveX control from the web browser, click on [Yes].
III) The interface will then appear. (Figure 4). Click the [Browse...] button and from the file selection dialog box select the Excel file that contains the data. Next, set the options in the interface.
IV) Finally, activate the Excel data sheet and click the [Run] button.

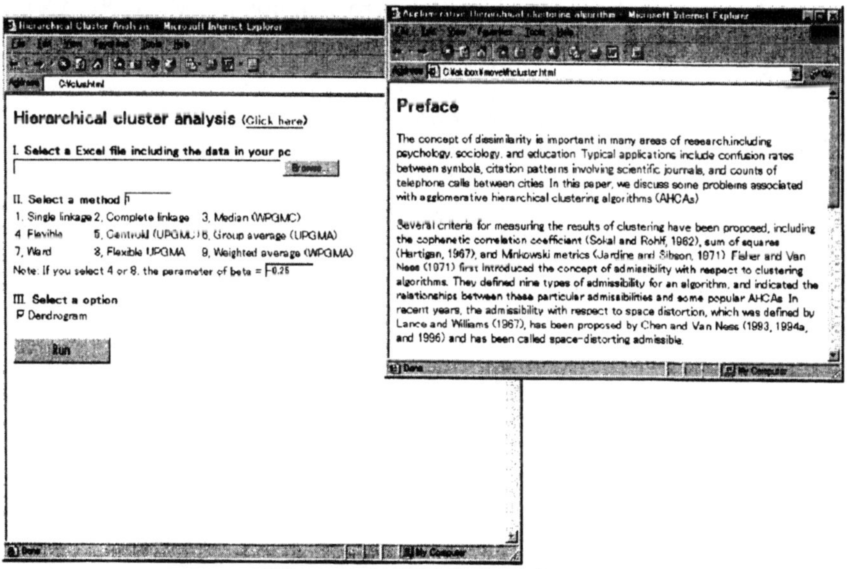

Figure 4. HTML interface

When the analysis is completed, the results (combined process of clusters, combined distances, dendrogram, and so on) are displayed in a new sheet. This shows that calculations that do not exist in Excel can be performed by using the DLL.

3 DLLSA/QC

The previous section proposed a system using DLL(s). In this section, we introduce a related project, the Dynamic Link Library for Statistical Analysis/Quality Control, DLLSA/QC.

The purpose of this project was to produce DLL versions of programs for statistical analysis written in various programming languages; the DLLs have been made available in a library for potential users. As a result, users can develop the input/output interface that best suits their needs and environment and still make use of the DLL as a statistical engine. The results of this project have been published on our web site. (http://www.sci.kagoshima-u.ac.jp/~dllsa/, http://www.sci.kagoshima-u.ac.jp/~dllqc/).

4 Concluding remarks

In this paper, we have proposed a new statistical computing system, SC-SWD, which includes a Web browser interface and DLL(s). By using a DLL, developers can reduce the time and effort needed to develop such software. In addition, by using an interface written in HTML, developers can add descriptions that correspond to the desired procedures and the data.

The current version of our system requires Excel, which allows users to input data easily and the results of the analysis to be shown conveniently. In our future work, we intend to improve the means by which data is input and output, and to change the format of the data and results files from Excel to text.

References

Asano, Ch. and Kimura, H. (1990). A study on system construction of general purpose statistical package for microcomputers — Micro-NISAN system, *Computational Statistics Quarterly*, **6**, 65-77.

Liu, L.-M., Chan, K.-K., Montgomery, A. L. and Muller, M. E. (1995). A system-independent graphical user interface for statistical software, *Computational Statistics and Data Analysis*, **19**, 23-44.

Nakano, J. (1998). Graphical user interface for statistical software using internet, in: Payne, R. and Green, P. (Eds.), *Proceedings of Computational Statistics 98*, 407-412: Heidelberg, Physica-Verlag.

Sato, S. (1997). Introduction to "Web Decomp" —Seasonal Adjustment System on WWW—, *Proceedings of the Institute of Statistical Mathematics*, **45**, 233-243 (in Japanese).

Takeuchi, A., Suenaga, K., Yadohisa, H., Yamaguchi, K., Watanabe, M., and Asano, Ch. (2001). Dynamic Link Library for Statistical Quality Control, *Proceedings of VII-th Intelligent Statistical Quality Control*, 317-330.

Takeuchi, A., Yadohisa, H., Yamaguchi, K., Watanabe, M., and Asano, Ch. (2000). DLLSA: Dynamic link library for statistical analysis, *Bulletin of The Computational Statistics of Japan*, **13**(2), 115-125 (in Japanese).

Thioulouse, J. and Chevent, F. (1996). NetMul, a world-wide web user interface for multivariate analysis software. *Computational Statistics and Data Analysis*, **21**, 369-372.

Yamaguchi, K., Inoue, T., Yadohisa, H. and Watanabe, M. (1998). Educational statistical software on the world wide web, *Proceedings of Computational Statistics 98: Short Communications and Posters*, 127-128.

Statistical Inference for a Robust Measure of Multiple Correlation

Catherine Dehon[1] and Christophe Croux[2]

[1] Ecares and Institut de Statistique, Université Libre de Bruxelles, CP 114, Av. F.D. Roosevelt 50, B-1050 Bruxelles, Belgium
[2] Departement of Applied Economics, Katholieke Universiteit Leuven, Naamsestraat 69, B-3000 Leuven, Belgium

Abstract. In regression analysis it is standard practice to report an R-squared statistic. This measure of multiple correlation is based on the Least Squares estimator, which is known to be extremely sensitive to outliers. The associated R-squared suffers from the same lack of robustness. Also the value of the F-statistic, used to see whether the explanatory variables have jointly a significant influence on the dependent variable, exhibits similar problems. Many robust estimation procedures are available now, but there has been less emphasis on the development of the inference part and in particular on the study of robust R-squared measures. It is however clear that applied statisticians desire to have the same tools to validate their model as when using the classical least squares estimator. In this note we will study the stability properties of robust measures of multiple correlation, and also investigate the size and power of F-statistics based on them.

Keywords. Confidence Interval, F-statistic, Multiple Correlation, R-Squared statistic, Regression analysis, Robustness

1 A robust measure of multiple correlation

The setting is the standard linear regression model with intercept:

$$y_i = \alpha + x_{i1}\beta_1 + \cdots + x_{ip}\beta_p + \varepsilon_i \qquad i = 1, \cdots, n$$

where n is the sample size, $x_i = (x_{i1}, \cdots, x_{ip})^t$ the vector containing the explanatory variables and y_i the response variable. Suppose that the errors ε_i are independent of the explanatory variables and i.i.d. according to a normal distribution $N(0, \sigma^2)$. For the estimation of β, normality is not really required, but we need this assumption for inference purposes.

With $r_i(\alpha, \beta) = y_i - x_i^t\beta - \alpha$ as residuals, the classical R^2 coefficient is defined as

$$R^2_{LS} = 1 - \frac{\sum_{i=1}^n r_i(\hat{\alpha}_{LS}, \hat{\beta}_{LS})^2}{\sum_{i=1}^n (y_i - \bar{y})^2}, \tag{1}$$

with $\bar{y}$ the sample average of the dependent variable. We see that the numerator in (1) equals the variance of the residuals in the full model, while the denominator is the variance of the residuals in the null model, i.e. the model with only an intercept term and no explicative variables.

Consider robust regression estimators which are defined by minimizing a robust measure of spread of the residuals. For example, when we use a trimmed standard deviation, we get the Least Trimmed Squares estimator (LTS) of

Rousseeuw and Leroy (1987). Taking the median of the absolute values of the residuals yields the Least Median of Squares regression estimator (LMS) of Rousseeuw (1984). When we take as measure of dispersion of the residuals an M-estimator of scale, we get the class of S-estimators of Rousseeuw and Yohai (1984). All these estimators have positive breakdown points, where the breakdown points measure the maximal fraction of contamination than an estimator can withstand. For LTS and LMS we take these versions of the estimator having the maximal breakdown point of 50%. For the S-estimators we take by default the 50% breakdown version using the Biweight loss function (BS), but we will also consider the 25% breakdown Biweight S-estimator (BS25).

According to Anderson-Sprecher (1994), a robust multiple correlation coefficient can now be obtained by as follows: let the regression estimators be defined by

$$(\hat{\alpha}, \hat{\beta}) = \underset{(\alpha,\beta)}{\operatorname{argmin}}\, S(r_1(\alpha,\beta), \ldots, r_n(\alpha,\beta))$$

with $r_i(\alpha,\beta) = y_i - x_i^t\beta - \alpha$ and S a robust residual scale estimator satisfying

$$S(ae_1, \ldots, ae_n) = |a| S(e_1, \ldots, e_n)$$

for all $e_1, \ldots, e_n$ and $a \in I\!R$. Estimating the null regression model, having only an intercept term, yields

$$\hat{\alpha}_0 = \underset{\alpha_0}{\operatorname{argmin}}\, S(y_1 - \alpha_0, \ldots, y_n - \alpha_0).$$

By comparing the estimated residual scales in the full model and in the null model, we define by analogy with the classical formula (1)

$$R_S^2 = 1 - \frac{S^2\left(r_1(\hat{\alpha},\hat{\beta}), \ldots, r_n(\hat{\alpha},\hat{\beta})\right)}{S^2\left(y_1 - \hat{\alpha}_0, \ldots, y_n - \hat{\alpha}_0\right)}. \tag{2}$$

For every residual scale estimator S we have another regression estimator and another measure of multiple correlation. Values of R_S^2 close to zero indicate that the explanatory variables are not explaining much of the dispersion of y, while a value of R^2 close to one indicates that the dispersion of the residuals is almost zero and so an excellent fit. This interpretation is standard, but the meaning of "dispersion" depends on the selected measure of scale S. Taking the standard deviation as measure of spread, results again in the usual definition. Taking the mean absolute deviation, corresponds with the L_1 regression estimator and gives the multiple correlation coefficients of McKean and Sievers (1987).

Under the assumption that the x-variables follow a multivariate normal distribution, all these different R_S will estimate the same population multiple correlation coefficient $\wp$. To allow a comparison between the different measures of multiple correlation, we will add this assumption.

2 Confidence intervals for multiple correlation

A theoretical study of robust R-squared measures was done by Croux and Dehon (2002). They looked at local robustness properties and found the most B-robust (minimizing the maximal influence that an observation can have on the estimator) R_S estimator in the class of all estimator of multiple correlation based on M-estimators of residual scale defined by a function ρ. This most robust estimator turned out to be the Least Quantile of Squares estimator LQS32, minimizing the 68% quantile of the absolute values of the residuals. Croux and Dehon (2002) proposed to use as 95% confidence interval for the Fisher transform of the population multiple correlation coefficient $\tanh^{-1}(\wp)$:

$$\tanh^{-1}(R_S) \pm 1.96\frac{W(R_S)}{\sqrt{n}} \tag{3}$$

with

$$W(R_S) = \sqrt{\frac{A(R_S)}{R_S^2 E_\Phi[Y\rho'(Y)]^2}},$$

and $A(R) = E[\{\rho(VR+U\sqrt{1-R^2})-\rho(U)\}^2]$ with U and V two independent standard normal variables. The constant $A(R_S)$ can easily be computed by bivariate numerical integration or by using the formulas given in Croux and Dehon (2002). The Fisher transformation is used as a variance stabilizing transformation: for the LS-estimator $W(R_S) \equiv 1$, but it also turned out that for many robust estimators $W(R_s)$ is a quite flat function of R_S. Of course, by back-transformation we get a confidence interval for $\wp$ itself.

Now we will present some numerical experiments for testing the stability of the R_S measures and the finite sample performance of the confidence intervals constructed according to (3). Observations were generated according to the model

$$y = \alpha + \beta_1 x_1 + \beta_2 x_2 + \varepsilon,$$

where $(x_1, x_2) \sim N_2(0, I)$ and $\varepsilon \sim N(0, \sigma^2)$ with $\sigma^2 = 0.5$. The regression parameters were set at $\alpha = 0$, $\beta_1 = \beta_2 = 0.5$, yielding a squared multiple correlation coefficient of $\wp^2 = 0.5$. For $m = 3000$ simulation runs, we generated each time 200 observations according to the above model, and replaced the first 10 observations by contaminated values. Four different contamination schemes are considered: (a) no contamination; (b) outliers placed at $x = (3,3)^t$ and y generated according to the model, giving us good leverage points; (c) outliers at $x = (3,3)^t$ and $y = 0$, a first type of bad leverage points; (d) outliers at $x = (3,3)^t$ and $y = -3$, a second type of bad leverage points. For each generated sample R_S^2 and the associated confidence interval were computed. All computations have been carried out using Splus build-in routines. The reported coverages in Table 1 give the percentage of runs for which the confidence interval covered the population value. Between parenthesis, the averages of the $m = 3000$ estimates of $\wp^2$ are reported.

From case (a) in Table 1 we see that in the case of no contamination the empirical coverages are close to the theoretical value of 95%. There is only a minor distortion for the LMS. In the presence of good leverage points (case b) the R^2 statistic increases for all estimators, but the most for LS and L_1. For the latter estimators, there is a complete implosion of the empirical coverage of the confidence interval. In case (c), where the most dangerous type

Table 1. Empirical coverages of the confidence interval for the multiple correlation coefficient under 4 different sampling schemes and for different estimators. The average value of the estimates of $\wp^2 = 0.5$ are reported in parentheses.

	LS	L_1	BS	BS25	LMS	LQS32	LTS
Case (a)	0.948	0.949	0.948	0.947	0.966	0.951	0.970
	(0.504)	(0.505)	(0.499)	(0.497)	(0.485)	(0.503)	(0.508)
Case (b)	0.063	0.337	0.862	0.541	0.934	0.827	0.931
	(0.647)	(0.617)	(0.562)	(0.595)	(0.546)	(0.572)	(0.564)
Case (c)	0.000	0.008	0.578	0.157	0.837	0.895	0.865
	(0.267)	(0.246)	(0.342)	(0.310)	(0.381)	(0.434)	(0.379)
Case (d)	0.000	0.000	0.953	0.958	0.960	0.892	0.954
	(0.008)	(0.201)	(0.497)	(0.494)	(0.523)	(0.551)	(0.544)

of bad leverage points were introduced in the sample, we see that the average R_S^2 values are all quite below the theoretical value of 0.5. Even for the robust procedures we have a non-negligible bias for R^2 and for the empirical coverage of the confidence interval. Here, the LQS32 has the best behavior. In fact, the maximal bias of R_S^2 over the four considered sampling schemes is minimal for this estimator, confirming that LQS32 is the Most B-Robust. For the confidence intervals, the LMS behaves at least as well as LQS32. It appears that the more efficient regression estimators BS and BS25 return robust estimators of $\wp^2$, but the associated confidence intervals are less reliable in the presence of contamination. Finally, the outliers introduced in case (d) are "harmless" bad leverage points, but only for the estimators with bounded influence function. Their effect on the LS and L_1 estimator is dramatic.

It turns out that although the robust regression estimators are highly resistant, this is less true for robust R-squared measures. Surely, they still outperform the classical measure, but it seems that robust estimation for correlation quantities is much harder than for regression parameters.

3 A robust F-test

Since we are comparing residual scales in two different models, it is natural to construct a robust test for the hypothesis that the true model equals the null model. The conventional F-test for the global validity of the model, testing whether all regression slopes parameters are jointly equal to zero is given by

$$F_{LS} = \frac{R_{LS}^2/p}{(1 - R_{LS}^2)/(n - (p + 1))}. \tag{4}$$

It is well-known that this statistic follows, under the null hypothesis, a Fisher distribution with $(p, n - (p + 1))$ degrees of freedom. But the power of this test turns out to be extremely low in the presence of outliers. Therefore we propose as a robust F-test simply

$$F_S = \frac{R_S^2/p}{(1 - R_S^2)/(n - (p + 1))}, \tag{5}$$

with R_S the robust estimate of multiple correlation defined in (2).

There is already existing a literature on robust testing for linear models (see e.g. Markatou and He 1994), but the test-statistic (5) is particularly simple and completely analogous to the classical F-test. The limiting distribution of (5) can be seen to be a Chi-Square distribution with p degree of freedom divided by p and multiplied by the asymptotic variance $\mathrm{ASV}(\hat{\beta}_S)$ of the estimator of regression based on minimizing S.

In this note, we are particularly interested in the finite-sample behavior of these robust F-tests. We focus on the LTS-estimator, since this estimator is probably the most often used high breakdown regression estimator, and gave reasonable results in the preceding section. Since the standard 50% breakdown LTS estimator has a very low efficiency, we will work with its 25% breakdown version LTS25 for which $\mathrm{ASV}(\hat{\beta}_S) = 3.618$

The first experiment aims at finding out which critical values should be used in practice. A first possibility would be to use the asymptotic critical value, given by $c_{\alpha,\infty,p} = \mathrm{ASV}(\hat{\beta}_S)\chi^2_{p,1-\alpha}/p$. By analogy with the classical case, we could also try $c_{\alpha,n,p} = \mathrm{ASV}(\hat{\beta}_S)F_{p,n-(p+1),1-\alpha}$, which is the exact finite sample critical value when using the LS estimator. Finally we can also simulate a critical value $\bar{c}_{\alpha,n,p}$ by computing F_S for $m = 5000$ generated samples under the null hypothesis and taking the $(1-\alpha)$ quantile of the empirical distribution of the simulated test values. For m tending to infinity this gives the exact finite sample critical value. From Table 2 we can compare these different critical values for $\alpha = 0.05$, for several values of the dimension p and different sample sizes n.

Convergence to the asymptotic value is very slow, and it is seen that $c_{\alpha,n,p}$ is, somewhat surprisingly, worse as an approximation as $c_{\alpha,\infty,p}$. It seems that we cannot rely on the approximations $c_{\alpha,n,p}$ or $c_{\alpha,\infty,p}$, and so we will use the critical values $\bar{c}_{\alpha,n,p}$ obtained by the Monte-Carlo method for the LTS25 estimator.

Table 2. Simulated $\bar{c}_{\alpha,n,p}$ and asymptotical critical values for robust F-test based on the LTS25 estimator at level $\alpha = 5\%$. The critical value $c_{\alpha,n,p}$ based on the Fisher distribution is reported in parentheses.

n	$p=1$	$p=2$	$p=3$
20	10.770 (15.969)	7.674 (12.994)	5.725 (11.718)
50	12.367 (14.626)	9.323 (11.560)	7.403 (10.155)
100	12.425 (14.248)	9.582 (11.180)	8.119 (9.766)
200	13.014 (14.070)	9.786 (11.005)	8.871 (9.590)
$c_{\alpha,\infty,p}$	13.898	10.839	9.425

In the last part, we study the level and power of the robust F-test. The test can be called robust if the level remains stable and has good power also in the presence of outliers. Observations were generated according to the model $y = \alpha + \beta x + \varepsilon$ where $x \sim N(0,1)$, $\varepsilon \sim N(0,1)$ and $\alpha = 1$. We considered four different values of β: 0 (corresponding to the null hypothesis), and 0.1, 0.5 and 1 (corresponding to the alternative hypothesis). To compute the empirical size and power of the test, we generated 500 samples according to the model and computed the percentage of times that the critical value was exceeded. We compared the LTS25 estimator with the classical LS-estimator. For the latter we used $F_{p,n-(p+1),1-\alpha}$ as the exact critical value. In Table 3, we have reported the frequency of rejections of the null hypothesis for the simulated data sets and for each value of β. The same experiment was then repeated but now 10% of the x-values in every data set were replaced by a random value from $N(10, 0.1)$, hereby creating leverage points. The empirical size (for $\beta = 0$) and power (for $\beta \neq 0$) under this type of contamination can also be retrieved from Table 3.

Table 3. Percentage of rejections of $H_0 : \beta = 0$ using F_{LS} and F_{LTS25} for different values of β, once without contamination and once in the presence of 10% outliers, for several values of the sample size n

		Without		With	
		LS	LTS25	LS	LTS25
	$n = 20$	0.056	0.078	0.020	0.032
$\beta = 0$	$n = 50$	0.042	0.040	0.052	0.012
	$n = 100$	0.042	0.040	0.080	0.044
	$n = 20$	0.060	0.114	0.072	0.052
$\beta = 0.1$	$n = 50$	0.080	0.064	0.040	0.036
	$n = 100$	0.164	0.062	0.052	0.032
	$n = 20$	0.682	0.408	0.056	0.072
$\beta = 0.5$	$n = 50$	0.928	0.530	0.028	0.112
	$n = 100$	0.986	0.720	0.196	0.148
	$n = 20$	0.996	0.916	0.008	0.368
$\beta = 1$	$n = 50$	1.000	0.990	0.008	0.804
	$n = 100$	1.000	0.996	0.388	0.896

¿From Table 3 we see that when no contamination is present, the LS estimator is the most powerful, which is of course no surprise. But we also see that the power completely implodes in the presence of the outliers. For the robust estimator we see that the procedure has rather low power for the smaller values of β (in both situations) and that only the results for $\beta = 1$ are satisfactory: here the robust procedure clearly picks up the alternative hypothesis. The overall picture is that the robust procedure gives better power in the presence of outliers than the classical procedure. But it also became apparent that we cannot say that the statistical inference procedure is highly robust, even when using the highly robust LTS25 estimator.

References

Anderson-Sprecher, R.A. (1994). Model Comparison and R-Squared. *The American Statistician*, **48**, 113-117.

Croux, C. and Dehon, C. (2002). Estimators of the Multiple Correlation Coefficient: local robustness and confidence intervals, to appear in *Statistical Papers*, www.econ.kuleuven.ac.be/christophe.croux

Markatou, M. and He, X. (1994). Bounded influence and high breakdown point testing procedures in linear models. *Journal of the American Statistical Association*, **89**, 543-549.

McKean, J.W. and Sievers, G.L. (1987). Coefficients of determination for least absolute deviation analysis. *Statistics and Probability Letters*, **5**, 49-54.

Rousseeuw, P.J. (1984). Least Median of Squares regression. *Journal of the American Statistical Association*, **79**, 871-880.

Rousseeuw, P. J., and Leroy, A. M. (1987). *Robust Regression and Outlier Detection*. New York: Wiley.

Rousseeuw, P.J. and Yohai, V.J. (1984). Robust regression by means of S-estimators. In: *Franke, J., Härdle W., Martin, R.D. (Eds.), Robust and Nonlinear Time Series Analysis, Lecture Notes in Statistics 26*. New York: Springer Verlag.

Statistical Software VASMM for Variable Selection in Multivariate Methods

Masaya Iizuka[1], Yuichi Mori[2], Tomoyuki Tarumi[3] and Yutaka Tanaka[3]

[1] Dept. of Law, Okayama University, 700-8530, Japan
[2] Dept. of Socio-Information, Okayama University of Science, Okayama 700-0005, Japan
[3] Dept. of Environmental and Mathematical Science, Okayama University, 700-8530, Japan

Abstract. A statistical software package VASMM (VAriable Selection in Multivariate Methods) has been developed for selecting a subset of variables in multivariate methods without external variables. The current version is fully implemented for variable selection in principal component analysis and factor analysis. The system has been constructed with interactive architecture on Internet. The users can not only use the system via a web browser but can also obtain information related to variable selection in multivariate techniques of their choice. It allows for us to perform variable selection easily in a variety of practical applications.

Keywords. web-based program, statistical tool, principal component analysis, factor analysis

1 Introduction

Selecting a subset of variables is useful in many practical applications where a multivariate method (MM) without external variables can be applied. Here the term "external variable" is used as a variable to be predicted or to be explained using the information derived from other variables. We sometimes meet the problem of variable selection in MMs such as principal component analysis (PCA) and factor analysis (FA) in various situations. For example, we wish to select a subset of items in a particular test so as to make a simplified version of the test without losing the amount of information of whole items, or to reduce the number of variables before a particular analytical technique in data mining when the number of variables is too large for the analysis. In practical data analysis various ad hoc methods are used for variable selection in MMs. Most of them are based on subjective judgement and the results usually differ depending on who analyses the data. So it will be valuable to provide objective methods and their software of variable selection in MMs.

In the past, however, we could not find suitable programs to perform any variable selection method easily except in multiple regression analysis, discriminant analysis and related methods. In these circumstances, we developed a statistical program called VASMM (VAriable Selection in Multivariate Methods) which contains a variety of selection methods and criteria to select a reasonable subset of variables in any MM without external variables.

2 Variable selection in multivariate methods without external variables

The problems of variable selection in MMs without external variables (PCA, FA, correspondence analysis among others) have been studied by many authors. Here we introduce ideas of variable selection briefly in two typical MMs: PCA and FA.

2.1 Variable selection in PCA

Several authors have studied variable selection in PCA (e.g. Jolliffe, 1972, 1973; Krzanowski, 1987; Tanaka and Mori, 1997). Most methods intend to select variables which represent the whole variables as well as possible in some sense. Here we show one of the ideas of variable selection in PCA. The criteria in Modified PCA (M.PCA) proposed by Tanaka and Mori (1997) are used as the selection criteria. M.PCA derives principal component (PC) scores which are computed using only a selected subset of variables but which represent all the variables including those not selected.

Suppose we obtain an $n \times p$ data matrix Y. Let Y be divided into an $n \times q$ submatrix Y_1 and an $n \times (p-q)$ submatrix Y_2 ($r \leq q \leq p$), and denote the covariance matrix of Y by S, that of Y_1 by S_{11} and that of Y_1 and Y_2 by S_{12} ($S_{21} = S_{12}'$). In M.PCA, Y is represented by r PCs as well as possible, where r is the number of PCs and the PCs are linear combinations of a submatrix Y_1, i.e. Y_1A. Then we can derive its maximization criteria to find A, using the idea of Rao (1964) and the approach of Robert and Escoufier (1976). These criteria are the proportion P and the RV-coefficient computed by

$$P = \sum_{j=1}^{r} \lambda_j / tr(S) \quad \text{and} \quad RV = \left\{ \sum_{j=1}^{r} \lambda_j^2 \Big/ tr(S^2) \right\}^{1/2},$$

respectively, where λ_j is the j-th eigenvalue of $[(S_{11}^2 + S_{12}S_{21}) - \lambda_j S_{11}]\boldsymbol{a}_j = 0$ in order of magnitude. They can be used to select the best subset Y_1 which provides the largest values of them among all subsets of size q.

As for PCA, we have already developed the system VASPCA (VAriable Selection in PCA, Mori et al., 2000), in which variable selection in PCA is easily performed.

2.2 Variable selection in FA

Variable selection in FA has been also studied by some authors, e.g. Tanaka and Kodake (1981), Tanaka (1983) and Kano and Harada (2000). Here we use the idea to find a subset of variables which has the optimum value of the closeness between the configurations of factor scores based on original variables and those based on selected variables. The closeness can be measured by D, Q (Tanaka. 1983) and the RV-coefficient (Robert and Escoufier, 1976).

Here we show these criteria briefly. Suppose an ordinary factor analysis model

$$Y = FL' + E,$$

where L is a $p \times r$ factor loading matrix, F is an $n \times r$ common factor score matrix and E is an $n \times p$ unique factor score matrix. Using an iterative procedure (principal factor analysis or maximum likelihood method), we can estimate L and a matrix of uniqueness D^2 and obtain F based on them. When we assign Y_1 to the factor analysis model, a factor score matrix based only on Y_1, denoted by $F^{(q)}$, can be estimated in the same way. Then we can derive the following criterion values to measure the closeness between F and $F^{(q)}$:

(1) Norm of the difference between F and $F^{(q)}T$

$$D\left(F, F^{(q)}T\right) = \frac{\min_T \left\| F - F^{(q)}T \right\|}{\|F\|},$$

(2) Norm of the difference between FF' and $F^{(q)}F^{(q)\prime}$

$$Q\left(FF', F^{(q)}F^{(q)\prime}\right) = \frac{\left\| FF' - F^{(q)}F^{(q)\prime} \right\|}{\|FF'\|},$$

(3) RV-coefficient between F and $F^{(q)}$

$$RV\left(F, F^{(q)}\right) = \frac{\left\| F'F^{(q)} \right\|^2}{\|F'F\| \cdot \left\| F^{(q)\prime}F^{(q)} \right\|},$$

where T is an appropriate orthogonal matrix. In selection step we compute one of these criteria and find a subset of variables which provides the optimum value of the criterion.

3 Statistical system VASMM

This system is constructed on the webs: http://mo161.soci.ous.ac.jp/VASMM/ and http://face.f7.ems.okayama-u.ac.jp/~masa/VASMM/. The user can perform variable selection with a web browser and can also obtain information of variable selection in any MM without external variables.

3.1 Implemented modules

The current version of VASMM implements two modules, VASpca (VAriable Selection in PCA) and VASfa (VAriable Selection in FA). Modules for other MMs are partially implemented. VASpca is a modified version of VASPCA. VASpca can perform variable selection in PCA using one of seven criteria based on the following approaches:

(P1) M.PCA (see 2.1) (P2) principal variables (P3) procrustes analysis
(P4) RV-coefficient (P5) influence analysis of variables
(P6) the idea of prediction error (P7) selection focusing loadings.

(see VASMM site for the details of these seven criteria.) We have prepared background information about variable selection in MMs without external variables for users who do not know variable selection very well and/or want to understand the underlying theories.

On the other hand, VASfa can select a subset of variables using

(F1) criterion to measure the closeness of configurations (see 2.2)

(F2) fitting function to evaluate the fitness between selected variables and original ones in the sense of model fitting.

Other new criteria and selection methods in other MMs will be built in when required.

3.2 Selection procedures

Using a particular criterion we can find the best subset which has the optimum value of the criterion among all possible subsets of size q. However, it is usually impractical due to high computational cost to compute criterion values for all possible subsets. As practical methods we implemented the following stepwise selection procedures:

Backward elimination	Backward-forward stepwise selection
Forward selection	Forward-backward stepwise selection

in which only one variable is removed or added sequentially (Mori et al., 1998). These procedures select a subset of variables for every q automatically.

3.3 Flow of VASMM

Fig 1 indicates the flow of VASMM. VASMM execution page consists of the following steps.

1) Specifying the MM of interest: The user can select an MM among PCA, FA, canonical correlation analysis and correspondence analysis.

[Initial step]

2) Data entry: VASMM reads a data file. Data file should be created with separation by comma, space or tab.
3) Determination of parameters in initial analysis: After reading a data file, VASMM provides some information (e.g. eigenvalue and its graph), then the user selects parameters such as the number of PCs or factors.

[Selection step]

4) Specifying a selection technique: The user specifies a criterion and a selection procedure.
5) Stepwise selection: Automatic and stepwise selection is performed to find a reasonable subset of variables based on 3) and 4).
6) Output of the results: The results of variable selection are displayed. The output consists of criterion values, their graph, selected variables and a guide to next selection.

After selection, the user can go back to 4) and can carry out variable selection again with other selection parameters.

4 An example

By way of illustrating the program, we use a MDOC (mild disturbance of consciousness) dataset (Sano et al., 1977) consisting of 85 observations and 25 variables. Here we show the execution of VASfa on VASMM using criterion to measure the closeness of configurations.

At first, the user accesses the VASMM site via Internet, then the top page of VASMM appears (Fig 2). Next the user selects the link to the VASfa. At the VASfa site, there are several links to description pages which explain criteria and to the execution page for selection of variables. To perform selection, the user selects one criterion in the execution page. Here suppose the user selects a criterion based on measurement of the closeness of configurations. The user can click the link to its description page to know the detail of the criterion.

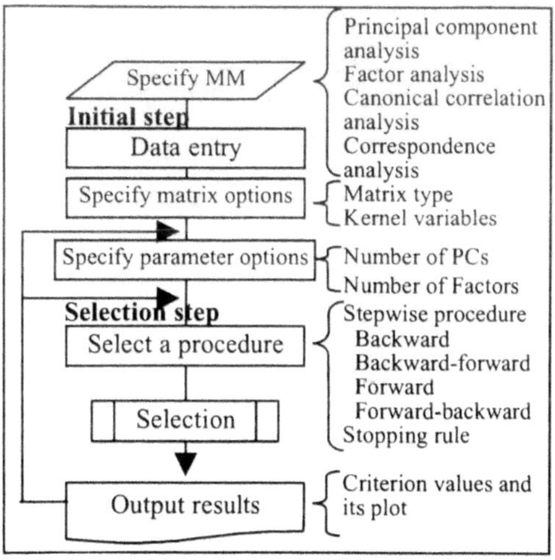

Fig 1. Flow of VASMM

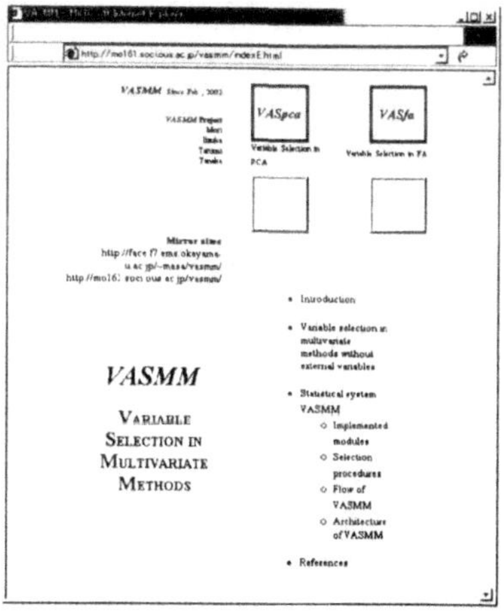

Fig 2. Top page of VASMM

Now the user proceeds to the initial step. In data entry step, the user open a data file which consists of comma, space or tab-separated values. If the user wants to demonstrate variable selection, a sample dataset is available in this page. Click the [execute] button, next page appears (Fig 3) to check whether the dataset is one which the user wants to analyze and whether the data format is correct. If the transmitted data is correct, the user specifies the number of factors looking at eigenvalues and their scree plot displayed on the next page.

Next the user proceeds to the selection step and then possible parameters are displayed. Here let specify parameters as follows: the number of factors is "two", the method of estimating the loading matrix L and uniqueness D^2 is "maximum likelihood estimate method", the method of estimating factor scores is "regression" and selection procedure is "backward-forward stepwise selection". Clicking the [calculate] button, the user obtains the results of variable selection (Fig 4) after computation. The results show a reasonable subset for every q. The graph of criterion values is also displayed, which is useful to observe the change of the values visually and helps the user to find the reasonable number of variables. The results in Fig 4 show that the RV-coefficient changes very slightly until the number of variables is 10 and it drops suddenly when the number of variables is less than 7. It can be stated that the selected 10 variables can provide close configuration of factor scores to original one. In such a way the user examines the results and/or repeats selection with another method, criterion or procedure if necessary.

5 Conclusion and future problems

We are developing the VASMM for variable selection in MMs without external variables. In particular, we showed a numerical example of variable selection in FA in this paper. Using the VASMM users can select a subset of variables easily inpractical situations. As for future works, we intend to add new variable selection methods for other MMs, and have plans to provide statistical functions written in other general statistical languages (e.g. XploRe) for users to perform variable

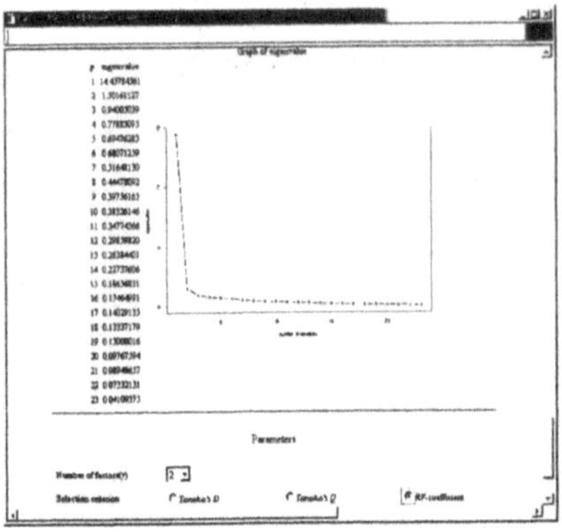

Fig 3. Prior information in VASfa

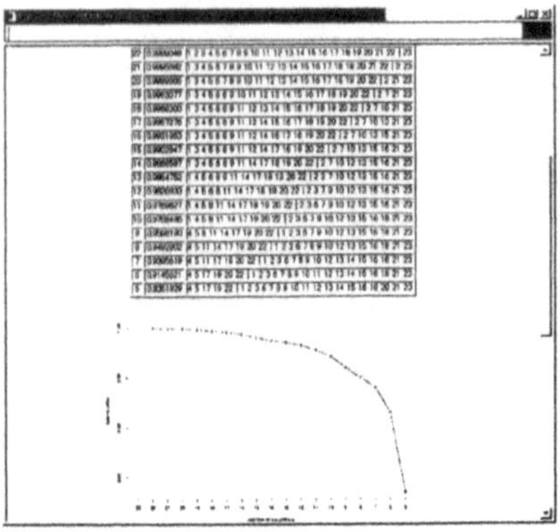

Fig 4. Outputted the result in VASfa

selection in various environments.

References

Jolliffe, I. T. (1972, 1973). Discarding variables in a principal component analysis. I, II. Appl. Statist., 21 (160-173), 22 (21-31).

Kano,Y and Harada,A.(2000). Stepwise variable selection in factor analysis, Psychometrika, 65(1), 7-22, URLs: SEFA http://koko15.hus.osaka-u.ac.jp /~harada/factor stepwise/, http://koko16.hus.osaka- u.ac.jp/~harada/scofa/input.html

Krzanowski, W. J. (1987). Selection of variables to preserve multivariate data structure, using principal components. Appl. Statist., 36, 22-33.

Mori, Y., Iizuka, M. Tarumi, T. and Tanaka, Y. (2000). Statistical Software "VASPCA" for Variable Selection in Principal Component Analysis, In: COMPSTAT2000 Proceedings in Computational Statistics (Short Communications) (Edited by Jansen, W. and Bethlehem, J.G.), 73-74.

Mori, Y., Tarumi, T and Tanaka, Y. (1998). Principal Component analysis based on a subset of variables -Numerical investigation on variable selection procedures -. Bulletin of the Computational Statistics of Japan, 11(1), 1-12. (in Japanese)

Rao, C. R. (1964). The use and interpretation of principal component analysis in applied research, Sankhya Ser. A, 26, 329-358.

Robert, P. and Escoufier, Y. (1976). A unifying tool for linear multivariate statistical methods: the RV-coefficient. Appl. Statist., 25, 257-265.

Sano, K., Manaka, S., Kitamura, K., Kagawa, M., Takeuchi, K., Ogashiwa, M., Kameyama, M., Tohgi, H. and Yamada, H. (1977). Statistical studies on evaluation of mind disturbance of consciousness -Abstraction of characteristic clinical pictures by cross-sectional investigation. Sinkei Kenkyu no Shinpo, 21, 1052-1065. (in Japanese)

Tanaka, Y (1983). Some criteria for variable selection in factor analysis, Behaviormetrika, 13, 31-45

Tanaka, Y. and Kodake, K. (1981). A method of variable selection in factor analysis and its numerical investigation. Behaviormetrika, 10, 49-61.

Tanaka, Y and Mori, Y. (1997): Principal component analysis based on a subset of variables: Variable selection and sensitivity analysis. American Journal of Mathematics and Management Sciences, 17, 1&2, 61-89.

Structural Equation Models for Finite Mixtures – Simulation Results and Empirical Applications

Dirk Temme[1], John Williams[2] and Lutz Hildebrandt[3]

[1] Institute of Marketing, Humboldt University of Berlin, 10178 Berlin, Germany
[2] Department of Marketing, Otago University, Dunedin, New Zealand
[3] Institute of Marketing, Humboldt University of Berlin, 10178 Berlin, Germany

Abstract. Unobserved heterogeneity is a serious, often neglected problem in structural equation modeling (SEM). Recently, a finite mixture approach to SEM has been proposed to solve this problem but until now only a few studies analyse the performance of this approach. The contribution of this paper is twofold: First, results from a Monte Carlo study into the properties of the software program MECOSA are presented. Second, an empirical application to data from a large-scale consumer survey in the fast moving consumer goods industry is described.

Keywords. Structural equation modeling, Unobserved heterogeneity, Model-based clustering, Finite mixtures, Monte Carlo simulation

1 Introduction

Structural equation modeling (SEM) is an established methodology in the social sciences. One of its key benefits is that it enables the estimation of relationships between unobservable theoretical constructs (e.g., attitudes) which are operationalised by multiple observed, albeit individually imperfect measures.

Empirical applications of SEM typically rest on the assumption that either the analysed sample is homogenous with respect to the underlying model or that heterogeneity can be adequately taken into account by forming sub-samples using one or two observed criteria (e.g., gender, age, brand loyalty). If unobserved heterogeneity is (still) substantial, parameter estimates might be seriously biased in this case (Jedidi, Jagpal and DeSarbo, 1997).

Meanwhile two alternative solutions to the problem of unobserved heterogeneity in SEM exist. Whereas the finite mixture approach presumes that heterogeneity in the population can be sufficiently captured by a limited number of homogenous segments, hierarchical Bayes methods (Ansari, Jedidi and Jagpal, 2000) rest upon the idea that individual-level parameters follow a continuous heterogeneity distribution. In this paper, we exclusively focus on finite mixture SEM. Since only a few studies exist which analyse the performance of this methodology we first describe the design and main results of a Monte Carlo study into the properties of the MECOSA approach to conditional finite mixture SEM. Second, the results of an empirical application of MECOSA to data from a large-scale consumer survey on attitudes towards a specific brand in the German fast moving consumer goods industry are reported.

2 Finite Mixture Structural Equation Models

2.1 Conditional and Unconditional Models

The finite mixture approach assumes that the sample is a discrete mixture of a limited, but generally unknown number of components each characterised by a specific distribution. Unconditional models (Yung, 1997; Jedidi, Jagpal and DeSarbo, 1997; Dolan and van der Maas, 1998) rely on the assumption that the variables are multivariate normal within the different components. For conditional models the more realistic assumption applies that the dependent variables are normally distributed given some exogenous regressors (e.g., demographics). To the best of our knowledge only two commercial software programs for the estimation of conditional finite mixture SEM are presently available (MECOSA, Arminger, Wittenberg and Schepers, 1996; Mplus, Muthén and Muthén, 1998).

2.2 The MECOSA Approach to Conditional Finite Mixture SEM

The following description of the MECOSA model is mainly based on Arminger, Stein and Wittenberg (1999; abbreviated as ASW below). Let $\mathbf{y}_i, i = 1,2,\ldots,n,$ be a $p \times 1$ vector of continuous dependent random variables and $\mathbf{x}_i$ a $q \times 1$ vector of continuous/dummy-type independent variables. The sample points are i.i.d. with density $h(\mathbf{y}_i, \mathbf{x}_i) = f(\mathbf{y}_i \mid \mathbf{x}_i) \cdot g(\mathbf{x}_i)$, where $g(\mathbf{x}_i)$ is the marginal density of the exogenous variables. Assuming multivariate normal dependent variables $\mathbf{y}_i$ in each component conditional on $\mathbf{x}_i$, the conditional density is given by the mixture

$$f(\mathbf{y}_i \mid \mathbf{x}_i) = \pi_1 \phi(\mathbf{y}_i; \mu_{i1}; \Sigma_1) + \pi_2 \phi(\mathbf{y}_i; \mu_{i2}; \Sigma_2) + \cdots + \pi_K \phi(\mathbf{y}_i; \mu_{iK}; \Sigma_K),$$

where $\pi_k, k = 1,2,\ldots,K,$ are the mixing proportions of the K mixture components ($\pi_k \geq 0$, $\sum_{k=1}^{K} \pi_k = 1$) and ϕ denotes the normal density with mean vector μ_{ik} and covariance matrix Σ_k. The conditional mean $E(\mathbf{y}_i \mid \mathbf{x}_i, k)$ is specified as a reduced form multivariate linear model

$$\mu_{ik} = \gamma_k + \Pi_k \mathbf{x}_i,$$

where γ_k is a $p \times 1$ vector of regression constants and Π_k is a $p \times q$ matrix of regression coefficients. The conditional covariance matrix Σ_k contains the variances and covariances of the regression residuals. The conditional means and covariances are parameterised by component-specific mean- and covariance structure models, for example a conditional LISREL model.

MECOSA currently offers three estimation methods: Minimum Distance Estimation (MDE), Direct EM (EM) and EM Gradient (EMG). The MDE method is a two-step procedure. In the first stage, an unrestricted multivariate regression model for finite mixtures is estimated by an EM algorithm. In the second stage, a minimum distance estimator is used to estimate the group-specific fundamental parameters based on the reduced form parameter estimates $\hat{\gamma}_k, \hat{\Pi}_k$ and $\hat{\Sigma}_k$ from step one.

In contrast, maximisation in both the EM and the EMG method is directly performed with respect to the fundamental parameters.

3 Simulation Study

3.1 Experimental Design

The experimental design of our simulation study into the properties of the MECOSA approach (for a more comprehensive description see Williams, 2002; Williams, Temme and Hildebrandt, 2002) extends the experimental factors 1. *Distribution of variables* (skewed or normal) and 2. *Estimation method* (MDE, EM and EMG) used in the study by ASW by four additional factors: 3. *Number of groups* (two or three), 4. *Group proportions* (equal or mixed), 5. *Knowledge about number of groups* (known or unknown) and 6. *Differences between the parameters in each group* (close or far). In total this design leads to 96 different conditions. For each condition 500 valid data sets with 2000 observations each have been simulated with GAUSS according to a parameterised structural equation model and subsequently analysed using MECOSA.

3.2 Main Results

For all conditions with an unknown number of groups we used the modified likelihood ratio test (ad hoc test) implemented in MECOSA to estimate the number of groups. Whereas for two groups the number of components was almost always estimated correctly, for those three group conditions with close parameters and mixed proportions more than 1100 data sets had to be simulated to achieve 500 replications with the number of groups estimated correctly (this is a prerequisite for meaningful comparisons between the results for different conditions).

The performance of MECOSA for the two group conditions in terms of estimator variance and bias is shown in Table 3.1. Both the highest mean MAD (mean absolute deviation) and the highest mean bias result for the MDE under conditions of mixed proportions and close parameters.

Table 3.1. Mean MAD and Mean Bias for the two group conditions

		Mean MAD				Mean Bias			
		Known		Unknown		Known		Unknown	
		Equal	Mixed	Equal	Mixed	Equal	Mixed	Equal	Mixed
		Close Parameters							
Skewed	MDE	.0339	.0443	.0338	.0452	-.0049	-.0116	-.0049	-.0127
	EM	.0316	.0409	.0319	.0400	-.0010	-.0018	-.0014	-.0012
	EMG	.0317	.0399	.0317	.0392	-.0021	-.0018	-.0013	-.0020
Normal	MDE	.0337	.0456	.0340	.0451	-.0047	-.0134	-.0061	-.0118
	EM	.0320	.0400	.0322	.0396	-.0015	-.0020	-.0002	-.0016
	EMG	.0323	.0398	.0319	.0402	-.0018	-.0019	-.0029	-.0033
		Far Parameters							
Skewed	MDE	.0230	.0283	.0227	.0282	-.0016	-.0033	-.0022	-.0028
	EM	.0221	.0272	.0226	.0268	-.0003	-.0004	-.0002	-.0007
	EMG	.0223	.0270	.0224	.0272	-.0004	-.0006	-.0007	-.0006
Normal	MDE	.0222	.0269	.0223	.0271	-.0018	-.0026	-.0019	-.0024
	EM	.0216	.0265	.0216	.0259	-.0007	-.0010	-.0004	-.0004
	EMG	.0215	.0265	.0217	.0257	-.0002	-.0005	-.0005	-.0006

Further analyses of individual parameter recovery using balanced ANOVA show that the largest effect on parameter recovery is due to the differences in parameter values between groups. In addition, the negative effect of mixed proportions on parameter recovery is quite pronounced, especially in conjunction with close parameters. Under most conditions the choice of the estimation method seems to have only a negligible effect but for mixed proportions combined with close parameters MDE performs considerably worse than Direct EM and EM Gradient.

4 Empirical Application

For marketing practice measuring customer based brand equity is of great importance in order to assess the effect of long-term investments in a brand. Marketing research companies and academics alike have proposed various approaches to measure brand equity. In this study we focus on the Brand Potential Index® (BPI) developed by the German company GfK, Nuremberg. The BPI consists of 9 items which form a unidimensional confirmatory factor model. The following items are measured on a 7-point scale with the endpoints (7) "I totally agree" and (1) "I totally disagree" (items 2 – 9 are all formulated in comparison to other brands): 1. will buy the brand in the future (*buy*), 2. will pay more for the brand (*pay*), 3. identify with the brand (*identify*), 4. trust the brand (*trust*), 5. will recommend the brand (*recom*), 6. brand positively differentiates itself (*positive*), 7. like the brand (*like*) 8. regret if brand is not available (*regret*) and 9. brand is of higher quality (*quality*). In a consumer survey 1048 subjects rated several competing brands in a specific convenience product category on the 9 BPI items. In addition, socio-demographic (e.g., age, household size) and behavioural (brand loyalty) information has been gathered, which are used as regressors. After eliminating several outliers with inconsistent answers a sample size of 1037 remained.

The data has been analysed using first the MDE procedure. Based on the finite mixture multivariate regression of the BPI items on the exogenous variables the ad hoc test ($LR_{ad\ hoc}$) as well as the parametric bootstrap test (LR_{Boot}) point to a three group solution although the BIC indicates a two group segmentation (see Table 4.1.) Since the entropy measure (Ramaswamy, DeSarbo, Reibstein and Robinson, 1993) for the three group estimation is .791, which indicates a good separation of the groups, we decided in favour of three groups.

Table 4.1. Summary statistics for model selection results

K	LL	$LR_{ad\ hoc}$	df	p	LR_{Boot}[1]	BIC
1	-13655.64	-	-	-	-	28248.73
2	-12932.54	1446.20	270	.000	331.15	27746.93
3	-12620.46	624.16	405	.000	450.29	28067.17
4	-12407.62	425.68	540	.999	450.47	28585.88

[1]Greatest Boot-strap LR value

With respect to the unconditional means of the BPI items (not shown here for space reasons) a clear order of the three groups emerges. Group 1 (N = 181) has the highest values (except for the item *buy* where the mean in group 2 is higher), followed closely by group 2 (N = 531). Group 3 (N = 325) clearly differentiates itself from the two other groups since all item means are considerably lower. Overall, mean differences for the regressors are only minor. The most pronounced difference concerns the variable *loyalty*. Whereas in the first group about 71% of the subjects report that most of the time they buy the focal brand for the two other groups this percentage is considerably lower (63% for group 2 and only 53% for group 3). Belonging to the western or eastern parts of Germany also makes a difference: In group 3 23% of the subjects live in East Germany whereas in group 1 this holds for only 17%. Loyalty has by far the strongest effect on the BPI items (positive and highly significant in all three groups). In group 2 respondents from those German federal states where the analysed food category is especially popular tend to react more positively to the items *recom* and *like* than inhabitants of other states. Only in group 3 *age* positively influences the items *positive* and *regret*.

In the next stage, we used the three available estimation methods to estimate the fundamental parameters of a conditional common regression model (Yung, 1997):

$$\mu_{ik} = \nu_k + \Lambda\alpha_k + \Lambda\Gamma_k \mathbf{x}_i,$$

$$\Sigma_k = \Lambda\Psi_k\Lambda' + \Theta_k.$$

Using MDE yields a Chi^2 statistic of 606.76 with 325 d.f. Although the model is rejected by this test, the RMSEA = .042 is below the cut-off value of .05, indicating a good fit. Because of space limitations we only report results for parameters with significant differences between the three estimation methods (see Table 4.2.).

Table 4.2. Parameter estimates for the conditional common regression model

	MDE			EM			EMG		
	Gr. 1	Gr. 2	Gr. 3	Gr. 1	Gr. 2	Gr. 3	Gr. 1	Gr. 2	Gr. 3
ν_1	**2.21**	**2.67**	**2.87**	**2.49**	**2.98**	**3.18**	**2.49**	**2.98**	**3.17**
ν_2	**1.92**	.93	1.20	**2.45**	**1.62**	**1.42**	**2.45**	**1.61**	**1.43**
ν_3		**1.64**			**2.15**			**2.15**	
ν_4		**2.14**			**2.42**			**2.42**	
ν_5		**2.15**			**2.51**			**2.51**	
ν_6		**2.23**			**2.64**			**2.64**	
ν_7		**1.90**			**2.42**			**2.42**	
ν_8		**1.85**			**2.33**			**2.33**	
ν_9		**2.41**			**2.73**			**2.74**	
α	.00[a]	**2.50**	.76	.00[a]	**1.79**	.41	.00[a]	**1.81**	.40
γ_1	.23	-.10	.09	.22	-.12	**.27**	.23	-.13	**.27**
γ_2	-.12	-.06	.08	**-.33**	-.03	.05	**-.34**	-.02	.05
γ_3	.25	.09	-.25	**.44**	.12	-.23	**.43**	.11	-.22
γ_4	**.09**	-.01	.02	.07	-.00	.01	.07	-.00	.01
γ_5	.16	.03	.09	**.45**	-.01	.15	**.47**	-.03	.16
γ_6	.14	-.03	.21	-.37	-.06	.09	-.37	-.05	.09
γ_7	**-1.20**	.09	.17	-.49	-.08	.19	-.52	-.07	.18
γ_8	-.21	**.23**	.21	.18	**.30**	.13	.18	**.29**	.15
γ_9	**2.68**	**.90**	**.70**	**2.87**	**1.08**	**.91**	**2.86**	**1.08**	**.92**

[a]fixed; significant parameters ($\alpha = .05$) are set in bold type

Whereas the direct EM and the EMG methods yielded almost identical estimates, the MDE estimations in part considerably deviate from those of the EM methods, leading to substantially different conclusions. For example, whereas *age* (γ_1) has a significant positive effect on BPI in group 3 when estimated using the EM methods (see also the results for the first step) this parameter is insignificant in the MDE solution. Also the MDE estimate for the dummy variably *West/East Germany* (γ_7) seems unreasonably high. Overall, in line with the simulation results, the EM estimates for this three groups/mixed proportions case seem to be more trustworthy than the MDE estimates.

5 Conclusion

Since both, the simulation results and the empirical application, have shown that in situations which typically occur in empirical studies (mixed proportions and/or close parameters) the MDE method performs worse than the EM algorithms, we suggest to use only the first stage of the MDE procedure and to estimate the fundamental parameters by using the EM Gradient method.

References

Ansari, A., Jedidi, K. & Jagpal, S. (2000). A hierarchical Bayesian methodology for treating heterogeneity in structural equation models. *Marketing Science*, **19 (4)**, 328-347.

Arminger, G., Stein, P. & Wittenberg, J. (1999). Mixtures of conditional mean- and covariance structure models. *Psychometrika*, **64**, 475-494.

Arminger, G., Wittenberg, J. & Schepers, A. (1996). *MECOSA 3 User Guide*. Friedrichsdorf, Germany: Additive.

Dolan, C.V. & van der Maas, H.L.J. (1998). Fitting multivariate normal finite mixtures subject to structural equation modeling. *Psychometrika*, **63**, 227-253.

Jedidi, K., Jagpal, H. & DeSarbo, W. (1997). Finite-mixture structural equation models for response-based segmentation and unobserved heterogeneity. *Marketing Science*, **16 (1)**, 39-59.

Muthén, L.K. & Muthén, B.O. (1998). *Mplus User's Guide*. Los Angeles, CA: Muthén & Muthén.

Ramaswamy, V., DeSarbo, W.S., Reibstein, D.J. & Robinson, W. T. (1993). An empirical pooling approach for estimating marketing mix elasticities with PIMS data. *Marketing Science*, **12 (1)**, 103-124.

Williams, J. (2002). *Mean and covariance structure models for finite mixtures*. Unpublished Ph.D. thesis, Otago University, Department of Marketing, Dunedin, New Zealand.

Williams, J, Temme, D. & Hildebrandt, L. (2002). A Monte Carlo study of structural equation models for finite mixtures. *Discussion Paper SFB 373*. Berlin: Humboldt Universität zu Berlin.

Yung, Y.F. (1997). Finite mixtures in confirmatory factor-analysis models. *Psychometrika*, **62**, 297-330.

Sweave: Dynamic Generation of Statistical Reports Using Literate Data Analysis

Friedrich Leisch

Institut für Statistik und Wahrscheinlichkeitstheorie, Technische Universität Wien, Wiedner Hauptstraße 8-10/1071, A-1040 Wien, Austria

Abstract. Sweave combines typesetting with LaTeX and data anlysis with S into integrated statistical documents. When run through R or Splus, all data analysis output (tables, graphs, ...) is created on the fly and inserted into a final LaTeX document. Options control which parts of the original S code are shown to or hidden from the reader, respectively. Many S users are also LaTeX users, hence no new software has to be learned. The report can be automatically updated if data or analysis change, which allows for truly reproducible research.

Keywords. R, S, literate statistical practice, integrated statistical documents, reproducible research

1 Introduction

The traditional way of writing a report as part of a statistical data analysis project uses two separate steps: First, the data are analyzed using one's favorite statistical software package, and afterwards the results of the analysis (numbers, graphs, ...) are used as the basis for a written report. In larger projects the two steps may be repeated alternately, but the basic procedure remains the same. Many statistical software packages try to support this process by generating pre-formatted tables and graphics that can easily be integrated into a final report using copy-and-paste from the data analysis system to the word processor. The basic paradigm is to write the report around the results of the analysis.

Another approach for integration of data analysis and document writing is to embed the analysis itself into the document, which reverses the traditional paradigm. Over the last decade a number of systems have been developed that integrate analysis and documentation and allow for *literate statistical practice*, see Rossini (2001) for a survey.

This new paradigm is probably most popular for creation of dynamic web pages and offers completely new possibilities for teaching statistics and delivering statistical methodology over the Internet. E.g., the ExploRe system (Härdle et al., 1999) provides means to embed statistical quantlets in web pages or electronic books to create interactive documents with direct access to a statistical data analysis package. Another example for a dynamic statistical analysis on a web page is given in Temple Lang (2001), by embedding R into

```
\documentclass[a4paper]{article}

\begin{document}

In this example we embed parts of the examples from the
\texttt{kruskal.test} help page into a \LaTeX{} document:

<<>>=
data(airquality)
kruskal.test(Ozone ~ Month, data = airquality)
@
which shows that the location parameter of the Ozone
distribution varies significantly from month to month.
Finally we include a boxplot of the data:

\begin{center}
<<fig=TRUE,echo=FALSE>>=
boxplot(Ozone ~ Month, data = airquality)
@
\end{center}

\end{document}
```

Fig. 1. A minimal Sweave file: `example-1.Snw`.

netscape as a plugin. Report rendering is performed using XML and XSL.

We introduce a new system, called Sweave, which combines ideas from both worlds described above using literate programming tools. The purpose is to create dynamic reports, which can be updated automatically if data or analysis change, while using standard tools for both data analysis and word processing. Sweave is written in the S language, either the open source R (`http://www.R-project.org`) or the commercial Splus (`http://www.insightful.com`) can be used for statistical data analysis.

2 Sweave files

Sweave source files are regular noweb files (Ramsey, 1998) with some additional syntax that allows control over the final output. Noweb is a simple literate-programming tool which allows to combine program source code and the corresponding documentation into a single file. These consist of a sequence of code and documentation segments, called *chunks*. Different command line programs are used to extract the code (*"tangle"*) or typeset documentation together with the code (*"weave"*).

A minimal Sweave file is shown in Figure 1, which contains two code chunks embedded in a simple LaTeX document. '`<<...>>=`' at the beginning of a line marks the start of a code chunk, while a '`@`' at the beginning of a line marks the start of a documentation chunk. Sweave translates this into the LaTeX document shown in Figures 2 and 3. The first difference between `example-1.Snw` and `example-1.tex` is that the LaTeX style file

```
\documentclass[a4paper]{article}

\usepackage{Sweave}
\begin{document}

In this example we embed parts of the examples from the
\texttt{kruskal.test} help page into a \LaTeX{} document:

\begin{Sinput}
R> data(airquality)
R> kruskal.test(Ozone ~ Month, data = airquality)
\end{Sinput}
\begin{Soutput}
        Kruskal-Wallis rank sum test

data: Ozone by Month
Kruskal-Wallis chi-squared = 29.2666, df = 4, p-value = 6.901e-06
\end{Soutput}
which shows that the location parameter of the Ozone
distribution varies significantly from month to month.
Finally we include a boxplot of the data:

\begin{center}
\includegraphics{example-1-002}
\end{center}

\end{document}
```

Fig. 2. The output of `Sweave("example-1.Snw")` is the file `example-1.tex`.

`Sweave.sty` is automatically loaded, which provides environments for typesetting S input and output (the LaTeX environments `Sinput` and `Soutput`). Otherwise, the documentation chunks are copied without any modification from `example-1.Snw` to `example-1.tex`.

The real work of Sweave is done on the code chunks: The first code chunk has no name, hence the default behavior of Sweave is used, which transfers both the S commands and their respective output to the LaTeX file, embedded in `Sinput` and `Soutput` environments, respectively. This emulates an R terminal session.

The second code chunk shows one of the Sweave extension to the noweb syntax: Code chunk names (the text within the double angle brackets) can be used to pass options to Sweave which control the final output: The chunk is marked as a figure chunk (`fig=TRUE`) such that Sweave creates EPS and PDF files corresponding to the plot created by the commands in the chunk. Furthermore, a `\includegraphics{example-1-002}` statement is inserted into the LaTeX file ('002' is the default for figures from the second code chunk). Option `echo=FALSE` indicates that the S input should not be included in the final document (no `Sinput` environment).

In this example we embed parts of the examples from the `kruskal.test` help page into a LaTeX document:

```
R> data(airquality)
R> kruskal.test(Ozone ~ Month, data = airquality)

        Kruskal-Wallis rank sum test

data:  Ozone by Month
Kruskal-Wallis chi-squared = 29.2666, df = 4, p-value = 6.901e-06
```

which shows that the location parameter of the Ozone distribution varies significantly from month to month. Finally we include a boxplot of the data:

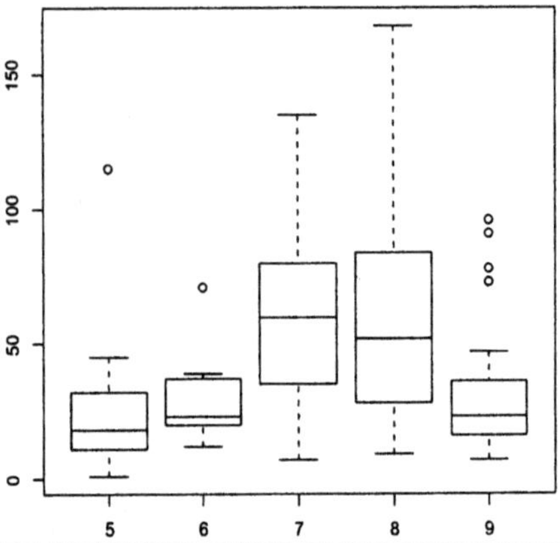

Fig. 3. The final document is created by running `latex` on `example-1.tex`.

3 Design principles and features

"One of the hardest tasks known is to convince a competent computer user to switch from a preferred computing environment to a new one" (Rossini, 2001). Hence, one of the major design principles of Sweave was to provide a system for literate statistical practice which S users will find familiar and convenient. This naturally leads to embedding the statistical analysis in LaTeX documents, because many S users already use LaTeX for typesetting.

The emacs text editor (Stallman, 1999) offers a perfect authoring environment for Sweave, especially for people who already use emacs for writing LaTeX documents and interacting with an S engine. We have chosen to use noweb as basis for the Sweave system because

1. the syntax is extremely simple and hence easy to learn
2. the ESS noweb mode for emacs already provides a perfect authoring environment

The importance of 2 should not be underestimated, a document format without convenient tools for authors will almost certainly be ignored by prospective users. However, it is not necessary to use emacs, Sweave is a standalone system, the noweb source files for Sweave can be written using any text editor.

ESS (*Emacs speaks statistics*, Rossini et al., 2001) allows to connect an Sweave file to a running R or Splus process while writing the document. Code chunks can be sent to the S engine and evaluated using simple keyboard shortcuts or popup menus. Syntax highlighting, automatic indentation and keyboard shortcuts depend on the location of the pointer: in documentation chunks emacs behaves as if editing a standard LaTeX file, when the pointer moves to a code chunk the mode switches automatically to S programming.

Sweave is not the first attempt to combine S and LaTeX using noweb: the RevWeb system (Lang and Wolf, 1997) uses the same principle. The major difference is that Sweave has been implemented in the S language itself, hence it works on all platforms where S is available (Linux, Mac, Unix, Windows). RevWeb uses system calls to the noweb command line utilities and therefore has not the same portability.

Another advantage of using S itself for all tangling and weaving steps is much finer control over the S output. Options that can be set either globally to modify the default behavior or separately for each code chunk control how the the output of the code chunks is inserted into the LaTeX file, e.g.:

- Hide or show S input code lines; include text output verbatim, as LaTeX code or hide it completely.
- Emulate an interactive R session: values of assignments are not printed, values of single objects are printed.
- Suppress evaluation of single code chunks.
- Text output can be written to separate files for each code chunk, EPS and PDF files of graphics can be created, and corresponding `\input{}` or `\includegraphics{}` statements inserted.

In addition the user can specify hook functions which are automatically executed before code chunks of certain types, e.g., to set graphical parameters. A detailed description of all current features can be found in Leisch (2002).

4 Summary and outlook

Sweave is already used for a wide variety of applications: Reports for medical statistical consulting that can be updated automatically when new data arrive or data change; lecture notes for statistics classes with executable examples; and manuals with embedded examples for bioinformatics software, especially DNA microarray analysis. Sweave is part of every R installation (version 1.5.0 or higher) and can be used to automatically create and check manuals in R packages. All code can be executed to verify consistency of code and documentation and is part of the quality assurance tests of the software.

A disadvantage of TeX-based file formats is that they are harder to parse than other formats, hence computation on documentation objects is not as easy as, e.g., with XML-based formats. Hence for the next generation of the R help page format an XML-based solution may be preferable. An obvious disadvantage of XML is that these formats tend to be very verbose, i.e., the markup occupies a larger percentage of the file making it harder to read for humans (simply compare an HTML file to a LaTeX file, especially more complicated markup like formulas). Hence, the ideal authoring environment is not obvious. These issues are currently under investigation, see also `http://www.omegahat.org`.

Sweave files are easy to write and offer the full power of LaTeX for high-quality typesetting. This makes them ideal for documents where quality is essential and which are mainly read in printed form, e.g., manuals, lecture notes or consulting reports. Using PDF even hyperlinked online documents can be created easily from LaTeX. Storing code and documentation in a single source file makes research completely reproducible, all results can easily be verified and regenerated.

Acknowledgments

This research was supported by the Austrian Science Foundation (FWF) under grant SFB#010 ('Adaptive Information Systems and Modeling in Economics and Management Science'). The author thanks Kurt Hornik for comments and ideas.

References

W. Härdle, S. Klinke, and M. Müller. *XploRe Learning Guide*. Springer Verlag, 1999.

L. Lang and H. P. Wolf. The REVWEB manual for Splus in Windows. URL `http://wiwi.uni-bielefeld.de/StatCompSci/software`. Universität Bielefeld, Germany, 1997.

F. Leisch. Sweave user manual. URL `http://www.ci.tuwien.ac.at/~leisch/Sweave`. Institut für Statistik und Wahrscheinlichkeitstheorie, Technische Universität Wien, Vienna, Austria, 2002.

N. Ramsey. *Noweb man page*. University of Virginia, USA, 1998. URL `http://www.cs.virginia.edu/~nr/noweb`. version 2.9a.

A. Rossini. Literate statistical analysis. In K. Hornik and F. Leisch, editors, *Proceedings of the 2nd International Workshop on Distributed Statistical Computing, March 15-17, 2001, Technische Universität Wien, Vienna, Austria*, 2001. URL `http://www.ci.tuwien.ac.at/Conferences/DSC-2001/Proceedings/`. ISSN 1609-395X.

A. J. Rossini, M. Mächler, K. Hornik, R. M. Heiberger, and R. Sparapani. Emacs speaks statistics: A universal interface for statistical analysis. Report 164, Department of Biostatistics, University of Washington, 2001. URL `http://software.biostat.washington.edu/statsoft/ess/ess-techrep.pdf`.

R. M. Stallman. *The Emacs Editor*. Free Software Foundation, Boston, MA, USA, 1999. URL `http://www.gnu.org`. version 20.7.

D. Temple Lang. Embedding S in other languages and environments. In K. Hornik and F. Leisch, editors, *Proceedings of the 2nd International Workshop on Distributed Statistical Computing, March 15-17, 2001, Technische Universität Wien, Vienna, Austria*, 2001. URL `http://www.ci.tuwien.ac.at/Conferences/DSC-2001/Proceedings/`. ISSN 1609-395X.

Testing for Simplification in Spatial Models

L. Scaccia[1] and R. J. Martin[2]

[1] Dipartimento di Scienze Statistiche, Università di Perugia, 06100 Perugia, Italy
[2] Department of Probability and Statistics, University of Sheffield, S3 7RH Sheffield, UK

Abstract. Data collected on a rectangular lattice occur frequently in many areas such as field trials, geostatistics, remotely sensed data, and image analysis. Models for the spatial process often make simplifying assumptions, including axial symmetry and separability. We consider methods for testing these assumptions and compare tests based on sample covariances, tests based on the sample spectrum, and model-based tests.

Keywords. Autoregressive process, axial symmetry, doubly-geometric process, lattice process, separability, spatial process.

1 Introduction

Data collected on a regular two-dimensional lattice arise in many areas. Most models used for such data assume axial symmetry, but can still be difficult to use. Many of the problems of two-dimensional modelling can be overcome by using separable processes. Under this assumption, time series methods can be used to analyze and model lattice data (Martin, 1990).

However, no tests for separability and axial symmetry are known apart from Guo and Billard (1998), who suggested the Wald test to test separability. This model-based test compares the fit of an AR(1) AR(1) process and the more general Pickard process to the data, and actually tests for axial symmetry and separability together.

We propose and compare some different model-free and model-based tests for testing axial symmetry and separability. The paper is structured as follows. In section 2 we give some notation and definitions, illustrate the simulation method and present the different models used. The different tests for axial symmetry and separability are discussed in section 3. In section 4 we compare the performances of the tests.

2 Definitions, models and simulation method

2.1 Notation

We assume that data occur on an n_1 by n_2 rectangular lattice, with rows indexed by $i_1 = 1, \dots, n_1$ and columns by $i_2 = 1, \dots, n_2$. Row and column lags are g_1 and g_2, with $g_j = -(n_j - 1), \dots, 0, \dots, (n_j - 1)$, for $j = 1, 2$. The sites are ordered lexicographically, so that (i_1, i_2) precedes $(i_1, i_2 + 1)$ for $i_2 < n_2$, and (i_1, n_2) precedes $(i_1 + 1, 1)$. Data can be considered as a realization of random variables Y_{i_1,i_2}. Let the vector Y contain the Y_{i_1,i_2} in site order and assume a constant mean. We assume $Y \sim \mathrm{N}(0, V\sigma^2)$ where $V = V(\alpha)$ is an n by n positive definite matrix depending on the q-vector of parameters α.

Assuming second-order stationarity, $C(g_1, g_2) = \text{Cov}\,(Y_{i_1,i_2}, Y_{i_1+g_1,i_2+g_2})$ is the covariance at lags g_1, g_2, with $C(0,0) = \sigma^2$. The spectrum is $f(\omega_1, \omega_2) = \sum_{g_1=-\infty}^{\infty} \sum_{g_2=-\infty}^{\infty} C(g_1, g_2) \cos(g_1\omega_1 + g_2\omega_2)/(2\pi)^2$ (Priestley, 1981, section 9.7). Note that $C(g_1, g_2) = C(-g_1, -g_2)$ and $f(\omega_1, \omega_2) = f(-\omega_1, -\omega_2)$ always.

2.2 Axial symmetry and separability

For an *axially* or *reflection symmetric* process $C(g_1, g_2) = C(g_1, -g_2), \forall g_1, g_2$ and, equivalently, $f(\omega_1, \omega_2) = f(\omega_1, -\omega_2), \forall \omega_1, \omega_2$. This means that the covariances and the spectrum are both symmetric about the axes.

For a process to be *separable*, $C(g_1, g_2) \propto C(g_1, 0) \cdot C(0, g_2), \forall\; g_1, g_2$ and $f(\omega_1, \omega_2) \propto f(\omega_1, 0) \cdot f(0, \omega_2), \forall\; \omega_1, \omega_2$ are also required. Therefore, the covariances and the spectrum are determined, up to a multiplicative constant, by the margins. Clearly separability implies axial symmetry.

It is often desirable in practice that a process should be axially symmetric, and that the covariances should have a simple form. Separable processes have both these properties, and are still flexible enough to provide a reasonable representation of many planar structures (Martin, 1990).

For a separable process, the V matrix can be expressed as $V = V_x \otimes V_z$, where V_x and V_z are two smaller matrices that arise from the two underlying one-dimensional processes, and this implies that its determinant and inverse (required for exact Gaussian maximum likelihood, generalised least-squares estimation and exact Gaussian simulation) are easily determinable, which is a major advantage of this subclass of processes (Martin, 1996).

2.3 Models simulated and simulation method

We compared the different tests proposed by simulation: separable (or axially symmetric) processes were used to simulate the null distribution of tests for separability (or axial symmetry), while non-separable (or non-symmetric) processes were used to simulate the distribution of the tests under the alternative hypothesis. We indicate tests for axial symmetry by S and tests for separability by R, using different superscripts for their basis: (c) sample covariances, (p) sample spectrum, and (m) model-based.

The separable process we used is the AR(1)·AR(1) process (Martin, 1979):

$$Y_{i_1,i_2} = \alpha_1 Y_{i_1-1,i_2} + \alpha_2 Y_{i_1,i_2-1} - \alpha_1\alpha_2 Y_{i_1-1,i_2-1} + \epsilon_{i_1,i_2}$$

where the ϵ_{i_1,i_2} are assumed to be independently distributed as $\text{N}(0, \sigma_\epsilon^2)$.

As an axially symmetric, non-separable model we used a particular case of a second-order conditional autoregressive process, the $\text{CAR}(2)_{SD}$ with symmetric diagonal term (Balram and Moura, 1993). It can be written as:

$$\begin{aligned} \text{E}(Y_{i_1,i_2}|\cdot) = {} & \beta_1(Y_{i_1-1,i_2} + Y_{i_1+1,i_2}) + \beta_2(Y_{i_1,i_2-1} + Y_{i_1,i_2+1}) \\ & + \beta_3(Y_{i_1-1,i_2-1} + Y_{i_1+1,i_2+1} + Y_{i_1-1,i_2+1} + Y_{i_1+1,i_2-1}) \end{aligned}$$

with constant conditional variance. Conditioning is on $Y_{l_1,l_2}, \forall (l_1, l_2) \neq (i_1, i_2)$. A $\text{CAR}(2)_{SD}$ with $\beta_3 = -\beta_1\beta_2$, $\beta_1 = \alpha_1/(1+\alpha_1^2)$ and $\beta_2 = \alpha_2/(1+\alpha_2^2)$ has the same covariance structure as an AR(1)·AR(1) with parameters α_1, α_2.

The non-separable, non-axially-symmetric model used, referred to as the Pickard process (Pickard, 1980, Tory and Pickard 1992) can be written as:

$$Y_{i_1,i_2} = \alpha_1 Y_{i_1-1,i_2} + \alpha_2 Y_{i_1,i_2-1} + \alpha_3 Y_{i_1-1,i_2-1} + \epsilon_{i_1,i_2}.$$

The AR(1)·AR(1) is a special case of the Pickard process with $\alpha_3 = -\alpha_1\alpha_2$.
Each process was simulated 1 000 times to estimate the distribution of the test statistics under the null and the alternative hypotheses. Choosing a matrix T such that $V = TT'$, the observation vector y can be simulated as $y = T\epsilon$ where ϵ is a random vector of n independent $\mathrm{N}(0,1)$ observations.

3 Testing axial symmetry and separability

3.1 Tests based on sample covariances

Since axial symmetry and separability are usually defined in terms of the covariance structure, it seems natural to try and test these two hypotheses using the sample covariances. The sample estimator of the covariances we used is $c(g_1,g_2) = \sum_{i_1=1}^{n_1-g_1}\sum_{i_2=1}^{n_2-g_2}(Y_{i_1,i_2} - \overline{Y})(Y_{i_1+g_1,i_2+g_2} - \overline{Y})/n$ and $c(g_1,-g_2) = \sum_{i_1=1}^{n_1-g_1}\sum_{i_2=1}^{n_2-g_2}(Y_{i_1,i_2+g_2} - \overline{Y})(Y_{i_1+g_1,i_2} - \overline{Y})/n$, for g_1, $g_2 \geq 0$, where $\overline{Y} = \sum_{i_1=1}^{n_1}\sum_{i_2=1}^{n_2} Y_{i_1,i_2}/n$. Writing $c(g_1,g_2)$ as a quadratic form in Y and using standard results (Cressie, 1993 section 2.4.2), it is easy, given V, to calculate $\mathrm{cov}[c(g_1,g_2), c(g_1^\star,g_2^\star)]$, from which it is clear that the $c(g_1,g_2)$ are highly dependent and have non-constant variance.

3.1.1 Testing axial symmetry

We can consider basing tests for axial symmetry on the sample differences $F(g_1,g_2) = c(g_1,g_2) - c(g_1,-g_2)$ for g_1, $g_2 \geq 1$. Under the null hypothesis $\mathrm{E}[F(g_1,g_2)] = 0\ \forall\, g_1,\ g_2$. However the $F(g_1,g_2)$ have non-constant variance and they are correlated. For $1 \leq g_j \leq a_j$, let F denote the vector containing the $F(g_1,g_2)$ in lexicographic order, and let $\mathrm{cov}(F) = \Sigma$. Since the $c(g_1,g_2)$ are asymptotically normally distributed, $F \sim \mathrm{N}(0,\Sigma)$ and a possible test uses

$$S^{(c)} = F'\hat{\Sigma}^{-1}F.$$

Under axial symmetry, $S^{(c)} \sim \chi^2_a$, asymptotically, with $a = a_1 \cdot a_2$. Notice that estimating Σ requires estimating V and, therefore, this test is not completely model-free. We fitted the AR(1)·AR(1) to estimate V.

3.1.2 Testing separability

Modjeska and Rawlings (1983) suggested testing separability by using the singular value decomposition of the matrix M of sample covariances $c(g_1,g_2)$ for $-(a_1-1) \leq g_1 \leq a_1 - 1$ and $0 \leq g_2 \leq a_2 - 1$. Since $C(g_1,g_2) \propto C(g_1,0)\cdot C(0,g_2)$ under the null hypothesis, the population M has rank one for a separable process. If λ_1 is the largest singular value of M, in absolute value, a test for separability could use:

$$R^{(c)} = \lambda_1^2 / \sum \lambda_i^2.$$

Its distribution needs to be simulated. Note that $R^{(c)} \in [0,1]$, low values leading to rejection of separability.

3.2 Tests based on sample spectrum

The main advantage of the sample spectrum is that it has better asymptotic properties than the sample covariances, being asymptotically independent at different frequencies. For ω_j a multiple of $2\pi/n_j$, the sample spectrum we used is $I(\omega_1,\omega_2) = \sum_{g_1=-n_1+1}^{n_1-1}\sum_{g_2=-n_2+1}^{n_2-1} c(g_1,g_2)\cos(g_1\omega_1 + g_2\omega_2)/(2\pi)^2$.

Then, for $\omega_j \neq 0, \pi$, we have $I(\omega_1,\omega_2)/f(\omega_1,\omega_2) \to$ i.i.d. Exp(1), $\mathrm{E}[I(\omega_1,\omega_2)] \to f(\omega_1,\omega_2)$ and $\mathrm{Var}[I(\omega_1,\omega_2)] \to f^2(\omega_1,\omega_2)$, as $n_1, n_2 \to \infty$. Exact formula for the mean and the variance of the one-dimensional sample spectrum are in Priestley (1981, section 6.1.3).

3.2.1 Testing axial symmetry

Consider the differences $G(\omega_1,\omega_2) = I(\omega_1,\omega_2) - I(\omega_1,-\omega_2)$ (with $\omega_j = 2\pi k_j/n_j$, $k_j = 1,2,\ldots,b_j$, and $b_j = (n_j-1)/2$ if n_j is odd, or $b_j = n_j/2-1$ if n_j is even). Clearly under axial symmetry $\mathrm{E}[G(\omega_1,\omega_2)] = 0$ and $\mathrm{Var}[G(\omega_1,\omega_2)] \to 2f^2(\omega_1,\omega_2)$ as $n_1, n_2 \to \infty$. Since the $G(\omega_1,\omega_2)$ do not have constant variance, we considered two modifications. Firstly, taking the logarithm of the $I(\omega_1,\omega_2)$ approximately stabilizes the variance and also reduces the non-normality. For $\omega_j \neq 0, \pi$ we then have, asymptotically $\log[I(\omega_1,\omega_2)] \sim$ Gumbel $(\log[f(\omega_1,\omega_2)], 1)$ from which $\mathrm{Var}\{\log[I(\omega_1,\omega_2)]\} \to \pi^2/6$ as $n_1, n_2 \to \infty$. Given $b = b_1 \cdot b_2$, a possible test uses:

$$S_1^{(p)} = \overline{D}\sqrt{b}/\sqrt{\pi^2/3},$$

where $\overline{D} = \sum_{\omega_1,\omega_2} D(\omega_1,\omega_2)/b$, $D(\omega_1,\omega_2) = \log[I(\omega_1,\omega_2)] - \log[I(\omega_1,-\omega_2)]$. Assuming b large enough for the central limit theorem to hold, under axial symmetry $S_1^{(p)} \sim \mathrm{N}(0,1)$. Other tests, for example, the sign test or the Wilcoxon test on the differences $D(\omega_1,\omega_2)$ or $G(\omega_1,\omega_2)$, can also be used.

A second modification is to standardize the differences $G(\omega_1,\omega_2)$, estimating $f(\omega_1,\omega_2)$ by $[I(\omega_1,\omega_2) + I(\omega_1,-\omega_2)]/2$. Thus, if $H(\omega_1,\omega_2)$ denotes $G(\omega_1,\omega_2)/[I(\omega_1,\omega_2) + I(\omega_1,-\omega_2)]$, then another possible test uses

$$S_2^{(p)} = \sqrt{3b}\ \overline{H}.$$

If the $I(\omega_1,\omega_2)$ are independent and exponentially distributed, $H(\omega_1,\omega_2)$ has a uniform distribution in the interval $[-1,1]$, and the sum converges rapidly to a $\mathrm{N}(0,b/3)$, so $S_2^{(p)} \sim \mathrm{N}(0,1)$ asymptotically under axial symmetry.

3.2.2 Testing separability

The same idea in section 3.1.2 can be applied to a matrix of $I(\omega_1,\omega_2)$ to test separability (we use $R_1^{(p)}$ to denote this test statistic).

As an alternative, under separability, $\log[I(\omega_1,\omega_2)]$ and $\log[I(\omega_1,-\omega_2)]$ can be regarded as two sample realizations of the same value of the log spectrum $\log[f(\omega_1,\omega_2)] = d + \log[f(\omega_1,0)] + \log[f(0,\omega_2)]$ where d is a constant. So testing for separability reduces to testing for lack of interaction in a two-way classification table with two realizations in each cell and, $\mathrm{var}(\log[I(\omega_1,\omega_2)])$ being approximately constant, this can be done using the statistic:

$$R_2^{(p)} = SS_{rc}/SS_e$$

where SS_{rc} is the mean of squares due to the interaction, and SS_e is the residual mean of squares. Under separability, and assuming the log differences are approximately normal, $R_2^{(p)} \sim F_{(b_1-1)(b_2-1),b}$, asymptotically.

3.3 Model-based tests

Assuming a specific model behind the data allows the use of model-based tests for axial symmetry and separability. The idea is to fit a particular unrestricted model (non-separable or non-symmetric) to the data, then to restrict it to be axially symmetric or separable, imposing constraints on its parameters and to test which model is more appropriate. The comparison can use the generalised likelihood ratio test (GLRT) or tests which are asymptotically equivalent, such as the Wald test or the Score test.

3.3.1 Testing axial symmetry and separability together

Under the assumption $Y \sim \mathrm{N}(0, \sigma^2)$, the log-likelihood can be written as $\ell(\alpha, \sigma^2; y) = -(n/2)\log(2\pi) - (n/2)\log(\sigma^2) - \log \mid V \mid /2 - y'V^{-1}y/(2\sigma^2)$. The GLRT statistic is then

$$S^{(m)} = 2[\ell(\hat{\alpha}, \hat{\sigma}^2; y) - \ell(\hat{\alpha}_0, \hat{\sigma}_0^2; y)].$$

Now, to test for axial symmetry and separability together we can consider the Pickard as the unrestricted model and the AR(1)·AR(1) as the restricted one, with the constraint $\alpha_3 = -\alpha_1\alpha_2$ (see section 2.3). Alternatively, the Wald (Guo and Billard, 1998) and the Score tests could be used, although we found that GLRT is preferred for smaller lattice sizes. Under the null text hypothesis, the three test statistics are asymptotically distributed as a χ_1^2.

3.3.2 Testing separability

Separability can be tested comparing the fit of a CAR$(2)_{SD}$ with the fit of an AR(1)·AR(1), applying the restriction $\beta_3 = -\beta_1\beta_2$ (see section 2.3) to the parameters of the CAR$(2)_{SD}$. The GLRT can be expressed this time as:

$$R^{(m)} = 2[\ell(\hat{\beta}, \hat{\sigma}^2; y) - \ell(\hat{\beta}_0, \hat{\sigma}_0^2; y)].$$

4 Results and discussion

For the tests on axial symmetry, a good correspondence between the simulated and theoretical distributions was found for $S_1^{(p)}$, $S_2^{(p)}$, $S^{(m)}$ on 11 by 11 or larger lattices. For $S^{(c)}$ with $a_1 = a_2 = 4$, a 15 by 15 lattice was required. Tests $S_1^{(p)}$, $S_2^{(p)}$ are completely general, the main assumption required being stationarity, while $S^{(c)}$, $S^{(m)}$ require partial or complete specification of a model. On the other hand, $S^{(m)}$ and $S^{(c)}$ are more powerful (see Table 1). Note that the power partially depends on how far the Pickard process is from the AR(1)·AR(1), measured by $|\alpha_3 + \alpha_1\alpha_2|$. Also note that $S^{(m)}$ is testing for separability at the same time.

Separability proved to be harder to test. Tests $R^{(c)}$ and $R_1^{(p)}$ are not invariant, their distributions strongly depending on the parameters of the process considered, and also can have values very close to 1 under the alternative hypothesis. Simulations show $R_2^{(p)}$ to be distributed as expected on 11 by 11 lattices, although this test does not seem to be very powerful. Also, it assumes axial symmetry, which therefore should be tested first. Test $R^{(m)}$ is very powerful (see Table 2), although it requires specification of a model.

As well as the tests mentioned, we have also considered some other tests, and considered variants of those mentioned - for example using different a_j. We have evaluated the tests on other parameter values than those mentioned, and on some other processes - for example some separable ARMA processes. We have some theoretical results to help explain the power for different parameters under the alternative hypothesis. We have tried the tests on a real data set - a SAR (synthetic aperture radar) image for which the postulated form of the point-spread function implies separability. The theoretical distribution of the data is Exponential, so further work is needed to see how the tests perform for theoretical SAR data. For further details, see Scaccia (2000) and Scaccia & Martin (2002).

Parameters	$S_1^{(p)}$	$S_2^{(p)}$	$S^{(c)}$	$S^{(m)}$	$\|\alpha_3 + \alpha_1\alpha_2\|$
$\alpha = (0.1, 0.2, 0.6)$	97.4%	97.1%	99.9%	100.0%	0.62
$\alpha = (0.6, 0.7, -0.8)$	87.0%	86.3%	99.5%	100.0%	0.38
$\alpha = (0.3, 0.4, 0.2)$	55.2%	59.0%	78.1%	99.8%	0.32
$\alpha = (0.1, 0.6, 0.2)$	44.6%	45.8%	61.2%	99.3%	0.26
$\alpha = (0.3, 0.6, -0.15)$	5.1%	5.3%	5.8%	6.8%	0.03

Table 1. *Simulated power of tests for axial symmetry for different Pickard processes, on a 15 by 15 lattice when the level of the tests is equal to 5%.*

Parameters	$R_2^{(p)}$	$R^{(m)}$	$\|\beta_3 + \beta_1\beta_2\|$
$\beta = (0, 0, 0.25)$	88.8%	100.0%	0.25
$\beta = (0.11, 0.07, 0.16)$	13.6%	98.7%	0.17
$\beta = (0.2, 0.2, 0.05)$	9.1%	92.7%	0.09
$\beta = (0.25, 0.25, 0)$	8.3%	86.8%	0.06
$\beta = (0.3, 0.2, -0.03)$	5.3%	12.3%	0.03

Table 2. *Simulated power of tests for separability for different $CAR(2)_{SD}$ processes, on a 15 by 15 lattice when the level of the tests is equal to 5%.*

References

Balram, N. and Moura, J.M.F. (1993). Noncausal Gauss Markov random fields: parameter structure and estimation. *IEEE Trans. Inform. Theory*, **39**, 1333-1355.

Cressie, N.A.C. (1993). *Statistics for Spatial Data.* New York: Wiley.

Guo, J.H. and Billard, L. (1998). Some inference results for causal autoregressive processes on a plane. *J. of Time Ser. An.*, **19**, 681-691.

Martin, R.J. (1979). A subclass of lattice processes applied to a problem in planar sampling. *Biometrika*, **66**, 209-217.

Martin, R.J. (1990). The use of time-series models and methods in the analysis of agricultural field trials. *Commun. Statist. -Theory Meth.*. **19**, 55-81.

Martin, R.J. (1996). Some results on unilateral ARMA lattice processes. *J. Statist. Plann. Inf.*. **50**, 395-411.

Modjeska, J.S. and Rawlings, J.O. (1983). Spatial correlation analysis of uniformity data. *Biometrics*, **39**, 373-384.

Pickard, D.K. (1980). Unilateral Markov fields. *Adv. Appl. Prob.*, **12**, 655-671.

Priestley, M.B. (1981). *Spectral Analysis and Time Series.* London: Academic Press.

Scaccia, L. (2000). *Testing for Simplification in Spatial Models.* Ph.D. thesis, University of Perugia, Italy.

Scaccia, L. and Martin R.J. (2002). Testing axial symmetry and separability of lattice processes. Preprint.

Tory, E.M. and Pickard, D.K. (1992). Unilateral Gaussian fields. *Adv. Appl. Prob.*, **24**, 95-112.

The Forward Search

Anthony Atkinson[1]

[1] Department of Statistics, London School of Economics, London, WC2A 2AE, UK
a.c.atkinson@lse.ac.uk

Keywords. Box-Cox transformation, fan plot, forward search, masked outliers, robustness

1 Introduction

This paper summarises joint research with Marco Riani on the forward search, a powerful general method for detecting multiple masked outliers and for determining their effect on models fitted to the data. Atkinson and Riani (2000) describe its use in linear and nonlinear regression, response transformation and in generalized linear models. These examples are here extended to include multivariate analysis. Riani and Atkinson (2001) describe an application to multivariate transformations and discriminant analysis.

The regression example is in §2. Section 3 is concerned with the effect of outliers on the Box-Cox transformation of the response in regression. The models considered are extended in §4 to include generalized linear models. Finally, in §5, an example is given of six measurements on Swiss bank notes. The observations are seen to fall into two groups, with some further structure.

2 Regression and the Surgical Unit Data

Let the standard linear regression model be written as

$$y = Q\theta + \epsilon = X\beta + w\gamma + \epsilon, \tag{1}$$

where the errors ϵ are independent and identically distributed with constant variance σ^2. The vector of p parameters θ is estimated by least squares applied to a subset of m observations to give the estimate $\hat{\beta}(m)$. We start the search with m small, usually p or perhaps $p+1$, and randomly select 1,000 subsamples. The initial subset $S^*(p)$ provides the least median of squares estimator $\hat{\beta}(p^*)$, that is it minimises the median squared residual (Rousseeuw 1984). We then order the residuals and augment the subset.

When m observations are used in fitting, the optimum subset $S^*(m)$ yields n residuals $e(m^*)$. We order the squared residuals $e^2(m^*)$ and take the observations corresponding to the $m+1$ smallest as the new subset $S^*(m+1)$. Usually this process augments the subset by one observation, but sometimes two or more observations enter as one or more leave. Due to the form of the search, outliers, if any, tend to enter as m approaches n.

In our example we look at a forward plot of the residuals $e(m^*)$, scaled by the final estimate of σ. We also look at forward plots of t tests from (1) where

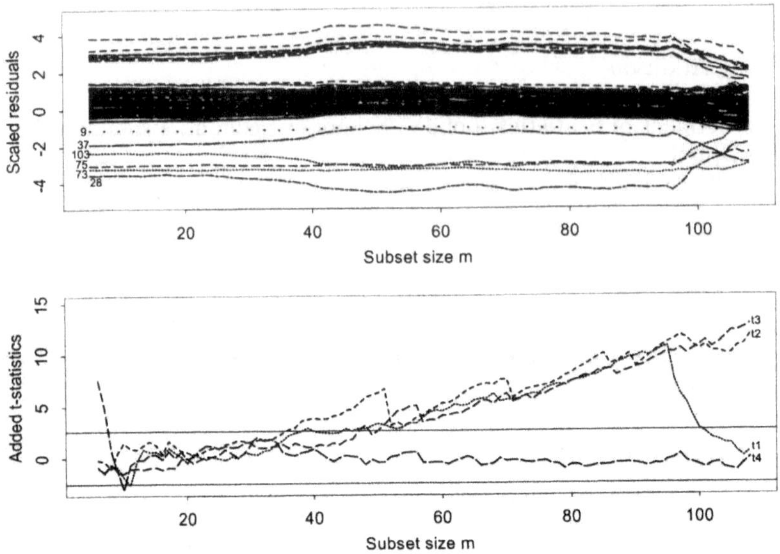

Fig. 1. Modified Surgical Unit data: upper panel, forward plot of scaled residuals; lower panel, added variable t test for each explanatory variable

we, in turn, take each of the columns of Q as the vector w (except the column corresponding to the constant term in the model). The well-established approach of added variables (for example §2.2 of Atkinson and Riani 2000) leads to an expression for the least squares estimate $\hat{\gamma}$ as a function of residuals from the regression of y and w on X. The added variable t test for w is

$$t_\gamma = \hat{\gamma}/\{s_w^2/(w^T Aw)\}^{1/2}, \tag{2}$$

where $A = I - H = I - X(X^TX)^{-1}X^T$ and s_w^2 is the residual mean square estimate of σ^2.

As an example we use the **Surgical Unit Data**, (Neter, Kutner, Nachtsheim, and Wasserman 1996, pp.334 & 438) with 108 observations on the time of survival of patients who had liver surgery. There are four explanatory variables. To illustrate the detection of masked outliers we changed the values of 12 observations. The details are in Atkinson and Riani (2003).

The effect of the outliers is clear. The upper panel of Figure 1 shows the forward plot of the standardised residuals. For most of the search the 12 outliers are clearly separated from the rest of the data. However, towards the end of the search, these outliers begin to be included in the subset of m observations used in fitting and become less apparent and so increasingly hard to detect.

The t statistics for the variables are plotted in the lower panel of Figure 1. The effect of the modification has been to make variable one non-significant: in the unmodified data it is the most important variable. The plot very clearly shows the effect of the outliers on the t tests for regression. Variable selection using t tests on all 108 observations would lead to the incorrect dropping of variable one. It is clear that a subset of observations are indicating a different model from the majority of the data. Which these observations are follows from the order in which the observations enter the search.

The plot of residuals in the upper panel of Figure 1 come from a single search fitting the model $E(Y) = A\theta$. The lower panel of the figure shows the results of four searches, one for each of the submodels $E(Y) = X\beta$. Exclusion from the model of the variable w which is to be tested ensures that the null distribution of t_γ is indeed Student's t (Atkinson and Riani 2003).

3 Transformations of the Univariate Response in Regression and the Doubly Modified Poison Data

We consider tests of the value of the transformation parameter λ in the Box and Cox (1964) family of normalized power transformations

$$z(\lambda) = \begin{cases} (y^\lambda - 1)/\lambda \dot{y}^{\lambda-1} & \lambda \neq 0 \\ \dot{y} \log y & \lambda = 0, \end{cases} \tag{3}$$

where the geometric mean of the observations is written as $\dot{y} = \exp(\Sigma \log y_i/n)$. The model is multiple regression with the response in (1) replaced by $z(\lambda)$. An approximate score test can be found by Taylor series expansion of $z(\lambda)$ about the null value λ_0 which adds an extra, constructed, variable to the regression model. To form an approximate t test for the significance of this variable, and so of the need for transformation, the variable w in (1) is replaced by the constructed variable which, provided X includes a constant, can be written

$$w(\lambda) = \begin{cases} y^\lambda \{\log(y/\dot{y}) - 1/\lambda\}/(\lambda \dot{y}^{\lambda-1}) & \lambda \neq 0 \\ \dot{y} \log y (0.5 \log y - \log \dot{y}) & \lambda = 0. \end{cases} \tag{4}$$

We combine calculation of the test statistic with the forward search. Since observations which are outlying on one scale may not be outlying for a different transformation, we conduct several searches for different values of λ. In most applications, including the example here, we use five searches for the values $\lambda = -1, -0.5, 0, 0.5$ and 1. If there are outliers for a particular λ they will enter the search last and influence the value of the test statistic.

As an example we use the **Poison Data** from Box and Cox (1964) in which two observations have been modified. For all the data the score statistic has a value of 0.64 when $\lambda = 0$ and the log transformation seems appropriate. The adjacent values of -0.5 and 0.5 are firmly rejected. However forward plots of the five t statistics, combined in a "fan" plot, reveal the effect of the two changed observations.

The effect of the two outliers is clearly seen in Figure 2. The effect is greatest for $\lambda = -1$, where addition of the two outliers at the end of the search causes the statistic to jump from an acceptable 1.08 to 10.11. The effect is similar, although smaller, for $\lambda = -0.5$. At the end of the search the log transformation is acceptable, but the forward plot shows that this conclusion is determined solely by the two outliers.

Although forward plots of the test statistic are easily interpreted, the statistic cannot have exactly a t distribution since the constructed variable (4) is a function of the response. The simulations reported in Atkinson and Riani (2002) indicate that the distribution is close to the t distribution of §2 if there is "strong" regression on X - values of R^2 around 80% or higher are necessary. A theoretical justification for these results is given by Atkinson and Riani (2003). A fuller analysis of the doubly modified Poison Data is in Atkinson and Riani (2000, §4.6).

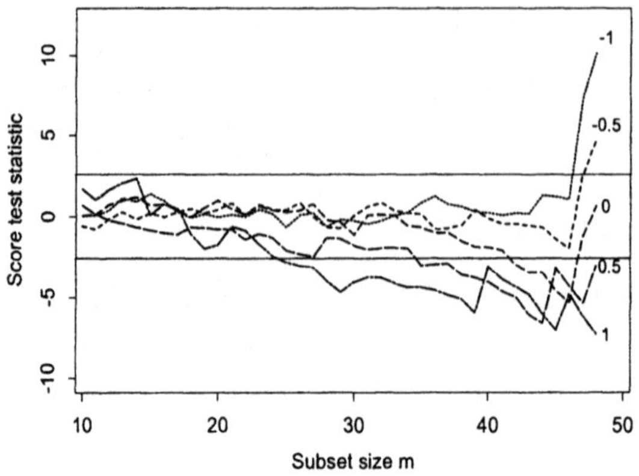

Fig. 2. Doubly modified poison data: fan plot – forward plot of $T_p(\lambda)$ for five values of λ. The curve for $\lambda = -1$ is uppermost; the effect of the two outliers is clear

4 Generalized Linear Models and Dielectric Breakdown Strength

The structure provided by the theory of generalized linear models allows us to apply the forward search to, particularly, gamma, Poisson and binomial data in a manner analogous to that used for regression. Chapter 6 of Atkinson and Riani (2000) contains theory and examples.

In generalized linear models we have a response y, a vector of linear predictors with elements $\eta = x^T\beta$ and a link function $g(\mu) = \eta$ connecting the two. As well as the customary problems about individual outliers and the correct form of the linear model, there is also a need to specify the correct form of link function, a need which is particulary acute with gamma data. One approach is to use a goodness of link test, which comes from a Taylor series expansion of the link function (for example, Atkinson and Riani 2000, §6.6.4). This results in a similar t test to those of the two previous sections, but now with constructed variable $\hat{\eta}^2$, where $\hat{\eta}$ is the fitted linear predictor from the model to be tested.

Atkinson and Riani (2000), §6.9, analyse data from Nelson (1981) on performance degradation from accelerated tests. The response, dielectric breakdown strength, is measured at the points of an 8×4 factorial with four observations per cell. The response is nonnegative with a skew distribution. The gamma distribution is appropriate.

The analysis starts by fitting a predictor with linear terms in temperature and log(time) and exploring the family of power (Box-Cox) links. Analysis of all 128 observations suggests a link with $\lambda = 2$. However the forward plot of the goodness of link statistic in the left-hand panel of Figure 3 reveals that, although the value of the statistic may be acceptable for all the data, it is not so earlier on, having a maximum absolute value of 5.58 when $m = 115$. The forward plot of deviance residuals, in the right-hand panel of the figure, shows that around this extreme value there is a rapid change in the values of some residuals. The figure also reveals that there are groups of residuals that form clusters. Much of the subsequent analysis is concerned with finding a structure for the linear predictor that accommodates these groups of observations from individual cells of the factorial arrangement.

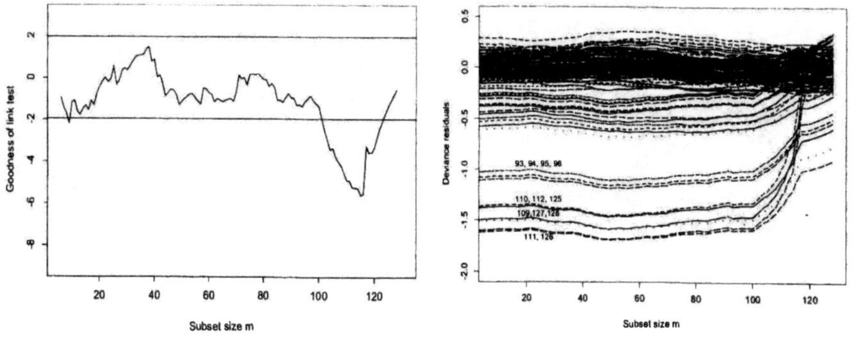

Fig. 3. Dielectric data, Box–Cox link with $\lambda = 2$: forward plots of goodness of link test and of deviance residuals

Here the forward search revealed the groups of observations that do not agree with that fitted to the rest of the data. The search also shows the effect of these observations on the proposed link.

5 Multivariate Observations and the Swiss Banknote Data

With multivariate observations we replace the squared residuals $e_i^2(m^*)$ with the squared Mahalanobis distances

$$d_i^2(m^*) = \{y_i - \hat{\mu}(m^*)\}^T \hat{\Sigma}^{-1}(m^*)\{y_i - \hat{\mu}(m^*)\}, \tag{5}$$

where $\hat{\mu}(m^*)$ and $\hat{\Sigma}(m^*)$ are estimates of the mean and covariance matrix of the observations based on the subset $S^*(m)$. These distances are used for ordering the observations and for determining how we move forward in the search. We use the robust bivariate boxplots of Zani, Riani, and Corbellini (1998) to determine an initial subset which is not outlying in any two-dimensional plot of the data. The content of the contours is adjusted to give an initial subset of the required size.

The data are readings on six dimensions of 200 Swiss bank notes, 100 of which may be genuine and 100 forged. All notes have been withdrawn from circulation, so some of the notes in either group may have been misclassified. Also, the forged notes may not form a homogeneous group. For example, there may be more than one forger at work. The data, and a reproduction of the bank note, are given by Flury and Riedwyl (1988, p.4-8).

The left-hand panel of Figure 4 shows the scaled Mahalanobis distances from a forward search starting with 20 observations on notes believed genuine. In the first part of the search, up to $m = 93$, the observations seem to fall into two groups. One has small distances and is composed of observations within or shortly to join the subset. Above these there are some outliers and then, higher still, a concentrated band of outliers, all of which are behaving similarly. The two groups are apparent.

The structure of the group of forgeries is also readily revealed by the forward search. The right-hand panel of Figure 4 is a forward plot of the scaled Mahalanobis distances just for the forgeries. In the centre of the plot, around $m = 70$ this shows a clear structure of a central group, one outlier from that

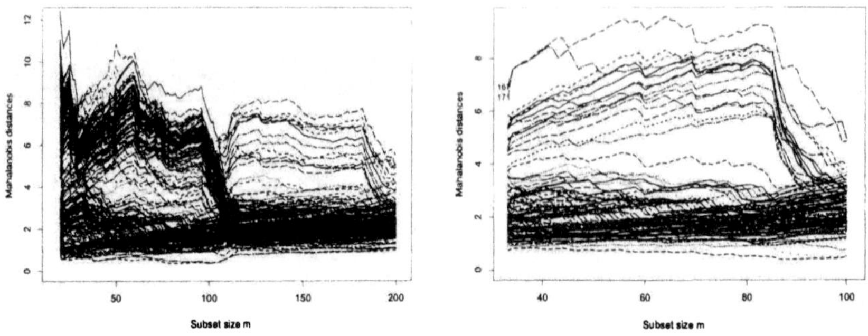

Fig. 4. Swiss Banknote Data, forward plots of Mahalanobis distances: both groups and, on the right, the forgeries, showing evidence of a third group

group and a second group of 15 outliers. As successive units from this cluster enter after $m = 85$, they become less remote and the distances decrease.

In this example the forward search clearly indicates not only the presence of two groups of notes, but also that the group of forgeries is not homogeneous, itself consisting of two subgroups.

References

Atkinson, A. C. and M. Riani (2000). *Robust Diagnostic Regression Analysis.* New York: Springer–Verlag.

Atkinson, A. C. and M. Riani (2002). Tests in the fan plot for robust, diagnostic transformations in regression. *Chemometrics and Intelligent Laboratory Systems 60*, 87–100.

Atkinson, A. C. and M. Riani (2003). Forward search added variable t tests and the effect of masked outliers on model selection and transformation. (Submitted).

Box, G. E. P. and D. R. Cox (1964). An analysis of transformations (with discussion). *Journal of the Royal Statistical Society, Series B 26*, 211–246.

Flury, B. and H. Riedwyl (1988). *Multivariate Statistics: A Practical Approach.* London: Chapman and Hall.

Nelson, W. (1981). The analysis of performance-degradation data. *IEEE Transactions on Reliability R-30*, 149–155.

Neter, J., M. H. Kutner, C. J. Nachtsheim, and W. Wasserman (1996). *Applied Linear Statistical Models, 4th edition.* New York: McGraw-Hill.

Riani, M. and A. C. Atkinson (2001). A unified approach to outliers, influence, and transformations in discriminant analysis. *Journal of Computational and Graphical Statistics 10*, 513–544.

Rousseeuw, P. J. (1984). Least median of squares regression. *Journal of the American Statistical Association 79*, 871–880.

Zani, S., M. Riani, and A. Corbellini (1998). Robust bivariate boxplots and multiple outlier detection. *Computational Statistics and Data Analysis 28*, 257–270.

The MISSION Client: Navigating Ontology Information for Query Formulation and Publication in Distributed Statistical Information Systems

Yaxin Bi

CES, The University of Edinburgh, St John's Land, Holyrood Campus, Edinburgh, EH8 8AQ, UK

Abstract. This paper describes the technologies developed for the summary and publication of large volumes of statistical information, such as produced by National Statistical Institutes (NSIs), in the European fifth framework project, MISSION (Multi-Agent Integration of Shared Statistical Information Over the [inter]Net). We review the MISSION system architecture that is built on a three-tier client / server architecture, and then focus on novel methods and techniques in designing and implementing different client components, and data model for ontologies within the library.

Keywords. Metadata, Ontology, XML, Integration of Distributed Statistical Information System, Graphical User Interface, Dissemination

1 Introduction

The effective access and use of data from distributed statistical information systems can be characterized as having a significant volume of data processing and output two major aspects, often with complex processing to produce statistical summaries, comparable figures and publications in various forms. These aspects involve rich interactions between users and statistical data residing on diverse locations, and can be treated as a process of distributed queries. To support various aspects of users' activities involved in statistical information systems, it not only requires an underpinning system architecture, but also requires a flexible data model for representing the content of statistical databases (ontology) and a smart user interface which provides users with the ability to browse ontology information, formulate complex queries, harmonize inconsistencies between variables, and output results with different layout.

A number of statistical information systems equipped with current access interfaces (client) and applications could be viewed as information workspaces, but in a variety of ways they limit the effectiveness of data access as well as the statistical summary and publication crossing the internet (Bi et al., 1999). In particular, conventional user interfaces situated in a two-tier client / server architecture are based on the view of menus or form fill-in as an isolated task in which the user formulates a query against a homogenous collection to obtain queried results. Many of statistical system clients, instead of giving users information about the contents of the data, requires users to fill out lengthy forms. A common assumption of such systems is that users are informed about the data that they are working on or they want to submit know-item queries. In fact, there are a number of areas where this view misses the reality of users doing real work.

- Iterative query refinement. Users are often unable to formulate information needs or express them effectively using the conventional user interfaces. Users often learn during the course of a session what they are trying to ask and how to ask it.
- Source heterogeneity. Users often access multiple data sources with differing characteristics of the same content, disparate content, or the same content with different forms. Understanding these characteristics is an important part of a user's activity.
- Parallel, interleave access. Users often switch among sources with slow or variable response. Through users may want to interleave access operation and track their progress, current systems are weak in their support for this process.
- Larger data processing. Data access is usually intertwined with other parts of the overall process, for example, formulating queries and defining output layout. Users switch among different techniques for searching or browsing sources and visualizing or utilizing results.

The MISSION system is built on a three-tier architecture composed of client, library and dataserver. The client is a presentation layer – a front-end interface, which is developed using the advanced technologies of Model-Controller-View (MVC) paradigm and Document Object Model (DOM). This front-end interface improves the previous drawbacks and provides users with a number of functions of metadata browsing, query construction, transformation, and publishing, and facilitates communication with the library.

The library is a mediation layer holding the metadata repository and acts as a workspace for software agents. It plays a mediation role between the client and the dataserver, that is, receiving queries from the client, sending the queries and getting the querying results back to the client by using agents. One of emerging standards for the World Wide Web XML (eXtensible Markup Language) is employed to encode statistical metadata required by the MISSION system in terms of ontology (Bi et al., 2001). This framework allows the system to extend simple ontologies for a range of uses and also provides mechanisms for the explicit representation of semantic knowledge to the content of statistical databases.

The dataserver is the local site where holds the physical data storage, management tools for data registering and data access.

In this paper, we concentrate on two main aspects of a data model which is incorporating XML with metadata – an ontology and the client interface. The MISSION client is aiming at making a more intuitive interface that allows users to construct complex statistical queries through understanding ontology information without knowledge of database structures.

2 MVC basics

The MVC architecture organizes an interactive application design by separating data presentation, data representation, and application behavior. The *Model* represents the structure of the data in the application, as well as application-specific operations on those data. A *View*(s) presents data in some form to a user, in the context of some application function. A *Controller* translates user actions (such as mouse motions, keystrokes, words spoken, etc.) and user input into method calls on the Model, and selects the appropriate View based on user preferences and Model state. Essentially, a Model abstracts application state and functionality, a View abstracts application presentation, and a Controller abstracts application behavior in response to user input (Gamma, et al., 1994).

3 The MISSION client architecture

The MISSION client is an MVC application. Figure 3.1 presents a high-level block diagram of the client in terms of MVC architecture. The client consists of four basic components which corresponding the functions of *browsing*, *query construction*, *transformation*, and *publishing*, which accommodate the functionality for the summary and publication of large volumes of statistical information required in the MISSION system. Each component is modeled to three distinct forms of functionality of model, view and controller, which inherit from the common MVC class and incorporate customizations for individual tasks. Such a configuration effectively takes advantages of MVC design paradigm, ensuring the explicit isolation of the various functions, abstract representation for each function, and interactions of each other.

It important to note that there are two types of interactions as illustrated in Figure 3.1: vertical and horizontal. The vertical interactions occur within each of the four components through event, registration and notification mechanisms, e.g. interactions between the distinct pieces of the model, view and controller. Let us see an example of how the BrowserModel, BrowserView, and BrwoserController communicate each other. Assume a user navigates ontology information presented as tree structure in the BrowserView, when the user clicks a node on the tree and the tree is expanded and more information is shown up. Such a process involves a series of interactions among the BrwoserView, BrowserController, and BrowserModel. In fact, each function on the BorwserView is defined as an action to which an event listener is registered in the BrowserController, the BrowserModel publishes its methods to the BrowserController and the actions (such as mouse motions, keystrokes, etc.) are regarded as event sources. Whenever a mouse button is clicked down on the BrowserView, an event occurs, the event source notifies all registered listeners in the BrowserController, and passes each an event object to perform the corresponding methods in the BrowserModel, which in turn triggers the BorwserView updates.

The horizontal interaction mainly occurs between the components when the metadata object is transferred from one model to another. Such a transfer is triggered when users drag a tree node from the BrwoserView and drop it onto the QueryView, or a transformation expressed is incorporated into a query formulation. There is an underlying object called MetadataObject which supports such communications between models. This object is embedded in the common MVC model and some of its attributes are rendered in the BorwserView. The object is derived from the ontology data which is simplified as subset containing essential attributes required by the QueryConstructor.

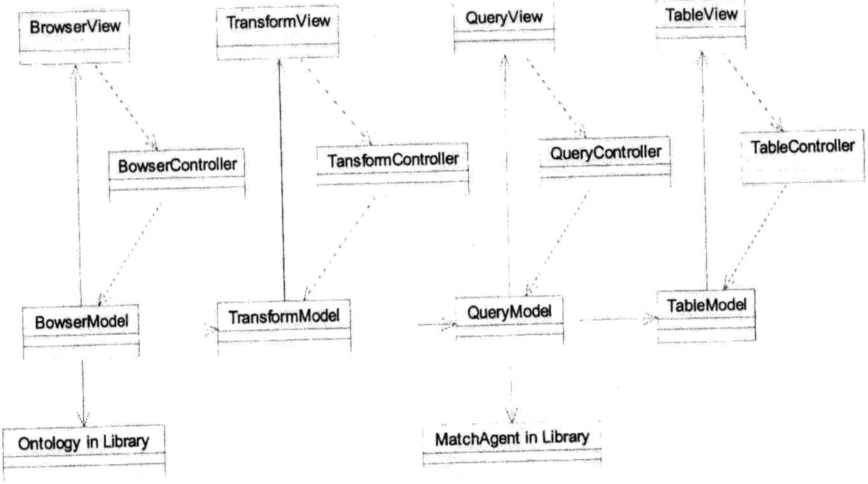

Figure 3.1: The MVC and tier partitioning of the MISSION client

4 From ontology to metadata object

An ontology is defined as shared formal conceptualization of a particular domain (Gruber, 1993). Practically, it can be regarded as a vocabulary and its definitions as well as relationships between concepts in it. In the MISSION system, an ontology is built on a range of statistical metadata such as theme, classification, coverage, etc. and used to facilitate query formulation and query processing. Looking at the characteristics of the domain of official statistics at which the MISSION is aiming, a data set is usually accompanied with metadata which interprets the meanings of the attributes and their values within the data set as illustrated in Table 4.1 and 4.2. Such metadta provides sufficient contextual information about attributes allowing users to understand what does each attribute ID stand for, for example, the attribute *schtype* represents *type of school,* and possible values captured for the attribute. However conceptual and background information associated with data sets is missing. For example, what subject the data set is related, and what is coverage of the data set. This missing makes the data set to lack semantic information, resulting in invalid queries to be formulated.

Therefore, metadata including conceptual information is not only to help users understand the content of databases, but also required in query processing.

LAND	YRLEFT	AGE	SEX	SCHTYPE
3	1994	18	1	2
3	1994	17	0	2
3	1993	18	1	2
3	1994	19	-9	2

Table 4.1: An example of data set

ID	Name	Type	Label	Value
SEX	Gender	Categorical	{ male, female, Not answered }	{0, 1, -9}
SCHTYPE	Type of school	Categorical	{Academic, Comprehensive Vocational, Not answered}	{1,2,3,-9}

Table 4.2: A piece of context information for Table 1

The issues of capturing and modelling metadata have drawn considerable efforts in the communities of official statistics and computational statistics, and a number of models in relation to modeling statistical metadata exist (Papageorgiou, et al., 2001). To address the issues in the context of query formulation and processing, we define a specific ontology based on metadata. It consists of conceptual frame and context information as shown Table 4.2. Figure 4.1 is a fragment of the DTD definition of the ontology.

```
<!ELEMENT ONTOLOGY (THEME , REFCLASSIFICTION? , CORRESTABLE? , COVERAGE? ,
DATASET+)>
```

Figure 4.1 is a fragment of the DTD definition of the ontology

In the definition, there are five high level elements, e.g. THEME, REFCLASSIFICTION, CORRESTABLE, COVERAGE, DATASET, the first four of them represent conceptual frame indicating what an ontology is related. The element DATASET is used to describe context information which is associated with the attributes of data sets, defined as Metadata Object. The hierarchy reflected in the elements is an important structure on the ontology because it depicts the explicit relationship between the elements situated in two distinct levels. In order to access and manipulate ontology information, we employ a standard model – DOM – to represent ontology information and metadata objects are directly derived from it.

5 The client interface

The underpinning client architecture and data model for ontology information are fundamental constituents of the client, which support the broader user activities with flexible ways. Figure 5.1 shows two of four basic functions of the client, e.g. metadata browsing and query construction, which are parallel to be held in the interface. Users can interleave access functions through their views and track their progress.

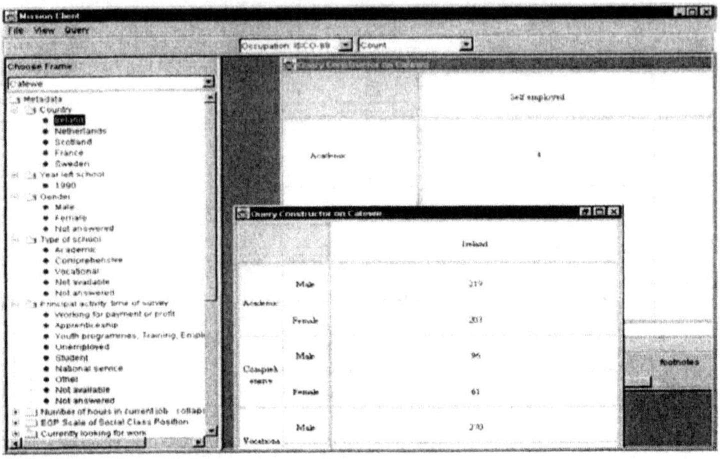

Figure 5.1: A screenshot of the MISSION client

The BrowserView presents an ontology as a tree. The user can switch from a view with general ontology information to another view with specific information – metadata objects. The query construction component employs "query by example" style as a way of query formulation, and incorporates the layout definition into query formulation process, which acts as the client publishing function. When a user starts to formulate the query, the user first navigates ontology information, and drag an individual node such as "Ireland" and group nodes, such as "Gender" and "Type of School", and then drop them on the query view. The layout will be automatically generated with the query results based on the properties of the metadata objects.

References

Bi, Y. and Lamb, J. (2001). Metadata Encoded in XML: Enabling Complex Query Formulation in Distributed Statistical Databases. *Proceedings of Third International Conference on Information Integration and Web-based Applications & Services*. Austria.

Gamma, E., Helm, R., Johnson, R., and Vlissides, J. (1994). *Design Patterns: Elements of Reusable Object-Oriented Software*. Addison-Wesley.

Gruber,T. (1993). A translation Approach to Portable Ontology Specifications. *Knowledge Acquisition*. Vol 5.

Papageorgiou, H., Pentaris, F. & Vardaki, M. (2001). A statistical Metadata Model for Simultaneous Manipulation of both Data and Metadata. *Journal of Intelligent Information System*. 17, 169-192. Kluwer Academic.

Time Series Modeling using Mobile Devices and Broadband Internet

Albert PRAT[1], Ignasi SOLÉ[1], David REMONDO[2], Daniel MARTINEZ[1] and Josep M. CATOT[1]

[1] Department of Statistics and Operational Research, UPC, Av. Diagonal 647, 08028 Barcelona, e-mail: albert.prat@upc.es
[2] I2CAT, Department of Telematics Engineering, UPC

Abstract. In this paper we present one application included in the Project @DAN, an AdvANced and high secure mobile platform to support the digital economy, that started on November 2001, and is financed by the *EU IST-2001-32634*.

This application will provide all types of services necessary to model and obtain forecasts of time series data by the professionals of any kind of organisations.

The user of this application will be able to obtain two different types of services: demand forecasts and consultancy using mobile terminals, broadband internet and internet payments.

The background system installed in a server, is FOREtess[+], which is an evolution of TESS (see Prat et al. (2000)).

Keywords: Forecasting, ARIMA models, Seasonal Adjustment, Smart Cards, UMTS, Internet2, Security.

1 Introduction to the @DAN project

The development of the new economy and the information society is being curbed due to both the insecurity that potential customers are feeling with regards to the communications and the lack of standards that could stimulate developments of this security.

@DAN, an aDvANced and high secure mobile platform to support the digital economy, seeks to remove these barriers in the mobile communications environment.

The main goals of the @DAN project can be summarised, but not limited to the following list:

[+] FOREtess is in process of being registered as a trademark by UPC.

Develop a secure mobile platform to support developments of m-commerce applications, specially these one pertaining to the field of the digital economy. Specifically, this platform will provide two basic functionalities: digital signature and secure payments based on a smart card. Each one of the components will be produced by a relevant expert in the field.

- ✓ The platform and related tools produced will be used for the development of two prototypes. These prototypes will shown a spectrum of functionalities that will prove the necessary capacities of the platform. The prototypes will be produced by two different universities which have showed in the past their skills in the subject.
- ✓ These two prototypes, will be tested and assessed by different industrial and services end users organizations. This assessment will be conducted by an SME that has interest in the commercialization of the results obtained.

2 Architecture of the platform to be developed

The @DAN platform will provide two basic functionalities: digital signature and electronic secure payment based on a cryptographic smart card. A relevant expert in the field will produce each one of the components.

This platform will be based in three components:

- ✓ UMTS component (wideband), with possibilities to read a smart card
- ✓ Smart card (cryptographic features)
- ✓ Digital certificates (X509.v3 based)

The next figure shows the architecture of the platform applied to the times series prototype.:

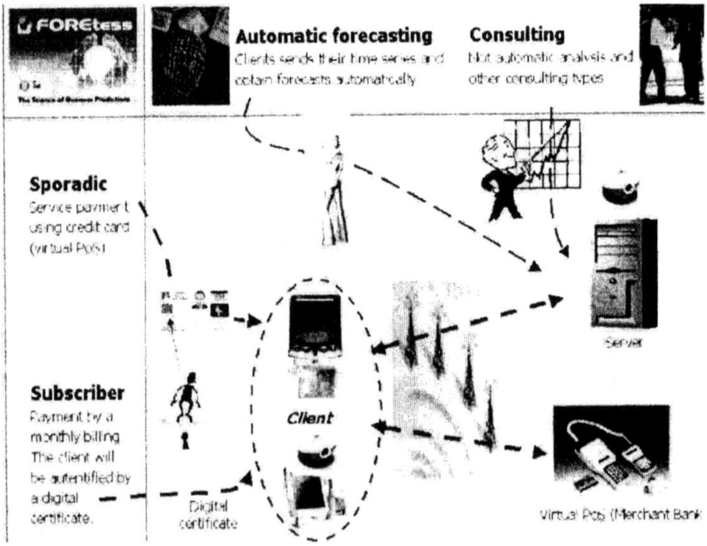

In what follows, we will comment on the requirements for the communication infrastructure and then we will see what is proposed as a solution for both the wireless and the fixed parts.

3 Solution adopted for the first phase of the @DAN project

The technology initially considered in the project, UMTS, is a cellular technology that provides a maximum throughput between 384 Kbit/s and 2 Mbit/s, depending on the speed of the terminal. This maximum throughput has to be shared among all users within the cell. Provided that there are few users in the UMTS cell, this technology can indeed be expected to be sufficient for the support of the requirements.

However, due to the unavailability of systems with this technology to date, and the uncertain perspectives, it has been decided to employ alternative wireless access technologies to fulfill the main objectives of the project.

Therefore, we resort to Wireless Local Area Networks technology (WLANs). In particular, we consider the systems based on the standards IEEE 802.11b. This technology has some attractive characteristics: it is widespread in the market and therefore relatively economical, it does not require spectrum licenses and, as we will see, it complies with the requirements of the project.

IEEE 802.11b provides radio communication between terminals at a nominal throughput of 11 Mbit/s. The maximum range is about 150 meters.

4 Fixed infrastructure

I2CAT is an experimental Internet platform that combines telecom companies, content service providers, technology providers, universities, governmental institutions and final users. The goal of this platform is to develop and to test new generations of applications, network services, network protocols and technologies, testing the acceptance and the social impact of the new generation of Internet (IPv6) by the final user.

I2CAT is based on an optical broadband network that interconnects all members and end users at high-speed rates. All the network providers are connected to the Gigapop, that resides in the UPC, using several technologies like Gigabit Ethernet, PoS (Packet over Sonet) or ATM (1.2 Gbps/622 Mbps/155Mbps).

The @DAN project will use the I2CAT platform that provides interconnection between the project partners UPF, UPC, SECARTIS and Caixa de Catalunya with high speed links above 2Mbps using wired and wireless LAN technologies. At the same time, @DAN takes advantage of the multiple services deployed on the I2CAT platform as VoD, streaming video and videoconference based on MPEG 1 /2/4 and H.323 standards.

These @DAN partners are transparently interconnected with the TCP/IP suite over Ethernet wire and wireless access technologies offering an adequate network platform to deploy secure @DAN applications.

ADSL is a technology that makes it possible to employ the existing telephone wires to transport data at relatively high data rates. In this way, we can build an access network that reaches a large part of the population without the need of installing new physical links. At this time, ADSL is being deployed commercially in Western European countries. With the commonly offered data rate of 256 Kbit/s downlink and 128 Kbit/s in the uplink we can expect ADSL to be suitable for our purposes. The client only needs to have a POTS (Plain Old Telephone Service) line, an ADSL modem and to contract the ADSL service from an operator.

5 Wireless terminals

The initial wireless terminals to be used in the project are Compaq iPAQ PDA's. These handheld computers have the Pocket PC operating system and can host up to two PCMCIA cards. There are IEEE802.11b PCMCIA cards commercially available and we expect that GPRS PCMCIA cards will be available soon. Other features of these devices are a Web browser, a Java virtual machine and Jdk2.0 for Java applets and applications development

In a second phase, SIEMENS handsets will be used.

6 Time series prototype

One of the two prototypes developed to prove and validate @DAN platform is the UPC Time Series application.

Based on the @DAN platform, UPC will develop an application that will allow individual persons or members of any kind of organisation to obtain forecasts and other useful information from their time series database in an easy, fast, accurate and secure manner.

The system is being designed to allow for two different types of users:

- ✓ Sporadic users that from time to time will use the application to obtain forecasts of some time series of interest.
- ✓ Subscription paying users that either from a desktop computer or from a mobile device will use the application for time series modeling and forecasting or for receiving consultancy services.

It will perform a fairly complete statistical treatment of time series, be that in routine large-scale applications or, for example, in careful analysis for short-term economic policy, control and monitoring. Although other frequencies are possible, our attention centres mainly on monthly and quarterly series.

UPC application prototype will be accessible by using workstations, laptops and mobile devices, which have a web browser.

The web based application performs all its operations on server side, and the only difference between diverse client devices is the user interface, specially designed for each one of them.

7 Functionalities of the application

The next table shows the different functionalities depending on the specific type of terminal used to access the prototype (the use of a mobile device imposes some restrictions on the functionalities to be implemented in such a handheld PC):

FUNCTIONALITIES	Desktop computer	MOBILE DEVICE
SECURE CONECTION	YES	YES
USER AUTENTIFICATION	YES	YES
VIRTUAL TPV	YES	YES
NEW USER'S REGISTRATION	YES	NO
TIME SERIES AND USERS DATABASE	YES	YES
DOMAINS ORGANISATION	YES	YES
IMPORT EXCEL, ACCES, ASCII files,	YES	POCKET EXCEL FILES
CREATE NEW TIME SERIES AND DOMAINS	YES	YES
UPDATING TIME SERIES MANUALLY	YES	YES
AUTOMATICALLY TIME SERIES UPDATING	YES	NO
EXPORT RESULTS	YES	POCKET EXCEL FILES
POSSIBILITY OF IMPORT TIME SERIES FROM PRIVATE DATABASE (EXTERNAL)	YES	YES
MODELLING		
AUTOMATIC AND MANUAL UNIVARIATE MODELLING	YES	YES
INTERVENTION ANALYSIS	YES	YES
INTERVENTION DATABASE	YES	YES
TRANFER FUNCTION	YES	YES
FORECASTING	YES	YES
AUTOMATIC SYSTEM	YES	NO
SEASONAL ADJUSTMENT	YES	YES
TREND AND CYCLE ESTIMATION	YES	YES
CONFIGURATION OF BACKEND SOFTWARE	YES	YES
THE PROBLEM OF AGREGATION	YES	YES
CONTROL OF FORECAST ERRORS		
EWMA GRAPHIC	YES	YES
CUSUM GRAPHIC	YES	YES
GRAPHICS AND PLOTS	YES	LIMITED
CONSULTANCY (video, multimedia)	YES	LIMITED

Finally, other relevant characteristics of this prototype are:

- ✓ The system will be provided with secure connection and user's authentication by electronic signature system, and secure Internet payments.
- ✓ The system will offer an online help system and also access to a Time Series Course.
- ✓ Possibility to ask questions to an expert (videoconference, forum, e-mail list,...)

8 Technical requirements estimated at the server side

The technical requirements estimated at this moment for the hardware and software to support the development at the server side are:

- ✓ Windows NT/2000 server
- ✓ Internet Information Server
- ✓ Scripting language installed (ASP, PHP, ...)
- ✓ Database server (MySQL)
- ✓ Videoconference Server (Microsoft Exchange 2000 Videoconference Server)

9 Conclusions

Technology is moving quicker than ever before, especially mobile technology. Global usage of the mobile device will continue to grow at a high rate. In fact, it could be said that the developments in technology are racing ahead of our imagination's capability to deal with it.

The past three years have seen an outbreak in the number of wireless communication devices. The emergence of these mobile delivery channels - digital phones, personal digital assistants (PDAs), etc. - presents an opportunity for many communities, e.g., academic and scientific communities looking to increase their efficiency and easiness to access the data and tools they needs to use.

It is probably. unlikely that individuals are confident in the use of wireless access to Internet until the right security solutions exist to support them. These include the delivery of identification and authentication, content privacy and non-repudiation support to wireless networks that are currently available for wired networks via public-key infrastructure (PKI) technology.

@DAN will try to provide a platform for the development of these new solutions. @DAN time series modelling and forecasting prototype will demonstrate the advantages of this new technology.

References

Prat, A., Gomez, V., Sole, I., and Catot, J.(2000). *TESS: system for automatic seasonal adjustment and forecasting of time series*, Proceedings of Compstat 2000, Physica-Verlag, pp. 391-396.

Acknowledgements

@DAN is partially funded by IST Program and the CICYT (TYC 2001-4716-E)

Unbiased Partial Spline Fitting under Autoregressive Errors

Michael G. Schimek
Karl-Franzens-University Graz, Institute for Medical Informatics, Statistics & Documentation, A-8010 Graz, Austria

Abstract. A special variant of semiparametric regression models are partial spline models. Schimek (2000) provides a cheap algorithm for unbiased estimation under an iid assumption. Here we introduce a similar algorithm for the case of autoregressive (AR) errors. Further we address the problem of smoothing parameter choice in this situation. Finally the results are illustrated on time series data from environmental epidemiology.

Keywords. AR errors, partial spline, semiparametric regression, smoothing parameter, time series, unbiased estimation

1 Introduction

Since the seminal paper of Speckman (1988) we have seen a strong interest in the further development of semiparametric regression models, mostly under an iid or normal error assumption. Monographs covering this field are Green and Silverman (1994), Eubank (1999), both focussing on spline smoothers, and Härdle et al. (2000), considering kernel smoothers. In this paper our interest is to study covariate effects on a time-dependent trend corrupted by autoregressive errors, hence a non-standard situation, however often seen in environmental, epidemiological and economic applications. Our approach is incorporating dependent errors into a partial spline concept. To obtain unbiased estimates we adapt ideas due to Speckman (1988). Apart from the numerical complexity of the estimation problem we face a number of problems stemming from the dependent error structure. All proposals we make should be seen against the background of computational efficiency.

2 The semiparametric model with independent errors

The predictor function of the semiparametric regression model consists of a parametric linear component and an arbitrary nonparametric component depending on a design variable t. The responses $y_1, \ldots, y_n$ are obtained at non-stochastic ordered values $t_1, \ldots, t_n$ of t. Let us then have

$$y_i = u_i^T \gamma + g(t_i) + \epsilon_i \qquad (1)$$

for $i = 1, \ldots, n$, where $u_1, \ldots, u_n$ are known k-dimensional covariate vectors, γ is a k-dimensional unknown coefficient vector, g is an unknown smooth function, and the $\epsilon_1, \ldots, \epsilon_n$ are independent, zero mean random variables with a common variance σ^2.

The sum of squares equation of model (1) is

$$SS(\gamma, g) = \sum_{i=1}^{n} (y_i - u_i^T \gamma - g(t_i))^2 + \lambda RP,$$

where

$$RP = \int_a^b [g''(t)]^2 \tag{2}$$

denotes the roughness penalty of a cubic smoothing spline with knots at $t_1, \ldots, t_n$ for a fixed smoothing parameter $\lambda > 0$.

The assumptions made above lead to partial splines. Green and Silverman (1994, chap. 4) suggest estimators which are asymptotically biased, however the bias is not severe for reasonable n (Schimek, 2000, p. 533ff). In using results due to Speckman (1988), Eubank et al. (1998) and Schimek (2000) could derive an unbiased risk estimator and $O(n)$ algorithms. Here, different from B-spline approaches, the full design information is relevant for knot placement. This extra effort - necessary for AR errors as introduced later, not for independent errors - is more than compensated by an efficient direct (non-iterative) algorithm, yielding unbiased curve and coefficient estimates.

We cannot emphasize enough how critical the choice of λ is in a semiparametric context: Apart from estimation aspects our task is to formally test covariate effects via the γ coefficients. Generalized cross-validation could be used here but unbiased risk estimation performs better. However the latter requires the consistent estimation of the error variance. Eubank et al. (1998) provide such an estimator based on pseudo-residuals which can be calculated in $O(n)$ steps.

As far as testing of the parametric component is concerned Speckman (1988) could prove that the coefficient vector γ is asymptotically NID without distributional assumption on the dependent variable respectively errors (for inferential tools see Schimek, 2000).

3 The semiparametric model with autoregressive errors

Let us now assume weak stationary autoregressive errors of order p (i.e. (AR(p)) for ϵ in model (1). The sum of squares equation is now

$$SS(\gamma, g) = \sum_{i,j} d_{ij} [y_i - \mathbf{u}_i^T \gamma - g(t_i)]^T [y_j - \mathbf{u}_j^T \gamma - g(t_j)] + \lambda RP \tag{3}$$

for $i = 1, ..., n$, $j = 1, ..., n$, where $\mathbf{u}_i = \mathbf{u}_j = (u_1, ..., u_k)^T$ are known covariate vectors and d_{ij} elements of the inverted covariance matrix (symmetric, banded with bandwidth $2p + 1$, and positive definite under the assumptions made). All other assumptions are the same as for (1) and the roughness penalty RP the same as in (2).

Regrettably, the iid results do not work for the AR case. This is even more true for the smoothing parameter selection. In the following we develop a similar non-iterative algorithm for dependent errors. Further we introduce a new concept which combines λ choice, estimation and inference about the coefficient vector γ.

3.1 Estimation, smoothing parameter selection and inference

Schimek (1988) developed a discrete roughness penalty approach (spline-like curve fitting) for known autoregressive and moving average errors. Here we adapt some of the ideas there for the AR case to work with partial (smoothing) splines. This is achieved by extending the methodology worked out in Schimek (2000).

Equation (3) for the partial spline in matrix notation is

$$SS(\gamma, \mathbf{g}) = (y - \mathbf{U}\gamma - \mathbf{g})^T D^{-1}(y - \mathbf{U}\gamma - \mathbf{g}) + \lambda \mathbf{g}^T K \mathbf{g} \tag{4}$$

where $K = QR^{-1}Q^T$ and $y = (y_1, \ldots, y_n)^T$, $\mathbf{U}^T = [\mathbf{u}_1, \ldots, \mathbf{u}_n]$ and $\mathbf{g} = (g(t_1), \ldots, g(t_n))^T$. Q and R are known band matrices from the value-second derivative representation of the cubic spline (Green and Silverman, 1994, p. 12f).

Let $h_i = t_{i+1} - t_i$ for $i = 1, \ldots, n-1$. Let Q be the $n \times (n-2)$ matrix with elements q_{ij} for $i = 1, \ldots, n$ and $j = 2, \ldots, n-1$, given by $q_{j-1,j} = h_{j-1}^{-1}$, $q_{j,j} = -h_{j-1}^{-1} - h_j^{-1}$, and $q_{j+1,j} = h_j^{-1}$ for $j = 2, \ldots, n-1$, and $q_{ij} = 0$ for $|i - j| \geq 2$. The columns of Q start with $j = 2$.

The symmetric matrix R is $(n-2) \times (n-2)$ with elements r_{ij} for i and j, running from 2 to $(n-1)$. They are defined as follows: $r_{ii} = \frac{1}{3}(h_{i-1} + h_i)$ for $i = 2, \ldots, n-1$, $r_{i,i+1} = \frac{1}{6}h_i$ for $i = 2, \ldots, n-2$, and $r_{ij} = 0$ for $|i - j| \geq 2$. R is positive definite.

Finally the $(n-2) \times (n-2)$ matrix K is symmetric and positive definite. For a weak stationary AR(p) process the inverted covariance matrix D^{-1} is symmetric, banded with bandwidth $2p+1$, and positive definite. Luckily its elements can be calculated directly (Schimek, 1988, p.38).

Our goal is to efficiently estimate the parameter vector γ, the function $\mathbf{g}$, and the mean vector μ. Minimizing (4) yields a penalized least squares solution. The estimators are

$$\gamma = (\tilde{\mathbf{U}}^T\tilde{\mathbf{U}})^{-1}\tilde{\mathbf{U}}^T(I - (D^{-1} + \lambda K)^{-1}D^{-1})y$$

and

$$\mathbf{g} = (D^{-1} + \lambda K)^{-1}D^{-1}(y - \mathbf{U}\gamma),$$

where $\tilde{\mathbf{U}} = (I - (D^{-1} + \lambda K)^{-1}D^{-1})\mathbf{U}$. Further we have $\mu = \mathbf{U}\gamma + \mathbf{g} = Hy$ with the hat matrix $H = (D^{-1} + \lambda K)^{-1}D^{-1} + \tilde{\mathbf{U}}(\tilde{\mathbf{U}}^T\tilde{\mathbf{U}})^{-1}\tilde{\mathbf{U}}^T(I - (D^{-1} + \lambda K)^{-1}D^{-1})$.

Compared to independent errors the algorithm for AR errors is only slightly more computationally demanding as long as $n \gg p$. This is a direct result from having banded matrices throughout. A similar algorithm cannot be designed for moving average errors (would imply full matrices). However the combination of smoothing splines and moving average errors is of limited value (see Schimek, 1992) and thus we can do without it. The estimators are asymptotically unbiased under the assumptions we have made.

The unbiased risk criterion for λ choice in combination with Eubank's et al. (1998) variance estimator (loss of consistency - Herrmann, 2000, p. 88) cannot be applied any more. Instead we apply an approach which efficiently combines smoothing parameter selection, estimation and inference about the coefficient vector γ. We are better off with Speckman's (1988, p. 426) variance estimator which can be corrected for dependent errors via H and is characterized by a positive but asymptotically negligible bias (results in slightly conservative tests),

$$\sigma^2 = \frac{RSS}{trace(I-H)^T(I-H)},$$

where $RSS = \frac{1}{n}||(I-H)y||^2$. Further for testing γ coefficients individually we need

$$Var(\gamma) = \sigma^2(\tilde{\mathbf{U}}^T\tilde{\mathbf{U}})^{-1}\tilde{\mathbf{U}}^T(I-(D^{-1}+\lambda K)^{-1}D^{-1})^2\tilde{\mathbf{U}}(\tilde{\mathbf{U}}^T\tilde{\mathbf{U}})^{-1}.$$

It is possible to derive tests without specific distributional assumptions (results from Speckman, 1988, still hold). For $k = 1$ there is an approximate z-distributed statistic, $z = \gamma/SE(\gamma)$, and for $k > 1$ an approximate F-distributed statistic. Let us define the parametric mean sum of squares $MSS_{par} = 1/n\gamma^T(P^TP)^{-1}\gamma$, where $P^T = (\tilde{\mathbf{U}}^T\tilde{\mathbf{U}})^{-1}\tilde{\mathbf{U}}^T(I-(D^{-1}+\lambda K)^{-1}D^{-1})$. We test $F_{df_1,df_2} = MSS_{par}/\sigma^2$ with degrees of freedom $df_1 = k$ and $df_2 = trace(I-H)^T(I-H)$.

Degree of smoothing and significance are connected via the significance trace. We take advantage of this concept and calculate our curve fits in combination with the γ estimates and model variance over a grid of typical λ values. Then we decide for a specific regression model under a prespecified AR error process.

4 An application

The practical value of our semiparametric partial spline model is demonstrated on an environmental epidemiology application. Particulate matter measurements ($PM_{2.5}$, i.e. airborne particles with aerodynamic diameter less equal 2.5 μm) have been taken for 214 days. Another time series is hospital admissions of a male adult population in an affected Austrian province for a certain pulmonary diagnosis. We analyze whether particle concentration effects the morbidity (admissions are an indicator for that).

First we fit a partial spline model under the inadequate assumption of iid errors under lag 1 (a one day delay is empirically known between particle inhalation and some health problems). In Figure 1 the fitted admission numbers (overall curve) under 60 nonparametric degrees of freedom (trace of the hat matrix) are displayed. For the covariate $PM_{2.5}$ $\hat{\gamma} = 0.041$ with $z = 1.621$ almost significant for $\alpha = 0.05$. When smoothing is increased the covariate effect gets more and more eliminated. Apart from that oversmoothing would not reduce the impact of the empirical AR(1) error with a positive autocorrelation coefficient in the admission series (makes the fitted curve look wiggly). For problems of smoothing under dependent errors see Kohn et al. (2000).

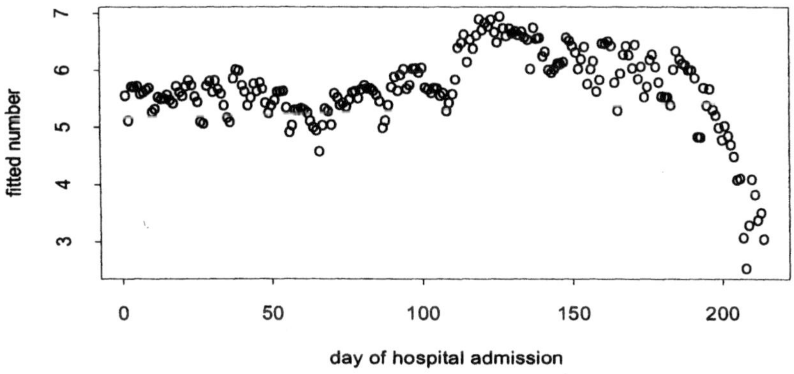

Fig. 1. Semiparametric fit assuming independent errors

In Figure 2 we see the partial spline fit (overall curve) under the correctly specified AR(1) error with a parameter $\phi = 0.83$. The nonparametric degrees of freedom remain fixed to 60 and are still adequate. However as can be clearly seen the curve is much smoother. More than that our γ estimate is with 0.047 slightly larger and now significant. What have we seen? Unbiased partial spline fitting under AR errors and the newly proposed smoothing parameter selection concept work reasonably well. The health risks in the above example are not very pronounced and not all statistical methods do well in this situation. What we can conclude is that the explicit modelling of dependent errors can improve our ability to assess even small covariate effects.

References

Eubank, R. L. (1999, 2nd ed.). Nonparametric regression and spline smoothing. Dekker, New York.

Eubank, R. L., Kambour, E. L., Kim, J. T., Klipple, K., Reese, C. S. and Schimek, M. (1998). Estimation in partially linear models. CSDA, 29, 27-34.

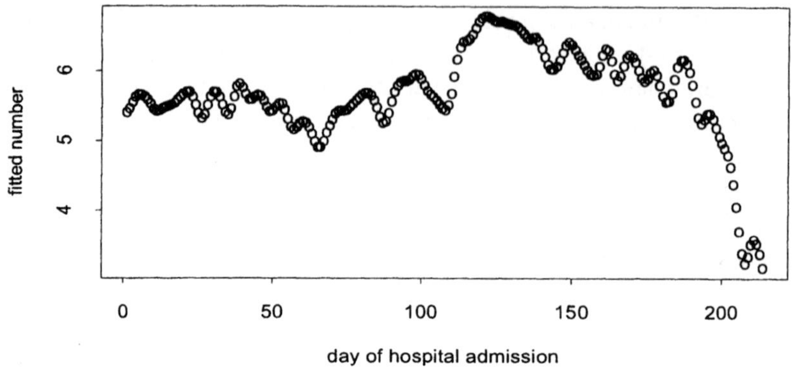

Fig. 2. Semiparametric fit assuming AR(1) errors

Green, P. J. and Silverman, B. W. (1994). Nonparametric regression and generalized linear models. A roughness penalty approach. Chapman & Hall, London.
Härdle, W., Liang, H. and Gao, J. (2000). Partially linear models. Springer, Berlin.
Herrmann, E. (2000). Variance estimation and bandwidth selection for kernel regression. In Schimek, M. G. (ed.) Smoothing and regression. Approaches, computation and application. John Wiley, New York, 71-107.
Kohn, R., Schimek, M. G. and Smith, M. (2000) Spline and kernel smoothing for dependent data. In Schimek, M. G. (ed.) Smoothing and regression. Approaches, computation and application. John Wiley, New York, 135-158.
Schimek, M. G. (1988). A roughness penalty approach for statistical graphics. In Edwards, D. and Raun, N. E. (ed.). Proceedings in Computational Statistics 1988. Physica, Heidelberg, 37-43.
Schimek, M. G. (1992) Serial correlation in spline smoothing: A simulation study. Computational Statistics, 7, 309-327.
Schimek, M. G. (2000). Estimation and inference in partially linear models with smoothing splines. JSPI, 91, 525-540.
Speckman, P. (1988). Kernel smoothing in partial linear models. JRSS, B, 50, 413-436.

Acknowledgment: The data set for the application comes from a research project which has been funded by the Austrian Ministry for the Environment, Youth and Family Affairs, the Austrian Ministry for Science and Traffic, and the Austrian Academy of Sciences.

Unobserved Heterogeneity in Store Choice Models

Rodríguez-del-Bosque, Ignacio[1], Suárez-Vázquez, Ana[2], Moral-Arce, Ignacio[3] and Rodríguez-Poo, Juan[4]

[1] Departament of Business Administration, University of Cantabria, Avda. De los Castros s/n, 39005, Santander, SPAIN
[2] Departament of Business Administration, University of Oviedo, Avda. Del Cristo s/n, 33071, Oviedo, SPAIN
[3] Departament of Economy, University of Cantabria, Avda. De los Castros s/n, 39005, Santander, SPAIN
[4] Departament of Economy, University of Cantabria, Avda. De los Castros s/n, 39005, Santander, SPAIN

Keywords. Shopping centre choice, heterogeneity, random effects models

1 Introduction

Discrete choice models are a tool which is shared by a large variety of scientific areas, and for this reason, different specifications have been developed. Of all these specifications, the so-called multinomial logit model (McFadden, 1974) is the one that has been used most to analyze situations in which there are multiple choice alternatives, specifically in retail studies (Arentze y Timmermans, 2001). The multinomial logit model has a strong theoretical base and is also easy to apply. However, a potential limitation of the studies that use this specification is the fact that consumer heterogeneity is not taken into account (Severin, Louviere and Finn, 2001). In this paper we present the results of a random effects model. The distribution of the heterogeneity is assumed to be unknown, and it is approximated through a piecewise constant function, as was proposed originally in Heckman and Singer (1984).

2 Treatment of heterogeneity in store choice models

Traditional procedures of accounting for heterogeneity, such as a priori segmentation and exogenous segmentation, do not assure intrasegment homogeneity and intersegment heterogeneity which, ideally, would take place during the segmentation process (for applications of these models in store choice see Finn and Louviere, 1990; Cliquet, 1995). In order to detect latent segments that are present in the market it is possible to resort to random effects models. Their use is more frequent for brand choice models than for store choice models. It is worth mentioning, however, research carried out by Dunn et al (1987) and Reader (1993).

2.1 Empirical application

The main source of data was gathered by interviewing a representative sample of shopping centre consumers in their homes between April and May 1998. The survey consists of shopping centre choices made by inhabitants of the three main cities in the region of Asturias and it covers the purchases made by 1099 households in six different shopping centres. Starting from previous studies, we have collected information about the following features: *Shopping Centre Image* (x_1); *Distance* (x_2) and *First visit* (x_3) —this variable takes into consideration the first shopping centre that the consumer visited in the region— and numbers of visits —For each household, the number of visits to any of the shopping centres during the year previous to the study. With this information the shopping centre which was most visited by each consumer was estimated (y)—. We also have collected information regarding demographic and attitudinal characteristics of the households.

Once we have defined our data set, we proceed to present our basic economic model. We will assume that the utility the i-th consumer obtains when visiting a specific shopping centre, j; U_{ij}, depends upon the set of three explanatory variables: Shopping Centre Image, x_1, Distance to the shopping centre, x_2, and First visit, x_3. This dependence is represented through the following structural model

$$U_{ij} = \mu_{ij} + \varepsilon_{ij} \qquad (1)$$

where μ_{ij} is a nonstochastic function of some explanatory variables and ε_{ij} is an unobservable random variable. More precisely, it is standard to assume $\mu_{ij} = \sum_{k=1}^{3} \beta_k x_{ijk}$,

Thus the i-th consumer will visit the k-th shopping centre if $U_{ik} > U_{ij}$ for all j. Under some crucial assumptions on the error term, McFadden (1974) showed that the multinomial logit model can be derived from a utility maximization process whenever the errors are independent and they follow a log-Weibull distribution. The multinomial logit model is defined by

$$P_{ij} = P(y = j | X = x_i) = \frac{exp\left(\sum_{k=1}^{3} \beta_k x_{ijk}\right)}{\sum_{l=1}^{6} exp\left(\sum_{k=1}^{3} \beta_k x_{ilk}\right)}$$

$$i=1,2,...,1099 \quad j=1,2,...,6 \qquad (2)$$

where $P(y=j|X=x_i)$ denotes the probability for the i-th household to choose shopping centre j given the set of explanatory values x_i. Related to the problem we are concerned with, in this model the parameters that appear in the nonstochastic function are constant. This means that heterogeneity in consumers' behavior is allowed only through the unobservable random variable ε_{ij} in (1). We tried to solve this problem by means of a random effects model.

2.2 Random effects model

In this subsection we start from the model originally proposed in Gupta and Chintagunta (1994). They base their analysis on the following logit formulation for studying consumer choice

$$P_{ij} = P(y=j|X=x_i) = \frac{\exp\left[\beta_{0ij} + \sum_{k=1}^{3} \beta_{ik} x_{ijk}\right]}{\sum_{l=1}^{6} \exp\left[\beta_{0il} + \sum_{k=1}^{3} \beta_{ik} x_{ilk}\right]} \qquad (3)$$

We note that the parameters β_{0ij} and β_{ik} are consumer specific. If there were several observations from each consumer, then it would be possible to estimate these parameters consistently. Unfortunately, in our data set we only have one observation for each. One solution to this problem is to assume that these coefficients vary randomly across consumers. Thus, the (conditional) probability of a randomly drawn household choosing shopping centre j will be given by

$$P_{j/s} = \frac{\exp\left[\tilde{\beta}_{0j} + \sum_{k=1}^{3} \tilde{\beta}_k x_{jk}\right]}{\sum_{l=1}^{6} \exp\left[\tilde{\beta}_{0l} + \sum_{k=1}^{3} \tilde{\beta}_k x_{lk}\right]} \qquad s = 1\ldots,S;\ j = 1,\ldots,6 \qquad (4)$$

For any randomly drawn consumer, the unconditional probability of choosing shopping centre j will be

$$P_j = \int_S P_{j/s} dG(s) \qquad (5)$$

To estimate the parameters of the model one can assume a particular parametric form for (5). An alternative solution is to approximate (5) by a finite number, S, of piecewise constant functions. This piecewise approximation is written as

$$P_j = \sum_{s=1}^{S} P_{j/s} P_s \tag{6}$$

In order to estimate the parameters of interest, Gupta and Chintagunta (1994) propose to construct the sample likelihood function as follows:

$$P_{i/s} = \prod_{j=1}^{6} \left[P_{ij/s}\right]^{\delta_{ij}} \quad \text{where:}$$

$$\delta_{ij} = \begin{cases} 1 \text{ if the ith household} \\ \quad \text{chooses hypermarke t } j \\ 0 \text{ otherwise} \end{cases} \tag{7}$$

$$L_i = \sum_{s=1}^{S} P_s P_{i/s} \quad (8) \qquad L = \sum_{i=1}^{995} L_i \quad (9)$$

And subsequently, calculate the parameters of interest by maximizing equation (10). In this paper, we propose an alternative two-step method. The main advantage of our approach is that P_s is allowed to vary across all households. With this method, we believe it is possible to account for the effect of heterogeneity.

In the first step we use household characteristics to estimate through maximum likelihood techniques the probabilities that a certain household belongs to a specific segment.

$$P_{is} = \frac{exp\left(\alpha_{0s} + \sum_{r=1}^{19} \alpha_{rs} D_{ri}\right)}{\sum_{l=1}^{S} exp\left[\alpha_{0l} + \sum_{r=1}^{19} \alpha_{ls} D_{ri}\right]} \tag{10}$$

where D_{ri} is the vector of 19 demographic variables of that household. Once we have estimated the parameters of interest, we predict the probabilities

$$\hat{P}_{is} = \frac{\exp\left(\hat{\alpha}_{0s} + \sum_{r=1}^{19} \hat{\alpha}_{rs} D_{ri}\right)}{\sum_{l=1}^{S} \exp\left[\hat{\alpha}_{0l} + \sum_{r=1}^{19} \hat{\alpha}_{ls} D_{ri}\right]} \tag{11}$$

In the second step, using the probabilities estimated in the first step, and the variables related to shopping centres (x_1, x_2, x_3), we estimate the probabilities of visiting the different shopping centers (results in Table 1).

$$L_i = \sum_{s=1}^{S} \hat{P}_{is} P_{i/s} \tag{12}$$

The proposed methodology enables us both to characterize segments and to assign households to segments on a demographic variables basis. The number of segments S is unknown and therefore needs to be inferred from the data. A common method to determine the appropriate number of segments is to select S which minimizes the CAIC criterion. Based on this method we selected 2 segments.

Table 1. Parameter estimates of the random effects model

	SEGMENT 1			SEGMENT 2		
V.	C.	S. E.	P[|Z|>z]	C.	S.E.	P[|Z|>z]
X2	-0.0599	0.0021	0.0000	-0.0818	0.0016	0.0000
X1	0.0579	0.0011	0.0000	0.0369	0.0006	0.0000
X3	1.0115	0.0443	0.0000	-0.5258	0.0358	0.0000
SC1	0.1856	0.0857	0.0152	-0.5576	0.0426	0.0000
SC2	0.8422	0.0833	0.0000	0.0492	0.0387	0.1015
SC3	-0.4181	0.0860	0.0000	-0.1552	0.0399	0.0001
SC4	0.8290	0.0643	0.0000	-1.2021	0.0374	0.0000
SC5	2.1831	0.0612	0.0000	-0.9466	0.0355	0.0000

Table 1 provides the parameter estimates for each of the two segments. Households in segment 2 have the highest sensitivity to travel time, whereas those in segment 1 are more sensitive to shopping centre image than those in segment 2. The interesting point of difference between the two segments, however, is in the quantity and sign of the "first visit" factor. The coefficient of this variable is positive in segment 1 and negative in segment 2. This makes intuitive sense because, as segment 1 is characterized by the lowest levels of travel time sensitivity, the households in this segment show more loyalty to the first centre visited, even if this does or does not imply a shorter shopping trip. Furthermore, there are differences in the intrinsic preferences for the centres in both segments.

The base centre is shopping centre 6 which is characterized by a low level of prices. Segment 1 has the highest preferences for all the centres except for shopping centre 3, which is also characterized by low prices.

3 Conclusions

The results of the empirical analysis in this paper demonstrate that the issue of unobserved heterogeneity in the context of shopping centre choice has significant substantive marketing implications. We have presented the estimation results of a random effect model which considers the existence of two market segments differing both in their intrinsic preference for the shopping centres and in their sensitivity to changes in the image, the distance and the first visit factor, which are the three explanatory variables considered in the model. Econometric models of this kind can be extremely useful tool for managerial decisions as they can provide a more realistic vision of consumer behavior.

References

Arentze, T.A. & Timmermans, H.J.P. (2001). Deriving Performance Indicators from Models of Multipurpose Shopping Behavior. Journal of Retailing and Consumer Services, 8, 325-334.

Cliquet, G. (1995). Implementing a Subjective MCI Model: An Application to the Furniture Market. European Journal of Operation Research, 84, 279-291.

Dunn, R., Reader, S. & Wrigley, N. (1987). A Non Parametric Approach to the Incorporation of Heterogeneity into Repeated Polytomous Choice Models of Urban Shopping Behaviour. Transportation Research A, 21, 327-343.

Finn, A. & Louviere, J.J. (1990). Shopping Centre Patronage Models: Fashioning a Consideration Set Solution. Journal of Business Research, 21, 3, 259-275.

Gupta, S. & Chintagunta, P.K. (1994). On Using Demographic Variables to Determine Segment Membership in Logit Mixture Models. Journal of Marketing Research, Vol. 31, 128-136.

Heckman, J.J. & Singer, B. (1984). A Method of Minimizing the Impact of Distributional Assumptions in Econometric Models for Duration Data. Econometrica, Vol. 52, nº 2, 271-320.

McFadden, D. (1974). Conditional Logit Analysis of Qualitative Choice Behavior, en P. Zarembka (ed.), Frontiers in Econometrics. Academic Press. New York. 105-142.

Reader, S. (1993). Unobserved Heterogeneity in Dynamic Discrete Choice Models. Environment and Planning A, 25/4, April, 495-519.

Severin, V., Louviere, J.J. & Finn, A. (2001). The Stability of Retail Shopping Choices Over Time and Across Countries. Journal of Retailing, 77, 185-202.

Using the Forward Library in S-plus

Kjell Konis[1]

[1] Statistician, Insightful Corporation, Seattle, WA 98109, USA
kkonis@insightful.com

Keywords. Forward Library, forward search, robustness

1 Introduction

The Forward Library is an S language implementation of the forward search described in Atkinson and Riani (2000). Software is provided for conducting the forward search in linear regression, in generalised linear models, and in analysing the effect of outliers on the Box-Cox transformation of the response in linear regression. The examples provided in this paper demonstrate how to conduct a forward search using both the graphical user interface (Windows) and the command line interface (Windows and Unix/Linux). S-plus version 6 is assumed.

Instructions for obtaining and installing the Forward Library are in section 2. Section 3 contains a detailed example of the forward search in linear regression using the Windows graphical user interface. A brief example of the Forward Search in generalised linear models is given in section 4. Section 4 also focuses on using the Forward Library from the S-plus command line interface. Finally, in section 5 we demonstrate how to create a user defined generalised linear model family in S-plus.

2 Installing the Forward Library

The Forward Library can be downloaded from the webpage at this url.

```
http://www.riani.it/ar/index.html
```

The file `forward.zip` contains the Windows version of the software and the file `forward.tar.gz` is for all Unix/Linux platforms. Download whichever is appropriate for your platform.

To install the Windows version of the software, right-click on the `forward.zip` file and select `expand to .../forward` then move the resulting forward folder to your S-plus library directory. On English computers this will often be `C:/Program Files/Insightful/Splus6/library`. To load the Forward Library into a running S-plus session select ▸ `Load Library...` from the S-plus File menu. This will display the Load Library dialog box shown in Figure 1. Choose forward from the list of library names and click OK to load the library. Notice that the Forward menu has been added to the S-plus menu bar.

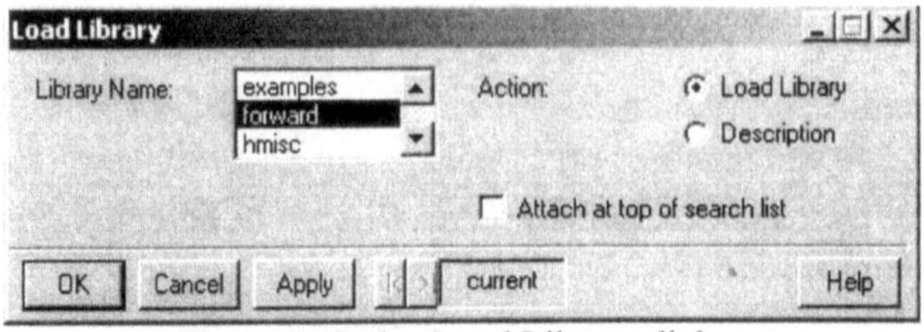

Figure 1: S-plus Load Library dialog

To install the Unix/Linux version of the software save `forward.tar.gz` in your home directory then use the following commands to unpack the archive and build the library.

```
% tar -zxpf forward.tar.gx
% cd forward
% Splus6 make
```

It may take a couple of minutes for the build process to complete depending on the system speed. System administrators can make the Forward Library available to all users by placing the forward directory in `SHOME/library/`. S-plus will look for the library first in `SHOME/library` then in the user's home directory. Use the following command to load the Forward Library into a running S-plus session.

```
> library(forward)
```

Note that this command can be used to load the library on all supported platforms, including Windows.

3 Forward Search in Linear Regression

A forward search in linear regression on the stack loss data is presented to demonstrate the Forward Library graphical user interface. The stack loss data have been analysed in several articles on robust statistics and it is known that observations 1, 3, 4, and 21 are outliers, Atkinson and Riani (2000). Start by opening the S-plus Object Explorer. In the left panel of the S-plus Object Explorer expand the `SearchPath` object then expand the `forward` object. The example data sets included in the Forward Library are displayed. Select `stack.dat` from this list.

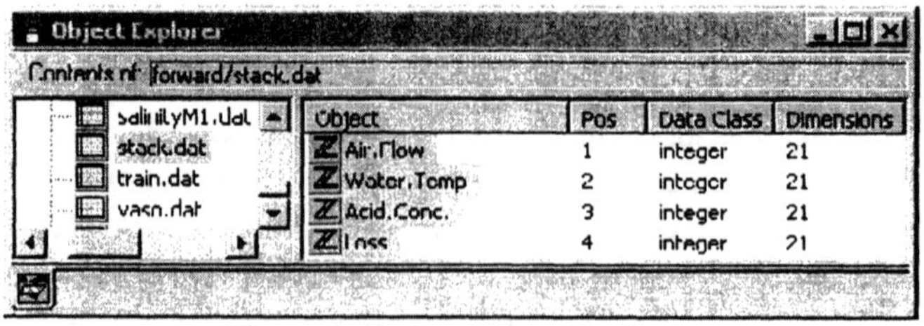

Figure 2: S-plus Object Explorer

The right panel of the Object Explorer displays the variables in `stack.dat`: the three explanatory variables `Acid.Conc.`, `Air.Flow`, and `Water.Temp`, and the response variable `Loss`. Select the response variable (`Loss`) first then control-click to select the explanatory variables. The Object Explorer should be similar to that in Figure 2.

Next, select Linear Regression from the Forward menu to display the `Forward Search: Linear Regression` dialog. Notice that the Data Set field has already been set to `stack.dat` and that the formula has been constructed from your selections in the Object Explorer. Switch to the Results page, deselect the `Long Output` check box and select `Short Output`. Also, take a look at the options on the Plots page; these can be used to generate the plots found on pages 50-62 of Atkinson and Riani (2000). Click `Apply` to run the forward search.

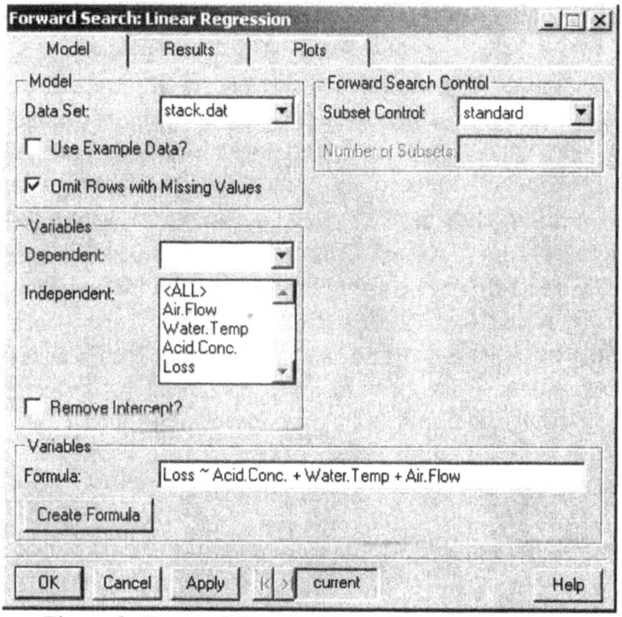

Figure 3: Forward Search: Linear Regression dialog

The example should finish quickly. The printed (short) output, which is shown on the next page, consists of the `Call`, the last 5 observations added in the search, and the coefficient estimates at the middle and end of the search. Note that the last four observations added during the search are the known outliers. Also, we see that the `intercept` and the coefficient estimate for `Air.Flow` remained relatively stable while the coefficient estimates for `Water.Temp` and `Acid.Conc.` changed considerably during the second half of the search. Selecting Long Output on the results page yields a more detailed summary of the forward search including the t statistics, the R^2 statistic, and an estimate of σ^2. The Long Output also provides the option to include 5% of the search, 10% of the search, or the entire search in the printed output.

```
*** Forward Search for Regression Models ***

Call:
fwdlm(formula = Loss ~ ., data = stack.dat, squared = F,
nsamp = "standard", na.action = na.omit)

Last 5 units included in the forward search:
m=17 m=18 m=19 m=20 m=21
   2    1    3    4   21

Estimated coefficients for 50% of the data and for the full
model:
     (Intercept)  Air.Flow  Water.Temp  Acid.Conc.
m=11   -35.00584 0.7175604   0.3299389  0.01363636
m=21   -39.91967 0.7156402   1.2952861 -0.15212252
```

The analysis of the stack loss data in section 3.2 of Atkinson and Riani (2000) suggests a first order model in `Air.Flow` and `Water.Temp` with response $\sqrt{\text{Loss}}$. To conduct a forward search on this model return to the `Forward Search: Linear Regression` dialog and click the `Create Formula` button. This will open the Formula dialog. Delete the terms on the right side of the ~ using the remove button. Next select Loss from the list in the Variables group. Since we want to fit the model using $\sqrt{\text{Loss}}$ press the Transformation button. In the Transformation dialog click the `Sq. Root: sqrt(x)` button then press Add. Verify that `sqrt(Loss)` has been added to the list of variables and click `OK`. In the Formula dialog select `sqrt(Loss)` from the variables list and click the response button. Finally, select `Air.Flow` and `Water.Temp` (control-click for multiple selection) and press the `Main Effect: (+)` button. The Formula dialog should be similar to Figure 4. Click `OK` to return to the `Forward Search: Linear Regression` dialog and then `Apply` to run the forward search.

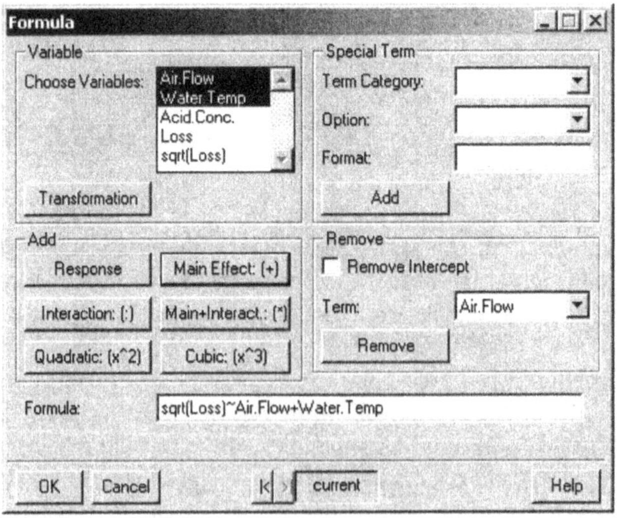

Figure 4: Formula dialog

4 Forward Search in Generalized Linear Models

We use the car insurance data analysed in section 6.8 of Atkinson and Riani (2000) to illustrate the forward search for generalized linear models. The three explanatory variables are the policy holder's age (PA, 8 levels), vehicle group (VG, 4 levels), and vehicle age (VA, 4 levels). The data frame carins.dat presents these data in an 8 x 4 x 4 factorial design. However, there are 5 cells for which there are no observations so we use the argument `na.action = na.omit` to tell S-plus to ignore the rows with missing values. The response y is the average settlement and the number of claims is also given.

By default, the S language uses the Helmert contrasts, however we wish to use the more intuitive sum contrasts. The default contrasts can be changed using the `options` function. The first entry (`contr.sum`) specifies that the sum contrasts will be used for factor variables and the second entry (`contr.poly`) specifies that polynomial contrasts will be used for ordered factor variables (this example does not include ordered factors but `options` expects two arguments for contrasts). Note that these settings will return to their original values upon exiting S-plus.

```
> options(contrasts=c("contr.sum", "contr.poly"))
```

In S-plus the model is specified with a formula object. The following command creates a formula object with response y and explanatory variables `PA`, `VG`, and `VA`.

```
> carins.formula <- y ~ PA + VG + VA
```

The Forward Library function implementing the forward search in generalized linear models is named `fwdglm`. The required arguments are the same as those for the S-plus function `glm`.

```
> cairns.fglm <- fwdglm(cairns.formula, family =
+ Gamma.BoxCox(-1), data = cairns.dat, na.action =
+ na.omit, weights = cairns.dat$Claims)
```

The function `fwdglm` fits the forward search and the `<-` operator assigns the output to the object `cairns.fglm`. The S-plus generic functions `print`, `summary`, and `plot` can now be used to examine `cairns.fglm`. The `print` function displays a brief summary of the contents of `cairns.fglm` and is invoked automatically when `cairns.fglm` is input at the command line. For example, both of these commands

```
> print(cairns.fglm)
< ... >
> cairns.fglm
< ... >
```

produce the same output. The `summary` function provides a more detailed summary of `cairns.fglm`. You can specify the number of steps to include in the summary with the optional argument `steps`. For instance, the following command will create a summary of the last 8 steps in the search.

```
> summary(cairns.fglm, steps = 8)
```

Lastly, the plot function produces a number of diagnostic plots useful for analysing the forward search. The plots are (1) Residuals, (2) Leverage, (3) Minimum and Maximum Deletion Residuals, (4) estimated β and t statistics, (5) Deviances, (6) Link Test, (7) Cook's Distance, and (8) Weights. By default, the plot function will produce all 8 plots. You can use the optional argument `which.plots` to specify (by number) which plot to create. For example, to see a plot of the deviances use this command.

```
> plot(cairns.fglm, which.plots = 5)
```

5 Adding a GLIM Family in S-plus

In the S language, the link function, its derivative, the deviance function and a method for obtaining initial values are stored in family objects. Given this information, the fitting process is essentially the same for all generalized linear models. S-plus includes the function `make.family` that allows users create their own GLIM family objects. The following example is the Forward Library function that creates the Gamma family object with Box Cox link used in section 4. For more information see `help(make.family)`. Also, a detailed example of a negative binomial family can be found on page 233 of Venables and Ripley (1999).

```
Gamma.BoxCox <- function(lambda = -1)
{
  if(lambda == 0)
    return(Gamma(link = "log"))

  gbc.lnk <- list(names = paste("BoxCox: (mu^(",
    lambda,  ") - 1) / (", lambda,")", sep = ""),
  link = substitute(function(mu)
    {(mu^lambda - 1) / lambda}),
  inverse = substitute(function(eta)
    {(1 + lambda * eta)^(1 / lambda)}),
  deriv = substitute(function(mu)
    {mu^(lambda - 1)}),
  initialize = expression(mu <- y + (y == 0) / 6) )

  make.family("Gamma.BoxCox", link = gbc.lnk,
    variance = glm.variances[, "mu^2"])
}
```

References

Atkinson, A. and Riani, M. (2000). *Robust Diagnostic Regression Analysis*. New York: Springer-Verlag.

Venables, W. and Ripley, B. (1999). *Modern Applied Statistics with S-PLUS: Third Edition*. New York: Springer-Verlag.

S-plus 6.0 (2001). Insightful Corporation: www.insightful.com.

Variance Stabilization and Robust Normalization for Microarray Gene Expression Data

Anja von Heydebreck[1], Wolfgang Huber[2], Annemarie Poustka[2], and Martin Vingron[1]

[1] Dept. of Computational Molecular Biology, Max-Planck-Institute for Molecular Genetics, 14195 Berlin, Germany, heydebre@molgen.mpg.de
[2] Div. of Molecular Genome Analysis, German Cancer Research Center, 69120 Heidelberg, Germany, w.huber@dkfz.de

Abstract. We introduce a statistical model for microarray gene expression data that comprises data calibration, the quantification of differential gene expression, and the quantification of measurement error. In particular, we derive a transformation h for intensity measurements, and a difference statistic Δh whose variance is approximately constant along the intensity range. The parametric form $h(x) = \operatorname{arsinh}(a+bx)$ is derived from a model of the variance-versus-mean dependence for microarray intensity data, using the method of variance stabilizing transformations. The parameters of h together with those of the calibration between experiments are estimated with a robust variant of maximum-likelihood estimation.

Keywords. gene expression, microarray, variance stabilization, robust regression

1 Introduction

Microarrays simultaneously measure mRNA abundances for thousands of genes in a cell population or tissue sample (The chipping forecast, 1999). Each gene is represented by a probe which is fixed to a spot on the array, and the measurement is performed by quantitating the fluorescence intensities from labeled sample cDNA that has hybridized to the probes. Multiple samples of interest are processed either by labeling them with different dyes, and letting them hybridize simultaneously against a single array, or by labeling them with a single dye, and letting them hybridize separately against multiple arrays. The relative abundance of a gene transcript in different samples can be quantified by comparing the fluorescence intensities of the corresponding probes. Due to variations in sample treatment, labeling, dye efficiency and detection across experiments, this requires appropriate calibration, which is often also called "normalization".

As the measurements are associated with non-negligible error, the usefulness of any measure of relative mRNA abundance depends on knowing its error distribution. The analysis of replicate microarray data shows that the variance of the measured spot intensities typically increases with their mean. For high intensities, the coefficient of variation is approximately constant (Chen et al. 1997). This is one of the reasons why microarray data are usually analysed on the logarithmic scale. However, this practice has limitations which particularly affect the data from weakly expressed genes, where typically a larger variance of the log-transformed intensities is observed. Furthermore, many image quantization methods produce a certain fraction of non-positive intensities, for which the (real-valued) logarithm is not defined. Often, measurements below a threshold are dismissed, but it is unclear where to set the threshold and what to do with the missing values in the subsequent analysis.

Our approach is based on a model for the distribution of measurement error that has been proposed by Rocke and Durbin (2001). From this model, we derive a parametric family of variance-stabilizing transformations of the measured intensities. Together with a class of affine-linear calibration transformations, these are incorporated into a statistical model which allows for maximum likelihood estimation of its parameters. We obtain a "difference statistic" Δh, whose variance is roughly constant along the intensity range, as a new measure of differential gene expression.

The estimation of the model parameters uses replicate data from either the different color channels of one array or from a series of one-color arrays. The different samples need not be exact biological replicates. Rather, they should be biologically related closely enough that the expression of most genes does not change. We use a robust estimation technique, which seeks to ignore differentially expressed genes and fits the model only to that subset of data points (typically, 50 to 90%) that is closest to the model mean.

2 The model

A microarray data set may be seen as a matrix (y_{ki}) of real numbers. The rows k correspond to the probes on the arrays, representing genes, and the columns i to the samples. The number n of probes may range from a few hundred to tens of thousands. The number of columns is $d = 2$ for the two-color glass chip technology, and may range up to a few hundred for series of one-color arrays. The values y_{ki}, with $k = 1, \ldots, n$ and $i = 1, \ldots, d$, are the intensity data as produced by the image quantization software. Many programs estimate local background intensities, which may be subtracted.

Due to variations in experimental factors such as amount of sample mRNA, or labeling and hybridization efficiencies, the values y_{ki} cannot directly be compared. We assume that the different columns (samples) can be brought onto the same scale through affine-linear mappings, parametrized by the real-valued parameters $o_2, \ldots, o_d$ and $s_2, \ldots, s_d > 0$:

$$y_{ki} \mapsto \tilde{y}_{ki} = o_i + s_i \, y_{ki}, \tag{1}$$

where $i = 1, \ldots, d$, and $o_1 = 0$, $s_1 = 1$ without loss of generality.

For a probe k representing a non-differentially expressed gene, we may regard the $\tilde{y}_{ki}$ as realizations of the random variable Y_k with mean $\mathrm{E}(Y_k) = u_k$ and variance $\mathrm{Var}(Y_k) = v_k$. We assume that v_k depends on k through a quadratic function of the mean u_k of the following form:

$$v_k = v(u_k) = (c_1 u_k + c_2)^2 + c_3, \qquad \text{with } c_3 > 0. \tag{2}$$

The motivation for this assumption will be discussed in Subsection 2.1. Now the method of variance stabilization can be used to derive a transformation h such that the variance $\mathrm{Var}(h(Y_k))$ is approximately independent of the mean $\mathrm{E}(h(Y_k))$ (Tibshirani 1988). We obtain

$$h(y) = \int^{y} 1/\sqrt{v(u)}\, du \tag{3}$$

from a linear approximation of $h(Y_k)$ around $h(u_k)$ ("delta method"). Inserting (2) into (3) yields

$$h(y) = \gamma \operatorname{arsinh}(a + by), \tag{4}$$

where the parameters of h are related to those of (2) through $\gamma = c_1^{-1}$, $a = c_2/\sqrt{c_3}$, and $b = c_1/\sqrt{c_3}$. The relation of the arsinh function to the logarithm is given by

$$\operatorname{arsinh}(x) = \log(x + \sqrt{x^2+1}), \quad x \in \mathbb{R}. \tag{5}$$

Hence, for large intensities the transformation (4) becomes equivalent to the usual logarithmic transformation. However, unlike the logarithm, it does not have a singularity at zero and is real-valued also for negative arguments.

Now we apply the variance stabilizing transformation (4) to the calibrated data $\tilde{y}_{ki}$ from Equation (1) to obtain the transformation $y_{ki} \mapsto h(\tilde{y}_{ki}) = \operatorname{arsinh}(a + b(o_i + s_i\, y_{ki}))$. The parameter γ may be omitted since it is merely an overall scaling factor. Setting $a_i = a + b\, o_i$, $b_i = b\, s_i$, and

$$h_i(y_{ki}) = \operatorname{arsinh}(a_i + b_i\, y_{ki}), \tag{6}$$

we can incorporate both the calibration transformation (1) and the variance-versus-mean dependence (2) of the y_{ki} in the following model:

$$h_i(Y_{ki}) = \mu_k + \varepsilon_{ki}, \qquad k \in K. \tag{7}$$

Here, K denotes the set of probes representing not differentially expressed genes, and the ϵ_{ki} are assumed to be i. i. d. $N(0, \sigma^2)$. The condition $\mathrm{E}(\varepsilon_{ki}) = 0$ reflects the goal of calibration, whereas the common variance σ^2 of the error term is aimed at by variance stabilization.

2.1 The variance-versus-mean dependence

The basic assumption underlying our statistical model is that the variance v_k depends on k as in Equation (2). First, this implies the practical assumption that v_k depends on k mainly through the mean intensity u_k. Second, we assume that this dependence is given by a quadratic function of the form (2). This is motivated by the following model for the measurement error of gene expression arrays proposed by Rocke and Durbin (2001):

$$Y = \alpha + \beta e^{\eta} + \nu, \tag{8}$$

where β is the expression level in arbitrary units, α is an offset, and ν and η are error terms which are assumed to be independent and normally distributed with mean zero. This leads to

$$\mathrm{E}(Y) = \alpha + m_\eta \beta \tag{9}$$

$$\mathrm{Var}(Y) = s_\eta^2 \beta^2 + s_\nu^2, \tag{10}$$

where m_η and s_η^2 are mean and variance of e^η, and s_ν^2 is the variance of ν. Thus we obtain a variance-versus-mean dependence as in Equation (2).

The quadratic relationship (2) can also be motivated by estimating $v(u)$ directly from microarray data. A typical example is shown in Figure 1. The estimated curves have a non-zero intercept, that is, $v(0) > 0$, and the convexity of the curve in the right plot is in agreement with the assumption that $c_3 > 0$ in Equation (2).

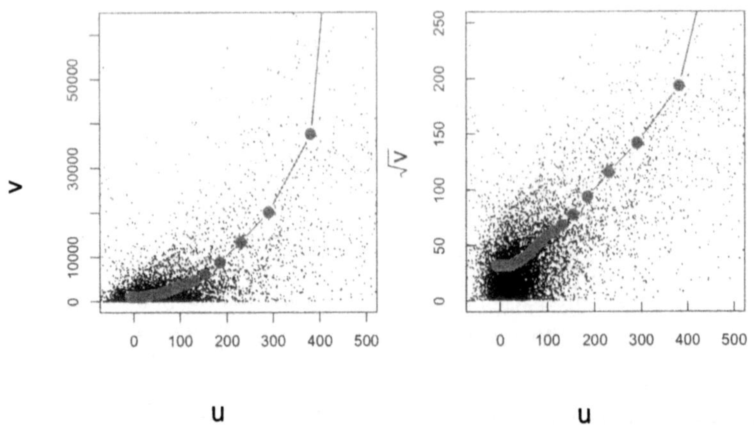

Fig. 1. *Variance-versus-mean dependence $v(u)$ in microarray data. Shown are the data from one mRNA sample, labeled both in red and green and hybridized against an 8400-element cDNA slide. The plots show the variance (left), and the standard deviation (right) versus the mean. The small dots correspond to single-spot estimates $\hat{v}_k = (y_{1k} - y_{2k})^2/2$, $\hat{u}_k = (y_{1k} + y_{2k})/2$, the lines show a moving average.*

2.2 Parameter estimation

The parameters of the model (7) are estimated from data with a robust variant of maximum likelihood estimation. The detailed derivation, as well as results on convergence and identifiability are described in Huber et al. (2002). Given the data (y_{ki}), $k \in K$, $i = 1, \ldots, d$, the profile log-likelihood (Murphy and van der Vaart 2000) for the parameters $a_1, b_1, \ldots, a_d, b_d$ is

$$-\frac{|K|d}{2} \log \left(\sum_{k \in K} \sum_{i=1}^{d} (h_i(y_{ki}) - \hat{\mu}_k)^2 \right) + \sum_{k \in K} \sum_{i=1}^{d} \log h_i'(y_{ki}), \tag{11}$$

with h_i as in (6). For a fixed set of probes K, we maximize (11) numerically under the constraints $b_i > 0$. The set K is determined iteratively by a version of *least trimmed sum of squares (LTS) regression* (Rousseeuw and Leroy 1987). Briefly, K consists of those probes for which $r_k = \sum_{i=1}^{d} (\hat{h}_i(y_{ki}) - \hat{\mu}_k)^2$ is smaller than an appropriate quantile of the r_k. The LTS regression addresses the fact that the distribution of the y_{ki} is produced by a mixture from genes that are differentially expressed, and ones that are not.

The *difference statistic* that quantifies the change in expression between samples i and j of a gene represented by probe k is now given as

$$\Delta\hat{h}_{k;ij} = \hat{h}_i(y_{ki}) - \hat{h}_j(y_{kj}), \qquad k = 1, \ldots, n. \tag{12}$$

It is easily seen that in the limit of large intensities, $\Delta\hat{h}_{k;ij}$ coincides with the log-ratio, whereas for near-zero intensities, it approaches the difference between the calibrated intensities.

3 Application

We now look at the effects of different transformations on data from a cDNA microarray experiment where samples from neighboring parts of a kidney tumor were labeled with green and red fluorescent dyes, respectively. The expression levels of most genes in the two samples are expected to be unchanged, so that observed differences should reflect experimental variability. To visualize the variance-versus-mean dependence, the differences of transformed

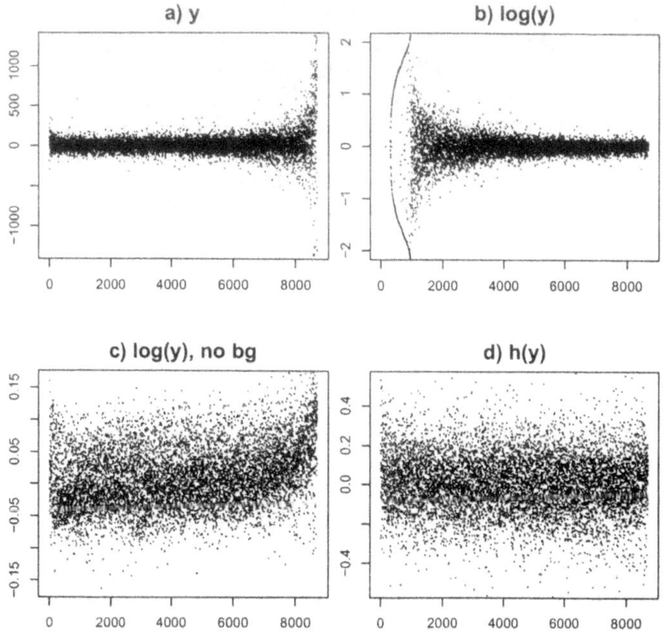

Fig. 2. *The difference between the two color channels of a cDNA microarray versus the rank of their average. The plots show the untransformed intensity data (***a***), log-transformation with/without background subtraction (***b** *and* **c***), and the transformation h proposed in this paper (***d***).*

intensities are plotted in Figure 2 against the rank of their mean (Sawitzki 1990). Figure 2a displays the untransformed data. The width of the difference distribution increases with the signal average. Taking logarithms after replacing all intensities below 1 by the value 1 yields Figure 2b. Logarithms of spot intensities without local background subtraction are shown in Figure 2c. The distribution of their differences has a roughly constant width; however, it does not follow a horizontal line. Finally, Figure 2d shows the difference statistic Δh as defined in (12). The distribution is centered around the x-axis, and its width is constant along the intensity range. We have observed similarly good results with microarray data from many different sources.

4 Discussion

A long-standing problem in the analysis of microarray gene expression data is how to take into account the dependence of the variance of a spot intensity of its mean. We propose the general approach of applying a variance stabilizing transformation. The resulting difference statistic Δh replaces the commonly used log-ratio as a measure of differential gene expression.

Our approach relies on the following assumptions: First, the variance of a spot intensity depends on its mean via a quadratic function with negative discriminant. This is grounded in the analysis of a large number of experiments and is in agreement with the model of Rocke and Durbin. Second, we assume that the relationship of measurements between samples is captured by an affine-linear transformation. While non-linear behavior may be observed under certain conditions, it has been demonstrated that microarray technology provides a large working range in which intensities increase linearly with mRNA concentrations. It appears to us that in many cases, non-linearities that have been observed in the logarithmic plot (as in Figure 2c) disappear when using the appropriate affine-linear calibration.

Other authors (Baggerly et al. 2001, Newton et al. 2001, Theilhaber et al. 2001) have also suggested quantitative models for the variance-versus-intensity dependence, and propose to augment log-ratio values by "significance values" derived from such models. However, these are only calculated after the calibration of data from different samples, which is complicated by the intensity-dependent variances of spot intensities in the first place. In our approach, the calibration is an integral part of the overall model fitting. The proposed statistical model has only few parameters, which are well identifiable according to our experience with simulated data. Finally, since the difference statistic Δh is simply obtained as the difference between the transformed intensity values of the individual samples resulting from simultaneous calibration, our approach not only allows for the comparison of *two* samples, but without any further effort can also be used for multivariate analyses comparing more than two samples.

Our method is implemented as an R package, which is freely available for academic use at http://www.dkfz.de/abt0840/whuber/.

Acknowledgements
We thank Günther Sawitzki and Dirk Buschmann for fruitful discussions, and Rainer Spang for critical reading of the manuscript.

References

The chipping forecast (1999). Supplement to *Nature Genetics*, **21**.

Baggerly, K.A., Coombes, K.R., Hess, K.R., Stivers, D.N., Abruzzo, L.V. and Zhang, W. (2001). Identifying differentially expressed genes in cDNA microarray experiments. *J. Comput. Biol.*, **8**, 639–659.

Chen, Y., Dougherty, E.R. and Bittner, M.L. (1997). Ratio-based decisions and the quantitave analysis of cDNA microarray images. *J. Biomed. Optics*, **2**, 364–374.

Huber, W., von Heydebreck, A., Sültmann, H., Poustka, A., and Vingron, M. (2002). A model for the calibration of microarray data and the quantification of differential expression. *Submitted.*

Murphy, S.A. and van der Vaart, A.W. (2000). On profile likelihood. *J. Amer. Stat. Assoc.*, **95**, 449–465.

Newton, M.A., Kendziorski, C.M., Richmond, C.S., Blattner, F.R. and Tsui, K.W. (2001). On differential variability of expression ratios: improving statistical inference about gene expression changes from microarray data. *J. Comput. Biol.*, **8**(1), 37-52.

Rocke, D.M. and Durbin, B. (2001). A model for measurement error for gene expression arrays. *J. Comput. Biol.*, **8**, 557-569.

Rousseuw, P.J. and Leroy, A.M. (1987). *Robust Regression and Outlier Detection.* John Wiley and Sons.

Sawitzki, G. (1990). Tools and Concepts in Data Analysis. In: *Softstat '89. Proceedings of the 5th Conference on Scientific Use of Statistical Software*, 237-248. Stuttgart, Gustav Fischer.

Theilhaber, J., Bushnell, S., Jackson, A. and Fuchs, R. (2001). Bayesian estimation of fold-changes in the analysis of gene expression: The PFOLD algorithm. *J. Comput. Biol.*, **8**, 585–614.

Tibshirani, R. (1988). Estimating transformations for regression via additivity and variance stabilization. *J. Amer. Stat. Assoc.*, **83**, 394-405.

Weights and Fragments

Stephan Morgenthaler[1]

[1] FSB - Institute of Mathematics, Swiss Federal Institute of Technology, 1015 Lausanne, Switzerland

Abstract. This talk explores two related ideas. The first one is to let the data guide us in downweighting observations in order to improve the fit of the model. Such procedures can, for example, be used to identify subsets of the data that closely fit a particular regression model and as such provide an alternative to robust methods. The second idea consists in fragmenting the data into pieces, analyzing the pieces separately and constructing a global analysis by putting the fragments back together. In analyzing a two-way table, for example, different interaction patterns may apply to different parts of the table.

Keywords. Robustness, parsimony, structural change, model break

1 Main Ideas

The description of a response variable in terms of explanatory variables and factors usually assumes either an additive or transformed-additive structure that holds over the whole observational range or – at the other extreme – estimates the structural relationship nonparametrically based only on smoothness assumptions. In this talk we discuss solutions that lie inbetween these two extremes. Most users of statistical fitting techniques are interested in parsimony, because this generally furthers understanding over mere description and is advantageous when one is interested in a predictive useage of the fit. They also want accuracy, which leads one to reject the simple additive methods in many situations, in particular when a large database is being analyzed.

The presence of isolated observations in the database that do not adher to an overall structure can be handled with weights. The main difficulty is in identifying such cases, which usually requires the sifting of trial fits but can sometimes be achieved with searches and iterations. If nothing is done, such isolated cases may wreck an analysis on many levels. Unimportant explanatory variables or factors may be given undue attention. Inversely, important factors may appear unimportant. Furthermore, the coefficients of important variables may be poorly estimated.

In some situations, such as in the case of two-way tables, more complex departures from a simple overall structure can be observed. Whole rows rather than individual cells may behave differently from the rest. In other situations, a subset of the cases may exhibit a common but different structure. This is the case, for example, if two regression lines rather than a single one are needed to explain the link between two variables. If this is the case, splitting the data set into more homogeneous fragments and analysing the fragments separately can provide a solution. In the remainder of this talk, these ideas will be discussed in the context of two examples.

2 Downweighting Outlying Observations

Experience shows that in a chemical industrial setting 1% to 10% of all measurements performed for the purpose of planned experiments result in grossly wrong data and should be treated as outliers. As an example, we consider a study that was performed to optimize the yield of a chemical synthesis (see Morgenthaler and Schumacher, 1999). A 5-Bromo-pyrimidine solution with a fixed concentration and n-Butyl-lithium (N.BULI) is continuously fed into a batch reactor (CSTR) at constant temperature. The reaction mixture is transfered into a tubular reactor with a static mixer where a ketone (KET) is continuously added yielding a carbinole. The temperature of this reactor (TR) and the temperature of the ketone (TKET), as well as the amount of n-Butyl-lithium and the ketone (in Mol-Eq.) were set according to the experimental design. The 5-Bromo-pyrimidine solution had to be prepared every day giving slightly different concentrations. In order to compensate for this a blocking factor (BLOCK) has been introduced. The amounts of n-Butyl-lithium and the ketone used in the experiment had minor deviations from the levels given by the statistical design (-1, 0, 1). The yield (in %) has been calculated with regard to the ketone (KET) as this is the most expensive reactant. Table 1 contains the data.

For the analysis we propose to fit a second order polynom in x_1, x_2, x_3 and x_4, the standardised versions of the raw variables KET, N.BULI, TR, and TKET. The blocking variable is taken into account by using it as an additive factor on three levels. This model has 17 terms, namely a global mean, two constants for the block factor, four linear and four quadratic terms and six interaction terms. The resulting fitted equation is:

$$\begin{aligned}\hat{y} = 94.7 &- 1.6\,x_1 + 1.6\,x_2 - 2.3\,x_3 - 0.6\,x_4 + 13.6\,x_1\,x_2 \\ &-5.9\,x_1\,x_3 - 0.2\,x_1\,x_4 + 6.8\,x_2\,x_3 - 1.9\,x_2\,x_4 + 0.3\,x_3\,x_4 \\ &-6.5\,x_1^2 - 12.6\,x_2^2 - 3.3\,x_3^2 - 1.6\,x_4^2 + \text{ block effect.}\end{aligned} \tag{1}$$

The underlying residual standard error is estimated as $\hat{\sigma} = 6.9$ and has 10 degrees of freedom. R^2 and R^2_{adj} are equal to 0.80 and 0.49, respectively.The residuals do not exhibit any pathological pattern and all stay well within the range of ± 2 standard deviations. One would not expect any problems. The residuals of runs 4 and 7 give only a weak hint to the presence of outliers. The value of $\hat{\sigma}$ of 6.9 is somewhat high for chemical data of this type but we do not have any external estimate of this quantity and so this problem cannot be further investigated.

In order to compensate for outlying values among measurements two actions are possible. We can either *include all the data into our computations but reduce the weight given to outlying values* or we can *modify or remove the outlying values and base our computations on cleaned data*. These ideas can be applied, whenever we have measurements $y_1, \ldots, y_n$ resulting from experimental runs, together with a structural model for their expected values,

$$E(y_i) = \mu_i(\beta_1, \ldots, \beta_p), \tag{2}$$

where μ_i is known up to the unknowns $\beta_1, \ldots, \beta_p$. If we have estimates $\hat{\beta}_1, \ldots, \hat{\beta}_p$ of the unknowns, reasonable weighting schemes, based on the residuals $r_i = y_i - \mu_i(\hat{\beta}_1, \ldots, \hat{\beta}_p)$, are possible. In many situations, it is of interest to include the possibility of a zero weight for observations with very large residuals. Let $d > 0$ denote the cutoff value. A simple polynomial function

Table 1. Data from a Box-Behnken Design.

	KET	N.BULI	TR	TKET	BLOCK	y	x_1	x_2	x_3	x_4
1	0.82	1.00	−55	0	1	96.49	−0.1	−0.1	0	0
2	0.82	1.00	−45	−25	1	93.22	−0.1	−0.1	1	−1
3	0.82	1.00	−65	−25	1	94.24	−0.1	−0.1	−1	−1
4	0.91	0.91	−55	0	1	77.20	0.8	−1.0	0	0
5	0.91	1.09	−55	0	1	89.86	0.8	0.8	0	0
6	0.73	0.91	−55	0	1	94.87	−1.0	−1.0	0	0
7	0.73	1.09	−55	0	1	63.46	−1.0	0.8	0	0
8	0.82	1.00	−65	25	1	91.88	−0.1	−0.1	−1	1
9	0.82	1.00	−45	25	1	91.88	−0.1	−0.1	1	1
10	0.83	1.02	−55	0	2	100.28	0.0	0.1	0	0
11	0.92	1.02	−55	−25	2	90.44	0.9	0.1	0	−1
12	0.74	1.02	−55	−25	2	92.53	−0.9	0.1	0	−1
13	0.83	0.93	−65	0	2	86.55	0.0	−0.8	−1	0
14	0.83	1.11	−65	0	2	82.12	0.0	1.0	−1	0
15	0.83	1.11	−45	0	2	90.85	0.0	1.0	1	0
16	0.83	0.93	−45	0	2	69.08	0.0	−0.8	1	0
17	0.92	1.02	−55	25	2	88.55	0.9	0.1	0	1
18	0.74	1.02	−55	25	2	91.55	−0.9	0.1	0	1
19	0.83	1.02	−55	0	3	90.96	0.0	0.1	0	0
20	0.83	0.92	−55	−25	3	71.99	0.0	−0.9	0	−1
21	0.83	1.11	−55	−25	3	80.21	0.0	1.0	0	−1
22	0.74	1.02	−45	0	3	86.80	−0.9	0.1	1	0
23	0.93	1.02	−65	0	3	86.89	1.0	0.1	−1	0
24	0.74	1.02	−65	0	3	82.86	−0.9	0.1	−1	0
25	0.93	1.02	−45	0	3	69.52	1.0	0.1	1	0
26	0.83	1.11	−55	25	3	76.14	0.0	1.0	0	1
27	0.83	0.92	−55	25	3	75.45	0.0	−0.9	0	1

KET: Educt 1 of the reaction, $x_1 = (\text{KET}-0.83)/0.1$
N.BULI: Educt 2 of the reaction, $x_2 = (\text{N.BULI}-1.01)/0.1$
TR: Temperature of the reactor, $x_3 = (\text{TR} + 55)/10$
TKET: Temperature TKET, $x_4 = \text{TKET}/25$
BLOCK: Day
y: yield

This table shows the setting of the four process variables (the first four columns) and the yields obtained (6th column). The explanation of the variables is given below the table. In the fitting stage we will make use of the standardized variables $x_1, \ldots, x_4$ also shown in the table.

which achieves the stated goal is **Tukey's biweight function** (see Mosteller & Tukey, 1977)

$$w_i = \begin{cases} \dfrac{(d^2 - r_i^2)^2}{d^4}, & \text{if } |r_i| < d\,; \\ 0\,, & \text{if } |r_i| > d\,. \end{cases} \tag{3}$$

It is clear how to make use of such weights. Let $(\hat{\beta}_1, \ldots, \hat{\beta}_p)$ be the current estimates. We then compute the residuals and weights and compute new estimates based on (3). We may then repeat this procedure, because the residuals and weights computed with the new estimates will in general be different from the previous ones. This process is known as iteratively reweighted least squares.

The weighting procedures that we have discussed cannot be used in practice. It is obviously necessary to adapt them to the scale of the residuals. To do this, let $S(r_1, \ldots, r_n)$ be a residual scale such as

$$S = S(r_1, \ldots, r_n) = \text{median}\Big\{|r_1|, \ldots, |r_n|\Big\} = \text{MAD} \qquad (4)$$

and compute the weights with the scaled residuals $(r_1/S, \ldots, r_n/S)$. Applying this robustness technology to measurements from a designed experiment should allow us to find **automatically** a model that fits a large majority of the observations and allows the occasional occurrence of outliers. In our example, the fitted equation using Tukey's biweight method with $d = 7.5$ is

$$\begin{aligned}\hat{y} = 95.0 &- 5.1\,x_1 + 5.2\,x_2 - 2.2\,x_3 - 0.6\,x_4 + 25.2\,x_1\,x_2 \\ &-5.9\,x_1\,x_3 - 0.2\,x_1\,x_4 + 6.4\,x_2\,x_3 - 1.9\,x_2\,x_4 + 0.3\,x_3\,x_4 \\ &-9.5\,x_1^2 - 17.4\,x_2^2 - 2.0\,x_3^2 - 0.3\,x_4^2 + \text{ block effect.}\end{aligned} \qquad (5)$$

If we fit simply by least squares, we obtain (1) and a fitted response surface that has its maximum close to the standard setting $x_i = 0$. Only a small gain in yield is possible. This fit reveals inconsistencies among the blocks with runs 4 and 7 being suspect. Deleting them from the analysis results in a very large residual at run 4 and a different judgment of the importance of the variables x_1 and x_2. A large gain in yield is predicted when x_1 is set to -1, x_2 is set to $-1/2$ and the variable x_3 remains at its central value of 0. The fourth variable TKET has a relatively small bearing on yield. The observation from run 7 conforms well to the fitted surface. This suggests that in fact a single outlier is present in the data, the measurement from run 4. The Tukey method with $d = 7.5$ gives practically identical results to least squares fitting after deletion of run 4 and does indeed find this solution automatically.

3 Fragmenting a Data Table

The data set under investigation comes from the International Maize and Wheat Improvement Centre (CIMMYT), an internationally funded, non-profit research organization aimed at improving the productivity, profitability, and sustainability of maize and wheat farmers in low-income countries. CIMMYT disseminates most of its elite germplasm through a system of international nurseries, where the term "nursery" means a collection of wheat lines assembled to meet a breeding objective (Fox, 1996). One of the major CIMMYT nurseries was the International Spring Wheat Yield Nursery (ISWYN), with the first, ISWYN_1, distributed in 1964 and the last, ISWYN_{30}, distributed in 1994 to co-operators around the world as pre-packaged seed, ready for sowing in an international multi-environment trial. The one chosen for analysis in this note is ISWYN_{11}, which contains the yield (in tonnes per hectare) of 50 genotypes grown in each of 65 environments. The 50 genotypes comprised 49 entries that were chosen by the CIMMYT breeders (from their lines and nominations from different parts of the world) and a local check. The 65 environments comprised various locations around the world, with co-operators sometimes using the same location under different conditions or time of planting (and hence labeling it as a different environment).

A simple non-additive decomposition of a two-way table is given by Tukey's one-degree-of-freedom-for-nonadditivity model (Tukey, 1949):

$$\hat{y}_{i,j} = m + a_i + b_j + g a_i b_j \,,$$

where $\hat{y}_{i,j}$ refers to the fitted value for the cell (i, j) and m, a_i, b_j, g are the fitted constants.

3.1 Splitting the Data Table

To prepare the splitting of the table, we first sort the rows and columns according to the values of a_i and b_j. In the reordered table, the rows and columns are such that $a_i \leq a_{i'}$ and $b_j \leq b_{j'}$ for $i \leq i'$ and $j \leq j'$, that is the north-east corner of the reordered table contains the smaller values, whereas the south-west corner contains the larger values. We then choose cut points for the rows, each of which splits the table into two subtables, one with row label smaller or equal to the cut and one with row label larger than the cut. Thus, $K - 1$ cut points lead to K subtables. Doing the same with the columns, leads to splits in the other direction and doing both results in subtables that contain both fewer rows and fewer columns that the original table. If $K - 1$ row cut points and $L - 1$ columns cut points are selected, a total of KL subtables are obtained. Let $\mathrm{Y} = (\mathrm{y_{i,j}}), \mathrm{i} = 1, \ldots, \mathrm{I}; \mathrm{j} = 1, \ldots, \mathrm{J}$ denote the original two-way table. Splitting it results in

$$\mathrm{Y} = \begin{bmatrix} \mathrm{Y_{1,1}} & \mathrm{Y_{1,2}} & \cdots & \mathrm{Y_{1,L}} \\ \vdots & \vdots & \vdots & \vdots \\ \mathrm{Y_{K,1}} & \mathrm{Y_{K,2}} & \cdots & \mathrm{Y_{K,L}} \end{bmatrix},$$

where the sizes of the subtables are determined by the location of the cut points. In our example, we chose the cut points as $7, 13, 25, 37, 43$ for the 50 rows and $9, 17, 33, 48, 56$ for the 65 columns. This corresponds to the median, the quartiles and the eighths and gives 36 subtables.

3.2 Analyzing the splits

Let a be the vector with components a_i and b the vector of the b_j. The odoffna decomposition in matrix notation is

$$\mathrm{Y} = \mathrm{m11^T} + \mathrm{a1^T} + \mathrm{1b^T} + \mathrm{gab^T},$$

where 1 denotes a vector of all ones of the approriate dimension. When we perform an odoffna analysis for each subtable separately, we have

$$\mathrm{Y_{k,l}} = \mathrm{m_{k,l}11^T} + \mathrm{a_{k,l}1^T} + \mathrm{1b_{k,l}^T} + \mathrm{g_{k,l}a_{k,l}b_{k,l}^T}.$$

Fitting by least squares lowers the residual sum of squares from 1542.17 to 1122.03 when passing from a global analysis to a fragmented analysis. The residual sum of squares in the 36 subtables are as shown in Table 2. A look at Table 2 shows that fragmentation leads to a sizable improvement only in a fraction of the subtables. We could therefore simplify the fit by allowing individual fits only when this seems necessary. Also, our method for constructing the splits corresponds to a simple first idea that can probably be improved, at least in particular cases.

References

Fox, P.N. (1996). The CIMMYT Wheat Program's International Multi-environment Trials. *Plant Adaptation and Crop Improvement.* M. Cooper and G.L. Hammer, eds. Oxford: CAB International.

Morgenthaler, S. and Schumacher, M. (1999). Robust Analysis of a Response Surface Design. *Chemometrics and Intelligent Laboratory Systems*, Amsterdam: Elsevier.

Table 2. The left-hand matrix shows the residual sum of squares of the global odoffna fit in the 36 subtables, whereas on the right-hand the fragmented residuals sums of squares are shown.

$$\begin{bmatrix} 28.93 & 33.9 & 74.5 & 92.5 & 65.3 & 126.0 \\ 5.44 & 29.1 & 42.4 & 43.0 & 19.4 & 30.5 \\ 12.43 & 55.4 & 57.5 & 78.2 & 52.7 & 62.9 \\ 8.53 & 83.7 & 58.9 & 57.3 & 38.0 & 51.6 \\ 7.34 & 24.4 & 33.0 & 33.0 & 30.9 & 29.0 \\ 7.78 & 37.7 & 34.7 & 46.1 & 20.3 & 29.8 \end{bmatrix} \begin{bmatrix} 7.42 & 20.3 & 49.4 & 63.0 & 38.9 & 73.3 \\ 3.70 & 24.7 & 36.4 & 38.2 & 14.0 & 21.2 \\ 10.08 & 38.0 & 46.5 & 63.1 & 42.1 & 49.8 \\ 6.85 & 65.8 & 50.4 & 50.2 & 28.2 & 40.6 \\ 5.91 & 15.5 & 28.2 & 26.6 & 24.8 & 23.7 \\ 4.50 & 20.8 & 27.5 & 29.0 & 15.5 & 17.9 \end{bmatrix}$$

Mosteller, F. and Tukey, J.W. (1977). *Data Analysis and Regression.* Reading: Addison-Wesley.

Tukey, J.W. (1949). One Degree of Freedom for Non-Additivity. *Biometrics.*

XQS/MD*Crypt as a Means of Education and Computation

Jörg Feuerhake

Institut für Statistik und Ökonometrie, wirtschaftswissenschaftliche Fakultät der Humboldt–Universität zu Berlin, Spandauerstr. 1, 10178 Berlin, Deutschland

Abstract. This paper introduces MD*Crypt as a means of distributing statistical methods and computing power. This paper explores MD*Crypt regarding topics of education. The role of MD*Crypt in projects associated with teaching statistics is examined. In an outlook, the potential of the MD*Serv/MD*Crypt architecture in frameworks of distributed computing and method providing is mentioned.

Keywords. education, XQS, MD*Crypt, distributed computing

1 Introduction

Nowadays statistics always goes with processing large amounts of data. The use of strong computers is absolutely necessary to handle this data. All considerations need to take this into account. When teaching statistics one faces particular demands. In the attempt to match those demands, it is necessary to consider how they can be met. This paper seeks to describe how computational power and a variety of methods are made available to students via the Internet. Distributing methods and computing power will be referred to as "distributed computing".

Since we are concerned with teaching statistics at the graduate level, we needed to provide students with a strong tool for computational statistics. Or more precisely, we need to provide computing power, as well as a wide selection of implemented statistical methods. The software to use in this case is XploRe, because of its well selected pool of statistical methods. This software has grown to be a strong standalone statistical environment. XploRe's developed language is highly applicable to meet student's needs.

With the emergence of Web technologies it became possible to acquire more students in a convenient way. It will be shown in this paper how attracting more students is achieved and the relationship to MD*Crypt will be discussed in further detail. Furthermore, particular features of the chosen architecture are discussed and evaluated.

Later, other approaches to deploy statistical methods and computing power are evaluated. The Rweb project is reviewed in more depth showing its quite remarkable approach to solving statistical problems.

Starting from the status quo of development and redefining the term "distributed computing", ways of improvement avaliable to the described architecture will then be discussed . If the MD*Crypt architecture is reviewed middleware centered, one can see these middlewares as core of a peer2peer web. The implications of this type of architecture will be described at the end of this paper.

2 MD*Crypt as an example

2.1 The Goal: Accessibility

As accessibility of information is a crucial question in teaching, students must be provided with the latest information about the issues they are concerned with. With 'paper-based' information, that task was not trivial. When the Internet emerged it became possible to publish new knowledge in almost an instant. Nonetheless, in teaching statistics, there is more than just published information. Furthermore, it is important to provide computing power and statistical algorithms, as well as the possibility to explore taught material and topics interactively in the classroom. To reach this goal, it becomes important to go beyond the means of pure Hyper-Text-Transportation-Protocol. In order to perform statistical computations, it is necessary to ensure that the computing machine is capable of actually computing at a certain level. One could achieve this by making sure that computing is done on a particular server machine. Client/Server computing appears to be relevant for these goals to be achieved.

In a subsequent step, it becomes crucial to provide an easy way to access server capacity. This has to be done in a way that can easily be carried out on a wide range of platforms. This requirement leads to the architecture described below.

2.2 Resulting Architecture

XploRe provides a large and well developed pool of statistical methods, as well as the ability to solve complex numerical problems. Hence XploRe becomes a promising candidate for serving as a statistical engine in teaching. Further, the methods implemented are well documented and the documentation is easy to obtain, or so called "browser enabled". The XploRe server implementation communicates on the stdio stream of it's underlying platform. Since this stdio communication is not appropriate for TCP/IP, it is important to have a middleware, providing the methods to connect via TCP/IP.

Using XploRe as a server for method providing and computational tasks, it is important to choose a suitable environment to implement the client. What technology guarantees most widespread platform access? Since Java's main argument was "write once, run anywhere", that seemed to be the language to implement the XQS communication package. Once using Java, there was only a short step made in sticking to using Java proposed approaches to programming.

Returning to the XQS, it suffices to state that MD*Serv does implement the communication between XQS and XQC. Since XQS is communicating on a stdio stream, it additionally needs to read from the TCP/IP stream and then write the same information to the stdio stream of the listening XQS. For the server to client communication, the same mechanism applies in reverse. Within the XQS/XQC architecture, MD*Serv captures a central role just as Figure 1 suggests.

The future client needs to implement the XQS Listener. By doing this the client is forced to implement the method *"handleServerReply()"*.This method is meant to handle incoming results from the XQS. It is crucial , that the handling of the result is left to the programmer of the future client, who has the knowledge of how to implement the representation of the result, such as graphic representation, most efficiently. In conclusion, we finally face an architecture as seen in Figure 1 according to the graphic. While MD*Serv and MD*Crypt handle all issues of TCP/IP communication, the XQS provides methods, data and computing power. The actual client needs to implement the user interface and handles the XQS results according to the technical environment on the client's machine.

The entire architecture relies on the ability of the future client programmer to interpret the XQS results properly and efficiently. That leads directly to the need for proper documentation and support. Another point that pushes Java to be the language of choice for MD*Crypt is the Javadoc tool which enables MD*Crypt to provide class documentation in an instant. Though this would not be sufficient to explain MD*Crypt to a third party developer it is a big step towards third party support.

For some projects, the whole architecture was brought to a single machine. So, with stronger processors in widespread use, the XQS/XQC-architecture can be de-

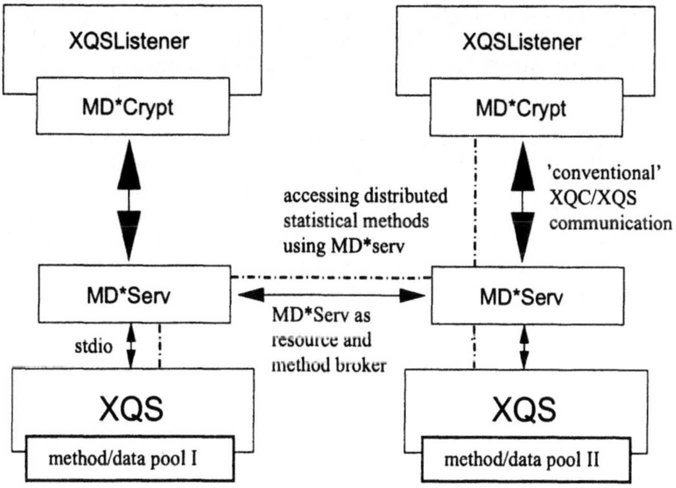

Fig. 1. the role of MD*Serv

ployed on closed single media, such as an electronic book. Creating media for teaching statistics on a XQS/XQC basis means writing net-based means of teaching as well as media ready for single machine use in the same step.

3 Related Projects

MD*Crypt is, as we have seen, a package to deploy computing power and statistical methods platform independent. Projects utilizing MD*Crypt to either use XploRe's statistical methods or reach accessibility via the Internet will now be introduced. This will be explained with regards to issues of education.

3.1 ReX

Recognizing the fact that Microsoft Excel is the most common software to analyze and process data, it is obvious to bring the methods one wants to deploy into the Excel interface. This can be done by offering an Excel add-in. Since it is possible to generate a com dll from Java code, MD*Crypt can be utilized to act as an Excel add-in. The Excel user is then able to use methods provided by XploRe.

Knowledgeable students will very likely be able to work with the Excel look and feel. Furthermore, students can concentrate on the contents taught if they do not need to familiarize themselves with an unknown user interface, ReX is a promising candidate as a means of teaching statistics. With the possibility to setup the XQS/XQC architecture on a single machine, it is possible to use the XQS/XQC architecture even in environments with impermanent Internet access. In situations, such as reviewing lectures at home, this becomes a major advantage.

3.2 XQC

XploRe's XQC provides a user interface very similar to XploRe's standalone user interface. Because one can run it as a Java applet, it becomes available on almost any machine connected to the Web. Students who are used to working within XploRe's user interface find it quite simple to use the XQC. Note that this is possible without having XploRe installed on the client machine. Therefore one can teach or study in environments poorly arranged for statistical teaching purposes.

3.3 e-books

A point that has not been emphasized so far is that teaching statistics should always come with the possibility to explore certain issues taught interactively. When using new formats of publishing contents, the teacher should go beyond presenting just the functionality of a book. One needs to find ways to extend the features of electronically enhanced books towards interactive examples. Here with MD*Crypt and the features of XploRe's XQC it is then possible to present interactive examples for statistical content just incorporated into an electronic book. There are already quite a developed amount of e-books using this feature.

These e-books present their content as a browser enabled e-book and provide the possibility to calculate examples, change their parameters or graphic representation all via the XQC.

4 Known Problems

There are of course some challenges to overcome with the MD*Crypt approach. Since MD*Crypt is a Java package, it relies on Java runtime on the client machine. If it is utilized in an applet, as it is done within XploRe's XQC, it is likely that on the client machine an installation effort has to be carried out. Within some environments this becomes a major drawback.

Another drawback is related to Java's applet technology. Since applets are regarded as potential security problems some crucial functionality is not present. For example confined server connecting and disabled copy/paste functionality are problems as long as applets are not signed by a trusted institution. Not every third-party programmer has access to assets necessary to get a trusted signature for his client. Since third party programmers are crucial for MD*Crypt this remains to be a problem. Pure html interfaces do not face these boundaries in functionality. See the Rweb section for further consideration.

5 Layout and Data-processing

As it was shown, with MD*Crypt it becomes possible to access all implemented algorithms provided by XploRe and handle the provided results via TCP/IP. This is a major improvement relative to XploRe's standalone version. The next section will show that other approaches can achieve this goal as well. They may even do this with less effort, like client programming, for the user. However, a main achievement is MD*Crypt's usability, adjusted to the use-case. With changing environments and working scenarios different clients with different user interfaces may connect XploRe Servers via MD*Crypt. This division of layout from data-processing does provide a lot of scalability. One may use XploRe's algorithms in a lot more occasions than XploRe's standalone could cover. Furthermore the clients may adjust to almost any use-case.

6 Different Approaches: Rweb

One remarkable project in deploying statistical methods via the Internet is Rweb. Rweb implements information transport with a common gateway interface. One great advantage to Rweb is its pure html interface. Since it does not rely on Java or a different runtime as its core environment, it runs without the installation of a plug-in. Rweb pays a large price for this convenience. Rweb cannot change its interface according to the underlying environment or as a reaction to different tasks.

Rweb's second advantage compared with the MD*Crypt approach is that it does not need further programming to implement a client. The browser runs again, but it is not adjustable to perform certain tasks.

Other Web-based data analysis approaches do not have the general approach which MD*Crypt or Rweb try to implement. In this context, one should mention Data Analysis Lab, WebStats and Statlet. The statistical methods in these programs are implemented as applets and run in Java enabled browsers. Unfortunately, they implement a very small range of methods and avoid Client/Server approaches. Client/Server architectures are most important to make sure that large scale problems can be solved as well.

7 Development Paths

The term distributed computing encompasses just more than making available methods and capacities geographically independent. Distributed computing means further on sharing capacities and methods throughout the Internet. Please see Figure 2 as an outline of a network of MD*Serv/XQS running on different machines:

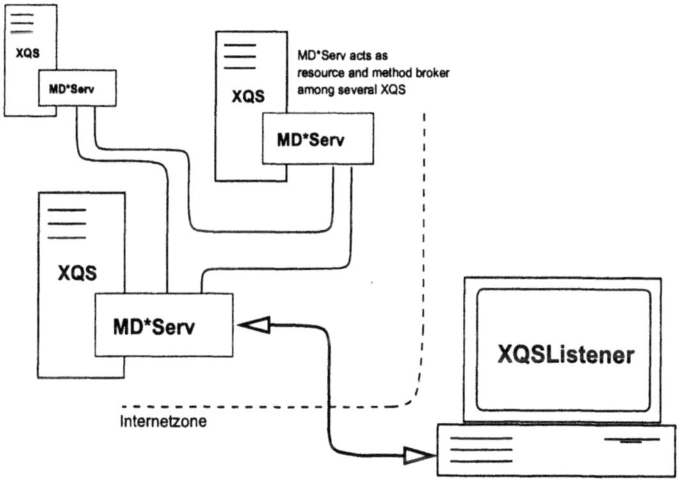

Fig. 2. XQS/MD*Serv systems connected

In returning to the described architecture, one can begin with the stated central role of MD*Serv. MD*Serv then, is easy to recognize as the pivotal point of a system of XQS's. MD*Serv would then be capable of providing functionality often referred to as distributed computing. MD*Serv is by now prepared to contact other instances of MD*Serv. When enabled to hold and pass information concerning its running servers, the servers' computing capacity and the methods available on their system it is possible to extend MD*Serv to serve as a resource and method broker. A XQC would then be able to contact one MD*Serv on a particular URL:Port and MD*Serv would dispatch this client's request. The result could be a grid of homogeneously working XQS and even more importantly, statistical methods avaliable on one server could be deployed to the client even if it connected to a different machine. With pure applet technology and in the absence of a middleware this architecture remains impossible. Please refer to Figure 2 to evaluate the impact.

The main task to be accomplished by MD*Serv will be resource broking. Resource broking would mean passing providing and evaluating information about XQS's avaliablity and about the methods these XQS's could provide to solve problems. Further, MD*Crypt/MD*Serv now provide communication patterns, such as graphic contexts and complex numeric arrays, ready for the use in statistical environments. This does not force MD*Serv to only connect to XploRe's servers. Provided, MD*Serv would map command structures of other languages and the engines could be connected, any of the mapped statistical systems could act as a XQS. Refer to the $\hat{\Omega}$ project's Java interface to find Java classes that provide methods to contact R. MD*Serv is not only able to dispatch requests among XploRe's servers, but could also access methods implementet within other statistical environments.

The improvements of distributed computing for students and teachers are straightforward. A grid is not as prone to failure as much as single servers are. Methods meant for distribution do not need to be stored on every server system. Method libraries can be shared as well as computing capacities.

8 Concluding Remarks

With MD*Crypt, a package is provided to deploy statistical methods and computing power to solve statistical problems. Programming skills provided, MD*Crypt is highly adjustable to a wide range of environments. Through this package a general approach in providing statistics can be utilized.

In developing MD*Crypt, it will be important to improve the underlying protocol. Furthermore, potentials of the achieved architecture need to be pushed forward to achieve new functionality and foolproof the recent architecture.

A different approach to the deployment of a large amount of statistical methods is Rweb, which follows a cgi approach in request handling. It is a project worth noticing for it's achievements in accessibility via the Internet.

The possibility of solving large scale statistical problems is one goal when teaching statistics. First steps are made to achieve wide accessibility to computing resources and statistical methods.

Resources on the Web

MD*Crypt, http://www.md-crypt.com

$\hat{\Omega}$ project's Java interface, http://www.omegahat.org/RSJava/

ReX MS Excel add-in, http://www.md-rex.com

Rweb html-interface, http://www.math.montana.edu/Rweb/

XQC XploRe's Java-interface, http://www.xplore-stat.de/java/java.html

References

Aydinli ,G. (2002) 'Netbased Spreadsheets in Quantitative Finance'. In: *"Applied Quantitative Finance"*, Berlin: Springer Verlag.

Aydinli ,G. , Härdle, W. , Kleinow, T. and Sofyan, H. (forthcoming 2002) 'Linking Office Applications with XploRe'. In: *Computational Statistics; special edition*, Heidelberg: Physica-Verlag.

Kleinow,T. and Lehmann,H. (forthcoming 2002) 'Client/Server based Statistical Computing'. In: *Computational Statistics;*, Heidelberg: Physica-Verlag.

Manzur, M.R. 'GridSim: A Toolkit for the Modeling and Simulation of Global Grids'. avaliable on the web:
http://www.csse.monash.edu.au/ rajkumar/papers/gridsim.pdf;

Author's Index

Keyword Index

GPSR Compliance

The European Union's (EU) General Product Safety Regulation (GPSR) is a set of rules that requires consumer products to be safe and our obligations to ensure this.

If you have any concerns about our products, you can contact us on ProductSafety@springernature.com

In case Publisher is established outside the EU, the EU authorized representative is:

Springer Nature Customer Service Center GmbH
Europaplatz 3
69115 Heidelberg, Germany

Batch number: 10725962

Printed by Printforce, the Netherlands